Gas Turbine Combined Cycle Power Plants

Gas Turbine Combined Cycle Power Plants

S. Can Gülen

CRC Press
Taylor & Francis Group
Boca Raton London New York

CRC Press is an imprint of the
Taylor & Francis Group, an **informa** business

CRC Press
Taylor & Francis Group
6000 Broken Sound Parkway NW, Suite 300
Boca Raton, FL 33487-2742

Printed on acid-free paper

International Standard Book Number-13: 978-0-367-19957-9 (Hardback)

Visit the Taylor & Francis Web site at
http://www.taylorandfrancis.com

and the CRC Press Web site at
http://www.crcpress.com

Contents

Preface

Either write something worth reading or do something worth writing.

Benjamin Franklin

In keeping with the sage advice of The First American, the author stuck to what he knows best from long experience in doing it. (Whether it has been worthwhile or not is for the reader to decide after reading this book.) Consequently, what you have in your hands is a monograph on natural gas-fired gas and steam turbine combined cycle power plants for electric power generation. It does not cover esoteric or failed variants long forgotten in dust-covered books and/or journals in library shelves (except in a historic context – e.g., Emmet's binary vapor cycle). The focus is fully on tried-and-tested machines presently chugging along in the field.

The goal of the author is to provide the reader with specialized knowledge that can be readily applied to the solution of day-to-day problems encountered in design, development, optimization, operation and maintenance of gas and steam turbine combined cycle power plants. Such knowledge comprises specific data (some hard to find – even on the internet, at least not in a compact and readily usable form), formulae, code snippets, charts and rules of thumbs. The book also includes a broad range of quantitative methods and tools that have been developed and used by the author over the course of more than two decades spent in the industry.

Having said that, it must be stressed that this volume is not exactly a handbook of generalized information for laypersons. At the same time, it would be ludicrous to claim that every piece of information included in the following pages is fully original and cannot be found elsewhere, i.e., online, other books, archival papers, etc. Nevertheless, it is safe to say that this book is a compendium of expert knowledge, some of which is either impossible to find in other resources or, even if found, is either too sketchy or too diluted or too obscure to be of immediate, practical use.

The intended readership is primarily professionals, i.e., engineers and researchers, who are working in the industry or research organizations on various aspects of electricity generation with gas and steam turbine combined cycle power plants. Although this is not a textbook per se, graduate and undergraduate students who are working on projects towards their degree with an ultimate goal of joining the industry can be added to that group as well.

At the time of writing (i.e., near the end of the second decade of the twenty-first century), almost all aspects of combined cycle power plant design are dominated by highly sophisticated and extremely expensive (i.e., not readily available to individual practitioners) computer software with steep learning curves. Most complicated combined cycle calculations for tens or hundreds of cases can be done in a matter of seconds by specialized software. (Commonly used design software products will be introduced in Chapter 2 – some of them are used in selected examples.) Their impact on productivity in power plant design and analysis is immense.

Consequently, such software is widely used by junior engineers and researchers as well as by practitioners with decades under their belt in the industry and/or academia in the course of their job-related activities. The author is one of them, too. The inconvenient truth is that there is no ghost in the machine. Even the most sophisticated software product is a tool developed by engineers and researchers who mastered the fundamental disciplines of mechanical and chemical engineering and translated them into calculators with user-friendly graphical interfaces for input and output. The old adage still holds; if the input is garbage, so is the output. This is so because there is no artificial intellect. There is only human intellect. But even the most sublime intellect is practically useless without a wealth of subject matter knowledge and experience. This monograph is intended to be just a building block in the reader's cumulative knowledge and experience base towards the attainment of full mastery in this particular field of power plant engineering.

Needless to say, quantitative and/or qualitative information (including opinions) that you will find in the following pages are based solely on the author's personal experience, observations, calculations, guesstimates, analysis of sample data culled from trade and archival publications (books, journals, magazines, standards, codes, etc.) and common engineering sense. Even though I do not think that I am way off-base (after all, pretty much everything that you will find herein is firmly grounded on first principles), to err is human. As such, I accept responsibility for any unintentional mistakes, typos and such.

Before moving on, there is a simple fact that needs to be stated unequivocally. As is the case with everything I have written before and will write in future, this book is dedicated to the memory of Mustafa Kemal Atatürk (1881–193∞) and his elite cadre of reformers. Without their vision, sacrifices and groundbreaking work, there would not have been a fertile ground where my parents, teachers, mentors, family and friends could shape me into the author of this book. Everything started with him; the rest was easy.

MATLAB® is a registered trademark of The MathWorks, Inc. For product information, please contact:

The MathWorks, Inc.
3 Apple Hill Drive
Natick, MA 01760-2098 USA
Tel: 508-647-7000
Fax: 508-647-7001
E-mail: info@mathworks.com
Web: www.mathworks.com

Author

Dr. S. Can Gülen (PhD 1992, Rensselaer Polytechnic Institute, Troy, NY), PE, ASME Fellow, Bechtel Fellow, has 25 years of mechanical engineering experience covering a wide spectrum of technology, system, and software design, development (GT PRO® PEACE®, THERMOFLEX®), assessment and analysis, primarily in the field of steam and gas turbine combined cycle (109FB-SS, IGCC 207FB, H-System) process and power plant turbomachinery and thermodynamics (in Thermoflow, Inc., General Electric and Bechtel). Dr. Gülen has authored/co-authored one book, two book chapters, numerous internal/external archival papers and articles (40+), design practices, technical assessment reports and US patents (20+) on gas turbine performance, cost, optimization, data reconciliation, analysis and modeling.

1 Introduction

Neither the gas turbine nor the steam turbine is the most efficient heat engine burning fossil fuels. At the time of writing, the *crème-de-la-crème* of either prime mover is at low 40s in thermal efficiency (as a percentage). That distinction belongs to the reciprocating, i.e., piston-and-cylinder, internal combustion engines burning natural gas. Presently, best-in-class representatives of that group can claim close to 50% in brake thermal efficiency. However, when combined through a heat recovery boiler, a gas and steam turbine power plant is by far the most efficient and cleanest technology to generate electric power from burning a fossil fuel, i.e., the natural gas. Field-proven performance of the best of the bunch is above 60%. Why this is so is going to become clear when we look at the thermodynamic principles governing the respective cycles of the two prime movers.

Using a system approach, there are four major subsystems in a combined cycle power plant system:

1. Gas turbine generator (operating in *Brayton* cycle)
2. Steam turbine generator (operating in *Rankine* cycle)
3. Heat recovery boiler or, using the more common terminology, heat recovery steam generator (HRSG)
4. Heat rejection system comprising either
   a. A water-cooled condenser (with or without a cooling tower) or
   b. An air-cooled condenser (ACC).

In addition to these four major subsystems, there is a maze of condensate, cooling water and steam piping with a large number of valves along with pumps (condensate, boiler feedwater and circulating cooling water pumps) and myriad heat exchangers, which are lumped under the term "balance of plant" (BOP). (In some sources, the BOP includes some or all of the equipment in the heat rejection system as well – not in this book.)

Each major subsystem can be and is the subject of a separate book in and of itself. Therefore, no attempt will be made herein to go into the nitty-gritty details of any of them. For such information, the reader's best source should be the original equipment manufacturer's (OEM) user manuals, technical information letters (TILs), white papers (such as General Electric's GER series freely available on its website) and information documents (such as GE's GEK series available to its customers). Another good source of pertinent technical information is OEM papers presented in technical conferences. Quite a few of them are available on OEM websites. Internally published technical journals such as *The Brown Boveri Review* are unfortunately a thing of the past. The only OEM continuing that tradition is Mitsubishi Hitachi Power Systems (MHPS); MHI (Mitsubishi Heavy Industries, the old name of the company) Technical Review articles are available online and provide in-depth information on MHI/MHPS steam and gas turbine technology.

The focus herein will be on the salient aspects of design and off-design performance and operability of each major subsystem within the combined cycle system. Furthermore, this is primarily a book of numerical/quantitative analysis. Aspects of performance and operability, which are amenable to numerical calculations, will be covered in detail; others will be mentioned in passing. High-level, physics-based but reasonably accurate calculation procedures will be described with two goals in mind:

1. Familiarize the reader with the "ghost in the machine", i.e., the basic fundamental principles of thermodynamics (by far, the most important discipline of mechanical engineering – at least in this author's opinion) governing the operation of the prime movers and heat exchange devices in the gas turbine combined cycle power plant.

2. Provide tools to do high-level analysis at different stages of design (conceptual, detailed, etc.) and operation (performance test, root cause analysis, etc.) and check the results of sophisticated computer simulation tools with steep learning curves and "black box" inner details.

This book is not intended for laypersons or dilettantes. There are books out there serving that segment quite adequately. This book is intended for serious practitioners and researchers in the field, who need practical "tools" to help them in their day-to-day activities. As such, no ink will be wasted on generalized information and textbook examples or esoteric variants, which have zero chance to see the daylight in the field in commercial operation. All information in this book, including charts, equations, examples and rules-of-thumb, is directly related to actual, state-of-the-art hardware in commercial operation. This is as close as you can get to the "real stuff" – one step removed from proprietary information – unless you are the employee of an OEM.

In the next chapter, the reader will be introduced to vital references (textbooks, industry codes and standards, etc.) and software tools requisite for the "job" because, to stress it once more, this book is intended for those who need to get the "job done" and do it properly. This requires access to commercial or in-house simulation software such as "cycle decks" and flowsheet simulators. Nowadays, complex cycle calculations are done using such tools, which are extremely expensive (e.g., commercial software) and proprietary (e.g., in-house software) so that only those employed by large organizations have access to them. All rigorous examples in the book are done with such specialized software so that the readers can familiarize themselves with their usage. It is an unavoidable fact that a lone engineer lacking specialty software knowledge cannot go too far in power generation industry (just like an amateur investor who, no matter how well he or she is versed in finance and economics, cannot hope to compete with institutional investors armed with millions of dollars' worth of computer software and hardware). It is hoped that this book will help inspire the reader in that direction (except those who already have hands-on experience in those tools).

1.1 NOTE ON UNITS

The entire professional career of the author was spent in US companies. While thoroughly conversant in SI units, his technical experience is primarily based on research studies, commercial projects and myriad assignments dealing with every conceivable aspect of gas and steam turbine combined cycle power plant technology where the *US customary system* (USCS), i.e., British units, was the standard unit system.[1] Therefore, this book primarily uses the USCS. Not doing so would require conversion of an immense body of work into SI units, which would introduce unnecessary errors into the treatise.

Despite this caveat, in many places in the text, SI units are provided in parentheses for convenience. Otherwise, the reader is asked to do his or her unit conversions. Note that the US companies still play a big role in power generation industry, and many of them still use the "archaic" pounds and degrees Fahrenheit and so on and so forth. Anyone who is serious on building a career in this particular industry would be well advised to be conversant in either unit system.

Ordinarily, no space and ink would be wasted to provide unit conversions herein. In this day and age when everybody is armed with a smartphone, even the most obscure units are one "click" away – one can simply *google* them. Nevertheless, in case your electronic arsenal is out of reach or your battery is dead or the power system is down because of Hurricane Zelda, below is a list of the most common unit conversions and some important caveats.

[1] The US customary system was developed from English units which were in use in the British Empire before the United States became an independent country. In the United Kingdom, the system of measures was overhauled in 1824 to create the imperial system, changing the definitions of some units. Nevertheless, as far as the most commonly used units in power generation calculations, both are the same.

Temperature

$°C = (°F - 32)/1.8$
$K = °C + 273.15$ (degrees Kelvin)
$R = °F + 459.67$ (degrees Rankine).

Unless a particular equation is an empirical curve-fit and the author specifically identifies which temperature units to use, one has to make sure that *absolute* temperatures (in degrees Kelvin or Rankine) are used in scientific formulae.

Pressure

1 psi = 0.0689476 bar (pounds per square inch)
1 bar = 100 kPa (kilopascal) = 0.1 MPa (megapascal)
1 bar = 1,000 mbar (millibar).

Note that one has to make sure whether a stated pressure is "absolute" or "gauge". If it is the latter, one has to add the atmospheric pressure to the *gauge* value to arrive at the *absolute* value. In many cases, this identification is made by appending an "a" for atmospheric or "g" for "gauge" to the base unit designation. For example,

1 psia = 1 psig + 14.696 or
1 bara = 1 barg + 1.01325.

Note the usage of standard atmospheric pressure in the equations above, which is defined as

1 atm = 1.01325 bar = 14.696 psi = 29.92 in. Hg (inches of mercury).

Steam pressures are expressed in psi, bar or MPa.

Condenser pressures are usually expressed in inches of Hg or millibar.

1 in. Hg = 33.8639 millibar
1 in. Hg = 0.491154 psi.

Gas turbine inlet or exhaust pressure losses are expressed in inches of water column (H_2O) or millibar.

1 in. H_2O = 2.4884 millibar
1 in. H_2O = 0.0360912 psi.

Mass or Mass Flow Rate

1 lb/s = 0.452592 kg/s
1 kg/s = 2.20462 lb/s.

Large steam flow rates are commonly expressed in kpph or tph.

1 kpph = 1,000 lb/h = 0.277778 lb/s
1 lb/s = (1/3.6) kpph
1 tph = 1,000 kg/h.

Note that, in the United States, one "short ton" is equal to 2,000 lbs.

In SI units, one "metric ton" is equal to 1,000 kg. Thus, a metric ton is slightly larger than a US ton – it is equal to 2,204.6 pounds (that is why the other one is labeled "short").

"Tonne" is an alternative spelling used to describe a metric ton. It is almost never used in American English, but it is widely used outside of the United States. Be careful!

Power and Energy

1 Btu (British thermal unit) = 1.05506 kJ = 1,055.06 J
1 Btu/s = 1.05506 kW = 1,055.06 W
1 MMBtu = 1,000,000 Btu (Million Btu)
1 MMBtu = 293.071 kWh
1 kW = 3,412.142 Btu/h
1 MW = 1,000 kW = 3.412142 MMBtu/h
1 GW = 1,000 MW
1 MMBtu/h = 293.071 kW.

When using kW or MW for heat transfer rate, it is customary to append "th" for thermal to the units, i.e., kWth or MWth. Similarly, when using kW or MW for electric power, it is customary to append "e" for electric to the units, i.e., kWe or MWe.

There are two very important constants in British units, which can lead to gross errors if not used properly or ignored altogether:

1 lbf = 32.174 lb-ft/s^2, where g_c = 32.174 ft/s^2
1 Btu = 778.17 ft-lbf, where J = 778.17 Btu/ft-lbf.

Of course, g_c is the standard acceleration of gravity, which is 9.80665 m/s^2 in SI units.

The third important constant is the universal gas constant:

R_{unv} = 8.314 kJ/kmol-K = 1,545 ft-lbf/lbmol-R = 1.986 Btu/lbmol-R.

The most common pitfall is in using pressure in formulae such as the ideal gas relationship. Note that

1 psi = 1 lbf/in.^{2} = 144 lbf/ft^2.

Example: What is the density of air at standard conditions of 14.7 psi and 59°F? Using the ideal gas relationship and assuming the molecular weight of air is 29 lb/lbmol.

$p = \rho RT$
$\rho = (14.7 \times 144)/(1{,}545/29)/(59 + 459.67) = 0.0766$ lb/ft^3.

1.1.1 Odds and Ends

There are quite a few industry oddities in terms of units used to express common quantities, which cannot be found in standard textbooks. Listed below are some of them:

cuft: Cubic feet (ft^3)
cuin: Cubic inch (in^3)
fps: Feet per second (ft/s)
ips: Inch per second (in/s)
kpph: 1,000 pounds per hour (1,000 lb/h)
ppb: Parts per billion (0.000000001)
pph: Pounds per hour (lb/h)

ppm: Parts per million (0.000001)
ppmv(d): Parts per million by volume (dry) – i.e., a volume fraction of 0.000001
ppmw(d): Parts per million by weight (dry) – i.e., a mass fraction of 0.000001
pps: Pounds per second (lb/s)
SCF/scf: Standard cubic feet (pronounced "scuff")
STPD/stpd: Standard tons per day.

2 Prerequisites

As stated in the Preface, this is not a book for beginners or laypersons. There is no introductory material including derivation of key equations from the first principles. The reader is expected to have at least a 4-year degree in engineering under his or her belt and be familiar with the fundamental disciplines of thermodynamic, heat transfer and fluid mechanics. If not, this is the time to dust off your favorite college textbooks and refresh your memory. If this is a chore that you are unwilling to tackle, there is a good chance that you will neither enjoy reading this book nor draw much benefit from it. If you need help with selecting appropriate references, the author is happy to oblige – so, read on.

2.1 BOOKS AND PERIODICALS

One does not have to be a math whiz to read this book. Nevertheless, a reasonably good knowledge of differential calculus would be very helpful. If your math skills are rusty, the excellent book by Zeldovich and Yaglom is the perfect resource for getting up to the speed [1].

For engineering thermodynamics, the author's preference is the first edition of the textbook by Moran and Shapiro [2]. In the remainder of this treatise, this particular book will be referred to as "Moran and Shapiro". Another favorite of the author is the book by Zemansky [3]. For heat transfer, the recommendation here is the textbook by Incropera and Dewitt [4]. For fluid mechanics, the author's choice is the textbook by White [5]. For more advanced readers, the one-and-only source is the timeless classic by Landau and Lifshitz [6]. You may not need it *per se*, but it is a good idea to have some basic understanding of the gas dynamics branch of fluid mechanics. In case you are so inclined, the go-to references are Liepmann and Roshko's classic treatise [7] and the superb book by P. A. Thompson [8] (full disclosure: late Dr. Thompson was the author's PhD adviser and mentor).

It is also a good idea to own a copy of a general reference book for typical engineering calculations. The handbooks recommended by the author are (the reader can, of course, use their own favorites):

1. *Marks' Standard Handbook for Mechanical Engineering* (any fairly recent edition will do).
2. *Perry's Chemical Engineer's Handbook* (any fairly recent edition will do).
3. For those readers who know the German language, *Dubbel – Taschenbuch für den Maschinenbau* (by Springer-Verlag, any fairly recent edition will do) is by far the best handbook known to the author.

The Professional Engineer (PE) exam reference manual (for USA) by Lindeburg is an excellent resource for many practical formulae in thermofluids, hydraulics and other disciplines of mechanical engineering [9]. It is concise, to the point and contains many examples. It is highly recommended that you obtain a recent edition if you do not already have one left over from your PE exam days.

An indispensable reference book that you *must* have if you are serious about doing your own calculations is *Numerical Recipes in FORTRAN* (or *C*) by William H. Press et al. [10]. The codes in those books (e.g., for finding the roots of a highly nonlinear function) are extremely robust, easily translatable into Visual Basic (VB) and very useful.

As far as the equipment is concerned, you should have a good book each on gas turbines and steam turbines. For the former, the author recommends the recent book written by him – not because it is the best in the field (that is a decision for the audience to make) but it is a good match to the treatise herein. ***This particular source will be referred to in the remainder of the current book***

as GTFEPG [11]. In it you can find an exhaustive list of excellent references on the subject. The author would strongly recommend to own a copy of a recent edition of the book by Saravanamuttoo et al. for the gas turbine fundamentals [12]. For the more application-oriented "hardcore" information, especially on land-based gas turbines for electric power production, the best source by far is the superb handbook edited by Lechner and Seume (if you know German) [13]. If English is your preference, the handbook edited by Jansohn is also a good source [14].

Steam turbines have been around for more than a century. There are hundreds of treatises going back to the monumental work of Stodola (which also covers the gas turbines of 1920s vintage). The author's recommendations for stocking the steam turbine technology shelf of your library is the two-volume treatise by Leyzerovich [15], the three-volume work by W.P. Sanders [16] and the two books, one by K.C. Cotton [17] and the other by P. Shlyakin [18], in addition to the ASME Performance Test Code PTC 6.2, which contains useful calculations.

For the heat recovery steam generator (HRSG), the best available source is the *HRSG Users Handbook*, which you can order directly from the HRSG User's Group (www.HRSGusers.org). At $390 a copy, it is probably too expensive for young professionals or students. Another good source, especially for practical calculations, is the ASME Performance Test Code PTC 4.4-2008 "Gas Turbine Heat Recovery Steam Generators". At $100 per copy, it is a good investment (www.techstreet.com).

For condensers, water or air-cooled, the best sources are the HEI (Heat Exchange Institute) Standards, which are available for purchase online at www.techstreet.com. For the cooling towers, please refer to CTI (Cooling Tower Institute) Standards (www.cti.org).

The best sources for new and emerging technologies are academic journals (requires subscription and/or membership to professional organizations), conference proceedings (ditto) and trade publications (almost all of them available online at no charge). The reader is encouraged to consult them for up-to-date information on new research and development. The list below, by no means comprehensive, is provided as a starting point.

1. *ASME Journal of Energy Resources Technology*
2. *ASME Journal of Engineering for Gas Turbines and Power*
3. *ASME Journal of Turbomachinery*
4. *Chemical Engineering* (www.che.com)
5. *Combined Cycle Journal* (www.ccj-online.com)
6. *Electric Light & Power* (www.elp.com)
7. *Energy, The International Journal*, Elsevier
8. *Gas Turbine World* (www.gasturbineworld.com)
9. *Hydrocarbon Processing* (www.HydrocarbonProcessing.com)
10. *Modern Power Systems* (www.modernpowersystems.com)
11. *POWER* (www.powermag.com)
12. *Power Engineering* (www.power-eng.com)
13. *Turbomachinery International* (www.turbomachinerymag.com)

The two most important sources of technology information are annual "handbooks" published by the **Gas Turbine World** (GTW) and **Turbomachinery International** (TMI). These publications will be referred to frequently in this book (e.g., GTW 2016 for the 2016 handbook).

2.2 SOFTWARE TOOLS

In order to derive the maximum benefit from this book to impact your daily work, you need to perform calculations. Some of the equations developed in the following chapters are amenable to quick implementation with a pen, a piece of paper and a $10 drugstore calculator (or the calculator app in your smartphone). Once you go beyond a few "back of the envelope" type estimates, you will probably require to implement quite a few of such equations in an Excel spreadsheet. Nevertheless, at a certain point, you will realize that it is a good idea to distill some of your calculations into compact snippets of computer *code*. This is where the real fun starts.

It is highly recommended that you are well versed at least in one of the programming languages[1]:

1. Visual Basic (especially VBA in Excel)
2. C or C++
3. FORTRAN
4. Matlab (from MathWorks®)

The last one is extremely popular among the younger generation – students *and* practitioners. Even though the author has not had extensive experience with it, he is aware that it is a very simple and powerful tool for computational purposes. The most beneficial one, in this author's opinion of course, which is also quite easy to learn and implement, is *Visual Basic*. In particular, VBA (Visual Basic for Applications) is the programming language of Excel and other Microsoft Office programs. (It is highly unlikely that the reader of this book does *not* have access to MS Office and Excel through his organization or on his or her own.) If you are not already familiar with VBA, you can get started by automating tasks in Excel by using the Macro feature, which, essentially records a sequence of your Excel spreadsheet actions in a VB function.

An alternative to Microsoft Office is *Apache OpenOffice* (AOO), which is an open-source office productivity software suite. It contains a word processor (**Writer** – equivalent to MS Word), a spreadsheet (**Calc** – equivalent to MS Excel) and a presentation application (**Impress** – equivalent to MS PowerPoint) among others. Readers who are more familiar with it can of course substitute the AOO variant for Excel, which is the software program most familiar to the author.

If you are already familiar with C++ and/or FORTRAN and, even better, if you already have old programs that you had developed earlier in your educational and/or professional career, you can easily translate them into VB. It is very intuitive and easy. Another way to capture such legacy code in Excel or similar software programs is via converting them into a *dynamic-link library* (DLL), i.e., an executable file encapsulated in a DLL.

In the Appendix of the **GTFEPG**, some examples of Excel VBA code based on the formulae developed for the subject-matter hand will be provided. You can use them in your own applications *as is* or as starting points. For the younger generation of readers, these code snippets (or formulae and methods contained therein) can be building blocks or starting points to design their own "apps".

However, at some point, for serious work, calculations requisite for accurate and reliable engineering design analysis requires specialty software. This is especially true for transient (dynamic) analysis of the power plant for operability analysis and control system development. The software used for the latter task is extremely complex with a steep learning curve (and very expensive, of course), and it requires a huge amount of man-hours to develop and execute fully functional power plant models. It is no exaggeration to state that some people spend an entire career on such programs and associated tasks. Unfortunately, there are no shortcut calculation methods to replace them (which require solution of combinations of partial differential equations) and even simplified approaches to estimate their outcome are quite complicated and require a deep engineering knowledge of the underlying phenomena.

Software for steady-state performance calculations is commonly known as "heat balance simulation tools". The name derives from the fact that the underlying fundamental principle is the first law of thermodynamics (also known as conservation of energy) along with mass continuity (also known as conservation of mass). These two laws of conservation are applied to the individual pieces of equipment comprising the power generation system in question. In the end, a "balance" is established between the two forms of energy transfer, work and heat, and mass transfer across the entire system (i.e., the "control volume").

[1] There are many new computer languages out there. Whatever works for you is just fine.

There are several steady-state, heat balance simulation tools widely used in the industry:

1. **GateCycle**. Gate, as it is commonly known, is a PC-based simulation software with a GUI that allows the user to build a system by interconnecting individual components, which are available as icons (e.g., compressor, combustor, expander). In 2000, Enter Software was fully acquired by General Electric. Since then, however, marketing and support activities for Gate continuously dropped until 2013 or 2014, when, for practical purposes, GE Enter all but ceased to exit. Some readers might be familiar with Gate and have access to its legacy (and functional) copies through their organizations.
2. **THERMOFLEX®**. A general-purpose tool, which allows the user to construct any real or conceptual power plant or thermal network model by connecting appropriate building blocks in a flexible and unconstrained manner. This was developed by Thermoflow, Inc. in 1995 and is part of Thermoflow's wider suite of power plant products.
3. **EBSILON® Professional** (developed by Steag Energy Services GmbH)
4. **AxCYCLE™** developed by the **SoftInWay, Inc.**
5. **IPSEpro** (offered by SimTech in Austria)
6. **PEPSE** (now offered by Curtiss-Wright Nuclear Division)

GateCycle, THERMOFLEX, EBSILON, PEPSE, all belong to a specialized family of engineering software commonly known as "heat balance programs" or "flow-sheet simulators", which are widely used in chemical process industry (CPI). Some of these tools, e.g., **HYSYS** (formerly known as **HYSIM**), **ASPEN** (at the time of writing, the company owning it is the owner of HYSYS as well) and **PRO/II**, are also used by engineers to develop fossil fuel-fired power plant performance modeling (especially for Integrated Gasification Combined Cycle (IGCC) applications, which require simulation of chemical process components). HYSYS and ASPEN also have transient (i.e., dynamic or unsteady-state) simulation capability[2].

In addition to the flow-sheet simulators mentioned above, there is Thermoflow's suite of dedicated Expert System programs for power plant design and simulation. This includes **GT PRO®** & **GT MASTER®** for combined cycle design and simulation, **STEAM PRO®** & **STEAM MASTER®** for conventional Rankine Cycle power plant design and simulation and their optionally integrated **PEACE®** module which creates preliminary engineering details, cost estimates and economic performance models. The PEACE module can also support THERMOFLEX.

For transient simulation and off-design, unsteady-state performance calculations, in addition to HYSYS, commercially available software products include **ProTRAX** (formerly known as **PC-TRAX**) and **EASY5**. ProTRAX was originally developed as a power plant operator training simulator. Basically, the software creates a "digital twin" of the entire fossil fuel-fired power plant, which is then run in real time with all pertinent controls and operator screens to create real-life scenarios. Thus, operators can be trained in all features and procedures of the power plant they will be working in. (In fact, first dynamic simulators were developed for safety analysis of nuclear power plants almost 50 years ago.) Such dynamic simulators are indispensable for plant controls design, testing and commissioning especially when they are combined with the actual control system hardware and software including operator consoles (e.g., General Electric's Mark series systems). Examples of other commercially available dynamic simulation software are Simcenter Amesim (now part of Siemens PLM Software), Apros (by Fortum and VTT Technical Research Centre of Finland Ltd. for modeling and dynamic simulation of power plants), ISAAC Dynamics (a dynamic simulation software by the TransientGroup – formerly Struttura Informatica – in Florence, Italy), 3KEYMASTER (modeling and simulation platform by WSC, Inc. in Maryland, USA) and Modelica (developed by the non-profit Modelica Association and is free).

[2] Note that this brief introduction is not intended as a commercial for particular software programs. In addition to the software tools discussed herein, *with which the author had hands-on experience*, there may be other commercially available software packages for steady-state and/or transient simulation, which the author has not been aware of.

There is also Simulink by Mathworks, developer of MATLAB®, which can be used for dynamic simulation of gas turbine power plants and controls. Similar to MATLAB, Simulink is widely used by the new generation of engineering students and practitioners and might come in handy especially in doing gas turbine transient calculations described in Chapter 15.

As mentioned earlier, dynamic simulation software is extremely complex and expensive and requires significant amount of training for an engineer to be well versed in its use. While their steady-state counterparts (especially Thermoflow software) are comparatively easier to use, they are also very expensive (initial purchase *and* annual maintenance fees for upgrades and customer support can run to tens of thousands of dollars) and unlikely to be affordable by individuals. In any event, all software tools require a certain experience level achievable via a sometimes steep and lengthy learning curve. While they make our lives easier and increase our productivity, they hide the fundamental principles that are at work from us – sometimes with disastrous results (i.e., the famous GIGO: garbage in, garbage out).

The material in this book is intended to make the inner works of such software tools "transparent" to the reader, who may already be a user of them. (Thermoflow software, especially GT PRO/MASTER and THERMOFLEX will be used in some of the coverage.) By the same token, methodology and fundamental principles, expressed as simple physics-based equations, introduced in the following chapters should provide a companion to or an alternative for carrying out certain *basic* design and analysis tasks.

2.3 CODES AND STANDARDS

Major combined cycle equipment, i.e., the prime movers (gas and steam turbines), synchronous ac generators, HRSGs and heat rejection system components (condenser, cooling tower) are designed, manufactured and tested to the original equipment manufacturer (OEM) drawings, procedures and processes that are in general compliance with applicable sections of the codes and standards in effect on the proposal date and enforced by respective organizations in the project site (e.g., European Union, USA).

Plant performance tests are performed to demonstrate that the plant meets the guaranteed performance (primarily power output and efficiency or heat rate) offered by the EPC Contractor to the Plant Owner. **American Society of Mechanical Engineers** (ASME) **Power Test Codes** (PTC) have been developed to provide guidance on how to conduct power plant performance testing. The codes provide comprehensive, practical information on calculations pertaining to major fossil fuel-fired power plant equipment and, as such, constitute very useful resources for information on the core subject matter of the chapter. Test codes most relevant to the reader of this book are listed below.

1. PTC 22 – Gas Turbines
2. PTC 4.4 – HRSGs
3. PTC 46 – Overall Plant Performance
4. PTC 47-2006 – Integrated Gasification Combined Cycle Power Generation Plants
5. PTC 6.2 – Steam Turbines (Combined Cycle)

Obviously, one can add to the list the **International Organization for Standardization** (ISO) standards (in all likelihood more familiar to readers based in Europe).

First published in 1914, the **ASME Boiler and Pressure Vessel Code**, especially its sections on the boiler and pressure vessel sections, is an indispensable resource in the USA as well as global electric power generation industry (in addition to other industries). A related and very important resource for pipe material selection is ASME B31.1 Power Piping Code.

The European equivalent of the ASME Boiler and Pressure Vessel Code is the **Pressure Equipment Directive** (PED) 2014/68/EU (formerly 97/23/EC) of the European Union (EU), which sets out the standards for the design and fabrication of pressure equipment (e.g., steam boilers, pressure vessels, piping, safety valves and other components and assemblies subject to pressure loading) generally over 1 L in volume and having a maximum pressure more than 0.5 bar (about 7.3 psi) gauge. The PED has been mandatory throughout the EU since May 30, 2002, with 2014 revision fully effective as of July 19, 2016.

Finally, **VGB PowerTech e.V.** is an international (voluntary) technical association for generation and storage of power and heat based in Germany. The organization brings together companies and supports their generation and internal utilization of electricity, heat and by-products resulting therefrom based on the most up-to-date technologies. Their publications and the *VGB PowerTech Journal* constitute a significant resource for engineers and researchers alike. The reader is strongly encouraged to consult their website and the lists of publications therein (many available in both German and English) when looking for hard-to-find technical information (www.vgb.org).

A selection of US codes and standards governing equipment design, specification and manufacturing are listed below. Note that, in addition to Europe or the USA, major equipment may be sourced from other regions, such as Asia (i.e., China, South Korea, Indonesia) or Middle East, using materials manufactured and graded to other material standards such as Chinese Q235/ Q345 carbon steel (which is used in place of A36/A572) or DIN, BS, JIS standards. The OEM designs the components to the applicable code based on the characteristics of the material being used. If the component is designed and certified to a specific code or standard (i.e., pressure parts, primary structural components) then the material specification used will only be those accepted by the applicable code. These standards are equivalent in intent to the design and operability requirements of equipment used in the USA.

- Major equipment
  - AGMA (American Gear Manufacturers Association)
  - ANSI (American National Standards Institute) B133.2 Basic Gas Turbine
  - ANSI C50.10 General Requirements for Synchronous Machines
  - ANSI C50.13 Requirements for Cylindrical Rotor Synchronous Generators
  - ANSI C50.14 Requirements for Gas Turbine Driven Cylindrical Rotor
  - Synchronous Generators
  - ANSI/NEMA (National Electrical Manufacturers Association) MG1 Motors and Generators
  - ANSI/NEMA MG2 Safety Standard for Construction and Guide for Selection
  - Installation and Use of Electric Motors and Generators
  - ASME Boiler and Pressure Vessel Code
  - IEEE Std. 421 IEEE Standard Criteria and Definitions for Excitation Systems for Synchronous Machines
  - ISO 7919-2 Vibration Standards for Large Rotating Apparatus (Steam Turbine)
  - ISO 7919-4 Vibration Standards for Large Rotating Apparatus (Gas Turbine)
- Industry standards
  - ABMA (American Boiler Manufacturers Association)
  - AFBMA (Anti-Friction Bearing Manufacture's Association)
  - ASNT (American Society for Nondestructive Testing)
  - ASTM (American Society for Testing and Materials)
  - EJMA (Expansion Joint Manufacturer's Association)
  - HEI (Heat Exchange Institute)
  - HI (Hydraulic Institute Standards)
  - TEMA (Tubular Exchanger Manufacturers Association)
  - IES Lighting Handbook
- Protection and control
  - ANSI B133.4 Gas Turbine Controls and Protection Systems
  - NFPA (National Fire Protection Association) 12 CO_2 Extinguishing System
  - NFPA 850 Recommended Practices for Fire Protection for Fossil Fueled Steam and Gas Turbine Electric Generator Plants
  - NFPA 2001 Standard on Clean Agent Fire Extinguishment Systems

- Electrical equipment
  - ANSI B133.5 Gas Turbine Electrical Equipment
  - ANSI C2 National Electrical Safety Code
  - ANSI C37.20 American Standard for Switchgear Assemblies Including Metal-Enclosed Bus
  - ANSI C37.90 Relays and Relay Systems Associated with Electric Power Apparatus
  - ANSI C57.13 Standard Requirements for Instrument Transformers
  - ANSI/IEEE C37.2 Electrical Power System Device Function Numbers
  - ANSI/IEEE 100 IEEE Standard Dictionary of Electrical and Electronic Terms
  - ANSI/NFPA 70 National Electrical Code
  - FM Factory Mutual (on selected devices – not complete assemblies)
  - IEEE 1 Standard General Principles for Temperature Limits in the Rating of Electrical Equipment
  - IEEE 32 Standard Requirements, Terminology and Test Procedures for Neutral Grounding Devices
  - IEEE 112 Test Procedures for Polyphase Motors and Generators
  - IEEE 115 Test Procedures for Synchronous Machines
  - IEEE 450 Recommended Practices for Maintenance, Testing and Replacement of Large Lead-Acid Storage Batteries for Generators
  - IEEE 484 Recommended Practices for Installation Design and Installation of Large Lead-Acid Storage Batteries for Generating Stations and Substations
  - NEMA AB1 Molded Case Circuit Breakers and Molded Case Switches
  - NEMA FU1 Low Voltage Cartridge Fuses
  - NEMA PE5 Constant-Potential-Type Electric Utility (Semiconductor Static Converter) Battery Chargers
  - NEMA ST20 Dry Type Transformers for General Application
  - UL Underwriters Laboratory (UL labels on selected devices – not completely assembled electrical enclosures)
- Enclosures, structures and acoustics
  - AISC Specification for Design, Fabrication and Erection of Structural Steel for Buildings
  - AISI American Iron and Steel Institute
  - ANSI B133.8 Gas Turbine Installation and Sound Emission
  - ANSI S1.4 Sound Level Standards and Specifications for Sound Level Meters
  - ANSI S1.13 Method for Measurement of Sound Pressure Levels
  - ANSI S6.1 Qualifying a Sound Data Acquisition System
  - ANSI S12.36 Survey Methods for the Determination of the Sound Power Levels of Noise Sources
  - AWS D1.1 Structural Welding Code – Steel
  - ISO 6190 Acoustics – Measurement of Sound Pressure Levels of Gas Turbine Installations for Evaluating Environmental Noise – Survey Method
  - NEMA IS1.1 Enclosures for Industrial Controls and Systems
  - NEMA PB1 Panel Boards
  - NEMA 250 Enclosures for Electrical Equipment
  - OSHA Federal Occupational Safety and Health Administration Standards
  - IBC International Building Code
  - ASTM American Society for Testing and Materials
- Quality
  - ISO 9001 Quality Systems – Model for Quality Assurance in Design/Development, Production, Installation and Servicing.

International standards include

- EN (Europe) European Standards
- JIS (Japan) Japanese Industrial Standards
- BS (UK) British Standards
- GB (China) Chinese National Standards
- DIN German Standards.

It is imperative to be familiar with applicable standards especially when installing equipment from OEMs, whose home countries have significantly different norms than the US ones. For example, the wall thickness of JIS pipe does not match ANSI pipe of the same diameter. A transition piece is thus required wherever piping connecting equipment manufactured in Japan to the rest of the plant is field-welded. (Flanged connections can be made to match.)

REFERENCES

1. Zel'dovich, Y.B., Yaglom, I.M., 1987, *Higher Math for Beginners* (Mostly Physicists and Engineers), Prentice-Hall, Inc., Englewood Cliffs, NJ.
2. Moran, M.J., Shapiro, H.N., 1988, *Fundamentals of Engineering Thermodynamics*, John Wiley & Sons, Inc., New York.
3. Zemansky, M.W., 1957, *Heat and Thermodynamics*, 4th Edition, McGraw-Hill Book Company, Inc., New York.
4. Incropera, F.P., Dewitt, D.P., 2002, *Introduction to Heat Transfer*, 4th Edition, John Wiley & Sons, Inc., New York.
5. White, F.M., 1979, *Fluid Mechanics*, McGraw-Hill Book Company, Inc., New York.
6. Landau, L.D., Lifshitz, E.M., 1987, *Fluid Mechanics*, 2nd Edition, Pergamon Press, Oxford, UK.
7. Liepmann, H.W., Roshko, A., 1957, *Elements of Gasdynamics*, John Wiley & Sons, New York.
8. Thompson, P.A., 1988, *Compressible Fluid Dynamics* (Advanced Engineering Series), McGraw-Hill Book Company, Inc., New York.
9. Lindeburg, M.R., 1998, *Mechanical Engineering Reference Manual for the PE Exam*, 10th Edition, Professional Publications, Inc., Belmont, CA.
10. Press, W.H., Teukolsky, S.A., Vetterling, W.T., Flannery, B.P., 1992, *Numerical Recipes in FORTRAN: The Art of Scientific Computing*, 2nd Edition, Cambridge University Press, New York.
11. Gülen, S.C., 2019, *Gas Turbines for Electric Power Generation*, Cambridge University Press, New York.
12. Saravanamuttoo, H.I.H., Rogers, G.F.C., Cohen, H., Straznicky, P.V., 2009, *Gas Turbine Theory*, 6th Edition, Pearson Education Limited, Harlow, Essex.
13. Lechner, C., Seume, J., 2010, *Stationäre Gasturbinen*, 2. neue bearbeitete Auflage, Springer, Heidelberg.
14. Jansohn, P. (Editor), 2013, *Modern Gas Turbine Systems: High Efficiency, Low Emission, Fuel Flexible Power Generation* (Woodhead Publishing Series in Energy), 1st Edition, Woodhead Publishing Ltd, Cambridge, UK.
15. Leyzerovich, A., 1997, *Large Steam Turbines: Design and Operation* (*2 Volumes*), PennWell Books, Tulsa, OK.
16. Sanders, W.P., 2004, *Turbine Steam Path: Maintenance & Repair* (*Volumes I and II*), *Mechanical Design and Manufacturing* (*Volume III*), PennWell Books, Tulsa, OK.
17. Cotton, K.C., 1998, *Evaluating and Improving Steam Turbine Performance*, 2nd Edition, Cotton Fact Inc., Rexford, NY.
18. Shlyakin, P., 2005 (Reprint), *Steam Turbines – Theory and Design*, Translated from Russian by A. Jaganmohan, University Press of the Pacific, Honolulu, HI.

3 Bare Necessities

This chapter is analogous to the 30-min warm-up before a basketball game. Key terminology, concepts and selected, basic estimation methods (charts, transfer functions, etc.) are covered in a single narrative so that the reader is "acclimated" to the subject matter at hand. It is highly recommended that you read this chapter in one sitting when you pick up this book for the first time. It is the author's hope that even those readers who spent decades in the field can find a few nuggets in the following paragraphs.

It should be noted that the name "combined cycle" is a generic term; it can refer to *any* two or three (or more – at least in theory) thermodynamic cycles connected to each other through their heat rejection and heat addition processes. A gas turbine combined cycle (GTCC) is a specific type of combined cycle comprising

- Gas turbine Brayton cycle
- Steam turbine Rankine cycle

Due to their respective positions in the temperature–entropy or T–s diagram (see the discussion below), the former is referred to as the "topping cycle"; the latter is referred to as the "bottoming cycle".

Strictly speaking, the bottoming cycle working fluid can be any pure fluid (or a *mixture* of pure fluids) other than steam (e.g., H_2O). Examples are

- Organic fluids such as pentane (organic Rankine cycle or ORC)
- Carbon dioxide (i.e., supercritical CO_2 cycle)
- Ammonia–water mixture (the Kalina cycle)

ORCs are available for "low-grade" waste heat recovery applications. Supercritical CO_2 Rankine cycles are also viable candidates for such applications. Neither are viable candidates for utility-scale (i.e., at least 500 MWe or more) electric power generation in combination with heavy-duty industrial gas turbines. Kalina cycle went through a period of serious research and development interest in the 1990s, but it did not present a serious challenge to the steam Rankine bottoming cycle due to myriad reasons. For general information on ORC and Kalina, refer to the chapter authored by Gülen and the references listed therein [1]. Ample material can be found online as well simply by googling them. Therefore, the focus in this book is specifically on the steam Rankine bottoming cycle. Without an explanatory moniker, the terms combined cycle and bottoming cycle refer to gas–steam turbine combined cycle and steam Rankine cycle, respectively.

3.1 WHY COMBINED CYCLE?

Gas and steam turbine combined cycle power plant for electric generation is based on two prime movers, which are heat engines whose operations are described by two fundamental thermodynamic cycles:

1. Gas turbine Brayton (or Joule) cycle
2. Steam turbine Rankine cycle

Gas and steam turbines are *prime movers*; they generate shaft power. Shaft power generated by the prime mover is converted into electric power by a synchronous alternating current (ac) generator.

Therefore, the correct terminology for the power generation equipment in a combined cycle power plant is "gas turbine generator" (GTG) and "steam turbine generator" (STG).

Brayton cycle is an *open* cycle; the working fluid mass flow rate and composition change from the compressor inlet (air) to the turbine exhaust (combustion products) whereas its phase, gas, does not change. It is the cycle governing the operation of an internal combustion engine.

Rankine cycle is a *closed* cycle; the working fluid mass flow rate and composition do not change. (In almost all cases of commercial interest, the working fluid is H_2O.) Its phase, however, *does* change – from water (liquid) to steam (gas). It is the cycle governing the operation of an external combustion engine.

There is a special class of gas turbine that operates in a *closed* Brayton cycle. However, at present, it is not used for electric power generation in a commercial power plant. Due to its unique nature (typically low cycle pressure ratio (PR) and low turbine inlet and exhaust temperatures), it is highly unlikely to be commercially deployed in a combined cycle configuration any time soon – if ever. At the time of writing, the focus on the closed Brayton cycle is within the context of supercritical CO_2 working fluid. Another niche for the closed cycle gas turbine is in future nuclear reactor power plant applications. For a full history of closed cycle gas turbines, the reader is referred to the monograph by Frutschi [2]. For supercritical CO_2 cycles, the most comprehensive source is the handbook edited by Brun et al. [3].

Basic thermodynamic concepts leading to the gas and steam turbine combined cycle are enumerated below:

- The most efficient heat engine cycle is the *Carnot cycle* operating between
  - A hot-temperature *reservoir*[1]
  - A low-temperature reservoir
- Hot and cold working fluid temperatures of a heat engine operating in a Carnot cycle are constant (i.e., they represent *isothermal* heat addition and rejection)
- The same may or may not be the case for a "real" (but still ideal) heat engine cycle
- A "real" (but still ideal) heat engine cycle can be compared to an *equivalent* Carnot cycle via its
  - Mean-effective heat addition temperature (METH)
  - Mean-effective heat rejection temperature (METL)
- METH is a proxy for the hot reservoir temperature
- METL is a proxy for the cold-reservoir temperature

In essence, a gas–steam turbine combined cycle plant is an attempt to "mimic" a Carnot cycle operating between two temperatures (see Figure 3.1):

- Hot reservoir temperature (i.e., gas turbine, turbine inlet temperature, TIT)
- Cold-reservoir temperature (i.e., the ambient temperature)

It is practically impossible (so far) to devise a machine with isothermal heat addition at a meaningfully high temperature. Thus, a gas turbine with a given cycle/compression PR and TIT cannot do better than a "Carnot-like" cycle with a hot reservoir temperature of METH (see Figure 3.1). The deviation from the Carnot ideal on the hot end is thus quantified by the upper-left triangular area on the T–s diagram.

The situation is much better for the cold end, where the deviation from the Carnot ideal is quantified by the lower-right triangle. Even a single-pressure, non-reheat steam Rankine cycle takes care of a big chunk of that problem. The key to that is constant pressure–temperature heat rejection via steam condensation (denoted by METL in Figure 3.1), which does a very good job of approximating the "true" cold-reservoir temperature.

---

[1] The term "reservoir" is the short-hand for thermal energy reservoir, which is a (hypothetical) thermodynamic system with infinitely large heat capacity so that, when it is in thermal contact with another system or its environment, its temperature does not change.

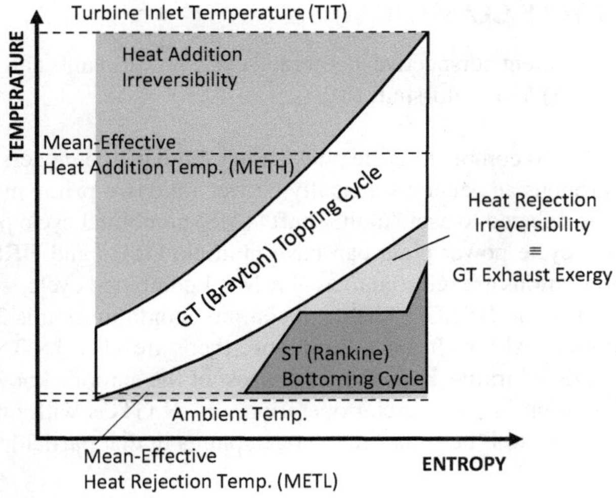

FIGURE 3.1 Temperature–entropy (T–s) diagram of GTCC.

Quantitatively, the lower-right triangle in Figure 3.1 is *exactly* equal to the gas turbine exhaust *exergy*, which is a thermodynamic property and, as such, can be calculated *exactly* with known gas composition, pressure, temperature and mass flow rate using an appropriate equation of state. A *perfect* bottoming cycle (i.e., a Carnot engine in and of itself) would convert that exergy into useful work fully and have an exergetic efficiency of 100%.

The topping (gas turbine) and the bottoming (steam turbine) cycles are connected to each other via a specially designed heat exchanger, which utilizes the exhaust gas of the gas turbine to convert the condensate coming from the steam turbine condenser to steam. It is referred to as a *heat recovery steam generator* (HRSG). Another term, less commonly used, is *heat recovery boiler* (HRB), which in this author's opinion is the more apt term because the term "generator" should really be exclusively reserved for the synchronous ac generator. In any event, HRSG or HRB is a special class of "waste heat recovery" boiler (WHRB). Note that

- The HRSG is the cycle "heat adder" for the steam Rankine (bottoming) cycle
- The gas turbine combustor is the cycle heat adder for the gas turbine Brayton cycle

The HRSG can be "fired" or "unfired". In the latter case, the only heat source for steam production is the gas turbine exhaust. In the former case, remaining oxygen (O_2) in the gas turbine exhaust gas, about 10%–12%, is utilized for a second combustion in a "duct burner". In those instances, the HRSG is referred to as "supplementary-fired" or "duct-fired". The terminology stems from the fact that, in the earlier applications, the burners were placed in the transition duct between the gas turbine exhaust and the HRSG inlet. Duct firing will be discussed in more detail in Chapter 6.

There are three major options for GTCC power plant "heat sink":

1. Once through (open loop) water-cooled condenser
2. Closed-loop, water-cooled condenser with a
   a. Mechanical draft cooling tower
   b. Natural draft cooling tower
3. Air-cooled "dry" condenser

There are other, less frequently encountered heat sink options such as the Heller unit. They will be discussed in Chapter 7.

3.2 COMBINED CYCLE CLASSIFICATION

From a powertrain arrangement perspective, combined cycle power plants can be classified into two categories: single-shaft (SS) and multi-shaft (MS).

- In a "single-shaft" (SS) combined cycle power plant, both prime movers can be connected to a single synchronous ac generator (usually between the two prime movers). Otherwise, the arrangement is referred to as a "multi-shaft" (MS) combined cycle power plant.
- An MS combined cycle power plant can have multiple GTGs and HRSGs with a single STG. Such configurations are referred to as N × N × 1 combined cycle, where N designates the number of GTGs and HRSGs. The most common configuration is 2 × 2 × 1 with two GTG-HRSG "trains". Although not as common, there are also 3 × 3 × 1 configurations with three GTG–HRSG trains. However, to the best of the author's knowledge, there is no N × 1 × 1 configuration in commercial operation, i.e., N GTGs with one HRSG and one STG. (There may very well be some small cogen plants in that particular arrangement.)

From a bottoming steam cycle perspective, combined cycle classification is made based on the number of the HRSG pressure levels and whether there is a *reheat* feature or not. In particular,

- State-of-the-art GTCC power plants with advanced class, heavy-duty industrial gas turbines have *three-pressure reheat* (3PRH) HRSG and steam cycle.
- The next most common variant is *two-pressure reheat* (2PRH). Small GTCC power plants with aeroderivative or small industrial GTGs have two-pressure no reheat (2PNR) or *one-pressure no reheat* (1PNR) steam cycles.
- The other two options are *three-pressure no reheat* (3PNR) and *one-pressure reheat* (1PRH); the former may be available in vintage power plants still in operation. The latter is probably only of academic interest at this point.

In a three-pressure HRSG, steam is generated at three pressure levels: high, intermediate (or "medium" in some references) and low, i.e., HP, IP and LP, respectively. In two-pressure HRSGs, there is HP and LP steam production.

Reheat refers to the rerouting of the steam exiting the HP turbine to the HRSG to bring up its temperature back to the level at the HP turbine inlet in a "reheat superheater". The exhaust steam from the HP turbine is referred to as "cold reheat" steam (CRH). The reheated steam at the exit of the reheat superheater, which is to be readmitted at the IP turbine inlet, is referred to as "hot reheat" steam (HRH).

The steam turbine has three sections, which may be in three different casings or in two casings, which are referred to as HP, IP and LP sections of turbines. The most common arrangement is the two-casing one, with one casing housing the HP and IP turbines in "opposed flow" configuration and the other casing housing the LP turbine, which generates roughly 50% of the total steam turbine output. Depending on the amount of steam flow, the LP turbine can be single-flow axial exhaust or double-flow down or side exhaust. All these aspects of steam turbine configuration will be covered in detail in Chapter 5.

A schematic diagram of a 3PRH 1 × 1 × 1 MS GTCC with a closed-loop water-cooled condenser and natural draft cooling tower is shown in Figure 3.2. The diagram identifies all the significant elements of a GTCC with the exception of boiler feedwater pumps and the gas turbine fuel gas performance heater:

1. Gas turbine
   a. Air inlet (flow rate, composition and temperature)
   b. Fuel flow rate (composition, flow rate and heating value)
   c. Exhaust gas (flow rate, composition and temperature)
   d. Generator output

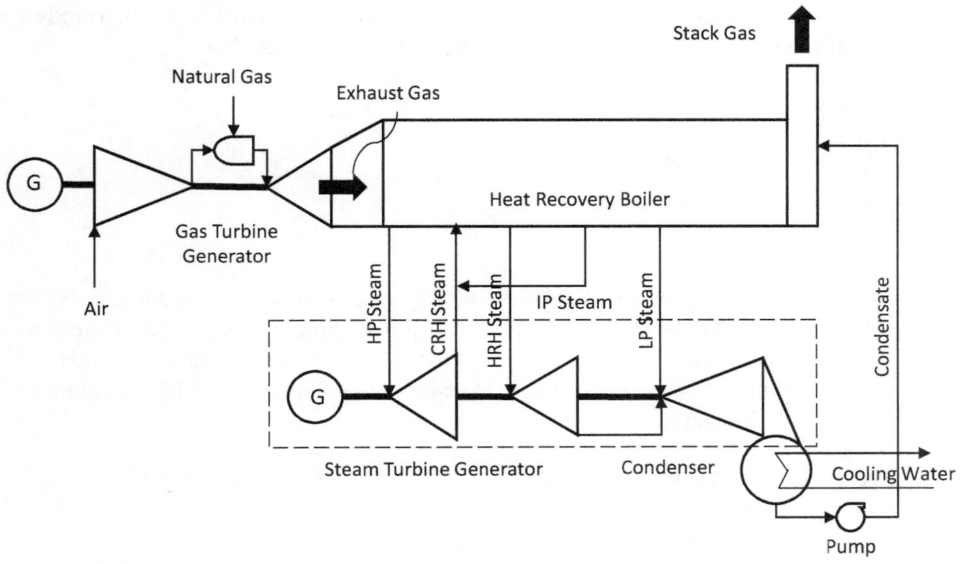

FIGURE 3.2 Generic GTCC diagram with 3PRH steam bottoming cycle.

2. HRSG
   a. Stack gas (temperature)
   b. HP, IP and LP steam (flow, pressure and temperature)
   c. Condensate inlet (flow, pressure and temperature)
3. Steam turbine
   a. Main (HP) admission steam (flow, pressure and temperature)
   b. CRH steam (flow, pressure and temperature)
   c. HRH steam (flow, pressure and temperature)
   d. LP admission steam (flow, pressure and temperature)
   e. LP exhaust steam (flow, pressure and temperature)
   f. Generator output
4. Condenser
   a. Steam pressure (temperature)
   b. Cooling water (flow rate and temperatures)

3.3 SIMPLE CALCULATIONS

Note that steam pressure *and* temperature are needed so that one can calculate steam enthalpy from the *ASME Steam Tables*. For air, fuel and exhaust gas, composition and temperature are enough to calculate the enthalpy and fuel LHV (lower heating value) using JANAF tables[2] and fuel gas component LHV data. Composition of gas turbine inlet air is found from the composition of dry air and the ambient humidity. Exhaust gas composition is found from the combustion reaction calculation. With known enthalpies, gas and STG outputs can be calculated via "control volume" heat and mass balance analysis of GTG, STG and HRSG. For rigorous design and off-design performance analysis, this can only be done using a heat and mass balance simulation software such as Thermoflow's GT PRO®/GT MASTER®. However, rather accurate estimates can be easily obtained by using first

[2] The JANAF (Joint Army Navy Air Force) thermochemical tables are compilations of the thermodynamic properties of substances over a wide temperature range. They are available from the American Chemical Society: 2nd Edition (1971); 1974 supplement; 1975 supplement; 1978 supplement.

principles and a few judicious rules of thumb from first law or second law (of thermodynamics) perspectives. All these aspects will be covered in the rest of the current book.

There are two types of quantitative analyses of interest:

1. Performance
   a. Design
   b. Off-design
2. Operability

In performance analysis, the primary goal is calculating net electric power output and thermal efficiency (or, equivalently, heat rate or HR) of the GTCC power plant at design and off-design conditions. The former is obvious; it is the *raison d'être* of the book you are reading. *Thermal efficiency* is the ratio of power output (net or gross, see below) to *heat (fuel) consumption*. Its importance cannot be overstated due to its impact on

- Conservation of limited fuel resources
- Minimization of operating costs
- Reduction of criteria pollutants and CO_2 emissions.

Another key performance parameter is heat consumption, which is the product of fuel mass flow rate into the combustor of the gas turbine (plus fuel supplied into the duct burners of the HRSG in a combined cycle power plant, if applicable) and fuel LHV. For example, a gas turbine burning 100% CH_4 fuel at a rate of 30 lb/s has a heat consumption of 30 × 21,515 = 645,450 Btu/s, which is equal to about 681 MWth. If the gas turbine in question generates 275 MWe of power, its efficiency is 275/681 = 40.4%.

3.3.1 Design Performance

A most basic GTCC performance calculation at a given design point (usually, ISO baseload) requires a knowledge of

1. Gas turbine efficiency
2. HRSG effectiveness
3. Steam cycle (caution: *not* the steam turbine) efficiency

For state-of-the-art, advanced class gas turbines, gas and steam turbine efficiencies are in low 40s (as a percentage). The effectiveness of a 3PRH HRSG is around 90%. Even this rudimentary information is adequate for generating useful information from bits and pieces of data (as will be demonstrated later). However, let us provide a "sneak preview" with a basic example. What is the estimated ISO baseload combined cycle performance of a gas turbine rated at 300 MWe and 40% efficiency? Here is a step-by-step calculation:

1. 300/0.4 = 750 MWth heat consumption
2. 750 × (1 − 0.4) = 450 MWth exhaust gas energy
3. 450 × 0.9 = 405 MWth recovered in the HRSG for steam generation
4. 405 × 0.4 = 162 MWe steam turbine output
5. 300 + 162 = 462 MWe gross CC output
6. 462 × (1 − 2%) = 453 MWe net CC output (2% auxiliary load)
7. 453/750 = 60.4% net CC efficiency.

Note that steam turbine-to-gas turbine output ratio is 162/300 = 0.54, which leads to a well-known rule of thumb that can be used for "conversational" estimates, i.e., in a typical combined cycle, steam turbine output is roughly half of that of the gas turbine(s). Much more accurate yet quite simple, physics-based methods will be introduced in the following chapters.

Another frequently asked question (and frequently provided "answer") pertains to the relationship between gas turbine efficiency and combined cycle efficiency: What is the impact of 1% change in gas turbine HR to the combined cycle HR? In order to answer this, let us translate the seven-step calculation above into a mathematical equation:

$$\eta_{CC} = \left(\eta_{GT} + (1-\eta_{GT})\eta_{HRSG}\eta_{ST}\right)(1-\alpha). \tag{3.1}$$

Taking the derivative

$$\frac{\partial \eta_{CC}}{\partial \eta_{GT}} = (1-\eta_{HRSG}\eta_{ST})(1-\alpha). \tag{3.2}$$

Using the values in the example

$$\frac{\partial \eta_{CC}}{\partial \eta_{GT}} = (1-0.9\cdot 0.4)(1-0.02) = 0.627.$$

In other words, each 1 percentage point in gas turbine efficiency is worth two-thirds of a percentage point in combined cycle efficiency.

Since $\eta = 3412/\text{HR}$, after a bit of algebra, we obtain

$$\frac{\Delta \text{HR}_{CC}}{\text{HR}_{CC}} = \frac{\Delta \text{HR}_{GT}}{\text{HR}_{GT}}(1-\eta_{HRSG}\eta_{ST})(1-\alpha)\frac{\eta_{GT}}{\eta_{CC}} \tag{3.3}$$

or

$$\frac{\Delta \text{HR}_{CC}}{\text{HR}_{CC}} = \frac{\Delta \text{HR}_{GT}}{\text{HR}_{GT}}\frac{\partial \eta_{CC}}{\partial \eta_{GT}}\frac{\eta_{GT}}{\eta_{CC}}. \tag{3.4}$$

Again, using the values in the example

$$\frac{\Delta \text{HR}_{CC}}{\text{HR}_{CC}} = 0.01\cdot 0.627\frac{0.4}{0.604} = 0.42\%.$$

Thus, as a rule of thumb, each 1% change in gas turbine HR is equal to 0.4% change in GTCC HR. (You can try the formula with values for a vintage F class GTCC; the answer is about the same.)

Taking the derivative of Equation 3.1 with respect to η_{ST}

$$\frac{\partial \eta_{CC,net}}{\partial \eta_{ST}} = (1-\eta_{GT})\eta_{HRSG}(1-\alpha). \tag{3.5}$$

Using the typical numbers from the previous example,

$$\frac{\partial \eta_{CC,net}}{\partial \eta_{ST}} = 0.531.$$

In other words, each 1 percentage point in steam turbine efficiency is worth about a half percentage point in combined cycle efficiency. In relative terms, assuming $\eta_{GT} \sim \eta_{ST}$

$$\frac{\partial \eta_{CC,net}}{\eta_{CC,net}} \cdot \frac{\eta_{ST}}{\partial \eta_{ST}} = \frac{1}{\left(\frac{1}{(1-\eta_{GT})\eta_{HRSG}} + 1 \right)}. \tag{3.6}$$

Once again, with the same typical numbers used earlier,

$$\frac{\partial \eta_{CC,net}}{\eta_{CC,net}} \cdot \frac{\eta_{ST}}{\partial \eta_{ST}} = 0.35.$$

This is another, widely used "rules of thumb" in combined cycle engineering. Each one percent (*not* percentage point) change in steam turbine efficiency (or HR) is worth about one-third of a percent change in combined cycle efficiency.

3.3.2 Off-Design Performance

Off-design calculations typically require rigorous simulation software (e.g., Thermoflow's GT MASTER®), which make use of hardware information such as component maps, pump curves and steam turbine exhaust loss curves. Nevertheless, one can still manage by judicious application of "correction curves". The recommended source for such curves is OEM-supplied user manuals and "thermal kits". In many cases, generic curves available in the literature can be used with reasonable accuracy. This will be discussed in more detail in the upcoming chapters.

For quick estimates, a typical part load curve that can be used reasonably accurately for state-of-the-art 3PRH systems with advanced class gas turbines is presented in Figure 3.3. The y-axis is combined cycle HR as a fraction of its value at 100% load (x = 1.0). Normalized efficiency can be found by using the inverse of the HR values (e.g., at 75% load, x = 0.75, normalized combined cycle efficiency is 1/1.042 = 0.96). The curve is applicable to 1 × 1 × 1 as well as multi-gas turbine

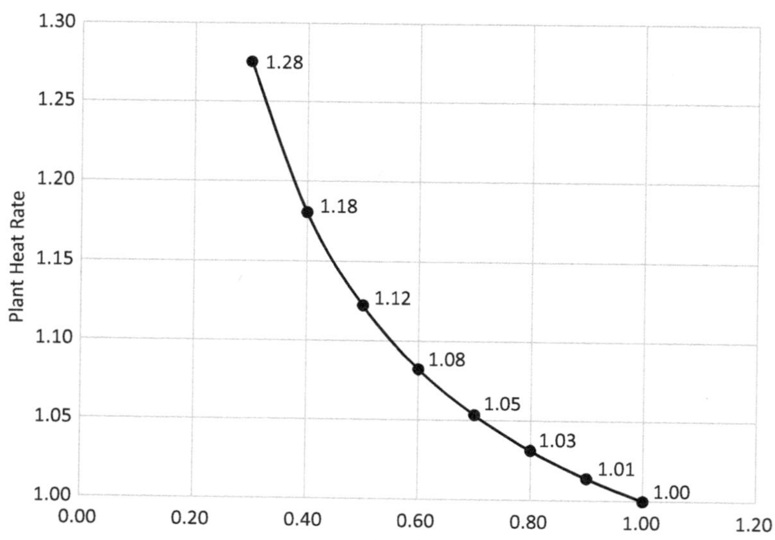

FIGURE 3.3 Typical GTCC part load curve (ISO conditions).

GTCCs when *all* gas turbines in the latter are loaded at the same level. (The data is from Figure 4 – *Frame 7 Automatic PL Curve* – in the paper by Gülen and Joseph [4].)

Obviously, in a 2 × 2 × 1 or 3 × 3 × 1 combined cycle, one can achieve very good part load HR by shutting down one gas turbine (or two gas turbines). Again, for a "quick and dirty" estimate, the curve in Figure 3.3 can be shifted to the left to start at 50% load (0.5 on the x-axis). For example, at 50% load, normalized combined cycle HR is 1.0. At 40% load, it is the value at x = 0.8 in Figure 3.3, which is 1.031 from the curve or using the following transfer function:

$$y = 0.8591 + \frac{19.85}{1 + \left(\frac{x}{0.004411}\right)^{0.911}} \tag{3.7}$$

where y is normalized combined cycle HR and x is combined cycle load as a fraction. In order to estimate combined cycle load as a function of gas turbine load, you can use the following transfer function

$$y = 0.8884 \cdot x + 0.1199 \tag{3.8}$$

where x is the gas turbine load and y is the combined cycle load.

Gas turbines are air-breathing engines with nearly constant volumetric flow because they run at constant rotational speed (3,000 or 3,600 rpm). An increase in ambient temperature and commensurate drop in air density causes significant lapse in gas turbine power output. Thus, performance correction with rising ambient temperature is of significant interest. (There are methods to mitigate the drop in power such as compressor inlet air cooling or HRSG duct firing, which will be covered later in the book.) For quick estimates, typical output and HR ambient lapse curves that can be used reasonably accurately for state-of-the-art 3PRH systems with advanced class gas turbines are presented in Figure 3.4. (The data is from the correction curves for 2 × 2 × 1 and 1 × 1 × 1 SCC6-5000F combined cycle with mechanical cooling tower-based heat rejection system [5].)

Note that the curves in Figure 3.4 are when the gas turbines (hence the GTCC) are run at "full load". This is a good place to take the opportunity to clarify the common confusion of the terms "base" and "full" load. First of all, the term *load* refers to a prime mover's synchronous ac generator output. When synchronized to the electric grid, the generator output is exactly balanced by the

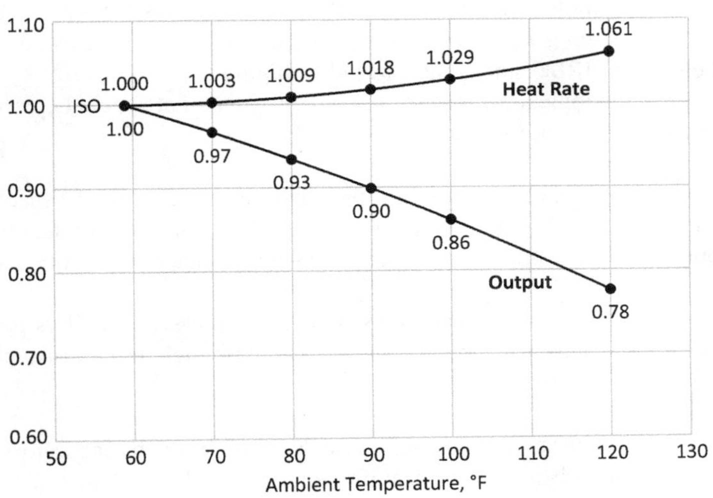

FIGURE 3.4 Typical GTCC ambient lapse curves (full load – ambient relative humidity is 60%).

electric load imposed by the grid – hence the term load. The term *full load* refers to 100% load. Strictly speaking, baseload refers to full load at particular site ambient conditions (e.g., ISO). Thus, the rating performance of a gas turbine at 100% load and ISO conditions can be referred to as "base-load" performance. In fact, typical "sticker performance" of a gas turbine in trade publications is referred to as "ISO baseload performance".

However, a gas turbine can run at full load (i.e., 100% load) at any given site ambient conditions, e.g., on a hot day at 95°F and 70% RH. In that case, when referenced to its ISO baseload, the gas turbine question output will be much less than 100%. Thus, specification of full load must be accompanied by the applicable site ambient conditions and it refers to the gas turbine loading with

- Inlet guide vanes (IGVs) at their nominal fully open position (i.e., nominal volume flow)
- Fuel flow per the specific control curve (i.e., not over – or underfired)

Therefore, whenever one comes across a gas turbine performance number quoted at "x percent" load, he or she *must* ascertain what exactly is meant by that "x percent".

From Figure 3.4, full load at ~80°F ambient (60% ambient RH) is 0.93. If the "ISO baseload rating" of a particular GTCC is listed as 500 MWe, at 80°F, its full load output would be $0.93 \times 500 = 465$ MWe. If it were to run at 70% load at 80°F ambient, its output would be $0.70 \times 465 = 325.5$ MWe. At this particular operating point, using the correction factors from Figures 3.3 and 3.4, its HR as a fraction of its ISO baseload rating value would be $1.009 \times 1.042 = 1.051378$. Its efficiency as a fraction of its ISO baseload rating value is $1/1.051378 = 0.95113$. If its ISO base-loading efficiency were 60%, its efficiency at 70% load at 80°F ambient would be $0.95113 \times 60 = 57.068$ or ~57%.

3.3.3 Lower or Higher Heating Value?

Note that HRs and efficiencies of gas turbines in simple and combined cycle are expressed on a LHV basis. In particular, natural gas, which is assumed to be 100% CH_4 (methane) has an LHV of 21,515 Btu/lb at 77°F (about 50 MJ/kg at 25°C). The *higher heating value* (HHV) is the "true" energy content of the fuel, which includes the latent heat of vaporization that is released by the gaseous H_2O in the combustion products when they are cooled to the room temperature. In other words, HHV is the value measured in a *calorimeter.* In a real application (e.g., a gas turbine), the combustion products, by the time they reach the exhaust gas stack (e.g., around 180°F (90°C) for a modern combined cycle power plant), are not cooled to a temperature to facilitate condensation, which, depending on the amount of H_2O vapor in the gas mixture, is around 110°F (~43°C). Thus, the latent heat of vaporization is *not* recovered and utilized. (It *can* be done by adding a condensing heat exchanger before the HRSG stack, but it would be highly uneconomical.)

This is why LHV is the logical measure of fuel energy input that should be used in engineering calculations. There are, however, two exceptions:

1. The efficiency of the IGCC power plant is usually (but not always) referenced to the HHV of the gasifier feedstock (typically, coal or petcoke). This goes back to the practice historically adopted by the US utilities when coal was the primary fossil fuel used in electric power generation.
2. The price of fuel in economic calculations is always expressed in dollars per million Btu (HHV). This is so because when plant operators purchase fuel, the contract is always based on the "true" energy content of the fuel.

Note that LHV is not measurable; it can only be calculated from the laboratory analysis of the fuel by subtracting the latent heat of vaporization. For 100% CH_4 (methane), the ratio of HHV to LHV is 1.109. Many handy formulae can be found in handbooks, e.g., in Marks or Dubbel cited in Chapter 2, to estimate LHV and HHV for various fuels and fuel gas compositions.

3.3.4 Gross or Net?

It is of utmost importance to specify whether a quoted efficiency or power output is on a "net" or "gross" basis. As far as this book is concerned,

- *gross* refers to the power measured at the low-voltage terminals of the gas and steam turbine generators and the sum total of them, which gives the GTCC gross output
- *net* refers to the power after subtracting all plant power consumers (e.g., pumps; compressors; lighting and heating, ventilating and air-conditioning or HVAC) from the gross
- The difference between the net and gross is commonly known as the *auxiliary load* or auxiliary power consumption
- Strictly speaking, auxiliary load must include the losses incurred in the step-up transformer because this is the power supplied by the power plant to the electric grid.

For a GTCC power plant, the auxiliary load can be anywhere between 1.6% (typical for *GTW Handbook* ratings) to more than 3% – mainly dictated by the steam turbine heat rejection system. The combined cycle power plant auxiliary load is a combination of

- Power consumption of major balance of plant (BOP) pumps and fans
- Power consumption of gas turbine and steam turbine auxiliary equipment (lube and hydraulic oil pumps, blowers, etc.)
- Power consumption of minor BOP pumps and compressors
- Miscellaneous plant power consumption (lighting, HVAC, etc.).

Auxiliary power consumption for 39 combined cycle power plants (65–860 MW) is extracted from data published in a trade publication and plotted in Figure 3.5. The data indicates that the auxiliary load ranges between 1.2% and 2.1% of the plant gross output. The average is 1.57% ± 0.24%. The curve-fit has a slope of 1.67%.

A breakdown of typical combined cycle plant auxiliary power consumption reveals that the boiler feedwater pumps account for 30%–40% of the total. The second biggest contributor is the heat rejection system, which accounts for 15%–25% of the total. Thus, the two major plant subsystems account for 50%–60% of the total.

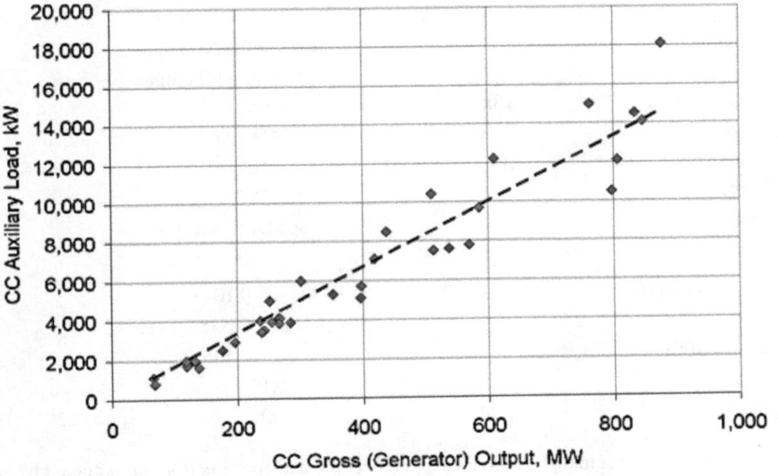

FIGURE 3.5 CC auxiliary load data from *Gas Turbine World Handbook 2007–2008*.

Major heat rejection system power consumers are the cooling water circulating pump and – if applicable – the cooling tower or air-cooled condenser forced draft fans. Estimation methodology for the power consumption of those items will be covered in Chapter 7. Other BOP equipment that consumes power includes myriad small pumps, compressors, miscellaneous plant HVAC and lighting. These items and generating equipment skids add up to a power consumption that is a small fraction of the combined cycle gross output (i.e., less than 0.5%). The exact value is very much dependent on site and customer requirements.

For quick and quite reliable estimation of combined cycle auxiliary power consumption, the reader is referred to the information in Table 3.1. For most practical purposes, using the total (as a fraction of combined cycle gross output) contribution is sufficiently accurate. Note that the numbers in Table 3.1 are ideal estimates to be used in conceptual studies. In an actual design exercise, based on requisite site ambient and loading conditions and economic criteria, a diligent cost-performance trade-off can result in higher percentages (e.g., as high as 3% of the plant gross output).

Rating performances listed in trade publications and commercial performance guarantees are on a "new and clean" basis. Even with good maintenance practices following OEM recommendations to the dot, plant equipment will degrade over time via normal wear and tear with resulting performance loss (lower output and efficiency). Performance loss due to compressor fouling can be (almost fully) recovered via online or offline "washes". Hot gas path degradation requires an overhaul of the machine (as prescribed by the OEM), which rarely results in full performance recovery. Typical degradation factors for a well-maintained GTCC are shown in Figure 3.6. A GTCC rated at 500 MWe (new and clean) is expected to lose 1.25% output after 5,000 h of fired operation (6,250 kWe). Its heat rate will increase by about 0.65% (or, equivalently, its efficiency will decrease by 0.65%).

TABLE 3.1
Standardized CC Power Plant Auxiliary Load Consumption

	Open Loop Water-Cooled (Cold)	Open Loop Water-Cooled (Normal)	Closed-Loop Water-Cooled	Air-Cooled
Ambient temperature		ISO		
Condenser pressure	1.00	1.20	1.50	2.50
Direct				
GT equipment auxiliaries		0.25% of GTG output		
ST equipment auxiliaries		0.10% of STG output		
Boiler feed pumps (% of STG)	1.90	1.95	1.90	2.00
Cooling tower fans (% of STG)	N/A	N/A	0.70	N/A
CW circ pump (% of STG)	0.95	1.10	1.00	N/A
ACC fans (% of STG)	N/A	N/A	N/A	1.50
Fuel compressor (PR > 20)		See Equation 10.4		
Indirect				
Auxiliary CW pump (% of STG)		0.10		
Other pumps		Number of GTGs × 175 kW		
Step-up transformer (% of CC)		0.50		
Miscellaneous (% of CC)		0.10		
Total (% of CC gross)	1.90	2.00	2.20	2.10

Note that these are derived from actual plant bid data and reflect the nameplate performance quoted by vendors of procured equipment such as pumps, fans and compressors.

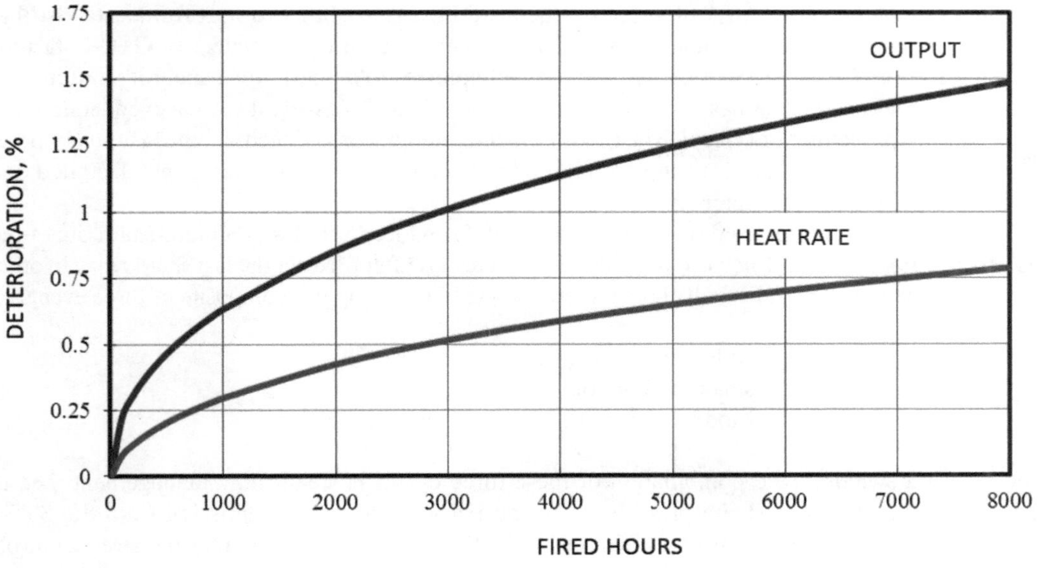

FIGURE 3.6 Typical GTCC degradation profiles.

Note that 0.65% decrease in efficiency due to degradation in the example above is on a *relative* basis. In other words, if the new and clean efficiency of the sample GTCC were 60%, after 5,000 h of operation, its efficiency would drop by $60 \times 0.0065 = 0.39$ "percentage points" to become $60 - 0.39 = 59.61\%$. Not paying attention to this distinction is a very common source of error.

3.4 OPERABILITY

In operability analysis, one is not after well defined "numbers" such as power output or efficiency *per se*. Key operability questions pertain to plant "events" such as

- Plant startup (hot, warm or cold depending on the length of layoff period)
- Plant shutdown (normal or trip)
- Load ramping (up or down)
- Minimum emissions-compliant turndown
- Response to grid events (e.g., frequency excursions).

Proper analysis of such events (and requisite equipment abilities) require sophisticated "dynamic" simulation models including control algorithms. Requisite software products are extremely expensive with steep learning curves. Building a proper model to develop plant controls from scratch can take man-years of investment. Only OEMs have the requisite resources and capability. Even models with much lower fidelity require significant amount of time and effort. Merely scratching the surface of such endeavors is well beyond the scope of the book (and the ability of the author). Nevertheless, a basic understanding of underlying physical fundamentals and ability to do simple numerical estimates is absolutely necessary so that one can have a firm understanding of operational problems. Chapter 15 will cover the pertinent details.

Combined cycle power plants are rarely operated as "baseload" units. On the other hand, they are too expensive to be deployed as "peakers". Thus, most modern GTCCs operate in a "cyclic duty" mode in between the two extremes. The exact nature of a GTCC's operating rhythm is dependent on the particular grid operator, the mix of generating assets on the grid and the existing demand.

One measure of the "flexibility" of a state-of-the-art GTCC is its ability to start from a standstill to its full load in the shortest amount of time. The "four-minute mile" equivalent of GTCC startups is "hot start" in 30 min (or less). OEMs develop technologies to achieve this capability and market them under trademarked names such as "Rapid Response" and "Flex-Plant". A basic understanding of the startup procedure of typical GTCC with modern equipment and technologies is thus imperative. Salient aspects of GTCC startup are highlighted below with a generic example. Detailed discussion can be found in Chapter 15.

Figure 3.7 shows normalized rotational speed (RPM) and load profiles of a multi-shaft, duct-fired GTCC in a "hot" startup, which is loosely defined as a restart after 8 h after the last shutdown. (In other words, the equipment metal is still "hot".) In essence, GTCC startup is a composite of three events:

1. Gas turbine startup and loading
2. HRSG warm-up and steam production
3. Steam turbine startup and loading via controlled steam admission.

The key to a synchronized combination of these three events is steam flow management, i.e., to strike a balance between steam production (in the HRSG) and steam admission (into the STG). The main actor behind the proverbial "screen" is HRSG HP drum and STG thermal stress control.

In terms of a prime mover startup, there are three key steps:

- Acceleration (roll-up) from turning gear speed (a few rpm) to synchronization speed (3,000 or 3,600 rpm)
- Speed match and synchronization to the grid – commonly referred to as *full speed, no load* (FSNL)
- Load ramp to 100% (full) load – commonly referred to as *full speed, full load* (FSFL).

In the case of a gas turbine, roll-up is started by "cranking" the unit with an external "driver" until the combustor is "lit up" and the turbine starts generating enough net power to accelerate the unit to the synchronization speed. In state-of-the-art, advanced class gas turbines, the external driver is the synchronous ac generator, which is run as a synchronous ac motor by the *Load Commutated*

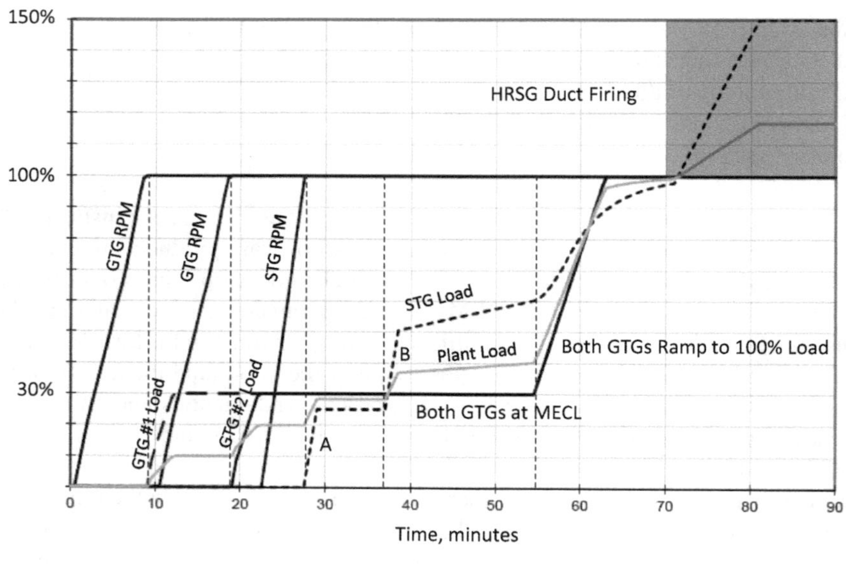

FIGURE 3.7 Typical GTCC hot startup (2 × 2 × 1 duct-fired).

Inverter (LCI). The steam turbine does not have an external driver. It is rolled by steam admitted into the HP or IP turbine in a controlled manner.

Let us look at the key components of the startup process in Figure 3.7:

1. The first (lead) gas turbine, GTG #1, starts and rolls up to full speed, synchronizes to the grid and loads up to its *minimum emissions-compliant load* (MECL), which in this case is 30% load.
2. Thereafter, the second (lag) gas turbine, GTG #2, starts and rolls up to full speed, synchronizes to the grid and loads up to its MECL.

   Modern gas turbines with dry-low-NOx (DLN) combustors cannot maintain their rated emissions (25 ppm typically) across the load range. In older vintage F class gas turbines, the load level below which NOx and CO emissions increased rapidly was 50%–60%. Advanced class machines can go to a lower load, e.g., 40% or 30%, while maintaining the rated NOx (and CO) emissions. In this example, the gas turbines are brought to the MECL and kept there for a while. The reason for that is to control the HRSG steam production at a level commensurate with STG admission requirements. Until the start of STG roll, all steam generated in the HRSG is bypassed into the condenser. Steam temperature is controlled by "terminal" attemperators via feedwater spray (extracted from the boiler feed pumps).
3. Steam turbine roll and acceleration starts via admission of HRH steam into the IP turbine.
4. Upon synchronization, the STG is ramped to 25% load by increasing the IP steam flow (labeled "A" in Figure 3.7).
5. After a wait period determined by the STG stress controller, HP steam admission starts and the STG is loaded to 50% load (labeled "B" in Figure 3.7).
6. With both GTGs at MECL, steam flow and temperature (via terminal attemperators) increase in a controlled manner until 60% load.
7. At that point, both GTGs start the load ramp to 100% load.
8. Steam production increases with increasing exhaust flow/energy as the GTGs are loaded.
9. STG load increases via admission of steam generated in the HRSG at a rate dictated by the heating rate of the HRSG metal.
10. Once the GTGs reach 100% load, HRSG duct burners kick in and the STG is loaded to 150% load with increased steam production in the HRSG.

The startup described above is not possible with the vintage F class gas turbines. Their MECL is too high so that steam generated in the HRSG with increased exhaust gas flow/energy is too much to bypass into the condenser. In older units with "conventional" startup procedure, HRSG steam production and temperature were controlled by the gas turbine exhaust energy, i.e., running the gas turbine at very low load (about 20%). This, of course, required a longer time for the HRSG to warm up. During that time, significant NO_X and CO were emitted through the stack. Considerably more stringent environmental requirements render obtaining an air permit with this type of obsolete technology practically impossible.

A typical daily operating profile of a combined cycle power plant is shown in Figure 3.8. This particular profile is sometimes described as the plant is "two-cycled" daily. Possible variations are myriad (e.g., the plant being brought down to a minimum load overnight instead of being shut down). This particular plant is operated 5 days a week, from Monday through Friday, in an annually recurring rhythm.

- Every morning, the plant provides baseload between 5 am (obviously the plant starts earlier to be at baseload at that time) and 11 am and is brought down to 40% load until 3 pm (presumably wind and/or solar power generation takes over).
- Between 3 pm and 8 pm, the plant is once again base-loaded to provide rush hour and "nightlife" power to a busy metropolitan area.

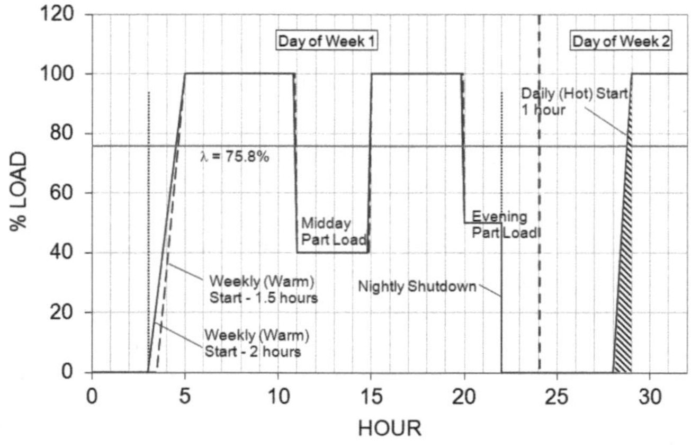

FIGURE 3.8 Example cyclic duty profile (1 day – Monday after the weekly shutdown) for a GTCC power plant. (Adapted from the paper by Aabadi, K. A. et al., "EDF Martigues Repowering Project Design and Construction of two Modern and Efficient CCGT Units," presented in POWER-GEN Europe, 26–28 May 2009.)

- After running at 50% load between 8 pm and 10 pm, the plant is shut down until 5 am next day.
- Every Friday, the plant is shut down at 10 pm for the weekend until next Monday 5 am. Twice a year, the plant is shut down for 1 week to carry out routine (scheduled) maintenance tasks.

Ignoring the time spent during startups, in terms of operating hours, the plant is (or planned to be) on line for

$$(52-2)\text{weeks/year} \times 5 \text{ days/week} \times 17 \text{ hours/day} = 4{,}250 \text{ hours/year}.$$

At a nominal 500 MWe baseload power requirement, total annual energy (megawatt-hours or MWh) production is

$$50 \times 5 \times 500 \text{ MW} \times (100\% \times 11 + 40\% \times 4 + 50\% \times 2) = 1{,}700{,}000 \text{ MWh}.$$

From the description above, it could be inferred that the plant start schedule is as follows:

- Two (2) *cold* starts (Monday morning after each scheduled maintenance shutdown)
- Forty-eight (48) *warm* starts (Monday mornings after each scheduled weekly shutdown)
- Two hundred (200) *hot* starts (every morning Tuesday through Friday).

Typically, **three** general types of plant operating regimes are possible:

- Peaking
- Cyclic
- Base-loaded.

These operating regimes can be characterized by the number of annual fired starts in conjunction with operating hours. A comparison based on generic definitions is shown in Tables 3.2 and 3.3. Maintenance factor in Table 3.2 is a multiplier applied to operating and maintenance costs. It should be stated that the operating regime definitions, especially the definition of the base load, in Tables 3.2 and 3.3 can be somewhat outdated. Modern gas turbines with state-of-the-art hot gas path technology

TABLE 3.2
Definition of Plant Operating Regimes

	Hours per Annum	Maintenance Factor
Peaking	600	10×
Cyclic	4,800	2.5×
Baseload	7,500	Baseline

TABLE 3.3
Definition of Plant Starts

	Cold Start	Warm Start	Hot Start
Downtime	≥72 h	≤48 to 72 h	≤8 to 12 h
Start time	1 h (conventional)	2 h (conventional)	3 h (conventional)
	30 min (fast)	90 min (fast)	2 h (fast)
Peaking (starts/annum)	40	100	10
Cyclic (starts/annum)	5	45	225
Baseload (starts/annum)	9	10	2

(nickel-based superalloys, advanced film cooling and thermal barrier coatings) operating at close to 3,000°F (close to 1,700°C) in the hottest part of the turbine are equipped with ample instrumentation and monitored by advanced diagnostic software. They are very expensive (unit purchase price and the maintenance agreement) and are operated closer "to the edge" to maximize owners' revenue and profit. The base load definition for those machines is typically based on the ratio of "fired" operating hours to number of starts (see Chapter 16). A gas turbine with a ratio of 50 or so is deemed to be a "base-loaded" unit.

REFERENCES

1. Gülen, S.C., 2017, Advanced fossil fuel power systems, Chapter 13 in *Energy Conversion*, 2nd Edition (Mechanical and Aerospace Engineering Series), Eds. D.Y. Goswami and F. Kreith, CRC Press, Boca Raton, FL.
2. Frutschi, H., 2005, *Closed-Cycle Gas Turbines*, ASME Press, New York.
3. Braun, K., Friedman, P., Dennis, R., 2017, *Fundamentals and Applications of Supercritical Carbon Dioxide (SCO_2) Based Power Cycles*, Woodhead Publishing, Duxford, UK.
4. Gülen, S.C., Joseph, J., 2012, Combined cycle off-design performance estimation: A second-law perspective, *J. Eng. Gas TurbinesPower*, Vol. 134, pp. 011801.
5. 2008, *Siemens Gas Turbine SGT6-5000F Application Overview*, Siemens Power Generation, Inc., Orlando, FL.

4 Gas Turbine

4.1 BRIEF OVERVIEW

This section is a quick tour through the landscape of "land-based" or "stationary" gas turbines, which will identify key hardware, actors (i.e., the original equipment manufacturers or OEMs) and terminology (e.g., "firing temperature").

There are two types of gas turbine applications outside aircraft and marine propulsion:

1. Mechanical drive
2. Electric power

In either case, the output of the gas turbine itself is the shaft power. In mechanical drive applications, the shaft power generated by the gas turbine is the driver of another turbomachine, e.g., a pipeline or process compressor. In electric power applications, shaft power generated by the gas turbine is the driver of a synchronous ac generator. The electric power generated by the ac generator is either supplied to the grid or used by the host industrial facility for internal power needs. In the latter case, the gas turbine is said to be in an "islanded operation". Note that, upon a grid disturbance, the gas turbine can be disconnected from the grid (i.e., it "trips" or "load-rejects") but still run at a low load to provide the plant auxiliary power in readiness for the restart. This operating mode is also referred to as islanded operation.

There are three types of gas turbines in electric power generation use:

1. Aeroderivative gas turbines
2. Small industrial gas turbines
3. Heavy-duty industrial gas turbines

Aeroderivative gas turbines are typically rated at less than 100 MWe; most common ones are rated at 50–60 MWe or less and can go down to about 25 MWe. The largest aeroderivative gas turbine (at the time of writing) is General Electric's intercooled LMS100 rated at around 100 MWe. Small industrial gas turbines are also rated below 100 MWe, but they can go down to about 5 MWe or lower (e.g., Solar gas turbine products).

As the name suggests, aeroderivative gas turbines are based on jet engine gas turbines, whose useful output is *thrust*. In the most basic aeroderivative configuration, the original jet engine is used as a "gas generator" and the exhaust gas is used in a "power turbine" for shaft power generation.

From a thermodynamic cycle perspective, aeroderivative gas turbines are characterized by high cycle pressure ratio (PR). For example, the cycle PR of GE's LM6000 rated at about 46 MWe is 30:1. In comparison, the cycle PR of GE's small industrial GT 6B.03, which is rated at 44 MWe, is 12.7:1. (The numbers cited are from *Gas Turbine World 2018 Handbook.*) Consequently, aeroderivative gas turbines are not ideal candidates for combined cycle applications because of their low exhaust temperature (824°F for LM6000 vis-à-vis 1,023°F for 6B.03).

Small industrial gas turbines can be similar in basic architecture to the aeroderivatives or the heavy-duty industrial gas turbines. General Electric's 6B.03 cited above is an example for the latter. Siemens's SGT-800, rated at 47.5–57 MWe, is another one. Siemens's SGT-600, rated at about 25 MWe, is an example for the former. It is a two-shaft gas turbine with a balanced gas generator shaft (with a two-stage turbine) and a power shaft (a two-stage power turbine) rotating at 7,700 rpm.

This brings us to a key distinction between "small" (i.e., less than 100 MWe rating) and "large" (rated at 100 MWe or higher) gas turbines. In order to fully understand the said distinction, we must turn to the rotational or shaft speed of the gas turbine, which is intimately tied to the particular "grid frequency".

There are two grid frequencies in the world, 50 Hz (e.g., in Europe) and 60 Hz (e.g., in the USA). Thus, a gas turbine generator operating in synchronous connection to the grid must rotate either at 3,000 rpm (50-Hz grid) or at 3,600 rpm (60-Hz grid). The reason is simple; for example, 60 Hz means 60 "cycles" or "revolutions" per second. This translates to $60 \times 60 = 3{,}600$ revolutions per minute (rpm) or $2\pi \times 60 = 120\pi$ radians per second (rad/s).

The problem is that a gas turbine's rotational speed cannot be set arbitrarily to any desired value. Aerothermodynamic and aeromechanical design considerations dictate what rotational speed (rpm) is optimal at what size. A gas turbine's size is quantified by its shaft output (in kW), which is a direct function of its air mass flow rate (in lb/s or kg/s). This in turn fixes the flow annulus area for the compressor and the turbine, which is dictated by the shaft diameter and compressor/turbine blade heights. The shaft diameter is dictated by the shaft torque, which, of course, is a function of the shaft power.

In order to have a rough idea about the relationship between power and airflow, for gas turbines rated at 25 MWe or smaller, to very good approximation,

$$\dot{W}[\text{MWe}] = 1.3\dot{m}[\text{lb/s}].$$

In other words, these "small" gas turbines have a power density of 130 kW/pps (~300 kJ/kg), which is the "specific work" output of the machine. It will be seen in Section 4.3 that large, advanced class gas turbines have a power density twice as much, which is driven by high turbine inlet or *firing* temperatures.

In order to get an idea about the shaft speed as a function of power rating, one can use the formula

$$N[\text{rpm}] = \frac{25{,}000}{\sqrt{\dot{W}[\text{MWe}]}}.$$

Beyond 25 MWe, aerodynamic and mechanical considerations would permit 3,600 rpm (or 3,000 rpm for 50-Hz applications) even though, say, 5,000 rpm or 6,000 rpm would be a more optimal design choice. This, however, would require a large, expensive and parasitic power-consuming gearbox. Thus, many industrial units are designed to run at 3,000 or 3,600 rpm.

There are, however, some notable exceptions. In particular,

1. Smaller aeroderivative gas turbines with self-contained gas generator and power turbine, e.g., some of General Electric's LM2500 units, which run at 6,100 rpm.
2. Smaller industrial gas turbines such as GE's Frame 6, which runs at 5,100 rpm.
3. Beyond 100 MWe, a gearbox becomes prohibitively large and expensive. Alstom (now owned by GE) GT11N2 is at 3,600 rpm for both 50- *and* 60-Hz versions (this is a not-so-small 115 MWe gas turbine). At the time of writing (2018), this machine is not offered new.

In general, 50- and 60-Hz gas turbines by a particular OEM are "speed-scaled" versions of the same basic design. The goal is to maintain the tangential speed at the blade tips to preclude undesirable shock losses. Accordingly,

$$\dot{m} \propto A = \pi\frac{D^2}{4}$$

$$U = \omega\frac{D}{2} = \pi\frac{ND}{60}$$

where N is rotational speed in rpm and D is a characteristic engine diameter. Thus,

$$\frac{\dot{m}_{60}}{\dot{m}_{50}} \propto \frac{D_{60}^2}{D_{50}^2} = \left(\frac{3{,}000}{3{,}600}\right)^2 = \frac{1}{1.44} = 0.694.$$

In other words, 50-Hz units are roughly 40% larger than corresponding 60-Hz units (or, equivalently, 60-Hz units are 30% smaller than corresponding 50-Hz units). Deviations usually stem from upgrades made by the OEM to certain units over the years. Since such upgrades are typically not done for 50- and 60-Hz machines in lock-step, the aerodynamically "correct" size ratio of 1.44 can get slightly off-kilter.

Heavy-duty industrial gas turbines have a very simple architecture: an axial compressor connected to an axial turbine via a solid shaft supported by two bearings. The synchronous ac generator is connected to the power shaft on the compressor end, which is known as a "cold end drive". One prominent exception is GE's older E class products such as the 7EA (60 Hz) gas turbine, whose generator is connected to the power turbine on the turbine end, which is known as a "hot end drive". (This machine is unique in its three-bearing arrangement as well with a third "hot" bearing between the compressor and the turbine.) The combustion section is placed between the compressor and the turbine. In state-of-the-art gas turbines, there are two basic combustor types:

1. Annular (e.g., Siemens' old "V machines", i.e., F class gas turbines)
2. Can-annular or "cannular" (e.g., GE gas turbines)

The former is an adaptation of aircraft jet engine gas turbine combustion chamber. The latter comprises identical, cylindrical combustion chambers (hence, "cans") in circumferential arrangement around the shaft, thus forming a *quasi*-flow annulus. In earlier days, the combustor chamber was a large cylindrical structure (referred to as a "silo"), which was outside the main gas turbine (e.g., older Siemens and ABB/Alstom gas turbines).

The only major variation in basic gas turbine architecture described above are former ABB/Alstom "sequential combustion" gas turbines GT24 (60 Hz) and GT26 (50 Hz). They are approximations of the textbook Brayton cycle improvement, i.e., *reheat*, into actual hardware.

Heavy-duty industrial gas turbines are classified according to their *nominal* turbine inlet temperature (TIT), which is the hot gas temperature at the exit of the combustion section just before entry into the turbine. There are four major classes:

- E class with nominal TIT of 1,300°C (2,372°F)
- F class with nominal TIT of 1,400°C (2,552°F)
- H class with nominal TIT of 1,500°C (2,732°F)
- J class with nominal TIT of 1,600°C (2,912°F)

There are some deviations from this neat classification rule. In particular,

- Mitsubishi's G class with nominal TIT of 1,500°C (2,732°F)[1]
- GE's H-System (a GE trademark) with nominal TIT of 1,500°C (2,732°F)

Unlike the conventional heavy-duty industrial gas turbines, whose "hot gas path" (HGP), i.e., the combustor and the turbine, is air-cooled in an "open loop" arrangement, these two classes of gas turbines utilize steam cooling in a "closed loop". The "open loop" term stems from the fact that the cooling air extracted from interstage locations in the axial compressor reenter the

1 Mitsubishi is the shorthand for Mitsubishi Hitachi Power Systems (MHPS), which used to be MHI (Mitsubishi Heavy Industries) before 2016.

gas flow-path after cooling the HGP components. In contrast, steam used for HGP component cooling, which is extracted from the steam (bottoming) Rankine cycle, does not interact with the HGP; it accomplishes its cooling duty within the channels inside the cooled component (via convection).

In the case of the G class gas turbines, closed-loop steam cooling is used for the combustor and the first turbine stage ring. In GE's H-System, the first two stages of the four-stage turbine are fully steam-cooled (i.e., both stationary nozzle vanes and rotating rotor blades). Thus, while the G class is essentially a *fortified* air-cooled gas turbine, the H-System is a *bona fide* steam-cooled gas turbine. While the G class has had a successful commercial operation history and is still offered new by Mitsubishi, the H-System, with six units in commercial operation since 2003, has been discontinued as a product offering by GE (see Section 8.6.5 for the demise of the H-System). GE now offers its HA class gas turbines (for "air-cooled H" to differentiate from the H-System) with nominal TIT of 1,600°C and, thus, equivalent to the J class (which is the designation used by Mitsubishi).

Note the designation "nominal" in TIT specification for heavy-duty industrial GT classes. The reason for that is that these are not "hard" numbers; they merely designate the introductory value of this key cycle design parameter. Over time, when new generations of a gas turbine with improved ratings are introduced, the TIT creeps up. For example, the 1,600°C-TIT J class was introduced in 2011. The recent offerings (in 2017–2018) have a TIT of around 1,650°C. The similar is true for the F class gas turbines; new F class gas turbines offered by the OEMs, e.g., GE's FA.04 class, have TITs pushing 1,500°C.

The TIT is frequently confused with two other temperatures, the so-called *firing temperature* (TFIRE) and the TIT per ISO-2314 standard. The former is a *real* temperature in the sense that it can (in theory at least) be measured, whereas the latter is a *hypothetical* number (see Figure 4.1). Also known as rotor inlet temperature (RIT), the firing temperature is arguably the most important gas turbine parameter (even more so than TIT) because it quantifies the true work generation ability of the cycle working fluid. The difference between TIT and TFIRE/RIT is a direct measure of the HGP component material durability (alloy and casting) and effectiveness of *thermal barrier coating* (TBC) and cooling technologies. The ultimate limit of TFIRE = TIT is the holy grail of the turbine designer (or, more precisely, the metallurgist). As it is, the lowest registered delta between the two is about 80°F, which has been achieved in the H-System deploying buckets made from single crystal alloy (durability) with TBC (protection) and closed-loop steam cooling (no hot gas temperature dilution). In air-cooled gas turbines, the TIT-TFIRE delta is around 200°F, somewhat lower for the most advanced H/J class machines and somewhat higher for the others.

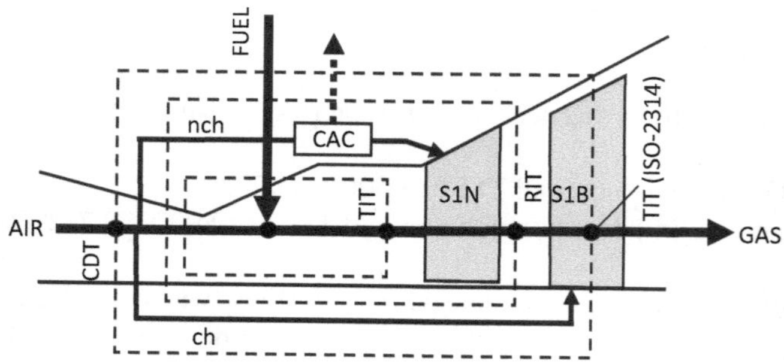

FIGURE 4.1 Gas turbine HGP. S1N, stage 1 nozzle (stator); S1B, stage 1 bucket (rotor); CAC, cooling air cooler; CDT, compressor discharge temperature; nch, nonchargeable (denotes compressor extraction air used to cool parts *upstream* of S1B inlet; ch, chargeable (denotes compressor extraction air used to cool parts *downstream* of S1B inlet).

Note that another classification of heavy-duty industrial gas turbines is also possible using TIT per ISO-2314. Accordingly,

- E class with nominal TIT (ISO-2314) of 1,100°C (2,012°F)
- F class with nominal TIT (ISO-2314) of 1,200°C (2,192°F)
- H class with nominal TIT (ISO-2314) of 1,300°C (2,372°F)
- J class with nominal TIT (ISO-2314) of 1,400°C (2,552°F)

At the time of writing (in 2018–2019), there are four major OEMs:

1. General Electric (including former Alstom, acquired at the end of 2015)
2. Siemens (formerly Siemens–Westinghouse, formerly separate companies, i.e., Kraftwerk Union (KWU) Siemens Power Generation and Westinghouse Electric Corporation)
3. Mitsubishi Hitachi Power Systems (MHPS, formerly Mitsubishi Heavy Industries, MHI)
4. Ansaldo Energia

General Electric designs, manufactures and sells gas turbines covering the entire spectrum, i.e., aeroderivatives, small industrial gas turbines and E, F and HA class heavy-duty industrial gas turbines. Its HA class is the equivalent of Mitsubishi's J class, which comes in steam-cooled and air-cooled (JAC) variants (see Section 4.3 to understand why this is so). Steam-cooled J class is essentially the (nominal) 1,600°C TIT upgrade of the G class. As mentioned earlier, GE's fully steam-cooled H-System is not offered anymore.

Siemens' gas turbine portfolio is similar to that of GE in terms of breadth. Unlike GE, which has its own aviation jet engine gas turbine technology, Siemens entered the aeroderivative segment via acquisition of Rolls-Royce. In 2014, Siemens bought the land-based gas turbine and compressor business of Rolls-Royce. As part of the deal, under a 25-year licensing agreement, Rolls-Royce also granted Siemens access to relevant Rolls-Royce aeroderivative technology for use in the 4–85 MW power output gas turbine range. Until the announcement of its HL class in 2017, Siemens did not have a gas turbine competing against GE's HA and MHPS's J/JAC class in its product portfolio. (The "HL" designation stands for "H technology on the way to 'L' technology for 65% combined cycle efficiency".)

In 2013, MHPS (then MHI) has completed its acquisition of Pratt & Whitney Power Systems, the small- and medium-sized, aeroderivative gas turbine business unit of P&W. Thus, the three major industrial gas turbine OEMs have their own aeroderivative (small) gas turbine divisions. This leaves Solar Turbines, owned by Caterpillar, as the other major small industrial gas turbine OEM, whose gas turbines, while not aeroderivative *per se*, are similar to aero units in design *spirit*.

Under an agreement that ran from 1991 through October 2004, Siemens had licensed its "V" class turbine technology to Ansaldo Energia ("V" for *Verbrennung*, combustion in German). Those gas turbines, i.e., V64.3, V94.2 and V94.3, were further developed, manufactured and sold by Ansaldo as AE64.3, AE94.2 and AE94.3, respectively. Following the acquisition of Alstom technology by Ansaldo, Siemens filed an arbitration action to block the agreements and terminate the license with Ansaldo Energia. Siemens claimed the agreements were outside the original license and allowing them to go forward would damage its gas turbine business and its position in the market. In 2016, however, Ansaldo Energia successfully defended the case in the International Chamber of Commerce (ICC) Arbitral Tribunal regarding its right to continue to use the "V" technology under the license agreement.

In 2015, Ansaldo Energia acquired Alstom's GT26 (50 Hz) and GT36 (50 and 60-Hz GT36-S5 and GT36-S6, respectively) gas turbine assets and technology, as required by European regulators. The acquisition followed GE's completion of a deal to acquire the remainder of Alstom's vast power and grid business for $10.6 billion. The European Commission had granted its approval to GE's merger with Alstom on the condition that Alstom divests "central parts" of its heavy-duty gas turbine business and key personnel to Ansaldo. According to the EC, the divestiture was intended to avoid the possibility of higher prices imposed by a *quasi*-monopoly resulting from the merger.

Future development, design and manufacturing of GT26 and GT36 are going to be undertaken by Ansaldo Gas Turbine Technology Co., Ltd, which is a joint venture between Ansaldo Energia SpA and Shanghai Electric Group Company Ltd established in Shanghai in 2014. Ansaldo Energia SpA is 59.9% owned by Cdp Equity in the Cassa Depositi e Prestiti Group, an Italian state-owned entity, and 40% by Shanghai Electric Group of China.

4.2 RATING PERFORMANCE

In this book, quite frequently, rating performances of gas turbines by different OEMs will be used as examples in calculations, discussions, etc. Rating (or rated) performances of gas turbines are annually published in two major trade publications:

- Gas Turbine World (GTW), http://www.gasturbineworld.com/
- Turbomachinery International (TMI), https://www.turbomachinerymag.com/

The data in these publications are directly obtained from the marketing departments of the OEMs. As such, they are identical; i.e., for a given year, say, 2016, whether you look up numbers from GTW or TMI is immaterial. Barring typos (quite rare), you will get the exact same information.

Gas turbine performance is defined by four "cardinal" parameters:

- Output (kWe or MWe)
- Heat rate (Btu/kWh or kJ/kWh)
- Exhaust gas mass flow rate (lb/s or kg/s)
- Exhaust gas temperature (°F or °C)

Rated/rating gas turbine performance is based on the following assumptions:

- ISO conditions (14.7 psia/1 atm (i.e., sea level), 59°F/15°C and 60% relative humidity)
- 100% (i.e., "full") load
- 100% methane (CH_4) fuel
- Gross output as measured at generator low-voltage terminals (i.e., no step-up transformer losses)
- Zero inlet and exhaust pressure losses (*unless* otherwise noted)
- No auxiliaries (*unless* otherwise noted – however, shaft-driven auxiliaries must have been accounted for in OEM-provided data)
- "New and clean" (i.e., no deduction for field deterioration)

For a precise evaluation of the performance, one needs the following information stated clearly:

- Inlet loss (if not stated as zero), how many inches of H_2O?
- Exhaust loss (if not stated as zero), how many inches of H_2O?
- Fuel temperature
- Any overboard air leaks?
- Exhaust/aft bearing blower? If yes, what is the flow rate and delivery temperature?
- Casing heat loss
- Shaft friction losses

This, of course, is not the case for ISO baseload rating data found in trade publications. Reasonable assumptions for missing information will be provided in the main body of the text.

In other words, missing or "fuzzy" information is not a big impediment for proper evaluation of a particular gas turbine's performance if one is really determined to carry it out with as much accuracy as possible.

There are two types of gas turbine rating performances: simple and combined cycle. Typically, combined cycle rating performance includes the following information:

- Gas turbine output (almost always)
- Steam turbine output (frequently but not always)
- Gross output for the plant (sometimes)
- Net plant output
- Net plant heat rate and efficiency

In rare instances, the bottoming cycle configuration (e.g., three-pressure reheat) and condenser pressure (e.g., 1.2 in. Hg) are also provided. Steam cycle conditions (main or HP steam pressure and temperature, hot reheat steam temperature) are never listed. Plant auxiliary load definition is not included either. Unfortunately, these omissions introduce a high-level uncertainty into the combined cycle rating performances.

There are two key changes to gas turbine simple cycle boundary conditions in combined cycle:

- Increase in exhaust pressure loss – it can be 10–12 inches of water column or more depending on the size of the HRSG and presence of an selective catalytic reduction (SCR)
- Fuel gas performance heating, typically up to 365°F–440°F (185°C–227°C).

The rules of thumb are:

- Each inch of H_2O extra exhaust loss is worth 0.1% in gas turbine output (lower) and heat rate (higher)
- Exhaust flow does not change
- Each 4 in. of H_2O extra exhaust loss is worth about 2°F in exhaust temperature (higher).

Impact of fuel heating on gas turbine output is negligible. Each 100°F (~56°C) is worth about 0.3% in heat rate (lower). Thus, increasing fuel gas temperature from 77°F (typical for simple cycle rating) to, say, 410°F, reduction in heat rate is $(410 - 77)/100 \times 0.3\% = 1.0\%$. Consequently, fuel gas performance heating compensates for the bulk of adverse impact of increased exhaust loss.

These generic derivatives can and should be used to check the thermodynamic consistency of simple and combined cycle performances published by the OEMs in trade publications and marketing brochures. This will be illustrated later in the book. For a critique of published rating performances, refer to the paper by Gülen [1].

4.3 TECHNOLOGY LANDSCAPE

The best way to get a handle on the gas turbine technology landscape is to look at major OEMs' 50-Hz product lineups. This should be enough to provide a full picture without a detailed listing of the 60-Hz machines. In most cases, they are the speed-scaled versions of their 50-Hz brethren (see Section 4.1) with lower airflow/output and similar efficiency and exhaust gas temperatures.

We start with the product lineup of General Electric, which illustrates the key differentiating features of E, F and HA/J class technologies. ISO baseload rating data obtained from *GTW 2018 Handbook* is summarized in Table 4.1 (YOI, Year of Introduction; OUT, ISO baseload in MW; EFF, Efficiency in %). TIT is a proprietary piece of information and is not disclosed by the OEMs. The TIT values in Table 4.1 are calculated using heat and mass balance analysis as described in Section 4.4.1.

TABLE 4.1
GTW 2018 GE 50-Hz Product Line Data (Calculated TIT)

	YOI	OUT MW	EFF %	TIT °C	PR	MEXH lb/s	TEXH °F
9E.03	1992	132	34.6	1,291	13.1	927	1,012
9E.04	2014	145	37.0	1,316	13.3	917	1,007
9F.03	1996	265	37.8	1,459	16.7	1,467	1,104
9F.04	2015	288	38.7	1,489	16.9	1,457	1,150
9F.05	2003	314	38.2	1,563	18.3	1,574	1,184
9HA.01	2011	446	43.1	1,598	23.5	1,891	1,164
9HA.02	2014	557	44.0	1,670	23.8	2,205	1,193

TABLE 4.2
GTW 2018 GE 50-Hz Product Line Data (Calculated)

	HR Btu/kWh	HC MMBtu/h	HC MWth	MFUEL lb/s	SP WK kW/pps
9E.03	9,862	1,302	382	16.82	145.0
9E.04	9,222	1,337	392	17.27	161.2
9F.03	9,027	2,392	701	30.90	184.5
9F.04	8,817	2,539	744	32.80	202.2
9F.05	8,932	2,805	822	36.23	204.2
9HA.01	7,917	3,531	1,035	45.61	241.7
9HA.02	7,755	4,319	1,266	55.80	259.2

Using simple algebra and basic correlations between performance parameters, one can extract additional information from the data listed in Table 4.1, which is summarized in Table 4.2 (HR, heat rate, which is 3,412.14/EFF in British units; HC, heat consumption, which is HR multiplied by output in kW and SP WK, specific output, which is output divided by airflow).

ISO baseload rating data obtained from *GTW 2018 Handbook* for MHPS 50-Hz product line is summarized in Tables 4.3 and 4.4.[2] (Note that M701G and M701J are steam-cooled gas turbines.)

ISO baseload rating data obtained from *GTW 2018 Handbook* for Siemens 50-Hz product line is summarized in Tables 4.5 and 4.6.

From the product line data presented above (from three major OEMs), state of the art in the largest and most efficient heavy-duty industrial gas turbines can be summarized as

- Airflows of 2,000 lb/s or higher
- Outputs of 500 MWe or higher (specific output ~250+ kW/pps)
- TITs pushing 1,700°C
- Cycle PRs above 20:1 – as high as 25:1
- Efficiencies of 42% or higher.

These "jumbo" gas turbines are optimized for combined cycle applications (note the very high exhaust gas temperatures) with advanced bottoming steam cycles (e.g., 600°C steam temperatures).

[2] Mitsubishi Hitachi Power Systems (MHPS) – formerly Mitsubishi Heavy Industries (MHI).

TABLE 4.3
GTW 2018 MHPS 50-Hz Product Line Data

	YOI	OUT MW	EFF %	PR	MEXH lb/s	TEXH °F
H-100	2013	118.08	38.3	20.1	695	1,025
M701DA	1981	144.09	34.8	14.0	999	1,008
M701G	1997	334	39.5	21.0	1,664	1,089
M701F	1992	385	41.9	21.0	1,650	1,167
M701JAC	2015	493	42.9	25.0	1,977	1,187
M701J	2014	478	42.3	25.0	1,977	1,166

TABLE 4.4
GTW 2018 MHPS 50-Hz Product Line Data (Calculated)

	HR Btu/kWh	HC MMBtu/h	MFUEL lb/s	MAIR lb/s	SP WK kW/pps
H-100	8,909	1,052	13.59	681	173.3
M701DA	9,805	1,413	18.25	981	146.9
M701G	8,638	2,885	37.27	1,627	205.3
M701F	8,144	3,135	40.50	1,609	239.2
M701JAC	7,954	3,921	50.65	1,926	255.9
M701J	8,067	3,856	49.81	1,927	248.0

TABLE 4.5
GTW 2018 Siemens 50-Hz Product Line Data

	YOI	OUT MW	EFF %	PR	MEXH lb/s	TEXH °F
2000E	1981	187	36.20	12.8	1,230	997
4000F	1995	329	41.00	20.1	1,596	1,110
8000H	2008	450	41.00	20.0	2,061	1,166
8000HL	2017	465	42.00	24.0	1,874	1,256
9000HL	2017	564	42.50	24.0	2,205	1,256

TABLE 4.6
GTW 2018 Siemens 50-Hz Product Line Data (Calculated)

	HR Btu/kWh	HC MMBtu/h	MFUEL lb/s	MAIR lb/s	SP WK kW/pps
2000E	9,426	1,763	22.77	1,207	154.9
4000F	8,322	2,738	35.37	1,561	210.8
8000H	8,322	3,745	48.38	2,013	223.6
8000HL	8,124	3,778	48.80	1,825	254.8
9000HL	8,029	4,528	58.49	2,147	262.8

A typical gas turbine combined cycle (GTCC) deploying these machines can have more than 1 GWe capacity in 2 × 2 × 1 to minimize installed cost ($/kW). High thermal efficiency (well above 60% net lower heating value, LHV, at ISO baseload) reduces variable costs and targets markets with expensive natural gas (or LNG) in need of baseload and clean (i.e., low emissions) electric power generation. Even in the USA, with cheap natural gas (due to the shale gas "boom"), high efficiency is advantageous for dispatch ranking in day-ahead markets like PJM (a regional transmission organization in the USA) and is sought after by independent power producer (IPP) developers (and their financers).

4.4 BASIC CALCULATIONS

There are two basic types of calculation methods in quantitative gas turbine analysis:

1. Heat and mass balance (HMB)
2. Stage-by-stage simulation

There are two basic motivations in this endeavor:

1. Predictive
2. Investigative

Whether predictive or investigative in nature, the ultimate goal in undertaking gas turbine (or steam turbine or any type of heat engine for that matter) calculations is *performance*, which is defined by shaft and/or generator output and thermal efficiency. Performance calculations can be done either in design mode or off-design mode.

4.4.1 Heat and Mass Balance Analysis (First Law)

Unless one is involved in actual turbine hardware design and development, a good grasp of HMB analysis is sufficient for predictive or investigative design performance calculations. For high-fidelity off-design calculations, one requires more in-depth information on the machine than an HMB model can provide (e.g., component maps) so that a stage-by-stage simulation model is requisite. For off-design performance *estimates*, however, generic or OEM-provided correction curves or simple derivatives are acceptable methods.

For HMB analysis, one can resort to a commercial software package such as Thermoflow's GT PRO® or to an in-house code (if one is employed by an OEM) such as a "cycle deck". Since such software products are highly expensive and proprietary, access to them is fairly limited to a select few. Luckily, one can do an equally good job with MS Excel and a good knowledge of fundamental thermodynamics.

In this book, the author will try to instill a good understanding of HMB analysis, including key details requiring utmost attention and/or pitfalls, in the reader. Stage-by-stage simulation of a gas turbine requires in-depth coverage of many different subjects, which can be found in **GTFEPG**. Herein, sample analyses will be provided and dissected in detail so that the reader knows what is required in modeling the "guts" of a state-of-the-art, heavy-duty industrial gas turbine in order to derive the most benefit out of it.

Let us take GE's 9HA.01 in Table 4.1 and subject it to a HMB test, which is summarized in Table 4.7. Generic gas turbine control volume for HMB test is shown in Figure 4.2. In order to do a rigorous HMB analysis, one has to have an accurate knowledge of material (i.e., air, fuel and, if applicable, dilution water or steam for NOx control) and energy streams crossing the control volume boundary identified in Figure 4.2. In case if such knowledge is not available, reasonable assumptions are also adequate.

TABLE 4.7
GE 9HA.01 Heat and Mass Balance

	m	p	T	h	Q	
	pps	psia	°F	Btu/lb	Btu/s	kW
Air in	1,835.8	14.7	59.0	–0.2	–443	–467
Fuel in	46.00	525.8	77.0	8.9	989,640	1,044,129
Compressor discharge	1,615.5	345.4	897.6	208.8	337,382	355,958
Overboard leak	1.8	328.1		208.8	383	404
Exhaust frame blower	11.0		110.0	12.07	133	140
Exhaust	**1,891.0**	14.9	**1,164.0**	294.9	557,697	588,404
Heat loss					2,473	2,609
Mechanical loss					1,350	1,424
Heat consumption					989,232	1,043,699
Shaft output					427,426	450,961
Generator output					422,725	**446,000**

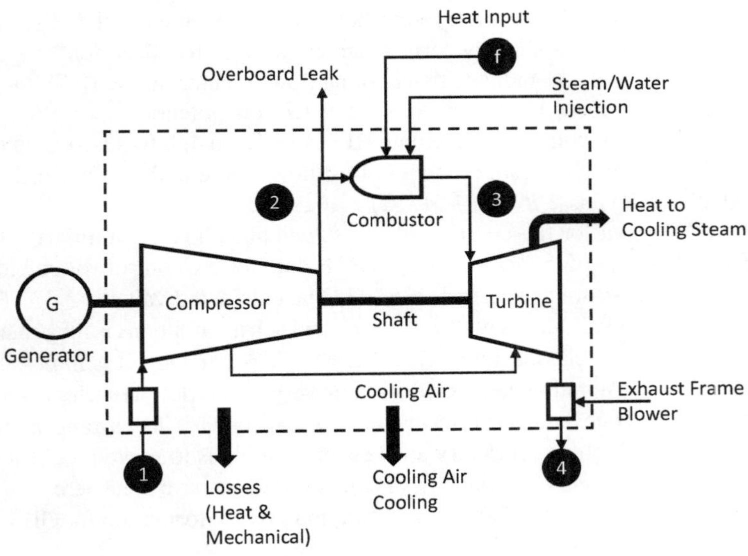

FIGURE 4.2 Generic gas turbine control volume.

The following assumptions are used in the HMB calculation summarized in Table 4.7:

- Overboard leak is 0.1% of airflow
- Exhaust frame blower airflow is 0.6% of airflow
- Exhaust frame blower exit temperature is 110°F
- Casing heat loss is 0.25% of heat consumption
- Mechanical losses are 0.35% of shaft output
- Generator efficiency is 98.9%
- Fuel gas is unheated at 77°F (25°C)
- Fuel is 100% CH_4 with an LHV of 21,503 Btu/lb
- Compressor polytropic efficiency is 93.8%

Note that compressor discharge information is not needed for heat and/or mass balance *per se* because it is within the gas turbine control volume. However, it is requisite for TIT and TFIRE/RIT estimates. State-of-the-art gas turbines use DLN (Dry-Low-NOx) combustors; thus, steam or water injection is not applicable (except for distillate operation, which is rarely used). Some gas turbines have the "cooling air cooling" (CAC) feature, which involves cooling the air extracted from the compressor discharge in an external heat exchanger before using it in the HGP for parts cooling. As mentioned earlier, MHPS G and J class gas turbines have steam-cooled parts, which necessitates to account for the heat transferred to the cooling steam in the HMB analysis.

HMB analysis of 9HA.01 gas turbine using published data in Table 4.1 (boldface numbers in Table 4.7) indicates a deviation of 0.37 percentage points in efficiency (75 Btus in heat rate, which is 0.95%) – see Table 4.8.

The only missing piece of information is TIT, which can be estimated from the HMB data in Table 4.7. In order to be able to do this, we need to estimate the *chargeable* and *nonchargeable* cooling flows. Referring to Figure 4.1, we can identify two control volumes to estimate the firing temperature, TFIRE or RIT, and the TIT. One crucial assumption that we have to make is that *all* HGP cooling air is extracted from the compressor discharge at CDT (compressor discharge temperature). (Typically, interstage extractions are used to cool parts further downstream in the HGP.) This simplification introduces a small error into the calculations, but it is unavoidable unless one is willing to build a complex stage-by-stage compressor and turbine model. Note that TFIRE (RIT) is not dependent on the value of nonchargeable cooling flows (nch), which enter the HGP upstream of the point where RIT is defined. In passing, the term "chargeable" is a shorthand for "chargeable to useful shaft work" – an old GE term – to indicate that cooling flows reentering the HGP *downstream* of the stage 1 blade row lose part of their mechanical work generating potential.

The HMB for the control volume to calculate RIT is in Table 4.9. It is based on the assumption that chargeable (ch) and nonchargeable (nch) cooling flows are equal to 12% each of compressor inlet airflow. Calculated RIT is 2,764°F (1,517°C).

Similarly, the HMB data for the control volume to calculate TIT is summarized in Table 4.10. Calculated TIT is 3,004°F (1,651°C). Finally, the HMB data for the control volume to calculate TIT per ISO-2314 definition is summarized in Table 4.11. The calculated value is 2,574°F (1,412°C).

The variation in TIT and RIT values with varying ch and nch assumptions is illustrated in Table 4.12. As advertised by GE, the HA class is 1,600°C (2,912°F) TIT technology. It is impossible to "reverse-engineer" the published GTW rating data exactly. It may very well reflect a machine with a TIT of, say, 2,895°F or 3,000°F. Unless one has exact information on the secondary flow arrangement and compressor performance, which is highly proprietary, the best one can do is to go with published information and data by using the first principles – which is, of course, what we have done here.

Using the HMB analysis described above, key Brayton cycle information for GE's 50-Hz product line is established as summarized in Table 4.13.

Using the published values of cycle PR and exhaust temperature as independent variables, the author had found that the firing temperature data reasonably was well represented by a

TABLE 4.8
Heat Balance Consistency of Published 9HA.01 Rating Data

		GTW 2018	HMB Check
OUT	MW	446	446
EFF	%	43.1	42.73
HR	Btu/kWh	7,910	7,985
MEXH	lb/s	1,891	1,891
TEXH	°F	1,164	1,164

TABLE 4.9
TFIRE (RIT) Control Volume HMB

	m	p	T	h	Q	
	pps	psia	°F	Btu/lb	Btu/s	kW
Air into combustor	1,613.7	345.4	897.6	208.8	336,998	355,554
Fuel into combustor	46.0	525.8	77.0	8.9	989,640	1,044,129
Heat loss					2,473	2,609
S1B inlet	1,659.7		**2,763.5**	797.84	1,324,165	1,397,073

TABLE 4.10
TIT Control Volume HMB

	m	p	T	h	Q	
	pps	psia	°F	Btu/lb	Btu/s	kW
Air into combustor	1,393.4	345.4	897.6	208.8	290,992	307,014
Fuel into combustor	46.0	525.8	77.0	8.9	989,640	1,044,129
Heat loss					2,473	2,609
Combustor exit	1,439.4		**3,003.6**	887.99	1,278,158	1,348,534

TABLE 4.11
TIT (per ISO-2314) Control Volume HMB

	m	p	T	h	Q	
	pps	psia	°F	Btu/lb	Btu/s	kW
Air into combustor	1,834.0	345.4	897.6	208.8	383,005	404,093
Fuel into combustor	46.0	525.8	77.0	8.9	989,640	1,044,129
Heat loss					2,473	2,609
TIT ISO 2314	1,880.0		**2,573.9**	728.82	1,370,172	1,445,613

TABLE 4.12
Sensitivity of TIT and RIT to Chargeable (ch) and Nonchargeable (nch) Cooling Flows

	ch = 10%	ch = 11%	ch = 12%
RIT, °F	2,729	2,746	2,763
nch		TIT, °F	
9%	2,896	2,917	2,938
10%	2,917	2,938	2,959
11%	2,938	2,959	2,981
12%	2,959	2,981	3,004

TABLE 4.13
Calculated TITs of Gas Turbines in Table 4.1

		TIT	TFIRE	TIT (ISO-3214)	TIT	TFIRE	TIT (ISO-3214)
	PR		°F			°C	
9E.03	13.0	2,316	2,149	2,008	1,269	1,176	1,098
9E.04	13.2	2,370	2,200	2,055	1,299	1,204	1,124
9F.03	16.8	2,602	2,418	2,261	1,428	1,326	1,238
9F.04	16.8	2,710	2,517	2,352	1,488	1,381	1,289
9F.05	18.3	2,773	2,577	2,410	1,523	1,414	1,321
9HA.01	21.8	2,949	2,746	2,574	1,621	1,508	1,412
9HA.02	23.5	3,037	2,827	2,649	1,670	1,553	1,454

formula derived from the turbine isentropic correlation (see **GTFEPG**, Ref. [11] in Chapter 2), i.e., (temperatures in °F)

$$\mathrm{TFIRE} = \frac{\mathrm{TEXH} + 460}{0.882\left(\mathrm{PR}^{-0.25} - 1\right) + 1} - 410. \tag{4.1}$$

Equation 4.1 fits the TFIRE values in Table 4.13 quite well for F and HA class but underestimates the TFIRE data for the E class. It is interesting to note that GE's advanced F class products, i.e., 9F.04 and 9F.05, are essentially H class gas turbines (TIT per ISO-2314 around 1,300°C). Similarly, the HA class is essentially the J class (TIT per ISO-2314 around 1,400°C).

In **GTFEPG**, it was shown that a generic linear correlation can be established between TIT and TFIRE. The same is true for the data in Table 4.13. The correlation is given by (temperatures in °F)

$$\mathrm{TFIRE} = 0.9405 \cdot \mathrm{TIT} - 29.912. \tag{4.2}$$

We would like to do the same HMB analysis with M701JAC in Table 4.3. This machine presents some difficulties due to its unique accessory systems and dearth of associated information. In fact, even the efficiency is listed as ">41%". According to the information obtained from the OEM in conferences, the key enabler of the next-generation JAC with 1,650°C TIT is the enhanced air-cooling (EAC) technology. In EAC system, compressor discharge air extracted from the combustor casing is cooled by two cooling air coolers (CAC) and pressurized by a booster compressor [2]. The compressor has a PR of 1.3:1 and volume flow of about 3,000 m³/h. After the pressurized air is used for cooling the combustor, it is returned back into to the combustor casing. The MT-FIN cooling structure of the combustor in the EAC system is the same as that in the steam-cooled G and J class combustors. The upstream side of the combustor is cooled by compressor discharge air and the downstream side is cooled by the EAC air in order to minimize the parasitic load of the booster compressor. (About 1,000 kW should be added to the GT auxiliary load to account for the EAC booster compressor.)

The HMB data for M701JAC is shown in Table 4.14. The following assumptions are made:

1. 12,500 Btu/s heat rejection from the CAC estimated as follows: cooling flow equal to 9% of airflow is cooled by 300°F (air c_p is assumed to be 0.24 Btu/lb-R);
2. No fuel gas heating (fuel supply at 77°F or 25°C);
3. Cycle PR is assumed to be 25:1.

The last item requires some explanation. All J/JAC gas turbine variants, 50- and 60-Hz, are listed with a cycle PR of 23 in *GTW 2018 Handbook*. According to Figure 10 in Ref. [2], "[s]imilar to

TABLE 4.14
MHPS M701JAC Heat and Mass Balance

	m	p	T	h	Q	
	pps	**psia**	**°F**	**Btu/lb**	**Btu/s**	**kW**
Air in	1,916.8	14.7	59.0	–0.2	–462	–488
Fuel in	50.57	558.4	77.0	0.0	1,087,469	1,147,345
Compressor discharge	1,686.8	367.4	923.4	215.6	363,679	383,703
Overboard leak	1.9	349.0		215.6	413	436
Exhaust frame blower	11.5		110.0	12.07	139	146
Exhaust	**1,977.0**	14.9	**1,187.0**	302.3	597,694	630,603
Heat loss					2,719	2,868
Cooling air cooling					12,500	13,188
Mechanical loss					1,350	1,424
Heat consumption					1,087,469	1,147,345
Shaft output					472,469	498,483
Generator output					467,272	**493,000**

M501H gas turbine, a PR of 25 is adopted to suppress the exhaust gas temperature rise". Detailed HMB analysis (described in **GTFEPG**) indicated that the latter is closer to the "true" value, which is the value used herein.

HMB analysis of M701JAC gas turbine using published data in Table 4.3 (boldface numbers in Table 4.14) indicates a cycle efficiency of 42.97% – see Table 4.15.

Using the same approach used above for 9HA.01, TIT for M701JAC is calculated as 1,680°C (3,056°F). Similarly, TFIRE (RIT) and TIT per ISO-2314 are calculated as 1,551°C (2,825°F) and 1,454°C (2,649°F), respectively.

The HMB data for M701J is shown in Table 4.16. In addition to 12,500 Btu/s of heat removed from the cooling air, steam cooling duty is set to 15,000 Btu/s. It should be emphasized that, from a purely energy balance perspective, a distinction between cycle heat rejection sources is not necessary. What really counts is the total amount of heat rejection, 27,500 Btu/s in this case, which is removed from the gas turbine control volume. The distinction will become important for bottoming cycle performance estimation, which will be discussed later in this section.

HMB analysis of M701JAC gas turbine using published data in Table 4.3 (boldface numbers in Table 4.16) indicates a cycle efficiency of 42.11% – see Table 4.17.

Using the same approach used above for 9HA.01, TIT for M701J is calculated as 1,653°C (3,008°F). Similarly, TFIRE (RIT) and TIT per ISO-2314 are calculated as 1,527°C (2,781°F) and 1,431°C (2,609°F), respectively.

TABLE 4.15
Heat Balance Consistency of Published M701JAC Rating Data

		GTW 2018	**HMB Check**
OUT	MW	493	493
EFF	%	42.9	42.97
HR	Btu/kWh	7,954	7,941
MEXH	lb/s	1,977	1,977
TEXH	°F	1,187	1,187

TABLE 4.16
MHPS M701J Heat and Mass Balance

	m	p	T	h	Q	
	pps	**psia**	**°F**	**Btu/lb**	**Btu/s**	**kW**
Air in	1,917.4	14.7	59.0	–0.2	–462	–488
Fuel in	50.03	558.4	77.0	0.0	1,075,784	1,135,017
Compressor discharge	1,725.6	367.4	923.4	215.6	372,049	392,534
Overboard leak	1.9	349.0		215.6	413	436
Exhaust frame blower	11.5		110.0	12.07	139	146
Exhaust	**1,977.0**	14.9	**1,166.0**	296.1	585,414	617,647
Heat loss					2,689	2,838
Steam cooling					15,000	15,826
Cooling air cooling					12,500	13,188
Mechanical loss					1,350	1,424
Heat consumption					1,075,784	1,135,017
Shaft output					458,094	483,316
Generator output					453,055	**478,000**

TABLE 4.17
Heat Balance Consistency of Published M701J Rating Data

		GTW 2018	HMB Check
OUT	MW	478	478
EFF	%	42.3	42.11
HR	Btu/kWh	8,067	8,102
MEXH	lb/s	1,977	1,977
TEXH	°F	1,166	1,166

4.4.2 Simplified Cycle Analysis

For most practical purposes, one can get by with even a simpler, parametric calculation than the rigorous HMB analysis presented above. We start by noting that, for an advanced class gas turbine, there are three significant material/energy streams:

1. Fuel (heat) consumption, which is the product of fuel gas mass flow rate and LHV
2. Exhaust gas
3. Shaft output

If one looks at the energy transfer values (the last two columns) in Table 4.7, if heat consumption is 100%, energy transfer out of the control volume with exhaust gas is 56.4% and shaft output is 43.2%. In other words, within a heat balance error of 0.5%,

$$\dot{Q}_{exh} + \dot{W}_{shft} \approx HC. \tag{4.3}$$

In terms of shaft efficiency,

$$\dot{Q}_{exh} \approx HC(1 - \eta_{shft}) \tag{4.4}$$

or

$$q_{exh} \approx 1 - \eta_{shft}. \tag{4.5}$$

The deceptively simple Equation 4.5 is actually the backbone of all combined cycle calculations. How it can be used to facilitate bottoming cycle calculations will be covered in Section 5.3. Herein, the focus will be on the "topping" cycle, i.e., the gas turbine.

Let us look at the gas turbine from the most basic perspective, i.e., the ideal, air-standard Brayton cycle, which is shown in Figure 4.3. There are *four* processes in the ideal cycle {1-2-3-4-1}:

1. Isentropic compression (proxy for the axial compressor)
2. Constant pressure heat addition (proxy for the combustor)
3. Isentropic expansion (proxy for the turbine)
4. Constant pressure heat rejection

The fourth process does not exist in a gas turbine, which operates in an "open" cycle, which is of interest to us in this book. (Closed cycle gas turbines have been tried in the past and dropped long time ago. For more information on them, the reader is referred to **GTFEPG** and the references cited therein.)

Before proceeding, it should be pointed out that the origins of the treatment of the Brayton cycle thermodynamics below can be traced back to the pioneering work done by Elmasri in the 1980s. The reader is encouraged to consult archival papers and seminar notes by Elmasri for a more in-depth look into the derivation of "mean-effective" cycle temperatures (from the first principles) and their importance in simple and combined cycle analysis [3–7].

From the cycle diagram in Figure 4.3, it is clear that the Brayton cycle is a very poor approximation of the Carnot cycle {1-2c-3-4c-1}, which comprises constant temperature (isothermal) heat addition and heat rejection processes. "How poor?" one might ask. In undergraduate thermodynamics, students learn that the Carnot cycle efficiency is given by

$$\eta_{Carnot} = 1 - \frac{T_1}{T_3}. \tag{4.6}$$

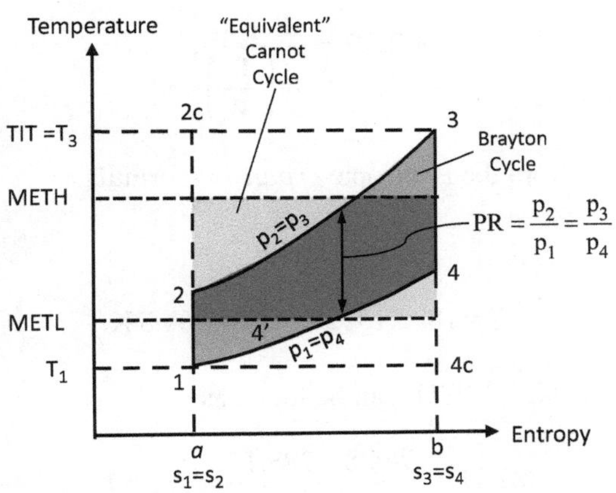

FIGURE 4.3 Ideal, air-standard Brayton cycle T–s diagram.

Thus, for a 1,600°C TIT class gas turbine

$$\eta_{\text{Carnot}} = 1 - \frac{15 + 273.15}{1600 + 273.15} = 0.846 \text{ or } 84.6\%.$$

In the Brayton cycle, the counterpart of the isothermal heat addition temperature is the *mean-effective heat addition temperature*, METH, which is the logarithmic mean of temperatures T_2 and T_3, i.e.,

$$\text{METH} = \frac{T_3 - T_2}{\ln\left(\frac{T_3}{T_2}\right)}. \tag{4.7}$$

(For full-blown derivation of METH and its significance, the reader is referred to **GTFEPG** and the references therein.) The temperature T_2 is found from the isentropic *compression* formula as

$$T_2 = T_1 \cdot \text{PR}^k \tag{4.8}$$

with

$$k = \frac{\gamma - 1}{\gamma}. \tag{4.9}$$

For air, $\gamma = 1.4$ so that the isentropic exponent $k = 0.2857$. For a gas turbine with PR of 23:1

$$T_2 = 288.15 \cdot 23^{0.2857} = 705.8 \text{ K}.$$

Substituting T_2 into Equation 4.7, METH can be found as

$$\text{METH} = \frac{1873.2 - 705.8}{\ln\left(\frac{1873.2}{705.8}\right)} = 1{,}196 \text{ K}.$$

The counterpart of the isothermal heat rejection temperature in the Brayton cycle is the *mean-effective heat rejection temperature*, METL, which is the logarithmic mean of temperatures T_4 and T_1, i.e.,

$$\text{METL} = \frac{T_4 - T_1}{\ln\left(\frac{T_4}{T_1}\right)}. \tag{4.10}$$

The temperature T_4 is found from the isentropic *expansion* formula as

$$T_4 = T_3 \cdot \text{PR}^{-k}$$

$$T_4 = 1873.15 \cdot 23^{-0.2857} = 764.8 \text{ K}.$$

Substituting T_4 into Equation 4.10, METL can be found as

$$\text{METL} = \frac{764.8 - 288.15}{\ln\left(\frac{764.8}{288.15}\right)} = 488.3 \text{ K}.$$

Consequently, one can easily see that the ideal (air-standard) Brayton cycle is equivalent to a Carnot cycle operating between two constant temperature reservoirs at METH and METL. Thus, its efficiency is given by

$$\eta_{Br,ID} = 1 - \frac{METL}{METH}. \tag{4.11}$$

For the example herein, we find that

$$\eta_{Br,ID} = 1 - \frac{488.3}{1196} = 0.592 \text{ or } 59.2\%.$$

This is the answer to the question of how poor the ideal Brayton cycle is vis-à-vis the Carnot cycle with the same T_3. Its efficiency is 84.6 – 59.2 = 25.4 percentage points lower. In other words, even if one could design and build a gas turbine with 1,600°C TIT and isentropic components with absolutely zero losses, one could not do better than 25 percentage points *lower* than the corresponding Carnot efficiency!

Note that, from Thermodynamics 101, it is known that the efficiency of an ideal, air-standard Brayton cycle is a function of the cycle PR *only*. In other words, it does not depend on the TIT (or T_3). One can easily confirm this from

$$\eta_{Br,ID} = 1 - PR^{-k}$$
$$\eta_{Br,ID} = 1 - 23^{-0.2857} = 0.592. \tag{4.12}$$

"Why did we go through all this rigmarole then?" one might be compelled to ask. The reason for going through the mean-effective temperature route is that Equation 4.12 does not provide us with sufficient information about the underlying thermodynamic drivers. In order to understand those drivers better, let us dig a bit deeper into the cycle diagram.

The triangular area {2-2c-3-2} in Figure 4.3 is the cycle *heat addition irreversibility* and is quantitatively equal to the rectangular area given by

$$i_H = (T_3 - METH) \cdot \Delta s \tag{4.13}$$

where the entropy delta is given by

$$\Delta s = s_3 - s_2 = s_4 - s_1.$$

The triangular area {1-4-4c-1} in Figure 4.3 is the cycle *heat rejection irreversibility* and is quantitatively equal to the rectangular area given by

$$i_L = (METL - T_1) \cdot \Delta s. \tag{4.14}$$

Sum of the two heat-exchange irreversibility terms in Equations 4.13 and 4.14 gives us the total cycle irreversibility as

$$i_{Cycle} = \left(1 - \frac{METH - METL}{T_3 - T_1}\right) \cdot (T_3 - T_1)\Delta s. \tag{4.15}$$

However, it is noted that

ΔT_B = METH – METL is the Brayton cycle's thermal "driving source"
$\Delta T_C = T_3 - T_1$ is the Carnot cycle's thermal driving source
$w_C = (T_3 - T_1)\cdot\Delta s$ is the Carnot cycle work.

Thus, Equation 4.15 quantifies the "lost" Carnot cycle work, i.e.,

$$w_{Lost} = \left(1 - \frac{\Delta T_B}{\Delta T_C}\right)\cdot w_C. \tag{4.16}$$

From the values calculated earlier, ΔT_B = 1,196 – 488.3 = 707.7 K, ΔT_C = 1,873.2 – 288.2 = 1,585 K and, thus, the term in the parentheses on the left-hand side of the lost work formula, Equation 4.16, is 1 – 707.7/1,585 = 0.5535. In other words, going from the Carnot entitlement to the *still-ideal* Brayton cycle obliterates more than one-half of the ultimate cycle work potential set by the second law of thermodynamics. The mechanism behind this severe reduction is the "truncation" of the original driving source, ΔT_C, down to ΔT_B. This truncation also reduces cycle heat input by almost one-third. This is why the reduction in ideal cycle *efficiency* potential is less than the reduction in ideal cycle *work* potential.

The bottom line is quantified by the *cycle factor*, which is found as

$$F_C = \frac{\eta_{Br,ID}}{\eta_{Carnot}} = \frac{0.592}{0.846} = 0.7.$$

From the rating performances covered in Section 4.3, we already know that the state of the art in gas turbine technology is still too far from the ideal Brayton cycle performance. Simple cycle efficiency of the latest generation of advanced class gas turbines with compressor PRs around 23:1 and TITs of 1,600°C or higher is in the 42% ballpark. Thus, the "technology factor" is

$$F_T = \frac{\eta_{SOA}}{\eta_{Br,ID}} = \frac{0.42}{0.592} = 0.71.$$

The technology factor is a very "solid" number, which can be used to great effect. For GE's 50-Hz product line summarized in Table 4.1, the technology factor perspective is summarized in Table 4.18.

TABLE 4.18
Cycle and Technology Factors (GE 50-Hz Product Line)

			TIT	EFF			
	YOI	**PR**	**°F (°C)**	**%**	**Carnot Efficiency**	**Cycle Factor**	**Technology Factor**
9E.03	1992	13.1	2,316 (1,269)	34.6	81.32	0.64	0.66
9E.04	2014	13.3	2,370 (1,299)	37.0	81.67	0.64	0.71
9F.03	1996	16.8	2,602 (1,428)	37.8	83.06	0.67	0.68
9F.04	2015	16.8	2,710 (1,488)	38.7	83.64	0.66	0.70
9F.05	2003	18.3	2,773 (1,523)	38.2	83.96	0.67	0.68
9HA.01	2011	21.8	2,950 (1,621)	43.1	84.79	0.70	0.73
9HA.02	2014	23.5	3,037 (1,670)	44.0	85.09	0.70	0.74

For a given T_3, i.e., TIT, there are two ways to increase METH, i.e., to reduce heat addition irreversibility and, hence, increase cycle efficiency:

1. Increase cycle PR
2. Reheat combustion

For a given T_1 (almost always the ambient temperature), there are two ways to reduce METL, i.e., to reduce heat rejection irreversibility and, hence, increase cycle efficiency

1. Increase cycle PR
2. Add a "bottoming" cycle (e.g., steam Rankine cycle).

On an ideal, air-standard cycle basis, increasing T_3 does not change cycle efficiency; METH and METL increase in lock-step. This is not the case for a *real* cycle; T_3 has a positive effect on Brayton cycle efficiency. However, more importantly

1. Increasing T_3 increases net cycle work
2. Increasing T_3 increases the "combined" cycle efficiency.

Appreciation of cycle PR effect is quite straightforward. For a given T_3, higher PR leads to higher T_2 and higher METH. In "real" machine terms, less fuel is burned in the combustor to achieve the specified T_3. The downside is reduction in the cycle's specific power output because of the increase in compressor power consumption. The second downside is lower T_4, which reduces the combined cycle potential. Therefore, PR alone is not an ideal "knob" for better combined cycle efficiency. On the other hand, increase in PR always goes hand in hand with increase in TIT, which requires some exploration.

First of all, from the ideal Brayton cycle calculations, it can be shown that the *optimum* PR for *maximum* net specific output is given by

$$PR_{opt} = \sqrt[2k]{\frac{TIT(=T_3)}{T_1}} = \left(\frac{T_3}{T_1}\right)^{\frac{1}{2k}}. \tag{4.17}$$

Thus, for a 1,600°C TIT class gas turbine, PR_{opt} can be calculated as 28:1. This is close enough to the value in the actual hardware, which is around 23:1. If one introduces the compressor and turbine isentropic efficiencies into the air-standard Brayton cycle calculations, Equation 4.17 becomes (the algebra is left to the reader as an exercise)

$$PR_{opt} = \left(\eta_c \eta_t \frac{T_3}{T_1}\right)^{\frac{1}{2k}}.$$

Assuming 0.9 for both efficiencies, PR_{opt} is calculated as 18.3:1; once again, close enough to the actual hardware value but on the lower side.

One could not expect more from a very simple equation with a constant value of γ. In **GTFEPG**, using the published rating data for 60-Hz machines, a semiempirical version of Equation 4.17 is derived as follows:

$$PR_{opt} = \tau_3^{\frac{1}{2k'}}$$

with

$$\tau_3 = \frac{T_3}{T_1} \tag{4.18}$$

and

$$k' = c_1 \cdot \left(\frac{TIT}{TIT_{ref}} \right)^{c2} \tag{4.19}$$

where c_1 and c_2 are curve-fit parameters and TIT_{ref} is a *reference* TIT, which is set to 1,600°C (2,912°F) for the 7HA.02 data point with PR of 23:1. The modified isentropic exponent is given by with the curve-fit constants

$$c_1 = 0.3093,$$

$$c_2 = -0.2939.$$

Equation 4.18 predicts the cycle PR of GE's 50-Hz product line rather well as demonstrated in Figure 4.4.

There is little doubt that the most cost-effective gas turbine design intuitively leads to maximizing the specific output. However, there is another consideration at play here, which is also implied by the ideal T–s diagram in Figure 4.3. In particular, we pay attention to the triangular area {1-4-4c-1} in Figure 4.3, which quantifies the cycle *heat rejection irreversibility*. What if we could place another power cycle in there, which has a mean-effective heat addition temperature of METL and mean-effective heat rejection temperature of T_1? This will be looked at in Chapter 8.

4.4.3 Stage-by-Stage Gas Turbine Model

HMB analysis (i.e., first law analysis) does not tell one much about the "inner workings" of a gas turbine. Summarizing the discussion and learning in this chapter until this point, the following points emerge:

1. Verify published rating data via energy and mass conservation
2. Define and calculate cycle and technology factors to establish the state of the art in gas turbine design

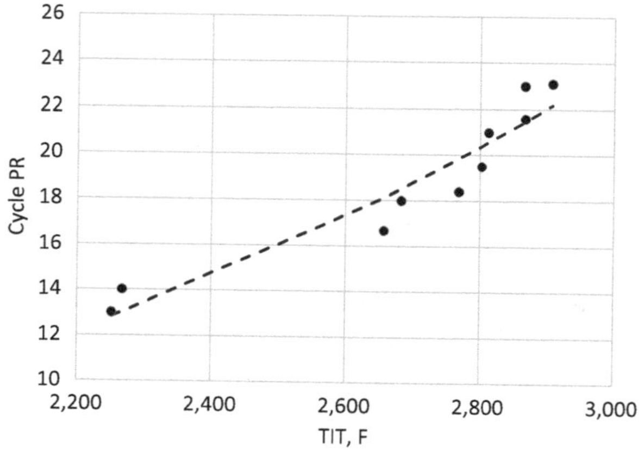

FIGURE 4.4 GE 50-Hz product line PR-TIT data (curve from Equations 4.17–4.19).

3. Estimate TIT and TFIRE (with reasonable assumption of chargeable and nonchargeable cooling flows)
4. Establish the relationship between cycle (or compressor) PR and TIT
5. Quantify exhaust gas energy and exergy
6. Estimate bottoming cycle power output from exhaust gas energy and exergy.

For most practical purposes, this is sufficient information to understand and evaluate combined cycle performance ratings. Unless one is involved in actual gas turbine design and development, to delve deeper into the aerothermodynamic and aeromechanical design principles of key components (compressor, combustor and turbine) can indeed be overkill. This would be well beyond the scope of the current book anyway. For an in-depth coverage, the reader is referred to **GTFEPG** and the references cited therein (i.e., Chapter 10 for turbine aero and Chapter 11 for compressor aero in Ref. [11] in Chapter 2).

Nevertheless, one has to be aware of the key hardware design considerations in order to make a realistic assessment of existing state of the art and claims of future improvements. Requisite understanding can be grouped into design objectives:

1. Cost-effective axial compressor design (number of stages, 3D-Aero design of stator vanes and rotor blades) to achieve requisite PR with maximum efficiency
2. Cost-effective combustor design (low pressure drop) to achieve requisite TIT with low emissions as dictated by applicable regulations
3. Cost-effective axial turbine design (number of stages, 3D-Aero design of stator vanes and rotor blades) with available materials, coatings and film cooling techniques with maximum efficiency.

As the reader no doubt noted, the key term is "cost-effective". From a conceptual cycle design perspective, one can go as high as, say, 3,500°F in TIT with 30+ compressor PR. As long as the conservation laws are not violated, on paper, one can come up with seemingly impressive cycle concepts. The difficulty arises when the designer, say, tries to come up with an aeromechanically sound compressor design with reasonable stage loading to enable high compressor efficiency without ending up with an exorbitantly large and expensive component requiring expensive alloys for manufacturing. A DLN combustor to achieve the requisite TIT with single-digit NOx and CO emissions may turn out to be impossible with the available cooling flow budget.

In order to assess all these aspects of gas turbine design, one has to resort to a few basic design parameters particular to each component, i.e.,

- Stage loading and flow parameters (compressor and turbine)
- Stage degree of reaction (compressor and turbine)
- Compressor polytropic efficiency
- Cooled turbine stage efficiency

In the following paragraphs, a very brief introduction into turbomachinery aerothermodynamics is provided. The goal is to acquaint the reader with the basic concepts. A recent revision of Logan's classic introductory text by Sultanian is highly recommended for self-study and learning [8].

4.4.3.1 Turbine Aero

Turbine aerothermodynamics (or turbine aero in the industry jargon) within the context of steam turbine steam-path design (analogous to the expander or "hot gas path" section of a gas turbine) will be covered in detail in Section 5.1. The treatment is very similar to that for a gas turbine (after all, steam *is* a gas) and the underlying principles are the same. In very simplistic terms, one is interested in coming up with a turbine design (hardware) that is going to translate the thermodynamic work from paper to machine. Let us call this exercise "hot gas expansion path layout". For simplicity, let us ignore cooling. At the beginning of the layout exercise, the following are known:

- Gas flow rate, $\dot{m}$
- Gas inlet pressure, p, and temperature, T
- Gas exit p and T
- Gas specific heat c_p (assume constant at an average value)
- Total work generated

$$\dot{W} = \dot{m}c_p\Delta T_{tot}. \tag{4.20}$$

Note that in analyzing gas flow in a turbomachine, one must keep track of "thermal" and "kinetic" energy of the fluid. The former is quantified by the *static* value of the enthalpy from the static values of pressure and temperature via a suitable *equation of state*. The equation of state can be as simple as the ideal gas equation of state, i.e., $p = \rho RT$, or quite cumbersome (e.g., JANAF tables or a "real gas" equation of state such as Peng–Robinson or Benedict–Webb–Rubin). The combination of thermal and kinetic components (the latter, per unit mass, is simply $v^2/2$ where v is the gas absolute velocity) gives the *total* or *stagnation* enthalpy of the gas in question. In simplified treatment of turbomachinery gas dynamics, as a result of "perfect gas" assumption with constant c_p, one frequently encounters the differentiation between static and total/stagnation temperatures. This convention simplifies the flow equations and is pretty accurate for most cases. In the discussion below, static values of p and T are notated without subscript, whereas their total/stagnation values are notated by the subscript tot. In literature, there are many different conventions (e.g., subscript 0 for total/stagnation values, subscript s for static values). In mathematical terms, in US customary system (USCS) units

$$T_{tot} = T + \frac{v^2}{2g_c J c_p}.$$

with $g_c = 32.174$ ft-lbm/lbf-s^2 and $J = 778.17$ ft-lbf/Btu (in SI units, the conversion factors are unity).

Equation 4.20 quantifies the *thermodynamic* power (i.e., rate of work done by the hot expanding gas); in the actual machine, it will be translated into *shaft* power and *torque*, τ. This will happen by the following mechanism (refer to Figure 4.5):

1. Static enthalpy of the hot gas entering the turbine stage will be translated into velocity while flowing through the stator vane.
2. This will be facilitated by the nozzle-like flow channels formed by adjacent stator vanes.
3. Note, however, since no work is done by the hot gas (yet), total enthalpy does not change (except a small friction loss).
4. Fast flowing hot gas (accelerated by the "nozzle" vanes) will flow across the rotor blades (typically with airfoil-like cross sections) attached to the shaft of the turbine. During this part of the process, two possibilities emerge:
   a. Hot gas impacts on the blades and transfers its momentum to them in analogy to water hitting the "buckets" of a Pelton (hydraulic) turbine. The blades/buckets are pushed by this momentum exchange and, thus, cause the rotation of the shaft to which they are attached. This is the theory behind the "impulse" turbine.
   b. Hot gas flows around the blades in a manner analogous to air flowing around the sail of a sailboat or air flowing around the wing of an airplane. In this version of events, the blades (they don't look like "buckets" anymore; they are shaped just like the wing of an airplane) are moved by the "lift" action of the gas on the pressure side. Again, their motion causes the rotation of the shaft to which they are attached. This is the theory behind the "reaction" turbine

Shaft power generated by this rotation is

$$\dot{W} = \frac{2\pi N}{60}\tau \tag{4.21}$$

where N is the rotational speed of the shaft in revolutions per minute (rpm). Ignoring the losses, this is equal to the thermodynamic power, i.e.,

$$\dot{W} = \dot{m}c_p\Delta T_{tot} = \frac{2\pi N}{60}\tau$$

or, in per unit mass terms

$$w = c_p\Delta T_{tot} = \frac{2\pi N}{60}\frac{\tau}{\dot{m}}. \tag{4.22}$$

But the torque generated by the hot gas action on the blades is also equal to the rate of change of angular momentum, i.e.,

$$\tau = \dot{m}r\Delta v \tag{4.23}$$

where r is the pitch-line radius of the blade (see below) and gives the distance between the centerline of the rotating shaft and the turbine pitch line and Δv is the change in the angular (tangential) component of the gas velocity. In Figure 4.5, only the axial (x) and radial (r) coordinates are shown. The third coordinate can be visualized as coming out of the page towards the reader. (Geometrically, it is a tangent to the circle formed by the radius r.) Angular momentum change is along this third coordinate. Combining Equations 4.22 and 4.23 and noting that $\omega = 2\pi N/60$ (the *angular velocity*, units are radians per second or rad/s), we obtain the famous *Euler turbine equation*

$$c_p\Delta T_{tot} = \omega r\Delta v$$

or in British units

$$c_p\Delta T_{tot} = \frac{\omega r}{g_c J}\Delta v. \tag{4.24}$$

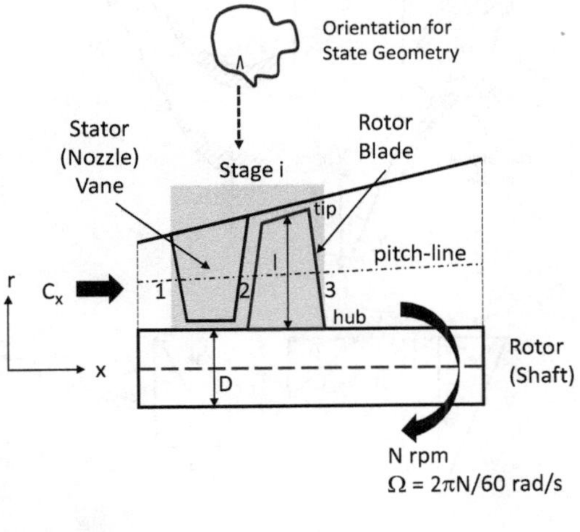

FIGURE 4.5 Turbine stage schematic.

Using the rotor speed (tangential speed at radius r) definition, $U = \omega r$ (in m/s or ft/s), the Euler turbine equation becomes (using SI version henceforth in order not to clutter the formulae)

$$c_p \Delta T_{tot} = U \Delta v. \tag{4.25}$$

Key turbine design parameters are calculated from the *velocity triangles* of the turbine stage in Figure 4.5, which are depicted in Figure 4.6. The schematic of the stage in Figure 4.6 is from the point of view of an observer looking at the gas flow-path from above (see the head cartoon in Figure 4.5). For the correct interpretation of important turbine aerothermodynamic formulae and the motion of the hot gas through the turbine stage, the nomenclature in Figure 4.6 should be well understood. This may sound trivial but rest assured that it is not. There is a significant variety in velocity and angle notations in different books and papers on the subject. The nomenclature adopted herein is from the superb treatise by Mattingly [9].

In Figure 4.6, roman numerals 1, 2 and 3 designate stage (stator) inlet, rotor inlet (stator exit) and stage exit (rotor exit), respectively. Subscript R designates "relative" velocities (see below). Furthermore,

- V is the velocity (of the gas) in a stationary coordinate system (*absolute* velocity)
- V_R is the velocity in a moving coordinate system (i.e., for an imaginary observer sitting on the rotor), i.e., *relative* velocity
- U, which is defined above, is the velocity of the moving coordinate system (i.e., the rotor)
- α is the angle between the axial coordinate and the absolute velocity
- β is the angle between the axial coordinate and the relative velocity
- u and v are axial and tangential components of the absolute velocity, respectively
- u_R and v_R are axial and tangential components of the relative velocity, respectively.

In addition,

- Axial components of the absolute and relative velocities are equal to each other, i.e., $u = u_R$
- Stage exit absolute flow angle, α_3, is known as the *swirl* angle
- Tangential components of absolute velocities v_2 and v_3 are known as *whirl* velocities.

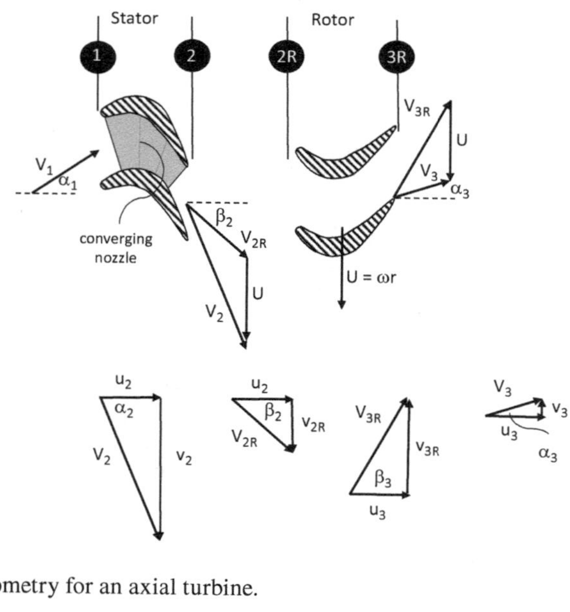

FIGURE 4.6 Stage geometry for an axial turbine.

Rotor speed, U, is a function of the radius and varies (increases) from the blade hub (root) to the blade tip. This will also change the gas velocity angles shown in Figure 4.6 along the rotor blades. A blade design that takes into account the variation in the velocity triangles is called the *vortex* design. The simplest treatment going back to the early days of turbine HGP layout when everything had to be done manually with pen, paper and slide rules is the *free vortex* design. The assumption in free vortex design is that the losses incurred by the variation of incidence angles along the radial direction are negligibly small as long as the blades are not too long. This assumption has proven to be reasonable for a long time when turbine mass flow rates (and power output) were low enough that the blade lengths were compatible with this approximation. In free vortex design, all gas velocity angles and speeds are evaluated at a mean diameter or the turbine *pitch line*.

Here, let us turn to the question of physical flow area, which is determined by the gas flow rate, blade height or length, ℓ, and the shaft diameter, D. From the mass continuity, we find that

$$\dot{m} = \rho u_2 \ell^2 \left(\pi \frac{D}{\ell} + 1 \right). \tag{4.26}$$

(For most practical purposes, $u_2 \approx u_3$). Since there is a strong correlation between the gas flow rate (which determines total power generation) and the turbine geometry, it is represented by the *flow coefficient*, which is defined as

$$\phi = \frac{u_2}{U}. \tag{4.27}$$

From the stage velocity triangles, using fundamental correlations of trigonometry, it can be shown that the change in the tangential component of the gas velocity, $\Delta v = v_2 + v_3$, can be related to the rotor inlet and exit angles of the hot gas, α_2 and α_3, respectively, i.e.,

$$\Delta v = \phi U (\tan \alpha_2 + \tan \alpha_3). \tag{4.28}$$

(For detailed derivation, see Chapter 10 in **GTFEPG**.) Substituting Δv from this equation into the Euler turbine equation, we find that

$$c_p \Delta T_{tot} = \phi U^2 (\tan \alpha_2 + \tan \alpha_3)$$

or

$$\frac{c_p \Delta T_{tot}}{U^2} = \phi (\tan \alpha_2 + \tan \alpha_3). \tag{4.29}$$

The term on the left-hand side of Equation 4.29 is the *stage loading coefficient*, ψ, which is the ratio of stage work to the square of the rotor speed. Let us first look into where the term "loading" comes from. The numerator in the definition of ψ, i.e., thermodynamic work per unit mass, is the "cause". The denominator in the definition of ψ, i.e., the rotational work per unit mass, is the "effect". *Prima facie*, this does not seem to make sense. The cause is "thermal" and the effect is "kinetic". How can this be possible one might wonder. Two points can clarify this apparent dichotomy.

1. As mentioned earlier, "total" or "stagnation" temperature, T_{tot}, combines thermal *and* kinetic components, i.e., at the inlet of the stator:

$$T_{tot,1} = T_1 + \frac{u_1^2}{2c_p}.$$

2. The stator ensures that the static temperature/enthalpy (i.e., thermal energy) of the hot gas at the stage/stator inlet is translated into kinetic energy (that is what a "nozzle" does), i.e.,

$$T_{tot,1} = T_{tot,2} = T_2 + \frac{u_2^2}{2c_p}$$

but

$$u_2 > u_1 \text{ and } T_2 < T_1.$$

Typically, u_1 is about 150 m/s and u_2 can be 800 m/s or higher. Note that 150 m/s is about 10 kJ/kg. At 1,500°C TIT (i.e., a modern H class gas turbine), at turbine stage 1 stator inlet, static/thermal enthalpy is about 2,000 kJ/kg (from $h = c_pT$ with $c_p = 1.148$ kJ/kg-K). This is why in HMB calculations, where one is not looking into the "guts" of the turbomachine, total/static dichotomy can be (and is) ignored with little loss in accuracy. On the other hand, noting that 800 m/s is 640 kJ/kg, precise bookkeeping is a *must* for stage aerothermodynamics.

In conclusion, without the stator vanes, which form effective nozzles through which the hot gas accelerates, there would indeed be very little "cause" (i.e., slow hot gas flow) and practically no "effect" (i.e., no discernible shaft rotation).

For a reasonable mechanical design, cause and effect should be comparatively close to each other (obviously the cause cannot be less than the effect, i.e., $\psi < 1$ is not possible). Ideally, with no losses, one would wish for the cause to be equal to the effect, i.e., $\psi = 1$. Indeed, this is what one gets for a 50% reaction design in an ideal turbine stage for maximum efficiency (see below and also Section 5.1). If the cause is much larger than the effect, i.e., $\psi \gg 1$, the implication is a waste of effort, i.e., low efficiency, which is indeed the case (see the Smith chart below). This is essentially the easiest way to think about making the transition from cycle thermodynamics to stage *aero*thermodynamics.

After the brief digression above on the stage loading, let us return to Equation 4.29. The exit swirl angle α_3 is the *third* important stage parameter. There are two *angles* to this importance (no pun intended):

- For the last turbine stage (usually, the *fourth* stage in almost all large industrial gas turbines nowadays), a low swirl angle is desirable so that relative and axial gas velocities are close to each other. This reduces the amount of kinetic energy left in the gas leaving the gas turbine.
- For the other stages, however, a higher exit swirl angle means larger stage work output (see Equation 4.29). In other words, the larger the gas flow "whirl" in the rotor, the larger is the torque imparted by the gas on the rotor.

The degree of reaction R is the *fourth* important stage parameter. It is defined as the ratio of the static enthalpy drop in the rotor to the total enthalpy drop across the stage (i.e., rotor and stator). From the stage geometry in Figure 4.6, assuming $u_2 \approx u_3$, it can be shown that

$$R = \frac{\phi}{2}\left(\tan\beta_3 - \tan\beta_2\right). \tag{4.30}$$

Zero reaction (R = 0) defines the *impulse* stage where the gas is accelerated through the stator vanes and impinges on the rotor blades imparting momentum on them. Since the angles β_2 and β_3 are equal to each other (for R = 0), the gas simply reverses direction with constant relative velocity (ignoring friction losses). It can be shown that the resulting stage loading parameter is

$$\psi = 2\left(\phi \tan\alpha_2 - 1\right). \tag{4.31}$$

There are several advantages and disadvantages associated with the impulse design:

- Zero static enthalpy drop implies little or no leakage across the rotor because of negligible pressure drop
- High value of ψ lowers the rotor-relative stagnation temperature, which lowers the cooling flow requirement

- However, higher values of ψ leads to poorer aerodynamic efficiencies
- Large values of α_2 can lead to supersonic stator (nozzle) exit velocity and shock waves, which reduces the efficiency
- Large turning angle increases the losses due to viscous boundary layer separation.

A 100% reaction stage (R = 1) is useless for a practical turbine. In order to visualize this condition, consider a lawn sprinkler with two opposing water jets coming out of each end causing the sprinkler to rotate around its axis and water a circular patch of lawn around it. Easiest way to see this in a turbine context is via the stage loading parameter. For a general, zero swirl case ($\alpha_3 = 0$), it can be shown that

$$\psi = 2(1 - R). \tag{4.32}$$

Thus, for R = 1, $\psi = 0$, and from Equation 4.29, stage work or torque is zero. For R = 0.5 or 50% reaction, from Equation 4.32, $\psi = 1$. The implication is equal enthalpy drop across stator and rotor and symmetrical velocity triangles. For R = 0.5, the counterpart of Equation 4.31 is

$$\psi = 2\phi \tan \alpha_2 - 1. \tag{4.33}$$

Low degree of stage reaction leads to higher stage loading, which reduces the required number of turbine stages for a given thermodynamic cycle design. This is beneficial from a cost (i.e., lower number of parts) as well as a performance perspective due to reduced cooling requirements (because of lower amount of space to be cooled). Furthermore, the reduction in the rotor-relative stagnation temperatures is another performance benefit via lower blade cooling flows (see Equation 4.40 below). Especially in the first turbine stage, a higher stage loading means a larger drop in temperature so that lower temperature gas enters the subsequent stage and lowers the chargeable cooling flow requirement. On the other hand, these positive attributes of high ψ are balanced by lower aerodynamic efficiencies. This can be best observed by the Smith chart in Figure 4.7, which is derived from efficiencies found from a wide range of Rolls-Royce axial turbine test results [10]. The data is corrected for tip clearance and leakage losses so that they are on the optimistic side. (A table showing the discrete chart data can be found in **GTFEPG**.)

As shown in the Smith chart, at any given flow coefficient, increasing stage loading leads to decreasing stage efficiency. Modern aircraft gas turbines have ψ and ϕ values in the range of 1.3–2.2 and 0.5–1.1, respectively [9]. Typical heavy-duty industrial gas turbines have similar values for the stage loading coefficient, whereas the flow coefficients are closer to the lower end of that range.

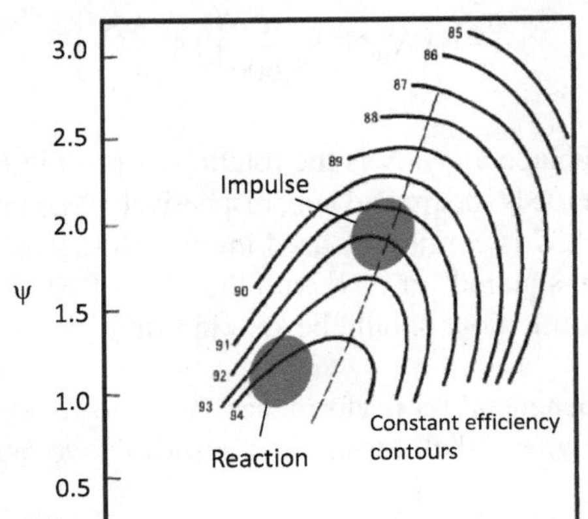

FIGURE 4.7 Smith chart [10].

The design problem is then one of optimization to strike the right balance between cost and performance. It is not difficult to see that in most cases, the optimum will fall somewhere around the middle of the two design choices, i.e., R = 0 and R = 0.5. This is indeed the case with optimum R in the range of 0.2–0.3 for most cooled turbines, i.e., a stage loading of 1.4.

Another important parameter is the annulus area, A_{an}. At any given point in the turbine HGP, A_{an} can be calculated as a function of the flow properties at the mean radius (i.e., the pitch-line radius) and the mass flow rate. For the purposes herein, the most important point is the turbine exit and the exit annulus area, which determines the length of the last stage blade:

$$A_{an} = \frac{\dot{m}\sqrt{T_{3,tot}}}{p_{3,tot}\cos\alpha_3 MFP}. \tag{4.34}$$

In Equation 4.34, the *mass flow parameter* (MFP) is a function of the Mach number and given by

$$MFP = \frac{Ma_3\sqrt{\dfrac{\gamma}{R_{gas}}}}{\left(1+\dfrac{\gamma-1}{2}Ma_3^2\right)^{\frac{\gamma+1}{2(\gamma-1)}}}. \tag{4.35}$$

The Mach number in Equation 4.35 is based on the absolute velocity of the gas and the speed of sound (which is based on the static temperature) as follows

$$Ma_3 = \frac{V_3}{\sqrt{\gamma R_{gas} T_3}}. \tag{4.36}$$

The length of the last stage blade is a critical parameter due to fact that it is the longest blade with the highest tip speed. The large centrifugal force acting on that component sets the critical mechanical design and manufacturing limits. A longer and heavier blade (as quantified by the large exit annulus area) has a large impact on the turbine cost via blade material selection.

The centrifugal stress acting on the last stage blade can be calculated as

$$\frac{\sigma_c}{\rho} = \left(A_{an}N^2\right)\frac{\pi}{3{,}600}\left(1+\frac{A_{x,t}}{A_{x,h}}\right) \tag{4.37}$$

where ρ is the density of the blade alloy, N is the rotational speed in rpm and $A_{x,t}$ and $A_{x,h}$ are the blade cross-sectional areas at the blade tip and hub, respectively. The first term in parentheses on the right-hand side of Equation 4.37 is commonly used in the industry as a measure of the centrifugal stress and referred to as "A-N-squared" or AN^2 with the units of 10^{-9}in.^{2}-rpm^2 in USCS.

In general, the following limitations should be kept in mind:

1. Although large A_{an} is beneficial for performance due to lower exit kinetic energy, for very low Ma_3, experience suggests that the size/cost disadvantage overcomes the performance advantage.
2. Mechanical limitations may force even higher Mach numbers and smaller A_{an} especially for uncooled last stage blades with high rotor-relative temperatures.
3. Last stage blade size (thus A_{an} and AN^2) is determined by the worst of the two operating conditions: (i) cold day performance (i.e., highest mass flow rate) and (ii) expected growth (i.e., airflow and output) capability.

4. In any event, exit Mach number should be safely below the sonic limit. Conceptual designs or reverse-engineered cases with turbine last stage exit Mach numbers much lower or higher than standard design practices should be reconsidered. (Usually, it should be kept below 0.8; typical initial design value is 0.5–0.6.)
5. Typical F class AN^2 values are around 70–80 (for 50-Hz units with about 1,400 lb/s airflow). Advanced class gas turbines with airflows pushing 2,000 lb/s (or exceeding it) have AN^2 values well above 100. Integrated Gasification Combined Cycle (IGCC) turbines with higher turbine flows would have higher values vis-à-vis their natural gas-fired cousins.

A numerical example using vintage F class gas turbine data should illustrate the salient aspects of the foregoing discussion. Based on the turbine inlet and exhaust temperatures, an overall ψ can be calculated as shown below. Note that nothing is known about the turbine geometry so that estimation of $U = \omega r$ via known pitch-line radius is not possible *a priori*. However, considering that the speed of sound within the HGP is typically greater than 2,000 ft/s and sonic or supersonic tangential velocities leading to shock waves are not desirable, assuming U = 1,500 ft/s is a quite reasonable choice for this exercise. This implies a pitch-line radius of about 4–5 ft at 3,000 rpm. Assuming that TIT is 2,600°F and exhaust temperature is 1,085°F (roughly a late 1990s Siemens V94.3A 50-Hz gas turbine), we find that

$$\psi = \frac{0.3\,\text{Btu/lb-R} \times 32.174\,\text{lbf-s}^2/\text{lb-ft} \times 778.17\,\text{Btu/ft-lbf} \times (2{,}600 - 1{,}085)\text{R}}{(1{,}500\,\text{ft/s})^2} = 5.1.$$

(Using USCS in this example is a deliberate choice; it is a very error-prone calculation so that a worked-out example should be beneficial from a learning perspective. The chore is much simpler in SI units – no need for the extra unit conversion factors.) The value of ψ we found above is clearly too high to be accommodated by a single-stage turbine design. Determining the actual number of stages is a process of optimization. For a four-stage turbine, one can infer the average value of $\psi = 5.1/4 \sim 1.28$. From the Smith chart in Figure 4.7, this is indeed a very reasonable average value from an aerodynamic efficiency perspective. From a cooling flow distribution perspective, a higher value is preferable for the front-end stages. In general,

- If the average stage loading is higher than 1.6–1.7, one should consider adding another stage
- If the average stage loading is less than ~1.2, one should consider removing a stage.

What we found in this example is rather close to the lower limit. This is why in 1990s, some OEMs (e.g., Siemens/Westinghouse and MHI) went with four-stage turbines whereas GE went with three-stage turbines with a highly loaded first stage. (You can repeat this exercise for an advanced class gas turbine with TIT of 2,900°F, and you will find out that a three-stage turbine is out of the question.)

Assuming a reasonable value of 0.6 for the last stage exit Mach number and a specific heat ratio of $\gamma = 1.3$ and $R_{gas} = 0.0685$ Btu/lb-R for the hot gas, from Equation 4.35, MFP is found as

$$\text{MFP} = \frac{0.6\sqrt{\dfrac{1.3 \times 32.174}{0.0685 \times 778.17}}}{\left(1 + \dfrac{1.3-1}{2}0.6^2\right)^{\frac{1.3+1}{2(1.3-1)}}} = 0.4345\ \text{lb}\sqrt{\text{R}}/\text{s-lbf}.$$

Consequently, assuming a reasonable 5° swirl angle for the last stage and 15 psia pressure, the exit stage annulus area can be calculated as

$$A_{an} = \frac{1{,}479\sqrt{1{,}085 + 460}}{(15 \times 144)\cos\left(\dfrac{5\pi}{180}\right)0.4345} = 62\ \text{ft}^2.$$

This annulus area translates into an AN^2 value of 81 for the rotational speed of 3,000 rpm. The stage exit axial velocity can be found from the mass flow rate definition as

$$u_3 = \frac{1,479}{0.025 \times 61} = 945 \text{ ft/s}$$

with hot gas density of 0.025 lb/ft³. Corresponding flow coefficient can be found as $\phi = 945/1,500 = 0.63$.

4.4.3.2 Turbine Cooling

Rotor-relative stagnation or total temperature is the gas temperature that determines the amount of cooling flow via the cooling effectiveness curve,

$$\frac{\dot{m}_{ca}}{\dot{m}_{gas}} \propto \left(\frac{\varphi}{\varphi_\infty - \varphi}\right)^X, \tag{4.38}$$

$$\varphi = \frac{T_{gas} - T_b}{T_{gas} - T_{ca}}. \tag{4.39}$$

Equation 4.38 represents the correlation between cooling air flow (as a fraction of stage gas flow) and the cooling effectiveness, φ, defined by Equation 4.39 (where subscripts b and ca denote bulk metal and cooling air temperatures, respectively). In Equation 4.39, which defines the cooling effectiveness, T_{gas} is the rotor-relative stagnation/total temperature, which is given by

$$T_{2R,tot} = T_{3R,tot} = T_{3,tot} + \frac{V_3^2}{\psi}\left(0.5 + \phi \tan \alpha_3\right). \tag{4.40}$$

Depending on the type of component cooling technology (i.e., convection, film or effusion – full-coverage film – cooling), the asymptotic value of φ, φ_∞, at infinitely high coolant flow rate and the exponent X have different values. Sample cooling effectiveness curves are plotted in Figure 4.8 [11]. For the convection and film cooling curves, a linear relationship (i.e., X = 1 in Equation 4.38) is inferred from the figure (φ_∞ is 0.8 for convection and 0.9 for film cooling).

In *convection cooling*, the coolant enters the blade from the root and goes through the hollow core of the blade (or channels within) to exit at the tip and mix with the main gas flow. In *film cooling*, the coolant exits through holes on the blade surface and mix with the main gas flow.

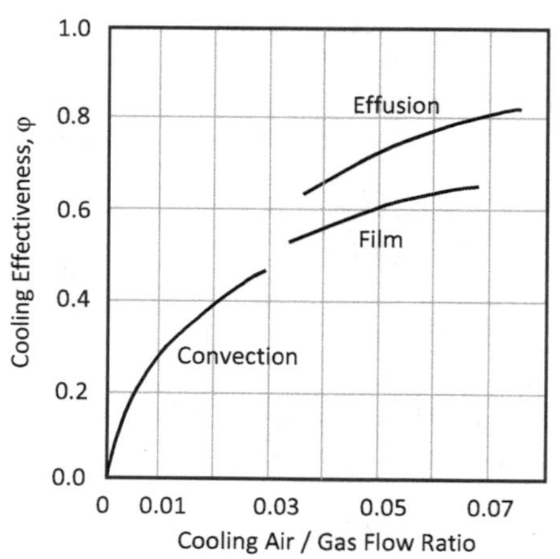

FIGURE 4.8 Typical cooling effectiveness curves [11].

In doing this, the coolant forms a boundary layer (i.e., film) on (preferably the hottest) parts of the blade surface. In the extreme upper limit, a very large number of small holes are electrochemically drilled into the blade wall so that the protective (i.e., insulating) film covers the entire blade surface. This is called as *transpiration*, *full-coverage film* or *effusion* cooling. Presently, this technology is not available due to difficulties and costs associated with the manufacturing of components with large number holes. This may change with advances made in *additive manufacturing* (commonly known as *3D printing*).[3] In the meantime, convective heat transfer coefficients are enhanced via different techniques such as ribbed channels with turbulators and impingement or pin-fin cooling to increase the effectiveness of convection and film cooling.

4.4.3.3 Compressor Aero

Development of efficient compressors has been critical in the coming of age of gas turbines as prime movers in aviation and reliable power producers. The compressor design itself has evolved hand in hand with the science of aerodynamics during the first quarter of the twentieth century. Design of axial, multi-stage compressors with aerodynamically optimized blades and airfoil profiles is crucial to gas turbine performance and cost.

The principles governing the aerothermodynamics of axial compressors are very similar to those for the axial turbines discussed earlier. The nomenclature and definition of key parameters are analogous or even identical to those for the turbines with a big difference:

- A turbine stage comprises a stator and a rotor in the direction of the gas flow
- A compressor stage comprises a rotor and a stator in the direction of the airflow.

In the compressor stage, air is accelerated by the rotor and decelerated in the stator, where the kinetic energy imparted on the air by the rotor is converted to static pressure. The compressor is driven by the turbine on the same shaft and roughly consumes one-half of the shaft power generated by the latter. The airflow in the compressor takes place in an adverse pressure gradient so that the design of a compressor for a given overall PR requires many stages with small pressure rises. In order to give the reader a sense of the significant development in axial compressor aerothermodynamics, consider that

- The 8-stage compressor of WWII jet engine Jumo-004 had a PR of about 3:1
- The 18-stage compressor of a 1990s vintage F class gas turbine had a PR of about 16:1
- Modern H class gas turbines with 14-stage compressors have cycle PRs of 20:1 or higher
- The next-generation advanced class gas turbine of Siemens (the HL class) is designed for a PR of 24:1 with only 12 stages.

The velocity triangles of a compressor stage are similar to those for a turbine stage (see Figure 4.6). The absolute velocity of the air in a compressor stage increases in the rotor while its velocity relative to the rotor decreases. This diffusing flow can only be stable for a small pressure rise in order to prevent the *stalling* or flow-reversing effect of the growing boundary layer on the annulus walls. This is why, unlike in turbines where the pressure gradient and the boundary layer growth are favorable in the flow direction, a compressor requires many more stages.

Compressor design algorithms involve empirical knowledge obtained from a large body of rig tests and 3D numerical computations using advanced CFD software. Fortunately, for the purposes of performance calculations in design and off-design conditions, one does not need to go into the aerodynamic design details of the compressor stages. The pertinent design parameters are

[3] At one time, metal foams with a naturally porous structure were considered for the manufacture of turbine blades to facilitate effusion cooling.

- The overall PR
- Number of stages
- Stage PR
- Polytropic efficiency.

The polytropic efficiency has the useful property to be constant for individual stages and the overall machine. Using a vintage F class gas turbine as an example (the same one used in Section 4.4.1), based on the published value of PR, 17.2:1 and an assumed polytropic efficiency of 92.5%, CDT can be estimated from (with inlet temperature of 59°F)

$$\text{CDT}[°\text{F}] = 519 \cdot \text{PR}^{\frac{\gamma-1}{\eta_p \gamma}} - 460$$

as 790°F with $\gamma = 1.4$. (For detailed discussion of polytropic and isentropic efficiencies, and other pertinent information, refer to Chapter 11 of **GTFEPG** and the references cited therein.)

The entitlement value for compressor polytropic efficiency is probably somewhere between 93% and 94%. It is not recommended to use values higher than that for conceptual and reverse-engineering studies. A scaling with number of stages and/or overall PR is a common way of adjusting the polytropic efficiency for design studies, e.g.,

$$\eta_p = \eta_{p,\max} - \frac{\text{PR}^{\alpha}}{N_{\text{stg}}}. \tag{4.41}$$

The stage loading coefficient for a compressor is analogous to that for the turbine, i.e.,

$$\psi = \frac{c_p \Delta T_{\text{tot}}}{U^2}.$$

Results of the cascade tests have shown that, for the most efficient operation, the design point value of ψ is limited to the range of 0.3–0.4. Compressor ψ is less than 1 because, in this case, the "cause" is the denominator of the ratio in its definition (i.e., rotating shaft) and the "effect" is the numerator, i.e., the rise in pressure (and temperature) of air flowing through the stage. Total temperature rise across the compressor stage is

$$\Delta T_{\text{tot}} = \frac{\psi U^2}{gJc_p} = \frac{\psi(\omega r)^2}{gJc_p}. \tag{4.42}$$

For the example herein, assuming a pitch-line radius of 4 ft and $\psi = 0.3$, we find that (50-Hz gas turbine, air-specific heat 0.25 Btu/lb-R)

$$\Delta T = \frac{0.3 \times \left(\frac{2 \times \pi \times 3{,}000}{60} \times 4\right)^2}{32.174 \times 778.17 \times 0.25} = 79°\text{F}.$$

The work capacity or temperature rise across a compressor stage is reduced by the work-done factor λ, which is about 0.85 for an axial compressor with 12 stages or more. Thus, the ideal average temperature rise calculated above becomes 67°F (about 37°C).

Dividing the temperature rise across the compressor, i.e., 790 − 59 = 711°F, by the estimated average temperature rise of 67°F, we arrive at 11 stages. This crude estimate is not too far off from the actual value of 15 (for the vintage Siemens V94.3A with Pratt–Whitney-designed compressor with 3D-Aero airfoils). This gives an average temperature rise of 47°F or 26°C and, everything else being the same, implies a ψ value of 0.18 (too low). From a different perspective, the (average) pitch-line radius is 3.4 ft with $\psi = 0.3$ and 15 stages, which is plausible. (In any event, to get more precise with simple calculations of this type is not possible.)

The limiting value of ψ can be gleaned from the velocity triangles for the compressor stage. Starting from the Euler equation for the compressor, i.e.,

$$c_p \Delta T_t = \omega r (v_2 + v_3) \tag{4.43}$$

it can be shown that ψ and ϕ (which is defined the same way for the compressor, i.e., via Equation 4.27) are correlated to each other via

$$\psi = 1 - k\phi \tag{4.44a}$$

$$k = \tan\alpha_1 + \tan\beta_2. \tag{4.44b}$$

In Equation 4.44b, α_1 is the angle between air *absolute* velocity and the stage centerline at the stage/rotor inlet; and β_2 is the angle between air *relative* velocity and the stage centerline at the stator inlet. Equation 4.44a implies that when there is no flow, i.e., $\phi = 0$, $\psi = 1$. Clearly, ψ should be less than one for stable and efficient compressor operation (just like ψ cannot be less than one for the turbine stage, ψ cannot be greater than one for the compressor stage; i.e., recall the cause–effect "thing").

Using the isentropic formula for the compressor stage and expanding it by means of the binomial theorem (since $\Delta T/T_{in} \ll 1$), one gets[4]

$$\frac{\Delta p_{tot}}{\rho U^2} = \eta_s \psi \tag{4.45}$$

where ρ is the air density and η_s is the stage isentropic efficiency. The term on the left-hand side of Equation 4.45 is the pressure coefficient π. Thus, the overall performance of a compressor stage can be expressed in terms of three "coefficients": stage loading coefficient ψ, flow coefficient ϕ and pressure coefficient π (or the isentropic efficiency, because $\pi = \eta_s \psi$).

When $\psi = 1$, the Euler equation for the compressor stage, which is equivalent to that for the turbine stage, tells us that $\psi = \Delta v/U$. In other words, the total change in the tangential air velocity (i.e., the *whirl*) is exactly equal to the peripheral speed of the rotor at the pitch-line diameter. This is equivalent to excessive diffusion in the stage and is unfavorable for efficiency as well as stability. A good range for ψ at design is 0.3–0.4. At off-design conditions, flow angles will change so that k cannot stay constant. Thus, the linear relationship between ψ and ϕ given by Equation 4.44a breaks down. It is more like a *concave* function. In particular,

- At high values of ϕ (i.e., α_1 approaching zero), choking will occur with a steep drop in efficiency and limited flow.
- At very low values of ϕ (i.e., α_1 approaching 90°), the flow will stall with a sharp drop in efficiency.

Thus, it is imperative that all stages operate in a region of high efficiency, safely removed from the choke or stall limits, at all possible off-design conditions. This is the big challenge in multi-stage axial compressor design. It is easy to figure out that this is nearly impossible to achieve, at least at operating conditions far from the design point, without forcing changes in the compressor geometry. This is the reason why almost all industrial gas turbine compressors are equipped with moveable *inlet guide vanes* (IGVs) and *variable stator vanes* (VSVs) in the front stages (usually, stages one, two and three are VSVs). These vanes are operated by the turbine controller to ensure stable operation at all ambient and loading conditions, including startup and shutdown, without damaging the compressor via stall or surge.

Design calculations for a gas turbine compressor do not pose any difficulty. Specifying the overall PR, number of stages and the polytropic efficiency is enough. In most cases, the number of stages

[4] The discussion follows the treatment in Section 5.11 of *Gas Turbine Theory* (3rd Edition) by Cohen, Rogers and Saravanamuttoo, Longman Scientific & Technical (1987).

is known and the individual stage PR can be calculated from the overall PR as the one-over-nth power of the latter, where n is the number of stages. If not, as demonstrated earlier, estimating them with sufficient accuracy is straightforward. Usually, using a number of "pseudo stages" that is large enough for extraction flows to supply the turbine HGP cooling air is good enough.

There are two ways to account for the off-design performance characteristics: *stage stacking* and performance maps. Stage stacking is the term for successive calculation of pressure and temperature changes in each stage separately, where the exit conditions from the preceding stage provide the inlet conditions to the next stage. Using the stage characteristics for each individual stage (exact relationships in lieu of the approximation in Equation 4.44), the overall compressor performance can be predicted analytically. The analytical predictions are later verified, validated and fine-tuned using data obtained from the rig testing.

While stage stacking is an extremely valuable tool for the compressor designer and the aerodynamicist, it is an overkill for the performance engineer. The most feasible approach for the latter is to use the compressor *performance maps*, which are graphical compilations of the dimensionless data gathered in the rig tests. Performance maps relate pressure rise, mass flow rate and rotational speed of the compressor at a broad range of ambient (i.e., suction) and loading conditions. A sample axial compressor performance map is shown in Figure 4.9 [12].

Performance map parameters are dimensionless "corrected" parameters that represent the pressure, temperature and mass flow characteristics of the compressor over a range of ambient and loading conditions for a fixed IGV setting. The horizontal axis of the performance map is the *corrected flow parameter*, CFP, normalized with respect to a reference condition (designated by the subscript 0), i.e.,

$$CFP = \frac{\dot{m}}{\dot{m}_0}\sqrt{\frac{T p_0}{p T_0}}. \tag{4.46}$$

The vertical axis of the performance map is the PR of the compressor as a fraction of the reference point PR. Each corrected speed line is defined by the *corrected speed parameter*, CSP, also normalized by its reference value, i.e.,

$$CSP = \frac{N}{N_0}\sqrt{\frac{T_0}{T}}. \tag{4.47}$$

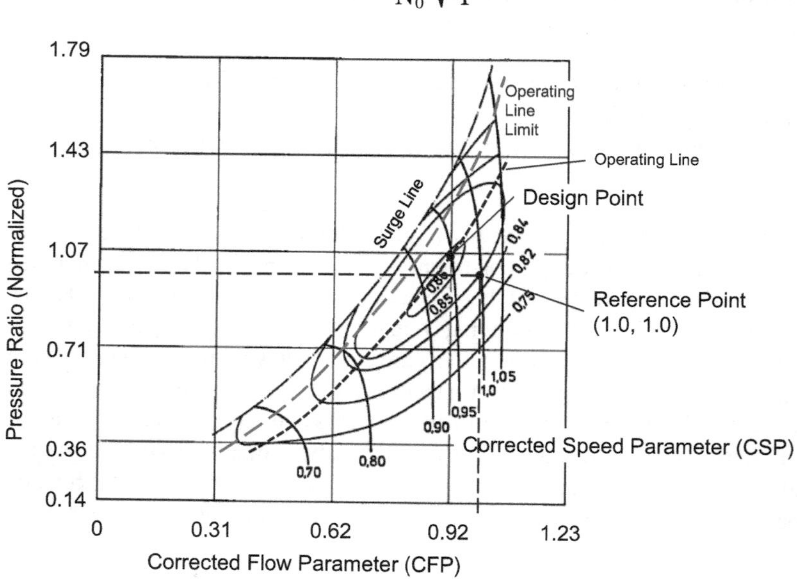

FIGURE 4.9 Axial compressor performance map [12].

The elliptical contours on the map are constant isentropic efficiency lines. Mathematically, it is straightforward to show that corrected flow and speed parameters are equivalent to flow (in the axial direction) and rotational (with reference to U) Mach numbers, respectively. In terms of flow geometry, all operating conditions with the same values of CSP and CFP should result in similar velocity triangles. This means that vane angles and airflow directions will match and the compressor will have the same performance in terms of PR, isentropic efficiency and, consequently, temperature ratio. This is the principle that underlies the generation of dimensionless maps to represent the entire operating envelope of the compressor.

For a heavy-duty industrial gas turbine synchronized to the grid, rotational speed N is constant. Thus, CSP varies only with compressor inlet temperature; i.e., CSP is inversely proportional to the inlet temperature. At high values of corrected speeds (i.e., on cold days), the speed lines are nearly vertical whereas the lower corrected speeds (i.e., on hot days) display a parabolic relationship (concave) between PR and airflow. The operating line (the "op-line") on the map in Figure 4.9 represents the effect of ambient temperature (via density, i.e., $\rho = p/RT$) on airflow and PR.

At any given mass flow rate, a compressor works against a pressure resistance, which is determined by the mass flow through a flow restriction or orifice downstream, e.g., a valve in process applications or the S1N throat area in a gas turbine, which acts as a *choked nozzle*. Thus, air pressure at the compressor discharge is proportional to the hot gas pressure at the turbine inlet, which goes up and down in proportion to the gas flow (airflow minus cooling flows plus fuel flow). At constant rotational speed, compressor is a constant volume flow machine. Consequently, airflow through the compressor goes up and down in inverse proportion to the ambient temperature. Since the ambient/suction pressure does not vary appreciably, at constant IGV setting, this dynamic leads to the op-line shown in Figure 4.9, i.e., from left to right or from high inlet/ambient temperature to low inlet/ambient temperature, increasing airflow and increasing PR.

Each speed line is bound by two limits: *stall/surge line* on the left/high end and *choke line* on the right/low end. At high CSP (i.e., low ambient, high airflow), the speed line is nearly vertical because of choking at the compressor inlet. At a given CSP, with increasing airflow and decreasing PR, choking in the rear stages dictates the compressor operation. With increasing PR and decreasing airflow, stable operation is limited by stall (in rear stages at low ambient temperatures and in front stages at high ambient temperatures). This is of importance for operation with low-Btu syngas and will be discussed in more detail in Section 18.1.

An *operating limit line* (OLL) is imposed on the compressor operation as an upper limit for stable operation, which is sufficiently/safely away from the surge line. The difference between the OLL and the surge line is a margin accounting for various instabilities arising from blade tip clearances and machining tolerances. This *surge or stall margin* (SM) is usually defined as follows for a given corrected speed (i.e., speed line)

$$SM = \frac{PR_{SL}}{PR_{OL}} - 1 \tag{4.48}$$

where the numerator of the fraction on the right-hand side is the surge pressure and the denominator is the op-line pressure. A typical value for SM is 0.15.

The first defense against compressor surge or stall is recirculation or bypass, which involves air bled from the compressor discharge that is sent back (i.e., recirculated) to the compressor inlet. Note that air recirculation to compressor inlet is also used to prevent inlet icing.

Opening and closing of the IGVs will increase and decrease the compressor mass flow rate. This can be approximated by a simple formula as follows

$$C = 1 + (K-1)\left(1 - \frac{Y_{IGV}}{100}\right) \tag{4.49}$$

where

C = Flow multiplier at different IGV settings

Y_{IGV} = 0–100 for fully closed to fully open

K = Fraction of mass flow when Y_{IGV} = 0 (fully closed).

Equation 4.49 will scale the base map to off-design conditions with different IGV settings. (The reader is referred to pp. 302–303 in **GTFEPG** for a more in-depth coverage.) However, it does not account for the change in the shapes of the speed lines, which may vary from those in the base (i.e., IGVs fully open) map. This will result in errors in stall/surge margin calculations and pressures.

The change in the speed lines via modulating IGVs and/or VSVs can be qualitatively estimated via the ideal relationship between ψ and ϕ given by Equation 4.44. Changing the position of the VSVs will change α_1 for the particular stage while β_2 stays constant. For example, increasing α_1 (i.e., increasing k) will reduce ψ at a given ϕ and the stage characteristic will be shifted to the left (on a $\psi - \phi$ surface). The reason is the reduction in the airflow increasing α_1. While the effect can be predicted reasonably accurately via detailed calculations, the actual settings are determined and optimized via comprehensive rig tests.

In a stage-by-stage HMB simulation model, the best way of modeling the compressor in its entire operating range is to divide the compressor into blocks, which can be done in two ways:

- Each block corresponds to a compressor stage
- Each block corresponds to a compressor section with the break points set to turbine cooling flow extraction locations (e.g., 9th stage, 11th stage and 13th stage exits for F class machines).

The first one is equivalent to stage-stacking model. This may be the best option for modeling the conceptual studies where the exact locations of cooling flow bleed points are subject to optimization. The second one is better suited to most conceptual and reverse-engineering studies with the focus on design and off-design performance.

Such a model structure is shown in Figure 4.10 for a gas turbine with three-stage axial turbine. Typical axial compressor for this type of unit (e.g., GE's vintage F class products) has up to 18 stages

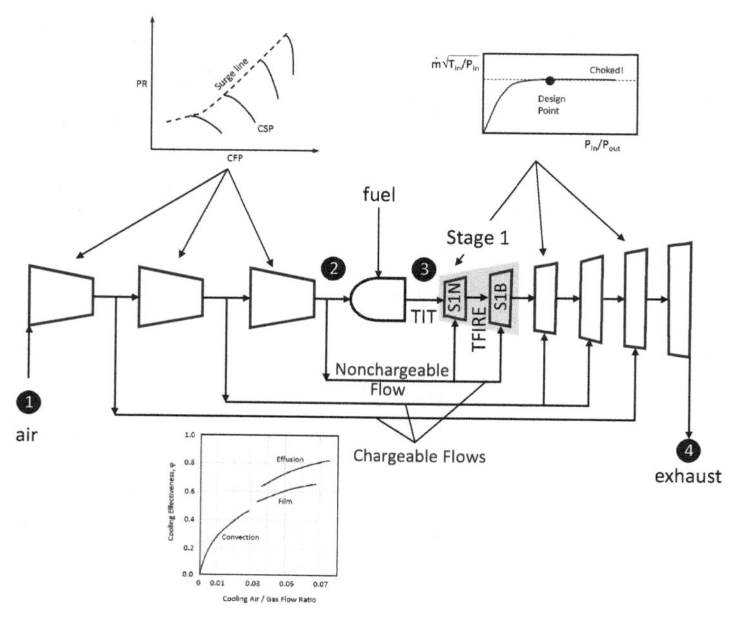

FIGURE 4.10 Stage-by-stage gas turbine model.

with two interstage bleed locations for chargeable cooling flows. Although tedious and iterative, requisite performance map, cooling flow and turbine stage geometry calculations can be done in MS Excel with VBA functions. A more convenient alternative is a software that can facilitate this, e.g., Thermoflow, Inc.'s THERMOFLEX with cooled turbine stage module (GASCAN).

4.5 FUEL FLEXIBILITY

Development of heavy-duty industrial gas turbine combustion systems was focused on natural gas (essentially methane) as the primary fuel with distillate oil as secondary fuel. However, the ability to burn a wide variety of fuels, known as fuel flexibility, is a critical gas turbine feature that provides a hedge against increasing fuel costs and reducing fossil fuel resources. For comprehensive coverage of gas turbine fuel flexibility, the reader is referred to the paper by Meher-Homji et al. [13]. The critical gas turbine system component from a fuel flexibility perspective is the combustion system. Apart from emissions, the key areas of combustor operability that are affected by fuel composition and LHV are lean blowout (LBO), flashback, combustion dynamics (noise) and autoignition. The key fuel characteristics that have an impact on the aforementioned operability problems are composition (specifically H_2 content), heating value (Btu/scf), stoichiometric flame temperature and *flammability*. Flammability of a fuel is expressed by upper and lower limits, UFL and LFL, respectively, which refer to the range of compositions, for fixed pressure and temperature, within which an explosive reaction is possible when an external source of ignition is introduced. A representative selection of fuel gases in the order of decreasing volumetric heating value is given in Table 4.19 [14]. Note that syngas in the table is carbureted water gas, which is produced by passing steam over red-hot coke. The calorific value is boosted by passing the gas through a heated retort into which oil is sprayed.

For stable diffusion flame operation, a *flammability ratio* in excess of 2.0 (ratio of UFL to LFL) is required. From Table 4.19, this value for natural gas or methane at 1 atm is around 3.0 (cf. close to 20 for hydrogen). Fuels with very low flammability ratios such as blast furnace gas (BFG) are usually blended with other fuels (e.g., coke oven gas, COG, in steel mill applications). In low-Btu fuel combustion systems, a conventional fuel such as natural gas or fuel oil is always utilized for startup and shutdown.

Flammability limits depend on the type and strength of the ignition source, type of atmosphere (e.g., in oxygen limits are much wider than in air) and pressure and temperature of atmosphere. Natural gas UFL at 20 atm is nearly 60% for a flammability ratio of ~12.0 (LFL dependence on pressure is much weaker). Increasing temperature reduces both UFL and LFL.

Autoignition temperature (AIT) is the temperature, at a given pressure, for which an explosive reaction at a fixed composition mixture within its flammable range is possible. The reaction is initiated by autocatalytic reaction without any external ignition source. The minimum AIT is a strong

TABLE 4.19
Selected Fuel Properties (UFL and LFL Are Upper and Lower Flammability Levels, Respectively, at 1 atm and 77°F and Given in Terms of Fuel Concentration by Volume) [14]

	Flame Temperature		NOx	H_2	CO	LHV	LFL	UFL
	°F	°C	rel.	%(v)	%(v)	Btu/scf	%(v)	%(v)
Natural gas	3,720	2,049	1.42	-	-	929.3	4.8	13.5
Methane	3,550	1,954	1.00	-	-	896.0	5.0	15.0
COG	3,590	1,977	1.10	53.2	5.9	579.7	4.8	33.5
Syngas (from coal)	3,690	2,032	1.34	32.5	24.1	431.8	5.3	40.7
CO	3,860	2,127	1.74	-	100.0	316.0	12.5	74.2
Hydrogen	3,930	2,166	1.91	100.0	-	270.0	4.0	75.0
BFG	2,600	1,427	0.10	3.5	26.5	95.0	35.0	73.5

function of fuel type, pressure and concentration. At 1 atm, minimum AITs for methane and hydrogen are, 1,350°F and 1,140°F, respectively. Increased pressures typically reduce the AIT.

Another important fuel characteristic is *flame speed* or burning velocity, which quantifies the rate of flame propagation. Flame speed is a function of fuel concentration and is less than 2 ft/s for most hydrocarbon fuels, whereas it can be more than 8 ft/s for hydrogen at 40% concentration. This is the primary reason for the difficulty in designing DLN combustors for fuels with large hydrogen content (due to flashback concerns). Therefore, conventional diffusion combustors are used for most hydrogen-containing fuels with diluent nitrogen or steam/water injection for NOx control. Another concern for combustor design is the tendency of the combustion flame to be "blown out" from its anchor point. For many low-Btu fuels, H_2 and CO are the two key components. Fuels or fuel mixtures with high H_2 content (very high flame speed) have much smaller reaction times than fuel mixtures with high CO content (low flame speeds). Thus, "fast" mixtures with higher H_2 content will blow out at leaner equivalence ratios than "slow" mixtures with high CO content.

Combustion dynamic instability, also referred to as combustion noise or "humming", is characterized by large-amplitude pressure oscillations driven by unsteady heat release. The problem is especially severe when pressure and heat release oscillations are nearly "in phase", i.e., the phase difference is less than 90°. Fuel composition and LHV variations (especially in fuels such as BFG) affect the combustion dynamics by altering the phase angle. DLN combustors are especially susceptible to this problem due to the fact that they operate near the LBO limit, where small perturbations are amplified to large effects. Combustion dynamic instabilities are potentially harmful to the combustor parts and downstream HGP components.

The key engineering parameter that is used to characterize gas composition is the *Wobbe index*. The basic WI is usually modified to incorporate the gas compressibility effects and known as the *modified Wobbe index* (MWI):

$$\text{MWI} = \frac{\text{LHV}}{\sqrt{T_{\text{fuel}} \dfrac{\text{MW}_{\text{fuel}}}{28.96}}}$$

where MW is the molecular weight of the fuel and T is its temperature in °R with the LHV in Btu/scf. In essence, MWI is a relative measure of the energy input to the combustor at a fixed PR and determines the ability of the fuel cleanup and injection system to accommodate the variations in composition and heating value.

There is an allowable range of MWI to ensure proper operation of the fuel injection system with the requisite fuel nozzle PRs at all modes of operation. That range is typically ±5%; however, OEMs for most modern H class machines claim allowable MWI ranges up to ±10%. For example, for the natural gas fuel with 1,000 Btu/scf at 365°F, the value of the MWI is about 47. Thus, the allowable variation can be up to about ±5. As the MWI formula indicates, in systems with performance fuel heating, some of the effects of composition variation can be moderated by changing the fuel gas temperature. Within certain MWI limits, the gas turbine can operate successfully without requiring combustion system hardware modification. Dual fuel systems require separate independent fuel handling trains, which include control valves, manifolds and fuel nozzles.

Composition and MWI variability in low-Btu fuels is a significant concern, which is amplified by fuel blending and "co-firing". For example, in steel mill applications, a fuel compressor is required to ensure delivery of the fuel to the gas turbine at sufficiently high pressure to ensure stable and reliable fuel flow control. In most cases, BFG is blended with COG to provide a higher LHV mixed gas fuel to the GT combustor. Fuel blending and compression are accommodated by improved controls and system design changes with allowable MWI limit of ±10%.

Due to its high flammability and lower ignition temperature, burning H_2 or any other syngas with high H_2 content in standard DLN combustors with fuel-air premixing is not possible. Such gaseous fuels are burned in diffusion flame combustors with steam (from the bottoming cycle) or nitrogen (from the air separation unit, ASU, if H_2 is generated in a gasification plant) diluent injection for NOx control. For those and other issues pertaining to H_2 combustion in heavy-duty industrial gas turbines (i.e., enthalpy drop difference in turbine expansion, turbine-compressor mismatch due to changing turbine mass flow rate and increased heat transfer rate with impact on HGP cooling), refer to the paper by Chiesa et al. [15]. For detailed information on combustion of syngas basics and specific problems including blowout and stability (so-called "humming"), consult the books edited by Lieuwen and Yang [16] for a general understanding and Ref. [17] for syngas-specific issues.

Nevertheless, due to its strong appeal from a sustainability perspective (i.e., zero CO_2 emissions), major gas turbine OEMs (supported by the governments, e.g., the US Department of Energy, DOE) have been actively developing DLN combustors capable of burning gaseous fuels with high hydrogen content. For example, General Electric recently announced that their DLN 2.6e combustor is capable of handling up to 50%(v) H_2 in fuel gas in their HA class heavy-duty industrial gas turbines. In January 2018, MHPS announced that they had successfully fired 30%(v) hydrogen mix in a large gas turbine combustor. Ansaldo Energia Group's PSM also claims that their FlameSheet™ DLN combustor technology can handle up to 45%(v) hydrogen in the fuel gas mix.

The aforementioned "mismatch" stems from the disturbance of the pressure-flow balance between the compressor and the turbine. For low or medium heating value syngas, the main driver for modification is the increased turbine mass flow rate due to higher fuel flow in the combustor. Some of the increase can be absorbed by opening up the turbine inlet flow area (e.g., via staggering the nozzle vanes). This can account for maybe 6%–8% of the increased mass flow. The rest is achieved by modifications to the compressor to accommodate the increased cycle PR and/or closing the IGVs to reduce the airflow. A detailed description can be found in the paper by Johnson [18].

REFERENCES

1. Gülen, S.C., 2019, Disappearing thermo-economic sanity in gas turbine combined cycle ratings – A critique, GT2019-90833, *ASME Turbo Expo 2019*, June 17–21, 2019, Phoenix, AZ.
2. Yuri, M., Masada, J., Hada, S., Wakazono, S., 2017, Operating results of J-series gas turbine and development of JAC, *Mitsubishi Heavy Ind. Tech. Rev.*, Vol. 54, No. 3, pp. 16–22.
3. El-Masri, M.A., 1984, On thermodynamics of gas turbines, part 1, *29th International Gas Turbine Conference and Exhibit*, June 4–7, 1984, The Hague, the Netherlands.
4. El-Masri, M.A., 1985, On thermodynamics of gas turbine cycles: Part 1-Second law analysis of combined cycles, *J. Eng. for GTs Power*, Vol. 107, pp. 880–889.
5. El-Masri, M.A., 1986, Exergy analysis of combined cycles part 1: Air-cooled Brayton-cycle gas turbines, 86-JPGC-GT-9, *Joint ASME/IEEE Power Generation Conference*, October 19–23, 1986, Portland, OR.
6. El-Masri, M.A., 1987, Exergy analysis of combined cycles: Part 1—Air-cooled Brayton-cycle gas turbines, *J. Eng. Gas Turbines Power*, Vol. 109, No. 2, pp. 228–236.
7. Elmasri, M.A., Design of Gas Turbine Combined Cycle and Cogeneration Systems Seminar, Chapter 2: Gas Turbine Cycle Fundamental Thermodynamics, October, 1990.
8. Sultanian, B.K., 2019, *Logan's Turbomachinery*, 3rd Edition, CRC Press, Taylor & Francis Group, Boca Raton, FL.
9. Mattingly, J.D., 2005, *Elements of Gas Turbine Propulsion*, Tata McGraw-Hill Edition, New Delhi.
10. Smith, S. F., 1965, A simple correlation of turbine efficiency, *J. Royal Aeronaut. Soc.*, Vol. 69, p. 467.
11. Kano, K., Matsuzaki, H., Aoyama, K., Aoki, S., Mandai, S., 1991, Development study of 1,500°C class high temperature gas Turbine, 91-GT-297, *ASME International Gas Turbine and Aeroengine Congress and Exposition*, June 3–6, 1991, Orlando, FL.
12. Münzberg, H.-G., Kurzke, J.T., 1977, *Gasturbinen – Betriebsverhalten und Optimierung*, Springer Verlag, Berlin abd Heidelberg.
13. Meher-Homji, C.B., Zachary, J., Bromley, A.F., 2010, Gas turbine fuels – System design, combustion and operability, *Proceedings of the 39th Turbomachinery Symposium*, October 4–7, Houston, TX.
14. Baumeister, T. (Editor-in-Chief), 1986, *Marks' Standard Handbook for Mechanical Engineers*, 8th Edition, McGraw-Hill, Book Company Inc., New York.

15. Chiesa, P., Lozza, G., Mazzocchi, L., 2005, Using hydrogen as gas turbine fuel, *J. Eng. GTs Power*, Vol. 127, No. 1, pp. 73–80.
16. Lieuwen, T.C., Yang, V. (Editors), 2005, *Combustion Instabilities in Gas Turbine Engines*, AIAA, Inc., Reston, VA.
17. Lieuwen, T.C., Yang, V., Yetter, R. (Editors), 2010, *Synthesis Gas Combustion – Fundamentals and Applications*, CRC Press, Boca Raton, FL.
18. Johnson, M.S., 1992, Prediction of gas turbine on- and off-design performance when firing coal-derived syngas, *J. Eng. Gas Turbines Power*, Vol. 114, No. 2, pp. 380–385.

5 Steam Turbine

Steam turbines do not come in well-defined product lines similar to the gas turbines. As we have seen in Chapter 4, a specific gas turbine product is fixed in its performance (i.e., generator output and efficiency) and key performance parameters (i.e., airflow, turbine inlet temperature or TIT and exhaust conditions) at baseload (i.e., inlet guide vanes or IGVs at their "fully open" position with normal firing according to the control curve) at a specific site condition (i.e., ISO ambient conditions). Original equipment manufacturer (OEM)-provided rating data is regularly published in trade publications such as the *Gas Turbine World*. This is a natural result of the fact that the gas turbine is an "air-breathing" heat engine with constant volumetric flow. In other words, at a constant rpm (3,000 or 3,600 rpm, depending on the grid frequency), the machine sucks in a well-defined volumetric quantity of air. Once the ambient conditions are specified (e.g., ISO conditions), the airflow is also fixed (via density). Thus, at its nominal TIT, say, 1,500°C, everything about the gas turbine is known precisely.

This is not the case for combined cycle steam turbines. Their output is a function of

- High-pressure (HP), intermediate-pressure (IP) and low-pressure (LP) steam admission pressures and temperatures
- HP, IP and LP steam flow rates
- Condenser pressure.

One can, of course, fix steam admission conditions and the condenser pressure precisely (note that there are *many* such combinations). However, steam flows are strongly dependent on design and performance characteristics of *two other pieces* of key plant equipment:

- Gas turbine exhaust flow and temperature
- Heat recovery steam generator (HRSG) design (e.g., three-pressure reheat, 3PRH, or two-pressure reheat, 2PRH) with or without supplementary firing.

Consequently, there is a wide range of steam flows and steam turbine performances. Another confounding factor is the availability of a suitable LP turbine "last-stage bucket" (LSB) that can derive the full benefit from the specified condenser pressure.

Steam turbines are thus classified in terms of "product lines". General Electric's steam turbine product line (after the 2015 acquisition of Alstom) is summarized in Figure 5.1. Note that

- The "A" series designate a separate HP and combined IP-LP casing configuration with axial LP turbine exhaust.
- The "D" series designate a multi-flow LP turbine with "down" or "side" exhaust and with
  - Separate HP and IP turbines in opposed flow configuration (D-650)
  - Combined HP and IP turbines in opposed flow configuration (D-400 and D-600)
  - Separate HP turbine (no IP turbine), e.g., D-200, for non-reheat applications.

The second tier of classification is based on the reheat feature, i.e., "200" designation for non-reheat and "400" and "600" for reheat applications.

The third tier of classification is based on HP and hot reheat (HRH) steam conditions:

- "400" series for HP pressures up to 1,800 psig (124 barg) – single shell configuration
- "600" series for HP pressures up to 2,680 psig (185 barg) – double shell configuration.

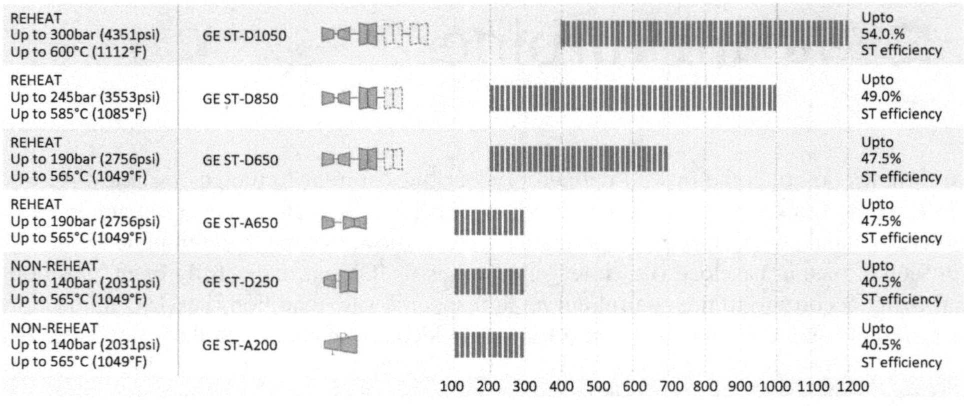

FIGURE 5.1 General Electric steam turbine product line (2016).

Note that 2,680 psig is the maximum HP or main steam throttle pressure (specified at the inlet of the throttle valve) for subcritical steam cycles with drum-type HRSGs.

Both "400" and "600" series are capable of 1,112°F (600°C) HP and/or HRH steam temperatures, whereas the "200" series temperature limit is 1,050°F (565°C). The break point is a function of available materials, manufacturing processes and, ultimately, cost. Materials used in combined cycle steam turbines and HRSG superheaters are summarized in Table 5.1 [1]. For more details on material selection, the reader is referred to the excellent book by Viswanathan [2]. Until the advanced class gas turbines with exhaust temperatures well above 1,100°F came into the market, state of the art in combined cycle steam cycle was 1,800 psig HP pressure and 1,050°F HP and HRH temperatures. Recently, steam cycles with 2,400 psig (or higher) HP pressure and 1,112°F steam temperatures have been adopted – especially for markets with expensive natural gas where thermal efficiency is at a premium.

There is extensive field experience with ferritic steels with up to 12% Cr, which allows for steam temperatures up to 1,112°F (600°C) at high pressures. There has been large-scale, coordinated (i.e., government, academia and industry) research activities to extend this to 625°C or even to 650°C (e.g., COST 522 in Europe or NIMS in Japan), especially for large components such as rotors, casings and valve bodies [3]. (Monoblock rotor forgings with austenitic steel are not possible.)

TABLE 5.1
Material Selection for HRSG Superheater Tubes and Steam Turbine Components

Component	Maximum 1,050°F (565°C)	Maximum 1,112°F (600°C)	625°C–650°C
Superheater tubes	1%–2% CrMo	Austenitic steel	Highly corrosion-resistant austenitic steel
Reheater tubes	9% Cr	9% Cr	
HP (main) steam pipes, headers, valve bodies	2¼ CrMo	9% Cr	Austenitic steel
High-temperature rotors (forging)	1%–2% CrMoV	9%–12% Cr	Austenitic steel, 9%–12% Cr
HP/IP turbine inner casing (casting)	1%–2% CrMoV	9%–12% Cr	Austenitic steel (casting or forging)
HP/IP turbine outer casing (casting)	1%–2% CrMoV	1%–2% CrMoV	9%–12% Cr
LP rotor (forging)		High-purity 3.5% NiCrMoV	

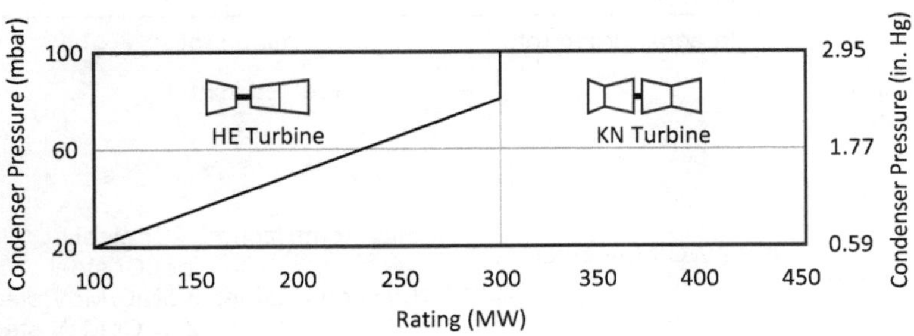

FIGURE 5.2 Siemens CC steam turbine product line application chart.

Material limits can be somewhat extended by steam cooling of the rotor to increase its creep strength. This, of course, has some impact on cycle efficiency in addition to rotor life limit, which may not be as high as that of one manufactured from higher-grade material.

Siemens has product line structuring scheme similar to that of GE in Figure 5.1. In particular,

- Siemens equivalent of GE "A" series is the "HE" turbine ("H" refers to the separate HP turbine and "E" to the combined IP-LP turbine with axial exhaust);
- Siemens equivalent of GE "D" series is the "KN" turbine ("K" refers to the combined HP-IP opposed flow turbine and "N" to the double-flow LP turbine).

The design space of Siemens combined cycle steam turbines is shown in Figure 5.2. As shown in the chart, which is from an early 2000s Siemens publication [4],

- HE turbines (now SST-3000 series) are mainly for lower power range, high condenser pressure (i.e., steam turbine "back pressure") applications[1].
- KN turbines (now SST-5000 series) are mainly for high power range, low back pressure applications.

In general, there are *four* basic types of casing (in some references, "cylinder") configurations:

1. Combined HP-IP-LP (one casing, axial exhaust)
2. Separate HP with combined IP-LP (two casings, axial exhaust)
3. Separate combined HP-IP and double-flow LP (two casings, down/side exhaust)
4. Separate HP, IP and double-flow LP (three casings, down/side exhaust).

Obviously, with increasing gas turbine size and/or number (i.e., two or three gas turbines in 2 × 2 × 1 or 3 × 3 × 1 configuration), one will have to switch from a single-flow, axial exhaust to double-flow, down/side exhaust configuration (or even to two double-flow LP turbines – especially with the largest 50-Hz machines rated at more than 500 MWe). This transition is dictated by the LSB size (i.e., exhaust annulus area). Differentiation within each LP turbine exhaust family depends on different configurations.

Let us refer to the configurations enumerated above from 1 to 4 as A1, A2, D2 and D3, respectively. In A1 configuration, one has to go with a monoblock rotor and two bearings (one at each end). The attractiveness of this compact design from a size/cost perspective (i.e., shorter and less expensive manufacturing, transportation, installation and startup) is countered by aerothermodynamic

[1] Note that, strictly speaking, in the earlier "E" designation, IP and LP flow paths in the combined IP-LP section are in opposed flow arrangement. In SST-3000, designated as "IL", IP and LP flow paths are arranged similar to GE's "A" series.

	Integral forged rotor	Welded rotor
Combined HP-IP-LP rotor	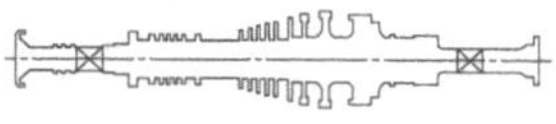 2¼CrMoV steel	Welded joint 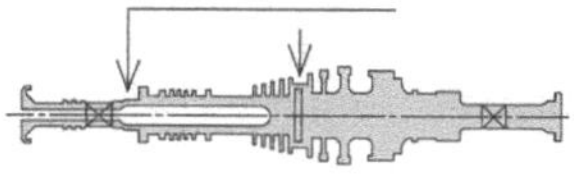 High-temp. zone: 12Cr steel or 9Cr steel Low-temp. zone: 3.5NiCrMoV steel, 2 ¼ Cr MoV steel

FIGURE 5.3 Welded versus monoblock (integral forged) rotors for single casing CC steam turbine [6].

design ramifications. The A2 design features an opposed flow arrangement with the HP turbine flow-path expanding in one direction and the IP-LP turbine in the other. The generator is driven at the HP turbine exhaust side. The monoblock rotor (forging) must have (i) creep rupture strength commensurate with HP steam conditions and (ii) fracture toughness commensurate with the centrifugal force acting on the LSB at the LP section. From Table 5.1, it is clear that this requires different materials. One manufacturer approached this by manufacturing the rotor from 2¼ CrMoV steel with dual tempering (i.e., different tempering and hardening methods were used on the HP and LP ends) and LP side diameter of 1,950 mm (~77 in.) [5]. For sufficient stability, the rotor diameter has to be commensurately large on the HP side, which does not provide the optimal dimensions for designing the HP flow-path. Furthermore, a monoblock rotor of large diameter and mass (usually with a small-diameter hole in the center) is not suitable to flexible operation including fast starts because of the large thermal stress generated by the large temperature gradient inside. The alternative is to use a welded rotor construction, which was done by the same OEM for the largest single casing combined cycle steam turbine (see Figure 5.3) [6].

The A2 configuration with two separate casings (and three bearings, one in the middle between the casings) allows for an optimal HP turbine rotor diameter. The two rotors are combined via a rigid, bolted coupling.

Beyond a certain unit rating, a single exhaust annulus cannot accommodate the requisite LP steam flow rate (either requisite LSB becomes too big or the exhaust losses become unacceptable) and one has to switch to a double-flow configuration. This results in the D2 configuration with two casings. Usually, there are two bearings for each rotor for a total of four bearings. Similar to the A2 configuration, the two rotors are combined via a rigid, bolted coupling.

5.1 IMPULSE VERSUS REACTION

In terms of aerodynamic design to extract useful shaft work from high pressure-and-temperature steam, steam turbines are classified into two categories: impulse and reaction. In the impulse design, what happens in the steam turbine is similar to what happens in a hydraulic turbine, specifically a Pelton wheel. High-energy steam (as quantified by its enthalpy) is accelerated through a converging nozzle and impinges on the bucket-shaped blades attached to the rotor "wheel" to create the rotational motion of the latter. (This is where the term "bucket" comes from – it is frequently used in lieu of the term "blade" in steam turbine jargon.) In thermodynamic terms, *static* enthalpy of steam at the turbine inlet is converted into kinetic energy in the nozzle and then into mechanical work in the rotor. (The nozzle in question is actually a series of nozzle shaped passages formed by stator vanes.) The process is graphically described in Figure 5.4 (on the left).

In the reaction design (on the right in Figure 5.4), static enthalpy drop in the stage is divided between the stator (i.e., the nozzles) and the rotor (i.e., the buckets). Therefore, the change in the kinetic energy of steam (i.e., its acceleration) in the stator is smaller than that in the impulse design

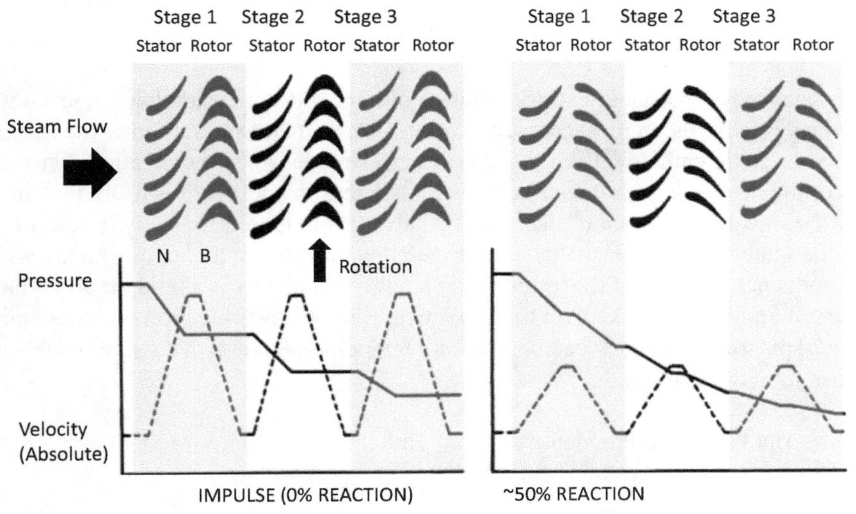

FIGURE 5.4 Impulse and reaction stages.

(compare the velocity profiles in Figure 5.4). Consequently, work generated in the rotor has two components: momentum exchange, i.e., transfer of kinetic energy to motion, and conversion of enthalpy to velocity and transfer of it to motion via reaction (i.e., analogous to gas turbine jet engine). This is why the flow channels between the rotor blades of a reaction turbine are also shaped as converging nozzles, i.e., in order to accelerate steam.

How is this explanation consistent with the decreasing velocity profile across the rotor blades in Figure 5.4, one might ask. The velocity profile in Figure 5.4 is the *absolute* velocity profile, i.e., the velocity measured by an observer standing outside the steam flow-path (i.e., *fixed* or *stationary* reference frame). In a *moving* reference frame, which applies to an (imaginary) observer "sitting" on a rotating blade, steam velocity *increases* while passing through the rotor blades. Before going into that further, however, we must spend some time on thermodynamics.

Steam enthalpy is a thermodynamic property; that is, with two other known properties, e.g., pressure and temperature, it can be determined exactly via an *equation of state* (EOS). In steam turbine calculations, the EOS is the *ASME Steam Tables*, from which the enthalpy can be found as h = f(p,T,x), where x is the *quality* of steam. For "dry" steam (saturated or superheated), x = 1.0. For saturated or compressed liquid phase of H_2O, x = 0. For "wet" steam, i.e., a mixture of dry steam and liquid H_2O ("moisture"), 1.0 > x > 0 and it is the mass fraction of dry steam in the two-phase mixture (wet steam). The counterpart of quality is *moisture fraction*, y, which is given by y = 1 − x.

The enthalpy found from the *ASME Steam Tables*, in Btu/lb or kJ/kg, is *static* enthalpy. The energy of steam flowing in a pipe or through a valve or through the turbine flow path is quantified by its *total* enthalpy, which is a combination of static enthalpy and kinetic energy of steam, i.e.,

$$h_0 = h(p,T,x) + \frac{1}{2}V^2 \tag{5.1a}$$

where V is the velocity of steam (ft/s in British units, m/s in SI units). This is where many headaches start. Let us focus on the units first. In SI units,

$$1\frac{kJ}{kg} = 1{,}000\frac{J}{kg} = 1{,}000\frac{Nm}{kg} = 1{,}000\frac{m^2}{s^2}.$$

Thus, if one uses J/kg for the static enthalpy in Equation 5.1a, simple substitution of values of h and V is sufficient. In British units, Equation 5.1a becomes (see Section 1.1 on units)

$$h_0 = h(p,T,x) + \frac{V^2}{2Jg_c}. \tag{5.1b}$$

The other headache is associated with the notation of total and static enthalpies or, even worse, lack thereof in many references. In most cases, it is impossible to understand whether the author refers to total enthalpy or static enthalpy. Even if a distinction is made, since the notation convention is not standardized, in one reference, total enthalpy is h_0, whereas in another, it is h_t, and yet in another, it is h_s (where s stands for "stagnation" because if you bring a fluid isentropically to rest, its stagnation enthalpy is the same as its total enthalpy). Static enthalpy may be just h or h_s or who knows what else.

This author is not a big fan of the term "total enthalpy". Enthalpy is a fluid thermodynamic property, and it is unambiguously fixed via the EOS. One cannot substitute total enthalpy into the EOS with, say, *total* pressure, and back-calculate a *total* temperature. Nevertheless, in order not to create more confusion, in this section

- the subscript "0" is used to identify "total" enthalpy.
- no subscript is used for "static" enthalpy.

(In Section 4.4.3, the subscript *tot* was used for total properties. This was an intuitive choice, and there were only a few equations where it appeared. Herein, there are many equations where total properties are used frequently, and the subscript 0 is chosen to reduce the notation "clutter".)

As much as possible, without creating undue clutter, h(p,T,x) will be used for static enthalpy to emphasize that it is a property calculated from the EOS (i.e., *ASME Steam Tables*) using two other properties (usually pressure and temperature). A distinction will *always* be made between total and static values. With that in mind, let us look at a generic stage. First, a note on notation:

- 0 designates stage (i.e., stator) inlet.
- 1 designates stator exit (i.e., rotor inlet).
- 2 designates stage (i.e., rotor) exit.
- V designates absolute velocity (fixed reference frame).
- W designates relative velocity (moving reference frame).
- U designates rotational speed at the "mean line" or "pitch" radius, r_m.

Mean line is a point in the flow annulus, which is $\ell/2$ away from the blade hub and blade tip (ℓ is the blade length). For a rotational speed of N rpm,

$$U = 2\pi \frac{N}{60} r_m = \omega \cdot r_m. \tag{5.2}$$

At the stage inlet, total enthalpy (i.e., total energy) of steam is (it is assumed to be superheated steam so that quality x is ignored)

$$h_{0,0} = h(p_0, T_0) + \frac{V_0^2}{2}. \tag{5.3}$$

Across the stator nozzle vanes, steam accelerates to a velocity V_1. Since no work is done in a stator (and ignoring friction losses), total enthalpy does *not* change:

$$h_{0,1} = h_{0,0},$$

i.e.,

$$h(p_1, T_1) + \frac{V_1^2}{2} = h(p_0, T_0) + \frac{V_0^2}{2} \tag{5.4a}$$

or

$$h_1 + \frac{V_1^2}{2} = h_0 + \frac{V_0^2}{2}. \tag{5.4b}$$

In the rotor, work is generated via energy transfer, i.e.,

$$w = \left(h_1 + \frac{V_1^2}{2}\right) - \left(h_2 + \frac{V_2^2}{2}\right) = (h_1 - h_2) + \frac{1}{2}\left(V_1^2 - V_2^2\right). \tag{5.5a}$$

One simplifying assumption for ease of discussion is a "repeating stage", i.e., $V_2 = V_0$. With that assumption, Equation 5.5a becomes

$$w = (h_1 - h_2) + \frac{1}{2}\left(V_1^2 - V_0^2\right). \tag{5.5b}$$

Combining Equations 5.4b and 5.5b, we obtain

$$w = h_0 - h_2. \tag{5.6}$$

In other words, shaft work generated in a steam turbine (per unit mass flow rate of steam, i.e., in Btu/lb or kJ/kg) is equal to the static enthalpy drop across the turbine. Note that one does *not* have to know whether the turbine is of impulse or reaction type! Mathematically, Equation 5.6 is exact via the assumption of the repeating stage, i.e., $V_2 = V_0$. Even if this equality does not hold exactly in the actual hardware, it is still true that $V_2 \sim V_0$ and furthermore,

$$\frac{V_0^2}{2} \ll h(p_0, T_0).$$

Let us consider steam at 166.5 bar and 565.6°C at the HP turbine inlet. From the *ASME Steam Tables*, the enthalpy is 3,473 kJ/kg. Typical steam velocity in the main steam pipe is about 200 ft/s (~60 m/s) or less. Let us be overzealous and say 100 m/s. Thus, the kinetic energy part is 5,000 m^2/s^2 vis-à-vis 3,473,000 for the static enthalpy or about ~0.15% of it! (Typically, at the stage inlet steam velocity is about 40 m/s in HP turbines, about 60 m/s in IP/LP turbines.)

This is why heat and mass balance simulation software does not worry about the aerothermodynamics in the "guts" of the steam turbine. It is based on purely thermodynamic calculations.

Let us go back to our turbine stage and look at the shaft work from a different perspective. Rate of energy transfer through the net torque applied by steam accelerating through the rotor blades gives the stage work (per unit steam mass flow rate)

$$w = \tau\omega = \omega r_m \left(V_{\theta,1} - V_{\theta,2}\right) = U\left(V_{\theta,1} - V_{\theta,2}\right), \tag{5.7}$$

where V_θ designates the tangential component of the absolute velocity, V. Where does the torque come from? Recall that steam enters the rotor, which is rotating at an angular speed of ω, at a speed of V_1. Consequently, for an imaginary observer sitting on a rotor blade at radius r_m, the "apparent" velocity of steam entering the passage between two blades is W_1, which is found via a vector subtraction of U (tangential rotor speed at radius r_m, see above) from V_1. The force imparted by the steam flow on the "pressure surface" of the rotor blade in its path with a momentum of $\dot{m} V_\theta$, is the key mechanism of exerting torque on the blade causing its rotation. (Note that only the tangential component, V_θ, of the absolute velocity, V, can exert a torque on the blade.) Steam then exits the rotor with absolute tangential velocity, $V_{\theta,2}$, so that the *net* torque exerted on the blade per unit mass flow rate is

$$\tau = r_m\left(V_{\theta,1} - V_{\theta,2}\right). \tag{5.8}$$

The actual calculation on the right-hand side of Equation 5.8 is a "vector subtraction", i.e., using a generic velocity vector:

$$\Delta\vec{v} = \vec{v}_1 - \vec{v}_2.$$

If both velocities are in the *same* direction,

$$\left|\Delta\vec{v}\right| = \left|\vec{v}_1\right| - \left|\vec{v}_2\right|.$$

If the second velocity is in the *opposite* direction,

$$\left|\Delta\vec{v}\right| = \left|\vec{v}_1\right| + \left|\vec{v}_2\right|.$$

Consequently, when substituting values into Equation 5.8, one has to be cognizant of the respective directions of $V_{\theta,1}$ and $V_{\theta,2}$. Velocity triangles and the expansion line on an enthalpy-entropy (h-s) surface are shown in Figure 5.5. Subscripts S and R designate stator and rotor, respectively. The subscript "is" denotes the isentropic expansion. The angle between the absolute velocity vector V and the flow path centerline (∃) is denoted α (not shown to minimize the clutter). The angle between the relative velocity vector W and the flow path centerline is denoted β.

From Figure 5.5, via basic trigonometric correlations, it is found that

$$V_{\theta,1} = V_1 \sin\alpha_1,$$

and

$$V_{\theta,2} = V_2 \sin\alpha_2.$$

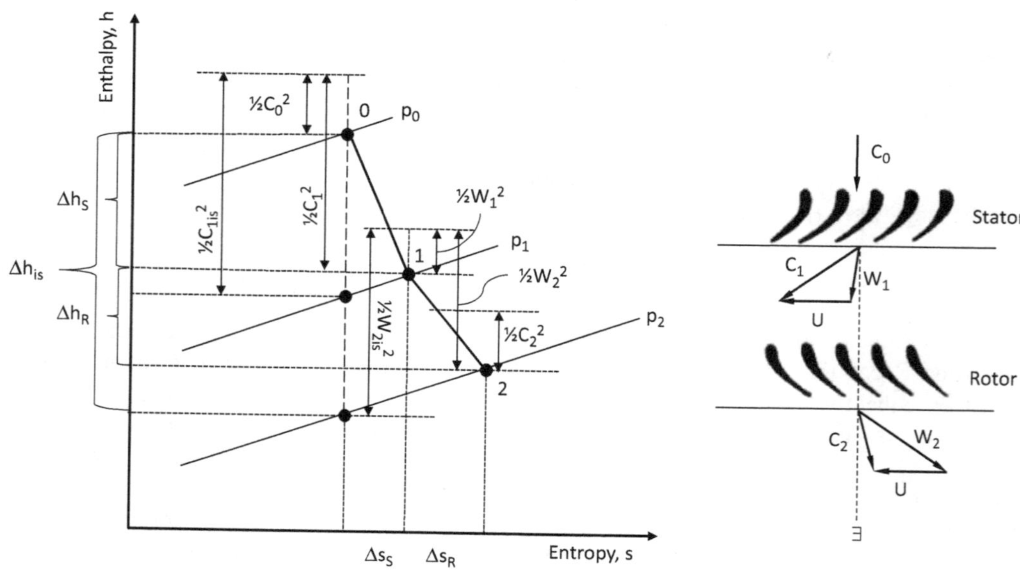

FIGURE 5.5 Generic steam turbine stage.

Combining Equations 5.2, 5.6 and 5.7, we obtain the famous *Euler's turbine equation*, i.e.,

$$h_0 - h_2 = U\left(V_{\theta,1} - V_{\theta,2}\right). \tag{5.9a}$$

(The two tangential velocities are in opposite directions so that we can also go with the "+" sign in Equation 5.9a with the implicit requirement that we substitute the absolute values.) Equation 5.9a is specifically for a repeating stage. For the most general case, one can use the total stage inlet and exit enthalpies, i.e.,

$$h_{0,0} - h_{0,2} = U\left(V_{\theta,1} - V_{\theta,2}\right). \tag{5.9b}$$

Stage reaction (or degree of reaction) is defined in several different ways. Herein, for ease of discussion, we will use the following definition (refer to Figure 5.5 for the terms in the equation):

$$R = \frac{\Delta h_R}{\Delta h_S + \Delta h_R}. \tag{5.10}$$

Note that the denominator of Equation 5.10 is approximately the stage work, w, for the general case (exactly for a repeating state). The numerator of Equation 5.10 is the static enthalpy change across the rotor. Thus, via substitution from Equation 5.5b, we obtain

$$R = 1 - \frac{\left(V_1^2 - V_0^2\right)}{2w}. \tag{5.11}$$

Substituting Equation 5.7, Equation 5.11 becomes

$$R = 1 - \frac{\left(V_1^2 - V_0^2\right)}{2U\left(V_{\theta,1} - V_{\theta,2}\right)}. \tag{5.12}$$

Note that, from vector algebra, we know that

$$V_0^2 = V_{\alpha,0}^2 + V_{\theta,0}^2$$

and

$$V_1^2 = V_{\alpha,1}^2 + V_{\theta,1}^2,$$

where the subscript α denotes the axial component of the absolute velocity, which does not change appreciably across the stage. Consequently,

$$R = 1 - \frac{\left(V_{\theta,1} + V_{\theta,2}\right)}{2U}. \tag{5.13}$$

The term in the numerator of Equation 5.13 is commonly known as the "change in whirl", where the term "whirl" refers to the tangential component of the absolute steam velocity.

This is where we should introduce the two most important parameters in turbine design: the *stage loading parameter*, ψ, and the *flow coefficient*, ϕ. Also known as the "work coefficient", stage loading factor is a non-dimensional measure of net torque exerted on the blade, as quantified by the ratio of "change in whirl" in the numerator of Equation 5.13 (i.e., the "cause"), to the rotational speed in the denominator (i.e., the "effect"). Thus, we have

$$\psi = \frac{w}{U^2} = \frac{h_0 - h_2}{U^2} = \frac{V_{\theta,1} - V_{\theta,2}}{U}. \quad (5.14)$$

The reason why ψ is called "loading parameter" is obvious when we combine Equation 5.14 with Equation 5.8, i.e.,

$$\psi = \frac{\tau}{Ur_m} = \frac{N_b F_b}{\dot{m}U},$$

where F_b is the tangential force acting on a single blade (N in SI units, lbf in British units) at the mean diameter and N_b is the total number of the blades in the stage rotor.

The flow coefficient is the ratio of the axial component of absolute velocity, V_1, of steam entering the rotor to the rotor speed (in some references, the "spouting velocity"), i.e.,

$$\phi = \frac{V_{\alpha,1}}{U} = \frac{V_1 \cos\alpha}{U}. \quad (5.15)$$

From the velocity triangles in Figure 5.5, we note that the axial components of absolute and relative velocities are identical. With that in mind, substituting Equation 5.15 in Equation 5.13 and going through a little algebra, we can write the stage reaction as

$$R = \phi \frac{\tan\beta_2 - \tan\beta_1}{2}. \quad (5.16)$$

For an impulse stage, $R = 0$ so that the change in whirl is given by

$$V_{\theta,1} + V_{\theta,2} = 2U$$

with

$$\beta_1 = \beta_2$$

and

$$\psi = 2 \quad \text{(from Equation 5.14).}$$

In other words, steam is accelerated across the stator nozzle vanes only, and it only changes direction across the rotor blades ($W_2 = W_1$). This can be seen from the airfoil profiles (crescent shape with no area change in the flow channel between adjacent blades) of the impulse blades (buckets) in Figure 5.4.

For a reaction stage with $R = 0.5$ (50% reaction), it can be shown that the stage loading parameter is simply

$$\psi = 2\phi \tan\alpha_1 - 1. \quad (5.17)$$

This design has lower ψ than the impulse stage, which leads to lower velocities and losses with better stage efficiency. For a zero exit swirl stage, i.e., $\alpha_2 = 0$, the stage loading parameter becomes

$$\psi = 2(1 - R). \quad (5.18)$$

Thus, for $R = 0.5$, $\psi = 1$. (Note that these are ideal results with simplifying assumptions and ignoring stator and rotor efficiencies.)

There is no 100% reaction stage (i.e., R = 1), which would be a reversed impulse stage. (Stage loading would be zero from Equation 5.18; in other words, no cause, no effect, no turbine – a pointless machine.) Steam would only change direction across the stator nozzle vanes and accelerate across the rotor blades. Consequently, the airfoil profiles of the stator nozzle vanes would look like the rotor buckets of an impulse stage. The rotor blade profiles, on the other hand, would be exactly like the stator nozzle vanes of the impulse stage; i.e., they would practically become "rotating nozzles". This is why physical examples used to illustrate 100% reaction stages are Hero's "aeolipile"[2] or common, rotating lawn sprinkler.

There is no pure impulse design, either. All practical turbines are a hybrid of both with impulse designs characterized by low degrees of reaction (about 5%) and high stage loading so that they can achieve the same work output in a relatively low number of stages. Reaction designs would be characterized with higher degrees of reaction and low stage loading so that one would need a larger number of stages to achieve the same work output as an impulse design with the same steam conditions. (If one goes with R = 0.5 and R = 0, for reaction and impulse designs, respectively, stage loading parameters become 1 and 2, respectively, so that a reaction turbine requires twice as many stages to convert the same available enthalpy.)

In reaction stages, there is a symmetry in steam flow relative to stator vanes and rotor blades (in fact, perfect symmetry for R = 0.5 under ideal assumptions). This means that same airfoil profiles can be used for both so that flow deflections along the flow channels between the blades are moderate. Consequently, profile losses (due to formation of boundary layers on the blade surface) and secondary losses (friction losses due to viscous effects in the end wall boundary layers at the blade hub and tip, i.e., the casing) are lower. These losses in nozzle vanes and rotor blades account for two-thirds of the stage efficiency losses. (By the same token, due to the significant difference in airfoil profiles of stator nozzle vanes and rotor blades in the impulse design, profile and secondary losses are higher.) On the other hand, due to the significant pressure drop across the rotor blades (see the 50% reaction stage pressure profile in Figure 5.4), there is a large axial thrust exerted on the rotor. In single-flow designs (e.g., the HP turbine of "A" series steam turbines), a dummy "balance piston" is required to compensate for this axial thrust. The leakage flow across the dummy balance piston has a negative impact on the turbine efficiency [7].

Typically in the HP turbine with the lowest volumetric flow rate (due to higher steam density), reaction design is favorable because it leads to longer blades, lower velocities and therefore lower secondary losses. In the IP and LP turbines with larger volume flows and larger enthalpy drops (e.g., about 50% higher vis-à-vis the HP turbine), impulse design with lower R and higher ψ is favorable because a lower number of stages are needed [7].

In earlier "A" and "D" series turbines, GE used impulse design (known as the "wheel-and-diaphragm" construction) in the HP turbine as well. In a wheel-and-diaphragm impulse design, stator nozzle vanes are carried in welded nozzle diaphragms (see Figure 5.6). The gap between the vane tip and the shaft is sealed by a rub-tolerant, spring-backed packing. (Note that due to the high pressure difference across the stator, leakage prevention is of utmost importance for stage efficiency.) Rotor blades are carried in the rims of wheels machined from a solid rotor forging [8].

In the reaction design, which was adopted by GE in HEAT™ product line, a "drum-type" rotor construction was used. In this design, rotor blades are inserted into the surface of the drum. The reason for that is obvious. Due to the pressure drop across the blade row, a very high thrust force would exist on the rotor if the blades were mounted on wheels with faces exposed to the pressure differential. Note the use of diaphragms supporting the nozzle vanes in Figure 5.7.

---

2 The ball of Aeolus, the Greek god of the air and wind, which is a steel ball with two protruding nozzles as described by the Greek mathematician Hero (or Heron) of Alexandria. The ball is filled with water, which, upon heating, becomes steam and exits through the nozzles, thereby causing the ball to turn around its axis in accordance with Newton's third law.

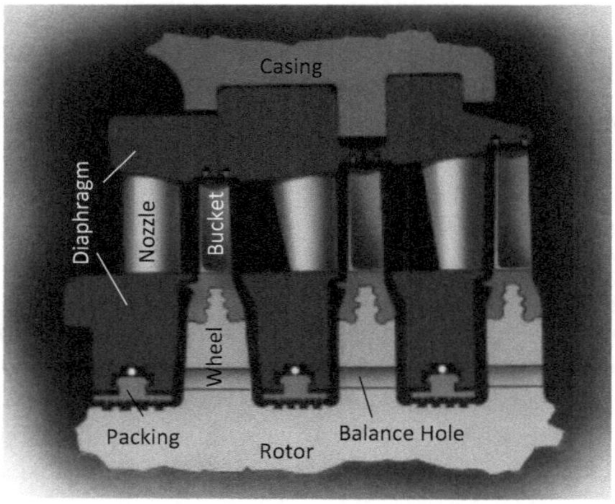

FIGURE 5.6 Typical wheel-and-diagram construction (impulse stage) [9].

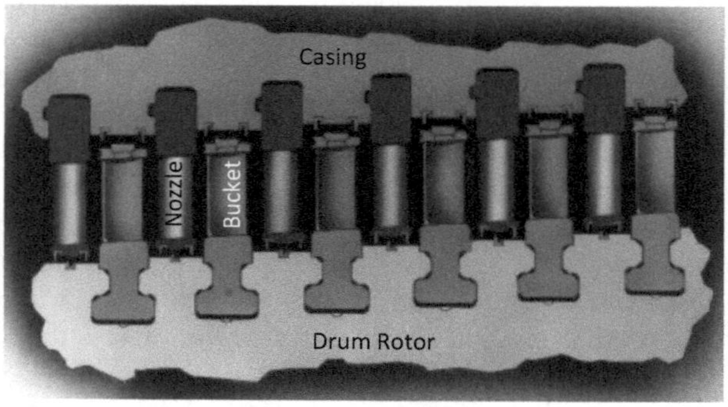

FIGURE 5.7 Typical drum rotor construction (reaction stage) [9].

For the ideal impulse stage, the efficiency correlation as a function of the velocity ratio is

$$\eta_I \cong 4\frac{U}{V_1}\left(\cos\theta - \frac{U}{V_1}\right), \tag{5.19}$$

where $\theta = 90 - \alpha_1$ (typical value for impulse stages is 20°, i.e., $\alpha_1 = 70°$). By differentiation, the optimum velocity ratio for maximum efficiency is

$$\frac{U}{V_1} = \frac{\cos\theta}{2} \approx 0.47.$$

Indeed, for GE's 1990s vintage impulse designs, similar to that shown in Figure 5.6 with four HP stages, the velocity ratio was 0.5. The "dense pack" variant (1995) with seven stages had a velocity ratio of 0.57. For the ideal 50% reaction stage, the efficiency correlation as a function of the velocity ratio is

$$\eta_R \cong 2\frac{U}{V_1}\left(\cos\theta - \frac{U}{2V_1}\right). \tag{5.20}$$

By differentiation, the optimum velocity ratio is found as

$$\frac{U}{V_1} = \cos\theta \approx 0.94,$$

which is twice that of the optimum velocity ratio for the pure impulse stage. This finding implies that, for the same tangential blade speeds, the work production per stage by a reaction turbine is one-half of that by an impulse turbine. Thus, for the same available energy (enthalpy), a reaction turbine requires roughly twice as many stages as an impulse turbine. This is the same conclusion arrived at via comparison of stage loading parameters for R = 0 and R = 0.5. This is not surprising and leads to another definition of the loading parameter. In comparison, GE's reaction HP design with a drum rotor and 12 stages had a velocity ratio of 0.62 (HEAT™ technology).

Note that, for a pure impulse stage with R = 0, from Equation 5.10 and 5.5b with $V_0 \approx 0$,

$$w \approx \frac{V_1^2}{2}$$

so that from Equation 5.14

$$\psi \approx \frac{1}{2}\frac{V_1^2}{U^2},$$

which leads to $\psi \sim 2$ at the optimal velocity ratio of ~0.5. From Equation 5.15 with $\alpha_1 = 70°$ and an optimum velocity ratio of 0.47, flow coefficient is found as $\phi = 0.73$.

For a reaction stage with R = 0.5, from Equations 5.10 and 5.5b with $V_0 \approx 0$,

$$w \approx V_1^2$$

so that from Equation 5.14,

$$\psi \approx \frac{V_1^2}{U^2},$$

which leads to $\psi \sim 1$ at the optimal velocity ratio of ~1.0. From Equation 5.15 with $\alpha_1 = 70°$ and an optimum velocity ratio of 0.94, flow coefficient is found as $\phi = 0.36$.

As a rough guideline, the following parameters from Traupel [10] (p. 492) can be used to choose between impulse or reaction design:

- Impulse
    - R = 0.05–0.25
    - $\alpha_1 = 74°–78°$
    - $\beta_2 = 60°–73°$
    - $\phi = 0.35–0.60$
- Reaction
    - R = 0.5
    - $\alpha_1 = \beta_2 = 68°–74°$
    - $\phi = 0.30–0.50$

These two key design parameters can be used as a guidance in the initial setting of the steam flow-path geometry (i.e., velocity triangles including the rotational speed) and key dimensions (e.g., rotor diameter). Stage efficiency estimates can be made using generalized charts such as the Smith chart (see Figure 4.7) or the Balje chart [11].

The best known OEM data for steam turbine section (not individual stage) efficiencies are provided in the paper by Spencer, Cotton and Cannon [12]. The *SCC method*, as it is widely known, is based on empirical data from fossil fuel-fired boiler steam turbines with correlations using steam conditions, volume flow and pressure ratio as the key parameters. The 1974 method is an extension of the ASME paper by the authors published in the early 1960s (the GE report cited above may be difficult to obtain but most of the information therein is available in Cotton's book – Ref. [17] in Chapter 2). Although still quite useful for performance calculations, it has to be updated to reflect the modern technology. (Such a correction to 1999 technology is provided in Thermoflow software.) See Section 5.3 for more on this method.

In any event, rigorous stage-by-stage steam turbine design and calculation is beyond the scope of this section. A basic knowledge of HP, IP and LP section efficiencies is sufficient for most "homemade" calculations. Commercial software tools have several built-in methods such as the 1974 SCC method and others with user inputs to modify them as needed. Steam turbine OEMs, of course, rely on their in-house design tools, which combine theoretical calculations with empirical data from their own hardware. (Needless to say, this information is highly proprietary and not available in the open literature.)

There are several assumptions implicit in steam turbine section efficiencies, i.e.,

- whether the quoted efficiency is "bowl to exhaust" or inclusive of inlet valves and exit losses (for HP and IP sections);
- whether the quoted efficiency is inclusive of exhaust losses (for the LP section).

Narula et al. published "cylinder" or steam-path efficiencies, measured from bowl to exhaust, for modern large combined cycle steam turbines [13], i.e.,

- 89.5%, 92.2% and 94.2% for HP, IP and LP turbines (cylinders) in units with no duct firing in the HRSG;
- 89.4%, 92.3% and 92.7% for HP, IP and LP turbines (cylinders) in units with duct firing in the HRSG.

Note that including the exhaust loss for the LP turbine can decrease the overall (effective) section efficiency by 3 percentage points or more. For example, an LP cylinder efficiency of 94.2% (i.e., exhaust enthalpy is h_{ELEP} – see Section 5.2) is probably equivalent to about 91% for the LP turbine (i.e., exhaust enthalpy is h_{UEEP}).

Representative section efficiencies for 2 × 2 × 1 and 1 × 1 × 1 F class gas turbine combined cycle power plants have been published by Gülen and Smith [14]. They are summarized in Table 5.2. Efficiencies in the table are inclusive of inlet valve losses (typically about 2% pressure drop) and exhaust loss in the LP section. Note that

- for the same gas turbine frame (i.e., 50 or 60 Hz), 2 × 2 × 1 steam turbine is more efficient than its 1 × 1 × 1 counterpart;
- for the same configuration, the 50-Hz steam turbine is more efficient than its 60-Hz counterpart.

TABLE 5.2
Typical F Class Combined Cycle Steam Turbine Efficiencies [14]

Configuration	Hz	HP (%)	IP (%)	LP (%)
2 × 2 × 1	50	90.1 ± 1.5 [86.9–91.9]	91.1 ± 1.5 [87.8–92.9]	91.4 ± 1.5 [88.1–93.2]
2 × 2 × 1	60	87.8 ± 2.5 [84.0–92.0]	88.8 ± 2.5 [85.0–93.1]	89.1 ± 2.5 [85.2–93.4]
1 × 1 × 1	50	88.7 ± 1.6 [85.9–90.7]	89.7 ± 1.6 [86.8–91.8]	90.0 ± 1.6 [87.1–92.1]
1 × 1 × 1	60	86.5 ± 2.6 [82.6–90.3]	87.5 ± 2.6 [83.6–91.3]	87.7 ± 2.6 [83.8–91.6]

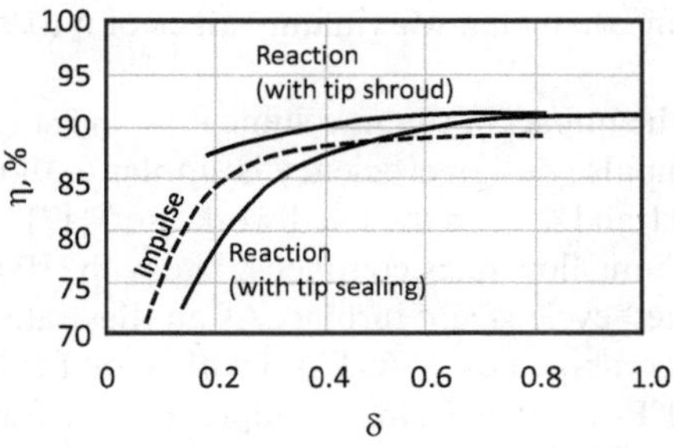

FIGURE 5.8 Stage efficiencies for impulse and reaction stages [15].

This is a reflection of the larger steam flow (i.e., larger volume flow) resulting from the size of the gas turbine via its speed (i.e., 3,000 rpm for 50 Hz vs. 3,600 rpm for 60 Hz, 44% more exhaust flow ergo 44% more steam production) and/or number of them in the plant (i.e., two vs. one, twice the exhaust flow ergo twice the steam production).

For the impact of volumetric steam flow on efficiency, please refer to the chart in Figure 5.8 (from Figure 12.3 on p. 208 in Bloch [15]). The x-axis is the volume coefficient

$$\delta = \frac{\dot{V}}{U \cdot r_m^2}, \tag{5.21}$$

where $\dot{V}$ is the volumetric flow rate of steam (ft³/s or m³/s), r_m is the mean or pitch-line radius and $U = \omega \cdot r_m$. Using the stage geometry, δ can be written as

$$\delta = \frac{V_{1,\alpha} 2\pi\ell}{\omega \cdot r_m^2} = \frac{V_1}{\omega}\cos\alpha \frac{2\pi\ell}{r_m^2}, \tag{5.22}$$

where ℓ is the blade length, and the spouting velocity is approximately.

$$V_1 \simeq \sqrt{2\phi R(h_0 - h_{2,i}) J g_c}, \tag{5.23}$$

where ϕ is the velocity coefficient for converging nozzles (0.97 is a good value for most practical purposes). (For convenience and generality, J and g_c applicable with British units are included in the equation; for SI units, they should be ignored.) Substituting the shaft speed N (in rpm) for ω

$$\delta = \frac{\sqrt{2\phi R(h_0 - h_{2,i}) J g_c}}{N} 60 \cos\alpha \frac{\ell}{r_m^2}. \tag{5.24}$$

Note that the pitch-line radius r_m is given by

$$r_m = \frac{1}{2}(D_{rot} + \ell), \tag{5.25}$$

where D_{rot} is the rotor diameter.

A chart similar to the one in Figure 5.8 is provided in Figure 9.5.4 on p. 494 of the first volume of Traupel's magnum opus [10]. The x-axis parameter is similar to δ, i.e., rewriting Equation 5.22 with substitution from Equation 5.15 and ignoring the constant 4π,

$$\delta' = \phi \frac{\ell}{D_m},$$

where $D_m = 2r_m$ is the pitch-line diameter. Maximum values of η (0.91–0.93) in Traupel's chart are at $\delta' = 0.09–0.1$.

Another chart illustrating the impact of steam volume flow and stage design on section efficiency is shown in Figure 5.9. The impulse design efficiency is similar to those published in Cotton's book (e.g., see charts in Figures 4.11 and 4.48 in the cited work, Ref. [17] in Chapter 2). Shaded areas in the chart designate typical volume flow rates commensurate with HP and IP (i.e., HRH steam at the IP bowl) sections of a combined cycle steam turbine. As an illustrative example, consider a typical 2 × 2 × 1 F class gas turbine combined cycle (GTCC) with a total exhaust flow rate of about 3,000 lb/s and temperature of 1,100°F–1,150°F. Using a simple transfer function that will be introduced later in the chapter, Equation 5.39 (Section 5.3.4, see Section 6.2.3 for an updated version), steam flow is estimated at about 475 lb/s (about 1.7 million lb/h). Assuming 1,815 psia and 1,050°F steam inlet conditions, steam volume flow rate at the HP turbine inlet is about 800,000 ft^3/h. Similarly, assuming 325 psia and 1,050°F steam inlet conditions, steam volume flow rate at the IP turbine inlet is about 4.5 million ft^3/h. From the chart in Figure 5.9, HP section efficiency range is 89.5%–91.5%; IP section efficiency range is 92.5%–93.5%. (In Table 5.2, corresponding ranges are 86.9%–91.9% and 87.8%–92.9%, respectively. The lower bound of the range in the table reflects vintage technologies with smaller units.)

For the HP section of a steam turbine, the thermodynamic design (i.e., the "heat balance") sets the inlet conditions T_0, p_0 and the exhaust pressure p_{exh}. Section pressure ratio is given by $PR = (p_0/p_{exh})$. Stage pressure ratio is approximately $PR_{stg} = PR^{1/n}$ where n is the number of stages. HP turbine shaft speed is set by the generator frequency, i.e., 3,000 or 3,600 rpm for 50 and 60 Hz, respectively. (Optimizing for speed is not feasible – except for very small units of less than 15 MWe or so – because the gearbox would be too expensive and introduce parasitic losses.) The output of the turbine from the heat balance and the shaft speed sets the shaft torque. Rotor diameter D_{rot} is thus a critical design parameter because it determines the root (hub) radius of the rotor blades and the hub–tip ratio:

$$\upsilon = \frac{D_{rot}}{D_{rot} + 2\ell} \tag{5.26}$$

which should not exceed 0.9. For HP stages, υ is about 0.85; for the longer LP stages, it can be as low as 0.5 or even lower. Since the blade length is set by the volumetric flow at the rotor inlet as

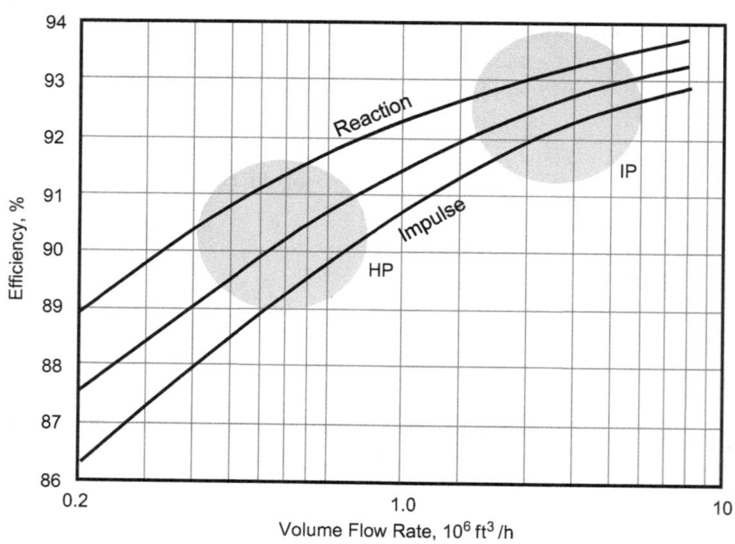

FIGURE 5.9 Section efficiencies for impulse and reaction designs.

$$\ell = \frac{D_{rot}}{2}\left(\sqrt{1+\frac{4A}{\pi D_{rot}^2}}-1\right) \tag{5.27}$$

with

$$A = \frac{\dot{V}}{V_{1,\alpha}} = \frac{\dot{m}}{\rho_1 V_1 \cos\alpha},$$

in addition to its importance from a mechanical design perspective (rotor length and stiffness from the shaft torque), rotor diameter is a critical factor for the flow-path design and efficiency (via ϕ and ψ). This is illustrated by the curves in Figure 5.10 [16], which is for the HP section of an ultra-supercritical (USC) steam turbine with steam conditions of 30 MPa (300 bar, 4,350 psi) and 600°C (1,112°F). The change in efficiency with increasing D_{rot} is in agreement with the predictions of Figure 5.8 along with Equations 5.24 and 5.25. The trend associated with the number of stages can be tied to the change in stage loading. Lower number of stages (i.e., higher loading) is detrimental to stage efficiency as predicted by the Smith chart (Figure 4.7). Similarly, stage efficiency is adversely affected if the number of stages is too high with low loading. There is an optimum number of stages and stage loading, which is more pronounced at higher rotor diameter.

5.1.1 Steam Turbine Irreversibility

The first and second laws of thermodynamics for the steam turbine can be combined into the following form:

$$\dot{I}_{ST} = \dot{m}_s \cdot T_0 \cdot (s_{out} - s_{in}), \tag{5.28}$$

where T_0 is the "dead-state" reference temperature (typically, 59°F), the term in the parentheses on the right-hand side is the entropy change between turbine inlet and exit, and the total steam mass flow rate is given by

$$\dot{m}_s = \mu \cdot \dot{m}_{exh}.$$

For the calculation of μ, see Section 6.2.3 (or Equation 5.39). Strictly speaking, Equation 5.28 is for a simple steam turbine without any extraction, admission and leakage streams between the inlet and exit. A more generalized version can be written by combining all inlet and exit streams. This, however, does not add more insight to the discussion of the subject so that the derivations below are

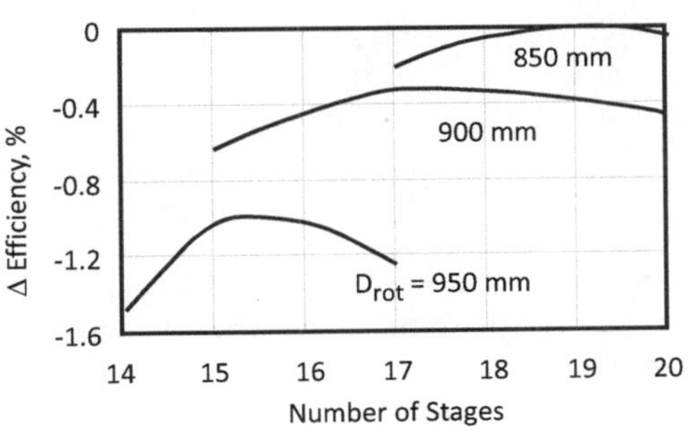

FIGURE 5.10 Effect of rotor diameter on stage efficiency for the HP section of an ultrasupercritical steam turbine [16].

based on Equation 5.28. However, even Equation 5.28 is not very amenable to a quick estimation of the steam turbine irreversibility without resorting to detailed heat balance calculations. A more suitable formulation of Equation 5.28 can be obtained by making use of the fundamental thermodynamic relationship

$$dh = Tds + vdp.$$

Integrating this exact differential between any two states and combining with the ideal gas relationship, the generic steam turbine irreversibility can be written as

$$\dot{I}_{ST} \propto \dot{m}_s \cdot T_0 \cdot (1-\eta) \cdot \frac{(PR-1)}{k \cdot (PR+1)}, \tag{5.29}$$

where PR is the ratio of turbine inlet and exit pressures. Equation 5.29 quantifies the steam turbine thermodynamic irreversibility with a suitable proportionality factor depending on the units used. Coefficients k_1 and k_2 for HP and IP sections are 0.33, and k_3 for the LP section is 0.2.

The efficiencies considered herein are total section efficiencies including associated piping and valves. For the last steam turbine section (the LP section) exhausting to the condenser, the efficiency is bowl to *Used Energy End Point* (UEEP) and implicitly contains the exhaust losses. It can be modeled as a base value with a correction for the exhaust loss, i.e.,

$$\eta_{LP} = \eta_{base} - \lambda \cdot TEL,$$

where λ denotes the decrease in the base, bowl-to-UEEP LP efficiency per unit increase in the total (dry) exhaust loss (TEL). A good value is $\lambda = 0.25\%$; that is, the η_{LP} decreases by 0.25 points for each Btu/lb (2.326 kJ/kg) increase in TEL, which is obtained from the particular exhaust loss curve. UEEP can be calculated from the *Expansion Line End Point* (ELEP) as follows:

$$UEEP = ELEP + TEL(1-0.01\,y)0.87(1-0.0065\,y).$$

ELEP is the enthalpy found from the definition of the isentropic efficiency for the LP stage with known LP bowl (i.e., LP turbine inlet) conditions and exhaust (condenser) pressure.

Steam turbine section pressure ratios can be calculated as follows (1, 2 and 3 refers to HP, IP and LP sections, respectively):

$$PR_1 = \frac{p_{HP}}{p_{CRH}} = (1-\Delta P_{rht})\frac{p_{HP}}{p_{IP}} \tag{5.30}$$

$$PR_2 = \frac{p_{ip}}{p_{lp}} \tag{5.31}$$

$$PR_3 = \frac{p_{lp}}{p_{cond}}. \tag{5.32}$$

In modern combined cycle power plants, the steam turbine consists of multiple sections, the current standard being three for modern three-pressure, reheat systems. For the entire steam turbine, Equation 5.29 can be expanded to

$$\dot{I}_{ST} = \sum_{i=1}^{3}\left\{c \cdot \dot{m}_{s,i} \cdot T_0 \cdot (1-\eta_i) \cdot \frac{(PR_i-1)}{k_i \cdot (PR_i+1)}\right\}. \tag{5.33}$$

The proportionality factor c in Equation 5.33 is 0.125 for a consistent set of English units. Conversion to a system in SI units is straightforward. For the example in Section 5.3.2, we have

$$PR_1 = 2{,}414.7/467.5 = 5.17,\ \dot{m}_{s,1} = 225\ \text{lb/s},\ \eta_1 = 0.8983$$

$$\dot{I}_{ST,1} = c \cdot \dot{m}_{s,1} \cdot T_0 \cdot (1-\eta_1) \cdot \frac{(PR_1 - 1)}{k_1 \cdot (PR_1 + 1)} = 0.125 \cdot 225 \cdot 519(1-0.8983)\frac{(5.17-1)}{0.33(5.17+1)} = 3{,}009\ \text{Btu/s}$$

$$PR_1 = 425/55 = 7.73,\ \dot{m}_{s,1} = 265\ \text{lb/s},\ \eta_1 = 0.9176$$

$$\dot{I}_{ST,2} = 0.125 \cdot 265 \cdot 519(1-0.9176)\frac{(7.73-1)}{0.33(7.73+1)} = 3{,}276\ \text{Btu/s}$$

$$PR_1 = 55/0.6817 = 80.7,\ \dot{m}_{s,1} = 295\ \text{lb/s},\ \eta_1 = 0.9036$$

$$\dot{I}_{ST,3} = 0.125 \cdot 295 \cdot 519(1-0.9036)\frac{(80.7-1)}{0.2(80.7+1)} = 8{,}999\ \text{Btu/s}.$$

Thus, the total steam turbine irreversibility is

$$\dot{I}_{ST} = 3{,}009 + 3{,}276 + 8{,}999 = 15{,}283\ \text{Btu/s}$$

$$\dot{I}_{ST} = 15{,}283\ \text{Btu/s} \times 1.05506 = 16{,}125\ \text{kW}.$$

In the beginning of Section 5.3, gas turbine exhaust exergy was calculated as 286 MW. Thus, steam turbine irreversibility is 16.125/286 = 0.0564 or less than 6% of it.

5.1.2 Supercritical Steam Turbine

A frequently proposed option to improve the bottoming steam cycle efficiency is supercritical (SC) steam generation in the HP evaporator (i.e., above 3,200 psia or roughly 220 bar). SC steam cycles and turbines are widely used in coal-fired boiler power plants. USC steam cycles with 600°C steam became commercial in the 2000s, and advanced USC (A-USC) steam cycles with up to 700°C steam have been in development. If realized, the latter is expected to result in coal-fired power plant efficiencies of 46% (HHV) or above 50% (LHV).

The feasibility of a SC steam bottoming cycle from a second law (exergy) perspective will be covered in detail in Section 6.4. Herein, the ramifications of SC HP steam in the design of the HP section of the combined cycle steam turbine will be covered utilizing the basic flow-path design correlations discussed earlier in the chapter.

One reason for this exercise is to highlight the basic design principles "lurking in the background" when looking at the thermodynamic cycle performance calculations. In other words, one must be aware of the hardware design impact of key cycle pressures and temperatures used in heat balance calculations. As an example, consider the following parameters for a hypothetical bottoming steam cycle:

- HP steam at the main combined stop and control valve (SCV) 3,600 psia and 1,112°F;
- HRH steam at the inlet of IP turbine SCV 600 psia and 1,112°F.

Assuming a reheater pressure loss of 15%, at the exhaust of the HP turbine, pressure is 600/(1–15%) ~ 700 psia. With a 50-Hz advanced class gas turbine, the amount of HP steam generated at specified conditions is about 1,000 kpph (about 278 lb/s or ~450 t/h). Assuming 88% section efficiency and 275 lb/s steam flow (ignoring leakages), the following performance is calculated:

	p, psia	T, °F	h, Btu/lb	s, Btu/lb-R
Inlet	3,600	1,112	1,501	1.5203
Exit (isentropic)	700	617	1,293	1.5203
Isentropic work			208	
Actual work (88% η)			183	
Exit (actual)	700	655	1,318	1.5431
Turbine output (275 lb/s), kW			53,105	

There is nothing wrong *per se* in this calculation. Let us look at whether one can actually build this turbine (3,000 rpm, pressure ratio of 5.14 and shaft output of 53.1 MW). Note that the HP turbine shaft transmits the power generated by the entire steam turbine, which is of the order of 250 MW. For 3,000 rpm, $\omega = 2\pi N/60 = 314.16\,s^{-1}$ so that shaft torque is found as

$$\tau = 236{,}953 \text{ Btu/s}/314.16\,s^{-1} \times 778.17 = 556{,}302 \text{ ft-lbf.}$$

Maximum allowable torque with a safety factor of 1.3 is 723,192 ft-lbf (1,034,507 N-m). Using the fundamental torsional stress theory (please check your machine design textbook), assuming a maximum allowable shear stress of 4,000 psi (27.6 MPa), requisite shaft diameter can be estimated from the following correlation:

$$\tau = \sigma \frac{\pi D_{rot}^3}{16}. \tag{5.34}$$

Substituting the estimated parameters into Equation 5.34, the result is

$$D_{rot} = \sqrt[3]{\frac{16 \cdot 1{,}034{,}507 \cdot 10^3}{27.6\pi}} = 575.8 \text{ mm} = 22.7 \text{ in.}$$

Using the equations developed earlier in the section, a series of cases are run to investigate the feasibility of the HP turbine heat balance from an aeromechanical design perspective. The results are summarized in Table 5.3. Clearly, whether impulse or reaction as defined by the stage reaction R, it is impossible to reconcile a feasible and efficient flow-path geometry with the mechanical stress requirement with such low volumetric flow rate (about 60 ft^3/s or 1.7 m^3/s) at SC steam conditions. (Note that, with a conventional steam cycle with 1,815 psia and 1,050°F HP steam, volumetric flow would be about 150 ft^3/s.) Even for a marginally acceptable blade height (~35 mm) and hub–tip ratio, exhaust pressure should be increased to 1,600 psia and the rotor diameter reduced by more than 10%. As a reference point, consider that, in a typical 700 MWe SC pulverized coal power plant, the HP turbine flow rate is about 5 million lb/h (about 1,350 lb/s) with a volumetric flow rate of 275 ft^3/s. This is more than twice that one would get in 2 × 2 × 1 combined cycle with an advanced class 50-Hz gas turbine.

In such a 2 × 2 × 1 combined cycle (with a 500 MW steam turbine) with twice the volumetric flow, i.e., about 125 ft^3/s, a feasible HP flow-path design with SC steam is possible. This would likely entail a "very high-pressure" (VHP) turbine with a pressure ratio of 2.25 (similar to the last column in Table 5.3) and five impulse stages (R = 0.1). The first stage of such a VHP would have a blade height of about 80 mm with a hub–tip ratio of 0.82 (for a rotor diameter of 29 in.).

TABLE 5.3
Supercritical HP Turbine First-Stage Design Parameters

P_in	psia	3,600	3,600	3,600	3,600	3,600
P_out	psia	700	700	700	1,200	1,600
T_in	°F	1,112	1,112	1,112	1,112	1,112
Rotor diameter	in.	22.7	22.7	19.2	19.4	19.8
	mm	577	577	488	492	503
R		0.50	0.20	0.37	0.37	0.27
Number of stages		16	7	12	12	6
α	degs	75	75	78	76	78
ℓ	in.	0.80	0.84	1.23	1.20	1.36
	mm	20.4	21.4	31.2	30.4	34.6
hub–tip ratio		0.93	0.93	0.89	0.89	0.88
r_m	in.	11.75	11.77	10.21	10.28	10.58
N	rpm	3,000	3,000	3,293	3,309	2,750
ω	1/s	314.16	314.16	344.87	346.48	287.93
δ		0.22	0.22	0.30	0.29	0.32
ψ		1.93	1.73	2.00	1.32	2.00
ϕ		0.51	0.48	0.40	0.40	0.40

5.2 LAST-STAGE BUCKET

Low condenser pressure, p_{cond}, is crucial to steam turbine performance. It sets the cycle heat rejection temperature, which is the saturated steam temperature at p_{cond}, whose impact on the cycle efficiency is as strong as cycle heat addition temperature, which is set by HP and HRH steam temperatures (and pressures to a lesser extent). On average,

- Each 25°F in HP/HRH steam temperature is worth 0.25 percentage point in steam cycle efficiency (typically, around 40%).
- Each 0.2 in. Hg (~7 mbar) in condenser pressure is worth 0.3 percentage point in steam cycle efficiency.

There are two confounding effects at play, which makes *achieving* and *exploiting* low p_{cond} tricky:

- Condenser cooling water temperature (achieving low p_{cond})
- LP turbine exhaust annulus area (exploiting low p_{cond}).

The first one is obvious enough. The lower the temperature of the condenser cooling water, the lower the saturated steam/water temperature in the condenser and the lower the p_{cond}. If cold water is naturally available, e.g., from a river or a lake or the sea, no parasitic power is consumed to cool the warm water returning from the condenser in a cooling tower, which is beneficial to the bottom line. Unfortunately, water scarcity and environmental regulations governing warm water discharge into natural water resources render this option increasingly difficult in many regions in the world. The alternative, "dry" air-cooled condensers, which make use of the ambient air as the heat "sink", are large parasitic power consumers (via electric motor drives of air circulation fans) and constitute large structures with a large footprint and costly equipment and construction materials and labor.

In terms of the second confounding effect, the key parameter is steam density at the LP turbine exhaust. The lower the p_{cond}, the lower the steam density and, for given steam mass flow rate, the higher the steam volumetric flow. If the flow annulus area, which is dictated by the length of

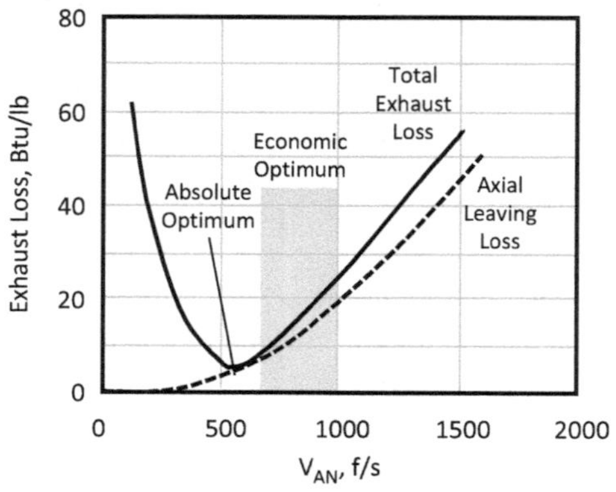

FIGURE 5.11 Exhaust loss curve and optimum V_{AN} selection.

the LSB, is not large enough, the exhaust velocity of steam can be very high, e.g., approaching sonic levels, with commensurately high exhaust losses. For a given steam flow rate and back pressure, an optimal exhaust annulus area can be determined by a combination of LSB size and LP turbine flow paths (e.g., two-flow, four-flow) The latter is typically an even number, i.e., 2, 4 or 6 (very rare in combined cycle applications – none so far to the best knowledge of the author), which can be achieved via a combination of 1, 2 or 3 two-flow LP turbine sections.

Optimum LSB selection is illustrated in Figure 5.11. If a given LSB length results in a small exhaust annulus area, steam exhaust velocity (V_{AN} in the figure) will be too high. Similarly, if the LSB length results in a large exhaust annulus area, V_{AN} will be too low. In either case, exhaust loss will be too high. By far the largest contributor to the *total exhaust loss* (TEL) is the kinetic energy of the steam at the exit of the last LP turbine stage (the dashed line labeled "axial leaving loss" in Figure 5.11). The other contributor is the exhaust hood loss, which is the extra loss incurred by having the specified exhaust or condenser pressure occur at the hood flange instead of at the wheel annulus. At very high values of V_{AN}, the flow will be choked (i.e., sonic velocity is reached).

On the left portion of the curve, at low steam flows and velocities, the last stage starts acting like a compressor; that is, pressure at the stage inlet is lower than the condenser pressure. Flow separation near the blade hub (root) and vortex formation are the key loss mechanisms and dominate the kinetic energy (leaving loss) at low values of V_{AN}. Similar effect is observed at extremely high condenser pressures during operation at hot ambient temperatures – especially with air-cooled condensers. Prolonged operation under such conditions can lead to blade flutter and shorten the part life.

The thermodynamic minimum is at around 550–600 ft/s. However, in order to provide a safety margin for low load operation and operation at high condenser pressures, the true optimum is at somewhat higher values of V_{AN} (700 ft/s or higher).

The TEL curve that comes with the steam turbine "thermal kit" is typically on a "dry" basis and must be corrected for the moisture content in the steam flow (about 8% in modern combined cycle units). The enthalpy correction given by Cotton (Ref. [17] in Chapter 2) is

$$h_{UEEP} = h_{ELEP} + TEL(1 - y)0.87(1 - 0.65y),$$

where

TEL = Total exhaust loss (Btu/lbm of dry steam) read from the exhaust loss curve with V_{AN}
UEEP = Used Energy End Point
ELEP = Expansion Line End Point (what one would get by using the stage isentropic efficiency)
y = Weighted average moisture (as a fraction) at ELEP.

For the details, the reader should consult the cited book by Cotton (pp. 88–98). Calculation of V_{AN} from cycle information is discussed in Section 5.3.4.

Available LSB sizes for combined cycle steam turbines by Siemens and GE for 50-Hz (3,000 rpm) and 60-Hz (3,600 rpm) applications are summarized in Tables 5.4–5.7. Note the difference in maximum available size between 50- and 60-Hz families of LSBs. Since 50-Hz gas turbines have larger exhaust energy than 60-Hz gas turbines (due to higher exhaust mass flow rate via nominal 1.44 size ratio stemming from the rotational speed difference), commensurate steam production in the HRSG and steam flow in the steam turbine are higher. Thus, a larger LSB is requisite to accommodate the largest 50-Hz gas turbine. Also note that a longer LSB translates into larger LSB mass and pitch diameter, which results in larger centrifugal stress acting on the blade. As a result, approximately above 48/40 in. of length, at 50/60 Hz, respectively, instead of steel, stronger but much more expensive titanium is used in manufacturing LSBs (e.g., Siemens 56 in. LSB, which was used in the steam turbine of the Irsching GTCC in Germany, which first broke the 60% net GTCC efficiency barrier – at 0.6 in. Hg back pressure; see Section 8.6.1).

TABLE 5.4
Siemens LSB (50 Hz)

LSB Length		Annulus Area	
in.	mm	ft²	m²
27.5	699	53.8	5.0
31.4	798	67.8	6.3
36.3	922	86.1	8.0
38.5	978	107.6	10.0
45.1	1,146	134.5	12.5
56.0 Ti	1,422	172.2	16.0

TABLE 5.5
Siemens LSB (60 Hz)

LSB Length		Annulus Area	
in.	mm	ft²	m²
22.9	582	37.7	3.5
26.2	665	47.4	4.4
30.2	767	60.3	5.6
32.0	813	74.3	6.9
37.6	955	93.6	8.7
47.0 Ti	1,194	119.5	11.1

TABLE 5.6
GE LSB (50 Hz)

LSB Length		Annulus Area	
in.	mm	ft²	m²
22.0	559	42.2	3.9
26.0	660	51.6	4.8
33.5	851	72.7	6.8
42.0	1,067	101.2	9.4
48.0	1,219	125.7	11.7

TABLE 5.7
GE LSB (60 Hz)

LSB Length		Annulus Area	
in.	mm	ft²	m²
23.0	584	32.9	3.1
26.0	660	41.1	3.8
30.0	762	55.5	5.2
33.5	851	66.1	6.1
40.0	1,016	87.3	8.1

TABLE 5.8
Exhaust Loss (Btu/lb of Dry Steam)

VAN, ft/s	26 in.-Btu/lb	30 in.-Btu/lb	33.5 in.-Btu/lb	40 in.-Btu/lb
128	64.09	99.21	106.70	114.58
150	56.15	88.90	95.43	102.18
175	47.58	77.88	84.28	91.36
200	40.55	67.65	74.38	82.53
250	29.75	50.85	56.72	64.24
300	22.48	38.38	43.00	49.00
350	17.28	29.83	32.82	36.45
400	13.55	23.15	25.40	28.10
450	10.82	17.80	19.40	21.25
500	8.85	13.60	14.66	15.80
550	8.30	11.20	11.93	12.62
600	8.50	10.45	10.80	10.88
650	9.50	10.75	10.75	10.21
700	11.25	12.00	11.65	10.45
800	15.82	16.20	15.20	12.68
900	21.13	21.17	19.80	16.39
1,000	26.82	26.42	25.22	21.84
1,100	32.82	32.10	31.06	27.67
1,200	38.50	37.55	36.70	33.37
1,300	43.80	42.70	42.05	38.80
1,400	48.30	47.35	46.69	43.19

Exhaust loss curves for several LSB sizes are provided in Table 5.8. The first three are from the curves in Figure 4.24 in Cotton's book cited earlier. The fourth one is an extrapolation from the first three and should be treated as a "guesstimate". Strictly speaking, they are for 60-Hz products by GE in Table 5.7. Nevertheless, they can be used in conceptual calculations for similar hardware from other OEMs. For quick estimates, assumption of 10 Btu/lb should be quite adequate, but care should be given to the boundary conditions and hardware size/cost ramifications (see Section 5.3 for more on this). It should also be mentioned that the exhaust losses from two different OEMs for similar LSB length and total annulus area can be quite different, i.e., by about 5 Btu/lb at minimum V_{AN}. (Refer to Figure 4 in Zachary [17].) Consequently, one *must* use the OEM-supplied exhaust loss curve if one is available.

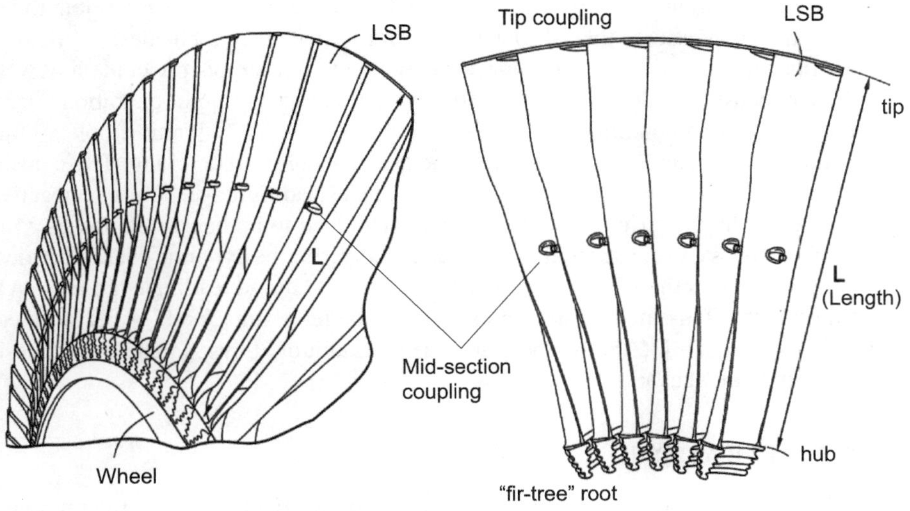

FIGURE 5.12 Combined cycle steam turbine last-stage rotor and blades. (From US 7,195,455 (Stonitsch, DeMania, Barb) "Application of high strength titanium alloys in last stage turbine buckets having longer vane lengths".)

In a combined cycle steam turbine, LP turbine exhaust steam flow is about 30% higher than the HP turbine inlet steam flow. The LP turbine generates close to 50% of the total unit power output. A typical LP turbine has 4–6 stages. Thus, the last stage can generate up to about 15% of the total unit power output. Combined with the blade length and commensurate change in stage velocity triangles in the radial direction, design of efficient three-dimensional (3D) LSB requires careful optimization of flow distribution and blade profiles. The resulting blade is of a "twisted and tapered" design with varying profiles from the blade hub to the blade tip (see Figure 5.12). As shown in the figure, individual blades are attached to the rotor wheel (disc) via "fir tree"-shaped roots or "dovetails". (There are other types of roots frequently used for connecting HP, IP or LP blades to their respective wheels. The reader can consult the books by Leyzerovich or Sanders cited in Chapter 2 for more information.)

As shown in Figure 5.12, the blades are continuously coupled at the tip and mid-section so that they cannot move independently. Tip coupling is accomplished via plates ("covers") with two sets of double holes, which fit onto the *tenons* protruding from the blade tips (two per blade). One plate thus combines two adjacent blades. There are two plates, one (*outer* cover) on top of the other (*inner* cover), so that inner and outer covers alternately couple adjacent buckets to one another. This arrangement forms a circumferential "shroud" at the last-stage rotor tip.

The mid-section connection is made with either (i) small nubs or "snubbers" machined from the blade forging or (ii) tie-wires through the holes in the blade. During turbine operation, the blades elastically "untwist" due to the centrifugal force acting on them so that tip shroud covers and snubbers contact. Such contact forces reduce blade vibration in response to stimuli from the steam flow by increasing the "stiffness" of the blade-wheel (disc) assembly as a whole. Another benefit is high tolerance to flutter occurring at low loads and high back pressures. One benefit of the tip shroud is to preserve the converging-diverging nozzle shape formed by the adjacent blades at the tip to accommodate supersonic Mach numbers without shock waves (extremely detrimental to expansion efficiency).

Unlike GE, some OEMs (e.g., Siemens and former Alstom) adopted "free-standing" LSB designs. Free-standing blades facilitate more efficient moisture separation than do shrouded blades. The snubbers in the flow channel disrupt the flow, creating additional losses and increased erosion due to local wetness concentration [18]. However, free-standing blades are more susceptible to flutter at low loads and high back pressures than shrouded and mid-span connected blades.

In modern units with advanced class gas turbine combined cycles, moisture content of wet steam at the LP turbine exhaust is typically less than 10% by mass at design point operation. Depending on site ambient and loading conditions, it can be as high as 10%–12%. Water droplets impinging on the LSB at high speeds can lead to erosion on the leading edge and cracking. The conventional method of protection is Stellite™ strips brazed to the blade leading edge.[3] On the negative side, discontinuities created by the strips in the blade profile can lead to higher profile losses. On top of that, breaking of the strips can cause local damage and changes in the dynamic characteristics of the blade [18]. A more recent method for protecting against droplet-caused leading edge erosion is laser hardening of the blades. This method delivers similar or better results for 17-4 PH stainless steel vis-à-vis flame-hardening [18]. (Another benefit of using titanium alloys is their greater resistance to wetness losses and damages.)

5.3 BASIC CALCULATIONS

The most common need for steam turbine calculations is performance-related. In combined cycle applications, the key question to be answered is this: What is the power output of the bottoming cycle steam turbine generator (STG) for a given gas turbine? For *bona fide* design calculations, one has to use a heat and mass balance simulation software such as Thermoflow, Inc.'s GT PRO® or an in-house proprietary software. (Each OEM has one, e.g., GE's legacy Fortran codes MHB and THB for HRSG and steam turbine thermodynamic calculations, respectively.)

There are several types of performance calculations:

- Design and off-design performance prediction
  - Rigorous models
  - Quick estimates (for sales and marketing).
- Performance test analysis
  - Rigorous models
  - Correction curves (manufacturer's "thermal kit").
- "Reverse engineering" of thermal kit data or competitor's equipment.

Design calculations require solution of *steady-state, steady-flow* (SSSF) formulation of the first law thermodynamics for key turbine control volumes, i.e., HP, IP and LP turbines. This can be easily boiled down to sequential solution of multiple enthalpy and mass balances. Very reasonable estimates can be made with pen, paper, steam property tables in the Appendix of one's undergraduate thermodynamics textbook and a calculator (most likely on a smartphone nowadays). Using an Excel spreadsheet with a steam property add-in or DLL can make this a routine exercise.

For higher fidelity, the following information is required:

- Section (cylinder) efficiencies (e.g., see Table 5.2)
- Packing leakage configuration and flow estimates
- LSB exhaust loss curve (e.g., see Table 5.8)
- Bearing losses
- Generator losses.

[3] Stellite alloys are a group of cobalt–chromium superalloys. Erosion shields for LSBs are typically made from Stellite alloy 6B or from cast Stellite materials. Stellite was a trademark of the Deloro Stellite Company, which is now a part of the Kennametal group.

This information is built into the proprietary software used by the OEMs. Commercial software such as GT PRO® and GateCycle also has built-in formulations, which are very good approximations of OEM methods.

Design calculations typically assume "rubber equipment". In other words, the designer stipulates that the particular plant hardware is not limited by size, materials and other physical features. Off-design calculations are based on "actual" hardware, which is sized to deliver the performance specified at the design point. Thus, while the design calculations are purely thermodynamic in nature (i.e., enthalpy and mass balances), off-design calculations require knowledge and application of heat transfer and fluid mechanical principles governing the fluid flow in heat exchangers, pipes, valves and other types of plant equipment. Also requisite is a solid knowledge of available materials and their relative cost. It is easy to do a design calculation with 300 bar and 700°C HP steam (which can be obtained by, say, supplementary firing in the HRSG). As long as the fundamental laws of thermodynamics are respected, one can come up with a performance estimate. Needless to say, heat exchangers, pipes, valves, etc., which can safely operate at such conditions, require expensive alloy materials (in some cases with limited commercial field experience), which would be prohibitively expensive even if they are available in the first place. (Whether the turbine flow-path to handle that steam can be designed is an altogether different question, which has been briefly touched upon in Section 5.1.1.)

Coverage of full-blown steam turbine design calculations is well beyond the scope of this book (and the abilities of the author). However, very accurate estimates can be made from the first principles with varying degrees of effort. Let us start with the simplest ones.

The starting point is always the gas turbine. In the following discussion, we will assume a 1 × 1 × 1 multi-shaft combined cycle configuration. The gas turbine is specified through its four cardinal parameters:

- 424 MWe generator output
- 42.17% LHV efficiency (8,092 Btu/kWh heat rate)
- 100% CH_4 (methane) fuel heated to 410°F
- 1,825 lb/s exhaust flow
- 1,187°F (642°C) exhaust temperature.

Gas turbine heat and mass balance analysis is summarized in Table 5.9.

Let us start with the simplest, "rule of thumb" approach. In a combined cycle power plant, STG output is about one half of the GTG output (as discussed in Chapter 3). Thus, in this example, STG output is 212 MWe. This is a very useful rule-of-thumb check for more detailed calculation results. Significant deviation from this number is a sure sign that something went awry in the process.

From the HMB in Table 5.9, gas turbine exhaust energy is 580,581 kW (with 60°F zero enthalpy reference). State-of-the-art 3PRH HRSG with advanced class gas turbines is about 90% effective. In another word, it converts about 90% of exhaust gas energy into steam production. Furthermore, a modern combined cycle bottoming steam turbine cycle efficiency is about 40%. Thus, STG output is

$$581 \times 0.9 \times 0.4 = 209.2 \text{ MWe}.$$

There is a "lurking" source of error or uncertainty in this type of calculation, namely, the reference temperature (enthalpy) used in the HRSG effectiveness definition and the zero enthalpy reference temperature of the exhaust gas enthalpy calculation. Using the same temperature for both is the most convenient choice. The said temperature can be the ISO temperature, 59°F (15°C), or 77°F (25°C), which is also commonly used in the literature. The difference between the two can result in an enthalpy difference of about 4 Btu/lb (see Appendix A). This translates to a difference of about 1.3%. Thus, the estimated value above, 209.2 MWe, can be 209.2 × (1–1.3%) = 206.5 MWe.

TABLE 5.9
Sample Gas Turbine Heat and Mass Balance (net η = 42.2%)

	$\dot{m}$	p	T	h	Q	
	pps	psia	°F	Btu/lb	Btu/s	kW
Air in	1,771.3	14.7	59.0	–0.2	–427	–451
Fuel in	44.29	488.8	410.0	206.7	961,539	1,014,481
Compressor discharge	1,558.7	320.4	866.9	200.8	313,033	330,268
Overboard leak	1.8	304.4		200.8	356	375
Exhaust frame blower	10.6		110.0	12.07	128	135
Exhaust	1,824.4	14.9	1,187.5	301.6	550,283	580,581
Heat loss					2,381	2,512
Mechanical loss					1,350	1,424
Heat consumption					952,383	1,004,821
Shaft output					406,871	429,273
Generator output					402,395	424,551
GT auxiliary					483	510
GT net output					401,912	424,041

In mathematical terms, the quick estimation above can be written as

$$\dot{W}_{STG} \approx HC \cdot (1 - \eta_{GTG}) \cdot \eta_{HRSG}\eta_{ST},$$

where HC is gas turbine heat consumption. Thus, combined cycle gross (generator) output becomes

$$\dot{W}_{CC,Gross} \approx HC \cdot \eta_{GTG} + HC \cdot (1 - \eta_{GTG}) \cdot \eta_{HRSG}\eta_{ST}.$$

The ratio of steam turbine output to that of the gas turbine is

$$\frac{\dot{W}_{STG}}{\dot{W}_{GTG}} \approx \frac{(1 - \eta_{GTG}) \cdot \eta_{HRSG}\eta_{ST}}{\eta_{GTG}}.$$

Thus, with typical values for state-of-the-art technology,

$$\frac{\dot{W}_{STG}}{\dot{W}_{GTG}} \approx \frac{(1 - 0.4) \cdot 0.9 \cdot 0.4}{0.4} = 0.54.$$

Hence, the one-to-two rule of thumb used above.

These were the "first law" estimates. From a second law (of thermodynamics, that is) perspective, the key input is the gas turbine exhaust gas *exergy*. The exergy of gas turbine exhaust gas burning natural gas fuel, to a good approximation, is given by Equation A.2, i.e.,

$$a_{exh} = 0.1961 \cdot 1187.5 - 86.918 = 146 \text{ Btu/lb}.$$

Thus, total exergy of the gas turbine exhaust gas is

$$a_{exh} = 1{,}824.4 \cdot 146 = 266{,}273 \text{ Btu/s}$$

or

$$A_{exh} = 1.05506 \cdot 266{,}273 = 281{,}000 \text{ kW}.$$

In other words, if the bottoming steam turbine cycle were substituted by an ideal, reversible heat engine, its output would be 281 MWe. In an earlier paper, Gülen and Smith showed that the bottoming cycle exergetic efficiency of well-designed systems with ~650°C gas turbine exhaust temperature is slightly lower than 75% [14]. This efficiency is based on bottoming cycle net output, which subtracts the boiler feed pump power consumption and miscellaneous losses from the STG output (roughly 2% [19]). Thus,

$$\dot{W}_{STG} \approx \frac{0.74 \cdot 281}{1 - 0.02} = 212 \text{ MWe}.$$

This is pretty much what can be done with high-level estimates. The range is between 209 and 212 MWe with an average of 211.5 MWe.

For a more rigorous calculation, one needs a bit more effort. The most important (and unavoidable) chore is estimation of isentropic "steam-path" efficiencies, which is discussed below in some detail.

5.3.1 Steam-Path Efficiency

Steam-path efficiency is the "expansion line" efficiency for the entire turbine section, HP, IP or LP, from bowl (inlet) to exhaust. For combined cycle steam turbines, steam expanding in the HP and IP sections is "dry" during normal operation. Therefore, the isentropic section efficiency, which is the ratio of the actual enthalpy drop to the isentropic enthalpy drop across the section, is adequate. For typical values, please refer to the discussion in Section 5.1 (e.g., see Table 5.2). The simple textbook formulation is

$$\eta_s = \frac{h(p_{in}, T_{in}) - h(p_{ex}, T_{ex})}{h(p_{in}, T_{in}) - h(p_{ex}, T(p_{ex}, s_{in}))}. \tag{5.35}$$

Known parameters are usually inlet pressure and temperature and section pressure ratio, which sets the exit pressure. Isentropic exit temperature and enthalpy are calculated using the exit pressure and the inlet entropy. Exit temperature and enthalpy are calculated from Equation 5.35 using the section efficiency.

Equation 5.35 is used for two purposes:

1. To calculate the section efficiency from the performance test data (inlet and exit pressures and temperatures) or
2. To predict performance from known section efficiency (i.e., exit enthalpy and temperature).

For the second purpose, one has to have a method to predict the section efficiency. There are two possibilities:

1. Use known efficiencies as a guide (e.g., Table 5.2) or
2. Predict the efficiency from known thermodynamic data.

The best-known (and quite archaic) predictive method was outlined in the aforementioned *Spencer–Cotton–Cannon* (SCC) paper originally published in 1963 [20]. The paper contains detailed calculation of isentropic "expansion line" efficiencies of "groups of stages" for electric utility steam turbines (both reheat and non-reheat) rated 16.5 MWe or higher. (At the time, they were all fossil fuel-fired boiler plants.) In addition, the paper also includes calculation methods for packing and valve steam leaks, mechanical and electrical (generator) losses and LP turbine exhaust loss. The method can be applied to combined cycle steam turbines by matching the HP, IP and LP sections with the corresponding reheat and condensing section charts in the paper. Key independent variables are section volumetric flow (inlet) and pressure ratio.

Obviously, significant progress in aerothermodynamic design of steam turbines in more than half a century since the publication of the paper rendered the efficiencies obtained from the original SCC method obsolete.[4] However, the basic thermodynamic drivers of isentropic expansion efficiency did not change, i.e., the expansion pressure ratio and inlet volumetric flow rate of steam. Thus, one can still use the revised 1974 SCC method with a calibration factor to update the raw efficiency number to a known technology level.

For the last turbine section discharging to the condenser, i.e., the LP section, the exhaust loss has to be separately evaluated as described in Section 5.2. For the overall section efficiency, one should use the UEEP as the final steam enthalpy (see the discussion accompanying Figure 5.11).

5.3.2 Steam Cycle Simple Calculation

There is nothing difficult about calculating the output and efficiency of a steam turbine Rankine cycle. In fact, any undergraduate textbook on engineering thermodynamics is full of examples of varying degrees in complexity. In a state-of-the-art 3PRH bottoming cycle, there are three key enthalpy balances for the HP, IP and the LP turbines. The information needed to do accurate heat and mass balance calculation includes

- Steam flow rates (HP, IP and LP)
- Steam turbine packing leakages
- Section efficiencies
- LP turbine exhaust loss.

Steam flow rates require heat and mass balance modeling of the HRSG. Reasonably accurate HP, IP and LP steam production rates in state-of-the-art 3PRH HRSGs are covered in Section 6.2.1. Typical section efficiencies can be found in Table 5.2. Exhaust loss estimation was covered in Section 5.2.

One can become overambitious and estimate packing leakages as well (e.g., see the SCC 1974 method [20]). Nevertheless, in this author's opinion this would be an overkill. If one is involved in steam-path design for an OEM (or steam turbine "thermal kit" development) or performance calculation for an EPC contractor, it is given that he or she has access to a proprietary in-house code (e.g., GE's MHB and THB) or a commercial software package (e.g., GT PRO® or GateCycle). If not, the accuracy requisite for the task at hand is guaranteed not to justify the diligent calculation of a packing leakage of several thousand pounds per hour (for a throttle steam flow of the order of hundreds of thousands).

Having said that, here is a method to estimate the steam turbine output quite accurately:

- Use the steam flow estimates in Section 6.2.1.
- Use the section efficiencies in Table 5.2.
- Calculate the output from simple heat and mass balance.
- Apply a correction factor for leakages and other losses.

---

[4] An updated version was published by the same authors in 1974 as a GE white paper (GER-2007C). Actually, this is the work cited in this book.

As an example, consider the following steam cycle:

- Pressures
  - 2,415 psia (HP throttle valve inlet);
  - 425 psia (HRH at the IP combined stop-control valve inlet);
  - 55 psia (LP bowl pressure).
- HP and HRH temperatures 1,050°F (commensurate with gas turbine exhaust temperature);
- IP and LP steam admission temperatures 600°F;
- HP (main), IP and LP admission steam flows from Section 6.2.1 for gas turbine exhaust mass flow rate of 1,500 lb/s;
- HP, IP and LP "dry step" efficiencies of 89%, 90% and 92%, respectively;
- Exhaust loss (dry) 0 Btu/lb;
- Condenser pressure 1.2 in. Hg.

Straightforward enthalpy and mass balance calculation in a software tool such as THERMOFLEX® or GateCycle (the same can be done easily in an Excel spreadsheet with cell-based calculations or simple VBA code) multiplied by a factor of 0.96 reproduces rigorous STG output calculated in GT PRO®. For the reader who wants to try his or her hand at this, sample HP, IP and LP turbine calculations are summarized in Figures 5.13–5.18.

As an example, referring to Figure 5.13,

$$\text{Expansion work} = \eta\left(h_1\left(p_1, T_1\right) - h_2\left(s_1, p_2\right)\right).$$

$$\text{Shaft work} = \text{Expansion work} - \text{Mechanical loss}(0.25\%).$$

Enthalpy, h, and entropy, s, are found from the *ASME Steam Tables*. (The calculations below are made with IFC-67.)

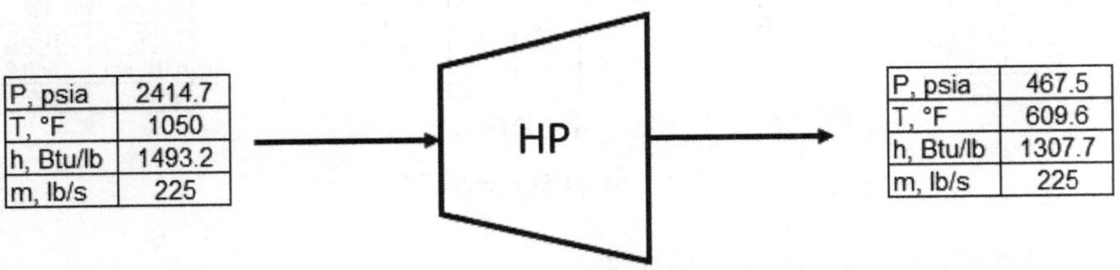

FIGURE 5.13 HP section heat balance; isentropic section efficiency 89.83%; expansion power 44,042 kW with 110 kW mechanical loss for shaft power of 43,932 kW. IFC-67 steam properties are used.

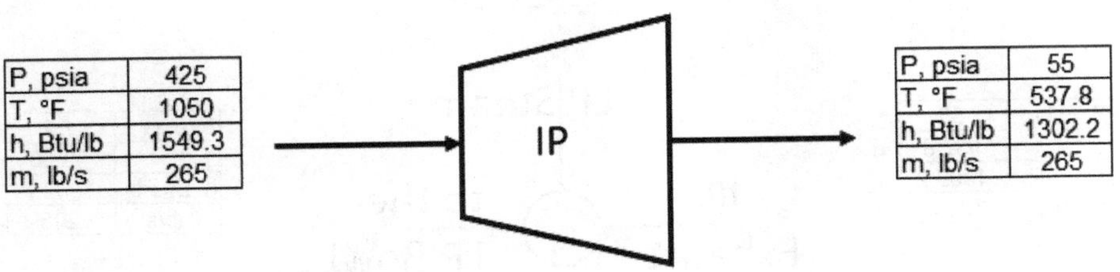

FIGURE 5.14 IP section heat balance; isentropic section efficiency 91.76%; expansion power 69,115 kW with 173 kW mechanical loss for shaft power of 68,942 kW. Steam flow is the sum total of 225 lb/s of cold reheat flow (HP exhaust) and 40 lb/s IP steam.

P, psia	55
T, °F	538.9
h, Btu/lb	1302.7
m, lb/s	295
s, Btu/lb-R	1.7976

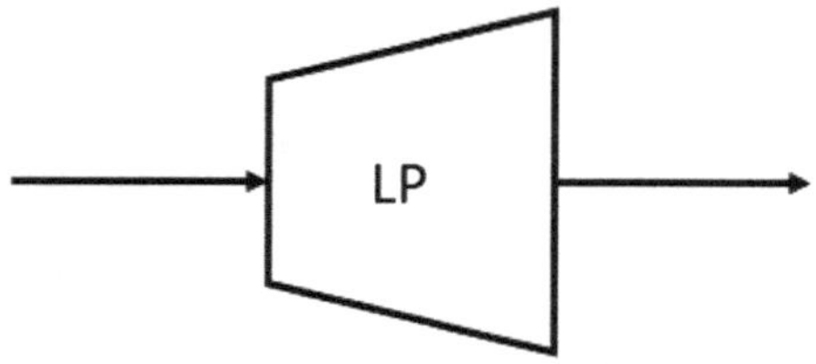

P, psia	0.6817
T, °F	89.24
x	0.9174
h, Btu/lb	1014.3
m, lb/s	295

FIGURE 5.15 LP section heat balance; overall section efficiency 90.36% (inclusive of exhaust losses); expansion power 89,776 kW with 224 kW mechanical loss for shaft power of 89,542 kW. Steam flow is the sum total of 265 lb/s of IP exhaust flow and 30 lb/s LP steam.

P, psia	467.5
T, °F	550
h, Btu/lb	1270.5
m, lb/s	40

P, psia	467.5
T, °F	609.6
h, Btu/lb	1307.7
m, lb/s	225

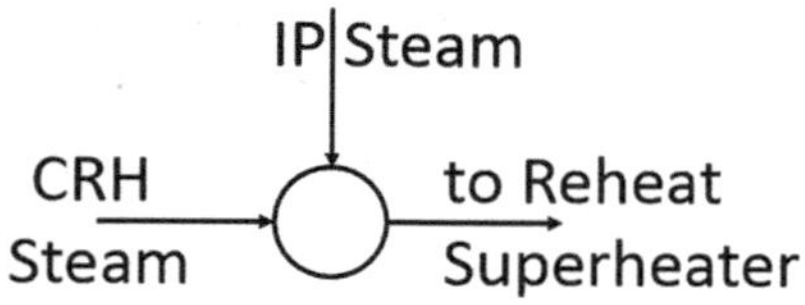

P, psia	467.5
T, °F	600.3
h, Btu/lb	1302.0
m, lb/s	265

FIGURE 5.16 Mixing of IP and CRH steams. Note that 467.5 psia includes 10% reheater pressure loss. Care should be taken that the IP steam temperature and CRH steam temperature difference does not exceed 75°F (~42°C) to prevent thermal stress problem. (In this case, the difference is 60°F and OK.)

P, psia	467.5
T, °F	600.3
h, Btu/lb	1302.1
m, lb/s	265

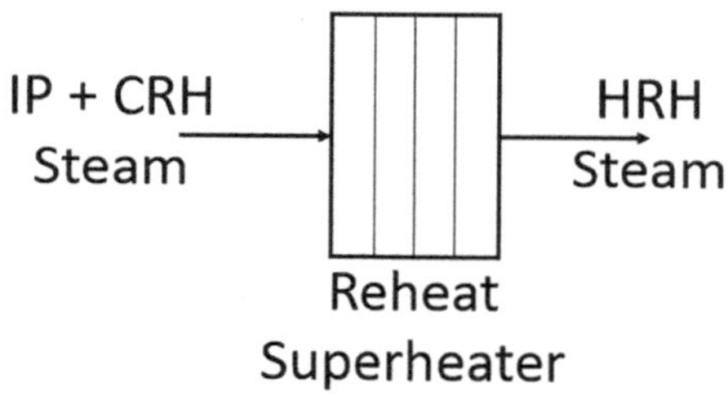

P, psia	425
T, °F	1050
h, Btu/lb	1549.3
m, lb/s	265

FIGURE 5.17 Reheater heat addition (65,530 Btu/s).

P, psia	55
T, °F	549.2
h, Btu/lb	1307.7
m, lb/s	30

P, psia	55
T, °F	537.8
h, Btu/lb	1302.2
m, lb/s	265

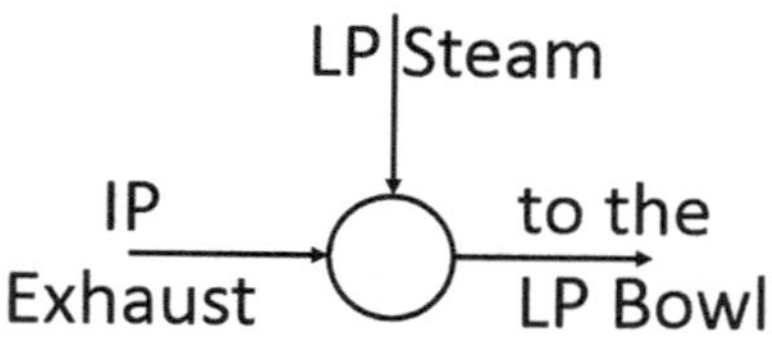

P, psia	55
T, °F	539.0
h, Btu/lb	1302.7
m, lb/s	295

FIGURE 5.18 Mixing of LP admission and IP exhaust streams. Care should be taken that the LP steam temperature and IP exhaust steam temperature difference does not exceed 75°F (~42°C) to prevent thermal stress problem. (In this case, the difference is less than 12°F and does not present a thermal stress issue.)

For mixing calculations, referring to Figure 5.16

$$\dot{m}_3 = \dot{m}_2 + \dot{m}_9;$$

$$h_3 = \frac{\dot{m}_2 h_2 + \dot{m}_9 h_9}{\dot{m}_3}; \; T_3 = f(p_3, h_3).$$

Total shaft output is

$$43{,}932 + 68{,}942 + 89{,}542 = 202{,}416 \text{ kW}.$$

Generator output is (assuming 98.7% efficiency including mechanical and electrical losses)

$$202{,}416 \times 0.987 = 199{,}785 \text{ kW}.$$

Applying the correction factor of 0.96 for leakages, etc., a "realistic" STG output is calculated as

$$199{,}785 \times 0.96 = 191{,}793 \text{ kW}.$$

Steam cycle efficiency is the ratio of STG output to the steam cycle heat input. The latter is calculated as follows:

$$\dot{Q}_{in} = \dot{m}_{HP}(h_{HP} - h_{fw}) + \dot{m}_{CRH}(h_{HRH} - h_{CRH}) + \dot{m}_{IP}(h_{HRH} - h_{fw}) + \dot{m}_{LP}(h_{LP} - h_{fw}). \tag{5.36}$$

In Equation 5.36, h_{fw} is the feedwater enthalpy at the HRSG inlet, which is pretty close to the condensate temperature (i.e., saturated liquid enthalpy at the condenser pressure). The terms on the right-hand side are, from left to right, heat transfer rates from the exhaust gas

- To make superheated HP steam
- To reheat cold reheat steam
- To make superheated IP steam
- To make superheated LP steam.

For this example h_{fw} can be taken as 54 Btu/lb. Thus, the calculation is summarized in a stepwise manner as follows:

$$\dot{m}_{HP}(h_{HP} - h_{fw}) = 225(1493.2 - 54) = 323{,}820 \text{ Btu/s}$$

$$\dot{m}_{CRH}(h_{HRH} - h_{CRH}) = 225(1549.3 - 1307.7) = 54{,}360 \text{ Btu/s}$$

$$\dot{m}_{IP}(h_{HRH} - h_{fw}) = 40(1549.3 - 54) = 59{,}812 \text{ Btu/s}$$

$$\dot{m}_{LP}(h_{LP} - h_{fw}) = 30(1307.7 - 54) = 37{,}611 \text{ Btu/s}$$

$$\dot{Q}_{in} = 323{,}820 + 54{,}360 + 59{,}812 + 37{,}611 = 475{,}603 \text{ Btu/s}$$

$$\eta = \frac{191{,}793}{475{,}603 \cdot 1.05506} = 0.3822.$$

Note that this calculation is slightly inaccurate because, in reality,

- cold reheat flow is less than the HP steam flow by the amount of packing leakages (about 2%);
- there is energy imparted by the boiler feed pumps (roughly 2% of the STG output).

Accounting for them, the steam cycle efficiency becomes 38.6%.

5.3.3 Steam Cycle Efficiency History

Key technologies driving the flange-to-flange steam *turbine* efficiency are

- Three-dimensional aerodynamic design of buckets and nozzles (for better isentropic efficiency)
- Advanced packing and gland seals to minimize leakages
- LP section LSB size to exploit low condenser pressures
- Exhaust diffuser design.

Key cycle design parameters driving the overall steam *cycle* efficiency are

- HP or *main* steam temperature
- HRH or IP admission steam temperature
- HP steam pressure
- Condenser pressure
- Volumetric flow rate of steam.

High steam pressures and temperatures as well as steam generation rates (leading to high volumetric steam flows) are driven by the gas turbine exhaust gas energy, primarily the gas turbine exhaust gas temperature. The relationship between the steam cycle efficiency and gas turbine exhaust gas temperature is demonstrated by the data plotted in Figure 5.19. The data is from the *Gas Turbine World 2006 Handbook*. The trend is very pronounced and interesting. For E class exhaust gas temperatures (i.e., less than ~1,050°F), there is a direct relationship between the two parameters. (The cluster below the trend line is for two-pressure or three-pressure, non-reheat steam cycles.) This is a clear manifestation of higher exhaust temperature's advantage in (i) higher steam temperatures and (ii) higher steam production in the HRSG. In F class range, the trend is quite flat, which reflects the steam temperature limit of 1,050°F (prevalent at the time).

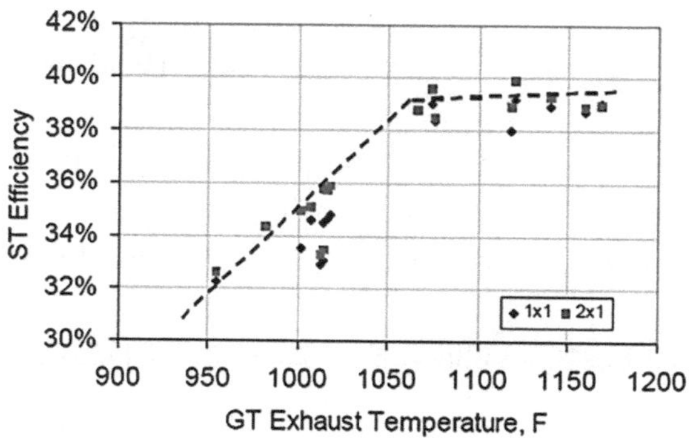

FIGURE 5.19 Steam cycle/turbine efficiency as a function of GT exhaust temperature (2006 vintage).

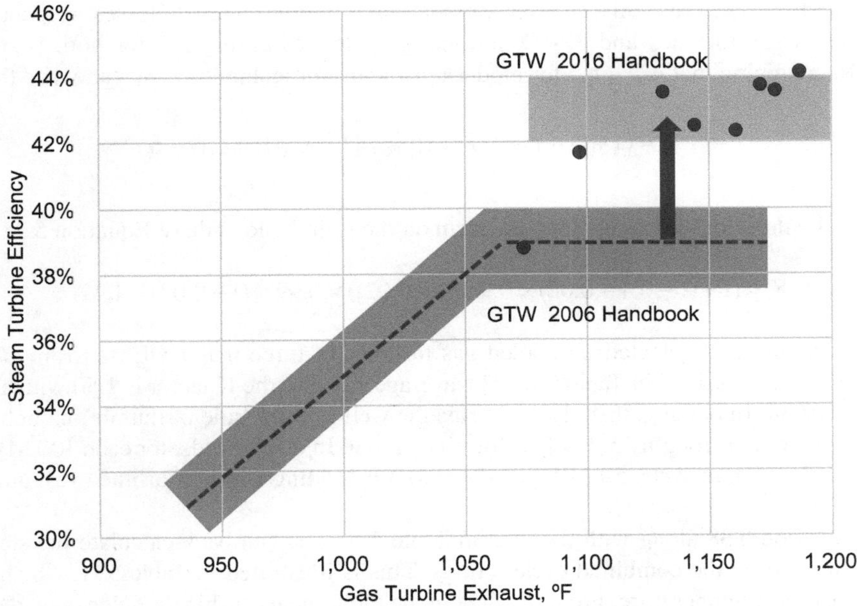

FIGURE 5.20 Steam cycle/turbine efficiency as a function of GT exhaust temperature (2016 vintage).

Data reflecting the 2016 state-of-the-art (from *GTW 2016 Handbook*) is plotted in Figure 5.20. The jump in advanced F, H and J class steam turbine cycle efficiencies by 3 percentage points (on average) between 2006 and 2016 is unmistakable. The main driver for that is the new standard in advanced bottoming cycle main (i.e., high pressure, HP) and HRH steam temperatures, which is 1,112°F (600°C).

Note that each percentage point in steam turbine cycle efficiency is worth 0.5 percentage point in combined cycle efficiency. Thus, a quite impressive 1.5 percentage points rise in combined cycle efficiency (at least as represented by OEM-supplied ISO baseload ratings in trade publications) between 2006 and 2016 is directly attributable to the bottoming cycle. Furthermore, each one percentage point in gas turbine efficiency is worth two-thirds of a percentage point in combined cycle efficiency (see Section 3.3).

According to the historical trends (see Figure 4.12 (for simple cycle gas turbines) and Figure 4.13 (for combined cycle gas turbines) in Chapter 4 of **GTFEPG**), between 2006 and 2016, the rise in gas turbine efficiency is 1.5–2 percentage points, whereas the rise in combined cycle efficiency is about 2–3 percentage points.

According to the derivative in Equation 3.5, the combined cycle efficiency impact of 1.5–2 percentage points in gas turbine efficiency is about 1–1.3 percentage points. This leaves about 1–1.7 percentage points attributable to the improvement in the bottoming, i.e., the steam turbine cycle efficiency. This confirms the finding from the data plotted in Figure 5.20.

Recall Equation 3.1 that establishes the relationship between the gas turbine, steam turbine and combined cycle efficiencies,

$$\eta_{CC} = \left(\eta_{GT} + \left(1 - \eta_{GT}\right)\eta_{HRSG}\eta_{ST}\right)(1-\alpha).$$

A slightly more involved but "exact" version is

$$\eta_{CC} = \left(\eta_{GT} + \left(1 - \eta_{GT} - \lambda_{GT}\right)\eta_{HRSG}\left(1 - \lambda_{BC}\right)\eta_{ST}\right)\eta_{Gen}(1-\alpha), \tag{5.37}$$

where λ_{GT} and λ_{BC} represent miscellaneous topping and bottoming cycle losses, respectively. For advanced class gas turbines and 3PRH bottoming cycles, assuming 2% for both is reasonably accurate. Recognizing that $\eta_{GT} \sim \eta_{ST}$ for modern prime mover technology, we can write that

$$\frac{\eta_{CC,net}}{\eta_{ST}} \approx \left(1+\left(1-\eta_{GT}-\lambda_{GT}\right)\eta_{HRSG}\left(1-\lambda_{BC}\right)\right)\eta_{Gen}(1-\alpha). \tag{5.38}$$

Using typical values for the parameters, the term on the right-hand- side of Equation 5.38 is

$$K=(1+(1-0.4-0.02)\times 0.9\times(1-0.02))\times 0.99\times(1-0.02)=1.47.$$

Rating performances of selected air-cooled gas turbines of three major OEMs (from *GTW 2016 Handbook*) are summarized in Table 5.10. The average value of the K factor is 1.50 with a standard deviation of 0.034. In passing, this also confirms the well-known "rule of thumb", namely, that the steam turbine output is roughly 50% of gas turbine output. In other words, for each 100 MWe of gas turbine output in *simple* cycle, 50 MWe is added to it by adding a steam turbine cycle and making a *combined* cycle.

Using Equation 5.38 along with the data in Table 5.10, one can back-calculate the steam cycle efficiency implied by the combined cycle ratings. This is illustrated in Table 5.11, which includes gas turbine exhaust temperature. From the data in the table, steam turbine efficiency is, on average, 42.5% (±1.7%) – cf. 40.6% (±1.4%) for the gas turbines in Table 5.10.

5.3.4 Exhaust End Analysis

From the basic thermodynamics, it is known that the heat sink temperature of a heat engine or power cycle is as important as its heat source temperature. For a Rankine steam turbine cycle, the heat sink is the condenser coolant, which is either water or ambient air. Thus, the closer one gets to the coolant's average or mean-effective temperature at the steam turbine exhaust, the better will be the steam cycle performance because of a reduction in exergy loss during heat rejection. The hardware design parameter controlling this behavior is the condenser pressure, which sets the (constant) temperature at which the exhaust steam condenses before being pumped back into the boiler or the HRSG.

On paper at least, one can get as low in condenser pressure as one wishes. In practice, however, aggressive designs come at the expense of asymptotic increase in condenser heat transfer area, i.e., size and cost, and LP turbine exhaust annulus flow area and LSB length, i.e., size and cost (notwithstanding exceeding mechanical design limits due to exorbitant centrifugal forces acting on the blades).

TABLE 5.10
Simple and Combined Cycle Efficiencies (*GTW 2016 Handbook* Data)

OEM	Model	GT Efficiency	CC Efficiency	K Factor
MHPS	M701F4	39.9%	60.10%	1.51
MHPS	M701F5	40.0%	61.10%	1.53
GE	9F-5	38.7%	60.65%	1.57
GE	9F-7	41.1%	61.25%	1.49
GE	9HA.01	42.4%	62.65%	1.48
GE	9HA.02	42.7%	62.75%	1.47
Siemens	SGT5-4000F	40.0%	58.70%	1.47
Siemens	SGT5-8000H	40.0%	60.50%	1.51

TABLE 5.11
Combined Cycle Steam Turbine Efficiency

OEM	Model	Exhaust Temperature, °F	Steam Cycle Efficiency
MHPS	M701F4	1,097	41.7%
MHPS	M701F5	1,131	43.5%
GE	9F-5	1,187	44.1%
GE	9F-7	1,144	42.5%
GE	9HA.01	1,171	43.7%
GE	9HA.02	1,177	43.6%
Siemens	SGT5-4000F	1,074	38.8%
Siemens	SGT5-8000H	1,161	42.3%

The starting point of steam turbine exhaust end analysis is gas turbine exhaust conditions, i.e., flow rate and temperature. The largest 50-Hz gas turbines rated at nearly 500 MWe have exhaust flows of the order of 2,000 lb/s (~900 kg/s) with temperatures close to 1,200°F (~650°C).

In an earlier paper, Gülen and Smith used the following correlation to estimate the steam turbine exhaust flow rate as a function of gas turbine exhaust parameters [14]:

$$\mu = \left(0.153 + 0.018\left(\frac{T_{exh}}{100} - 11\right)\right) - 0.002\frac{T_{stm} - 1050}{25}. \tag{5.39}$$

where μ is the ratio of steam and exhaust gas flow rates. Ignoring the adjustment for the steam temperature, for the large gas turbine with 2,000 lb/s and 1,200°F exhaust gas flow, the steam flow rate at the LP turbine exhaust is found as

$$\dot{m}_{stm} = 2{,}000 \times (0.153 + 0.018 \times (12 - 11)) = 342 \text{ lb/s} = 1{,}231{,}200 \text{ lb/h},$$

which is equal to 342 lb/s. Economically optimum design requires an annulus velocity of ~700 ft/s, which is given by [19]

$$V_{AN} = \frac{\dot{m}_{stm} v_g (1 - y)}{3{,}600 \, A_{an}}, \tag{5.40}$$

where v_g is the saturated steam specific volume at the condenser pressure, y is the exhaust steam moisture mass fraction, and A_{an} is the flow area of the exhaust annulus. Note that the volumetric flow rate of steam in ft^3/s is given by

$$\dot{V}_{stm} = \frac{\dot{m}_{stm} v_g (1 - y)}{3{,}600} \tag{5.41}$$

so that Equation 5.40 can be rewritten as

$$V_{AN} = \frac{\dot{V}_{stm}}{A_{an}}. \tag{5.42}$$

For reader's convenience, data from the *ASME Steam Tables* are summarized in Table 5.12. For conceptual calculations of this type, y = 0.08 (i.e., steam quality, x = 0.92) is a very reasonable assumption.

TABLE 5.12
Saturated Steam Data for Condenser Conditions (1 in. Hg = 33.9 mbar)

Condenser Pressure		Saturated Temperature	v_g
in. Hg	psi	°F	ft³/lb
0.50	0.25	58.8	1,256.6
1.00	0.49	79.0	652.4
1.50	0.74	91.7	445.0
2.00	0.98	101.1	339.3
2.50	1.23	108.7	275.0
3.00	1.47	115.1	231.6
3.50	1.72	120.6	200.3
4.00	1.96	125.4	176.7
4.50	2.21	129.8	158.2
5.00	2.46	133.8	143.3

The following transfer function represents the saturated steam specific volume accurately:

$$v_g\left[\text{in ft}^3/\text{lb}\right] = 652.77 \times P_{cond}\left[\text{in. Hg}\right]^{-0.943}. \qquad (5.43)$$

Rating performances for combined cycles are usually evaluated at a condenser pressure of 1.2 in. Hg. Using Equations 5.40 and 5.43, one can estimate the annulus area requisite for the calculated steam flow rate at this condenser pressure. First, from Equation 5.43, v_g is 549.2 ft³/lb. From Equation 5.40 with $y = 0.08$ and $V_{AN} = 700$ ft/s,

$$A_{an} = (342 \times 549.2 \times 0.92)/700 = 246.9 \text{ ft}^2.$$

The annulus area in square-feet can also be estimated from a curve-fit to exhaust end dimensions in Tables 5.4–5.7 as a function of the LSB length l_b in inches and the number of LP flow-paths N_{LP} as follows [19]:

$$A_{an} = N_{LP} 0.129 l_b^{1.7722}. \qquad (5.44)$$

From Equation 5.44, for $N_{LP} = 1$ (i.e., one LP flow-path)

$$l_b = (246.9/0.129)^{(1/1.7722)} = 71.1 \text{ in.}$$

This is a mechanically infeasible result. Thus, it is clear that this steam turbine must have at least two LP flow-paths, i.e., $N_{LP} = 2$, which gives $l_b = 48$ in. This is equal to the largest 50-Hz LSB length from GE (see Table 5.6).

REFERENCES

1. Zörner, W., 1994, Steam turbines for power plants employing advanced steam conditions, *10th CEPSI*, September 19–23, 1994, Christchurch, New Zealand.
2. Viswanathan, R., 1989, *Damage Mechanisms and Life Assessment of High-Temperature Components*, ASM International, Metals Park, OH.
3. Technology Status Report 018, Review of Status of Advanced Materials for Power Generation, Cleaner Coal Technology Programme, October 2002, DTI/Pub URN 02/1267.

4. Parvizinia, M., Wechsung, M., Düweler, M.G., Standardized Steam Turbines for Combined Cycle Power Plants, Siemens paper.
5. Matsushima, T., Nishimura, S., 1999, Single casing reheat turbine, PWR-Vol. 34, *1999 Joint Power Generation Conference*, Burlingame, CA, pp. 565–572.
6. Nakano, T., Tanaka, K.E., Nakazawa, T.A., Nishimoto, S., Takeda, K., Miyawaki, T.O., 2005, Development of large-capacity single-casing reheat steam turbines for single-shaft combined cycle plant, *Mitsubishi Heavy Ind. Ltd.*, Vol. 42, No. 3, pp. 108.
7. Simon, V., Stephan, I., Bell, R.M., Capelle, U., Deckers, M., Schnaus, J., Simkine, M., 1997, Axial steam turbines with variable-reaction blading, advances in turbine materials design and manufacturing, *Inst. Mater.*, Vol. 689, pp. 46–60.
8. Boss, M.J., 1994, Steam Turbines for STAG Combined Cycle Power Systems, General Electric GER–3582D.
9. Reinker, J.K., Mason, P.B., 1994, Steam Turbines for Large Power Applications, General Electric, GER–3636D.
10. Traupel, W., 2001, *Thermische Turbomaschinen, Erster Band*, Springer Verlag, Berlin, Heidelberg, GmbH.
11. Balje, O.E., A Study on Design Criteria and Matching of Turbomachines, Part A -- Similarity and Design Criteria of Turbines, ASME Paper No. 60-WA-230.
12. Spencer, R.C., Cotton, K.C., Cannon, C.N., A Method for Predicting the Performance of Steam Turbine-Generators 16500 KW and Larger, GE paper # GER-2007C, July 1974, revised version of ASME paper no. 62-WA-209.
13. Narula, R., Zachary, J., Olson, J., 2004, *Matching Steam Turbines with the New Generation of Gas Turbines*, PowerGen Europe, Barcelona.
14. Gülen, S.C., Smith, R.W., 2010, Second law efficiency of the rankine bottoming cycle of a combined cycle power plant, *J. Eng. GTs Power*, Vol. 132, p. 011801.
15. Bloch, H.P., 1995, *A Practical Guide to Steam Turbine Technology*, McGraw-Hill, New York.
16. Mikhailov, V.E., Khomenok, L.A., Pichugin, I.I., Kovalev, I.A., Bozhko, V.V., Vladimirskii, O.A., Zaitsev, I.V., Kachuriner, Y.Y., Nosovitskii, I.A., Orlik, V.G., 2017, Concept of turbines for ultrasuper-critical, supercritical, and subcritical steam conditions, *Therm. Eng.*, Vol. 64, No. 11, pp. 787–793.
17. Zachary, J., 2007, Strategies for integration of advanced gas and steam turbines in power generation applications, GT2007-27978, *ASME Turbo Expo 2007*, May 14–17, 2007, Montreal, Canada.
18. Zachary, J., Koza, D.J., 2006, The long and short of last-stage blades, *Power*, Vol. 150, No. 9, 40.
19. Gülen, S.C., 2011, Importance of auxiliary power consumption on combined cycle performance, *J. Eng. Gas Turbines Power*, Vol. 133, 041801.
20. Spencer, R.C., Cotton, K.C., Cannon, C.N., A Method for Predicting the Performance of Steam Turbine Generators …, 16500 kW and Larger, 1974 revised version of ASME Paper No. 62-WA-209.

6 Heat Recovery Steam Generator (HRSG)

The heat recovery steam generator (HRSG) is the vital connection between the topping Brayton cycle (gas turbine) and the bottoming Rankine cycle (steam turbine). It serves as the "heat sink" for the former and as the "heat source" for the latter. It is a "waste heat recovery boiler", which utilizes the exhaust gas of the gas turbine to convert condensate coming from the steam turbine condenser into steam at one, two or three different pressures, which are admitted into the steam turbine to generate useful shaft work. Typically, an HRSG is designed in accordance with the ASME Boiler & Pressure Vessel Code, Section 1.

The HRSG has three major types of "heat transfer sections":

1. Economizer
2. Evaporator (boiler)
3. Superheater

Depending on the pressure levels present in the HRSG, there are high-, intermediate- and low-pressure (HP, IP and LP) economizers, evaporators and superheaters. A fourth type of heat transfer section comprises "reheat" superheaters, which receive "cold reheat" (CRH) steam coming from the HP turbine exhaust and heat it back to its original temperature (i.e., main or HP steam admission temperature) to be readmitted into the IP turbine. The reheated steam is referred to as the "hot reheat" (HRH) steam.

There are three main parts in an HRSG (see Figure 6.1):

1. Inlet (transition) duct
2. Casing
3. Stack.

The inlet duct facilitates the smooth transition of gas turbine exhaust gas from the exhaust diffuser exit to the inlet of first *tube bank* (or *tube bundle*, which is the more common term in the USA) housed in the casing (typically the HP superheater). Since the flow area at the latter is much larger than that at the former, the geometry of the inlet duct is crucial to the flow uniformity. The reason for that is simple: temperature uniformity and the amount of heat transfer in the tube banks are directly proportional to the flow uniformity.

The casing houses the HP, IP and LP heat transfer sections. In addition, it also contains the selective catalytic reduction (SCR) system, which is included to reduce the nitrogen oxide (NOx) emissions in the gas turbine exhaust gas. The standard SCR design includes an aqueous ammonia (NH_3) evaporation and injection system. The catalyst modules are field-installed into the structural steel reactor housing which is typically placed in between the HP evaporator tube banks (for temperature requirements). Typically, an oxidizing catalyst system is also included in the SCR package to reduce carbon monoxide (CO) and non-methane, non-ethane, unburned hydrocarbons (VOC[1]) to the emission guarantees specified in the contract.

[1] Volatile Organic Compounds.

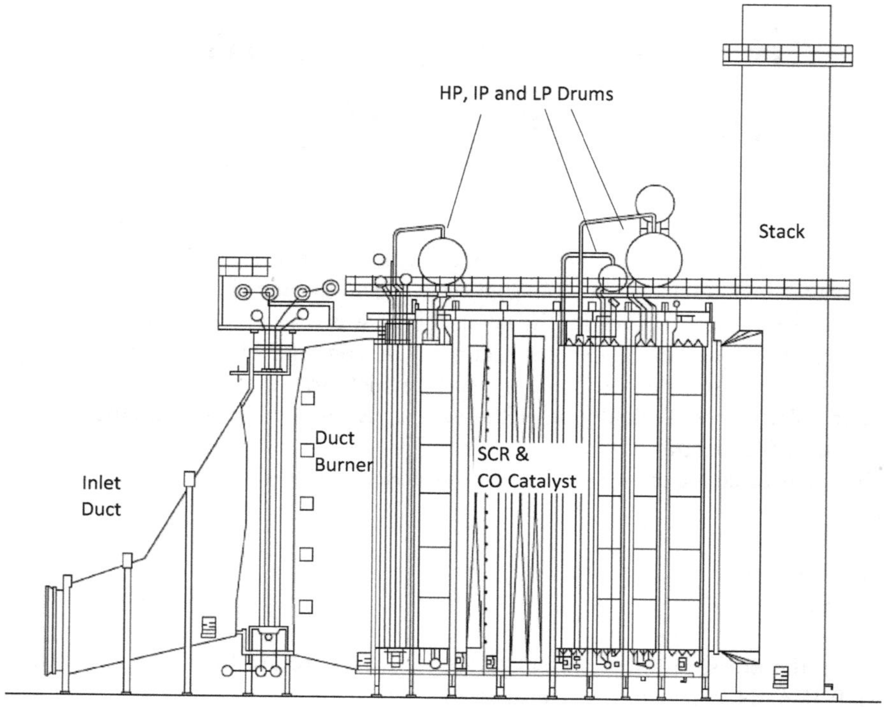

FIGURE 6.1 Horizontal HRSG schematic diagram.

Especially in the USA, a duct burner system is also included in the HRSG casing for supplementary firing in hot ambient conditions in order to make up for lost gas turbine output. The system includes a natural gas fuel burner skid, a burner management system, burner elements inside the HRSG ducting and interconnecting piping between the skid and the elements. The duct burner skid (contrary to its name) is usually located between the reheat superheater tube banks.

The HRSG stack is a self-supporting carbon steel stack. Its height is determined for adequate plume dispersion commensurate with applicable environmental regulations. Typically, it has a damper to reduce heat loss when the unit is offline. It has external insulation on the outlet duct from the HRSG to the stack and up to the damper. If required, a silencer is provided to meet the regulatory requirements (e.g., 115 dBA at the stack exit).

According to the orientation of exhaust gas flow and the arrangement of heat exchange sections in its path, HRSGs are classified into "horizontal" and "vertical" designs. The one shown in Figure 6.1 reflects a horizontal design, which is by far the most common version (especially in the USA). A vertical design would look as shown in Figure 6.2. This type of design is more common in Europe and Japan (due to a somewhat smaller footprint). Note the location of the steam drums at the top of the casing in Figure 6.2. The reason for that is to create sufficient head between the drum and the (horizontal) evaporator tube bank at the bottom. This ensures "natural circulation" of boiling water similar to that in the horizontal design with the drum located atop the (vertical) evaporator tube banks. This is possible in state-of-the-art triple-pressure (with reheat) designs with advanced class gas turbines, which led to a higher casing frame by stacking multiple pressure sections. In older, single-pressure systems, vertical design required "forced circulation". In such designs with shorter frames, the lack of sufficient head led to the substitution of "gravity" by a small pump, which, of course, is an added source of parasitic power consumption (albeit small). For a detailed discussion of pros and cons of either design, the reader is referred to Ref. [1].

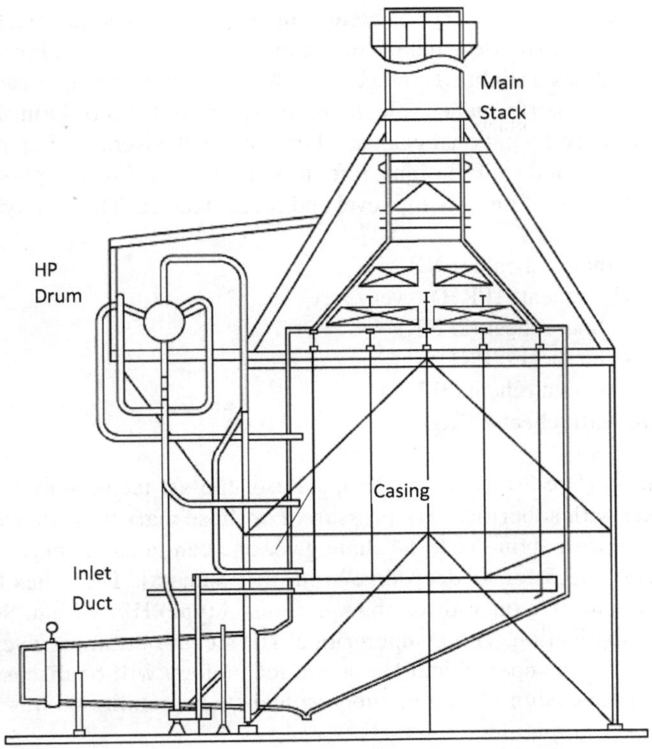

FIGURE 6.2 Vertical HRSG schematic diagram.

From a performance analysis perspective, the difference between the two designs is insignificant. However, there may be a case to be made in favor of the vertical HRSG from the perspective of better suitability to cyclic operation. The reader is referred to the paper by Fontaine for a detailed discussion of this interesting perspective [2].

There are two types of HRSGs, differentiated by their HP section design, which emerged in recent years (this book is written in late 2018 and early 2019) due to a demand for "fast start" power plants. Both are patented technologies of one original equipment manufacturer (OEM) (Siemens): DrumPlus™ and Benson®. Each design attacks the major limiting factor during thermal transients imposed by plant startup, namely, the thick walls of the HP drum. Thermal stresses created in the drum walls due to rapid heating can lead to low-cycle fatigue (LCF) and component failure. This puts a limit on the allowable pressure and temperature ramp rates and extends the time from initial roll to full load.

Originally developed by NEM (now owned by Siemens), DrumPlus technology retains the HP drum but with a smaller diameter and reduced wall thickness, which acts as a primary separator. Water–steam separation is performed in two stages with the secondary stage comprising a row of separator bottles external to the main drum.

Benson technology eliminates the HP drum altogether [3]. Thus, the HP section of the HRSG is essentially a *once-through boiler* (OTB). One additional advantage of the OTB and elimination of the thick-walled HP drum is the possibility of supercritical steam generation. When the steam pressure is above the critical value (22.064 MPa, which is 3,208 psia or 218 atm), the distinction between liquid and vapor phases disappear. This eliminates the need for a drum where the two phases are separated. (Originally developed more than 70 years ago for coal-fired boilers, the Benson/OTB technology is so far applied to subcritical HP pressures in HRSGs.)

From an operability perspective, OTB is better suited to cycling and fast startup (no thick-walled drum susceptible to thermal stresses) and it can run dry for longer periods. From a steady-state performance perspective, it has a slight advantage because the economizer approach temperature delta is essentially zero. From a performance calculation perspective, OTB or DrumPlus variants do not introduce any additional computational baggage. For a detailed coverage of operability characteristics of horizontal, vertical and OTB designs, refer to Ref. [4].

HRSGs are classified based on pressure level and reheat feature. The acronyms are as follows:

1. One pressure without reheat (1PNR)
2. One pressure with reheat (1PRH) – very rare
3. Two pressure without reheat (2PNR)
4. Two pressure with reheat (2PRH)
5. Three pressure without reheat (3PNR)
6. Three pressure with reheat (3PRH)

At the time of writing (late 2018, early 2019), state-of-the-art technology with advanced class gas turbines is 3PRH with subcritical HP pressure. For those markets with access to cheap natural gas (e.g., the USA with about $3 to $4 shale gas), one can make a very good economic case for 2PRH (see Section 13.5 for a "deep dive" into this subject). There has been discussion of supercritical HP pressure in a once-through (e.g., Benson type) HP section. So far, there has not been a commercial application with a supercritical HP steam bottoming cycle (BC). The practicality or feasibility of the supercritical HP steam technology will be discussed in Section 6.4. Steam turbine flow-path design impact of supercritical steam conditions was already covered in Section 5.1.1.

The arrangement of the heat transfer sections of a state-of-the-art 3PRH HRSG is shown schematically in Figure 6.3. In the figure, gas turbine exhaust gas flow is from the left (HRSG inlet) to the right (HRSG outlet). Consequently, the gas temperature decreases from left (highest) to right (lowest) whereas feedwater and steam temperatures increase from right to left. This basic heat transfer "fact" dictates the order of

i. Pressure sections (i.e., HP, IP and LP from left to right)
ii. Individual superheater, evaporator and economizer tube banks/bundles, SH, EVP and EC, respectively, in each pressure section from left to right).

The interlaced arrangement of the tube banks/bundles is dictated by the fundamental driver of optimum thermal system design: closest match of "source" (i.e., in this case, gas turbine exhaust gas) and "sink" (i.e., feedwater and steam) temperatures. Closest possible (and cost-effective one should add) match of gas/steam and gas/feedwater temperatures minimizes the "exergy destruction" in the HRSG. This will be explored in more detail in Section 6.1 below.

The schematic in Figure 6.3 is generic in the sense that it is based on a horizontal HRSG with drum-type evaporators. Nevertheless, from a thermodynamic steady-state design point calculation perspective, it is also representative of a vertical design (i.e., just replace "from left to right" with "from bottom to top") as well as a once-through (e.g., Benson type) evaporator design (e.g., just delete the circle representing the drum from the HP EVP section). Many practical details (e.g., level control valves (LCVs) between the economizer-evaporator sections) are omitted. Different interlacing arrangements as well as parallel arrangement of certain sections can be (and usually are) adopted by different HRSG manufacturers. Still, the schematic in Figure 6.3 is a very useful guide in order to understand the governing design principles as well as making sense of the off-design performance of the system at different site ambient and loading conditions (e.g., see Chapter 15). In that sense, to make Xerox copy of Figure 6.3 and keep it handy while reading different parts of the book might be a good idea.

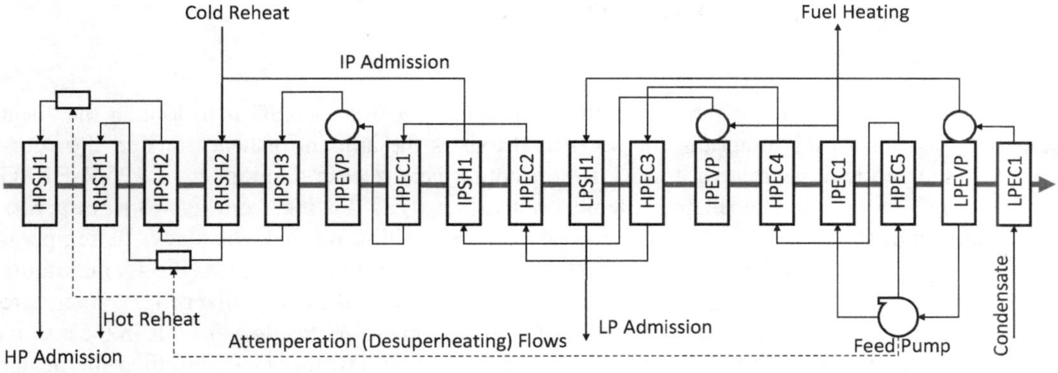

FIGURE 6.3 Schematic arrangement of heat transfer sections (i.e., tube banks/bundles) of a typical 3PRH HRSG.

The heat release diagram corresponding to the arrangement in Figure 6.3 is shown in Figure 6.4. This diagram shows the cooling of gas turbine exhaust gas (from left to right) while transferring heat to feedwater at three different pressure levels to evaporate and superheat it (from right to left). The heat release diagram is the most important graphical tool in HRSG analysis. This will be illustrated in detail in Section 6.1. The basic thermodynamic principle governing the HRSG heat transfer section arrangement, similar to that shown in Figure 6.3, is to minimize the "gap" between the gas cooling line and the evaporating feedwater heating lines.

There are some purely "visual" but quite important takeaways in Figure 6.4. This does not require lengthy calculations and/or profound thermodynamic explanations. Each evaporator in the HRSG is represented by a horizontal line, which quantifies the heat transferred from the flue gas to the boiling feedwater at constant (saturation) temperature (and pressure). In graphical terms, each horizontal line's length is proportional to steam production in HP, IP and LP evaporators. Thus, it is straightforward to make the following deductions just by "looking at" the diagram in Figure 6.4:

1. There is practically no room left to "squeeze in" a fourth pressure level
2. Eliminating the IP level would not make a big impact.

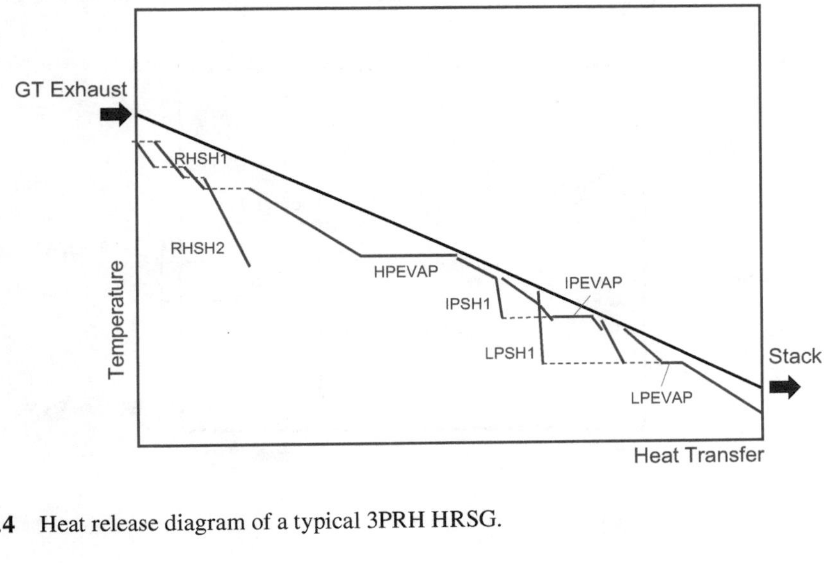

FIGURE 6.4 Heat release diagram of a typical 3PRH HRSG.

6.1 FUNDAMENTALS OF HEAT RECOVERY

6.1.1 Heat Release Diagram

The best way to illustrate the fundamental thermodynamics in an HRSG is to look at the "heat release" diagram. In this diagram, the horizontal axis is the amount of heat transfer; the vertical axis is steam and gas temperature. The conceptual heat release diagram for a 1PNR HRSG is shown in Figure 6.5. The diagram depicts hot exhaust gas flowing from left to right across the superheater, the evaporator and the economizer and exiting through the stack. Its temperature drops from TEXH to TSTACK. Heat transfer increases from zero at TEXH to its maximum value at TSTACK. Feedwater enters from the right and heated in the economizer to a temperature slightly below the saturation temperature at the evaporator pressure. Feedwater boils in the evaporator at constant pressure and temperature to become steam and further heated in the superheater before exiting the HRSG on the left.

From a first law perspective, what happens in this HRSG is described by a simple enthalpy balance, i.e.,

$$\dot{m}_{exh}\left(h_{exh} - h_{stck}\right) = \dot{m}_{stm}\left(h_{sup} - h_{fw}\right) = \dot{Q}_{rec}, \tag{6.1}$$

which quantifies heat energy recovered from the gas turbine exhaust gas (ignoring losses for simplicity). For specified gas turbine exhaust gas conditions, the design is fixed by the following four parameters:

1. Superheater *approach* temperature delta, which is the difference between final steam temperature and the entering gas temperature
2. Evaporator steam pressure
3. Evaporator *pinch*, which is the difference between the saturated steam temperature in the evaporator and the gas temperature at the evaporator exit
4. Economizer *approach subcool*, which is the difference between the saturated steam temperature in the evaporator and the feedwater temperature at the economizer exit.

Gas turbine exhaust gas is the "hot side/fluid" of the heat exchanger; its T-Q line (slightly curved in reality because of the change in c_p with gas temperature) is the "cooling line". Feedwater steam is the "cold side/fluid" of the heat exchanger; its three-piece T-Q line is the "heating line". In theory, for an infinitely large heat exchanger with both single-phase cold and hot fluids, cooling and heating lines

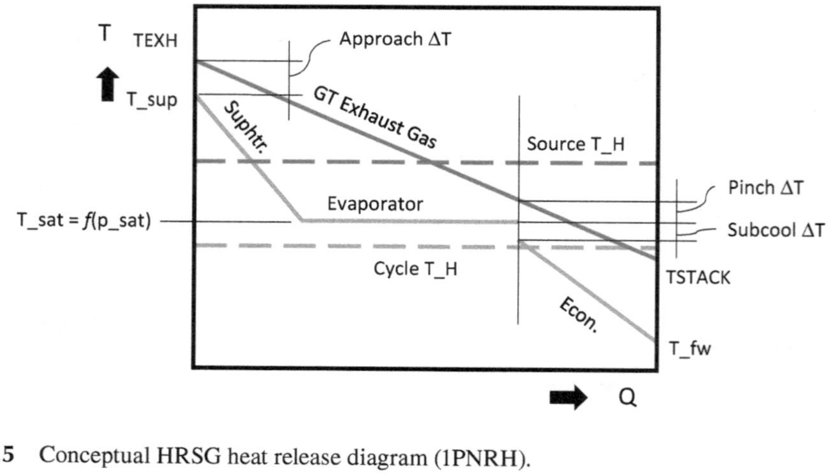

FIGURE 6.5 Conceptual HRSG heat release diagram (1PNRH).

coincide. In other words, at each point in the heat exchanger, hot and cold fluids are at the same temperature. This is when the heat transfer irreversibility in the heat exchanger is exactly equal to zero.

In the HRSG in Figure 6.5, even when pinch, subcool and approach temperature deltas are set to zero, this ideal condition is impossible to achieve because of the constant pressure (and temperature) phase change, from liquid to vapor, in the evaporator. In other words, even if one could build an infinitely large HRSG, the irreversibility would not be zero. This is the second law "curse" of the HRSG.

6.1.2 HRSG Irreversibility

The irreversibility in the HRSG is quantified by the mean-effective gas and steam/water temperatures in the HRSG. The former is given by

$$\overline{T}_{gas} = \frac{h_{exh} - h_{stck}}{s_{exh} - s_{stck}} \approx \frac{T_{exh} - T_{stck}}{\ln\left(\frac{T_{exh}}{T_{stck}}\right)}. \tag{6.2}$$

For the steam/water, we have

$$\overline{T}_{stm} = \frac{h_{sup} - h_{fw}}{s_{sup} - s_{fw}}. \tag{6.3}$$

If one could build a Carnot engine operating between two temperature reservoirs at $\overline{T}_{gas}$ and T_0, the work output would be

$$\dot{W}_{id,gas} = \left(1 - \frac{T_0}{\overline{T}_{gas}}\right)\dot{Q}_{rec}. \tag{6.4}$$

Similarly, if one could build a Carnot engine operating between two temperature reservoirs at $\overline{T}_{stm}$ and T_0, the work output would be

$$\dot{W}_{id,stm} = \left(1 - \frac{T_0}{\overline{T}_{stm}}\right)\dot{Q}_{rec}. \tag{6.5}$$

Even without a rigorous quantitative evaluation, just by looking at the heating and cooling lines in Figure 6.5, one could deduce that

$$\overline{T}_{stm} < \overline{T}_{gas}$$

so that

$$\dot{W}_{id,stm} < \dot{W}_{id,gas}.$$

Therefore, the irreversibility in the simple HRSG depicted in Figure 6.5 is

$$\dot{I}_{HRSG} = \dot{W}_{id,gas} - \dot{W}_{id,stm},$$

$$\dot{I}_{HRSG} = \left(1 - \frac{T_0}{\overline{T}_{gas}}\right)\dot{Q}_{rec} - \left(1 - \frac{T_0}{\overline{T}_{stm}}\right)\dot{Q}_{rec} = \left(\frac{T_0}{\overline{T}_{stm}} - \frac{T_0}{\overline{T}_{gas}}\right)\dot{Q}_{rec},$$

$$\dot{I}_{HRSG} = \frac{T_0}{\overline{T}_{stm}}\left(1 - \frac{\overline{T}_{stm}}{\overline{T}_{gas}}\right)\dot{Q}_{rec}. \tag{6.6}$$

6.1.3 HRSG Effectiveness

HRSG effectiveness can be defined in several ways. Each different formulation hinges on the definition of the maximum possible heat transfer in the HRSG. Logically, it is easy to see that the exhaust gas cannot be cooled below the temperature of the feedwater entering the economizer, T_{fw}. Thus, the maximum possible heat transfer is

$$\dot{Q}_{max} = \dot{m}_{exh}\left(h_{exh} - h_{fw}\right).$$

The problem with this definition is that it requires *a priori* knowledge of T_{fw}, which is a function of steam turbine back pressure (i.e., the pressure of steam condensing in the condenser). For a truly universal definition, the best option is to define a "reference" temperature. The ideal candidate is the ambient temperature, which is 59°F per the ISO definition. Another candidate is the zero enthalpy reference for the gas enthalpy, which can be set to 59°F or 77°F (see Appendix B). Going with this convention is the preferred choice herein so that

$$\eta_{HRSG} = \frac{h_{exh} - h_{stck}}{h_{exh}} = 1 - \frac{h_{stck}}{h_{exh}}. \tag{6.7}$$

In analyzing the heat transfer process in the HRSG and its impact on BC performance, the best approach is graphical. Consider the cooling line in Figure 6.5 as a solid beam or lever and the pinch point as the *fulcrum* about which the said beam pivots. For a fixed pinch point, which is defined by the pressure (and temperature) of the evaporating steam, the pivoting motion is controlled by the gas turbine exhaust temperature. In particular,

1. When the exhaust gas temperature increases, the "beam" pivots clockwise and the stack temperature goes down (see Figure 6.6a).
2. When the exhaust gas temperature decreases, the "beam" pivots counter-clockwise and the stack temperature goes up (see Figure 6.6b).

From Equation 6.7, it is easy to deduce that the HRSG effectiveness increases with increasing exhaust gas temperature and vice versa.

The second parameter the designer can control is the steam pressure in the evaporator, which can be used to move the pinch point up/left or down/right. For given gas turbine exhaust gas temperature,

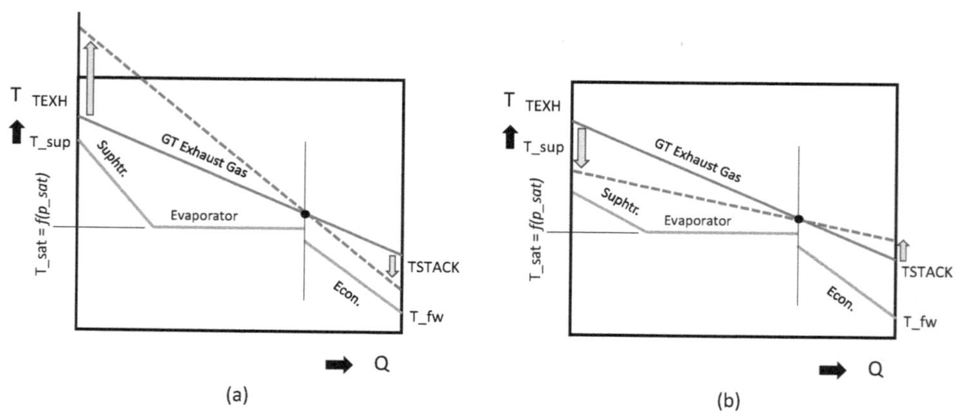

FIGURE 6.6 (a and b) Impact of GT exhaust temperature on HRSG stack temperature.

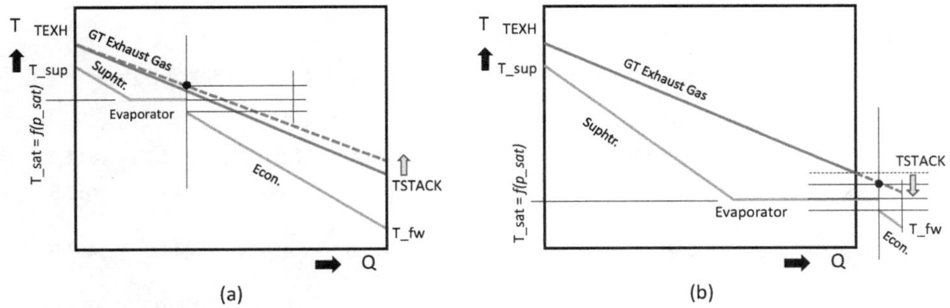

FIGURE 6.7 (a and b) Impact of evaporator steam pressure on HRSG stack temperature.

1. When the evaporator steam pressure increases, the pinch point moves to the left and up and the stack temperature goes up (see Figure 6.7a).
2. When the evaporator steam pressure decreases, the pinch point moves to the right and down and the stack temperature goes down (see Figure 6.7b).

From Equation 6.7, it is once again easy to deduce that the HRSG effectiveness increases with decreasing evaporator steam pressure and vice versa.

6.1.4 Simplest Possible HRSG: One-Pressure, No Reheat

In terms of the second part of the puzzle, one has to look at the impact of gas turbine exhaust temperature and steam pressure on the Rankine cycle efficiency. Note that the term used is "cycle efficiency", i.e., *not* "steam turbine efficiency", which is a *contributor* to the cycle efficiency. It is easy to figure out the beneficial impact of increasing gas turbine exhaust temperature on cycle efficiency *as long as the final steam temperature increases in lock-step with it.* At given steam pressure, higher steam temperature translates into higher steam exergy, i.e., work producing potential. It is quantified by the rise in $\overline{T}_{stm}$ and its favorable impact on steam cycle work via Equation 6.5. By the same token, we expect the steam pressure to have the same effect on cycle efficiency, but to a lesser degree vis-à-vis the steam temperature, also via higher exergy.

In order to put the aforementioned qualitative assertions to test, a series of cycle runs are made in Thermoflow's GT PRO® software. A 1PNR bottoming steam cycle using exhaust gas at 400 lb/s and 900°F, which is typical of small industrial or aeroderivative gas turbines (most likely candidates for this type of cycle), is run for a series of steam pressures, from 40 psig to 1,400 psig. Exhaust gas data is summarized in Table 6.1. Superheated steam temperature is fixed at 850°F (about 454°C). Condenser pressure is set to 1.2 in. Hg (41 mbar), which results in about 86°F (30°C) feedwater temperature at the economizer inlet. The results are summarized in Table 6.2.

TABLE 6.1
Gas Turbine Exhaust Gas (Hypothethical; Typical of a Small Industrial GT)

Exhaust gas composition	Mole %	O_2:11.28, CO_2: 4.313, H_2O: 9.535, Ar: 0.891, N_2: 73.98
Exhaust flow rate	lb/s (kg/s)	400 (181.4)
Exhaust temperature	°F (°C)	900 (482.2)
Exhaust enthalpy	Btu/lb	214.7 (Equation A.4)
Exhaust exergy	Btu/lb	89.6 (Equation A.2)
	Btu/s (kWth)	36,360 (38,362)

TABLE 6.2
Bottoming Cycle Steam Pressure (At the Steam Turbine inlet – 5%–6% Higher in the Evaporator)

Steam Pressure	**psia**	**1,414.7**	**1,214.7**	**1,014.7**	**814.7**	**614.7**	**414.7**
Mean-effective gas temperature	°F	658.5	649.1	638.3	625.7	610.8	592.1
Mean-effective steam temperature	°F	478.5	470.5	460.7	448.6	432.9	411.1
Heat recovery	Btu/s	52,541	54,094	55,866	57,914	60,340	63,361
Steam flow rate	lb/s	38.7	39.5	40.5	41.7	43.2	45.1
STG output	MWe	18.1	18.8	19.5	20.2	20.9	21.5
Stack temperature	°F	410.0	395.0	377.9	358.1	334.6	305.2
HRSG effectiveness		59.67%	61.37%	63.32%	65.57%	68.25%	71.59%
Steam cycle efficiency		32.69%	32.94%	33.08%	33.06%	32.78%	32.09%
Bottoming cycle efficiency		19.51%	20.22%	20.95%	21.68%	22.37%	22.98%
HRSG irreversibility	Btu/s	4,677	4,859	5,091	5,397	5,826	6,493
	% of GT exhaust exergy	12.86%	13.36%	14.00%	14.84%	16.02%	17.86%
Steam Pressure	**psia**	**314.7**	**214.7**	**114.7**	**84.7**	**54.7**	
Mean-effective gas temperature	°F	580.3	565.6	548.3	543.8	524.5	
Mean-effective steam temperature	°F	396.1	375.7	343.8	329.0	308.6	
Heat recovery	Btu/s	65,245	67,575	70,299	71,006	73,994	
Steam flow rate	lb/s	46.3	47.8	49.6	50.0	52.1	
STG output	MWe	21.6	21.7	21.1	20.5	20.1	
Stack temperature	°F	286.8	264.1	237.4	230.5	201.2	
HRSG effectiveness		73.68%	76.27%	79.30%	80.09%	83.43%	
Steam cycle efficiency		31.44%	30.38%	28.41%	27.35%	25.78%	
Bottoming cycle efficiency		23.16%	23.17%	22.53%	21.91%	21.51%	
HRSG irreversibility	Btu/s	7,006	7,772	9,210	9,994	10,961	
	% of GT exhaust exergy	19.27%	21.38%	25.33%	27.49%	30.14%	

Let us summarize the "as expected" findings first. With increasing steam pressure,

- There is a monotonic decrease in total heat recovery (due to higher boiling temperature of steam at higher pressures, which results in lower steam generation rate – also listed in the table – with fixed source temperature).
- There is a monotonic increase in HRSG stack gas temperature, which is reflected in decreasing HRSG effectiveness and increasing mean-effective gas temperature.
- Even though superheated steam and economizer feedwater inlet temperatures are fixed, there is a monotonic increase in mean-effective steam temperature (via positive impact of steam pressure on steam enthalpy).

Now let us look at "somewhat surprising" findings. With increasing steam pressure,

- Steam turbine output goes through a maximum at 200 psig and declines thereafter.
- Steam cycle efficiency first increases but becomes quite flat after about 600 psig with a slight maximum at 1,000 psig.

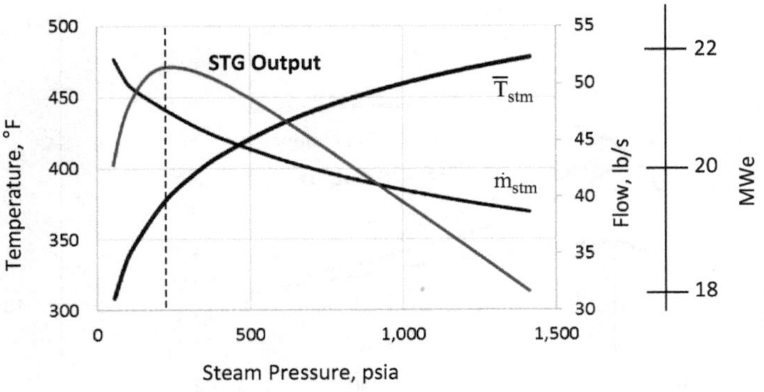

FIGURE 6.8 Competing drivers of steam cycle performance (STG).

The presence of a maximum in steam turbine output is the result of a "tug of war" between mean-effective steam temperature (i.e., the key driver of the steam cycle efficiency) and steam flow rate, which is depicted graphically in Figure 6.8. These two parameters are representative of the two parts of the expression on the right-hand side of Equation 6.5. Even though the rise in steam pressure is beneficial in driving the cycle efficiency upwards, its adverse impact on heat transfer and steam flow rate (i.e., a reduction in both) counters this benefit and steam turbine output starts declining after reaching a maximum.

Explaining the trend in steam cycle efficiency is more complicated. On an ideal basis, it only depends on $\overline{T}_{stm}$, which keeps increasing with increasing steam pressure. The culprit here is the detrimental impact of steam pressure on section efficiencies via lower volumetric flow rates (because of increasing density). For a quantification of this effect, refer to Figure 5.9. Significant reduction in volumetric flow (from 740 ft^3/s at 40 psig to only about 19 ft^3/s at 1,400 psig) puts a big damper on cycle efficiency. Another contributor to lower steam turbine efficiency is higher exhaust moisture in the LP stages. In a non-reheat cycle, higher steam inlet pressure results in deeper penetration into the two-phase region, which is depicted conceptually on a Mollier diagram Figure 6.9. (This is partially countered via lower exhaust loss because of decreasing volume flow at the LP exhaust.)

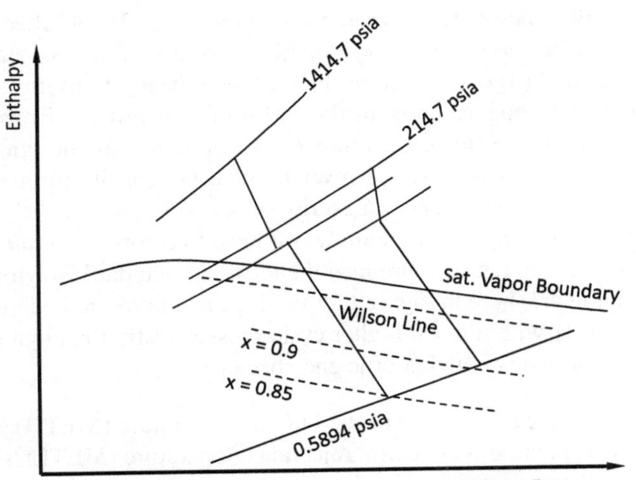

FIGURE 6.9 Steam turbine expansion path on Mollier chart.

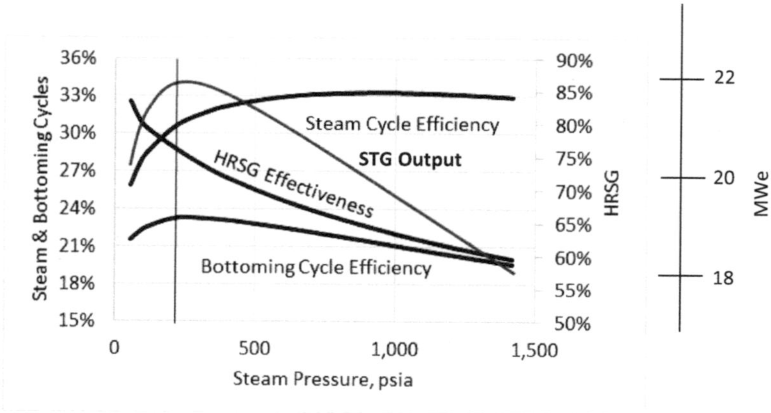

FIGURE 6.10 Key bottoming cycle performance parameters.

The final picture that emerges from the steam pressure sweep is summarized in Figure 6.10. In a nutshell, opposing trends in HRSG effectiveness and steam cycle efficiency result in a maximum in BC efficiency (which is the product of the two) at 200 psig, where the steam turbine output is maximum at 21.7 MWe.

The foregoing analysis using the simplest possible HRSG and steam BC highlights the fundamental dynamic present in more complex HRSG and BCs such as 3PRH. The dynamic in question boils down to the "exact nature" of the bottoming steam Rankine cycle in a gas turbine combined cycle (GTCC) power plant, i.e.,

- Is it a heat engine cycle in its own right, i.e., the "leading actor"
- Is it a "supporting actor", i.e., helps in achieving the highest "combined" cycle efficiency?

If it were the former, one would be aiming to achieve the highest possible steam cycle efficiency, which is at 800–1,000 psig steam pressure in the foregoing example. This, however, would harm the combined cycle efficiency because its contribution to the total would be about 2 MWe less than the maximum achieved at 200 psig. Clearly, the exact nature of the BC can be best described as a "supporting actor". In order to be able to do that, it must extract as much heat from the gas turbine exhaust gas as possible. Note that the maximum BC efficiency is quite close to the point where HRSG effectiveness is maximum.

This dichotomy between the two primary "missions" of a BC, i.e., (i) heat recovery from the gas turbine exhaust gas and (ii) conversion of the recovered heat into useful shaft work with high efficiency, is worthwhile to ponder upon further. Minimizing process irreversibility is (almost) an *axiom*[2] of thermal design. In this case, however, the optimum BC design (i.e., 200 psig steam pressure – see Table 6.2) is where the HRSG irreversibility is far from its minimum (i.e., at 1,400 psig steam pressure), ~21% of gas turbine exhaust exergy versus ~13%, respectively.

In order to understand this apparent anomaly, we have to resort to an *analogy* and go back to the Brayton gas turbine cycle. Let us reproduce the ideal, air-standard Brayton cycle T–s diagram used in Chapter 4 (Figure 4.3) here in Figure 6.11 (with an addition). In the figure, there is a second Brayton cycle {1-2′-3′-4′-1} with a much higher cycle pressure ratio than that of the original cycle {1-2-3-4-1}. This cycle has a much higher efficiency because

- It has a much higher mean-effective heat addition temperature (METH) (see Equation 4.7)
- It has a much lower mean-effective heat rejection temperature (METL) (see Equation 4.10)
- Which both help the cycle efficiency go up simultaneously (see Equation 4.11).

[2] An "axiom" is a statement or proposition that is regarded as being established, accepted, or self-evidently true.

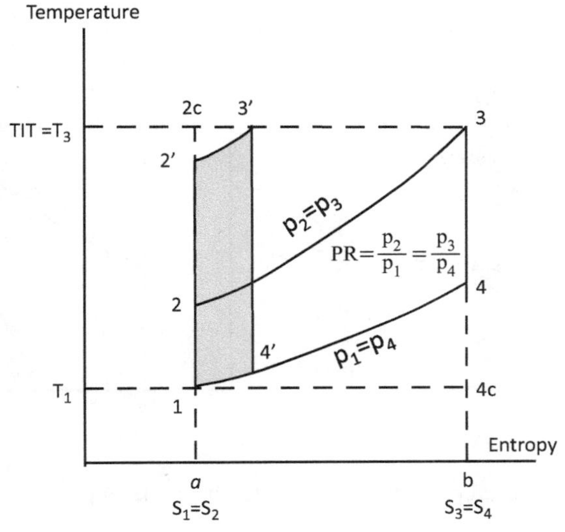

FIGURE 6.11 Ideal, air-standard Brayton gas turbine with high cycle PR.

(This effect is also concisely described by Equation 4.12.)

In fact, graphically (using Figure 6.11), it is easy to see that, at a very high cycle pressure ratio, METH ~ T_3 and METL ~ T_1 and the cycle efficiency is very close to the ultimate maximum set by the second law of thermodynamics, i.e., the Carnot efficiency. Yet it is also easy to see that this cycle would be a totally worthless contraption because its work output would be essentially zero (as quantified by the area encompassed by the cycle T–s diagram – shaded in Figure 6.11 for the second Brayton cycle with high cycle pressure ratio).

This is a striking demonstration of the pitfalls one might fall in when losing sight of the design objective. When designing a land-based (stationary) gas turbine for electric power generation, the objective is not maximum possible efficiency at all costs; the objective is highest specific power output with maximum possible efficiency for a cost-effective product. This results in a cycle pressure ratio where specific output is maximized (which also maximizes combined cycle efficiency as demonstrated in Chapter 4).

This "tug of war" between cycle work and efficiency is analogous to what is observed in 1PNR bottoming steam cycle design exercise described above. Going after high cycle efficiency and minimum HRSG irreversibility leads one to higher steam pressures with higher stack gas temperatures, which translate into high METHs (both $\overline{T}_{gas}$ and $\overline{T}_{stm}$). Unfortunately, this is not the design objective here. Let us look at this a bit more closely by using the diagrams in Figure 6.12.

In Figure 6.12a, the shaded triangular area {1-4-4c-1} quantifies the maximum possible (theoretically of course) work that can be produced by utilizing the heat energy in the gas turbine exhaust (state-point 4). Numerically, it is equal to the exergy at state-point 4, i.e., gas turbine exhaust gas exergy. (This was covered in detail in Chapter 4.) In that case, the BC "squeezed" into that triangular area is equivalent to a Carnot cycle operating between METL, mean-effective temperature (MET) between state-points 1 and 4, which is quantified by Equation 4.10 for a calorically perfect gas, and T_1.

In the 1PNR example presented earlier, we looked at what can be done in practice with a realistic piece of equipment, which can be cost-effectively manufactured and deployed in the field. The high steam pressure case with 1,400 psig is (conceptually) equivalent to the smaller trapezoidal darker-shaded area in Figure 6.12b. It is a poor approximation of the ideal in Figure 6.12a. In contrast, the low steam pressure case with 200 psig is (again, conceptually) equivalent to the large trapezoidal lighter-shaded area in Figure 6.12b. Thus, it is a better approximation of the ideal in Figure 6.12a.

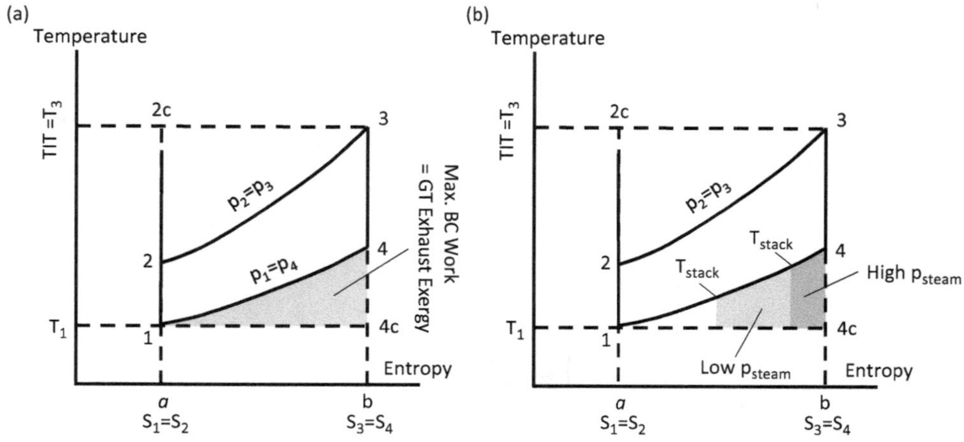

FIGURE 6.12 (a and b) Graphical representation of bottoming cycle output (shaded areas).

This "goodness" of approximation is quantified by the second law effectiveness, which is the ratio of steam turbine output to gas turbine exhaust gas exergy (on a gross basis). Second law effectiveness for the cases listed in Table 6.2 is plotted in Figure 6.13.

As shown in Figure 6.13, maximum second law effectiveness coincides with maximum BC efficiency and steam turbine output in Figure 6.10. In other words, total cycle irreversibility is minimized. Even though the HRSG irreversibility is not minimized at 200 psig, it is balanced by lower steam turbine and condenser irreversibility (low steam flow and low heat rejection) and stack exergy loss (lower stack temperature). Before looking at this in more detail, let us reemphasize the analogy between BC steam pressure selection and Brayton cycle pressure ratio selection.

Just like going after maximum Brayton cycle pressure leads the designer to an efficient but not much useful product (i.e., very low specific output), going after maximum steam pressure leads one to an efficient steam cycle, which, alas, is quite suboptimal in its contribution to the "common goal", i.e., combined cycle total power output.

BC exergy distribution as calculated by GT PRO® is summarized in Table 6.3. Major loss buckets are expressed as a fraction of gas turbine exhaust gas exergy. Note that GT PRO® calculates the

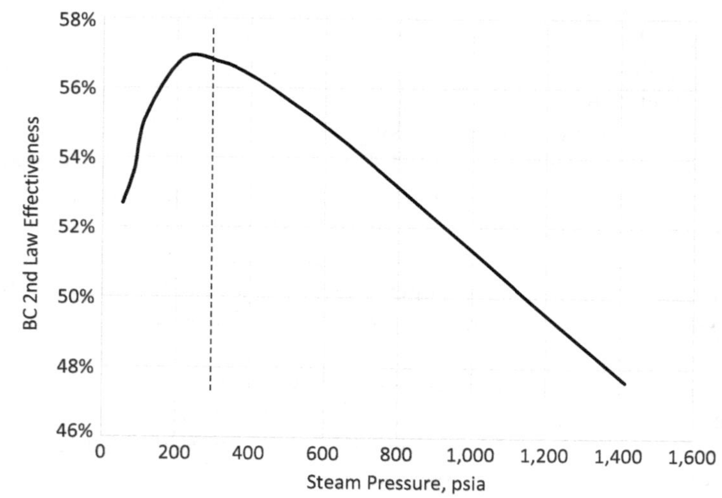

FIGURE 6.13 1PNRH bottoming steam cycle second law effectiveness.

TABLE 6.3
Exergy Accounting for the Cases in Table 6.2

Steam Pressure, psig	**1,400**	**1,200**	**1,000**	**800**	**600**	**400**
BC net output	47.6%	49.6%	51.5%	53.5%	55.4%	57.1%
Stack exergy loss	21.7%	20.1%	18.2%	16.1%	13.8%	11.1%
HRSG exergy loss	14.7%	15.2%	15.8%	16.7%	17.9%	19.7%
STG exergy loss	12.7%	11.9%	11.1%	10.2%	9.4%	8.4%
Condenser exergy loss	1.4%	1.4%	1.5%	1.5%	1.6%	1.7%
Miscellaneous losses	1.8%	1.8%	1.9%	1.9%	1.9%	2.0%
Steam Pressure, psig	**300**	**200**	**100**	**70**	**40**	
BC net output	57.7%	57.8%	56.1%	54.6%	53.6%	
Stack exergy loss	9.6%	7.8%	5.9%	5.4%	3.6%	
HRSG exergy loss	21.2%	23.4%	27.6%	29.8%	32.7%	
STG exergy loss	7.8%	7.1%	6.4%	6.1%	5.9%	
Condenser exergy loss	1.7%	1.8%	2.0%	2.0%	2.2%	
Miscellaneous losses	2.0%	2.0%	2.1%	2.1%	2.1%	

exergy for a dead state of 14.696 psia and 77°F with water as vapor. Therefore, the values are different from what would be obtained using a dead state temperature of 59°F, which is the convention adopted in most calculations in this book. Nevertheless, this does not impact the key takeaways from Table 6.3. BC net output listed in Table 6.3 is the second law effectiveness of the BC. Another term for it frequently encountered in the literature is "exergetic efficiency" or "rational efficiency".

Before delving into the exergy distribution in Table 6.3, note that exergy "loss" has two distinct components:

1. Exergy *destruction* (i.e., irreversibility)
2. Exergy *transfer* with heat transfer (out of the cycle).

For example, stack exergy loss refers to exergy transferred out of the BC control volume with the gas exiting the HRSG stack. However, condenser exergy loss contains both components: (i) exergy destruction or irreversibility stemming from the temperature difference between the condensing steam and the cooling water and (ii) exergy transfer out of the BC control volume with heat rejected to the cooling water.[3]

Steam turbine exergy loss in Table 6.3 is a combination of irreversibilities associated with non-isentropic expansion in the steam flow-path, friction losses and electrical losses (in the generator). This was explored in detail in Section 5.3.5. HRSG exergy loss is mainly irreversibility described earlier with small contributions from the casing and transition duct heat losses. Miscellaneous losses in the table include myriad small losses associated with valves and pipes in the balance of plant (BOP).

6.1.5 Next Level: Two-Pressure HRSG

The main takeaway from Table 6.3, which also provides the clue to BC improvement is the trade-off between low stack exergy loss (i.e., low stack gas temperature) and low HRSG exergy loss (i.e., high steam pressure). Clearly, in a single-pressure HRSG/steam cycle design, satisfying both objectives

[3] When the "dead state" is at 77°F, however, condenser exergy transfer is zero because the mean-effective temperature of the cooling water is lower than 77°F (unless, of course, at a very hot ambient and cooling water temperature). Thus, the condenser exergy loss in Table 6.3 comprises only exergy destruction (irreversibility). This will be covered in more detail in Chapter 7.

TABLE 6.4
Exergy Accounting for 2PNR Cases

HP Steam, psig	**1,400**	**1,300**	**1,200**	**1,100**	**1,000**	**800**	**600**
LP Steam, psia	**55**	**55**	**55**	**55**	**55**	**55**	**55**
BC net output	62.1%	62.5%	62.8%	63.1%	63.3%	63.7%	63.8%
Stack exergy loss	3.9%	3.9%	3.9%	4.0%	4.0%	4.1%	4.2%
HRSG exergy loss	16.8%	16.8%	16.9%	17.0%	17.1%	17.6%	18.4%
STG exergy loss	13.2%	12.8%	12.3%	12.0%	11.5%	10.6%	9.6%
Condenser exergy loss	2.0%	2.0%	2.0%	2.0%	1.9%	1.9%	1.9%
Miscellaneous losses	2.1%	2.1%	2.1%	2.1%	2.1%	2.1%	2.1%

is not possible; a compromise is necessary. This brings up the following question: What if we make steam in the HRSG at *two* different pressures such that

1. One pressure level is as low as possible to set the stack temperature low
2. The other pressure level is as high as possible to minimize HRSG exergy loss?

This brings one to the concept of a 2PNR BC, which will be investigated next. Exhaust gas conditions are set to values similar to that of an old BBC Type 13 gas turbine, 800 lb/s (~363 kg/s) and 925°F (~495°C). Steam cycle low pressure is set to 55 psia and the HP pressure range is set from 600 to 1,400 psig. (Note that steam pressures are admission pressures; HRSG drum pressures are slightly higher.) Steam temperature is set to 850°F. The results of GT PRO® runs are summarized in Table 6.4.

The improvement in BC pressure vis-à-vis the 1PNR configuration in Table 6.3 is unmistakable. The best performance point for the 2PNR steam cycle is at 600 psig – 40 psig, which is 63.8 – 57.8 = 6 percentage points better than the optimum 1PNR steam cycle (at 200 psig steam pressure). A side-by-side comparison of the two optima is presented in Table 6.5, which shows that the improvement is driven by two mechanisms:

- Reduction in HRSG irreversibility (via steam production at two pressure levels)
- Reduction in stack exergy loss (i.e., lower stack temperature via lower LP steam pressure).

TABLE 6.5
Comparison of Exergy Accounting, 2PNRH versus 1PNRH

	2PNRH	**1PNRH**
HP Steam, psig	**600**	**200**
LP Steam, psia	**55**	**NA**
Stack temperature, °F	214.9	264.1
BC net output	63.8%	57.8%
Stack exergy loss	4.2%	7.8%
HRSG exergy loss	18.4%	23.4%
STG exergy loss	9.6%	7.1%
Condenser exergy loss	1.9%	1.8%
Miscellaneous losses	2.1%	2.0%

TABLE 6.6
Steam Turbine Comparison, 2PNR versus 1PNR

	2PNR	1PNR
HP Steam, psig	**600**	**200**
LP Steam, psia	**55**	**NA**
Steam cycle heat input, Btu/s	150,754	67,575
STG output, kWe	50,588	21,804
Steam cycle efficiency, %	31.81	30.58
HP efficiency, %	92.75	95.10
LP efficiency, %	86.09	89.01
Exhaust loss, Btu/lb	6.16	3.86
Exhaust quality	0.87	0.91

The two beneficial effects cited above are dampened by the increase in steam turbine exergy loss, which is dissected in Table 6.6. There are two mechanisms leading to poorer steam turbine performance:

- Lower HP efficiency due to lower volumetric flow rate at higher throttle pressure
- Lower LP efficiency via higher exhaust loss (lower exhaust steam quality, i.e., higher moisture).

One might be compelled to raise an objection here: "Wait a minute! What do you mean 'poorer' steam turbine performance? 2PNR steam cycle efficiency is better than that for the 1PNR cycle". This is indeed true. But the better steam cycle first law efficiency is simply a reflection of the fact that the exergy input to the 2PNR steam turbine is higher, i.e., from a work production perspective, it has better "quality". Let us quantify this by using the admission steam properties for the 2PNR and 1PNR steam turbines, which are summarized in Table 6.7. As shown in the table, the mass-averaged admission steam exergy for the 2PNR steamer is 2.3% higher than that for its 1PNR counterpart. Consequently, although its "internal efficiency" is lower as indicated by the data in Tables 6.6 and 6.7, from a "cycle efficiency" perspective, it still comes ahead by 1.23 percentage points.

TABLE 6.7
Steam Turbine Admission Exergy Comparison, 2PNR versus 1PNR

	2PNR	1PNR
HP Steam, psig	**600**	**200**
LP Steam, psia	**55**	**NA**
HPT admission exergy, Btu/lb	549.77	498.33
HPT admission flow rate, lb/s	92.11	47.38
LPT admission exergy, Btu/lb	311.59	NA
LPT admission flow rate, lb/s	18.51	NA
Mass-averaged STG admission exergy, Btu/lb	509.91	498.33
Total exergy input, Btu/s	56,406	23,809
STG output, kWe	50,588	21,804
Exergetic efficiency, %	85.01	86.80

To be comfortable with this somewhat puzzling observation is crucial for understanding the fundamental thermodynamics of the bottoming steam Rankine cycle of the GTCC. Let us use a (admittedly less-than-perfect) financial analogy to make the distinction between first law and second efficiencies a bit more clear.

Let us assume that there are two neighboring countries, Upper and Lower Slobovia. They are both prosperous countries with stable political and economic systems but Upper Slobovia is slightly more so. Both have their own currencies, i.e., Upper Slobovian and Lower Slobovian Pound, US£ and LS£, respectively. Let us consider two investors, A and B, in Upper and Lower Slobovia, respectively, each with £10,000 of their native currency in their portfolios at the beginning of the year. Investor B is the better investor, and at the end of the year, he has 10% return whereas investor A's return is 9%. However, 1 US£ is worth 1 US Dollar whereas 1 LS£ is worth only 0.85 US Dollar. Since the US Dollar is a "reserve currency", the two investors' ultimate performance is measured in that currency. The comparison of the two investors' performances is summarized in Table 6.8.

The analogy is as follows: Just like the steam turbine in 1PNR cycle, investor B has better "internal efficiency"; it makes more with what she has vis-à-vis investor A. Unfortunately, in the "real world" where the "reserve currency" is the king, she comes up short vis-à-vis investor A. This is similar to the performance of the 2PNR steam cycle vis-à-vis the 1PNR cycle as measured by the steam cycle efficiency. This is so because investor A's currency has "higher exergy", i.e., is worth more in reserve currency.

Let us now look at the case with improving "topping cycle", i.e., the gas turbine, which is quantified by higher exhaust gas temperatures. We will consider two scenarios:

- Steam temperatures stay constant at 825°F
- Steam temperatures keep up with increasing exhaust gas temperatures (up to 1,112°F or 600°C, which is pretty much the state of the art at the time of writing).

The results of the two scenarios are summarized in Tables 6.9 and 6.10, respectively. The reader is reminded that the percentages in the tables are with respect to the gas turbine exhaust gas exergy. Thus, the BC net output is equal to the BC second law effectiveness (or BC exergetic efficiency).

Table 6.9 clearly illustrates the futility of pumping "high quality" (read: high exergy) exhaust gas into a BC when steam temperatures (i.e., HRSG, BOP and steam turbine materials, technology) cannot keep up with it. The telltale signs are the increasing HRSG exergy loss and stagnant BC exergetic efficiency (second law effectiveness).

Table 6.10 shows the dramatic improvement in performance when steam temperatures are commensurate with exhaust gas temperatures. This is a picture-perfect illustration of the most important thermal design principle: close matching of "source" and "sink" temperatures to minimize irreversibility (i.e., exergy destruction).

TABLE 6.8
Investment Analogy to 2PNRH versus 1PNRH Steam Cycle

	Investor A	Investor B
Analogous to	2PNR	1PNR
Portfolio value	US£ 10,000	LS£ 10,000
Portfolio rate of return	9%	10%
Return in native currency	US£ 900	LS£ 1,000
Exchange rate	1.00	0.85
Return in US dollars	$900	$850

TABLE 6.9
Gas Turbine Exhaust Temperature Increase – Steam Temperature Same (2PNR)

Exhaust Gas Flow, lb/s	800	800	800	800
Exhaust Gas Temperature, °F	**925**	**1,000**	**1,100**	**1,200**
HP steam, psig	615	615	615	615
HP steam temperature, °F	825	825	825	825
LP steam, psia	55	55	55	55
BC net output	63.8%	64.2%	64.2%	64.0%
Stack exergy loss	4.2%	3.2%	2.2%	1.5%
HRSG exergy loss	18.4%	19.4%	20.7%	22.0%
STG exergy loss	9.6%	9.4%	9.1%	8.9%
Condenser exergy loss	1.9%	1.9%	1.8%	1.8%
Miscellaneous losses	2.1%	2.0%	1.9%	1.9%

TABLE 6.10
Gas Turbine Exhaust Temperature Increase – Steam Temperature Keeps Up (2PNR)

Exhaust Gas Flow, lb/s	800	800	800	800
Exhaust Gas Temperature, °F	**925**	**1,000**	**1,100**	**1,200**
HP steam, psig	615	615	615	615
HP steam temperature, °F	825	950	1,050	1,112
LP steam, psia	55	55	55	55
BC net output	63.8%	65.6%	67.1%	67.8%
Stack exergy loss	4.2%	3.4%	2.5%	1.9%
HRSG exergy loss	18.4%	18.4%	18.7%	19.3%
STG exergy loss	9.6%	8.8%	8.0%	7.5%
Condenser exergy loss	1.9%	1.8%	1.8%	1.7%
Miscellaneous losses	2.1%	2.0%	1.9%	1.9%

6.1.6 The "Ultimate" HRSG: Three-Pressure with Reheat

There are two paths to further improvement in BC performance as quantified by the second law effectiveness:

- Introduce reheat, i.e., 2PRH steam cycle or
- Introduce the third pressure level, i.e., 3PNR steam cycle.

Let us start with 2PRH. For this exercise, an "F class-like" exhaust gas specification is adopted: 1,000 lb/s (454 kg/s) and 1,150°F (620°C). Main and HRH steam temperatures are set to 1,050°F (565°C). Steam pressure sweep results are presented in Table 6.11. Note that the table's last row includes the specific cost of the BC, which is obtained by dividing "Contractor's Price" from GT PRO's PEACE[4] add-in by the net output. Details of plant installed cost estimation will be covered in Chapter 13. For now, the reader is asked to accept the numbers presented herein at their face value. Suffice to say that "Contractor's Price" is what is known in the industry as the engineering, procurement and construction (EPC) contractor's price, which includes all "bare" equipment costs

[4] Plant Engineering And Construction Estimator.

TABLE 6.11
Exergy Accounting for 2PRH Cases

HP Steam, psia	1,515	1,415	1,315	1,215	1,115	1,015	815	615
LP Steam, psia	**55**	**55**	**55**	**55**	**55**	**55**	**55**	**55**
BC net output	69.8%	70.0%	70.2%	70.3%	70.4%	70.5%	70.5%	70.1%
Stack exergy loss	2.5%	2.5%	2.5%	2.5%	2.5%	2.5%	2.5%	2.5%
HRSG exergy loss	16.6%	16.5%	16.4%	16.4%	16.4%	16.4%	16.7%	17.2%
STG exergy loss	7.1%	7.0%	6.9%	6.8%	6.7%	6.5%	6.3%	6.1%
Condenser exergy loss	1.8%	1.8%	1.8%	1.7%	1.7%	1.7%	1.7%	1.8%
Miscellaneous losses	2.2%	2.2%	2.2%	2.2%	2.3%	2.3%	2.3%	2.3%
Capex, $/kW	$1,060	$1,050	$1,041	$1,036	$1,030	$1,022	$1,027	$1,024

plus materials and labor involved in erecting the power plant with piping, foundations (concrete and reinforced steel) and all the other bells and whistles as well as the "indirects" such as fees, engineering and supervision.

As can be seen in Table 6.11, although there is an optimum at 1,000 psig HP steam pressure (highest exergetic efficiency and lowest specific cost), the trend is very flat. Within the uncertainty of cost estimate and fine-tuning the cycle design and performance, one could easily go with, say, a 1,400 psig HP steam pressure.

Steam pressure sweep results for the 3PNR bottoming steam cycle are presented in Table 6.12. Main steam temperature is set to 1,050°F (565°C). LP steam pressure is set to 40 psig. IP and LP steam temperatures are calculated based on the minimum approach temperature delta with the limit of maximum 75°F difference between admission steam and steam at the admission point in the steam turbine (to prevent thermal stress problems). Once again, although there is an optimum at 2,200 psig HP steam pressure (highest exergetic efficiency but not the lowest specific cost), the trend is very flat.

It is quite clear that choosing between the 2PRH and 3PNR BCs is a question of cost-performance trade-off. From a thermodynamic perspective, however, there are very important takeaways, which will be helpful in guiding the designer in more advanced gas turbine applications. In particular,

1. In a reheat steam cycle, going to higher pressures does not pay off.
2. In a non-reheat cycle, adding a third pressure level matches the reheat cycle performance but at a higher cost.

TABLE 6.12
Exergy Accounting for 3PNR Cases

HP Steam, psia	2,415	2,315	2,215	2,115	2,015	1,915	1,815	1,715
IP Steam, psia	**483**	**463**	**443**	**423**	**403**	**383**	**363**	**343**
BC net output	70.72%	70.75%	70.76%	70.75%	70.74%	70.73%	70.71%	70.65%
Stack exergy loss	1.4%	1.4%	1.4%	1.4%	1.4%	1.4%	1.4%	1.4%
HRSG exergy loss	11.6%	11.8%	12.0%	12.2%	12.4%	12.6%	12.9%	13.1%
STG exergy loss	12.5%	12.3%	12.1%	11.9%	11.7%	11.5%	11.2%	11.0%
Condenser exergy loss	1.7%	1.7%	1.7%	1.7%	1.7%	1.7%	1.7%	1.7%
Miscellaneous losses	2.1%	2.1%	2.0%	2.0%	2.0%	2.0%	2.0%	2.0%
Capex, $/kW	$1,094	$1,081	$1,072	$1,057	$1,052	$1,042	$1,029	$1,019

TABLE 6.13
Exergy Accounting for 3PRH Cases

HP Steam, psia	2,415	2,315	2,215	2,115	2,015	1,915	1,815	1,715
IP Steam, psia	**483**	**463**	**443**	**423**	**403**	**383**	**363**	**343**
BC net output	73.65%	73.63%	73.5%	73.6%	73.6%	73.4%	73.4%	73.3%
Stack exergy loss	1.9%	1.9%	1.9%	1.9%	1.9%	1.9%	1.9%	1.9%
HRSG exergy loss	11.8%	11.9%	12.0%	12.2%	12.3%	12.4%	12.6%	12.7%
STG exergy loss	8.6%	8.5%	8.5%	8.3%	8.2%	8.2%	8.1%	8.0%
Condenser exergy loss	1.6%	1.6%	1.6%	1.6%	1.6%	1.6%	1.6%	1.6%
Miscellaneous losses	2.4%	2.4%	2.4%	2.4%	2.4%	2.4%	2.4%	2.4%
Capex, $/kW	$1,121	$1,107	$1,101	$1,092	$1,081	$1,078	$1,072	$1,064

3. In the 2PRH cycle, the benefit vis-à-vis the 2PNR comes from reduced steam turbine exergy loss (higher mass-averaged exergy input to the steam cycle).
4. In the 3PNR cycle, the benefit vis-à-vis the 2PNR comes from reduced HRSG exergy loss (via the third pressure level).
5. Overall, steam temperature (and reheat) is a stronger "effector" than steam pressure from an exergetic design perspective.

Let us finalize this analysis by doing a steam pressure sweep for the 3PRH cycle with 1,050°F main and HRH steam temperatures. Exhaust gas conditions are the same as those used for 2PRH and 3PNR cycles. The results are summarized in Table 6.13. The maximum is at 2,400 psig HP steam pressure (cf. 2,200 psig for the non-reheat version). It is also the most expensive design in terms of specific cost. Whether it pays to go from, say, 2,000 to 2,400 psig is economically feasible requires a careful cost-performance trade-off analysis, which will be discussed in Chapter 13.

A summary of what we have looked at in this section is provided in Table 6.14. For the 1PNR, maximum exergetic efficiency and minimum cost variants are not the same. Both are included in the table. In fact, as will be seen in Chapter 8, first combined cycles in the late 1960s and early 1970s had 1PNR BCs with ~500 psig steam pressures.

TABLE 6.14
Exergy Accounting for Bottoming Cycle Configurations

	1PNR	**1PNR**	**2PNR**	**2PNR**	**3PNR**	**3PNR**	**2PRH**	**3PRH**	**3PRH**
Exhaust gas flow, lb/s	400	400	800	800	1,000	1,000	1,000	1,000	1,000
Exhaust gas temperature, °F	900	900	925	1,100	1,150	1,150	1,150	1,150	1,150
HP steam, psia	215	415	615	615	2,215	1,715	1,015	2,415	1,715
IP steam, psia	NA	NA	NA	NA	443	343	NA	483	343
LP steam, psia	NA	NA	55	55	55	55	55	55	55
BC net output	57.8%	57.1%	63.8%	67.1%	70.76%	70.65%	70.5%	73.65%	73.3%
Stack exergy loss	7.8%	11.1%	4.2%	2.5%	1.4%	1.4%	2.5%	1.9%	1.9%
HRSG exergy loss	23.4%	19.7%	18.4%	18.7%	12.0%	13.1%	16.4%	11.8%	12.7%
STG exergy loss	7.1%	8.4%	9.6%	8.0%	12.1%	11.0%	6.5%	8.6%	8.0%
Condenser exergy loss	1.8%	1.7%	1.9%	1.8%	1.7%	1.7%	1.7%	1.6%	1.6%
Miscellaneous losses	2.0%	2.0%	2.1%	1.9%	2.0%	2.0%	2.3%	2.4%	2.4%
Capex, $/kW	$1,583	$1,556	$1,190	$1,033	$1,072	$1,019	$1,022	$1,121	$1,064

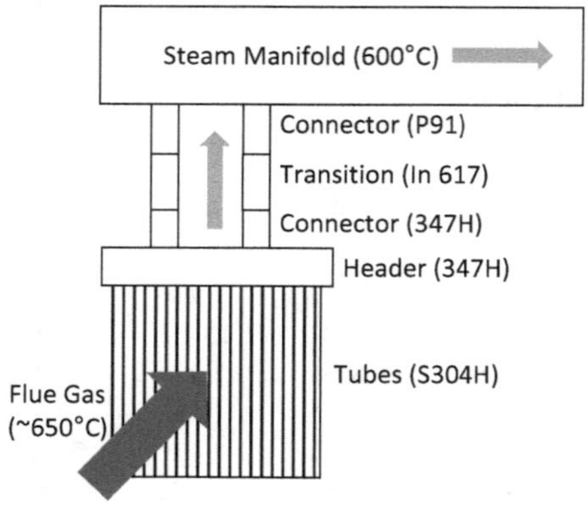

FIGURE 6.14 HRSG high-temperature section material selection (CMI Groupe).

6.1.7 Advanced Steam Conditions

Latest generation of "super-heavy" industrial gas turbines (see Section 8.5) push exhaust gas temperatures above the 1,200°F (~650°C) mark. From an exergetically optimum steam BC perspective, at least 600°C (1,112°F) steam temperature along with high main (HP) steam pressures (e.g., 180 bar or ~2,600 psia) is desirable. This pushes the high-temperature superheater section tube material selection from ferritic (T91) to austenitic (T92). This is considered a step change in technology by the HRSG vendors. Major concerns are LCF impact, low thermal conductivity (increased heating surface) and weldability (especially with dissimilar welded joints). Similar concerns exist from HP steam piping perspective as well (e.g., see Section 10.2). While it is pretty clear that major gas turbine OEMs quote their combined cycle ratings with advanced steam conditions, commensurate field experience and HRSG/BOP vendor and EPC contractor confidence are not widespread.

An alternative material for 600°C steam is Super 304H (fine grain structure) for superheater and reheater tubes, which has been applied as standard material for ultra-supercritical (USC) boiler tubes. Super 304H has the highest strength among 18Cr-8Ni austenitic stainless steels. Its fine-grained microstructure contributes to good steam oxidation resistance. For steam pipes, TP347H stainless steel (coarse grain structure) is a suitable solution. For even higher steam temperatures, i.e., 650°C or higher, Inconel 617 should be considered (it is ten times as expensive as T/P91).

Superheater and reheater materials should possess the following qualities: high creep strength, good weldability, resistance to corrosion on the gas side and oxidation on the steam side. For the heavier parts not directly exposed to the flue gas, i.e., headers and steam pipes/manifolds, resistance to thermal fatigue (LCF) is essential. This is where austenitic materials are at a disadvantage vis-à-vis ferritic materials due to their lower thermal conductivities and higher thermal expansion coefficients. Thus, mixing and matching materials based on their capabilities and weldability characteristics is essential. One HRSG vendor's solution (CMI Groupe), which was applied to the HRSG in GE's Bouchain HA class single-shaft combined cycle (see Section 8.6.2), is shown in Figure 6.14. (Note that the thermal expansion coefficient of Inconel 617 is between P91 and TP347H.)

6.2 HRSG PERFORMANCE CALCULATIONS

The in-depth coverage in the preceding section should have made clear that, as far as state-of-the-art combined cycle power plants with heavy-duty industrial gas turbines are concerned, HRSG configurations of interest are 2PRH and 3PRH. In general, the following rough guidelines apply:

1. Advanced class gas turbines with exhaust conditions of 1,500+ lb/s and 1,150+°F, 3PRH HRSG with 2,200+ psia HP steam generation and 1,085°F–1,112°F steam (HP and HRH)

2. F class gas turbines with exhaust conditions of 1,000–1,400 lb/s and 1,100+°F, 3PRH HRSG with 1,400–2,200 psia HP steam and 1,050–1,065°F steam
3. Vintage E and F class gas turbines with exhaust conditions of 500+ lb/s and up to 1,100°F, 2PRH HRSG with 900–1,400 psia HP steam generation and 950°F–1,050°F steam.

For economical design, pinch point delta should be set to 15°F–25°F with 7°F–10°F in economizer approach subcool. For expensive designs, pinch point delta can be set to 10°F–22°F with 5°F in economizer approach subcool.

The performance figure of merit for the HRSG that is pertinent to combined cycle performance calculation is the HRSG effectiveness given by Equation 6.7, which is reproduced below for convenience:

$$\eta_{HRSG} = 1 - \frac{h_{stck}}{h_{exh}}.$$

Thus, knowing the exhaust and stack gas temperatures is enough to calculate the HRSG effectiveness. Exhaust gas enthalpies can be found from Equations A.5 and A.6 in the Appendix. Similarly, stack gas enthalpies can be found from Equations A.7 and A.8.

Since the gas turbine exhaust temperature is known, the parameter that defines the HRSG effectiveness is the stack temperature. For typical HRSG designs with two-pressure or three-pressure reheat, HRSG effectiveness as a function of gas turbine exhaust gas temperature can be found using the following linear correlations:

$$\text{3PRH}: \eta_{HRSG}[\%] = 74.851 + 0.0144 T_{exh}[°F] \left(1{,}000°F - 1{,}200°F\right) \quad (6.8)$$

$$\text{2PRH}: \eta_{HRSG}[\%] = 59.477 + 0.0244 T_{exh}[°F] \, (1{,}000°F - 1{,}125°F). \quad (6.9)$$

The correlations are obtained from detailed heat and mass balance simulations (using GateCycle Version 5.61.2) with the following assumptions:

- 1.2 in. Hg (about 41 mbar) condenser pressure (i.e., the condensate reaches the HRSG at less than 90°F)
- Gas enthalpies are evaluated with 77°F zero enthalpy reference
- 100% methane-fired gas turbine.

Once the HRSG effectiveness is obtained from Equations 6.8 and 6.9, one can calculate the HRSG stack temperature from Equation 6.7 using the enthalpy formula, Equation A.7. Another way to approach the problem is from the stack temperature angle. Instead of asking "what is the HRSG effectiveness?" one can ask "what is the stack temperature?" As discussed earlier, in an LP pinch-controlled design, the higher the exhaust temperature, the lower the stack temperature and, thus, the better the heat recovery. As a guideline, for well-designed modern 3PRH and 2PRH BC systems with gas turbine exhaust temperatures ranging from 1,050°F to 1,200°F, stack temperatures ranging from 160°F to 200°F can be expected.

6.2.1 HRSG Pressure Loss

The presence of the HRSG between the gas turbine exhaust and the ambient increases the exhaust pressure loss of the gas turbine from its base (i.e., simple cycle) value. The latter is typically assumed to be 5 in. of *water column* (wc or H_2O), which is about 12.5 mbar. The exhaust gas at the gas turbine diffuser exit must go through additional hurdles before reaching the atmosphere:

1. The transition duct between the gas turbine and the HRSG
2. HRSG heat exchanger tube bundles

3. HRSG exhaust duct and silencer (if applicable)
4. CO catalyst (if applicable)
5. The SCR (if applicable)
6. The HRSG stack.

Obviously, the larger the HRSG, the larger is the pressure loss, which negatively impacts the combined cycle performance via reduced gas turbine output. Typically, each additional 1 in. of pressure at the gas turbine exit is worth about 0.1% of power output.

An exact calculation of HRSG pressure loss requires quite complicated analysis. The usual method is to use "allowances" for design calculations, which the HRSG manufacturer is reasonably expected to satisfy. For off-design calculations, simple scaling should suffice; e.g.,

$$\Delta p \propto \rho V_f^2, \tag{6.10}$$

where ρ is the gas density and V_f is the "face velocity". This basic correlation is translated into gas turbine exhaust gas flow parameters as

$$\Delta p \propto \dot{m}_{exh}^2 T_{exh}. \tag{6.11}$$

For typical calculations, it suffices to assume 15 in. wc total pressure loss allowance for 3PRH and 12 in. wc for 2PRH designs. For "aggressive" performance estimates, 2 in. can be subtracted from these values as long as the gas turbine exhaust mass flow rate is less than 1,500 lb/s. Two in. wc should be added if an SCR catalyst system is present.

Note that

$$V_f = \frac{\dot{m}_{exh}}{\rho A} \approx \frac{\dot{m}_{exh}}{\rho(w\ell)}, \tag{6.12}$$

where w is the HRSG "box width" and ℓ is the tube length so that

$$\Delta p \propto (w\ell)^{-2}. \tag{6.13}$$

Consequently, in order to reduce the pressure loss, one has to increase either w or ℓ or a combination thereof, which, of course, has an adverse impact on size and cost. However, this is not a simple optimization problem with a continuous objective function, i.e., $\Delta p = f(w, \ell)$. One should be cognizant of the fact that HRSGs are usually constructed using quasi-standard modules. Each module contains several "harps", which are individual tube bundles without casings or roof. For small HRSGs, one module containing several harps in series (in the direction of gas flow) is a fully cased "O-Section". In that case, the HRSG is said to be "one module wide". Larger HRSGs with larger gas turbines (and exhaust gas flows) require two modules side by side. Each module is partially cased in a "C-Section". (Installing them together in the field, i.e., one module a left-looking C and the other a right-looking C, results in an "O-Section".) An HRSG constructed in this way is said to be "two modules wide". Another method of modularization, the so-called "G-Fast" design by Siemens, involves first erecting the HRSG's main steel components (referred to as goal posts) – like two rows of soldiers facing each other. Subsequently, the modules are positioned between the goal posts and secured in place. (Note that each C-Section contains its own goal post attached to it.)

6.2.1.1 Stack Effect

The stack effect is a pressure difference caused by the difference in ambient air and exhaust gas densities. Modern combined cycle power plant HRSG stacks are 100–200 ft (30–60 m) tall. For round stacks, the design limit for gas velocity inside the stack is 4,000 ft/m or ~ 67 ft/s (~ 20 m/s).

Depending on the stack height and diameter, the stack effect should overcome the friction loss so that the gas has enough momentum left for an exit velocity of at least 30–40 ft/s. Depending on the particular site ambient conditions and governing environmental regulations, higher velocities can be needed to satisfy "plume abatement" requirement. This would also limit the stack gas temperature. A good answer can only be provided by the actual dispersion study. An illustrative calculation is given below.

Let us assume that the exhaust gas at the LP economizer exit is 1,450 lb/s at 175°F. The stack height H is 150 ft with a diameter of D = 22 ft. The ambient temperature is 59°F. Assuming that the average stack gas temperature is (59 + 175)/2 ~ 117°F, the stack effect SE is calculated as

$$SE = H \times (\rho_{air} - \rho_{gas}) = 150 \times (0.0762 - 0.0673)/144 \times 27.7076 = 0.26 \text{ in. wc} \qquad (6.14)$$

from Equation 22.28 in Ref. [9] in Chapter 2. (Densities are calculated using the JANAF tables. Gas density is evaluated at 117°F.) The friction loss FL along the stack is approximately given by

$$FL = 48\frac{H\rho\dot{Q}^2}{D^5} = 48\frac{150 \cdot 0.0673\left(\dfrac{1450}{0.0673}\right)^2}{(22 \cdot 12)^5} = 0.18 \text{ in. wc} \qquad (6.15)$$

from Equation 22.34b in the reference cited above. Thus, the stack effect is sufficient to overcome the friction loss and provide a gas exit velocity of

$$V = \frac{\dot{Q}}{\dfrac{\pi}{4}D^2} = \frac{\dfrac{1,450}{0.0673}}{0.25 \cdot 3.1416 \cdot 22^2} = 56.7 \text{ ft/s.} \qquad (6.16)$$

Note that another option is to find the ideal stack velocity and use that to calculate the flow area and stack diameter. Ideal stack velocity is calculated from the stack effect head as (with a velocity coefficient of $C_v = 0.5$)

$$V = C_v\sqrt{2g \cdot H \cdot \left(1 - \frac{T_{amb}}{\overline{T}_{gas}}\right)} = 0.5\sqrt{2 \cdot 32.174 \cdot 150 \cdot \left(1 - \frac{59 + 460}{117 + 460}\right)} \sim 50 \text{ ft/s.}$$

There is a minimum stack velocity requisite for proper dispersion of the gas discharging from the power plant stack into the atmosphere in vertical and horizontal directions (primarily determined by the prevailing wind speed and direction) and its dilution therein. This is schematically described in Figure 6.15. The vertical rise ΔH is controlled by the stack gas exit velocity and is referenced to a hypothetical point (denoted as "virtual source" in the figure). Its addition to the stack height results in the *effective* height. There are several methods to calculate ΔH, which are described in Ref. [5]. They include the stack gas exit velocity as an independent variable. Depending on the local topography and wind patterns, the effective stack height should be determined to comply with the applicable environmental regulations. A stack design resulting in an insufficient value of V can result in the accumulation of the plume around the power plant site.

Coming back to the example covered herein, if the stack exit temperature were, say, 135°F, then the average stack temperature would be 97°F. This would give a density of 0.0697 lb/ft^3 and the stack effect would be only 0.19 in. wc. It is highly unlikely that this would have enough margin to overcome the frictional loss in the stack. Thus, even if the thermodynamic design would be able to achieve such a low stack temperature, it is unlikely to be feasible from a stack exhaust plume dispersion point of view. As a general rule, stack exit temperatures around 150°F–160°F or lower should not be used without further investigation into the feasibility from different aspects of the plant design.

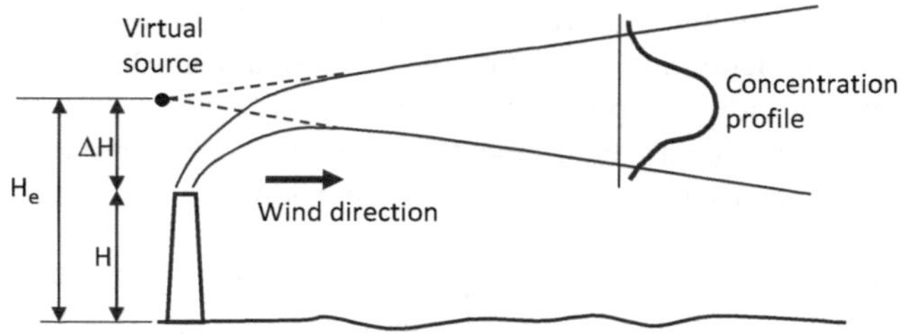

FIGURE 6.15 Schematic description of power plant stack gas dispersion (adapted from Figure 3.21 in El-Wakil [5]).

6.2.2 Heat Transfer in the HRSG

The first HRSG for a gas turbine was built in 1957 or 1958. This was facilitated by the development of equipment to continuously weld steel strips around a bare tube to manufacture spirally (or helically) wound finned tubes. In the late 1950s, HRSGs with finned-tube heat exchangers were deployed in chemical process and refinery plants in cogeneration (combined heat and power) applications to provide power and process steam to the host facility [6]. The significance of finned-tube heat exchangers can be best understood by considering the heat transfer process in the waste heat economizer or boiler. Ignoring the heat conduction across the relatively thin tube wall, the heat transfer between the hot exhaust gas (outside the tubes) and the water/steam (inside the tubes) can be summarized by the following formula

$$\dot{Q} = (UA)_{eff} \cdot LMTD, \tag{6.17}$$

where U is the overall heat transfer coefficient (Btu/ft^2-h-R or W/m^2-K), A is the heat transfer surface area (ft^2 or m^2) and LMTD is the log mean temperature difference defined as

$$LMTD = \frac{\Delta T_A - \Delta T_B}{\ln\left(\dfrac{\Delta T_A}{\Delta T_B}\right)}. \tag{6.18}$$

For rigorous derivation of Equations 6.17 and 6.18 for a simple counterflow (single pass) heat exchanger, refer to Weltyet al. [7] (specifically, pp. 406–407). Equations 6.17 and 6.18 can be applied to more complex geometries and configurations (e.g., crossflow, multi pass) via application of “correction factors”, CF, i.e.,

$$\dot{Q} = (UA)_{eff} \cdot (CF \cdot LMTD). \tag{6.19}$$

Values of CF (F is the symbol used in many references) for many common configurations are available in the literature (e.g., see pp. 412–413 in Ref. [7]).

In the LMTD formula,

- ΔT_A is the temperature difference between hot gas entering the “tube bank” and heated steam or water exiting it (also known as the *approach temperature difference*).
- ΔT_B is the temperature difference between the hot gas exiting the tube bank and steam or water entering it (also known as *terminal temperature difference*, TTD).

From the analogy to resistances in series in an electric circuit, the “effective” UA is found as

$$(UA)_{eff} = \frac{1}{(hA)_{gas}} + \frac{1}{(hA)_{stm/wat}}. \tag{6.20}$$

where ħ is the convective heat transfer coefficient. For a thin-walled tube, $A_{gas} \sim A_{stm/wat}$ and $ħ_{gas} \ll ħ_{wat/stm}$. Thus,

$$(UA)_{eff} \cong (hA)_{gas} \tag{6.21}$$

and

$$\dot{Q} \cong (hA)_{gas} \cdot (CF \cdot LMTD). \tag{6.22}$$

In other words, heat transfer in the economizer or superheater is controlled by the gas-side heat transfer (mostly convective with minor contribution from radiation). When gas turbine exhaust gas at close to 1,000°F was used for feedwater heating and/or economizing in a conventional boiler, high value of LMTD (several hundred degrees F) ensured that bare tube design was economically feasible.

In an HRSG where gas turbine exhaust gas is utilized to produce and superheat steam, LMTD values are much smaller. Using bare tubes would require to maximize the heat transfer surface area by utilizing a very large number of tubes and/or large tube diameters. This would result in excessive gas-side pressure loss (detrimental to gas turbine performance) and expensive heat exchangers. Increasing the gas-side UA by increasing the total heat transfer surface area by using fins made HRSGs for power generation combined cycles feasible, i.e.,

$$\dot{Q} \cong (hA)_{gas} (1+F) \cdot (CF \cdot LMTD), \tag{6.23}$$

where F is the ratio of total fin surface area to the bare tube surface area (F = 0 for bare tubes). It should be noted that, in the case of heavily duct-fired HRSGs, LMTD increases again so that bare tube superheaters can be used.

For a 1PNR system, HRSG energy balance and heat transfer calculations can be done reasonably easily in an Excel spreadsheet or a simple FORTRAN (or C++) code. Beyond that, the calculations become quite cumbersome and one has to resort to a commercial (or in-house) software product such as Thermoflow's GT PRO® (design) and GT MASTER® (off-design). For those who would like to try their hand at it, design calculations involve applying Equation 6.1 to each heat exchange section (economizer, evaporator and superheater) individually, i.e., for section i in an HRSG with N sections (i = 1 at the HRSG inlet and i = N at the HRSG stack)

$$\dot{m}_{gas} \left(h_{gas\text{-}in,i} - h_{gas\text{-}out,i} \right) = \dot{m}_{w/s} \left(h_{s/w\text{-}out,i} - h_{s/w\text{-}in,i} \right) = \dot{Q}_i, \tag{6.24}$$

$$h_{gas\text{-}in,i} = h_{gas\text{-}out,i-1}, \tag{6.25}$$

$$h_{s/w\text{-}in,i} = h_{s/w\text{-}out,i+1}. \tag{6.26}$$

$$h_{s/w\text{-}in,N} = h_{cond}. \tag{6.27}$$

$$h_{gas\text{-}in,1} = h_{exh}, \tag{6.28}$$

$$h_{gas\text{-}out,N} = h_{stack}, \tag{6.29}$$

For each heat exchange section i, using Equation 6.23,

$$\dot{Q}_1 \cong (hA)_{gas,i} (1+F) \cdot LMTD_i = \dot{m}_{gas} \left(h_{gas\text{-}in,i} - h_{gas\text{-}out,i} \right) \tag{6.30}$$

or

$$\dot{Q}_I \cong (hA)_{gas,i}(1+F)\cdot LMTD_i = \dot{m}_{w/s}\left(h_{s/w\text{-}out,i} - h_{s/w\text{-}in,i}\right) \tag{6.31}$$

with

$$LMTD_i = \frac{\Delta T_{A,i} - \Delta T_{B,i}}{\ln\left(\dfrac{\Delta T_{A,i}}{\Delta T_{B,i}}\right)}, \tag{6.32}$$

where

$$\Delta T_{A,i} = T_{gas\text{-}in,i} - T_{s/w\text{-}out,i} = \Delta T_{APP,i} \tag{6.33}$$

and

$$\Delta T_{B,i} = T_{gas\text{-}out,i} - T_{s/w\text{-}in,i} = TTD_i. \tag{6.34}$$

(For the evaporator, TTD is the same as the pinch delta. For the economizer, the approach temperature delta, ΔT_{APP}, is the sum total of pinch delta and approach subcool.)

During the design calculations, $(\hbar A)_{gas,i}$ for each heat exchange section is determined from Equation 6.31. Solution of Equations 6.24 through 6.29 for i = 1, 2, …, N determines the temperature profiles for the exhaust gas and water/steam (i.e., the heat release curves). The profile data is used to calculate $LMTD_i$ for each section using Equations 6.30 through 6.34. As stated earlier, this procedure is relatively easy to implement in and Excel spreadsheet for a 1PNR design (N = 3). While this has very little practical significance in terms of modern, utility-scale power generation applications, it can be very useful from a self-learning perspective.

The other important parameter in HRSG calculations is the gas-side pressure drop, which is important because of its impact on gas turbine performance. As a rule of thumb, each inch of water column in gas turbine exhaust pressure loss is worth about 0.1% in power output (reduction) and heat rate (increase). An aggressive HRSG design with low approach and pinch temperature deltas (to maximize steam production) results in large heat transfer surfaces, which translates into higher number of tube rows and larger gas-side pressure drop. Pressure drop for flow of gases over a bank of tubes is given by the following generic relationship [8]:

$$\Delta p \propto N \cdot \frac{\dot{m}_{gas}}{\rho_{gas} A_{xs}}, \tag{6.35}$$

where N is the number of tube rows and A_{xs} is the cross-sectional flow area (i.e., roughly equal to HRSG box width multiplied by the tube length). In design calculations, Δp is usually specified, e.g., 14 in. H_2O between the transition duct inlet (gas turbine exhaust diffuser exit) and the ambient.

Clearly, design calculations, while especially cumbersome for multi-pressure, reheat configurations, are relatively straightforward. For simplified off-design calculations, knowledge of $(\hbar A)_{gas,i}$ is adequate. Basically, Equations 6.30 and 6.31 are used in combination with Equations 6.24 through 6.29 in an iterative procedure using the boundary conditions imposed on the HRSG. Equation 6.35 suggests that a rough estimate of gas-side pressure loss for the entire HRSG can be estimated via proportionality to gas-side volume flow.

The main difficulty in HRSG modeling lies in translating $(\hbar A)_{gas,i}$ to a practical hardware design with exact dimensions and quantities. Calculation of $\hbar$ requires flow velocities on both the gas side and the water/steam side for each heat exchanger. The flow velocities, however, depend on the exact geometry of each heat exchanger. Furthermore, flow velocities and heat exchanger geometries affect the sectional pressure drops as well, on both the gas side and the water/steam side (typically via the famous Dittus–Boelter equation, see Ref. [8]). Since pressure drops on both sides had been assumed to solve the heat balances via Equation 6.35, an exact match requires a highly iterative calculation routine while simultaneously tweaking a large number of hardware design parameters (e.g., tube length, tube diameter, number of rows, number of tubes in a row, fin type, fin geometry). This is a job that can be properly done only by HRSG manufacturers with their in-house software and decades' worth experience distilled into myriad empirical correlations, selection algorithms, etc. Typically, the following information should be provided to the HRSG supplier:

- Gas turbine exhaust flow, temperature and composition (from the gas turbine OEM)
- Required steam flows, pressures and temperatures (e.g., from GT PRO® or GateCycle runs)
- Maximum allowable gas-side pressure loss (impact on gas turbine performance)
- Maximum allowable reheater-side pressure drop (impact on BC performance)
- Auxiliary equipment (e.g. duct burner, SCR/CO catalysts, deaerators)
- Noise emission requirements
- Air emission requirements.

This is why no attempt is made herein to delve into such calculations in nitty-gritty detail. A reader who wants to develop his or her own models for learning or other purposes can start with Equations 6.24 through 6.34. These equations have to be supplemented by heat transfer coefficient correlations and hardware information. A universally accepted reference for the latter, including gas-side heat transfer and pressure drop correlations, is the ESCOA Rating Method (which is used in GT PRO® and GT MASTER® as well). ESCOA is the acronym for *Extended Surface Corporation of America*, which is the trademark of the ESCOA Fintube Corporation in Oklahoma, USA. The original rating method can be found in the manual issued by the company [9]. For water and steam flowing in the tubes, there are many well-known correlations, which can be found in the literature. Simple pressure loss and heat transfer coefficients for feedwater in the economizers and steam in the superheaters are given below correlations (also used in GT PRO® and GT MASTER®):

$$\Delta p = \left(f \frac{L}{D_i} + \lambda \right) \frac{\rho \bar{U}^2}{2} \tag{6.36}$$

$$\text{Friction factor}: f = f\left(\text{Re}, \frac{\varepsilon}{D_i} \right) \tag{6.37}$$

$$\text{Nusselt number}: \text{Nu} = \frac{\hbar D_i}{k} \propto \text{Re}^{0.8}\,\text{Pr}^n \tag{6.38}$$

$$\text{Reynolds number}: \text{Re} = \frac{\rho \bar{U} D_i}{\mu} \tag{6.39}$$

$$\text{Prandtl number}: \text{Pr} = \frac{\rho c_p}{k}, \tag{6.40}$$

where Δp is the pressure drop of steam/water flowing in the heat exchanger tube (per pass), μ is the viscosity, ρ is the density, c_p is the specific heat, D_i is the inner diameter of the tube and L is its length with k for conductivity and $\hbar$ for convective heat transfer coefficient. (Note that for natural circulation evaporators, Δp is zero.) In Equation 6.36, which is a modified version of the Darcy–Weisbach equation, λ accounts for minor losses (can be set to 2, i.e., two "velocity heads") and $\bar{U}$ is the average flow velocity. There are many empirical relationships for the friction factor, f, which is inversely proportional to the Reynolds number (ε/D in Equation 6.37 is the pipe roughness in the Moody chart), e.g., Blasius, Nikuradse or Colebrook. The Blasius equation is valid for smooth pipes ($\varepsilon/D = 0$) and Reynolds numbers up to 100,000 and easy to use, i.e., $f = 0.036/\text{Re}^{0.25}$.

For fully developed turbulent flow in smooth and straight tubes, the commonly used form of Equation 6.38 is the Dittus–Boelter equation with the proportionality factor of 0.023 and n = 1/3 [8]. These equations should be supplemented by the *ASME Steam Tables* (to find density and specific heat) and property relationships for μ and k. The best source for the latter is the book by Poling et al. [10] (pretty much a "must have" book for chemical and mechanical engineers).

Front-end heat transfer sections of the HRSG comprise HP superheater and reheat superheater tube bundles. For optimal heat transfer, superheater/reheater tubes are "finned tubes". There are two types of fins:

- Solid – typically used for the first two rows of heat transfer surface (superheaters)
- Serrated – for the rest of the HRSG.

Typically, high-frequency, resistance-welded fins are used in finned tube construction. Even though serrated fins provide better heat transfer, solid fins are preferred because they are easier to clean (especially when burning "dirty" fuels like number two distillate).

Typical fin sizes are

- Fin pitch – from one (1) to seven (7) fins per inch
- Fin height – from 0.125 to 0.875 in. (typical values 0.5 and 0.75 in.)
- Fin thickness – from 0.04 to 0.06 in. (typical value 0.05 in.).

Typical finned tube sizes are

- 1.5 or 2.0 in. diameter tubes with
- 0.75 in. high fins (0.04 in. thickness) spaced at six fins per inch.

Typical tube materials used in the HRSG heat transfer sections are

- SA-178A, SA-210 (economizers and evaporators)
- SA-516-70 or SA-299 (evaporator drums)
- SA-213 T11, T12, T22 or T91 (superheaters and reheaters)
- Duplex 2205 stainless steel or SA-178A (condensate preheaters).

6.2.3 HRSG Steam Production

The *raison d'etre* of a modern 3PRH HRSG is steam production. In Section 5.3.3, the transfer function below (Equation 5.39) was used to estimate the steam turbine exhaust flow rate as a function of gas turbine exhaust parameters (Ref. [14] in Chapter 5):

$$\mu = \left(0.153 + 0.018\left(\frac{T_{exh}}{100} - 11\right)\right) - 0.002\frac{T_{stm} - 1,050}{25}, \tag{6.41}$$

where μ is the ratio of steam and exhaust gas flow rates, T_{exh} is the exhaust gas temperature and T_{stm} is the HP and HRH steam temperature (assumed to be the same as is the case in practically all real cycles). In this section, a generic version of Equation 6.41 is introduced:

$$\mu = \left[a + b\left(\frac{T_{exh}}{100} - 11\right)\right] - c\frac{T_{stm} - 1,050}{25}, \tag{6.42}$$

where the constants a, b and c are determined for three different HP pressures. The results are summarized in Table 6.15. (Note that GateCycle Version 5.62.1 was used to determine the transfer

TABLE 6.15
Curve-Fit Constants for Equation 6.42

	Original Equation 6.41	Equation 6.42 Main Steam 1,600 psig	Main Steam 2,000 psig	Main Steam 2,400
a	0.15300	0.152446	0.154021	0.155593
b	0.01800	0.017922	0.017935	0.017845
c	0.00200	0.001520	0.001558	0.001643

TABLE 6.16
Curve-Fit Constants for Equation 6.43

	1,600 psig	2,000 psig	2,400 psig
HP Admission			
a	0.7499	0.8161	0.8707
b	–0.0982	–0.1891	–0.2650
IP Admission			
a	–0.4762	–0.5003	–0.5256
b	0.6840	0.7122	0.7428
LP Admission			
a	–0.2657	–0.3076	–0.3371
b	0.4072	0.4697	0.5152

function coefficients for the original Equation 5.39. Updated coefficients are determined with GT PRO® Version 26 runs.)

Equation 6.42 gives the sum total of HP, IP and LP steam flow rates. In order to find the individual admission steam flow rates, the following transfer function is used:

$$\mu_i = a\frac{T_{exh}}{1{,}000} + b, \tag{6.43}$$

where i = 1, 2 and 3 correspond to HP, IP and LP, respectively, and μ_i is the fraction of the respective admission steam flow rate to the steam turbine exhaust flow rate. Values of the curve-fit constants *a* and b for three different HP pressures are provided in Table 6.16.

As an example, consider an advanced class 50-Hz gas turbine with 2,000 lb/s exhaust flow rate and 1,175°F exhaust temperature. The steam cycle is 2,400 psig/1,065°F/1,065°F. Using Equation 6.42 with the constants in Table 6.15,

$$\mu = \left[0.155593 + 0.017845\left(\frac{1{,}175}{100} - 11\right)\right] - 0.001643\frac{1{,}065 - 1{,}050}{25} = 0.167991.$$

Thus, steam turbine exhaust flow rate is

$$2{,}000\ \text{lb/s} \times 0.167991 = 336\ \text{lb/s}.$$

Using Equation 6.42 with the constants in Table 6.16,

$$\mu_1 = 0.8707\frac{1175}{1000} - 0.265 = 0.75807,$$

$$\mu_2 = -0.5256\frac{1175}{1000} + 0.7428 = 0.12522,$$

$$\mu_3 = -0.3371\frac{1175}{1000} + 0.5152 = 0.11911.$$

Note that the sum of the three fractions is

$$0.75807 + 0.12522 + 0.11911 = 1.0024 \sim 1.0,$$

which is satisfactory. Consequently, HP or main steam flow rate is

$$336 \times 0.75807 = 254.7 \text{ lb/s}.$$

IP admission steam flow rate is

$$336 \times 0.12522 = 42.1 \text{ lb/s}.$$

LP admission steam flow rate is

$$336 \times 0.11911 = 40 \text{ lb/s}.$$

Using the same assumptions as in the example in Section 5.3.2, steam turbine generator (STG) output is found as 234,309 kWe from the simple heat and mass balance. Using the correction factor of 0.96, actual STG output is 224,936 kWe or 225 MWe. Allocating 2% for condensate and feed pumps, BC net output is 225 × 0.98 = 220.4 MWe. From Equation A.2 in the Appendix, gas turbine exhaust exergy is

$$a_{exh} = 0.1961 \cdot 1{,}175 - 86.918 = 143.5 \text{ Btu/lb}.$$

Multiplying by the exhaust flow rate, total exhaust gas exergy is

$$2{,}000 \text{ lb/s} \times 1.05506 \text{ kWs/Btu} \times 143.5 = 302{,}801 \text{ kW}.$$

Thus, the implied BC exergetic efficiency is 220.4/302.8 = 0.728 or 72.8%.

6.2.4 Stack Temperature

As mentioned at the beginning of the section, once the HRSG effectiveness is obtained from Equations 6.8 and 6.9, one can calculate the HRSG stack temperature from Equation 6.7 using the enthalpy formula, Equation 6.8. Another way to estimate the stack temperature takes into consideration several key BC design parameters. In order to investigate this, let us look at the tail section of a typical 3PRH HRSG in Figure 6.16.

Let us define the following parameters:

DTP = Temperature difference between the LP steam generated in the LP evaporator and the exhaust gas leaving the LP evaporator (LP pinch)

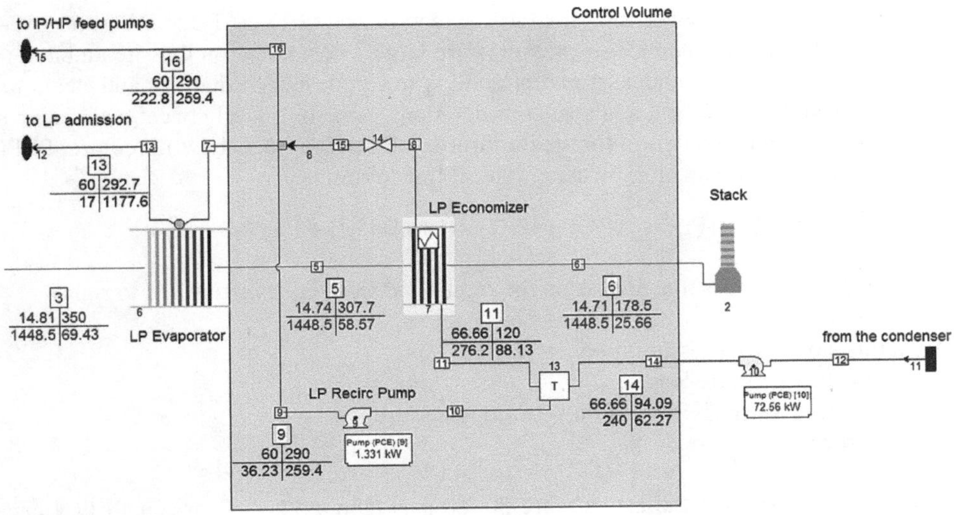

FIGURE 6.16 Typical 3PRH HRSG LP section (THERMOFLEX® model).

DTSUBC = Temperature difference between the LP steam generated in the LP evaporator and the LP feedwater leaving the LP economizer (LP approach subcool)

TCOND = Temperature of the condensate that is approximately equal to the saturation temperature at the condenser pressure, PCOND

TSAT = Temperature of the LP steam that is equal to the saturation temperature corresponding to the LP drum pressure, PLPD.

From the heat and mass balance for the control volume in Figure 6.16, ignoring the small contribution of the LP recirculation pump power, it can be shown that

$$\text{TSTACK} = \text{TAVE} + \text{DTAPP}, \tag{6.44}$$

where

$$\text{TAVE} = (1-\text{E}) \cdot \text{TSAT} + \text{E} \cdot \text{TCOND} \tag{6.45}$$

is a weighted average of LP steam and condensate temperatures and

$$\text{DTAPP} = \text{DTP} + \text{E} \cdot \text{DTSUBC} \tag{6.46}$$

is a modified approach temperature difference for the LP economizer. The heat capacitance ratio of the condensate to the gas turbine exhaust gas, E, can be found from

$$\text{E} = \frac{\dot{m}_{cond}}{\dot{m}_{exh}} \frac{c_{p,cond}}{c_{p,exh}} = \frac{\mu}{\chi}. \tag{6.47}$$

Condensate-to-gas flow ratio, μ, can be estimated from gas turbine exhaust parameters as described in Section 6.2.3. Gas-to-condensate specific heat ratio, χ, can be taken as 0.25 with reasonably good accuracy. Combining Equations 6.44 through 6.47, stack temperature can be estimated as

$$\text{TSTACK} = (\text{TSAT} + \text{DTP}) - \frac{\mu}{\chi}(\text{TSAT} - \text{TCOND} - \text{DTSUBC}). \tag{6.48}$$

State-of-the-art BC design parameters for the LP section are 12°F–20°F for DTP and 0°F–10°F for DTSUBC. LP drum and condenser pressures are largely dependent on the site ambient conditions, heat rejection system (condenser and/or cooling tower) design limitations and steam turbine technology as well as pipe sizing and cost considerations. At a high-level conceptual design stage, LP drum pressure can be related to the steam turbine LP section inlet (bowl) pressure, PLPIN, (a readily available design parameter) by using typical line losses, i.e.,

$$\text{PLPD} = (\text{PLPIN} + 3) \times (1 + 2\%) \times (1 + 4\%) \times (1 + 1\%) \times (1 + 2\%). \quad (6.49)$$

In Equation 6.49, the factors that appear on the right-hand side signify (from left to right)

- LP admission valve pressure drop (in psi)
- LP steam line pressure loss
- LP non-return valve (NRV) and flow nozzle pressure loss
- LP superheater pressure drop.

Typical numbers for state-of-the-art BCs are 50–80 psia (about 3.5–5.5 bar) for PLPIN and 0.8–2.0 in. Hg for PCOND (about 25–70 mbar). LP turbine admission pressures lower than 50 psia are unlikely to be encountered due to increasing steam pipe diameters at exceedingly low densities (or high specific volumes). Values higher than 80 psia are also unlikely in 3PRH systems because of efficiency (i.e., low stack temperature) requirements.

In the sample system shown in Figure 6.16, DTP = 15°F, DTSUBC = 2.7°F, TCOND = 94°F, TSAT = 292.7°F and $\mu = 0.1657$. From Equation 6.48,

$$\text{TSTACK} = (292.7 + 15) - \frac{0.1657}{0.25}(292.7 - 94 - 2.7) = 176.8°\text{F},$$

which is only 1.7°F lower than the "exact" value in Figure 6.16.

In theory, the HRSG can be designed with enough heat transfer surface to bring the exhaust gas temperature at the LP economizer exit within a few degrees of the feedwater entering the HRSG. Apart from the size and cost considerations that would prohibit such a large design, there are limitations imposed by water and sulfuric acid condensation on the LP economizer tubes.

This is the reason why all HRSG designs incorporate a recirculating water system that pumps the feedwater from the LP economizer discharge back into the economizer inlet to maintain an inlet temperature of at least 120°F (~50°C). This will be sufficiently above the water dew point temperature of the exhaust gas for most cases and will keep the gas-side heat transfer surfaces dry. This is the convention regardless of the fuel sulfur content, which is typically very low for natural gas.

In HRSGs with the SCR systems, ammonia (NH_3) that is injected into the gas path upstream of the catalyst reacts with sulfur to form ammonium bisulfate (NH_4HSO_4), which can form corrosive deposits on the heat exchanger tubes at temperatures well above the sulfuric acid dew point, e.g., up to 392°F (200°C). This is pretty much unavoidable. Other problems associated with the SCR are ammonia chloride plume from the stack and emission of gaseous ammonia. In general, SCR as a NOx emissions technology in combined cycle power plants is not recommended for gas turbines firing sulfur-bearing fuels as the primary fuel. For secondary fuels such as number 2 distillate, SCR should be considered on a case-by-case basis and with the satisfaction of the following criteria such as

- Use of fuel oils with very low sulfur content (<1% by weight)
- Low ammonia slip design (<10 ppmvd)
- Limit to less than 240 h/year operation.

Sulfuric acid dew point is a function of fuel sulfur content, fuel–air ratio and conversion rate of SO_2 to SO_3 (necessary for the formation of the sulfuric acid, H_2SO_4). There are several empirical calculation methods for sulfur dew point calculation. One such method is the formulation by Verhoff and Banchero [11], which is available in simulation software such as GateCycle and THERMOFLEX®. The formula is as follows:

$$T_{ADP} = \frac{1,000}{1.7842 + 0.0269\log p_{H_2O} - 0.1029\log(C \cdot p_{SO_2}) + 0.0329\log(C \cdot p_{SO_2})\log p_{H_2O}}, \quad (6.50)$$

where

T_{ADP} = Sulfuric acid dew point temperature [K]
p_{H_2O} = Partial pressure of water (in atm) with the mixture of gases at standard atmospheric pressure (101.325 kPa)
p_{SO_2} = Partial pressure of SO_2 (in atm) with the mixture of gases at standard atmospheric pressure
C = Conversion rate of SO_3 from SO_2 by volume.

For typical gas turbine fuels, sulfuric acid dew point is around 250°F with fuel sulfur weight fraction of 0.1% (with 5% conversion to SO_3). With higher conversion, T_{ADP} increases; with lower conversion, T_{ADP} decreases (see Figure 6.17). This explains why in 1960s and 1970s HRSG stack temperatures were at 300°F or above, which precluded the deployment of more efficient two- or three-pressure cycles with lower stack temperatures. Back then, gas turbines were fired with high sulfur-containing liquid fuels with high acid dew points. (Fuel sulfur content was limited to 0.8% by weight by the US DOE's New Source Performance Standards (NSPS) by 1979.) In the USA, the Kalaeloa Cogeneration Plant (in Hawaii) is probably the only combined cycle power plant designed to operate with low-sulfur (0.5% by weight) heavy fuel oil as the primary fuel [12]. Nominal HRSG stack temperature is 350°F. Significant difficulties presented by the fuel in terms of gas turbine and HRSG deposits can be found in an article published in 2Q-2009 issue of *Combined Cycle Journal* ("Kalaeloa Cogeneration Plant: Continual improvement defines black-oil-fired combined cycle").

Natural gas has very low sulfur. Including the sulfur in the odorant (mercaptan) added to gas for safety reasons, typical values are less than 10 ppm (0.00001 weight fraction) – mostly as hydrogen sulfide (H_2S). Based on the curves in Figure 6.17, depending on the conversion rate and considering

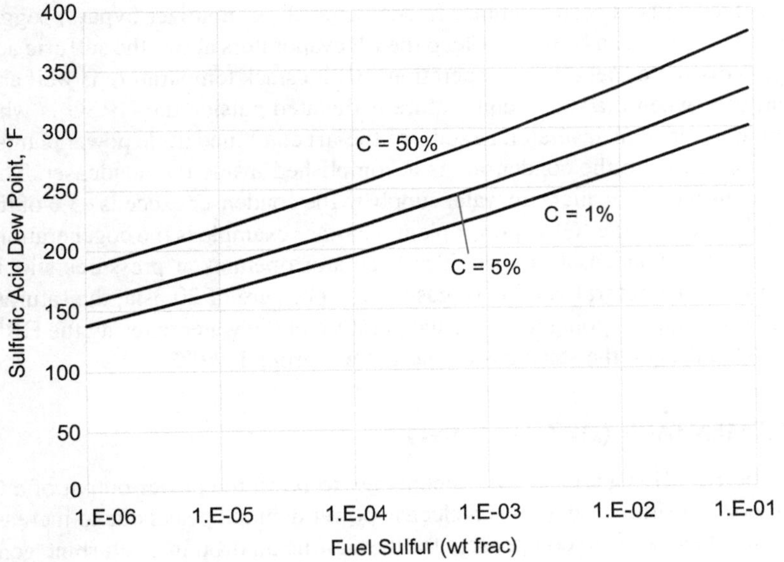

FIGURE 6.17 Stack gas sulfuric acid dew point variation (Equation 6.49).

the uncertainty involved in empirical correlations used in estimation, even such a minuscule sulfur concentration can be problematic. Fortunately, the reality is better than this worst-case scenario. Detailed field studies have shown that [13]

1. Acid dew point corrosion is not expected to be a significant problem in HRSGs with natural gas-fired gas turbines. The quantity of sulfur, in general, and sulfuric acid, in particular, is too low to cause major corrosion attack of carbon steel materials *as long as the tube metal temperature is maintained above the water dew point.* Otherwise, more expensive stainless steel tubes should be considered in LP economizer construction.
2. Gas turbine inlet fogging or water injection will have an impact on flue gas moisture content. The added moisture will raise the water dew point temperature and lead to corrosion if the system is not designed to raise tube temperatures sufficiently above that.
3. Installation of an SCR will be a strong factor. Any ammonia slippage leading to formation of ammonium sulfate or bisulfate will promote corrosion at temperatures typical of the LP section (300°F–350°F at the LP evaporator inlet). Unless the SCR is performing properly, in which case the impact of the ammonium sulfate formation should be negligible as in the sulfuric acid case, proper design accommodations should be made.

Thus, for low-sulfur fuels, LP feedwater recirculation to 120°F feedwater inlet temperature as shown in Figure 6.16 is used as a preventive design feature. Without this, LP economizer tubes will be at the temperature of the condensate, which is typically at 115°F or lower (at condenser pressures of 3 in. Hg or lower – see Table 5.12.), which will lead to condensation of water vapor in the exhaust gas on the tubes. (Equation A.1 can be used for water dew point calculation as a function of moisture mole fraction of the exhaust gas.)

Number 2 or distillate fuel oil is a standard backup fuel with high sulfur content. During distillate fuel oil-fired operation, water is injected into the combustor for NOx control. This increases the H_2O content in the exhaust gas and pushes the water dew point upwards. LP economizer tube metal temperature to accommodate this rise can be achieved via increased recirculation. Typically, the rule of thumb is to limit the total economizer feedwater flow rate to 200% of the design flow rate to prevent high-velocity, high-pressure drop operation, which can accelerate tube erosion. Thus, a common way to control the tube temperature in distillate fuel oil-fired operation is to bypass the LP economizer (with or without recirculation). In complete LP economizer bypass, pegging of the LP drum with IP or HP steam can be used to keep the LP evaporators above the sulfuric acid dew point. Consequently, in distillate fuel oil-fired operation, HRSG stack temperature is well above 200°F.

One special case when the stack temperature is dictated outside the HRSG is when the power plant incorporates a separate deaerator. In state-of-the-art combined cycle power plants, the removal of noncondensable gases in the condensate is accomplished inside the condenser. This is not possible in applications where the makeup water supply to the condenser exceeds ~3% of the total steam flow. An example of that is the IGCC power plant. Another example is the cogeneration applications with large amounts of condensate return. The deaerator operates at pressures slightly above the atmospheric to facilitate natural venting of gases. At a pressure of 20 psia, the saturation temperature is about 228°F, which is going to be the temperature of the water entering the HRSG. Thus, for a reasonable HRSG design, the stack gas temperature is around 250°F.

6.3 SUPPLEMENTARY (DUCT) FIRING

Duct firing in the HRSG is a widely used technology to boost the power output of a GTCC power plant especially on hot summer days when electric power demand peaks due to increased use of air conditioners. The objective is to compensate for the significant drop in gas turbine generator power output due to lower inlet airflow (i.e., high ambient temperature → low ambient air density) by burning additional fuel in the HRSG and exploit high power prices for additional revenue. The "duct" in question is the transition duct between the gas turbine exhaust and the HRSG inlet even though in most cases the burners are located behind the front-end superheater or reheater tube banks. Therefore, the alternative term "supplementary firing", also widely used in the industry, is probably a more apt description of the generic technology. According to a study done in 2014, between 2002 and 2014, 56 new GTCC power plants were installed or in development in the USA (total installed capacity of nearly 36 GWe with an average 640 MWe per plant) with average duct firing capacity of 77 MWe (i.e., 77/(640 – 77) ~ 14%).[5]

It is almost a foregone conclusion that duct firing is detrimental to the combined cycle heat rate. Although this is certainly true based on the field deployment of duct firing in utility-scale power generation, it is not a universally valid statement. There are several published references illustrating the possibility of combined cycle efficiency improvement via duct firing under certain boundary conditions. One of the earliest ones known to the author is the paper by Finckh and Pfost, which illustrates graphically that there is a limited "window of opportunity" where specific output and efficiency of a GTCC with a one-pressure reheat (1PRH) steam "bottoming" cycle *both* increase with supplementary firing [14]. A nice discussion and graphic illustration of the underlying thermodynamic "driving force" can be found in the second edition of the excellent book by Kehlhofer et al. [15] (i.e., pp. 98–103 and Figure 4.47 therein). A more recent paper by Jordal et al. includes references to the cited works above (and many more) and also includes a mathematical derivation of the necessary condition for efficiency improvement with duct firing [16] (i.e., Equation (7) therein). A peculiar application of HRSG duct firing is in a "split boiler" arrangement originally proposed by Ivan Rice [17] and closely examined in Ref. [16]. Especially when combined with steam cooling of the first stage of the gas turbine hot gas path (ideally, stage 1 nozzle stators, i.e., stationary components), a split boiler, where only a portion of the gas turbine exhaust gas is duct-fired, is shown by Jordal et al. to lead to a significant efficiency improvement in a two-pressure no reheat (2PNR) steam BC configuration with small (typically less than 100 MWe) industrial gas turbines [16].

At this point, it is appropriate to take a short break and reexamine the aforementioned Equation (7) in Ref. [16], which will shed light on the small "window" for combined cycle efficiency improvement with HRSG duct burning and why it can only work with "small" gas turbines (industrial or aeroderivative). Ignoring small gas turbine and steam cycle loss terms, on a shaft output basis, the formula for duct-fired GTCC efficiency can be written as follows (for derivation of the combined cycle efficiency formula, revisit Chapter 4):

$$\eta_{cc} \approx \frac{\eta_{gt} + (q_x + d)\eta_{bc}}{1+d} \tag{6.51}$$

$$\eta_{bc} = \eta_b \eta_{st}, \tag{6.52}$$

where
η_{cc} = Combined cycle efficiency
η_{gt} = Gas turbine efficiency
η_{bc} = Steam "bottoming" cycle efficiency
η_b = HRSG (i.e., "boiler") effectiveness
η_{st} = Steam turbine efficiency
F_{db} = Duct burner fuel consumption
F_{gt} = Gas turbine fuel consumption
d = Duct burner fuel consumption as a fraction of F_{gt}
q_x = Gas turbine exhaust energy as a fraction of F_{gt}.

[5] From "Cost of New Entry Estimates for Combustion Turbine and Combined Cycle Plants in PJM," prepared for PJM by The Brattle Group and Sargent & Lundy (May 15, 2014).

Taking the derivative of both sides of Equation 6.51 with respect to d, the result is

$$\frac{\partial \eta_{cc}}{\partial d} = \frac{(q_x + d)\frac{\partial \eta_{bc}}{\partial d} + \eta_{bc} - \eta_{cc}}{1+d}. \tag{6.53}$$

In order for the combined cycle efficiency to improve with duct firing, from Equation 6.53 it follows that the inequality below must hold:

$$\frac{\partial \eta_{bc}}{\partial d} > \frac{\eta_{cc} - \eta_{bc}}{(q_x + d)}. \tag{6.54}$$

Ignoring the contribution of minor energy/mass streams, normalized gas turbine exhaust energy is given by

$$q_x = (1 - \eta_{gt}) \tag{6.55}$$

so that Equation 6.54 can be reformulated as

$$\frac{\partial \eta_{bc}}{\partial d} > \frac{\eta_{cc} - \eta_{bc}}{(1 - \eta_{gt} + d)}. \tag{6.56}$$

As already noted by Jordal et al. [16], Equation 6.56 was derived by Kehlhofer as well (see Equation (12) on p. 11 in Ref. [18][6]).

The left-hand side of Equation 6.56 is the improvement in steam BC efficiency with duct firing. It can be related to several effects caused by extra energy input at the HRSG inlet, which raises the gas temperature. In particular, the said effects include

- Introduction of reheat (or even "double reheat")
- Higher main and reheat steam temperatures
- Higher steam pressures
- Higher steam mass and volumetric flow
- Higher steam turbine efficiency (due to higher volume flow).

The right-hand side of Equation 6.56 can be rewritten as follows

$$\frac{\eta_{gt}(1 - \eta_{bc})}{(1 - \eta_{gt} + d)}.$$

For a modern combined cycle power plant with an advanced heavy-duty industrial gas turbine and a three-pressure reheat (3PRH) steam BC, it can be shown that (e.g., see Chapters 4 and 5)

$$\eta_{gt} \sim \eta_{bc} \sim 0.4 (\text{i.e.}, 40\%)$$

so that, to a very good approximation, the right-hand side of Equation 6.56 can be written as

$$\frac{\eta}{1 + \frac{d}{1-\eta}},$$

[6] Even if you own the second or third editions of the book by Kehlhofer et al., the first edition with Kehlhofer the sole author is significantly different, albeit from a historical perspective mostly, and worth obtaining a copy for the wealth of hard-to-find material in it.

where $\eta \sim 0.4$. Thus, in order for duct firing to be beneficial to the combined cycle efficiency, for each incremental duct burner fuel energy input equivalent to 1% of gas turbine fuel consumption (i.e., 0.01), the BC efficiency should increase by 0.25–0.4 percentage points (i.e., 0.0025–0.004). The actual derivative for advanced 3PRH combined cycle systems is approximately 0.1 percentage point (i.e., 0.001).

Equation 6.56 suggests that the "hurdle" is less severe at high levels of duct firing. This, of course, brings up the question of "how high" one can go in duct firing. This is not a question amenable to a "one size fits all" generic answer. First of all, consider that duct firing is an "off-design" operation, which is strongly dependent on the hardware design limitations and system control philosophy. The controlling design limit is the "swallowing capacity" of the HP steam turbine. Without going into the fluid and thermodynamic details, duct firing increases HP steam production and, thus, steam flow through the HP turbine inlet. Since the combined cycle steam turbines operate in a "valves wide open" (VWO) mode, there is no mechanism to adjust the inlet flow area. Consequently, in accordance with fundamental compressible fluid flow principles, higher duct firing and higher steam flow lead to higher steam pressures.

The other limiting factor is the gas temperature downstream of the duct burner grid. Exhaust gas temperature rise across the duct burner can be related to duct burner fuel consumption as

$$\frac{\Delta T_{db}}{T_{exh}} \approx 1.4\,d, \tag{6.57}$$

where T_{exh} is the gas turbine exhaust temperature. State-of-the-art F, G, H and J class gas turbines have exhaust temperatures approaching 1,200°F (~650°C). Thus, at d ~ 0.2 (i.e., duct burner fuel consumption is equal to one-fifth of the gas turbine fuel consumption), gas temperature downstream of the duct burner is around 1,450°F (~800°C). Above 1,650°F, water-cooled "membrane" wall construction with unfinned ("bare") tube rows downstream is requisite (i.e., increased installed equipment cost). In order to reach that temperature with an advanced gas turbine, one would need duct firing above d ~ 0.3.

Typical steam cycle impact of duct firing is depicted by the linear curves in Figure 6.18. With duct firing level of d = 0.2, HP steam pressure increase is about 0.65; i.e., if the unfired HP steam pressure is P, fired HP steam pressure is 1.65 × P. This is the key dynamic in fired GTCC design.

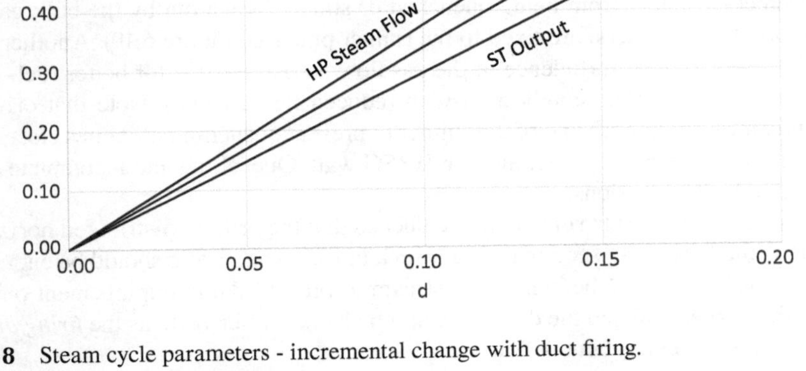

FIGURE 6.18 Steam cycle parameters - incremental change with duct firing.

In order to achieve maximum output gain from duct firing, unfired steam cycle design should be held back significantly. For example, state of the art in steam cycle design is about 2,500 psia HP steam pressure. If this is achieved via maximum firing, it follows that the unfired HP steam pressure is 2,500/1.65 ~ 1,500 psia.

In conclusion, duct firing beyond d = 0.2–0.25 is probably not feasible to strike a good balance between fired (typically on hot summer days) and unfired performance for combined cycles with advanced gas turbines and 3PRH (or maybe 2PRH) steam cycles. This unequivocally precludes the possibility of heat rate improvement with duct firing in large-scale combined cycle power generation.

6.3.1 Practical Considerations

When duct/supplementary firing is to be implemented in an HRSG for power augmentation, the following items should be carefully evaluated:

- Increased CO and NOx emissions (and their impact on SCR size/cost and pressure loss)
- Increased HRSG cost due to superheater tube alloys to handle high gas temperatures
- Maximum allowable gas temperature
- Location of the duct burner grid
- Fuel to be used in the duct burner.

Internal HRSG wall lining with insulation and metallic alloy sheet retainers is adequate up to 1,400°F–1,500°F (gas temperature downstream of the burners). Temperatures above 1,500°F would require that the HRSG walls are lined with refractory materials. Above 1,800°F, a screen section and different type of insulation may be used by the OEMs (i.e., increased installed equipment cost). Another option at such high duct firing temperatures is water-cooled membrane walls, which would be appropriate for small-size cogen applications.

The most common duct burner site is upstream of the HP evaporator, which will boost the HP steam production at the expense of IP and LP steam production. Typically, the duct burners are located in between the reheat superheater tube banks to ensure maximum steam production in the HP evaporator while requiring minimal attemperation to maintain maximum allowable HP and HRH steam temperatures.

A flow distribution device is needed upstream of the duct burner grid to facilitate uniform flow distribution. This and the duct burner itself will add to the gas pressure drop in the HRSG (i.e., gas turbine exhaust loss). Another advantage of placing the duct burner grid between reheater tube banks is that the superheater tube banks upstream of the duct burner act as flow distribution devices.

The duct burner "grid" is formed by "runners" arranged in vertical direction perpendicular to the gas turbine exhaust gas flow. Each runner can be thought of as a pipe with orifices evenly spaced along its length. Fuel gas is distributed through the runner to the orifices, which are essentially the "burners". Fuel ejected from the orifices mixes with the exhaust gas to form a diffusion flame. In order to protect the flame from being quenched or simply blown out by the bulk gas flow, "bat wing" protectors (flame holders) are fixed to the runner pipe (see Figure 6.19). Another function of the flame holders is to create turbulence in the gas flow across its edge for better fuel–gas mixing. This helps to achieve complete combustion with reduced flame length. Note that orifices should start sufficiently away from each end of the runner to prevent quenching of flames closer to the flow by "cold" gas flow between the runner and the HRSG wall. Quenching and incomplete combustion increases CO and UHC emissions.

There are typically five or six runners in the duct so that the orifices distributed horizontally and vertically across the flow cross-sectional area form a *burner grid*. There should be enough distance between the runners and the tube bank downstream to prevent flame impingement on tubes. The area between the burner grid and the downstream tube bank is referred to as the *firing duct*. Typical firing duct length is 12–18 ft (about 4–6 m).

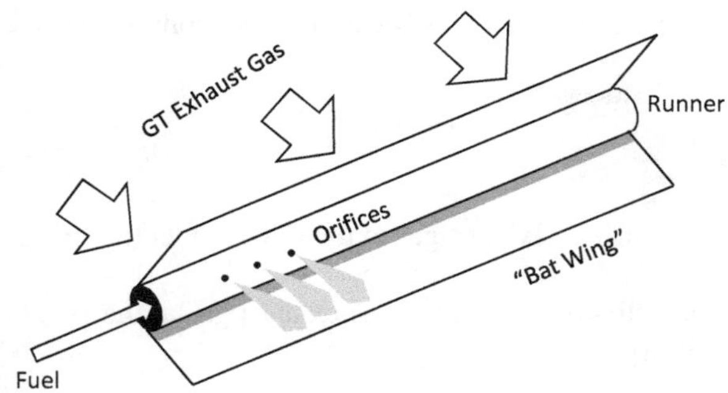

FIGURE 6.19 Duct burner runner schematic.

The operation of the duct burner is controlled by the *burner management system*. Unlike the DLN combustor of the gas turbine, the duct burner is not a sophisticated piece of equipment. Nevertheless, due to the complex nature of the gas flow inside the HRSG (i.e., large volume flow across a large cross-sectional area), the system is designed for a well-defined set of input parameters, i.e., exhaust gas flow, temperature, composition (especially O_2 and H_2O content of the gas) and velocity. In addition to the absolute values, even distribution of the input parameters across the flow area is of utmost importance for complete combustion and thermal performance (quantified by additional steam production per unit duct burner heat input). If the duct burner is designed to accommodate gas turbine loads below 100%, change in exhaust gas conditions with load makes the burner management even more complicated. Just to give an idea, consider that a duct burner designed to 1,425°F firing temperature at 100% gas turbine load would fire to 1,600°F at 50% load. At the same time, however, lower exhaust gas flow rate (by about 30%) and gas velocity would lower the convective heat transfer coefficient on the gas side and hamper the heat transfer downstream (leading to higher metal temperatures). In addition, the design of the duct burner system must comply with NFPA 85,[7] which defines the required permissives, safety precautions, parameters to be monitored, alarms, etc.

Typical duct burner problems identified by Electric Power Research Institute (EPRI) are[8]

- Condensation and corrosion in fuel lines
- Exhaust gas flow distribution
- Fuel gas flow maldistribution
- Flame impingement on downstream tube bank.

Prevention of such problems requires visual inspection of flames during operation (through ports provided for this purpose), monitoring of gas temperatures and drainage of fuel lines. During the HRSG outage, the system is checked for warped flame holders, oxidation/corrosion of the parts, sagging of the runners, signs of flame impingement on downstream superheater tubes and other damage.

6.3.2 Aeroderivative Gas Turbine Combined Cycle

In this section, a quantitative investigation of the claim made above is undertaken (i.e., duct firing in the HRSG can indeed *improve* the combined cycle efficiency).

[7] National Fire Prevention Association (NFPA) 85, "Boiler and Combustion System Hazard Code," is the main standard for fire/explosion protection of HRSGs and duct burners.

[8] Bill Carson (EPRI), 2015, "Duct burner operation beyond original design," Energy-Tech, August 2015 issue, pp. 5–8.

Ignoring losses and auxiliary loads, for the steam BC of a combined cycle, it can be shown that

$$\eta_{bc} = \varepsilon_{bc} \frac{\dot{A}_x}{\dot{Q}_x}, \tag{6.58}$$

where

$\dot{A}_x$ = Total *exergy* of gas turbine exhaust gas at the HRSG inlet (no firing) or duct burner exit (kWth)

$\dot{Q}_x$ = Total *energy* (i.e., enthalpy) of gas turbine exhaust gas at the HRSG inlet (no firing) or duct burner exit (kWth)

Note that specific exergy (per unit mass of fluid) is a thermodynamic property defined as

$$a = (h - h_0) - T_0 (s - s_0) \tag{6.59}$$

where h and s denote specific enthalpy and specific entropy, respectively, and the subscript 0 denotes the so-called "dead state" (i.e., when the fluid in question is at equilibrium with its surroundings with zero useful work generation potential). This was covered in more detail in Chapter 4.

Selecting the same thermodynamic state (e.g., ISO conditions) for "dead state" and zero enthalpy and ignoring the small change in mass flow rate and composition in the presence of a duct burner, Equation 6.58 can be written as

$$\eta_{bc} = \varepsilon_{bc} \left(1 - \frac{T_0 (s_x - s_0)}{(h_x - h_0)} \right). \tag{6.60}$$

It can be easily verified by a heat balance simulation software (e.g., Thermoflow, Inc.'s THERMOFLEX®) that the second term is practically constant with changing d. Consequently, by taking the derivative of both sides of Equation 6.60,

$$\frac{\partial \eta_{bc}}{\partial d} \propto \frac{\partial \varepsilon_{bc}}{\partial d}. \tag{6.61}$$

In other words, the determinant of the impact of duct firing on combined cycle steam BC efficiency is its impact on BC *exergetic* efficiency. A formula for the BC exergetic efficiency is given in the Appendix as a function of gas turbine exhaust temperature (Equation C.1).

It can be easily verified by the left-hand portion of the curve in Figure C.1 that, with aeroderivative gas turbines, i.e., high cycle pressure ratio and low exhaust temperature, there is a strong improvement in ε_{bc} with duct firing. In order to demonstrate the veracity of this observation, a hypothetical example is constructed using Thermoflow's GT PRO® heat balance simulation software.

The selected gas turbine is Siemens (Rolls-Royce) Trent 60D with the following cycle parameters and performance (at ISO baseload):

- Cycle pressure ratio of 34.6 with 422°C exhaust
- Dry low emission (DLE) combustor
- 54.6 MWe and 43.2% LHV efficiency.

A cost-effective, unfired BC for this gas turbine is

- 1PNR with 22.5 bara, 396°C steam
- Air-cooled condenser (ACC) at 68 mbar
- 12.5 MWe STG output.

Net combined cycle performance is 63.9 MWe and 51.38% LHV efficiency. The fired BC is defined as follows:

- Same heat rejection system
- Duct firing to 816°C
- 1PRH BC with 166.5 bara/600°C/600°C steam conditions.

Net combined cycle performance for the fired case is 103.5 MWe and 51.74% LHV efficiency. A comparison of the cases highlighting key parameters of interest is given in Table 6.17.

Based on the data in Table 6.17, $\partial\eta_{bc}/\partial d$ is calculated as 0.315 and satisfies the inequality in Equation 6.54. Consequently, there is 0.85 percentage points improvement in combined cycle gross efficiency but drops to 0.36 percentage points on a net basis. The significant difference between gross and net efficiencies is mainly due to the large increase boiler feed pump power consumption, which is driven by a combination of higher pumping head and higher feedwater/steam flow (see Table 6.18).

Improvement in aeroderivative combined cycle performance (~40 MWe and 0.36 percentage points) via duct firing is unassailable. It is predicted by fundamental thermodynamics and verified by detailed heat and mass balance simulation. The key question, of course, is whether it is cost-effective or not. As expected, there is a significant cost increase in major steam cycle equipment, i.e., STG, HRSG and ACC. Numbers obtained from GT PRO's cost estimation add-in PEACE® are summarized in Table 6.19.

It is also expected that installation and erection of larger and heavier equipment, foundations, piping and BOP equipment (especially feed pumps) will be higher for the fired GTCC. Estimates obtained from PEACE® are summarized in Table 6.20, which projects an installed cost increase of ~$40 million or about $1,000 per additional kilowatt of electric power.

Whether $1,000/kWe (incremental) is acceptable or not depends on the financial and economic criteria on a site and project basis. Without a rigorous (i.e., AACE Class 4 level) cost estimate and

TABLE 6.17
Fired and Unfired Aeroderivative GTCC Performance Comparison

	Unfired	Fired
GT efficiency	42.45%	42.45%
GT fuel consumption, kWth	124,266	124,266
Duct burner fuel consumption, kWth	NA	75,751
Exhaust gas into HRSG, °F	802	1,500
BC exergetic efficiency	47.23%	73.88%
GT exhaust energy, kWth	69,387	69,387
GTG output, kWe	52,751	52,751
STG output, kWe	12,472	53,940
HRSG effectiveness	62.09%	90.84%
ST efficiency	28.95%	40.90%
BC first-law efficiency	17.97%	37.15%
CC gross efficiency	52.49%	53.34%
RHS of Equation (7)	NA	0.291
ΔT_{db}	NA	701
$\Delta T_{db}/T_{exh}$	NA	0.87
d	NA	0.61

TABLE 6.18
Fired and Unfired Aeroderivative GTCC Auxiliary Load Comparison

	Fired	Unfired	Δ
GT fuel compressor(s)	287	287	0
HRSG feed pump(s)	1,157	64	1,092
Condensate pump(s)	21	9	12
ACC fans	569	222	347
Motors	451	335	116
Miscellaneous (GTG)	109	109	0
Miscellaneous (STG)	29	7	22
Miscellaneous (plant)	53	33	21
Total	**2,675**	**1,065**	1,610
Transformer losses	534	326	207
Total	**3,209**	**1,391**	1,817
Percent of gross output	3.0%	2.1%	

TABLE 6.19
Fired and Unfired Aeroderivative GTCC Major Equipment Cost Comparison (2015 Dollars)

	Unfired	Fired	Δ
GTG	21,350,000	21,350,000	0
STG	5,150,000	16,950,000	11,800,000
HRSG	4,200,000	12,350,000	8,150,000
ACC	3,150,000	7,000,000	3,850,000
Fuel Gas Compressor	1,150,000	1,150,000	0
CEMS	400,000	400,000	0
DCS	250,000	400,000	150,000
Electrical (Transformer)	1,750,000	2,550,000	800,000
Electrical (Generator)	500,000	800,000	300,000
Total	37,850,000	62,900,000	25,050,000

TABLE 6.20
Fired and Unfired Aeroderivative GTCC Total Installed Cost Comparison (2015 Dollars)

	Unfired	Fired	Δ
Major equipment	37,850,000	62,900,000	25,050,000
Pumps, tanks, etc.	1,850,000	3,550,000	1,700,000
Civil	4,200,000	6,950,000	2,750,000
Mechanical and piping	4,700,000	9,900,000	5,200,000
Electrical	1,825,000	3,150,000	1,325,000
Buildings and structures	3,000,000	4,000,000	1,000,000
Engineering and startup	6,000,000	8,300,000	2,300,000
Total	59,425,000	98,750,000	39,325,000

financial pro forma, the answer is simply unknowable. What one can do at a conceptual study phase as presented herein is to get a rough idea whether the concept has a realistic chance. For that, "maximum acceptable additional capital cost" (MACC) is estimated using the method described in detail in Chapter 13 (see Section 13.4). The assumptions used to calculate MACC in this particular example are:

- 4,000 h/year of operation
- Capital charge factor of 16% (typical of an independent power producer or IPP)
- $5 per million Btu (higher heating value or HHV) natural gas fuel
- Base specific cost of $1,200/kWe
- Levelization factor of 1.169.

Δ Capex	$39,325,000	
Δ Net efficiency	0.37%	
Δ Net output	39,650	kWe
Δ Heat consumption	75,738	kWth
MACC	$48,384,968	

Calculated MACC is $48.4 million, whereas estimated increase in capex is $39.3 million. In other words, the concept of highly duct-fired aeroderivative combined cycle with nearly 40 MWe extra output is not a "pipe dream" and deserves a closer scrutiny with the proverbial "sharper pencil".

6.4 SUPERCRITICAL BOTTOMING CYCLE

In a state-of-the-art 3PRH HRSG with an advanced class gas turbine, the largest BC exergy loss is the HRSG irreversibility, equivalent to about 10% of the gas turbine exhaust exergy, mainly driven by the constant pressure, p, and temperature, T, boiling characteristic of the cycle working fluid (H_2O). One possible way to reduce the HRSG irreversibility is a variable composition working fluid, e.g., ammonia–water as proposed by Kalina [19], which did not pan out as a feasible technology so far. Another possibility is a supercritical HP section (*once-through* instead of drum-type boiler) with steam pressure well above the critical pressure of H_2O, 221.2 bar [20]. Supercritical steam technology is widely used in fossil power plants with pulverized coal-fired boilers and feedwater-heated steam turbines (e.g., 250 bar, 540°C–560°C). USC steam technology with steam pressures up to 300 bar and temperatures up to 700°C is in development. In a coal-fired boiler with flue gas temperatures approaching 2,000°C, generation of 250 bar steam in large quantities commensurate with efficient steam turbine steam-path design (i.e., high volumetric flow of steam) is thermodynamically straightforward. In a 3PRH HRSG with 600°C–650°C exhaust gas, achieving the same with reasonable heat exchanger surface area (size and cost) is a considerable challenge.

The performance impact that can be obtained from a supercritical steam BC is relatively straightforward to estimate from an entitlement perspective. Using the definition of BC exergetic efficiency introduced earlier, the relationship between unfired combined cycle efficiency, η_{CC}, and BC *exergetic* efficiency, ε_{BC}, is obtained via partial differentiation:

$$\frac{\partial \eta_{CC}}{\partial \varepsilon_{BC}} = \frac{\dot{m}_{exh} \cdot a_{exh}}{HC} \cdot (1-\alpha) = \frac{\dot{A}_{exh}}{HC} \cdot (1-\alpha). \tag{6.62}$$

In Equation 6.62, $\dot{m}_{exh}$ and a_{exh} denote gas turbine exhaust mass flow and specific exergy, respectively, whereas α is the total plant auxiliary load as a percentage of the net BC output (about 5%–6% based on 2% of combined cycle gross output) and HC is the heat (fuel) consumption of the gas turbine. STG output is related to the net BC output via

$$\dot{W}_{ST} = \frac{\varepsilon_{BC} \cdot \dot{A}_{exh}}{(1-\lambda_{BFP})}, \tag{6.63}$$

where λ_{BFP} is the boiler feed pump power consumption as a fraction of the steam turbine output, $\dot{W}_{ST}$. Ignoring the small effect of α and λ_{BFP} (~2%), Equation 6.62 can be rewritten as

$$\frac{\partial\eta_{CC}/\eta_{CC}}{\partial\varepsilon_{BC}/\varepsilon_{BC}} \approx \frac{1}{1+\dot{W}_{GT}/\dot{W}_{ST}}. \tag{6.64}$$

The relationship between exhaust exergy and total exhaust energy (i.e., enthalpy) is given by

$$\frac{a_{exh}}{h_{exh}} = 1 - \frac{T_0 \cdot (s_{exh} - s_0)}{(h_{exh} - h_0)} \approx 1 - \frac{T_0 \cdot \ln(T_{exh}/T_0)}{(T_{exh} - T_0)} \tag{6.65}$$

where the subscript 0 denotes a reference state (typically ISO ambient). Using standard Brayton cycle state-point notation, Equation 6.65 becomes

$$\frac{a_{exh}}{h_{exh}} = 1 - \frac{T_0 \cdot \ln(T_4/T_1)}{(T_4 - T_1)}.$$

Substituting Equation 4.10 into Equation 6.65 yields

$$\frac{a_{exh}}{h_{exh}} = 1 - \frac{T_1}{METL}, \tag{6.66}$$

where a and h represent exhaust gas exergy and enthalpy, respectively, with zero enthalpy and dead state reference at 59°F. From Equations A.2 and A.5 with T_{exh} = 650°C (1,202°F) and T_0=15°C (59°F),

$$a_{exh} = 0.1961 \cdot 1202 - 86.918 = 148.8 \text{ Btu/lb}$$

$$h_{exh} = 0.3003 \cdot 1202 - 55.576 = 305.4 \text{ Btu/lb}$$

the ratio of exhaust exergy to exhaust enthalpy is 0.4872. Using Equation 6.65 with temperatures, the ratio is found as 0.4717. If the dead state is chosen as 77°F as in some studies published in the literature (also in Thermoflow's GT PRO® software), from Equation 6.65, the ratio is found as 0.4609.

For advanced class gas turbines, exhaust energy is about 60% of the gas turbine heat consumption (i.e., ~$1 - \eta_{GT}$) and the gas turbine output, $\dot{W}_{GT}$, is approximately twice that of the steam turbine, $\dot{W}_{ST}$. Thus, per Equation 6.64, each additional percentage point in ε_{BC} is worth approximately 0.25 η_{CC} percentage points. For older generation E and F class units (i.e., gas turbine efficiency in low to mid-thirties), gas-to-steam turbine output ratio is closer to 1.5 and 1.75, respectively, so that each additional percentage point in ε_{BC} can be as high as about 0.3 η_{CC} percentage point.

A supercritical HP "steam" section in the 3PRH HRSG improves the ε_{BC} via reduction of the HRSG irreversibility, which is driven by the mismatch in gas and steam temperatures. Other exergy loss mechanisms are unaffected with the exception of the total steam turbine exergy loss, which will be discussed later. For combined cycle power plants with advanced class gas turbines and state-of-the-art 3PRH BCs, the HRSG irreversibility is ~10% of the gas turbine exhaust energy. The heat release diagram in Figure 6.20 shows that only a fraction of that total is attributable to the HP evaporator. The "lost" work (i.e., irreversibility) in the HP evaporator is given by Equation 6.6, which is repeated here for convenience,

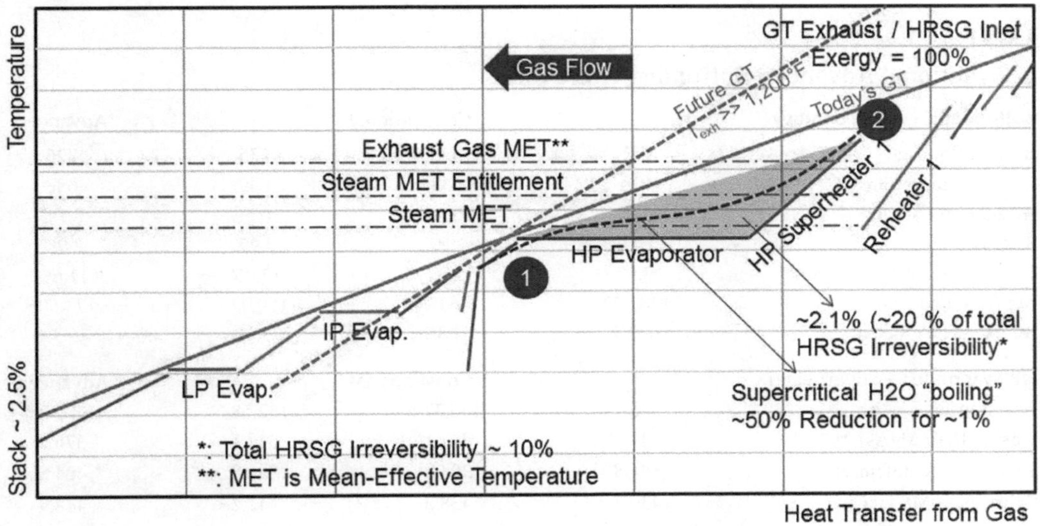

FIGURE 6.20 Combined cycle bottoming cycle heat release diagram for a typical 3PRH system.

$$\dot{I}_{HRSG} = \frac{T_0}{\overline{T}_{stm}}\left(1 - \frac{\overline{T}_{stm}}{\overline{T}_{gas}}\right)\dot{Q}_{rec}.$$

As demonstrated earlier in Chapter 4, METs can be calculated using the ratio of enthalpy, h, and entropy, s, changes between the inlet and exit conditions of each heat exchanger stream, i.e. ignoring the relatively small drop in pressure

$$\overline{T} = \frac{(h_2 - h_1)}{(s_2 - s_1)}. \tag{6.67}$$

The entitlement value for the MET of steam between state-points 1 and 2, i.e., entrance and exit of the combined evaporator + superheater section of Figure 6.20, respectively, is given by

$$\overline{T}_{stm,\,max} = \frac{T_2 - T_1}{\ln(T_2/T_1)}. \tag{6.68}$$

Equation 6.68 is the LMTD between water/steam temperatures T_1 and T_2 for a *hypothetical* heat transfer process that can be represented as a straight line between states 1 and 2 on the heat release diagram (i.e., constant specific heat, c_p, and no change in pressure). In order to put some numbers into the discussion, two 3PRH BC systems are considered:

1. *Conventional*: 125 bar – 565°C HP steam conditions, 8.3/5.6°C HRSG pinch/subcool temperature deltas and 50 mbar condenser pressure.
2. *Advanced*: 165 bar – 600°C HP steam conditions, 6.7/2.8°C HRSG pinch/subcool temperature deltas and 40 mbar condenser pressure.

Several cases are run using Thermoflow's GT PRO® software. Conventional BC is evaluated for a range of exhaust temperatures (590°C–650°C). The results from selected cases are summarized in Table 6.21.

As illustrated by the data presented in Table 6.21, out of a total of ~13.5% heat recovery lost work (of which ~10% is due to HRSG irreversibility), only a fraction can be attributed to the HP

TABLE 6.21
Conventional Advanced Bottoming Cycle Cases

Bottoming Cycle Technology		Conventional		Advanced
Exhaust flow, kg/s	635	635	635	726
Exhaust temperature, °C	616	638	649	638
HRSG irreversibility	10.0%	10.3%	10.5%	8.7%
Stack exergy	3.4%	3.1%	2.9%	2.9%
Total heat recovery lost work	13.5%	13.4%	13.4%	11.6%
Net BC output, kW	138,522	146,694	150,779	174,506
BC exergetic efficiency	73.2%	73.3%	73.4%	76.3%
HP Evaporator + Superheater		**Conventional**		**Advanced**
Heat transfer, kWth	134,119	146,379	152,583	148,721
Steam MET ($\Delta h/\Delta s$), °C	347.1	347.1	347.1	376.6
Steam MET entitlement, °C	394.8	394.8	394.8	424.8
Exhaust gas MET, °C	430.5	438.3	442.2	448.9
Lost work, kW (%)	7,379 (3.71%)	8,678 (4.13%)	9,364 (4.33%)	6,695 (2.78%)
Lost work entitlement, kW (%)	3,178 (1.6%)	4,094 (1.95%)	4,585 (2.12%)	2,398 (1.0%)
Maximum gain	2.11%	2.18%	2.21%	1.79%

evaporator. Therefore, if one could imagine a special working fluid and an extremely large heat exchanger surface area so that at each point during the heat transfer process, the Q-T lines of the working fluid and the hot exhaust gas are approximately parallel to each other, the additional net BC power generation would be equal to about 2.2% of $\dot{A}_{exh}$. This signifies the theoretical upper limit of the benefit due to a reduction of HP evaporator irreversibility mechanism, which cannot be achieved in a realistically designed heat exchanger. Thus, the conclusion is that the upper theoretical limit to the benefit of switching to a supercritical HP section in a conventional 3PRH HRSG is approximately 15% reduction in the total HRSG "lost work", which is equivalent to a net η_{CC} benefit of slightly higher than 0.5 percentage points ($0.25 \cdot 2.2 = 0.55$).

While this may seem a very meager benefit for such a drastic change in pressure (by a delta of ~85 to 125 bar), one should keep in mind that this is the only exergy loss/destruction mechanism *solely attributable* to a supercritical HP "boiler" with all other design parameters and criteria kept constant. In other words, the system retains same final steam temperatures, same heat exchanger design criteria (i.e., pinch, approach and subcool temperature deltas), same heat exchanger arrangement and same "everything else" as in the base, conventional state-of-the-art 3PRH HRSG design. With the same gas turbine (631°C–825 kg/s exhaust) in a 2 × 2 × 1 combined cycle configuration, Zhang et al. [20] reported that the efficiency delta between a subcritical and supercritical BC was 0.4 η_{CC} points. However, steam temperature difference between the supercritical and subcritical cases in the cited paper, i.e., 593°C versus 565°C, respectively, accounts for +0.20 η_{CC} points and is equally applicable to a subcritical BC. Thus, when this delta is excluded, the benefit from a supercritical BC (and *nothing else*) as reported in Zhang et al. [20] is only 0.20 net η_{CC} percentage points. Needless to say, some earlier claims of nearly one full percentage point improvement in 3PRH combined cycle efficiency due to a supercritical HP section are simply not in the realm of physical possibility [21].

A noteworthy exception is the study published by Bolland, which confirms the findings herein [22]. His 3PRH supercritical BC with an older generation F class unit (Siemens V94.3A) improved ε_{BC} by 1.8 percentage points, which was solely attributable to HRSG irreversibility reduction (see Figure 11 in Bolland [22]). This is about 60% of the lost work recovery entitlement per Equations 6.6 and 6.67 (or 6.68) using data extracted from Figure 2j of Bolland [22]. The HRSG irreversibility reduction translated into ~0.50 percentage points of improvement in η_{CC} in accordance with

Equation 6.64 and GT/ST output ratio of ~1.75. The requisite increase in HRSG heat transfer surface area was calculated as 25% or about 50% per incremental η_{CC} percentage point.

6.4.1 Feasibility of Supercritical Bottoming Steam Cycle

For a conventional (subcritical) 3PRH BC, maximum improvement potential afforded by a supercritical HP section is shown to be 2.2 ε_{BC} or 0.55 η_{CC} points. Graphically, this is represented by the shaded triangular area in Figure 6.20. The dashed line represents the H_2O "boiling" curve at a high supercritical pressure (~300 bar). From a technological and economic feasibility perspective, perhaps ~50%–60% of that potential is realistically achievable (certainly, not even close to 100%). Thus, the maximum practical gain from a supercritical HP section is expected to be 0.25–0.35 net η_{CC} points for state-of-the-art 3PRH systems.

Note, however, a much more significant impact is easily achievable by an advanced steam cycle (still subcritical) and larger HRSG along with lower condenser pressure (as permitted by the site conditions). From the second and fourth columns of Table 6.21, the increase in ε_{BC} between a conventional and advanced BC is 2.9 points (0.73 points in η_{CC}), i.e., *larger than the entitlement for the supercritical HP section.* (Of that increase, nearly half or 0.35 points in η_{CC} is attributable to the condenser pressure.) Considering that the advanced steam cycle conditions (165 bar, 600°C) are currently used with the new advanced class large gas turbines, all one needs is to increase the HRSG (and condenser) size as long as it pays for itself. (GT PRO-estimated HRSG heat transfer surface area increase was 38%, which translates into ~100% per incremental η_{CC} percentage point.) Furthermore, once the advanced steam cycle becomes the norm, the entitlement for a supercritical BC becomes even smaller, i.e., only 1.8 ε_{BC} points (0.45 points in η_{CC}), in which case it will be even less palatable.

One should also consider the impact of supercritical HP steam on the steam turbine efficiency due to reduced volume flow (due to much higher steam density) and its detrimental effect on steam-path efficiency. This can be overcome either by large multiple-gas turbine configurations (e.g., $2 \times 2 \times 1$, $3 \times 3 \times 1$) to generate enough steam, supplementary firing in the HRSG (which, obviously, decreases the combined cycle efficiency and defeats the purpose of introducing the supercritical HP feature in the first place – see section 5.1.2) or designing a geared, high-speed HP turbine section (i.e., a "topping" HP turbine) [23]. Unless proper design modifications are undertaken, supercritical HP steam will in fact increase the steam turbine exergy loss and limit the aforementioned benefit entitlement.

Another detrimental impact of a supercritical HP section is the increase in parasitic power loss associated with feedwater pumping. This can be significant and tends to obfuscate the bottom line in performance evaluation using complex models. Since the BC exergetic efficiency is based on the *net* power (i.e., STG output *minus* boiler feed pumps) of the BC control volume, the physics-based approach described herein circumvents this difficulty altogether.

It must be added that a supercritical HRSG can be padded with other design features such as [23]

- Double reheat, i.e., reheating of the topping HP exhaust steam to conventional HRH steam temperature
- CRH steam entry below HP pinch [24]
- Parallel reheat-superheater and HP economizer sections
- Parallel IP–LP superheater sections.

These design features will further improve the HRSG exergy utilization significantly, albeit at a significant increase in heat transfer surface area and cost. In that sense, it is possible to interpret them as a "brute force" approach to HRSG performance improvement by simply pouring more money into the design. Furthermore, as already emphasized several times herein, with the exception of the double reheat, they are equally applicable to a conventional, subcritical BC and should be properly discounted when making apples to apples comparisons between the two technologies.

REFERENCES

1. Volpi, G., Penati, M., Silva, G., 2005, Heat Recovery Steam Generators for large combined cycle plants (250 MWe GT output): Experiences with different design options and promising improvements by once-through technology development, *Power Gen Europe 2005*, June 28–30, 2005, Milan.
2. Fontaine, P., 2003, Cycling tolerance: Natural circulation vertical HRSGs, *2003 International Joint Power Generation Conference*, June 16–19, 2003, Atlanta, GA.
3. Brückner, J., Schlund, G., 2011, Pego experience confirms BENSON as proven HRSG technology, *Mod. Power Syst.*, Vol. 31, No. 6, pp. 21–24.
4. TETRA Engineering Group, 2003, *Guidelines for the Operation and Maintenance of HRSGs*, Ed. R. Swanekamp, The HRSG Users Group, Weatogue, CT.
5. El-Wakil, M.M., 1984, *Powerplant Technology*, International Edition, McGraw-Hill Book Co., New York.
6. Balling, L., Termühlen, H., Baumgartner, R., 2002, Forty years of combined cycle power plants, IJPG2001-26111, *2002 International Joint Power Generation Conference*, June 24–26, 2002, Phoenix, AZ.
7. Welty, J.R, Wicks, C.E., Wilson, R.E., 1984, *Fundamentals of Momentum, Heat and Mass Transfer*, 3rd Edition, James Wiley & Sons, New York.
8. Holman, J.P., 1981, Heat Transfer, 5th Edition, Mc-Graw-Hill Company, New York.
9. ESCOA, 1979, *Turb-X HF and Soldfin HF Rating Instructions*, ESCOA, Pryor, OK.
10. Poling, B.E., Prausnitz, J.M., Reid, R., 1989, *The Properties of Gases and Liquids*, McGraw-Hill Company, New York.
11. Verhoff, F.H., Banchero, J.T., 1974, Predicting dew point of flue gases, *Chem. Eng. Prog.*, Vol. 70, No. 8, pp. 71–72.
12. Khalaf, Z.R., Basler, B., 2002, Kalaeloa: Combined cycle power station burning low sulfur fuel oil in its ninth successful year, *J. Eng. Gas Turbines Power*, 124(3), pp. 534–541.
13. Gabrielli, F., Goodstine, S., Mastronarde, T., 2002, Cold-end corrosion in HRSGs, *Power Plant Chem.*, Vol. 4, No. 3, pp, 148–153.
14. Finckh, H.H., Pfost, H., 1991, Development Potential of Combined-Cycle (GUD) power plants with and without supplementary firing, ASME Paper 91-GT-227, *International Gas Turbine and Aeroengine Congress and Exposition*, June 3–6, 1991, Orlando, FL.
15. Kehlhofer, R.H., Warner, J., Nielsen, H., Bachmann, R., 1999, *Combined-Cycle Gas & Steam Turbine Power Plants*, 2nd Edition, PennWell, Tulsa, OK.
16. Jordal, K., Fridh, J., Hunyadi, L., Linder, U., 2002, New possibilities for combined cycles through advanced steam cycles, ASME Paper GT-2002-30151, *ASME Turbo Expo 2002*, June 3–6, 2002, Amsterdam, the Netherlands.
17. Rice, I.G., 1997, Split Stream Boiler for Combined Cycle Plants, US Patent 5,628,183.
18. Kehlhofer, R., 1991, *Combined-Cycle Gas & Steam Turbine Power Plants*, The Fairmont Press, Inc., Lilburn, GA.
19. Kalina, A.I., 1984, Combined cycle system with novel bottoming cycle, *J. Eng. Gas Turbines Power*, Vol. 106, pp. 737–742.
20. Zhang, W., Magee, J., Singh, H., Ruchti, C., Selby, G., 2012, HRSG development for the future, *PowerGen Europe 2012*, June 12–14, 2012, Köln, Germany.
21. Galopin, J.F., 1998, Going supercritical: Once-through is the key, *Mod. Power Syst.*, Vol. 18, p. 39.
22. Bolland, O., 1991, A comparative evaluation of advanced combined cycle alternatives, *J. Eng. Gas Turbines Power*, Vol. 113, pp. 191–197.
23. Tomlinson, L.O., Smith, R.W., Ranasinghe, J., Gülen, S.C., Rancruel, D.F., 2009, Supercritical Steam Combined Cycle and Method, US Patent Application Publication 2009/0090111 A1.
24. Smith, R.W., 2001, Multi-Pressure Reheat Combined Cycle with Multiple Reheaters, US Patent 6,220,013.

7 Heat Sink Options

What is a "heat sink"? For the simple cycle gas turbine, it is the ambient air. For the combined cycle gas turbine, it is the heat recovery steam generator (HRSG). Consequently, the heat sink for the gas turbine Brayton cycle becomes the heat source for the steam turbine Rankine cycle. The heat sink for the Rankine cycle is either the ambient air or an ambient water body (i.e., lake, river or ocean). In a simple cycle gas turbine, there is no intermediary between the heat ejection (i.e., the gas turbine exhaust) and the heat sink. In the Rankine cycle steam turbine, there is an intermediary between the heat ejection (i.e., the steam turbine exhaust) and the heat sink. The intermediary in question is the steam turbine condenser. Depending on the site ambient conditions and the prevailing environmental regulations, the condenser may be assisted by a cooling tower. These two types of power plant equipment are the subject matter of this chapter.

The reader is cautioned that this is not an equipment "tour guide", i.e., pictures of sample equipment and accompanying "this is X", "this is Y" type narrative. There are many "handbooks" where this information is available. For indispensable technical design, operation and maintenance information on heat rejection equipment, the best resources are the standards issued by respective technical institutes cited earlier in Chapter 2. Readers employed by industrial organizations are highly likely to have access to those standards and/or already familiar with them. All practitioners in the field of combined cycle design and optimization, including academia and research organizations, are encouraged to consult them for mission-critical information.

In terms of combined cycle design performance, one is primarily interested in the condenser pressure and the power consumption requisite to achieve it (in addition to size and cost of course). This can be accomplished by application of relatively straightforward thermodynamic correlations.

In terms of off-design performance, obtaining the same information requires considerably more effort including a basic knowledge of equipment operation. Especially with air-cooled condensers (ACCs) and cooling towers, this can be quite tricky unless one is using a software package like GT PRO® or GateCycle.

Steam turbine condensers are classified into two groups based on the cooling medium: (i) water-cooled or (ii) air-cooled. Water-cooled condensers can be "open loop" or "closed loop". In the former, also referred to as a "surface condenser" in the trade literature, cooling water is taken from a natural reservoir to accomplish the cooling duty in the condenser, and afterwards, warm water is returned to the same reservoir. The only power consumer is the circulating water pump ("circ pump" in short). In a closed-loop system, condenser cooling water itself is cooled in a mechanical- or natural-draft cooling tower. In the case of a mechanical cooling tower, parasitic power consumption by the air circulation fans should be added to that of the circ pump.

From a performance perspective, there is no doubt that an open-loop water-cooled condenser is the best option. Unfortunately, returning warm condenser water back to the source can negatively impact marine life via elevated water temperatures near the plant discharge (e.g., killing fish and/or their eggs). Consequently, environmental protection organizations such as the US Environmental Protection Agency (EPA) strictly control and limit the location, design, construction and capacity of cooling water intake/discharge structures, which curtail the deployment of this option in the new-build power plants (e.g., Section 316(b) of the Clean Water Act). In fact, a study of Form EIA-860 data for 2008–2012 found that all gas turbine combined cycle (GTCC) power plants under construction or built in PJM with a specified cooling system (about 4 GW installed capacity out of a total of 6.1 GW) had a cooling tower installed (including forced/

induced and natural draft units).[1,2] It is not clear how many out of the remaining GTCC power plants with 2.1 GW installed capacity has an open-loop heat sink system. It is, however, safe to assume that the number is very low. (One possibility is a "brownfield" project on a former fossil fuel-fired power plant site with existing permits.)

Natural draft cooling towers are very large and expensive structures (they became – incorrectly, of course – symbol of large nuclear power plants) and thus unlikely to be used in new-build ("green-field") combined cycle power plants. One notable exception is the combined cycle power plant in Bouchain, France, with GE's advanced HA class 50-Hz gas turbine (went commercial in 2016) where the natural draft cooling tower of the retired coal-fired power plant is used for cooling the condenser cooling water of the combined cycle steam turbine (see Section 8.6.2).

Airflow in this type of tower is created by the same mechanism as in the HRSG stack. Warm, moist air (less dense) rises due to the density differential compared to the dry, cooler outside air (denser). The resulting *buoyancy* produces an upward current of air through the tower. The upward airflow is assisted by the hyperboloid shape of the tower.

Mechanical cooling towers operating on the same principle, i.e., evaporative cooling, as their natural-draft cousins are the next best option in terms of parasitic power consumption. In this tower, water flows down in the form of a thin film over the packing (or fill) and the air flows upward

i. *Pulled* by the suction action of the electric motor-driven fans at the top (induced draft)
ii. *Pushed* by the blower action of the electric motor-driven fans at the intake (forced draft).

Practically all combined cycle power plant cooling towers are of the induced draft type. There is a further subdivision of mechanical cooling towers: crossflow and counterflow. In counterflow designs, the water flows vertically downward through the fill as the air flows upward. In crossflow designs, airflow is horizontal. With the adoption of "film-type" fills, most prevalent design used in GTCC applications is counterflow due to its smaller footprint and lower pumping head.

A common variation of mechanical cooling towers for *plume abatement* is the hybrid wet/dry design. The plume is caused when the warm moist air discharged from the tower is cooled by the ambient air and the water vapor condenses. While not toxic or harmful to human health, it is undesirable because it can affect visibility and safety as well as public perception. In a hybrid wet/dry design, the circulating water first flows through a finned-tube section before distribution over the tower fill. The airflow from the dry and wet sections mix in the fan plenum. The mixed air is warmer and no longer saturated so that it does not create a plume when it mixes with the colder ambient air. Depending on the design point, the hybrid wet/dry cooling tower is significantly more expensive than a conventional cooling tower.

In areas with scarce water resources and/or drastic environmental regulations, the ACC is the most widely used option. The ACC is a once-through (or open-loop) system with atmospheric air replacing water as the heat sink. Its advantages, vis-à-vis its water-cooled cousin with a wet cooling tower, are elimination of makeup water and blowdown (BD) requirements as well as water freezing and water vapor plume issues. Its main disadvantage is the fact that air is inferior to water as a heat transfer medium. Therefore, a large volume of air has to be moved across finned tubes (in which the condensing steam is flowing) by the action of electric motor-driven fans. The result is high parasitic power consumption at comparatively high condensing steam pressures (about 3 in. Hg or much higher in hot climates). The most common configuration is the "A-Frame" with finned-tube

[1] The survey Form EIA-860 issued by the U.S. Energy Information Administration collects generator-level specific information about existing and planned generators and associated environmental equipment at electric power plants with combined nameplate capacity of 1 MW or larger (see www.eia.gov/electricity/data/eia860/).

[2] From "Cost of New Entry Estimates for Combustion Turbine and Combined Cycle Plants in PJM", prepared for PJM by The Brattle Group and Sargent & Lundy (May 15, 2014).

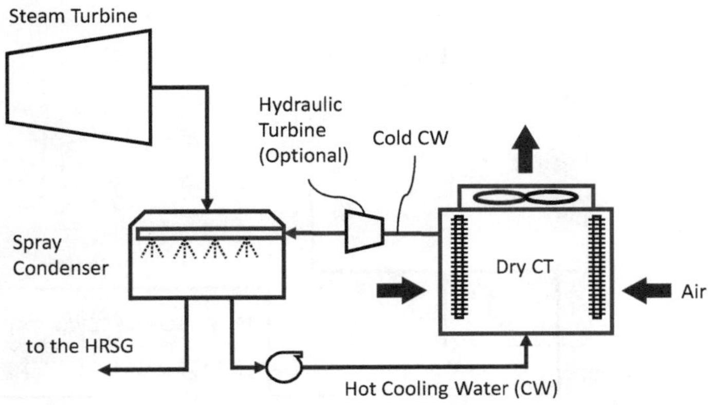

FIGURE 7.1 Heller system (hybrid cooling tower).

bundles extending at roughly 45° angle (i.e., the two legs of "A") from both sides of the steam header at the top.

One alternative system, which is far less common than the three basic types discussed above, is the Heller system, which was developed in Hungary in 1950s. The Heller system has two components: (i) a direct-contact spray condenser and (ii) a large air–water heat exchanger, i.e., "dry" cooling tower. The hot condensate is pumped from the spray condenser to the dry cooling water (except a small extraction diverted to the HRSG). Cold condensate is returned to the spray condenser as spray water. The system is shown in Figure 7.1.

The dry cooling tower and all of the piping operate at a positive pressure so that the air leak risk is minimal. In order to recover the pressure head, a hydraulic turbine is usually incorporated into the spray water return line (at additional cost of course). In comparison with an ACC, the Heller system is more expensive (installed equipment) but has lower parasitic power consumption by about 40% and 40% lower "area occupied by limit noise". It also has a lower minimum back pressure, i.e., about 1 in. Hg vis-à-vis about 2 in. Hg for the ACC. Thus, a life-cycle cost evaluation (plant net efficiency, installed cost, maintenance costs, etc.) is requisite to determine the appropriate heat sink for specific site ambient and loading conditions as well as operating profile (e.g., winter operating hours) and land availability.

7.1 WATER-COOLED SURFACE CONDENSER

The one and only source reference to be consulted for the surface condensers is HEI, Inc.'s *Steam Surface Condensers* [1]. While the theory is quite simple, actual calculations, especially for the off-design operation, can be tedious and iterative. Using a heat balance simulation software such as GT PRO® eliminates this difficulty for day-to-day tasks. Nevertheless, a fundamental understanding of the system is essential and can be accomplished by the use of quite simple formulae with a few simplifying assumptions.

The surface condenser is a special type of shell-and-tube heat exchanger. Steam exhausted from the steam turbine condenses at constant pressure (which is significantly below the atmospheric, i.e., the shell is practically a "vacuum chamber") and temperature on the cold tubes through which the cooling water flows. The latent heat of condensation of the condensing steam is picked up by the cooling water circulating through the tubes. In an *open-loop* system, the inlet (cold) water is pumped from a natural source such as a river or a lake, and the warm water is returned back to the source. In a *closed-loop* system, the warm outlet water is sent to a cooling tower where it is cooled again via direct contact with air before returning to the condenser inlet.

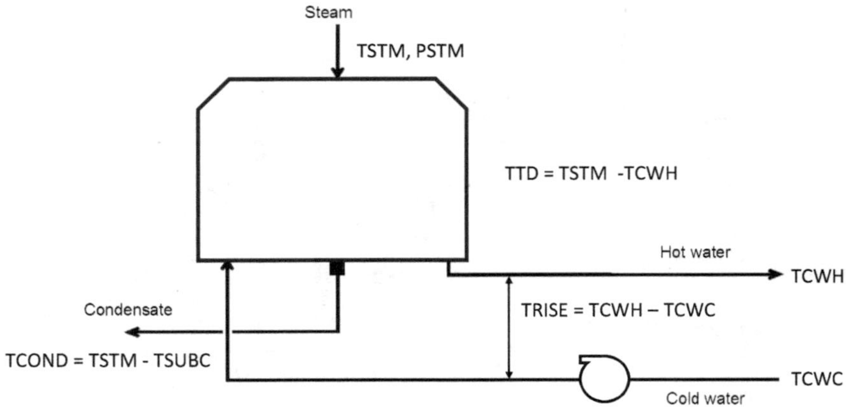

FIGURE 7.2 Schematic diagram of water-cooled steam surface condenser.

A simple schematic diagram of the surface condenser is provided in Figure 7.2. There are three important design/sizing parameters:

1. Terminal temperature difference (TTD)
2. Cooling water temperature rise (TRISE)
3. Condensate subcool (TSUBC).

The heat balance of the system is simple and can be written from three perspectives:

$$\text{Condensing steam}: \dot{Q}_C = \dot{m}_{stm}(1-y)h_{fg} \tag{7.1}$$

$$\text{Cooling water}: \dot{Q}_C = \dot{m}_{cw}c_{p,cw}\text{TRISE} \approx \dot{m}_{cw}\text{TRISE} \tag{7.2}$$

$$\text{Condenser design}: \dot{Q}_C = UA \cdot \Delta T_{LM}, \tag{7.3}$$

where y is the moisture fraction of steam, h_{fg} is the latent heat of evaporation at TSTM/PSTM, U is the condenser heat transfer coefficient (Btu/ft^2-h-F) and A is the condenser heat transfer surface area (ft^2). (The approximation in Equation 7.2 is valid when using the British units.) The overall heat transfer coefficient is UA in Btu/h-F.

Steam flow rate can be estimated as

$$\dot{m}_{stm} = \mu N_{GT}\dot{m}_{exh}, \tag{7.4}$$

where N_{GT} is the number of gas turbines in the combined cycle power plant. For estimation of μ as a function of gas turbine exhaust parameters, please refer to Section 6.2.3. The log-mean temperature difference is defined as

$$\Delta T_{LM} = \frac{\text{TRISE}}{\ln\left(\dfrac{\text{TSTM}-\text{TCWC}}{\text{TTD}}\right)}. \tag{7.5}$$

Cooling water "circ pump" power is given by

$$\dot{W}_{cwp} = \frac{\dot{m}_{cw}\Delta p}{\rho_{cw}\eta_{is}\eta_{mot}}, \tag{7.6}$$

where ρ_{cw} is the density of cooling water (62.4 lb/ft^3), η_{is} is the pump isentropic efficiency (75%–80%) and η_{mot} is the driver motor efficiency (90%–95%). Requisite pump head Δp is a function of the pressure drop in the condenser tube banks and the cooling water pipes. (When using US customary units with flow in lb/s, include the factor 0.195 on the right-hand side of Equation 7.6 to obtain pump power in kW.) Thus, in essence $\dot{W}_{cwp} \propto \dot{m}_{cw}$. From Equation 7.2, it is easy to see that, for a given bottoming cycle, parasitic power consumption of the cooling water circ pump is inversely proportional to the TRISE in the condenser. From Equations 7.3 and 7.5,

$$UA = \frac{\dot{Q}_C}{\Delta T_{LM}} = \frac{\dot{Q}_C}{TRISE} \ln\left(\frac{TSTM - TCWC}{TTD}\right) \quad (7.7)$$

which, recognizing the connection between TTD and TRISE for given TSTM and TCWC, can be rewritten as

$$\frac{UA}{\dot{Q}_C} = \frac{-1}{TRISE} \ln\left(1 - \frac{TRISE}{TSTM - TCWC}\right). \quad (7.8)$$

Plotting Equation 7.8 for typical values (TSTM = 101.7°F, TCWC = 65°F), one can see that at TTD values of 5°F or lower, the rise in $\frac{UA}{\dot{Q}_C}$ becomes exponential with adverse impact on condenser size and cost (see Figure 7.3). This should be considered as an upper limit for TCWH and TRISE (for given TCWC). It should also be noted that applicable environmental regulations can also impose an upper limit on TRISE, which increases the cooling water flow rate for given $\dot{Q}_C$ and leads to higher circ pump power consumption.

A typical condenser *heat release* diagram is shown in Figure 7.4. On the diagram, TCWM is mean-effective cooling water temperature, which is the logarithmic average of TCWC and TCWH. It can be shown that, however, for typical values of TCWC and TCWH,

$$TCWM = \frac{TCWH - TCWC}{\ln\left(\frac{TCWH}{TCWC}\right)} \cong \frac{1}{2}(TCWH + TCWC);$$

i.e., TCWM is equal to the arithmetic average of TCWH and TCWC (because cooling water specific heat is practically constant).

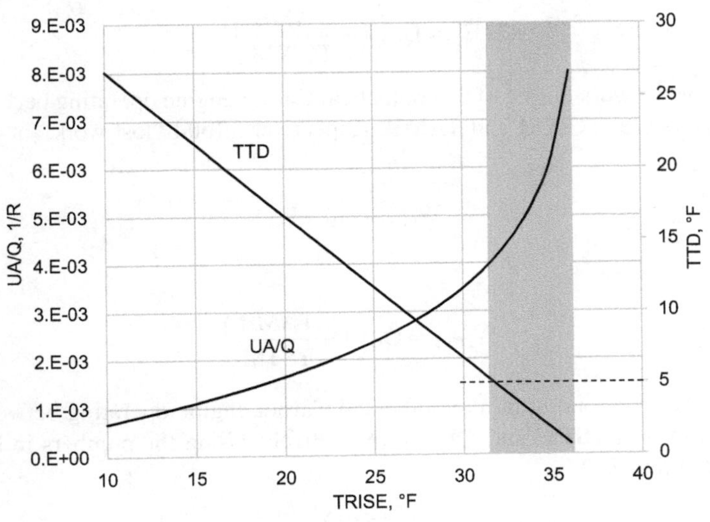

FIGURE 7.3 Plot of Equation 7.8 as a function of TRISE.

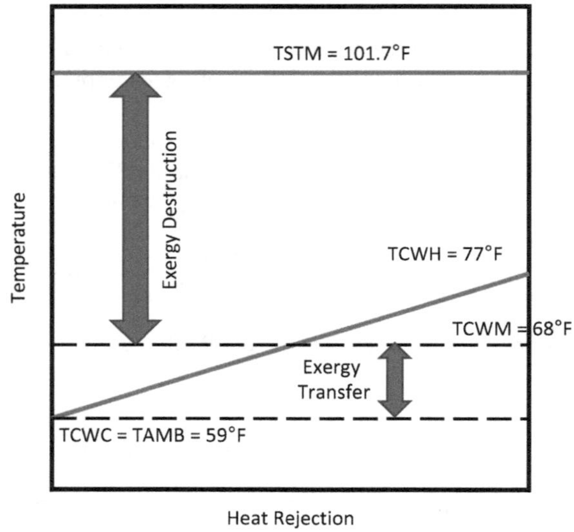

FIGURE 7.4 Heat release diagram of water-cooled steam surface condenser.

In Figure 7.4, two second law loss mechanisms in the condenser are identified:

1. *Exergy destruction* (irreversibility) due to temperature mismatch between "hot" and "cold" streams at TSTM and TCWM, respectively.
2. *Exergy transfer* out of the Rankine steam cycle via heat transfer to the cooling water.

Using the same approach in Section 6.1, exergy destruction in the condenser is quantified as

$$\dot{I}_C = \dot{Q}_C \frac{TAMB}{TCWM}\left(1 - \frac{TCWM}{TSTM}\right). \tag{7.9}$$

The exergy transfer via heat transfer to the cooling water is simply

$$\dot{E}_C = \dot{Q}_C \left(1 - \frac{TAMB}{TCWM}\right), \tag{7.10}$$

which, of course, is the work output of a hypothetical Carnot engine operating between hot and cold temperature reservoirs at TCWM and TAMB, respectively. Total "lost work" in the condenser is thus

$$\dot{W}_{LOST} = \dot{I}_C + \dot{E}_C \tag{7.11}$$

or

$$\dot{W}_{LOST} = \dot{Q}_C \left(1 - \frac{TAMB}{TSTM}\right), \tag{7.12}$$

which quantifies the work output of a hypothetical Carnot engine operating between hot and cold temperature reservoirs at TSTM and TAMB, respectively. Using the numbers in Figure 7.4, from Equations 7.10, 7.11 and 7.12

$$\dot{E}_C = \dot{Q}_C \left(1 - \frac{519}{528}\right) = 0.017\dot{Q}_C,$$

$$\dot{W}_{LOST} = \dot{Q}_C\left(1 - \frac{519}{562}\right) = 0.076\dot{Q}_C$$

so that

$$\dot{I}_C = \dot{W}_{LOST} - \dot{E}_C = (0.076 - 0.017)\dot{Q}_C = 0.059\dot{Q}_C.$$

From the heat release diagram in Figure 7.4 and the second law loss correlations, the most obvious conclusion is that the condenser losses can be brought down by reducing PSTM (i.e., TSTM). This will work in two ways:

1. It will close the temperature gap and reduce the irreversibility.
2. It will reduce the amount of heat rejection.

Limitations imposed by increasingly larger (i.e., costly) equipment and parasitic losses (via circ pump power consumption) put a brake on excessively aggressive designs. Another factor to consider is the increase in exhaust loss if a suitable last-stage bucket (LSB) is not available (e.g., see Section 5.2).

For the water-cooled surface condenser, the performance calculation of interest is the estimation of the condenser pressure at an off-design operating condition with a fixed condenser design, i.e., constant heat transfer surface area. What needs to be done is quite straightforward:

1. Calculate A from the design data.
2. Use the heat transfer equation (Equation 7.3) to find the condenser pressure at any off-design condition.

The following assumptions will simplify the calculations and will give quite accurate results for most combined cycle power plant cases:

1. Heat transfer coefficient $U = 400$ Btu/ft^2-h-F
2. Cooling water specific heat 1 Btu/lb-F
3. Exhaust steam moisture fraction $y = 0.08$.

Note that condensers are designed and constructed in accordance with the *Heat Exchange Institute* (HEI) standards for steam surface condensers [1]. The heat transfer coefficient is a function of tube diameter and cooling water velocity along with correction factors for TCWC, tube material and gauge and design cleanliness factor. For rigorous calculations, you should refer to those specifications or use a software package incorporating them.

Furthermore, condenser sizing should also account for the high-energy steam dump during steam turbine bypass operation (e.g., during startup or trip). In such an operation, steam flow rate can be as high as 150% of the normal full load value (thermal duty can be as high as 170%). In particular, condenser pressure should be less than the alarm point for the particular steam turbine during 100% bypass operation (see EPRI Report CS-2251 (1982) *Recommended Guidelines for the Admission of High Energy Fluids to Steam Surface Condensers* for more information).

Equations 7.1–7.3 can be combined into two equations:

$$1 = \frac{\mu N_{GT}\dot{m}_{exh}(1-y)h_{fg}}{\dot{m}_{cw}(TCWH - TCWC)} \tag{7.13}$$

and

$$1 = \frac{UA}{\dot{m}_{cw}\ln\left(\dfrac{TSTM - TCWC}{TSTM - TCWH}\right)}. \tag{7.14}$$

TABLE 7.1
Latent Heat of Condensation

Condenser P		Saturation Temperature	h_{fg}
in. Hg	psi	°F	Btu/lb
0.50	0.25	58.8	1,060.3
1.00	0.49	79.0	1,048.9
1.20	0.59	84.7	1,045.8
2.00	0.98	101.1	1,036.4
2.50	1.23	108.7	1,032.1
3.00	1.47	115.1	1,028.5
3.50	1.72	120.6	1,025.3
4.00	1.96	125.4	1,022.5
4.50	2.21	129.8	1,020.0
5.00	2.46	133.8	1,017.7

In addition to assumed values of U, y and water specific heat, the following information is known:

1. Condenser heat transfer surface area A from the design case
2. Cooling water inlet temperature
3. Cooling water flow rate (approximately constant at constant circ pump speed).

Therefore, for any given gas turbine exhaust gas and site ambient conditions, Equations 7.13 and 7.14 can be solved for TCWH and the condenser pressure. (Note that TSTM and h_{fg} are functions of the condenser pressure and can be read from the *ASME Steam Tables* – also see Table 7.1.) The calculation can be done easily in an Excel spreadsheet.

Evacuation of air and other noncondensable gases from the condenser is imperative for proper heat transfer from steam to cooling water and maintain high vacuum in the condenser. The three most common types of equipment for maintaining the vacuum in the condenser are steam-jet air ejectors (SJAE), liquid-ring vacuum pumps (LRVP) and a hybrid of the two. In modern combined cycle power plants (no cogeneration, condenser makeup is limited to 3%), LRVP is the preferred technology. In addition to maintaining a high vacuum in the condenser, the other important goal is to limit the dissolved oxygen in the condensate to 7 ppb (parts per billion – equal to 0.005 cc/L). In cogeneration plants with steam export or power plants with steam-injected gas turbines (quite rare in modern machines with dry-low-NOx, DLN, combustors), increased makeup water flow rates necessitate a separate deaerator.

7.2 WET COOLING TOWER

Cooling tower is a direct-contact heat and mass exchanger. In areas where a natural source for condenser cooling water is not available (or too scarce to be used for industrial purposes in large quantities) or where environmental regulations prohibit dumping of the warm cooling water back into the natural source (river, lake, ocean, etc.), the cooling tower facilitates the heat transfer from the warm cooling water returning from the surface condenser so that the cold water is sent back to the condenser. Thus, the condenser cooling water flow forms a "closed loop".

The type of the cooling tower that is commonly encountered in combined cycle power plants is the mechanical (induced) draft, counterflow, *wet* cooling tower. The water that comes into contact with air while flowing downwards through the tower fill is cooled via two mechanisms:

1. Sensible heat transfer
2. Latent heat of the water evaporating into air (i.e., via mass transfer).

The solution of the physics-based, partial differential system of equations governing the heat and mass transfer processes requires complex computer codes. Simplified methods are available for tower design and off-design calculations. The most widely used simplified cooling tower theory is that formulated by Merkel (published in 1926), which is based upon the assumption that an enthalpy potential difference is the driving force [2]. The method is also used in commercial heat balance simulation software, e.g., GateCycle. A simplified method that is amenable to manual calculations can be found in the paper by Kelly [3]. A critique of the Merkel method and a more precise approach is provided in the paper by Webb [4].

There are three thermal design parameters of interest in the cooling tower performance assessment and sizing (costing):

1. *Range*, which is equivalent to TRISE in the condenser;
2. *Approach* (TAPP), which is the temperature difference between the cold water leaving the cooling tower, TCWC, and
3. *Ambient wet-bulb temperature*, TWB (closely approximated by the adiabatic saturation temperature).

These parameters are indicated in the simple diagram of the wet cooling tower in Figure 7.5.

Cooling tower heat load (in Btu/h) is a product of TRISE and water flow rate in gallons per minute (GPM) and given by

$$\dot{Q}_{CT} = 500 \cdot \text{TRISE} \cdot \text{GPM}. \quad (7.15)$$

In passing, note that the factor of 500 is for fresh water. For once-through (open-loop) system using seawater (i.e., no cooling tower), it is 512. The entire theory covered by rigorous heat and mass transfer equations that govern the complex process inside the tower can be collapsed into a single number called the "rating factor", RF. From a publication by a major cooling tower original

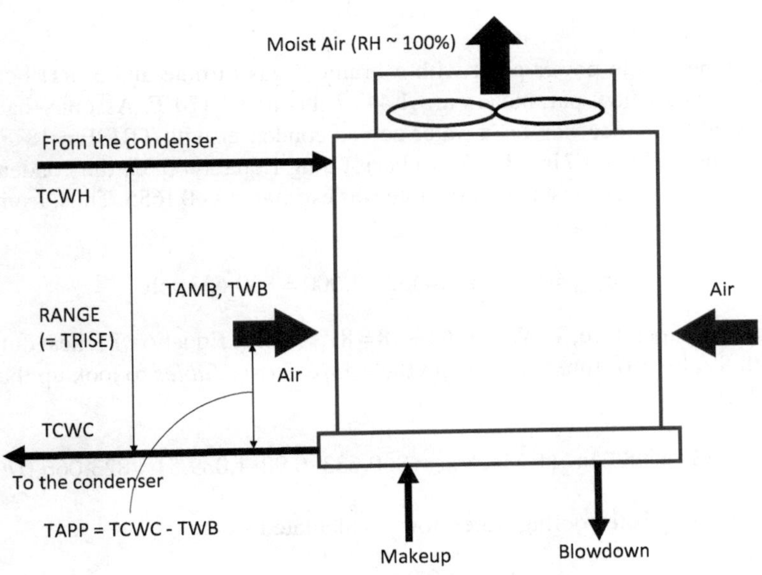

FIGURE 7.5 Schematic diagram of mechanical (induced draft) wet cooling tower.

equipment manufacturer (OEM), the following relationship for RF (*Tower Sizing Factor* or TSF in the cited reference) can be developed:

$$TSF = RF = \left(\frac{TRISE}{TRISE_0}\right)^{\beta}\left(\frac{TAPP}{TAPP_0}\right)^{\chi}\left(\frac{TWB}{TWB_0}\right)^{\delta}. \tag{7.16}$$

The exponents β, χ and δ in Equation 7.16 can be extracted from Figures 5, 6 and 7, respectively, in Ref. [5] as 0.5, –1.0 and –2.0 (with TRISE, TAPP and TWB in degrees Fahrenheit). The subscript 0 denotes a reference value. Specifically, TSF/RF is a numerical measure of the "difficulty" of the cooling duty that has to be performed by a particular cooling tower. Representative reference values from Ref. [3] are

$TRISE_0 = 20°F$
$TAPP_0 = 10°F$
$TWB_0 = 80$
$RF_0 = 1.0$

Manufacturers have developed the concept of the "Tower Units" or TU, which is a hypothetical measure of the cooling tower size and is defined as [3]

$$TU = RF \cdot GPM. \tag{7.17}$$

The parameter TU represents a lumped measure of the cooling tower design including the fill height, cross-sectional flow area and fan horsepower. Curves of TU with RF as the independent variable have been developed for different values of TWB and published [3]. Using these curves in their original graphical form or digitizing them into a computer program (e.g., using VBA in Excel), sizing and costing calculations can be performed.

The critical value from a combined cycle performance point of view is the power consumption of the fans that pull the air through the cooling tower fill (packing) to accomplish the desired cooling. Fan power consumption in horsepower (1 hp = 0.7457 kW) can be estimated as [3]

$$\dot{W}_{CTFAN} = 0.012 \cdot TU. \tag{7.18}$$

Consider a combined cycle power plant with a Frame 9 gas turbine and 3PRH bottoming steam Rankine cycle. Exhaust flow parameters are 1,443.7 lb/s and 1,170°F. Assume that the heat sink design at ISO conditions incorporates a water-cooled condenser with TRISE = 18°F and TCWC = 63°F. Condenser pressure is 1.7 in. Hg (57.6 mbar). Using Equation 5.33, the condensate flow as a fraction of the gas turbine exhaust mass flow rate was estimated as 0.1656. Thus, from Equation 7.4, the condensate flow is found as

$$\dot{m}_{stm} = 0.1656 \cdot 1{,}443.7 \cdot 3{,}600 = 860{,}633\,\text{pph}.$$

Cooling water exit temperature, TCWH, is 63 + 18 = 81°F. Using Equation 7.1, heat duty is calculated as follows (with 8% for moisture fraction and the *ASME Steam Tables* to look up the latent heat of condensation, h_{fg})

$$\dot{Q}_C = 860{,}633 \cdot (1 - 8\%) h_{fg}(1.7\text{ in. Hg}) = 860{,}633 \cdot 0.92 \cdot 1{,}039.51 = 823{,}066{,}070\text{ Btu/h}.$$

From Equation 7.2, requisite cooling water flow is calculated as

$$\dot{m}_{cw} = \frac{823{,}066{,}070}{18} = 45{,}725{,}893\text{ lb/h}.$$

The log-mean temperature is calculated as 22.5°F from Equation 7.5 so that the heat transfer surface can be found from Equation 7.3 as (with U = 400 Btu/ft²-h-F)

$$A = \frac{823{,}066{,}070}{400 \cdot 22.5} = 91{,}408\ \text{ft}^2.$$

Now consider that the gas turbine is running at 80°F ambient (40% relative humidity so that the TWB is 63.5°F). Exhaust flow parameters are 1,375 lb/s and 1,189.5°F. The available cooling water temperature TCWC is 75°F. What is the condenser pressure? First, one has to find the condensate (or steam) flow with μ found as 0.1691 from Equation 5.33:

$$\dot{m}_{stm} = 0.1691 \cdot 1{,}375 \cdot 3{,}600 = 837{,}076\ \text{lb/h}.$$

With the following known parameters

$\dot{m}_{cw} = 45{,}725{,}893$ lb/h
$A = 91{,}408$ ft²
$U = 400$ Btu/ft²-h-F
$y = 8\%$

from Equations 7.13 and 7.14, cooling water outlet temperature and the condenser pressure are found as TCWH = 106.6°F and PCOND = 2.2824 in. Hg. (Note that PCOND enters Equation 7.13 through the latent heat of condensation h_{fg}.)

Now let us assume that the cooling water is supplied by a wet cooling tower. The first task is to convert the cooling water flow rate to GPM, i.e.

$$\text{GPM} = \frac{45{,}725{,}893}{62.4 \cdot 8} = 91{,}600\ \text{gpm}.$$

At ISO conditions, the TWB is 51.5°F so that the approach TAPP = 63 – 51.5 = 11.5°F. From Equation 7.16, the RF is found as

$$\text{RF} = \left(\frac{18}{20}\right)^{0.5}\left(\frac{11.5}{10}\right)^{-1}\left(\frac{51.5}{80}\right)^{-2} = 1.991.$$

From Equation 7.17, the TU is calculated as

$$\text{TU} = 1.991 \cdot 91{,}600 = 182{,}341.$$

Finally, fan power consumption is estimated from Equation 7.18 as

$$\dot{W}_{CTFAN} = 0.012 \cdot 182{,}341 = 2{,}188\ \text{hp},$$

which is about 1,630 kW.

In the off-design case at 80°F ambient, the calculation is an iterative process between the condenser and cooling tower calculations. Recall that the condenser solution required the simultaneous solution of Equations 7.13 and 7.14 for TCWH and PCOND. In the open-loop case, TCWC was a known parameter, which in the case with a wet cooling tower is an output of Equation 7.17 with known TU (found from the design calculation at ISO conditions above) and constant cooling water flow rate GPM, i.e.,

$$\text{RF} = f(\text{TRISE} = \text{TCWH} - \text{TCWC}, \text{TAPP}, \text{TWB}) = \frac{\text{TU}}{\text{GPM}}$$

or

$$\left(\frac{TCWH - TCWC}{TRISE_0}\right)^{\beta}\left(\frac{TAPP}{TAPP_0}\right)^{\chi}\left(\frac{TWB}{TWB_0}\right)^{\partial} = \frac{TU}{GPM},$$

which results in

$$TCWH = TCWC + TRISE_0\left(\frac{TU}{GPM}\left(\frac{TAPP}{TAPP_0}\right)^{-\chi}\left(\frac{TWB}{TWB_0}\right)^{-\partial}\right)^{\frac{1}{\beta}}$$

or

$$TCWH = TWB + TAPP + TRISE_0\left(\frac{TU}{GPM}\left(\frac{TAPP}{TAPP_0}\right)^{-\chi}\left(\frac{TWB}{TWB_0}\right)^{-\partial}\right)^{\frac{1}{\beta}}. \tag{7.19}$$

Steam properties requisite for calculations are summarized in Table 7.1. Useful transfer functions for spreadsheet calculations are listed below.

$$PCOND[\text{in psia}] = 0.0434\exp(0.0305 \cdot TSTM[\text{in}°F]), \tag{7.20}$$

$$h_{fg}[\text{in Btu/lb}] = 1{,}093.9 - 0.5696 \cdot TSTM[\text{in}°F]. \tag{7.21}$$

Simultaneous solution of Equations 7.13, 7.14 and 7.19 with Excel Solver by varying TSTM, TRISE and TAPP results in TCWC = 70.95°F (i.e., TAPP is 7.45°F), TCWH = 88.38°F (i.e., TRISE is 17.44°F) and PCOND = 2.0212 in. Hg. The results show that the original assumption of 75°F for TCWC (presumably from a lake or river) was quite pessimistic. The cooling tower, as designed, can deliver 4°F colder cooling water at the cost of about 1.6 MW in fan power consumption for about 0.25 in. Hg. lower condenser pressure. The exact value of the steam turbine power output benefit is a function of the LP turbine configuration and the available LSB size.

For a mechanical-draft (wet) cooling tower, the heat sink is defined by the ambient TWB. As such, the ambient relative humidity is a very important factor. For the example above, consider that the ambient relative humidity is 70% instead of 40%. This would push TWB to about 72.5°F, i.e., 9°F higher than at 40% relative humidity. The new solution is TCWC = 78.17°F (i.e., TAPP is 5.71°F), TCWH = 95.54°F (i.e., TRISE is 17.37°F) and PCOND = 2.5097 in. Hg. The dramatic increase of nearly 0.5 in. Hg in condenser pressure is worth about 3.5 MW in steam turbine output.

7.3 CIRCULATING WATER PUMPS AND PIPING

"Circ" water pumps ensure the circulation of the condenser cooling water between the condenser and (i) the coolant source or (ii) the cooling tower. They are high specific-speed pumps with high flow rate and relatively low head. For a typical GTCC (no cogeneration), the flow range is 200–400 gpm per net plant output in MW at heads ranging from 30 to 90 ft. The higher flow and lower head are characteristic of the open-loop systems.

The most common type of circ water pump is vertical turbine wet-pit (VTP). Using VTP pumps reduces the size and complexity of the intake structure vis-à-vis that of a horizontal split-case pump in a dry-well-type installation. Circ water pumps typically operate at a fixed speed, and they are designed to permit reverse flow. In open-loop systems with a large head variation due to changes in the water level of the particular reservoir (e.g., lake or river), variable speed drive or variable pitch impeller pump can be more appropriate.

Typical circ water pump configurations are 2 × 100%, 2 × 50% or 3 × 50%. The most cost-effective solution is 2 × 50%. Here is why: In the case of one 50% pump being out of service, the plant can be run at baseload with reduced steam turbine output due to higher back pressure (because the condenser is operating with 50% of requisite cooling water flow). For example, for a 2 × 2 × 1 GTCC rated at 515 MW, the output loss is about 5.5 MW or 1.1% with a corresponding 1.1% increase in heat rate due to the increase in condenser pressure from 2.2 to 3.1 in. Hg. However, one must ensure that the rise in condenser pressure is well below the alarm point (typically 5 in. Hg) and that the resulting pump runout[3] of the operating pump is within allowable limits.

Cooling water piping is designed to *American Water Works Association* (AWWA) standards. Typical pipe materials are coated and lined carbon steel (for freshwater applications) and concrete (both for freshwater and seawater applications). Another option in lieu of concrete is fiberglass-reinforced plastic (FRP). Due to the large diameters (high flow rate) and long distances at relatively low design pressures, damage from *water hammer* is a concern that must be addressed in design of circ water piping.[4] This is the reason why the circ water pumps do not use discharge check valves. They are connected to the supply header via motor-operated butterfly valves interlocked with pump operation. Valve opening and closing times are chosen to reduce the water hammer effects. Other means of protection from water hammer are standpipes (a vertical vent pipe) and vacuum breakers. The latter is a valve installed at the outlet water box of the condenser. During transient operating conditions or shutdown of circ pumps, the valve opens and atmospheric air flows into the circulating water piping to prevent water hammer.

In order to minimize the pumping head and power, open-loop system piping layout is configured for maximum "siphonic recovery" of static head. The hydraulic gradient is below the actual elevation of some parts of the system so that the operating pressure of those segments is subatmospheric. Furthermore, the layout should ensure that the free water surface at the outlet is as close as possible to the water level at the intake. The theoretical lower limit of siphonic recovery is the vapor pressure of the water, which is less than 1 psia at the prevalent cooling water temperatures. A typical system design operating vacuum is 2.6 psia (28 ft of water column).

The intake structure for the open-loop system comprises an intake sump and a mechanical screening equipment (e.g., trash racks with mechanical rakes). Stop logs or sluice gates are provided to allow dewatering of the intake structure. The *American National Standard Institute* (ANSI) standard *ANSI/HI 9.8 Pump Intake Design* provides design guidelines for pump intake structures. In addition, online condenser tube cleaning systems and condenser inlet debris filters are important to maintain design heat transfer coefficients and eliminate the need for plant outages for condenser tube cleaning.

7.4 AIR-COOLED (DRY) CONDENSER

ACCs are once-through systems that use the ambient air as the heat sink. The ACC systems are advantageous in dry climates and other locations where water scarcity and environmental regulations such as water pollution restrictions prohibit the water-cooled systems. The elimination of makeup water supply, BD disposal and water vapor plumes makes ACC an environment-friendly option at the expense of lower steam turbine performance.

Heat transfer from condensing steam to the cooling medium, i.e., air in an ACC, takes place via convection instead of evaporation (mostly) in the wet cooling towers. Due to the lower specific heat of air vis-à-vis water (about one-fourth of it), achieving low condenser pressures require a large heat transfer surface area with large airflows – i.e., large fan power. The heat sink temperature is the ambient *dry bulb temperature* (TDB), which is higher than the TWB by up to 10°F (the heat sink temperature in a wet cooling tower). A schematic description of an "A-Frame" ACC is shown in

[3] Pump runout, or runout flow, is the maximum flowrate that can be developed by a pump. The runout conditions correspond to the lowest pump head at the highest flow rate as specified by the pump vendor.

[4] Water hammer or hydraulic shock is a pressure surge or wave caused when the water in motion in a pipe is forced to stop or change direction suddenly (e.g., due to sudden closure of a valve at the end of the pipeline).

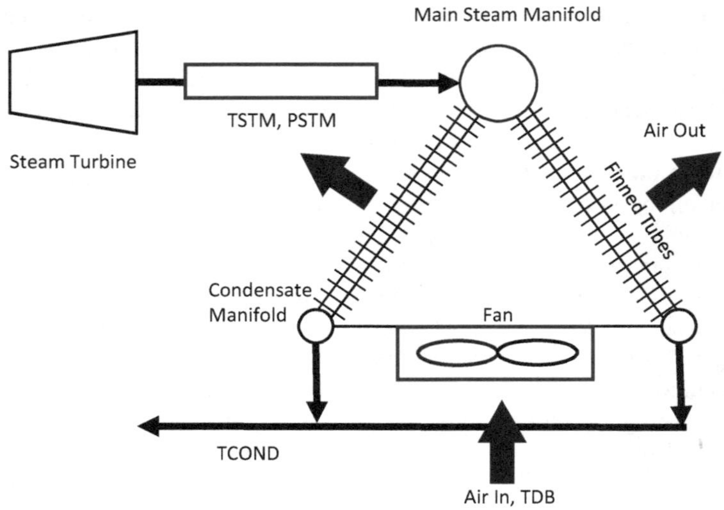

FIGURE 7.6 A-frame air-cooled (dry) steam condenser.

Figure 7.6. As shown in the figure, finned-tube bundles are sloped downwards to reduce the plot area. Forced-draft fans located under the A-frame push the ambient air across the finned-tube bundles, causing the steam ducted to the apex of the frame to condense as it flows down the inclined tubes.

Steam exhausting the LP turbine is piped to the condenser by a large-diameter duct (approximately 10% pressure drop). The A-frame structure is mounted on a steel structural support. Noncondensable gases are drawn off by the air ejection equipment. The condensate drains into a collection tank and is sent back to the HRSG via the condensate pump. A 3D view of an A-Frame ACC cell is shown in Figure 7.7. A typical 2 × 2 × 1 GTCC with two F class gas turbines (about 750 MW ISO baseload rating) can have an ACC with 25 such cells in five rows with five cells per row. The size of an ACC is determined as a result of the cost-performance trade-off between heat transfer surface area and fan power consumption.

Due to the size/cost restrictions and large parasitic power consumption, ACC systems are not feasible for condenser pressures below 68 mbar (2 in. Hg). A reasonable estimate of the ACC fan power consumption is given by

$$\dot{W}_{FAN} = \frac{\dot{m}_{stm} h_{fg}}{\varepsilon(\rho c_p)_{air}(TSTM - TAMB)} \frac{\Delta p_{air}}{140\eta_{fan}}. \tag{7.22}$$

Steam flow rate in Equation 7.22 is found from Equation 7.4. Typical values are as follows:

- ACC effectiveness, ε, is 55%:
- Air pressure drop Δp_{air} is 0.5 in. Hg.
- Air specific heat is 0.24 Btu/lb-°F.
- Air density is 0.0765 lb/ft^3.
- Fan efficiency, η_{fan}, is 80%.

With consistent British units, Equation 7.22 returns the fan power in kW.

Normal ACC design is based on a 40°F–50°F difference between TSTM and TAMB. For a plant with an ambient design condition of 90°F and 45% relative humidity (73°F TWB), a feasible back pressure with an ACC is about 4.5 in. Hg (cf. 2.1 in. Hg with a mechanical wet cooling tower at the same ambient conditions).

FIGURE 7.7 Typical A-Frame cell (SPX Dry Cooling).

For ISO ambient conditions and PCOND values between 2 and 5 in. Hg (68 mbar to 170 mbar), the following power law is found to represent the ACC power consumption reasonably well:

$$\frac{\dot{W}_{FAN}}{\dot{W}_{STG}} = \frac{1.58\%}{(PCOND - 1.036)^{0.365}}. \tag{7.23}$$

For different site conditions, Equation 7.22 should be used. A reasonable approximate estimate is 1.7% of steam turbine generator output. Equation 7.23 clearly shows that below PCOND of about 2 in. Hg (68 mbar), the ACC fan power consumption increases exponentially and the ACC becomes an unfeasible alternative to the wet cooling systems with or without a cooling tower in terms of net electric power maximization.

Wind causes air temperature at the fan inlet to rise due to hot air recirculation and results in the deterioration of the heat transfer via lower airflow. The effect is negligible below wind speeds of about 5 m/s (about 11 mph). Above that limit, each mile-per-hour increase in wind speed causes about 3% increase in condenser pressure. Each one inch of mercury (34 mbar) increase in condenser pressure is about 3% loss in steam turbine generator output.

A typical off-design performance curve for an ACC is shown in Figure 7.8. The design conditions are 1,556 kpph (196 kg/s) steam flow (y = 6%), 2.5 in. Hg (84 mbar) condenser pressure at TAMB = 49.1°F (9.5°C). Power consumption of the ACC fans is 2,183 kW. The curves in Figure 7.8 can be used as a guide to estimate off-design performance for different applications. Typical degradation in ACC performance is about 4% drop in condenser pressure after 10 years of operation (with regular cleaning of the finned tubes as prescribed by the vendor).

For ACC fan performance, vendor-provided fan curves are the best source. Those curves describe the fan performance at design inlet conditions as a function of fan blade pitch angle. The horizontal axis is airflow (say, in m^3/s), and the vertical axis is the static pressure (say, in Pa). Curves for each pitch angle setting are similar to the speed curves on a centrifugal compressor or pump performance map. The fan performance map also includes static efficiency contours similar to the efficiency contours for axial or centrifugal compressors. ACC performance (i.e., lowest possible steam turbine back pressure) is strongly dependent on the ability of the fans to drive sufficiently high airflow against pressure resistance of the tube bundles. Verification of fan performance during performance

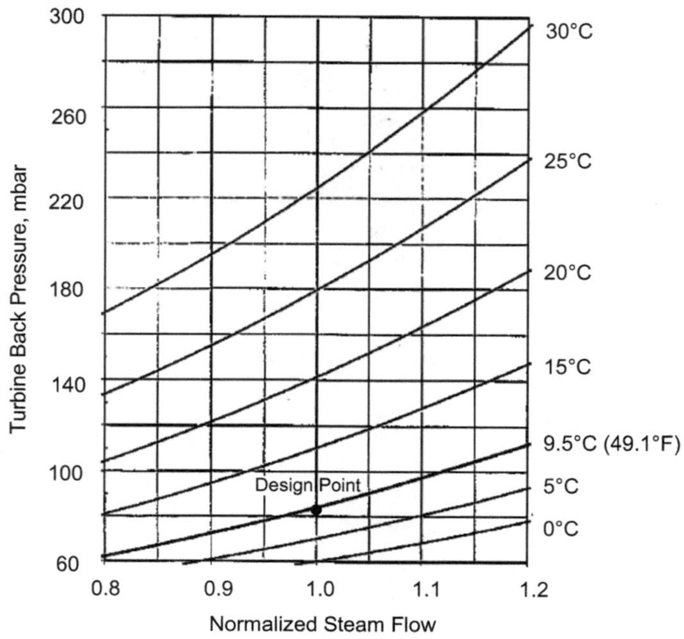

FIGURE 7.8 Performance curves of a typical ACC.

tests and field operation requires careful analysis because the only available measurement is the driver motor power.

For proper operation of an ACC, continuous and complete removal of the noncondensables in the system is imperative. Noncondensables are gases that enter the steam duct at vacuum via air leaks or from the chemicals used for feedwater treatment. Such gases can be trapped inside the tubes between the inlet steam from the turbine exhaust and the "backflow" steam at the tube outlet. The reader is referred to the paper by Larinoff et al. for a detailed description of the trapping mechanism [6]. During the winter, the trapped noncondensables can cause freezing of the condensate. In the summer, they blanket the heat exchange surfaces and impede the heat transfer between condensing steam and air. The result is an increase in steam turbine back pressure (via higher condensing steam temperature) and loss of steam turbine power output.

ACCs typically use SJAE for noncondensable removal in lieu of vacuum pumps due to the large steam-side volume of the ACC and the unavailability of cooling water at a temperature much lower than that of the condensate.

It is important to match the ACC design to the steam turbine back pressure limits, which are set by the OEM to minimize the possibility of LSB fatigue damage and LSB-exhaust hood damage from overheating. Overheating can cause bucket structural problems as well as exhaust hood distortion. Fatigue damage is caused by the buffeting of the LSB vanes due to steam flow at extreme off-design entrance angles (particularly at low loads). Typical alarm limit is 5 in. Hg. Due to the high design point back pressures with ACC, there is increased likelihood of reaching this limit on hot days.

One option to protect against this problem is the inclusion of a *fogger* to precool the inlet air stream. This, however, defeats the purpose of having an ACC instead of a wet cooling system in the first place, i.e., water scarcity or water conservation. Steam turbine load reduction requires significant reduction in gas turbine load and exhaust flow and temperature with a significant hit on GTCC output when it is typically most valuable (i.e., hot summer days with peak demand). Another option is special steam turbine back-end design for operation at high back pressures.

The lowest practical operating back pressure for an ACC is about 2 in. Hg. Unlike its water-cooled counterpart, the ACC has steam in the tubes, and there is a significant internal pressure drop therein and in the exhaust duct connecting the steam turbine to the condenser. At lower pressures, nearly exponential rise in steam specific volume increases the internal pressure drop, and the back pressure of the steam turbine does not drop much below 2 in. Hg at full load (i.e., full thermal duty).

7.5 HEAT SINK SYSTEM SELECTION

Selection of the "right" steam turbine heat sink is a function of availability and quality of water, disposal of the water (e.g., from the cooling tower BD) and site characteristics including the discharge permits issued by the regulating authorities. Once a system is selected, its performance is optimized based on the historical site weather data (i.e., dry bulb and TWBs). This will be discussed in more detail below.

If there is a naturally available water source such as a lake or river and the environmental regulations do not preclude the return of warm water to the source, open-loop water-cooled condenser is the ideal choice from a cost and performance perspective. Alas, discharge of warm water back to the source affects the full spectrum of organisms in the aquatic ecosystem in the said source, at all life stages from tiny photosynthetic organisms to fish, shrimp, crabs, birds and marine mammals, including threatened and endangered species. The result is destruction of billions of fish and destabilization of aquatic populations.

Furthermore, intake structures of the heat sink system can kill aquatic organisms by "entraining" them through the plant's heat exchangers where they succumb to physical, thermal and toxic stresses and "impinging" them on the intake screens (i.e., trapping them against the screens by the pressure of the intake flow), and the discharge of heated water from the cooling system may also harm wildlife.

Consequently, *Section 316(b) of the Clean Water Act* requires all power plants – new and old – to install the "best technology available" (BTA) for minimizing the adverse environmental impacts of cooling water intake structures. Congress included section 316(b) in the 1972 federal Clean Water Act, yet the implementation of section 316(b) has been stalled for decades. In December 2001, the EPA issued regulations for cooling water use at new facilities, identifying closed-loop cooling (i.e., with a cooling tower) as the BTA. The final rule under Section 316(b) was issued in 2014. Consequently, the option of open-loop water-cooled condenser in the USA is pretty much unavailable due to severe permit limitations on discharge temperature and quantity.

In closed-loop systems with a cooling tower, the quality of makeup water is the critical parameter. Cooling tower makeup water substitutes the losses due to evaporation (from cooling water to the air), BD, drift (water escaping the cooling tower as fine moisture droplets in the fan exhaust) and miscellaneous leaks. Modern mist eliminators can reduce the drift to 0.0005% of the circulating water flow. Thus, in a well-maintained and properly operated system, the last two items are negligible.

In addition to the quantity and quality of the makeup water, seasonal variations in flow rate and water characteristics must be considered in choosing the cooling tower makeup water source. For example, usage of agricultural or industrial wastewater can cost 2–2.5 times more than low total dissolved solid (TDS) water [7]. Recovering gray[5] or black[6] water for heat sink use is up to twice the cost of low TDS water because of the complex purification equipment required.

Another consideration is the cost of wastewater disposal – primarily BD from the HRSG evaporator drums and the wet cooling tower. The wastewater in question here is of course the latter but let us explain the reason for both BDs.

5 Relatively clean waste water from baths, sinks, washing machines and other kitchen appliances.

6 Wastewater from toilets.

In the drum of the evaporator, after steam leaves it, solids in the feedwater are concentrated in the water left behind. After a limiting value of solid concentration is reached, the feedwater in the drum is bled off at such a rate that the amount of solids entering the drum with the economizer feedwater is exactly balanced by that removed in the bleed stream.

The water loss mechanism in the wet cooling tower is evaporation of a portion of the circulating cooling water to the air stream. Recall that the relative humidity of air leaving the cooling tower is nearly 100%, whereas the ambient air relative humidity is typically 40%–70%. A simple formula to estimate the rate of evaporation is

$$E[\text{gpm}] = \frac{f \cdot \text{GPM} \cdot \text{TRISE}}{1{,}000}, \tag{7.24}$$

where f is a correction factor to account for sensible heat transfer between water and air [8]. A typical range for f is 0.65–0.90 (higher in the summer and lower in the winter). Note that the factor 1,000 in the denominator of Equation 7.24 is an approximation for the latent heat of evaporation of water in Btu/lb (e.g., see Table 7.1). Using a value of f = 0.78 [8], for the example above with GPM = 91,600 gpm and TRISE = 18°F, the evaporation rate is found as E = 0.78 × 91.6 × 18 = 1,286 gpm or 1,286/91,600 = 1.4%.

Similar to what happens in the evaporator drum, evaporation causes dissolved and suspended solids in the cooling water in the tower basin (or sump) increase in concentration. The remedy is of course the same, i.e., bleeding off the cooling water in the tower basin at a rate to balance the amount of solids coming in with the makeup water. With BD and makeup water (MU) in gpm, ignoring drift and leakage losses, we have

$$\text{MU} = \text{E} + \text{BD}, \tag{7.25a}$$

$$C = \frac{\text{MU}}{\text{BD}} = \frac{\text{E}}{\text{BD}} + 1, \tag{7.25b}$$

where C is the BD ratio, which is referred to as "cycles of concentration". Consequently,

$$\text{BD} = \frac{\text{E}}{C - 1}. \tag{7.26}$$

Note that dissolved/suspended solids (e.g., chlorides, calcium, magnesium, sometimes sulfates and even mercury – e.g., in the Ohio river) balance requires that

$$\text{MU} \cdot \text{TS}_{\text{MU}} = \text{BD} \cdot \text{TS}_{\text{BD}}, \tag{7.27}$$

where TS_{MU} is the total solids concentration in the makeup water (in parts per million, ppm), and TS_{BD} is the total solids concentration in the BD. It follows that

$$C = \frac{\text{TS}_{\text{BD}}}{\text{TS}_{\text{MU}}}. \tag{7.28}$$

Thus, BD ratio or cycles of concentration is a function of the makeup water chemistry, effectiveness of chemical process used to treat the makeup water and specific permit limits. Higher cycles result in lower BD rates (see Equation 7.26) but at the expense of more concentrated discharge (see Equation 7.28).

Wastewater (i.e., BD) discharge options include discharge to the source (for the open-loop systems), sewer discharge, evaporation ponds and *zero liquid discharge* (ZLD). Tawney et al. present relative costs of several wastewater disposal options using the cooling pond as the basis for comparison (i.e., 1.0) [7]. By far the most expensive one is the ZLD, which is used only where stringent permitting requirements need it. In addition to adding significantly to the total installed cost of the combined cycle power plant, ZLD systems can add significantly to the parasitic power burden of a power plant. There are several types of ZLD systems, but the most common is the thermal

TABLE 7.2
Comparison of Three Key Heat Sink Technologies

	Once-Through (Open-Loop)	Mechanical-Draft (Wet) Cooling Tower	Air-Cooled (Dry) Condenser
Installed cost	0.64	1.00	3.80
Water usage	46.00	1.00	0.00
Plant heat rate	0.98	1.00	1.05
Plant output	1.02	1.00	0.96
Condenser pressure	0.63	1.00	2.12
Noise	Lower	Base	Higher
Land usage	Lower	Base	Higher
Wastewater disposal requirements	Higher	Base	Lower
Treatment chemical usage	Lower	Base	Lower

Source: From Ref. [7].

evaporation system with a brine concentrator and crystallizer. The reader is referred to the water treatment section of the chapter on the advanced fossil fuel power systems by Gülen in Ref. [9] or to Section 10.9.5 in this book.

As an example consider that the cooling tower makeup water requirement of a typical GTCC is 220 gallons per MWh. For a 500 MWe power plant, this comes to about 1,850 gpm. For a cycle of concentration of $C = 5$, E and BD can be determined from Equations 7.25a and 7.26 as $E = 1{,}480$ gpm and $BD = 370$ gpm. At 90 kWh per 1,000 gallons of feed, the *mechanical vapor compressor* of the brine concentrator would consume nearly 2 MWe of power. One should also add the lost steam turbine output via steam provided to the crystallizer to that amount (if one is present). The impact of the ZLD system operation on plant performance is significant. Assuming that the GTCC in question is an advanced H or J class unit rated at 60% net efficiency with auxiliary power consumption at 2% of gross, addition of the ZLD reduces the net efficiency to $498/500 \times 60 = 59.76\%$. Consequently, a careful cost-performance study is imperative to make a selection between a ZLD system and an ACC.

Site characteristics also play a significant role in heat sink system selection. For mechanical-draft (wet) cooling towers, location, orientation and spacing of the system vis-à-vis the power island can impact the tower performance due to recirculation and wind direction. For example, rectangular cooling towers are oriented parallel to the prevailing wind direction so that the warm and moist air exiting the tower is blown away from the inlet [7].

Once-through (open-loop) systems must obviously be located near a large body of water. Otherwise, the extended length of circulating water pipes (typically, very-large-diameter concrete pipes) and larger circ pumps will adversely impact both total installed cost and performance (via increased parasitic power consumption) [7].

ACCs must be located close to the steam turbine in order to minimize the pressure drop between the steam turbine and the condenser. This will be favorable to both total installed cost (via shorter steam duct) and plant performance (via lower back pressure). In general, ACCs require a large, symmetrical plot space to accommodate their rectangular shape. Sites with irregular lots or limited acreage can result in excessive cost and reduced performance (Table 7.2).

7.6 HEAT SINK OPTIMIZATION

As stated earlier in Section 7.1.3, ACC optimization is pretty straightforward: evaluate the cost-performance trade-off between heat transfer surface area and fan power consumption. The same is true for the open-loop system with the added complexity introduced by the availability of a suitable LSB (to exploit low back pressures for maximizing steam turbine power output). For the mechanical-draft (wet) cooling tower systems, the optimization process is somewhat more involved. Key equations for optimization are Equations 7.9–7.12. For any given design ambient conditions (e.g., ISO), all parameters in those equations can be uniquely determined from four design parameters:

1. Condenser pressure, PCOND
2. Condenser TTD
3. Cooling tower approach to wet bulb, TAPP (in a closed-loop system)
4. Cooling water inlet temperature TCWC (in an open-loop system).

Equation 7.12 is unambiguous in showing that there is *one and only one* parameter that controls the total lost work, which is the condenser steam temperature, i.e., the condenser pressure. Obviously, the aim of an optimal GTCC bottoming cycle design is to minimize the lost work associated with steam turbine condenser heat rejection process. Unavoidably, the improvement via reducing the total lost work comes at a price, which is the power consumed by the cooling water circulating pump and if applicable, the cooling tower fans. The cooling water circulating pump ("circ pump") power can be calculated from the condenser duty $\dot{Q}_C$, TRISE, and circ pump head H, as follows:

$$\frac{\dot{W}_{CP}}{\dot{Q}_C} = \frac{H}{\eta_p \left(\rho c_p\right)_{cw} \text{TRISE}}. \tag{7.29}$$

For a reasonable design, pump head and overall efficiency of 65 ft (about 20 m) and 75%, respectively, can be assumed. Equation 7.29 is self-consistent in SI units when using kPa for H and kJ/kg-K for c_p. Note that

- 1 m of pump head is 9.81 kPa (1 kPa = 1,000 N/m^2)
- 1 ft of pump head is 0.43 psi.

For cooling water, it is sufficiently accurate to use

- Density, ρ, of 997 kg/m^3 or 62.4 lbm/ft^3
- Specific heat, c_p, of 4.186 kJ/kg-K or 1 Btu/lb-R.

Thus, in US customary or British units, Equation 7.29 becomes

$$\frac{\dot{W}_{CP}[\text{kWe}]}{\dot{Q}_C[\text{kWth}]} = \frac{0.001275 \cdot H[\text{ft}]}{\eta_p \text{TRISE}[°\text{F}]}.$$

For cooling tower fan power consumption, combining Equations 7.15–7.18, we can write that

$$\frac{\dot{W}_{CTFAN}}{\dot{Q}_{CT}} = \left(\frac{0.012}{500 \cdot \text{TRISE}}\right) \cdot \left(\frac{\text{TRISE}}{\text{TRISE}_0}\right)^{\beta} \left(\frac{\text{TAPP}}{\text{TAPP}_0}\right)^{\chi} \left(\frac{\text{TWB}}{\text{TWB}_0}\right)^{\partial},$$

$$\frac{\dot{W}_{CTFAN}[\text{kWe}]}{\dot{Q}_{CT}[\text{kWth}]} = \frac{0.082}{\text{TRISE}^{1-\beta}} \cdot \text{TRISE}_0^{-\beta} \left(\frac{\text{TAPP}}{\text{TAPP}_0}\right)^{\chi} \left(\frac{\text{TWB}}{\text{TWB}_0}\right)^{\partial}. \tag{7.30}$$

Equations 7.29 and 7.30 show that circ pump and fan power consumption are bot inversely proportional to TRISE. As such, the principle of performance improvement for a GTCC heat rejection system with a water-cooled condenser is rather simple:

Reduce the condenser pressure (i.e., condensing steam temperature) while keeping the TRISE constant to the extent possible.

This will reduce the lost work associated with heat rejection while keeping the increase in parasitic power consumption at a minimum. There are certain practical design guidelines to consider while doing that. Economically and mechanically feasible designs for condenser and cooling tower dictate the following limits:

1. TTD 5°F or 3°C (ideally 7°F or 4°C) so that the condenser size, weight (i.e., foundation concrete and reinforced steel) and cost remain reasonably low [1].
2. TAPP 10°F or 6°C (ideally 15°F or 8°C) so that the cooling tower size, weight and cost remain reasonably low. Decreasing the approach from 8°C to 6°C results in a size increase of about 30%. Decreasing it from 6°C to 3°C results in a size increase of 65% at which point the size-approach curve becomes nearly vertical [10].

Typically, cost considerations dictate that cooling towers are sized for warm weather operation by using the ASHRAE[7] 2.5% frequency summer TWB as the design basis. On average, in a given year, this TWB is exceeded for slightly more than 50 h (i.e., 2.5% × 90 days × 24 h/day = 54 h). A 1°F recirculation allowance above TWB is a typical design allowance. As already indicated above, cooling tower cost is most sensitive to TAPP, which should be 10°F or higher. A cost-effective design would have 30°F–40°F delta between TSTM and TWB. In other words, from the relationships above,

$$\text{DELTA} = \text{TAPP} + \text{TRISE} + \text{TTD}$$

and

$$30°\text{F} \leq \text{DELTA} \leq 40°\text{F}.$$

This rule of thumb is based on a summertime design because TAPP is inversely proportional to TWB. At lower values of TWB, DELTA should be increased. These considerations usually result in condenser pressures of 2.0–2.5 in. Hg.

7.6.1 Two-Step Condensation

The area between the two heat transfer curves in Figure 7.4 can be calculated as

$$\text{AREA} = \dot{Q}_c \text{TSTM}\left(1 - \frac{\text{TCWM}}{\text{TSTM}}\right). \tag{7.31}$$

Comparing Equation 7.31 with Equation 7.9, it is found that

$$\text{AREA} = \dot{I}_c \text{TSTM}\left(1 + \frac{\text{TRISE} + \text{TTD}}{2\text{TAMB}}\right). \tag{7.32}$$

Since TSTM (i.e., saturated steam temperature at the condenser pressure, PCOND) is a thermodynamic design parameter and TRISE and TTD are condenser hardware design parameters,

$$\text{AREA} \propto \dot{I}_c;$$

i.e., the geometric area between the heat transfer curves is a measure of the exergy destruction in the condenser. From purely graphical considerations,

[7] American Society of Heating Refrigerating and Air-Conditioning Engineers.

- Reduction of exergy destruction is equivalent to the reduction of the area between the heat transfer curves.
- This requires bringing the two T-Q curves (pretty much straight lines) closer to each other.
- Since condensation of steam takes place at constant temperature (TSTM), however, this is a quite difficult proposition.
- One possible approach (conceptually) is breaking the condensation curve into smaller segments in a manner analogous to reheat combustion

It is easy to visualize that, when this approach is taken to the extreme (i.e., infinitesimally small segments), both heat transfer curves can coincide and the exergy destruction becomes exactly equal to zero. This, of course, is impossible in practice. The only feasible implementation of this concept is a double-condenser configuration, which is schematically depicted in Figure 7.9.

Heat transfer for each condenser in Figure 7.9 can be written as follows:

$$\frac{1}{2}\dot{m}_{stm}(1-y_2)h_{fg}(TSTM2) = \dot{m}_{cw}c_p(TCWMED - TCWC), \tag{7.33}$$

$$\frac{1}{2}\dot{m}_{stm}(1-y_1)h_{fg}(TSTM1) = \dot{m}_{cw}c_p(TCWH - TCWMED). \tag{7.34}$$

Combining Equations 7.33 and 7.34, it is found that

$$TCWMED = TCWC + \frac{\lambda}{1+\lambda}TRISE \tag{7.35}$$

with

$$\lambda = \frac{(1-y_2)h_{fg}(TSTM2)}{(1-y_1)h_{fg}(TSTM1)}. \tag{7.36}$$

Since the moisture fraction of the exhaust steam in either LP turbine is very close to each other, i.e., $y_1 \sim y_2$, λ is simply the ratio of latent heats of condensation, i.e.,

$$\lambda \approx \frac{h_{fg}(TSTM2)}{h_{fg}(TSTM1)}. \tag{7.37}$$

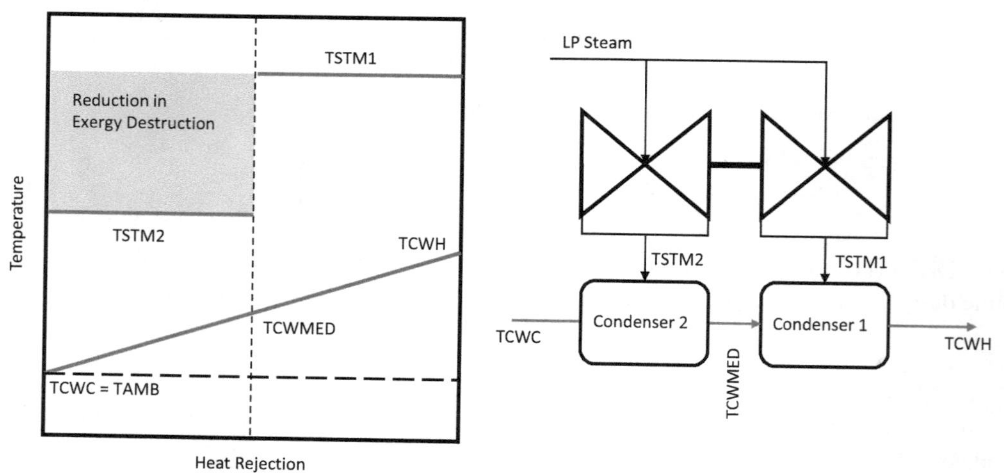

FIGURE 7.9 Four-flow LP steam turbine with two condensers.

Consequently, for given system boundary conditions, i.e., TCWC (~TAMB for once-through or open-loop systems), TRISE and TTD, i.e.,

$$\text{TSTM1} = \text{TCWC} + \text{TRISE} + \text{TTD}$$

and

$$\text{TCWH} = \text{TCWC} = \text{TRISE}.$$

The design problem is reduced to the solution of

$$\text{TSTM2} = f(\text{TCWMED}),$$

$$\text{TSTM1} > \text{TSTM2} > \text{TCWC}.$$

(It can be shown that TCWMED is very close to the average of TCWC and TCWH.) A sample calculation is summarized in Table 7.3. Note that the last row in the table is the non-dimensional exergy destruction, i.e., from Equation 7.9,

$$i = \frac{\dot{I}_C}{\dot{Q}_C} = \frac{\text{TAMB}}{\text{TCWM}}\left(1 - \frac{\text{TCWM}}{\text{TSTM}}\right). \tag{7.38}$$

The reduction in exergy destruction is

$$\left(\frac{1}{2}(0.0099 + 0.0194) - 0.0194\right)\dot{Q}_c = -0.0074\dot{Q}_c$$

or

$$\frac{-0.0074\dot{Q}_c}{0.0194\dot{Q}_c} = -0.3798 \approx -38\%.$$

TABLE 7.3
Two-Step or "Double" Condenser Example

Parameter	Units	Double Condenser	COND2	COND1
TCWC (~TAMB)	°F	84	84	90.0
TRISE	°F	12	6.0	6.0
TTD	°F	5	2.5	5.0
TCWH	°F	96	90.0	96
TCWM	°F	90	87.0	93.0
TSTM1	°F	101	92.5	101
PCOND1	in. Hg	2.0		
TSTM2	°F	92.5		
PCOND1	in. Hg	1.5		
HFG1	Btu/lb	1036.5		
HFG2	Btu/lb	1041.3		
λ	-	1.005		
TCWMED	°F	90.0		
i	-	0.0194	0.0099	0.0142

REFERENCES

1. Heat Exchange Institute, 1995, *Standards for Steam Surface Condensers*, 9th Edition, Heat Exchange Institute, Cleveland, OH.
2. Baker, D.A., Shryock, H.A., 1961, A comprehensive approach to the analysis of cooling tower performance, *J. Heat Trans.*, Vol. 83, No. 3, pp. 339–350.
3. Kelly, G.M., 1975, Cooling tower design and evaluation parameters, ASME paper 75-IPWR-9, *Industrial Power Conference*, May 19–20, 1975, Pittsburgh, PA.
4. Webb, R.L., 1988, A critical evaluation of cooling tower design methodology, pp. 547–557 in *Heat Transfer Equipment Design*, Hemisphere Publishing Co. Washington, DC.
5. Marley Corporation, 2009, Cooling tower performance – basic theory and practice, This vintage document is now superseded by "*Cooling Tower Fundamentals*, 2nd Edition", Eds. J.C. Hensley, SPX Cooling Technologies, Inc., Overland Park, KS.
6. Larinoff, M., Moles, W., Reichhelm, R., 1978, Design and specification of air-cooled steam condensers, *Chem. Eng.*, Vol. 85, pp. 86–94.
7. Tawney, R., Khan, Z., Zachary, J., 2003, Economic and Performance Evaluation of Heat Sink Options in Combined Cycle Applications, ASME Paper No. GT2003-38834.
8. Buecker, B., 2018, Cooling tower water conservation conundrums, Energy-Tech, August 2018, pp. 26–30.
9. Gülen, S.C., 2017, Advanced fossil fuel power systems, Chapter 13 in *Energy Conversion*, 2nd Edition, Eds. D.Y. Goswami and F. Kreight, CRC Press, Boca Raton, FL.
10. Hensley, J., 1992, Maximize tower power, *Chem. Eng.*, Access Intelligence, New York, Vol. 99, No. 2, pp. 74–82.

8 Combining the Pieces

In Chapters 4–7, major combined cycle components were covered in considerable depth. Obviously, the term "depth" should be understood within the limits imposed by a single book of finite number of pages. The reader should consult references cited in relevant chapters for a more expansive grounding in gas turbines, steam turbines, heat recovery boilers, condensers and cooling towers.

Especially in Chapters 4–6 on the two prime movers and the heat recovery boiler (or heat recovery steam generator, HRSG, in generally accepted industry jargon), the thermodynamic principles leading to the combined cycle concept have been identified and elaborated upon. Using those principles, reasonably rigorous but still simplified (i.e., amenable to pen-and-paper or spreadsheet calculations) combined cycle performance calculation methods have been introduced. At the risk of being repetitive, the key underlying thermodynamic principle is elaborated upon below (for the last time in the book). The goal in doing that is to demonstrate as clearly as possible (for the less experienced reader's benefit) that the concept of a "combined" cycle can materialize from the first principles *on its own*, i.e., without having to go through endless trial-and-error "tinkering" with different ideas and technologies. The latter, however, was the path to the development of combined cycle technology from its *fin de siècle* start to its coming of age in the late 1980s and early 1990s. This quite tortuous path will be described in some detail in Section 8.4. In fact, it is recommended that the reader reads Sections 8.1–8.4 in one sitting to make the connection with theory and reality.

8.1 TOPPING CYCLE

The gas turbine is an internal combustion engine. As such, it is a member of the larger family of heat engines. The starting point of any heat engine is the particular thermodynamic "cycle" that describes its basic operation. For the gas turbine, the cycle in question is the Brayton cycle. Like all heat engine cycles, the gas turbine Brayton cycle is a very poor approximation of the thermodynamic ideal quantified by the Carnot cycle. The cycle hierarchy is illustrated graphically on the temperature-entropy (T-s) diagram in Figure 8.1; in essence, from the top

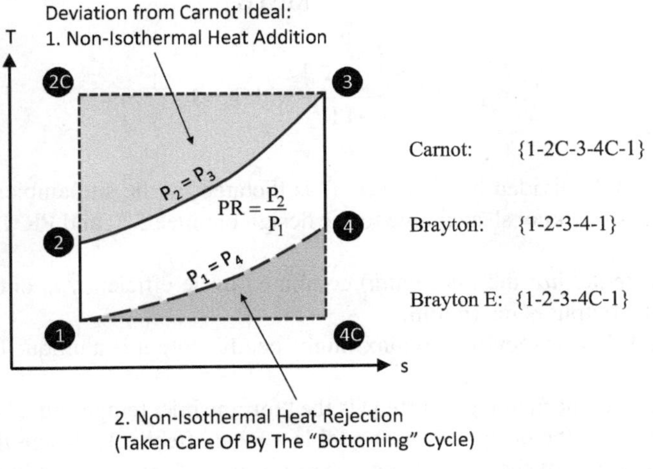

FIGURE 8.1 Ideal cycle hierarchy (Brayton "E" stands for "enhanced" Brayton – see the text).

- Carnot cycle is the theoretical maximum imposed by heat "source" and "sink" temperatures, T_3 and T_1 in Figure 8.1, respectively;
- Next comes the "air-standard", ideal Brayton cycle, whose performance is controlled solely by the cycle pressure ratio, PR;
- At the bottom lies the "real" cycle (i.e., the actual gas turbine) beset by hardware imperfections, which, strictly speaking, cannot be depicted on a T-s diagram.

The deviation of the ideal, air-standard Brayton cycle from the Carnot cycle can be quantified by the two, roughly triangular areas in the T-s diagram of Figure 8.1:

- Heat addition "imperfection" quantified by the area {2-2C-3-2}
- Heat rejection imperfection quantified by the area {1-4-4C-1}.

The second imperfection can be rectified to a great extent by a second thermodynamic cycle via "waste heat recovery". The waste heat in question is the energy content of the gas turbine exhaust (state point 4). This is the thermodynamic driver behind the Brayton–Rankine combined cycle, wherein the Rankine "bottoming" cycle, via the steam turbine, makes use of the gas turbine exhaust heat to generate additional power. The reason that a steam *Rankine* cycle is an ideal candidate for the bottoming cycle is twofold:

1. State-of-the-art gas turbine technology with exhaust temperatures around 1,150°F or higher constitutes a readily available heat source for generating high-pressure (HP) steam at 1,000°F–1,100°F in a "waste heat recovery" boiler, i.e., the HRSG.
2. Heat rejection from the steam turbine condenser at constant pressure and temperature is the perfect embodiment of the fourth ideal Carnot cycle process, i.e., *isothermal* cycle heat rejection.

The ideal combined cycle is essentially an "enhanced" Brayton cycle represented by the area {1-2-3-4C-1} in Figure 8.1. It is a function of heat "source" and "sink" temperatures, T_3 and T_1 and the cycle PR. From the cycle T-s diagram in Figure 8.1, the ideal "combined" efficiency with ideal Brayton (topping) and ideal bottoming cycles can be deduced to depend on the ratio of METH (Equation 4.7) and the ambient temperature T_1, which leads one to the combined cycle efficiency:

$$\eta_{CC} = 1 - \frac{T_1}{\text{METH}} \tag{8.1}$$

$$\eta_{CC} = 1 - \frac{\ln\left(\dfrac{\tau_3}{PR^k}\right)}{\tau_3 - PR^k} \text{ with } k = 1 - \frac{1}{\gamma}. \tag{8.2}$$

In Equation 8.2., τ_3 is T_3 divided by T_1. Since T_1 is dictated by the site ambient conditions, there are only two cycle design "knobs" available to the design engineer, T_3 and PR. It can be shown that

- The optimum (note: *not* the maximum) combined cycle efficiency is obtained when gas turbine specific output is maximum.
- Brayton cycle PR corresponding to maximum specific output is a unique function of T_3.

(The counterpart of T_3 in the actual gas turbine is the "turbine inlet temperature" or TIT.) Once these two parameters are fixed, the designer's next goal is to cost-effectively design the *bottoming cycle* to make the maximum use of the gas turbine exhaust exergy. This design effort hinges around the HRSG stack temperature, which is a very important parameter that does not appear in Equation 8.2.

8.2 BOTTOMING CYCLE

Superficially, one can dismiss the bottoming cycle of a combined cycle as another fossil fuel-fired steam turbine Rankine cycle and could not be more wrong. From a hardware perspective, the only (and immediately apparent) difference is the replacement of the fired boiler by a "heat recovery" boiler or, as commonly referred to, HRSG. However, it is the other, less apparent key difference, i.e., elimination of the feedwater heaters, that holds the key to understanding the unique thermodynamic nature of the bottoming cycle. This will be looked at below in detail.

8.2.1 THEORY

The two cardinal performance parameters in the bottoming cycle are the HRSG effectiveness and the steam turbine/cycle efficiency. The product of the two (ignoring a small bottoming cycle loss term) is the *overall* bottoming cycle efficiency, i.e.,

$$\eta_{BC} = \eta_{HRSG} \cdot \eta_{ST}. \tag{8.3}$$

For the combined cycle, the simple formula for net thermal efficiency is

$$\eta_{CC,net} = \left(\eta_{GT} + (1-\eta_{GT}) \cdot \eta_{BC}\right) \cdot \eta_{Gen} \cdot (1-\alpha). \tag{8.4}$$

To simplify the discussion, let us use the combined cycle *gross* efficiency and rewrite Equation 8.4 as

$$\frac{\eta_{CC,gross}}{\eta_{GT}} = \left(1 + (1-\eta_{GT})\frac{\eta_{BC}}{\eta_{GT}}\right). \tag{8.5}$$

From Equation 8.5, it is obvious that, for a given gas turbine, maximizing combined cycle efficiency is equivalent to maximizing the bottoming cycle efficiency. Maximizing bottoming cycle efficiency, as evidenced by Equation 8.3, requires to strike a balance between HRSG effectiveness and steam turbine efficiency. It can be easily shown, graphically as well as numerically, that, everything else being equal, maximum combined cycle thermal efficiency is achieved at maximum heat recovery, at which point the bottoming cycle exergetic efficiency is also a maximum – as it should (but the bottoming cycle *thermal* efficiency is not!). While this might seem counterintuitive, it is a reflection of the dichotomy in the nature of a bottoming cycle, i.e..

- "Assistant" to the topping cycle (exhaust heat recovery)
- "Solo player" in its own right (bottoming cycle overall efficiency).

As an "assistant", the worth of the bottoming cycle is measured by the lowness of the exhaust gas stack temperature (maximum heat recovery effectiveness). For the cycle itself, however, as a "solo player", low stack temperature means low mean-effective cycle heat addition temperature, which, of course, hurts the cycle efficiency. Indeed, this was the reason in the first place for introduction of the feedwater heaters into the conventional fossil steam turbine cycle. Since the ultimate goal is the best possible *combined* cycle efficiency, exhaust gas heat recovery is more important than the bottoming cycle thermal efficiency.

This is best illustrated by a graphical comparison as shown in Figure 8.2, which has three different bottoming cycles for the same gas turbine topping cycle:

- Bottoming cycle **A** is a **low** heat recovery cycle with high stack temperature.
- Bottoming cycle **B** is a **medium** heat recovery cycle with lower stack temperature.;
- Bottoming cycle **C** is a **high** heat recovery cycle with lowest stack temperature – e.g., a 3PRH steam Rankine cycle.

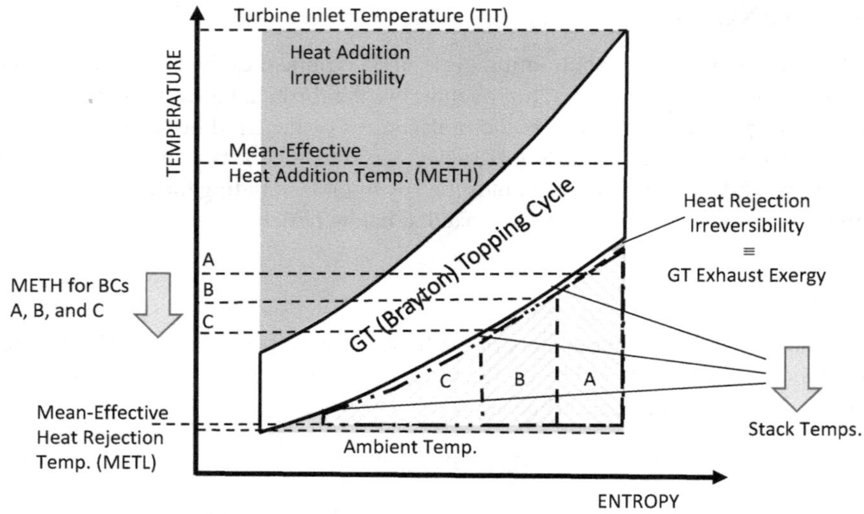

FIGURE 8.2 Illustration of the importance of exhaust gas heat recovery. In order not to clutter the graphic, state points are not numbered. Thus, be aware that cycle B comprises the area segments labeled A and B; cycle C comprises the area segments labeled A, B and C.

As shown in Figure 8.2, all three bottoming cycles have the *same* mean-effective *heat rejection* temperature, METL. However, in terms of mean-effective *heat addition* temperature, METH,

$$\mathrm{METH}_A > \mathrm{METH}_B > \mathrm{METH}_C.$$

This is so because

$$\mathrm{METH}_{A,B,C} \propto \frac{\mathrm{TEXH} - \mathrm{TSTCK}_{A,B,C}}{\ln\left(\dfrac{\mathrm{TEXH}}{\mathrm{TSTCK}_{A,B,C}}\right)},$$

where TEXH is the gas turbine exhaust temperature (same for all three cycles), and TSTCK is the HRSG stack temperature such that

$$\mathrm{TSTCK}_A > \mathrm{TSTCK}_B > \mathrm{TSTCK}_C.$$

At this point, the reader is reminded that

1. Bottoming cycle METH is the same as topping cycle (i.e., gas turbine) METL given by Equation 4.10.
2. Bottoming cycle METL is the saturated steam temperature in the steam turbine condenser, which is dictated by the ambient temperature T_1.
3. TEXH is T_4 in standard cycle notation.

Bottoming cycle *thermal* efficiency can be expressed as (noting that METL refers to the topping cycle)

$$\eta_{A,B,C} \propto 1 - \frac{T_1}{\mathrm{METL}_{A,B,C}}.$$

Therefore,

$$\eta_A > \eta_B > \eta_C.$$

Even with a high conversion (thermal) efficiency, however, a low heat recovery bottoming cycle, such as A, can only achieve a low fraction of the theoretically possible maximum quantified by the lower-right triangular area on the T-s diagram (i.e., low bottoming exergetic efficiency).[1] Only a high heat recovery bottoming cycle with the lowest possible stack gas temperature can approach the theoretical maximum (i.e., highest bottoming cycle exergetic efficiency) even with a more modest thermal efficiency. In the graphical example in Figure 8.2, that cycle is C.

In order to confirm this algebraically, consider that the HRSG effectiveness, let us call it ε for this exercise to prevent confusion with thermal efficiency, is

$$\varepsilon_{A,B,C} \propto 1 - \frac{TSTCK_{A,B,C}}{T_4},$$

and therefore,

$$\varepsilon_A > \varepsilon_B > \varepsilon_C.$$

The *overall* bottoming cycle efficiency, let us call it β for this exercise, is then

$$\beta = \varepsilon \cdot \eta.$$

Let us define that $\tau = TSTCK/T_4$ and $\mu = T_1/T_4$, so that the generic relationship for the overall bottoming cycle efficiency is

$$\beta \propto (1-\tau)\cdot\left(1-\mu\cdot\frac{\ln(\tau)}{\tau-1}\right).$$

Taking the derivative of β with respect to τ, we obtain

$$\frac{\partial\beta}{\partial\tau} \propto -1 + \frac{\mu}{\tau}.$$

Since $T_1 < TSTCK$, it follows that μ is less than τ; therefore, $\partial\beta/\partial\tau$ is *always less than zero.* In other words,

- Overall bottoming cycle efficiency β *always increases with decreasing* TSTCK.
- Overall bottoming cycle efficiency β *always increases with increasing* HRSG effectiveness even when bottoming cycle *thermal* efficiency *decreases.*

Consequently, since

$$TSTCK_A > TSTCK_B > TSTCK_C.$$

HRSG effectiveness ranking of bottoming cycles A, B and C is

$$\varepsilon_A < \varepsilon_B < \varepsilon_C$$

and *overall* bottoming cycle efficiency ranking is

$$\beta_A < \beta_B < \beta_C$$

---

[1] Theoretical maximum of combined cycle efficiency is when the gas turbine exhaust exergy (lower-right triangular area below the topping cycle) is completely converted into useful work – a practical impossibility.

even when bottoming cycle *thermal* efficiency ranking is

$$\eta_A > \eta_B > \eta_C.$$

8.2.2 Practice

Fundamental considerations outlined in Section 8.2.1 can be translated into practice by designing an HRSG to deliver the lowest possible stack temperature cost-effectively with minimum "harm" to the first-law efficiency of the steam cycle. How to accomplish that was described in detail in Section 6.1.

The state of the art in bottoming cycle design today is a three-pressure steam cycle with reheat. The first pressure level (i.e., HP) is chosen as high as possible to maximize the steam cycle first-law efficiency. Main (HP) steam temperature is chosen as high as possible to maximize HP steam exergy within the limits of the alloys used HP superheater tubes, and HP steam pipes and valves. The current maximum is set at 600°C (1,112°F).

Advanced class gas turbine exhaust conditions allow for single-reheat, i.e., reheating of HP turbine exhaust steam in the HRSG to the same temperature as the HP steam prior to the IP turbine inlet. The goal herein is the same as in conventional boiler-turbine power plants, i.e., increasing the METH of the Rankine cycle. Unlike some (mostly on paper) high-fired boiler plants, however, there is insufficient "driving energy" to allow for a second reheat. (In theory, this can be accomplished with supplementary firing, but the impact on the overall combined cycle efficiency would be detrimental.)

The stack gas temperature is dictated by the low-pressure (LP) section of the HRSG, specifically the LP evaporator pressure (see Section 6.2.4). There are three key considerations preventing the stack temperature from being set as low as possible (e.g., only a few degrees above the condensate temperature, which is dictated by the steam condenser pressure) even if the size/cost of the economizer is ignored:

- Sulfuric acid dew point
- Plume dispersion
- LP steam pipe size.

The first two are covered in detail in Chapter 6 (see Section 6.2.1 for plume dispersion and Section 6.2.4 for sulfuric acid dew point). The third one requires a closer look at steam velocity inside the LP steam pipe.

As a rule, very high fluid velocity inside a pipe will cause excessive friction and lead to erosion of the pipe from the inside, which can lead to very costly system failure. The density of the steam is proportional to the pressure so that the required pipe size increases with decreasing pressure. These factors will play against each other and impact bottoming steam cycle pipe sizing and cost.

The maximum recommended by the ASME boiler codes is 10,000 ft/min (~165 ft/s) for high-pressure saturated steam and 14,000 ft/min (~235 ft/s) for high-pressure superheated steam. Let us do a simple calculation. LP steam conditions at ISO baseload design point for a typical 50-Hz F class combined cycle with 65 psia LP evaporator pressure are summarized below:

	LP Superheater Exit	LP Admission Valve Inlet	LP Bowl
Pressure, psia	63.7	59.5	57.5
Temperature, °F	520	519.2	518.9
Flow, kpph	83	83	83
Specific volume, ft³/lb	9.04	9.68	10.02

The average specific volume of the steam in this pipe is 9.3 ft^3/lb, which translates into steam volumetric flow of about 12,950 cfm. The required diameter of the pipe can be found from

$$D = 12\sqrt{\frac{C}{0.7854 \cdot F}},$$

where C is the steam flow in cfm, D is the pipe diameter in inches and F is the permissible steam velocity in ft/min. Thus, the diameter for the example above can be calculated as

$$D = 12\sqrt{\frac{12{,}950}{0.7854 \cdot 14{,}000}} = 13.4 \text{ in.}$$

This would require a fairly large pipe with 14 in. outer diameter (OD). In order to reduce the HRSG stack temperature and improve the combined cycle performance, let us assume that the LP evaporator pressure is set to 55 psia. This increases the average volumetric flow to about 15,300 cfm, and the pipe diameter is calculated as 14.5 in. This value is high enough to drive the pipe selection to the next available size, which is 16 in. OD, and results in higher plant cost. *Prima facie*, another option is to specify a custom pipe size to fit the exact system requirements. However, designing pipe systems sized outside the standard availability (i.e., pipe schedules) is a very costly prospect. (The reader is cautioned that, in actual engineering design of the steam piping, the sample calculation above is carried out for steam conditions over the normal operating envelope of the plant and proper design margins are applied.) Typically, LP admission valve floor pressure (the lowest pressure upstream of the valve set by the valve stroke) is set to ensure that steam pressures and velocities at all operating conditions are commensurate with cost-effective specification of pipe sizing.

Intermediate pressure (IP) selection is subject to cycle optimization within certain limits set by other considerations, e.g.,

- Steam turbine mechanical design (in particular, HP-IP turbine thrust balance)
- Feedwater requisite for performance fuel heating
- HP turbine exhaust (cold reheat or CRH) temperature and CRH steam pipe size/material and cost
- LP turbine exhaust moisture content.

For example, Figure 8.3 shows the maximum possible fuel gas temperature with a shell-and-tube type heat exchanger (with a design approach temperature difference of 30°F (16.6°C)) where natural gas from the pipeline at 80°F (26.6°C) is heated using IP economizer water. For vintage F class gas turbines in late 1990s and early 2000s, fuel heating to 365°F (185°C) was the standard; current practice with advanced class gas turbines is to go as high as 440°F (226.6°C), which pushes the IP pressure upward.

For a sample optimization of an F class combined cycle bottoming cycle, please refer to the paper by Gülen and Jacobs [1]. Figure 8.4 shows the relative impact of bottoming cycle design parameters on combined cycle net output. These sensitivities have been calculated assuming a base bottoming cycle design with main steam conditions of 1,400 psig/1,050°F/1,050°F (96.2 barg/565°C/565°C). Main (HP) and hot reheat (HRH) steam temperatures are by far the dominant parameters; 1% increase in throttle steam and reheat steam temperature increases combined cycle net output by 0.035% and 0.045%, respectively. As an example, consider that a 50°F (27.7°C) or 4.55% increase in HP and reheat steam temperatures (separately) results in increases in net combined cycle output of approximately 0.16% and 0.21% respectively. Since gas turbine heat consumption is constant, this translates into 0.2–0.25 percentage points in net combined cycle efficiency. Throttle steam pressure is the third most significant parameter and yields a net increase in combined cycle output of

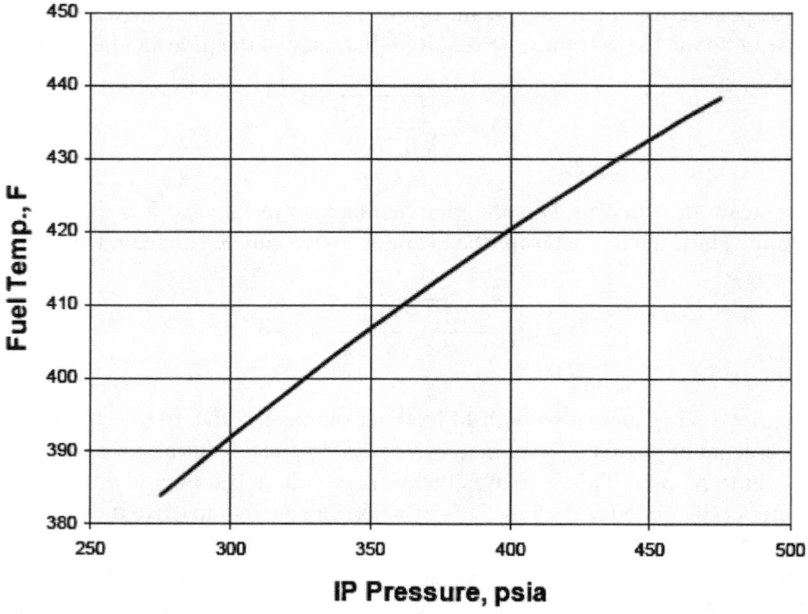

FIGURE 8.3 Fuel gas temperature heated with IP economizer bleed in a 30°F (16.6°C) approach heat exchanger [1].

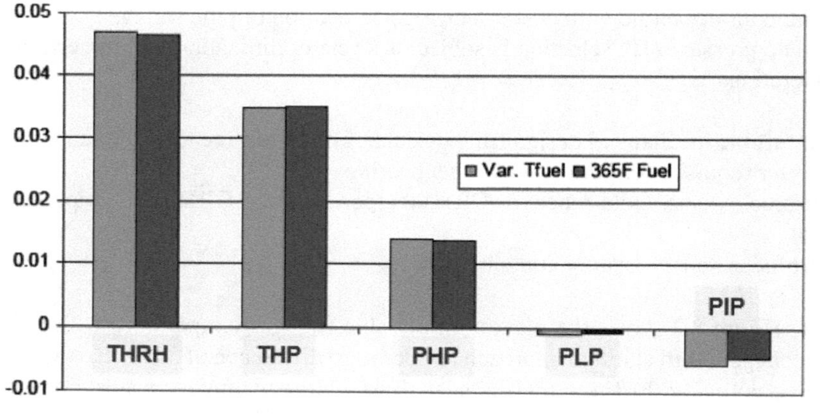

FIGURE 8.4 Combined cycle net output sensitivity to bottoming cycle design parameters (THRH, hot reheat steam temperature; THP, HP steam temperature; PHP, HP steam pressure; PIP, IP steam pressure; PLP, LP steam pressure) [1].

approximately 0.014% per percentage point increase in its value. It should be noted that this effect diminishes at higher throttle pressure levels, which will be explained below.

In particular, further increases in HP steam pressure beyond 2,200 psig have diminishing impact on combined cycle performance. Increasing HP pressure at given IP and LP pressures results in decreasing HP steam flow (less steam produced at higher HP evaporator temperature) and increasing IP and LP steam flows. The net effect is a decrease in IP and LP section flows and outputs, while HP section output increases for an increase in overall steam turbine shaft output. The favorable trade-off between increasing available energy and decreasing steam flow in HP section tapers off

TABLE 8.1
GTCC Bottoming Cycle Design Parameters ("Ball Park" Numbers; Even More Aggressive Designs Have Been Offered)

	Units	Aggressive	Moderate
HP/IP/LP steam pressures	barg	165/40/5	125/30/5
Main/reheat steam	°C	600/600	565/565
IP/LP steam	°C	315/315	260/260
GT fuel temp	°C	225	185
HRSG stack	°C	75	82
HP/IP/LP evaporator pinch	°C	7/7/7	10/12/10
Economizer approach	°C	3/6	4/6
ST HP/IP/LP efficiencies	% pts	+2.5/+1/+2.5	Base
ST LPT exhaust annulus	m^2	17.3	11.0

at higher pressures. Diminishing returns with higher HP pressure are made worse by increasing power consumption of the boiler feed pump. Especially, in markets with low fuel prices (e.g., the USA with $3 to $4 natural gas after the shale gas boon), justifying higher steam pressures and the accompanying cost is quite difficult.

In general, anytime a steam cycle conceptual design moves outside the standard design experience by a significant margin, i.e., 10% or more, steam pipe size and cost impacts (among other things) should be considered. (See Section 10.2 for more on this subject.)

State-of-the-art design parameters are summarized in Table 8.1, which has two columns: moderate and aggressive. Admittedly, these qualitative monikers are somewhat arbitrary (e.g., why 125 barg for the "moderate" HP steam pressure and not, say, 115 barg?). Unfortunately, there is no purely physics-based, clear-cut delineation that one can use as a yardstick. This difficulty can be traced back to the fact that there is no fixed product family classification for the bottoming cycles analogous to the "class hierarchy" for heavy-duty industrial gas turbines (i.e., the "topping cycle" of the combined cycle). Thus, some amount of fuzziness in labeling bottoming steam cycle designs is unavoidable.

8.3 COMBINED CYCLE

8.3.1 Second Law Analysis

Heat and mass balance analysis is based on direct application of the two conservation laws to the gas turbine control volume:

1. Conservation of energy (first law of thermodynamics)
2. Conservation of mass.

In this section, we look at the problem from a second law (of thermodynamics) perspective. The discussion below is rigorous in terms of the fundamental underlying concept and its significance, but the derivation is pedagogically oriented to practitioners in the field and makes use of graphical aids. For a more scholarly treatment of the subject, specifically aimed at gas turbine combined cycle (GTCC) power plants, the reader is referred to the paper by Elmasri [2]. Another scholarly treatment from a generic power plant perspective can be found in the paper by Sir Horlock [3]. The concept of "rational efficiency" used by Sir Horlock in his treatment is equivalent to the second law effectiveness used herein (the derivation is below). While Elmasri derives the formula for the "maximum possible work output" for the bottoming cycle, Horlock looks at the entire combined cycle using the maximum possible work output from the fuel supplied to the gas turbine combustor. The origins of rational efficiency can be traced to Haywood [4]. A more extensive treatment of exergy analysis of thermal power plants can be found in the book by Kotas [5]. For the history of the exergy concept, the reader is referred to the review paper by Sciubba and Wall [6].

As discussed earlier in Chapter 4 and Section 8.1, the triangular area {1-4-4c-1} in Figure 4.3. (or Figure 8.1) quantifies the Brayton cycle *heat rejection irreversibility.* If one could place another power cycle, which has a METH of METL (Equation 4.10) and METL of T_1, in there, the efficiency of that ideal *bottoming* cycle would be given by

$$\eta_{BC} = 1 - \frac{T_1}{METL}. \tag{8.6}$$

The work of the ideal cycle in question would be found as

$$w_{BC} = c_p(T_4 - T_1)\left(1 - \frac{T_1}{METL}\right). \tag{8.7}$$

Using Equation 4.10 for METL,

$$w_{BC} = c_p(T_4 - T_1) - T_1 c_p \ln\left(\frac{T_4}{T_1}\right). \tag{8.8}$$

For an ideal gas with constant specific heat (i.e., calorically "perfect" gas) undergoing an isobaric process (i.e., $p_4 = p_1$) between state points 4 and 1, Equation 8.8 is equivalent to

$$w_{BC} = (h_4 - h_1) - T_1(s_4 - s_1). \tag{8.9}$$

(You can refer to Moran & Shapiro or your favorite thermodynamics textbook for the requisite background.) A combination of any two (or more) thermodynamic properties is another thermodynamic property. Therefore, the right-hand side of Equation 8.9 quantifies the *change* in a thermodynamic property, which is known as *exergy* (or *availability* in US textbooks – more specifically, *flow* exergy or availability), between state points 4 and 1, i.e.,

$$w_{BC} = a_4 - a_1, \tag{8.10}$$

where

$$a_4 = h_4 - T_1 s_4$$

and

$$a_1 = h_1 - T_1 s_1.$$

Exergy is typically defined with respect to a reference state or "dead state", at which the working fluid is at equilibrium with its surroundings with zero work potential. Thus, the full definition of flow exergy (ignoring the contribution of kinetic, potential and chemical – for nonreacting systems – energy terms) is

$$a = (h - h_{ref}) - T_{ref}(s - s_{ref}).$$

In heat engine cycle analysis, the logical choice for the reference or dead state is the ambient, which is defined by T_1 and p_1. Consequently, Equation 8.10 can be simplified to

$$w_{BC} = a_4, \tag{8.11}$$

which is the most crucial equation in the thermodynamic analysis of GTCCs. Equation 8.11 tells us that

1. Maximum work of a heat engine, whose energy source is the gas turbine exhaust gas stream with temperature T_4 (and pressure p_4, to be exact) is exactly equal to the exergy of the said exhaust stream, which can be calculated from a suitable equation of state (say, JANAF tables) with T_4, p_4 and the gas composition.
2. The ideal heat engine, whose work output is given by the exergy of the gas turbine exhaust gas stream, is equivalent to a Carnot cycle operating between the two temperature reservoirs at METL and T_1.

For a gas stream at T_4, p_4 and composition $\{x_4\}$, METL can be calculated *exactly* as

$$\text{METL} = \frac{(h_4 - h_1)}{(s_4 - s_1)} \tag{8.12}$$

using a suitable equation of state. For a calorically perfect gas, Equation 4.10 is a reasonably good approximation.

In the foregoing analysis, we have derived the "engineering version" of the Kelvin–Planck statement of the second law of thermodynamics when applied to a heat engine operating on a power cycle (e.g., Brayton or Rankine)

> The maximum possible efficiency of a heat engine operating on a power cycle is given by the efficiency of a Carnot cycle operating between two thermal reservoirs, one at the mean-effective heat addition temperature and the other at the mean-effective heat rejection temperature of the said engine.

By the same fundamental principle,

1. The maximum possible efficiency of a gas turbine operating on a Brayton cycle is given by the efficiency of a Carnot cycle operating between two thermal reservoirs, one at the METH (from Equation 4.7) and the other at the METL (from Equation 4.10) of the said cycle.
2. The maximum possible efficiency of a gas turbine operating on a combined Brayton–Rankine cycle is given by the efficiency of a Carnot cycle operating between two thermal reservoirs, one at the METH (from Equation 4.7) and the other at the METL (T_1) of the said cycle.

The advantage of the second law analysis is the presence of an unambiguous upper limit to process efficiency, which is defined by a law of physics. For example, when asked about the maximum possible efficiency of a bottoming steam Rankine cycle, it is impossible to provide a definitive answer, say, 43% or 44%, without going through rigorous calculations. The answer to "What is the maximum possible *second law* efficiency (or, more aptly, *effectiveness*) of a bottoming steam Rankine cycle?", on the other hand, is obvious: 100%. This is so because the maximum amount of work that can be produced by the said cycle is exactly equal to the gas turbine exhaust gas exergy. Thus, when one calculates, say, 41.7% for the first law efficiency, it is difficult to gauge the "goodness" of the cycle in question. However, 75% second law effectiveness, e.g., tells us how good a design is by its proximity to the ultimate goal, i.e., 100% signifying a Carnot-equivalent ideal cycle.

Note that the term "efficiency" is appropriate when applied to the first law analysis, i.e., "the ratio of the useful work performed by a machine or in a process to the total energy expended or heat taken in". The second law analysis results in a ratio, which best fits the definition of the term "effectiveness", i.e., "the degree to which something is successful in producing a desired result". In the case of the bottoming steam Rankine cycle of a GTCC power plant, the "desired result" is perfect heat engine work (i.e., Carnot engine work). What is successfully achieved via state-of-the-art engineering design is the steam turbine generator (STG) output minus the auxiliary power consumption, which is the net bottoming cycle power output. The ratio of the two is the *second law effectiveness* (or "rational efficiency" in some references). Nevertheless, going with the established practice and, to a lesser extent, for pedagogical reasons, the term "bottoming cycle *exergetic efficiency*" will be used herein.

Let us put some numbers into this discussion. From a first law perspective, bottoming cycle gross output is given by

$$w_{BC} = h_4\big|_{T_{ref}} \eta_{HRSG}\eta_{ST}, \tag{8.13}$$

where the enthalpy of gas turbine exhaust gas (state point 4 in Brayton cycle) is with a zero-enthalpy temperature reference of T_{ref}. From a second law perspective, bottoming cycle work is given by

$$w_{BC} = a_4\big|_{T_{ref}} \varepsilon_{BC}, \tag{8.14}$$

where the gas turbine exhaust gas exergy is calculated with the same temperature reference, and ε_{BC} denotes the bottoming cycle exergetic efficiency. Thus,

$$\varepsilon_{BC} = \frac{h_4\big|_{T_{ref}}}{a_4\big|_{T_{ref}}} \eta_{HRSG}\eta_{ST}. \tag{8.15}$$

For a calorically perfect gas and assuming $T_{ref} = T_1$,

$$\varepsilon_{BC} = \frac{(T_4 - T_1)}{(T_4 - T_1) - T_1 \ln\left(\dfrac{T_4}{T_1}\right)} \eta_{HRSG}\eta_{ST}, \tag{8.16}$$

$$\varepsilon_{BC} = \frac{\eta_{HRSG}\eta_{ST}}{1 - \dfrac{T_1}{METL}} = \frac{\eta_{BC}}{1 - \dfrac{T_1}{METL}} \tag{8.17}$$

or

$$\eta_{BC} = \left(1 - \frac{T_1}{METL}\right)\varepsilon_{BC}. \tag{8.18}$$

In Equation 8.18, the term in parentheses on the right-hand side is easy to recognize as the ideal bottoming cycle efficiency. (Note that Equation 8.18 is identical to Equation 51 in Elmasri [2].) Before moving on, it should be emphasized that ε_{BC} is the counterpart of the technology factor, F_T, developed earlier for the Brayton gas turbine cycle.

For modern gas turbines, T_4 is between 1,100°F and 1,200°F and for ISO conditions T_1 = 59°F. A reasonable range for the second law effectiveness is 0.7–0.8. Once again, note that 1.0 means a perfect heat engine (essentially a Carnot engine), i.e., a practical impossibility. Equation 8.18 is plotted in Figure 8.5 in order to get an idea about what to expect (realistically) in terms of first law efficiency of the bottoming steam Rankine cycle.

Based on the picture depicted in Figure 8.5, thermal efficiency of the bottoming cycle in a state-of-the-art GTCC is about the same as that of a typical coal-fired boiler-steam turbine power plant.

For the combined cycle, the simple formula for net thermal efficiency was given by Equation 8.4. From a second law perspective, a similar simple formula can be derived as well. Translating the foregoing discussion into a straightforward performance roll-up (unfired GTCC) yields

$$\dot{W}_{CC,net} = \left(\dot{W}_{GT} + \dot{m}_{exh} a_{exh} \varepsilon_{BC}\right) \cdot \eta_{Gen} \cdot (1-\alpha). \tag{8.19}$$

Dividing by gas turbine heat consumption, we obtain

$$\eta_{CC,net} = \left(\eta_{GT} + \frac{\dot{m}_{exh} a_{exh}}{HC} \varepsilon_{BC}\right) \cdot \eta_{Gen} \cdot (1-\alpha). \tag{8.20}$$

Exhaust exergy, i.e., $a_4|_{T_{ref}}$ in Equation 8.14, can be estimated using a simple transfer function as a function of exhaust gas temperature (e.g., see Equations A.2–A.4 in the Appendix).

Bottoming cycle exergetic efficiency can be estimated using Equation C.1 in the Appendix. Note that Equation C.1 is based on "net" bottoming cycle output, which is the difference between the steam generator output and the power consumed by the cycle feed pumps. Thus, when using Equation C.1 for bottoming cycle exergetic efficiency, Equation 8.20 requires a small correction, i.e.,

$$\eta_{CC,net} = \left(\eta_{GT} + \frac{\dot{m}_{exh} a_{exh}}{HC} \frac{\varepsilon_{BC}}{1-\lambda_{fp}}\right) \cdot \eta_{Gen} \cdot (1-\alpha), \tag{8.21}$$

where λ_{fp} represents the feed pump power consumption as a fraction of steam generator output (from Table 3.1, a typical value is about 2%).

8.3.2 Optimum Combined Cycle Efficiency

Ignoring the miscellaneous topping and bottoming cycle losses and minor inputs, a simplified version for the combined cycle efficiency in Equation 8.4 can also be written as

$$\eta_{CC} = \eta_{TC} + (1-\eta_{TC}) \cdot \eta_{BC}, \tag{8.22}$$

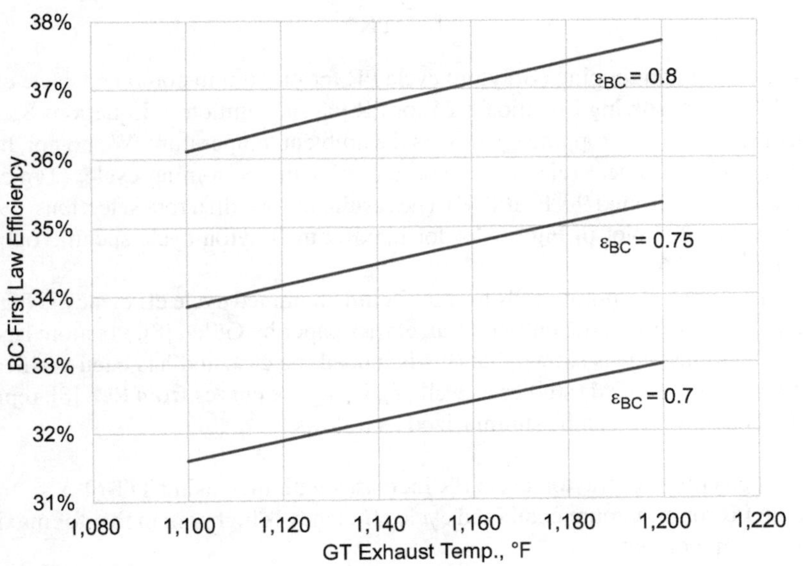

FIGURE 8.5 Bottoming cycle first law efficiency (Equation 8.18).

where subscripts TC and BC denote topping and bottoming cycles, respectively. Taking the derivative of both sides of Equation 8.22 with respect to the PR and setting the combined cycle efficiency derivative to zero to find its maximum, and noting that the bottoming and topping cycle efficiencies are approximately the same in magnitude, we find that

$$\frac{\partial \eta_{TC}}{\partial PR} \approx -\frac{\partial \eta_{BC}}{\partial PR}. \tag{8.23}$$

This correlation states that the maximum combined cycle efficiency occurs at the point where the rate of increase of the topping cycle efficiency with PR is the same as the rate of decrease of the bottoming cycle efficiency. The rate of increase of the topping cycle efficiency with increasing PR can be found from the Brayton cycle efficiency as

$$\frac{\partial \eta_{TC}}{\partial PR} = \frac{k}{PR^{k+1}}. \tag{8.24}$$

Similarly, the rate of decrease of the bottoming cycle efficiency with increasing PR is found as

$$\frac{\partial \eta_{BC}}{\partial PR} = \frac{k}{PR} \cdot \frac{\left(1 - \frac{T_4}{METL}\right)}{\left(\frac{T_4}{T_1} - 1\right)} \tag{8.25}$$

with

$$\frac{T_4}{METL} = \frac{T_4}{T_1} \cdot \frac{\ln\left(\frac{T_4}{T_1}\right)}{\left(\frac{T_4}{T_1} - 1\right)} \tag{8.26}$$

and

$$\frac{T_4}{T_1} = \frac{\tau_3}{PR^k}. \tag{8.27}$$

One can find the optimum topping (Brayton) cycle PR for maximum combined cycle efficiency as a function of TIT (T_3) by solving Equation 8.23 for PR via substitution of Equations 8.24–8.27. Note that the logical choice for the topping cycle T_1 is the ambient temperature. We do not, however, have to use it as the isothermal heat rejection temperature for the bottoming cycle. (Typical condenser steam temperatures are around 90°F–100°F.) The results for two different selections, 59°F and 95°F, are superimposed on the plot in Figure 4.4 for maximum Brayton cycle specific output, which is reproduced in Figure 8.6.

As shown in Figure 8.6, optimum PR for maximum combined cycle efficiency is higher than that for maximum gas turbine specific output. In an earlier paper by Gülen [8], via more rigorous cooled gas turbine calculations, it was shown that this is indeed the case and has been verified by original equipment manufacturer (OEM) studies as well [7]. Using the curves from Ref. [8], reproduced here in Figure 8.7, general trends can be summarized as follows:

1. Optimum gas turbine Brayton cycle PR increases with increasing TIT (T_3).
2. Optimum PR for maximum combined cycle efficiency is higher than that for maximum gas turbine specific output.
3. The difference in combined cycle efficiency (i) at optimum PR for gas turbine specific output and (ii) at the "true" optimum is negligible.

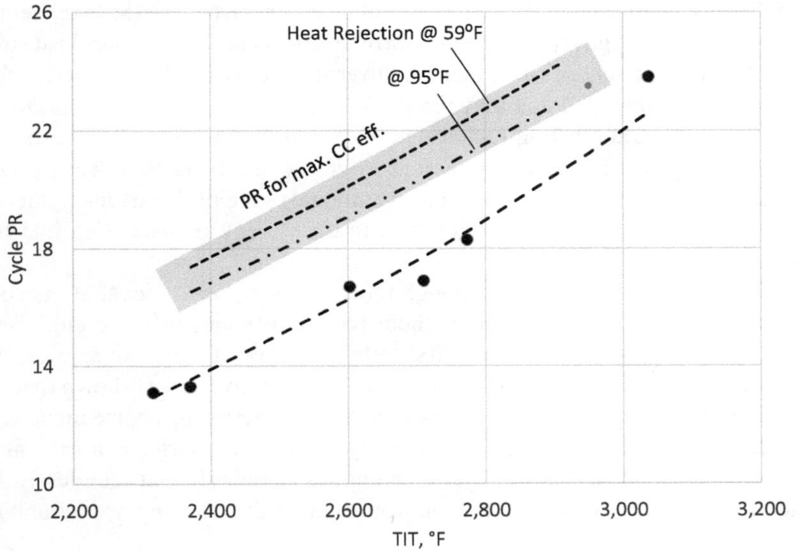

FIGURE 8.6 GE 50-Hz product line PR-TIT data (maximum CC efficiency curves from Equations 8.23–8.27).

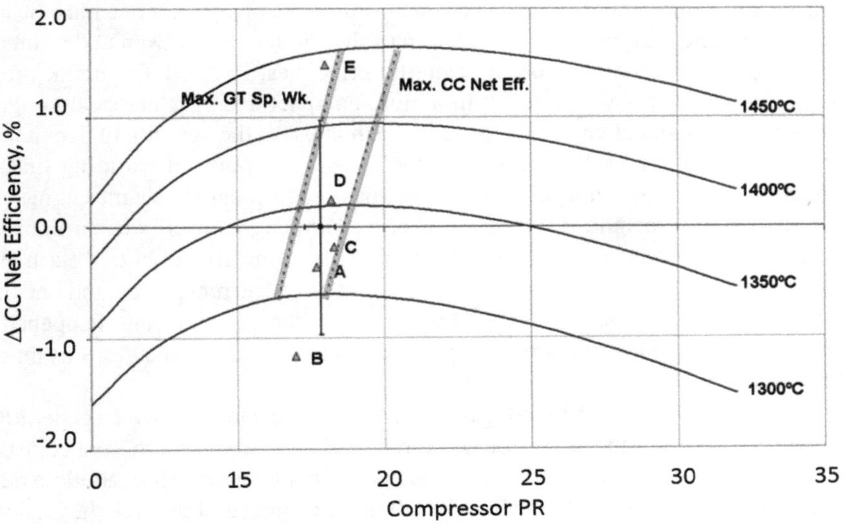

FIGURE 8.7 Relative CC net efficiency as a function of GT Brayton cycle PR and RIT. The published CC rating data is from Turbomachinery International 2009 Handbook. The origin is for the "average" state-of-the-art (2009 vintage!) F class air-cooled GT [8].

4. Since the cost-effective choice leads to a gas turbine design at a lower PR, this is where OEMs set their product designs.

8.4 HISTORY

The reader may wonder "why the need for a history?" or "why here in the middle of the book and not in the beginning?" The reason for presenting a concise history of the evolution of the combined cycle concept at this particular point in the treatise is to establish the symbiotic relationship between the science of thermodynamics and thermal power plant engineering. For those readers who already went through the chapters on gas turbine, steam turbine, HRSG and condenser (heat sink) and found the second law-based treatment of fundamental, governing equations too "esoteric" for a practical book, it is strongly recommended that they read this chapter with an open mind. This is where the author will try to demonstrate that, unless armed with a strong grasp of those "esoteric" scientific principles, even an engineering genius is bound to go awry. At the very least, especially when hampered by the "horse blinders" of existing (conventional) state of the art in engineering technology, one cannot envision the possibilities inherent in the system to chart a feasible future path to technology growth.

At this point, the reader should have enough tools in his or her "arsenal" to accomplish basic combined cycle performance calculations without having to resort to a pre-engineered software tool. The biggest help in this endeavor is the first and second law of thermodynamics. In the preceding sections of this chapter, in a self-contained albeit brief narrative, it was shown that the combined Brayton–Rankine cycle concept naturally followed from a desire to approximate a Carnot heat engine as closely as possible. Alas, the combined cycle did not emerge as a natural extension of the Brayton cycle via the second law of thermodynamics as quickly and seamlessly as one would expect. In fact, this has been the case since the beginning of the heat engine technology as will be discussed (briefly) below.

Thomas Newcomen and James Watt came up with their steam engine concepts and jump-started the Industrial Revolution in the nineteenth century decades before the science of thermodynamics matured into the vital branch of physics as we know it today. The steam engine of James Watt, for all its historical significance and the master craftsmanship of its inventor, if one must be honest, was an abomination from the perspective of the science of thermodynamics. Even at the time, albeit via intuition instead of application of rigorous scientific principles, Richard Trevitchik proposed that using steam at high pressures would result in a more compact and efficient steam engine. Today, even a second-year mechanical engineering student can attest to the veracity of Trevitchik's claim. Watt, on the other hand, was fully opposed to the idea to the point of claiming that Trevitchik deserved hanging for bringing into use the "dangerous" high-pressure steam engine. As is well known, of course, Watt's vacuum-driven atmospheric steam engine itself was an improvement on Thomas Newcomen's engine via a separate condenser.[2] (The same kind of bitter "clash of technologies" between a genius inventor, i.e., Edison on the side of direct current power, and one of the greatest scientific minds of all times, i.e., Tesla on the side of alternating current, happened a century later. Edison even went so far as pointing to the electric chair used in executions as an example of dangers of the ac technology.)

Let us now return to the combined gas-steam turbine cycle for power generation. In his pioneering work on steam and gas turbines, Aurel Stodola introduced the concept but did not spend a lot of time on it [9]. On page 1194 of Volume 2 of his *magnum opus*, Stodola describes an "explosion turbine utilizing the exhaust heat in a steam turbine" and derives the basic combined cycle efficiency relationship. It is highly likely that Stodola was inspired by the "explosion turbine" of Holzwarth, which he describes on pp. 1237–1251 at length and credits him for "having built the first economically practical gas turbine." This unique machine was the combination of an "explosion chamber" and a velocity-compounded two-stage turbine (also known as a *Curtis wheel*). In the former, constant-volume combustion, analogous to that in the cylinder of an internal combustion engine, added heat to the cycle while simultaneously increasing the pressure of the working fluid. The exhaust gas from the turbine was used in a "waste heat economizer" to

[2] James Watt maintained that high-pressure steam was "criminally insane" although his initial concepts recognized the possibility and superiority of it vis-à-vis the atmospheric variant, which depended on very low pressure (i.e., vacuum) on the other side of the piston. It is not clear whether this was a genuine concern of his or he was simply acting to protect his and his partner's (Matthew Boulton) commercial interests.

make steam for a steam turbine driving the turbo-blower, which provided the charge air to the explosion chamber (see Figure 8.8). This was a *bona fide* GTCC with a bottoming Rankine steam cycle. (Note that, in theory at least, a modern combined cycle can be constructed as shown in Figure 8.8 because the steam turbine output is roughly equal to the power consumed by the gas turbine compressor.)

Another inventive heat engine concept that made its way into Stodola's book was Emmet's mercury-steam "binary cycle". In this concept, both the topping and bottoming cycles are Rankine cycles. The working fluid in the topping cycle is mercury; the working fluid in the bottoming cycle is water/steam. The first commercial unit of 10 MWe rating went into commercial operation in 1929. Other plants with the binary mercury-steam cycle were built in several states. The last one was built in 1950 in New Hampshire (Schiller Station in Portsmouth). The mercury boiler operated at 200 psia and 1,020°F. Heat from the condensing mercury at 505°F was used to boil water at 540 psia and 475°F, which was then heated to 800°F in a coal-fired superheater [10]. Nominal rating of the power plant was 53 MWe (23 MWe from the mercury turbine) at 39% efficiency (higher heating value or HHV basis). Significant complexity associated with handling poisonous mercury inventory and the additional equipment, combined with the improving efficiency of coal-fired steam power plants with supercritical steam conditions, rendered the binary cycle obsolete.

Until 1960s, the term "combined cycle" was used for "greenfield" or "repowered" fossil fuel-fired conventional or power boiler-steam turbine power plants. In fact, the first commercial application of a gas turbine for electric power generation in the United States was GE's Frame 3 at the Belle Isle Station of Oklahoma Gas & Electric (1949). The exhaust gas from the 3.5 MWe gas turbine was used for boiler feedwater heating in an economizer with bare (unfinned) tubes. The power boiler unit was rated 35 MWe. A similar system was added to the same station in 1952.

There were basically three types of gas turbine-assisted steam power plant configurations:

1. Boiler feedwater heating
2. Supercharged (Velox) boiler
3. Exhaust boiler ("hot wind box").

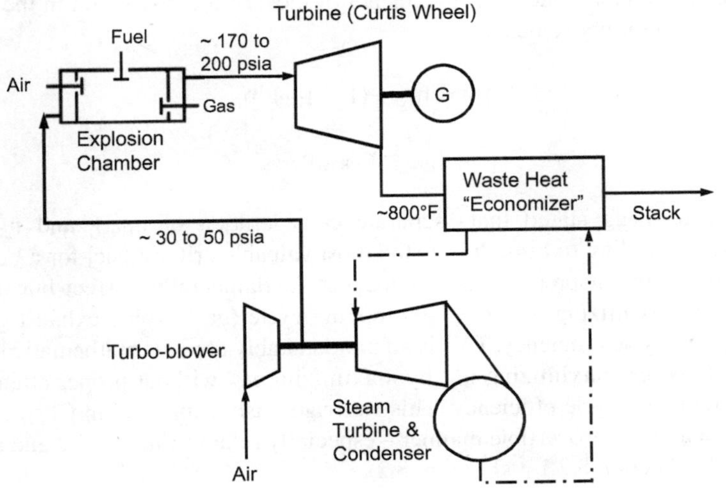

FIGURE 8.8 Schematic description of Holzwarth's "explosion" turbine [9].

These three options including several variants of feedwater heating are described in detail by late Dick Foster-Pegg [11]. In the supercharged boiler application, the gas turbine combustor is replaced by the coal-fired boiler, which utilizes compressed air (hence the term "supercharged") as combustion air and the combustion products are expanded in the turbine after steam production. Earlier versions did not extract enough net shaft power from the gas turbine to drive a generator; turbine output was just sufficient to drive the supercharger compressor.

The most common type of gas turbine-steam power plant combination is the "hot wind box" application, where the gas turbine exhaust gas is the combustion air for the power boiler. An enlarged economizer replaces the air preheater, and HP steam extraction for feedwater heating is eliminated. The heat rate of this type of combined cycle was only 5%–6% better than that of the base steam power plant with a conventional boiler. As reported by Foster-Pegg, up until 1966, there were only eight such units in the USA with a total capacity of about 660 MW, less than 1% of total installed capacity at the time. Clearly, the "combined" cycle concept was not a big hit among the utilities at the time [11].

In their 1960 paper on the theory of gas-steam turbine combined cycles, Seippel and Bereuter cited Stodola and Schröder (in 1936) as early proponents[3] of "harnessing gas and steam turbines together" and explained the reason for the dearth of combined cycle power plants partly due to a "dislike of complication in operation" [12]. The other obvious reason, not specifically mentioned by the authors, was that the added cost and complexity did not result in a dramatic increase in efficiency due to the "weak" exhaust gas temperatures of gas turbines of the time (about 700°F or 370°C), which was too low for HP steam production or reheat (about 1,000°F or 540°C for the then-prevalent state of the art). Most of the fuel burn still took place in the power boiler (mainly coal) with a much smaller amount of distillate firing in the gas turbine.

The beginning of the rise of combined cycle power plants was tied to the cost-effective manufacturing of finned-tube HRSGs in the late 1950s (see Section 6.2 for details). As mentioned above, gas turbine firing and exhaust gas temperatures at the time were still too low for high pressure and temperature steam production (let alone reheat) for efficient power generation. Thus, cogeneration units in chemical process and refinery plants formed the majority of commercial applications. It was not until late 1960s and early 1970s that the combined gas-steam turbine power plants came to their own in terms of number of plants installed and total installed capacity [13]. Even as late as 1978, however, a proper thermodynamic understanding of the role of the bottoming cycle in a combined cycle was lacking. In his 1978 paper, Foster-Pegg used the basic combined cycle formula, i.e., Equation 8.24, which was the anchor of most of the discussion in the earlier chapters and is repeated below for convenience:

$$\eta_{CC} = \eta_{TC} + (1 - \eta_{TC}) \cdot \eta_{BC}$$

$$\eta_{BC} = \eta_{HRSG}\eta_{ST}.$$

In his paper, Foster-Pegg stated that "separate consideration of η_{HRSG} and η_{ST} can be most misleading" [14]. According to him, "η_{BC} is the most valuable criteria [sic] for a bottoming cycle". As shown rigorously in Chapter 6, this statement is fundamentally correct but incomplete. It is certainly true that maximizing η_{BC} for given topping cycle (gas turbine exhaust gas energy) will maximize combined cycle efficiency. This is an unmistakably obvious mathematical certainty from Equation 8.24. However, maximizing η_{BC} by maximizing η_{ST} without proper attention to η_{HRSG} is detrimental to combined cycle efficiency. This was rigorously explained in Chapter 6 numerically and graphically in an incontrovertible manner – especially refer to Figure 6.12 and the accompanying discussion (also Section 8.2.1 and Figure 8.2).

[3] Seippel also mentions that "numerous patents, many of which would not stand up to a novelty search, describe different variants".

Still influenced by conventional fossil-fired boiler-steam turbine power plant thermodynamics, which aims to maximize cycle METH by maximizing boiler feedwater inlet temperature (utilizing steam extracted from the steam turbine), Foster-Pegg came up with bottoming cycle configurations with feedwater heaters [14]. This, of course, is an absolute GTCC efficiency killer because it prevents the designer from achieving the lowest possible HRSG stack temperature. This "mortal sin" in bottoming cycle thermodynamics was a direct result of the fact that Foster-Pegg exclusively focused on single-pressure bottoming cycles. (In that limited design space, of course, he indeed achieved maximum possible efficiencies.) As discussed in detail in Chapter 6, the proper route to maximizing η_{BC} is a two-pronged attack, i.e., a multi-pressure HRSG design (unfired and no feedwater heating – both detrimental to efficiency) with one pressure level low enough to minimize stack temperature and maximize η_{HRSG} and one pressure level high enough to maximize η_{ST}.

In all fairness, however, it should also be mentioned that in the 1970s when Foster-Pegg did his pioneering work, liquid fuels of varying quality, from number 2 distillate to heavy residuals, were the primary fuels used in gas turbines. This put a cap on minimum possible stack temperature in order to prevent sulfuric acid condensation on economizer tubes (about 300°F or 150°C). This was another impediment to designing efficient bottoming cycles even if the designer were aware of the need to maximize heat recovery and minimize the stack temperature.

At the end of 1970s, GE had twenty combined cycle power plants (STAG for *Steam-And-Gas* in GE parlance) in operation – most of them in the USA and Puerto Rico with two power plants in Korea. A large number of them were B and E class Frame 7-based single-shaft 1 × 1 × 1 units (STAG 107B and 107E). There were several STAG 407s (4 × 4 × 1) and one 607 (6 × 6 × 1 in Portland, Oregon, commissioned in 1977). ISO base rating of 107E was 99.3 MWe and 44.5% lower heating value (LHV) with a 1PNR bottoming cycle and deaerating condenser, which was the state of the art in 1979 [15]. The 107B was rated at 81.7 MWe and 42.7% LHV.

These STAG bottoming cycles featured vertical HRSGs with forced circulation and horizontal finned tubes. Typical steam conditions were 500 psia (~35 bar) and 810°F–835°F (430°C–445°C). Since GE's Frame 7B/E gas turbines were "hot end-driven" (i.e., the ac generator was connected to the gas turbine shaft on the turbine side), the HRSG was connected to the gas turbine exhaust via an L-shaped duct. The straight-condensing, axial exhaust steam turbine was connected to the collector end of the ac generator rotor from the HP section through a flexible coupling. The simplicity and robustness of the design was confirmed by the fact that nearly four decades after commissioning, several of these combined cycle power plants were still in operation.[4]

Before the F class came on the scene, gas turbine exhaust gas energy did not have enough "oomph" to justify multi-pressure reheat bottoming cycles. The two basic cycle improvements offered by GE, provided that the economic criteria of the project at hand justified the added complexity and cost, were feedwater heating and a "poor man's" two-pressure cycle with a *deaerating steam supply heater* (DASSH). The former is a well-known steam cycle performance improvement method via increasing the bottoming cycle METH. This is conceptually easily explainable via the T-s diagram in Figure 6.12b, which is reproduced here in Figure 8.9. By increasing the feedwater inlet temperature to the HRSG, heat transfer to the steam cycle takes places at a higher mean-effective temperature. The drawback is the increase in the stack gas temperature and concomitant reduction in the HRSG effectiveness. As long as the improvement in steam cycle efficiency is large enough to more than compensate for the reduction in HRSG effectiveness, the net effect is an improvement in bottoming cycle efficiency.

The two-pressure steam cycle with DASSH was a quite ingenious scheme to improve the bottoming cycle efficiency by decreasing the stack gas temperature and increasing the HRSG effectiveness without harming the steam cycle efficiency. The cycle diagram is shown in Figure 8.10. A final economizer section right before the stack extracts additional energy from the exhaust gas and reduces the stack temperature for improved HRSG effectiveness. Pressurized feedwater heated

[4] The reader is referred to an excellent article in *Combined Cycle Journal's* Q2 2009 issue, pp. 146–152, for more details.

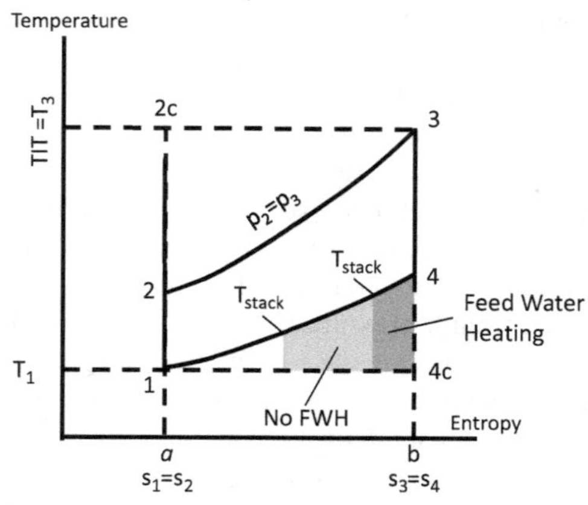

FIGURE 8.9 Impact of feedwater heating on bottoming cycle performance.

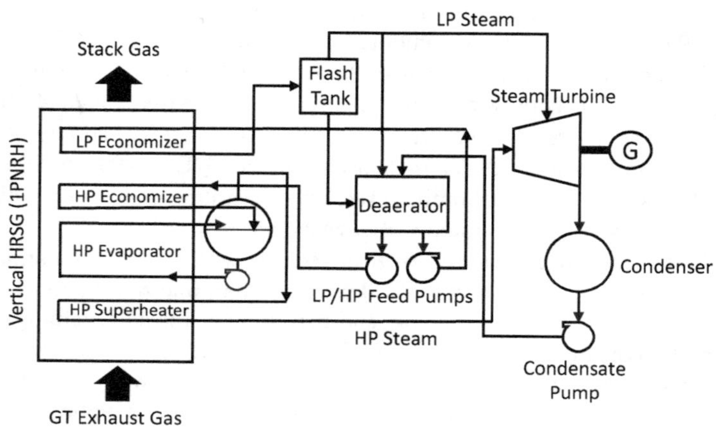

FIGURE 8.10 "Poor man's" two-pressure steam bottoming cycle.

in this economizer is flashed in a separate tank with the resulting LP steam sent to the steam turbine. Put another way, the flash tank constitutes a cheaper LP evaporator (without a superheater section), which would be present in a *bona fide* 2PNR HRSG and steam cycle. Doing this results in about 100°F drop in stack gas temperature and 6% increase in steam turbine output.

Korneuburg B combined cycle power plant near Vienna, Austria, was commissioned in 1980, and it represented the state of the art in bottoming steam cycle technology at the time. It was a 125 MWe unit with a BBC (Brown, Boveri & Cie) Type 13 gas turbine and 2PNR steam bottoming cycle. Its salient performance data is listed below:

- Natural gas-fired 81 MWe GTG with 29.4% net LHV efficiency
- 363 kg/s, 491°C exhaust gas conditions
- 33.2 bar, 433°C main (HP) steam
- 4.4 bar, 180°C LP steam
- 36 millibar (1.06 in. Hg), open-loop water-cooled condenser
- STG output 48,670 kW at generator terminals.

Stack temperature was 95°C (203°F), which did not present economizer corrosion and fouling problems with the low-sulfur natural gas fuel. Net LHV efficiency of the combined cycle power plant was 46.62%. A big contributor to the plant efficiency was the low condenser pressure with 8°C (~46°F) cooling water drawn from the river Danube.

In late 1970s, the US Department of Energy (DOE) initiated the *High-Temperature Turbine Technology* (HTTT) *Program* with the goal of 2,600°F (1,425°C) firing temperature (caution: not TIT) gas turbine. In an air-cooled gas turbine, this is equivalent to 1,600°C TIT, which was first achieved three decades later. General Electric's (GE) HTTT gas turbine technology included closed-loop water-cooled first-stage stator nozzle vanes [16]. The resulting gas turbine was envisioned to utilize syngas produced from coal gasification with a cycle PR of 12:1 and exhaust gas temperature of 1,300°F. In a 1982 paper, Rice and Jenkins proposed a reheat gas turbine with a cycle PR of 38:1 and compared its performance with the non-reheat HTTT variant with a PR of 12:1 [17]. The reheat gas turbine firing temperatures were 2,600°F and 2,200°F for the HP and LP turbines, respectively. The bottoming cycle for both gas turbines (exhaust temperature around 1,300°F) was a single-pressure reheat design with feedwater heating (2,400 psig main steam pressure with 1,000°F main and reheat steam temperatures). The authors calculated 57.4% net LHV efficiency for the reheat gas turbine vis-à-vis barely 51% net LHV for the non-reheat gas turbine [17].

The convoluted heat balance tables and diagrams in the Rice and Jenkins paper make a simple analysis difficult. The bottom line is that the steam-to-gas turbine output ratio of 0.4 with the single-pressure reheat bottoming cycle (with feedwater heating) was extremely poor for an advanced gas turbine with 1,300°F exhaust gas temperature. As a benchmark, it should be pointed out that with a modern 3PRH bottoming cycle, the ratio would be more like 0.55 and the combined cycle efficiency would be 63.5% net LHV. Note that the steam cycle itself had a first law efficiency of 40.43%, which is comparable to that of a 3PRH steam bottoming Rankine cycle. The culprit was (no surprise!) the low heat recovery effectiveness with 300°F stack gas temperature.

An OEM's perspective on the future of heavy-duty gas turbine technology in a paper published in 1983 is quite interesting in its modesty [18]. Both air-cooled and water-cooled technologies were considered with TITs up to 1,600°C (with water-cooled stator nozzle vanes and rotor blades). Alas, the highest combined cycle efficiency envisioned by the authors was about 52% LHV by the year 2000 (see Figure 8.11). This is a full 10 percentage points below of what is possible with advanced 3PRH steam cycle technology. Admittedly, the focus of the paper was gas turbine technology, and it is obvious that the authors did not bother to investigate bottoming cycle technology improvements. Nevertheless, this is a good illustration of the dangers inherent in ignoring the fundamental thermodynamics in favor of existing state of the art. Eventually, as noted on the chart in Figure 8.11, the projected efficiency levels (with 1,600°C, no less) were attained a decade earlier with air-cooled gas turbines with TITs well below 1,260°C. The key was, of course, the bottoming cycle.

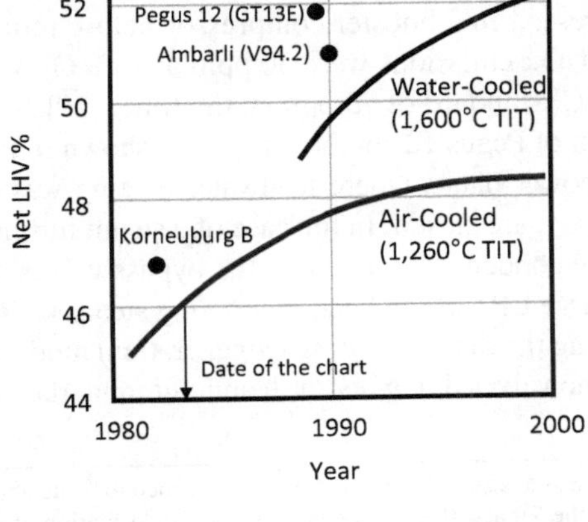

FIGURE 8.11 Projected combined cycle efficiency in 1983 (after Figure 6 in Ref. [18]).

In fact, the 50% benchmark was exceeded by 1990 with F class gas turbines. The F class, first introduced by GE in 1986 (Frame 7F), was as close as one could get to a "step change" in gas turbine technology. With 2,400°F (1,315°C) firing temperature and a cycle PR of 14:1 (with an 18-stage axial compressor), the F class was designed for optimum combined cycle efficiency (rather than simple cycle efficiency). Around the same time, other OEMs introduced their own F class gas turbines, i.e., Siemens V94.3 and Westinghouse and Mitsubishi W501F. Brown, Boveri & Cie (and its successor ABB) was also at work on their F class gas turbine, GT 15 (60 Hz, 150 MW) and GT 17 (50 Hz, 210 MWe) with a cycle PR of about 16:1 and TIT (per ISO-2314) of 1,160°C (2,120°F), which corresponds to a firing temperature of about 1,260°C (2,300°F). Alas, the development program, spanning a period of 15 years between 1976 and 1991, came to naught and was eventually dropped. The reader is referred to the fascinating book by Eckardt for the details of this interesting saga [19] (pp. 270–278).

There are two noteworthy combined cycle power plants, which established record efficiencies (for the time) at the end of the 1980s:

- Pegus 12 (Utrecht, Netherlands, 1989) with ABB GT13E and a three-pressure reheat bottoming cycle (223.5 MWe–51.78% LHV net)
- Ambarli (Istanbul, Turkey, 1990) with Siemens V94.2 and a two-pressure (no reheat) bottoming cycle (three 450 MWe blocks with 51% net LHV).

At the time, GT13E was the largest gas turbine in operation (147.2 MWe, 38%) with 491 kg/s (1,082 lb/s) airflow and 1,070°C TIT (per ISO-2314, about 1,170°C (~2,150°F) firing temperature). Cycle PR of GT13E was about 14:1 with exhaust gas temperature of 525°C (977°F). Steam cycle was 68.3 bar/494°C/448°C (990 psia/921°F/838°F) with 81.6 MWe steam turbine output [20]. Using Equation A.2, exhaust gas exergy is calculated as (adjusting for fuel flow and exhaust loss imposed by the HRSG)

$$A_{exh} = 1,082 \cdot (1+0.02) \cdot \left(0.1961 \cdot (977+6) - 86.918\right) \cdot 1.05506/1,000 \approx 123 \text{ MW}.$$

Thus, the bottoming cycle gross exergetic efficiency is 81.6/123 = 0.66, which is pretty good. It should be mentioned that Pegus 12 was a cogeneration plant supplying steam to a district steam heating facility. The efficiency cited above was when the plant was in straight condensing mode.

This impressive (for its time of course) efficiency was primarily driven by the gas turbine, which had a 5-stage turbine (still unique among all heavy-duty industrial gas turbines without reheat/sequential combustion). Increased compressor discharge temperature due to the cycle PR of 14.3:1 (increased from 11.8:1 for the earlier generation unit rated at 90 MWe) was accommodated by a complicated cooling air cooler scheme with heat rejection to LP feedwater and gas turbine fuel (in an indirect loop with an intermediate heat transfer medium – see Figure 8.12). The cooled air from the fuel heater was recompressed in a booster compressor before returning to the combustor [20]. As an interesting side note, NOx emissions were 50 ppm at 15% O_2 without diluent water or steam injection (well below the EEC[5] standard of 73 ppm at the time) [20].

A simplified cycle diagram of Pegus 12 combined cycle is shown in Figure 8.12 [20]. This was one of the first combined cycle power plants where feedwater heating was done in the HRSG. With no district heating, valves a, b and c are closed. In the case of a steam turbine trip, all steam generated in the HRSG is dumped into the condenser by opening the bypass valves d, e and f. In order to prevent the windage heating of the large LP turbine buckets, cooling steam is discharged to it via the valve j. If the steam turbine trips during the district heating (cogeneration) mode, steam can still be supplied to the steam users by opening the valves d, i, g and/or h and dumping the excess steam to the condenser

[5] European Economic Community was a regional organization which aimed to bring about economic integration among its member states. It was created by the Treaty of Rome of 1957. Upon the formation of the European Union (EU) in 1993, the EEC was incorporated and renamed as the European Community (EC).

through the valves e and f. Gas turbine can be run at part load to regulate steam generation without generating excess steam with the steam turbine offline in a regular cogeneration operation mode.

A reader with modern industry experience would immediately recognize that Pegus 12 bottoming cycle is effectively a 2PRH cycle. As shown in the cycle diagram in Figure 8.12, there is no third steam admission and no third HRSG evaporator section. Lowest pressure steam generation takes place in a flash tank in an arrangement very similar to that in GE's DASSH described earlier in the chapter (see Figure 8.10). Steam generated in the LP drum is sent to the deaerator. The IP turbine admission steam is the HRH steam from the HRSG. Clearly, the designers had to resort to a quite complex (albeit ingenious) steam cycle in order to squeeze every little bit of energy from the gas turbine exhaust at well below 1,000°F. At ISO baseload with no district heating, steam turbine to gas turbine output ratio is 81.6/144.5 = 0.565, which is very impressive. At maximum heat load of 182.7 MWth, the electric output is 209.1 MWe (157.8 MWe for the gas turbine at 0°C ambient).

Turning to the other noteworthy combined cycle power plant around 1990, its anchor, Siemens V94.2 (new designation SGT5–2000E), is also an E class engine with a cycle PR of about 11:1 and TIT (per ISO-2314) of 1,135°C. The bottoming steam cycle in Ambarli was straightforward 2PNR with

- 76 bar – 524°C HP steam (462 t/h)
- 6 bar – 197°C LP steam (93 t/h).

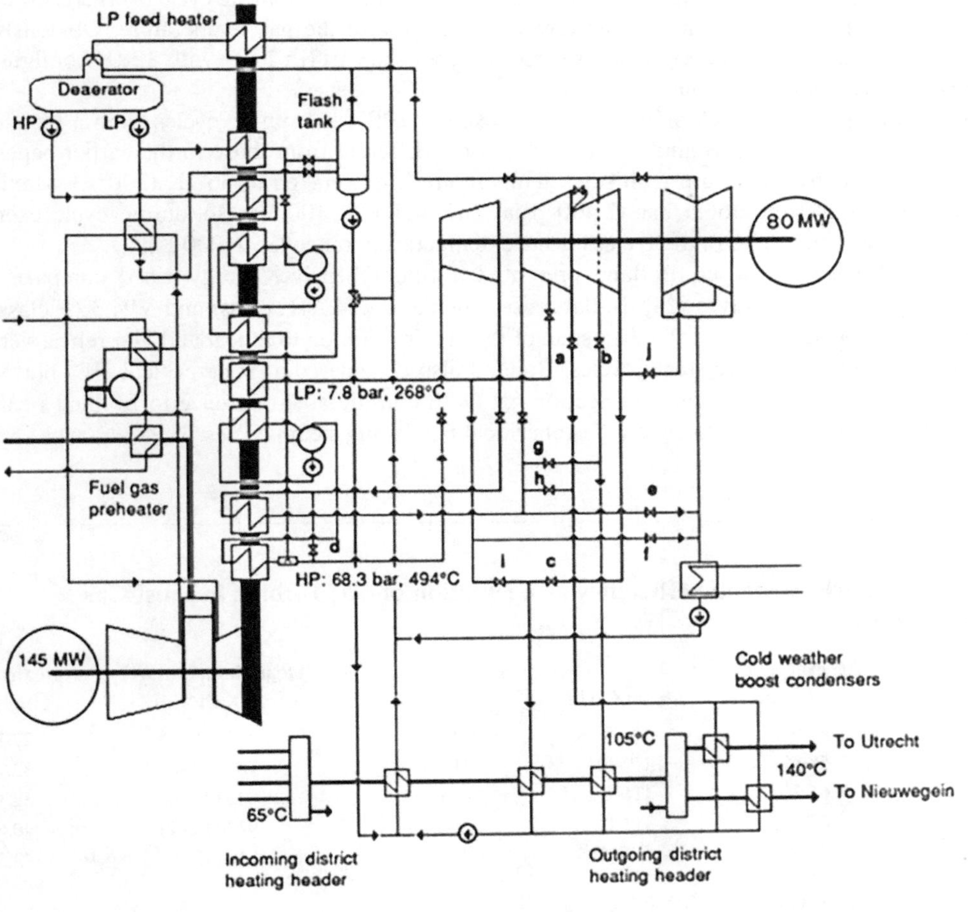

FIGURE 8.12 Simplified cycle diagram of Pegus 12 with ABB's GT13E [20].

Steam turbine output was 172.7 MW with a double-flow LP section and seawater-cooled condenser with titanium tubes. Using Equation A.2, exhaust gas exergy was (exhaust flow and temperature estimated using GT PRO® library engine #43)

$$A_{exh} = 2,241 \cdot (0.1961 \cdot 1,018 - 86.918) \cdot 1.05506/1,000 \approx 266 \text{ MW}.$$

Thus, the bottoming cycle gross exergetic efficiency is 172.7/266 = 0.648.

The first comprehensive look at the gas turbine (topping) and steam turbine (bottoming) cycles using the second law approach was by Elmasri in 1985–1987 (Ref. [6] in Chapter 4 and Refs. [1,21,22] in this chapter). Elmasri considered a single-pressure steam cycle without reheat and/or feedwater heating with a stack temperature of 420 K (300°F). The data in Figure 8 of his 1987 paper is summarized in Table 8.2. Elmasri's bottoming steam cycle exergy balance for the T_{exh} = 1,106°F led to the following breakdown (from Figure 9 in his cited paper):

1. Stack loss ~8%
2. HRSG heat transfer irreversibility ~14%
3. Condenser irreversibility and heat rejection ~6% (45 mbar condenser pressure)
4. Steam turbine and feed pump irreversibility ~7%.

With the exception of the last case in Table 8.2 with reheat (T_{exh} = 1,315°F, which is still – in 2018 – not a reality without duct firing in the HRSG), improvement in bottoming cycle exergetic efficiency is driven by the rise in steam temperature in lockstep with the gas temperature. Obviously, the performances are quite modest vis-à-vis existing state of the art (in 2018) with 3PRH configuration and advanced steam conditions.

In a subsequent paper, Chin and Elmasri looked at 2PRH bottoming cycles from a second law (exergy) perspective [23]. Boundary conditions were different from those in the earlier paper, i.e., 67.7 mbar condenser pressure with stack temperature limit relaxed to 367 K (200°F). Maximum steam conditions were 166.5 bar (2,400 psia) and 867 K (1,100°F). Bottoming cycle exergetic efficiency was about 65% in the F class exhaust temperature range (850–900 K).

In a 1990 ASME paper, Bolland presented a comprehensive, exergy-based comparison of bottoming cycle alternatives [24]. Bolland used Siemens V94.2 (E class) and V94.3 (F class) gas turbines as representatives of the then state of the art. In addition to reheat and non-reheat versions of two- and three-pressure steam cycles, Bolland also considered a "supercritical HP" bottoming cycle, which is covered in detail in Section 6.4 (with considerable reference to Bolland's calculations). Herein, the focus is solely on the subcritical bottoming steam cycles.

TABLE 8.2
Bottoming Cycle Exergetic Efficiency as a Function of Gas Turbine Exhaust Gas Temperature

TEXH, °F	BC Exergetic Efficiency	Steam Cycle	Steam Pressure, bar (psia)	Steam Temperature, K (°F)	Steam Cycle Efficiency
584	48.5%	1PNR	9.1 (132)	563 (554)	27.1%
845	58.8%	1PNR	24 (348)	703 (806)	32.7%
1,106	65.3%	1PNR	70 (1,015)	843 (1,058)	38.1%
1,237	68.0%	1PNR	120 (1,740)	923 (1,202)	40.6%
1,315	70.0%	1PRH	160 (2,321)	810 (998)	42.3%

Source: From Figure 8 in Elmasri, Ref. [6] in Chapter 4.

Based on the data in Tables 1 and 2 in Ref. [24], Bolland's findings can be summarized as shown in Table 8.3. Here is how this table is constructed:

1. Heat consumption is found from gas turbine output and efficiency.
2. Combined cycle output is found from heat consumption and efficiency.
3. Steam turbine output is found by subtracting gas turbine output from gross plant output.
4. Bottoming cycle pump power consumption is assumed to be 2% of steam turbine output to arrive at net bottoming cycle output.

The "reverse-engineered" bottoming cycle net exergetic efficiencies for V94.2 in Table 8.3 are on average 4 percentage points lower than those in Figure 11 in Ref. [24].

Bolland's V94.2 analysis has been repeated with GT PRO® using the software's library engine #43 (exhaust flow 1,117.5 lb/s, exhaust temperature, 1,008.5°F). The findings are summarized in Table 8.4 (bottoming cycle exergetic efficiency) and Table 8.5 (combined cycle efficiency).

In 1988, Elmasri formed Thermoflow, Inc. and unveiled his gas turbine simple and combined cycle modeling software GT PRO® [25]. In its initial release, GT PRO® supported two-pressure bottoming steam cycles (with reheat option) and a third (optional) LP drum for feedwater heating. In 1988, Enter Software, Inc. announced GATE (later GateCycle[6]) software (initially developed

TABLE 8.3
Bolland's "Advanced" Combined Cycle Analysis [24]

	V94.2	V94.3
GT output, MWe	144.30	189.50
GT efficiency, %	32.50	35.20
Heat consumption, MWth	444	538
Exhaust flow, kg/s (lb/s)	504 (1,111)	565 (1,246)
Exhaust temperature, °C (°F)	553 (1,028)	563 (1,046)
a, Btu/lb (Equation A.2)	114.59	118.23
A, MW	134.34	155.43
CC efficiency (2PNR)	51.53	53.61
CC output (2PNR), MWe	228.79	288.61
ST output, MWe	84.49	99.11
BC net exergetic efficiency (from Ref. [24])	61.64% (65.7%)	62.91%
CC efficiency (2PRH)	51.95	54.06
CC output (2PRH), MW	230.66	291.03
ST output, MW	86.36	101.53
BC net exergetic efficiency (from Ref. [24])	63.0% (66.8%)	64.46%
CC efficiency (3PNR)	52.08	54.12
CC output (3PNR), MW	231.24	291.36
ST output, MW	86.94	101.86
BC net exergetic efficiency (from Ref. [24])	63.42% (68.3%)	64.27%
CC efficiency (3PRH)	52.52	54.57
CC output (3PRH), MW	233.19	293.78
ST output, MW	88.89	104.28
BC Net exergetic efficiency (from Ref. [24])	64.85% (68.7%)	65.92%

[6] GATE (GAs Turbine Evaluation) was the gas turbine code; CYCLE was the code for the steam cycle analysis. Later, the software name changed to GateCycle.

TABLE 8.4
Bolland's "Advanced" Combined Cycle Analysis [24] – GT PRO Check

	Table 8.3		Figure 11 [24]		GT PRO	
	BC Exergetic Efficiency	Δ	BC Exergetic Efficiency	Δ	BC Exergetic Efficiency	Δ
2PNR	61.64%		65.70%		65.64%	
2PRH	63.00%	1.36%	66.80%	1.10%	66.78%	1.14%
3PNR	63.42%	0.42%	68.30%	1.50%	67.76%	0.98%
3PRH	64.85%	1.43%	68.70%	0.40%	70.44%	2.69%

TABLE 8.5
Bolland's "Advanced" Combined Cycle Analysis [24] – GT PRO Check

	Table 2 [24]		GT PRO	
	CC Efficiency	Δ	CC Net Efficiency	Δ
2PNR	51.53%		50.27%	
2PRH	51.95%	0.42%	50.59%	0.32%
3PNR	52.08%	0.13%	50.83%	0.24%
3PRH	52.52%	0.44%	51.58%	0.75%

under the sponsorship of EPRI[7]) for gas turbine simple and combined cycle performance simulation [26]. Similar to the early GT PRO®, the early GateCycle was also limited to the 2PRH steam cycles with the third (LP) evaporator supplying the steam necessary for deaeration of the boiler feedwater. These two software products enabled researchers and developers to carry out high-fidelity design and off-design performance simulation of GTCC power plants. They were thus instrumental in the rapid development of the technology in the 1990s.

In 1992, Sir John Horlock published his book on combined cycles in which he laid down the key aspects of governing thermodynamics from both first and second law perspectives. (A reprint edition with corrections and additional material was published in 2002 [27].) Sir Horlock drew heavily upon the pioneering work by Seippel–Bereuter and Elmasri (cited above). The first edition of an important book on combined gas-and-steam turbine combined cycles (by Kehlhofer in 1991, later editions with coauthors) also appeared around the same time [28]. Thus, by 1992 all the pieces, tools and theory, were in place for rapid development in gas-steam turbine combined cycle power plant technology.

The next breakthrough in combined cycle performance was GE's Frame 9FA single-shaft power block in the Netherlands. The 1.7 GWe power plant on the Atlantic coast near Eemshaven comprised five such blocks rated at 54.5% net LHV efficiency with a 3PRH bottoming cycle (53.6% with a 2PNR bottoming cycle) [29]. The gas turbine (PG9331FA in GE parlance at the time) was rated at 226.5 MWe and 35.7% at ISO baseload with 1,288°C (2,350°F) firing temperature at a cycle PR of 15:1. The exhaust gas temperature was 589°C (1,093°F). The first three blocks were commissioned in 1995. With a 50-mbar (1.4765 in. Hg) seawater-cooled condenser, steam turbine out was 128.3

[7] The Electric Power Research Institute, Inc., is an independent, nonprofit organization in the USA that conducts research and development related to the generation, delivery and use of electricity to help address challenges in electricity, including reliability, efficiency, affordability, health, safety and the environment.

MWe. Using Equation A.2, exhaust gas exergy was (exhaust flow and temperature estimated using GT PRO library engine #105)

$$A_{exh} = 1,355 \cdot (0.1961 \cdot 1,100 - 86.918) \cdot 1.05506 / 1,000 \approx 184 \text{ MW}.$$

Thus, the bottoming cycle gross exergetic efficiency is 128.3/184 = 0.697.

In 1992, the US DOE, through the Office of Fossil Energy and the Office of Energy Efficiency & Renewable Energy, initiated the *Advanced Turbine Systems* (ATS) Program to break the 2,300°F firing temperature barrier (then state-of-the-art F class firing temperature) towards 60% net LHV combined cycle efficiency. With considerable effort and investment by the DOE and its private partners (mainly major OEMs), the ATS program opened up the door to 2,600°F in firing temperature and 60% net efficiency (which was actually achieved in the field in 2011) by the end of the decade. Key enabling technologies that emerged from the ATS program (or accelerated by it) were single-crystal turbine blades, thermal barrier coatings (TBC), advanced brush seals, dry-low-NOx (DLN) combustors, closed-loop steam cooling (SC) and 3D aerodynamic design of compressor and turbine components.

In terms of an all-around gas turbine-based power plant system, the ATS program resulted in the H-System (a GE trademark). While described as "60% capable" by the OEM, none of the six H-System units actually performed at that level (four 50 Hz and two 60 Hz, all in single-shaft 1 × 1 × 1 configuration). However, the technologies incorporated into the H-System achieved design benchmarks in a field-proven, reliable platform (see Section 8.6.3). The gas turbine itself (about 40% simple cycle efficiency) should have been able to achieve 60% net LHV efficiency easily with a reasonably well-designed bottoming steam cycle. As explained in detail in Section 8.6.3, this was not the case. In fact, it was an air-cooled H class (1,500°C TIT) gas turbine that achieved 60+% combined cycle efficiency in a power plant in Germany in 2011 [30] (see Section 8.6.1).

8.5 STATE OF THE ART

Siemens STG outputs from the rating data in *GTW 2018 Handbook* is summarized in Table 8.6. Also included in the table are exhaust enthalpy (59°F zero-enthalpy temperature) and exergy (59°F dead state temperature) and bottoming cycle first law efficiency and second law effectiveness. With 77°F reference/dead state temperature, enthalpy and exergy values are reduced with commensurate rise in efficiency numbers (see Table 8.7).

From a high-level, qualitative analysis perspective, bottoming cycle performance parameters in Tables 8.6 and 8.7 do not differ significantly. Nevertheless, it is imperative to have a solid handle on property calculation, especially on reference/dead state temperatures, when performing engineering calculations. Not doing so can lead to significant errors. (In this book, 59°F is the preferred reference/dead state temperature.)

GE STG outputs from the rating data in *GTW 2018 Handbook* is summarized in Table 8.8. (Note that STG output for 2 × 2 × 1 configuration with higher efficiency is used in the table by dividing it by two, i.e., per gas turbine.) Note that the ε_{BC} values in Table 8.8 for the HA class machines are significantly higher than those in Table 8.6 and the others in Table 8.8.

In Tables 8.6–8.8, listed ε_{BC} and η_{BC} values are on a "gross" basis, i.e., based on the STG output. In order to arrive at a "net" value, one has to subtract the power consumption of boiler feed pumps and the condensate pump (i.e., exergy "input" to the bottoming cycle) from the STG output. To a very good approximation, these two items account for about 2% of STG power output. Thus on a net basis, listed ε_{BC} and η_{BC} values should be multiplied by a factor of 0.98. The results are summarized in Tables 8.9 and 8.10.

There is an implicit assumption in simplified combined cycle efficiency equations, Equations 8.4 and 8.6. Heat energy "pumped" into the bottoming cycle with the gas turbine exhaust gas was simply the difference between gas turbine heat consumption and useful work output. Thus,

TABLE 8.6
Siemens Bottoming Cycle STG Output and First/Second Law Efficiencies (59°F Reference)

		MEXH	TEXH	STG	A_{exh}	ε_{BC}	H_{exh}	η_{BC}
	YOI	lb/s	°F	MW	MW	(Gross)	MW	(Gross)
2000E	1981	1,230	997	93	142.2	0.654	316.4	29.4%
4000F	1995	1,596	1,110	160	223.5	0.716	467.7	34.2%
8000H	2008	2,061	1,166	217.5	313.5	0.694	640.5	34.0%
8000HL	2017	1,874	1,256	232	321.5	0.722	635.9	36.5%
9000HL	2017	2,205	1,256	272	378.3	0.719	748.2	36.4%

TABLE 8.7
Siemens Bottoming Cycle STG Output and First/Second Law Efficiencies (77°F Reference)

		STG	A_{exh}	ε_{BC}	H_{exh}	ε_{BC}
	YOI	MW	MW	(Gross)	MW	(Gross)
2000E	1981	93	135.3	0.688	310.8	29.9%
4000F	1995	160	213.4	0.750	460.4	34.7%
8000H	2008	217.5	299.8	0.726	631.1	34.5%
8000HL	2017	232	308.0	0.753	627.3	37.0%
9000HL	2017	272	362.4	0.751	738.1	36.9%

TABLE 8.8
GE Bottoming Cycle STG Output and First/Second Law Efficiencies (59°F Reference)

		MEXH	TEXH	STG	A_{exh}	ε_{BC}	H_{exh}	η_{BC}
	YOI	lb/s	°F	MW	MW	(Gross)	MW	(Gross)
9E.03	1992	927	1,012	75.5	110.2	0.686	242.9	31.1%
9E.04	2014	917	1,007	75.3	108.0	0.697	238.8	31.5%
9F.03	1996	1,467	1,104	152.6	203.5	0.750	427.1	35.7%
9F.04	2015	1,457	1,150	163.1	216.6	0.753	445.4	36.6%
9F.05	2003	1,574	1,184	186.0	245.6	0.757	498.2	37.3%
9HA.01	2011	1,891	1,164	230.4	286.9	0.803	586.5	39.3%
9HA.02	2014	2,205	1,193	274.6	348.3	0.788	704.2	39.0%

TABLE 8.9
Siemens 50-Hz GTCC Bottoming Cycle Performance (*GTW Handbook 2018*)

		ε_{BC}	η_{BC}
	YOI	(Net)	(Net)
2000E	1981	0.641	28.8%
4000F	1995	0.702	33.5%
8000H	2008	0.680	33.3%
8000HL	2017	0.707	35.8%
9000HL	2017	0.705	35.6%

TABLE 8.10
GE 50-Hz GTCC Bottoming Cycle Performance (*GTW Handbook 2018*)

	YOI	ε_{BC} (Net)	η_{BC} (Net)
9E.03	1992	0.672	30.5%
9E.04	2014	0.683	30.9%
9F.03	1996	0.735	35.0%
9F.04	2015	0.738	35.9%
9F.05	2003	0.742	36.6%
9HA.01	2011	0.787	38.5%
9HA.02	2014	0.773	38.2%

$$\eta_{CC} = \eta_{GT} + (1 - \eta_{GT}) \cdot \eta_{BC}.$$

We can now test the accuracy of this assumption. Using the data for GE and Siemens 50-Hz product line, unaccounted-for gas turbine heat balance items are calculated and summarized in Tables 8.11 and 8.12 (as a fraction of heat consumption, λ). Exhaust energy is calculated with 59°F zero-enthalpy reference.

The numbers in Tables 4.24 and 4.25 indicate that, to a very good approximation, $\lambda = 0.025$, can be used in Equation 8.22, i.e.,

$$\eta_{CC} = \eta_{GT} + (1 - \eta_{GT} - \lambda) \cdot \eta_{BC}.$$

For the 50-Hz product line of GE, 2 × 2 × 1 combined cycle ratings listed in the *Gas Turbine World 2018 Handbook*, are summarized in Table 8.13.

Using Equation 8.2, which is the ideal (air-standard) combined cycle (or "enhanced" Brayton cycle) efficiency, technology factors are calculated and summarized in Table 8.14. For comparison, the same analysis is done for Siemens 50-Hz product line, and the results are summarized in Table 8.15. In both OEM's published ratings, implied combined cycle technology factors are quite impressive at 0.8 or even higher. (Note that $F_T = 1.00$ means that the combined cycle is essentially a Carnot engine!)

Steam cycle efficiencies implied by the rating data Tables 8.13 and 8.15 are summarized in Tables 8.16 and 8.17. Calculations are based on

TABLE 8.11
Siemens 50-Hz Gas Turbine Heat Balance

	YOI	W MWe	HC MWth	HC-W MWth	H_{exh} MWth	λ -
2000E	1981	187	517	329.6	316.4	0.040
4000F	1995	329	802	473.4	467.7	0.012
8000H	2008	450	1,098	647.6	640.5	0.011
8000HL	2017	465	1,107	642.1	635.9	0.010
9000HL	2017	564	1,327	763.1	748.2	0.020

TABLE 8.12
GE 50-Hz Gas Turbine Heat Balance

		W	HC	HC-W	H_{exh}	λ
	YOI	MWe	MWth	MWth	MWth	-
9E.03	1992	132	382	249.5	242.9	0.027
9E.04	2014	145	392	246.9	238.8	0.033
9F.03	1996	265	701	436.1	427.1	0.021
9F.04	2015	288	744	456.2	445.4	0.024
9F.05	2003	314	822	508.0	498.2	0.019
9HA.01	2011	446	1,035	588.8	586.5	0.004
9HA.02	2014	557	1,266	708.9	704.2	0.007

TABLE 8.13
GE 50-Hz Product Line CC Efficiencies

		OUT	TIT	PR	GT Efficiency	CC Efficiency
	YOI	MW	°C		%	%
9E.03	1992	132	1,269	13.1	34.6	53.7
9E.04	2014	145	1,299	13.3	37.0	55.2
9F.03	1996	265	1,428	16.7	37.8	59.0
9F.04	2015	288	1,488	16.9	38.7	60.4
9F.05	2003	314	1,523	18.3	38.2	60.9
9HA.01	2011	446	1,621	23.5	43.1	63.6
9HA.02	2014	557	1,670	23.8	44.0	64.2

TABLE 8.14
GE 50-Hz Product Line Technology Factors

	Rating CC Efficiency	Ideal CC Efficiency	Technology Factor
	%	%	
9E.03	53.7	71.15	0.75
9E.04	55.2	71.52	0.77
9F.03	59.0	73.52	0.80
9F.04	60.4	74.09	0.82
9F.05	60.9	74.63	0.82
9HA.01	63.6	76.12	0.84
9HA.02	64.2	76.51	0.84

- Equation 5.37 assuming 2% each for miscellaneous topping and bottoming cycle losses and 1.6% for auxiliary load
- Equation 6.8 for HRSG effectiveness.

In the foregoing exercise, we used GE and Siemens 50-Hz product lines as examples. The reason for that is the fact that their gas turbines interact with the bottoming cycle only via their exhaust gas stream (fuel gas performance heating can be safely ignored without introducing a big error into the

TABLE 8.15
Siemens 50-Hz Product Line

	YOI	OUT MW	GT Efficiency %	PR	TIT °C	Rating CC Efficiency %	Ideal CC Efficiency %	Technology Factor
2000E	1981	187	36.20	12.8	1,297	53.3	71.4	0.75
4000F	1995	329	41.00	20.1	1,539	59.7	75.0	0.80
8000H	2008	450	> 41.0	20.0	1,592	61.0	75.4	0.81
8000HL	2017	465	42.00	24.0	1,702	>63.0	76.8	>0.82
9000HL	2017	564	> 42.0	24.0	1,703	>63.0	76.8	>0.82

TABLE 8.16
GE 50-Hz Product Line – Estimated Bottoming Cycle Performance (GTW 2018)

	YOI	GT Efficiency %	CC Efficiency %	TEXH F	HRSG Efficiency %	StmC Efficiency %
9E.03	1992	34.6	53.7	1,012	89.4	36.94
9E.04	2014	37.0	55.2	1,007	89.4	36.81
9F.03	1996	37.8	59.0	1,104	90.7	42.52
9F.04	2015	38.7	60.4	1,150	91.4	43.86
9F.05	2003	38.2	60.9	1,184	91.9	45.15
9HA.01	2011	43.1	63.6	1,164	91.6	45.01
9HA.02	2014	44.0	64.2	1,193	92.0	44.97

TABLE 8.17
Siemens 50-Hz Product Line Estimated Bottoming Cycle Performance (GTW 2018)

	YOI	GT Efficiency %	CC Efficiency %	TEXH F	HRSG Efficiency %	StmC Efficiency %
2000E	1981	36.2	53.3	997	89.2	34.27
4000F	1995	41.0	59.7	1,110	90.8	39.98
8000H	2008	41.0	61.0	1,166	91.6	42.23
8000HL	2017	42.0	63.0	1,256	92.9	44.45
9000HL	2017	42.5	63.0	1,256	92.9	43.86

analysis). In sequential (reheat) combustion gas turbines such as GT 24/26 with extensive *cooling air cooling* (CAC) and in steam-cooled gas turbines such as MHPS G and J class, there is significant interaction between topping and bottoming cycles. Handling such cases using a reductive first law approach requires a bit more algebra. Before going into the "how", let us first provide some introductory information.

In certain gas turbines, heat recovered from cooling air and/or heat rejected to cooling steam are utilized to make steam (or heat feedwater or steam) in the bottoming cycle. Especially in gas turbines with high compressor PRs such as ABB/Alstom (now shared between GE and Ansaldo) reheat gas turbines (PR > 30), compressor discharge temperature becomes close to 1,000°F. Air extracted from the compressor discharge is the hot gas path coolant for the first and hottest turbine stage (due to pressure requirements). Cooling the extracted compressor discharge air down to a more reasonable 600°F–700°F limits the amount of extraction, which otherwise would render the reheat combustion advantage insignificant (or worse, turn it into a *disadvantage*). CAC has also been used in GE's steam-cooled H-System (PR of 23:1) and some of the MHPS and Siemens/Westinghouse gas turbines (albeit to a lesser extent vis-à-vis GT24/26) – in order to use less expensive rotor alloys in the latter cases.

Heat removed from the cooling air (usually in a kettle-type evaporator, common in the chemical process industry (CPI) as reboiler of stripper columns) is utilized to make LP or IP steam. In steam-cooled gas turbines, typically IP or cold reheat steam from the bottoming cycle is used to cool hot gas path components and returned to the bottoming cycle at the HRH circuit. In order to account for the impact of these additional energy inputs to the bottoming cycle, Equation 8.22 is modified as follows:

$$\eta_{CC} = \eta_{GT} + (1 - \eta_{GT}) \cdot \eta_{HRSG}\eta_{STG} + q_{CAC}\eta' + q_{SC}\eta'' \tag{8.28}$$

with

$$q_{CAC} = \frac{\dot{Q}_{CAC}}{HC}, \, q_{SC} = \frac{\dot{Q}_{SC}}{HC}.$$

In Equation 8.28, η' and η'' are first-law conversion factors for heat supplied to the bottoming cycle via CAC and SC, respectively. In other words, each Btu/s of heat supplied to the bottoming cycle from source results in either η' or η'' kWe of additional STG output. To a very good approximation, the following values can be used in Equation 8.28:

- If CAC heat rejection is used for LP steam generation, $\eta' = 0.073$–0.075.
- If CAC heat rejection is used for IP steam generation, $\eta' = 0.11$–0.12.
- If CAC heat rejection is used for HP steam generation, $\eta' = 0.15$–0.16.
- For SC heat rejection, $\eta'' = 0.15$–0.16.

Armed with this information, let us take a peek at the combined cycle performance of MHPS J class using the HMB data in Table 4.16. From the data in the table

- $q_{SC} = 0.014$
- $q_{CAC} = 0.007$
- $\eta_{GT} = 0.423$.

Let us assume that $\eta_{HRSG} = 0.91$ and $\eta_{ST} = 0.40$, which are appropriate for state of the art in 3PRH bottoming cycle technology (as already discussed). Substituting these values into Equation 8.28 and assuming that CAC heat rejection is used to make IP steam, we have

$$\eta_{cc} = 0.423 + (1 - 0.423) \cdot 0.91 \cdot 0.4 + 0.007 \cdot 0.12 + 0.014 \cdot 0.16 = 0.636,$$

i.e., combined cycle *gross* efficiency is 63.6%. Applying the bare minimum 1.6% for auxiliary load (typical for GTW ratings), we find 62.6% for *net* combined cycle efficiency. In GTW 2018 Handbook, combined cycle efficiency for 50-Hz M701J is published as 62.3%. In conclusion, this performance rating is in line with expectations derived from the first principles.

8.6 THE HALL OF FAME

8.6.1 Irsching

Irsching 4 is a 1 × 1 × 1 single-shaft GTCC power plant with Siemens SGT5-8000H gas turbine. The plant was originally owned by the German utility E.ON (the current owner is Uniper). It is located next to a 40-year-old thermal power plant with three blocks. Until recently (superseded by the HL class), SGT5-8000H was the largest gas turbine in Siemens portfolio with close to 400 MW output in its latest version. The first unit was manufactured in the Berlin factory, but its size precluded its being tested in the factory test bed discussed earlier. At the time (mid-2000s) the highest load made possible by the water brake in the factory was 220 MW. (Today, the capacity of the water brake is higher. In fact, Siemens tested the 60-Hz version of 8000H in the factory.) Thus, Siemens found a suitable location and partner in Irsching and E.ON, respectively (e.g., access to natural gas via a pipeline from Russia, connection to the grid and space). The project started in early 2000s and took about seven years to complete. In mid-2008, first synchronization to the grid and achieving the baseload took place. The gas turbine was tested in simple cycle mode for about 18 months when the first phase of its validation was completed in August 2009. The extension to a combined cycle power plant (single-shaft 1 × 1 × 1) was completed at the end of 2010. In 2001, the power plant clocked 60.75% net efficiency.

At the time of writing (2019), the plant is in "cold standby" (i.e., one step removed from being completely "mothballed") – Section 8.6.5. Even at the time when the author visited the power plant in the Fall of 2013, the plant was operating at only one-fifth of its planned capacity. Apparently, the generation from coal and renewables (mostly wind from what the author could see) was such that there is really not much need for power from a plant (even a world-record holding one at that!) burning very expensive natural gas imported from Russia. (Side note: Per *Turbomachinery International* at the time, the share of coal in Germany's power generation mix had risen to 52% in the first half of 2013. Gas-fired power production was about at the same level as the wind, with both around 9% each.)

A key characteristic of the Irsching GTCC is its very low condenser pressure. On the day of the author's visit, at an ambient temperature 10°C (50°F), the condenser pressure was 23 mbar (~0.67 in. Hg!). The main steam condenser is an open-loop unit drawing cooling water from the river Danube. The LP section of the steam turbine is a dual-flow, side-exhaust design with 16 m^2 of exhaust annulus area with titanium LSBs (56 in. long – coupled with snubbers).

The HRSG is of Benson type with once-through HP section. Nominal main and HRH steam temperatures are 600°C (1,112°F); on the day of the visit, they were quite close to that (about 1,090°F each). Steam pressures in the HRSG are about 180/35/4 bar HP/IP/LP. Again, on the day of the author's visit, HRSG stack temperature was 87°C (189°F) with 192 MWe steam turbine output.

SGT5-8000H gas turbine is a nominal 1,500°C-TIT unit with 40+% simple cycle efficiency. Its performance on the day of the visit (360 MWe output and exhaust temperature 610°C (1,130°F) at 50°F ambient) suggested the following (via HMB analysis outlined in Chapter 4):

1. 40.2% net full load gas turbine efficiency
2. 18,694 Btu/lb natural gas LHV (77°F with 91.3%(v) methane content, author's assumption)
3. 1,500°C–1,550°C turbine inlet temperature (>2,800°F!)
4. About 2,600°F firing temperature (stage 1 rotor inlet)
5. Steam bottoming cycle second law efficiency ~75% (nearly 2 percentage points above the common state of the art per Equation 8.22, which is worth nearly 0.5 percentage points in net combined cycle efficiency)
6. 60.3% net combined cycle efficiency.

Irsching GTCC is (was?) indeed a phenomenal power plant bringing the best features of state-of-the-art combined cycle gas turbine technology together under one proverbial roof. In particular,

- Extremely effective noise abatement.[8]
- Inside the building, prime movers and the generator have their own enclosures.
- Extremely clean plant building with well-organized layout of myriad accessories.
- All condensate, feedwater and cooling water circulation pumps as well as lube oil pumps are either 2 × 100% or 3 × 50%.
- Condenser cooling water suction and discharge pipes are fiberglass reinforced plastic with 2.5 m diameter and 30 mm wall thickness.
- The condenser is located next to the steam turbine (with side exhaust).
- The HRSG is three-pressure reheat with once-through Benson-type HP section. There is a condensate polishing system.
- SSS clutch[9] between the steam turbine and generator.
- Hydraulic Clearance Optimization (HCO).[10]

8.6.2 Bouchain

Électricité de France (EdF) Bouchain is a 1 × 1 × 1 single-shaft GTCC with 9HA.01 gas turbine and D-650 steam turbine (2 × 48″ with side exhaust) with W23 generator in the middle and SSS clutch. It is the first GE 50-Hz HA class commercial unit. It is also the first GE *FlexEfficiency50* (FE50) GTCC.

EdF's fossil-fired fleet represents about 11% of its installed capacity in France (the bulk being nuclear) but accounts for only about 2%–3% of electricity production, reflecting its grid balancing role. The modernization program includes closure of around ten old coal units (the last coal firing unit at the Bouchain site closed down in April 2015[11]). The construction of the GTCC began in 2012 and the commissioning took place mid-2016. The plant is built by ENWESA under an EdF-GE consortium. The 3PRH HRSG is designed (and supplied) by CMI Energy to meet the operability requirements of the FE50 concept.

The plant has an installed (nameplate) capacity of 575 MWe, the equivalent of the electricity consumption of 680,000 French households. The claimed electrical efficiency is 61%. The expected output is slightly higher at 590 MWe and can go even beyond that. However, the summer output is limited by the cooling tower. The powertrain is single-shaft with the generator in the middle and separated from the steam turbine by a SSS clutch. It can reach full load in less than 28 min after an overnight shutdown ("hot" start).

The nominal bottoming steam cycle is 170 bar/565°C/565°C. The natural draft cooling tower of the existing coal-fired plant (decommissioned) was retained for the GTCC. At the time of the author's site visit in the summer of 2017, condenser pressure was 43 mbar (~1.25 in. Hg) on a 19°C day.

Similar to Irsching with Siemens SGT5-8000H, Bouchain with GE's 9HA.01 was also a world-record holder (certified by Guinness). The world-record performance, 605+ MWe and 62.22%, was measured and certified on 28 April 2016 shortly before the Commercial Operation Date (COD) (25 July 2016). The cited performance excludes 2.5 MWe EdF loads (essentially the circ-water pumps and miscellaneous plant consumption) and step-up transformer. (The latter should be between 2 and 3 MWe.)

Thus, the "true" net performance of the Bouchain GTCC is somewhere around 61.7% (ISO baseload), which is still impressive and in-line with *GTW Handbook* ratings. In fact, EdF Bouchain is the

---

8 Until one enters the plant building, it is nearly impossible to tell from the outside whether the plant is running or not.

9 The SSS (Synchro-Self-Shifting, pronounced as "Triple S") clutch, which automatically engages and disengages through shaft speed control only, is a mechanical device widely used in combined cycle plants to disconnect the steam turbine in a single shaft configuration, allowing the gas turbine/generator to be operated separately.

10 HCO shifts the gas turbine rotor against flow direction to optimize turbine clearances during steady state operation. The axial shift of the rotor is performed automatically by hydraulic pistons behind the compressor thrust bearing.

11 Bouchain is one of France's historical mining regions, which is north of Paris, between Douai and Valenciennes. The GTCC plant heat sink includes the 410 ft high natural draft cooling tower of the old coal-fired power plant.

most efficient fossil-fired power plant in France, which explains why it clocked ~7,000 h in its first 12 months. (The plant is operated in a load-following mode between 75% and full load.)

8.6.3 Inland Empire Energy Center

Inland Empire Energy Center (IEEC) was GE's first (and only) 60-Hz H class GTCC, which came online in 2008. The power plant is located in southern California. It comprises two single-shaft 107H powertrains (nominal rating 400 MWe each). The commercial rating of the 107H in IEEC was about 395 MWe and 57.6% net (LHV) at site conditions with inlet evap coolers. The IEEC power plant is unique in two respects: fully steam-cooled turbine hot gas path and fuel gas moisturization. Before focusing on the IEEC, let us first take a look at the steam-cooled H technology of GE.

8.6.3.1 Steam-Cooled H Technology

In a 20-year span covering the late 1970s and late 1990s, in Japan, Europe and USA, three separate advanced gas turbine projects were established with the aim to reach 60% efficiency (net LHV) in a combined cycle configuration. These were

- Japanese Ministry of International Trade and Industry (MITI) *Moonlight* Project (1978–1987)
- European Union's COST-522 Project (launched in 1998)
- US DOE's ATS Project (1992–2001).

As dictated by fundamental thermodynamics, all three programs were cognizant of the need to increase the hot gas temperature at the inlet of the first-stage turbine rotor to at least 1,450°C (2,642°F) or, preferably, 1,500°C (2,732°F) in order to achieve this goal. Three major OEMs, MHI (Japan), Westinghouse and GE (USA), focused on the closed-loop steam cooling as a key enabler. The reason was that, even with single-crystal components in the HGP, extreme dilution of the hot gas between the combustor exit and S1B inlet with open-loop air cooling pushed the flame zone temperature beyond the thermal NOx limit.

The 1,450°C–1,500°C firing temperature capable advanced gas turbine with closed-loop steam cooling was labeled as the *H class*. While all three OEMs tried to develop H class machines with fully steam-cooled first- and second-stage components (both stationary *and* rotating parts), ultimately GE succeeded in bringing the true H class gas turbines to the market in the early 2000s under the trademark *H-System*. Both MHI and Westinghouse (later acquired by Siemens) eventually settled at the intermediate *G class* (between the air-cooled F and fully steam-cooled H class machines) with closed-loop steam cooling limited to the combustor transition piece (and later to first-stage turbine blade ring).

Note that GE always referred to an "H-System" rather than an H class gas turbine. Indeed, GE has never published simple cycle performance for either of its H class units, 7H and 9H. A venerable industry veteran Ivan Rice had once proposed that steam-cooled gas turbine "combined cycle" systems should have been renamed "integrated cycle" systems. This is due to the fact that the gas turbine Brayton and steam turbine Rankine cycles of the H-System are intertwined via cooling steam exchange. In contrast, in standard air-cooled gas turbine systems the two cycles are merely "attached" to each other. (For practical purposes, performance fuel heating utilizing IP economizer feedwater extraction can be ignored.)

The steam-cooled gas turbine of the H-System is quite simply the most advanced commercially deployed machine in operation with the possible exception of latest J class gas turbines. Its main characteristics are as follows:

- Eighteen-stage aircraft-derivative compressor with IGVs and multiple stages of VGVs to achieve a PR of 23:1 (1,500+ pps airflow).
- Advanced DLN 2.5 combustor (14 cans in 9H, 12 cans in 7H).

- Four-stage turbine (hot gas path).
- First two turbine stages fully steam-cooled (stators *and* rotors).
- *Second*-generation single-crystal first-stage vanes/nozzles and buckets/blades (René N5) with DVC[12] TBC.
- GTD-222 second-stage vanes and directionally solidified GTD-111 second-stage buckets, both with DVC TBC.
- 1,500°C turbine inlet temperature (1,450°+C firing temperature).

During the GE media blitz of the early 2000s to promote the H-System, many trade publication articles were published. Generic technical information can be found in those, which carefully state that the H-System is "capable of 60% efficiency". The only two GE technical documents discussing the H-System are Refs. [31,32].[13]

There are six H-System GTCC power plants that went in commercial operation:

1. 109H in Baglan Bay, Wales, UK
2. 3 × 109H in TEPCO Futtsu, Japan
3. 2 × 107H in IEEC, California, USA.

8.6.3.2 IEEC 107H

No "world-record" performance was officially announced by GE for the H-System. Even though the technology was a commercial flop, its field performance was quite remarkable. A solid piece of evidence for successful field deployment of the H-System is given by *Electric Light & Power* magazine's annual "Operating Performance Ratings – Top 20 Power Plants" articles. In 2010, 2011 and 2012, IEEC with two 107H units made the Top 20 list in heat rate (Table 8.18).

Even though the capacity factor hovered around 60%, this is not a bad feat. Furthermore, IEEC ranked *number one* in 2011 and 2012 in terms of NOx emissions rate[14] (0.00385 lbs/MMBtu in 2012). The units were successfully tested in the summer of 2008 (the author was there). Both units have the unique fuel moisturization system for efficiency and NOx control (see below). Unfortunately, near the end of the testing in 2008 Unit 2 was shut down due to a compressor failure. Although GE did not publicly disclose the cause for the failure, the scuttlebutt inside the company at the time was a manufacturing defect in the compressor's last stages. The restart was delayed until 2010 due to difficulties encountered in procuring the spare parts.

TABLE 8.18
IEEC 2 x 107H statistics from *Electric Light & Power* magazine

Year	Rating	GWh	CF, %	MMBtu (HHV)	Average HR, Btu/kWh	Average Efficiency (LHV)	Rank
2010	690	3,597	~62	24,467,248	6,801	55.6%	2
2011	690	3,119	~54	21,557,687	6,912	54.7%	5
2012	690	3,530	~60	24,649,199	6,983	54.2%	16

[12] Dense vertically cracked thermal barrier coating (TBC). It is a GE proprietary plasma spray process, which aims to achieve a durable high-performance coating in terms of strain tolerance, spallation resistance and component life.

[13] Exceptions are specific articles/papers on components such as the combustion system and the fuel moisturization skid of 107H.

[14] One reason for that is an oversized Selective Catalytic Reduction (SCR) on those units because the combustion team had indicated they needed a year of running to get down to 9 ppm. The SCR system was sized for 15 ppm gas turbine exit and 2.0 ppm stack. In the field, gas turbine exit was 9 ppm so that stack emission was about 1.0 ppm.

8.6.3.3 Fuel Gas Moisturization

Moisturizing natural gas fuel from the pipeline by utilizing the lowest-exergy heat source in the HRSG, i.e., warm feedwater from the LP economizer just upstream of the stack, is beneficial to combined cycle efficiency and emissions. The fuel moisturization system in IEEC 107HA GTCC is shown schematically in Figure 8.13. The fuel moisturizer (or the fuel "saturator" as referred to in GE publications) is a packed column similar in construction to the distillation columns widely used in the chemical process industry. Saturator water is supplied from a dedicated water heating section in the HRSG. The location of the saturator water heater ensures that the lowest grade of energy is utilized at the highest possible efficiency. Detailed design information can be found in Smith et al. [33].

The performance impact can be numerically gauged by about 15°F reduction in HRSG stack temperature (from about 180°F to 165°F). Moisturized and warm fuel (containing about 15%(v) H_2O at about 300°F) from the saturator is further heated in a shell-tube-type heat exchanger using the IP feedwater from the HRSG. The performance impact of fuel gas saturation as reported by GE is about 1.7% on output and 0.16 percentage points in efficiency [34]. In addition to the performance improvement, the water vapor in the fuel also reduces combustor peak flame temperatures and NOx emissions [34].

A very rough estimate can be made as follows:

- Combined cycle efficiency from Equation 8.4 with 40% for η_{GT} and η_{ST}.
- HRSG effectiveness from Equation 6.7 with
  - 1,150°F exhaust gas temperature
  - 180°F stack temperature (no fuel moisturization)
  - 165°F stack temperature (with fuel moisturization)
  - exhaust and stack enthalpies from Equations A.5 and A.7.

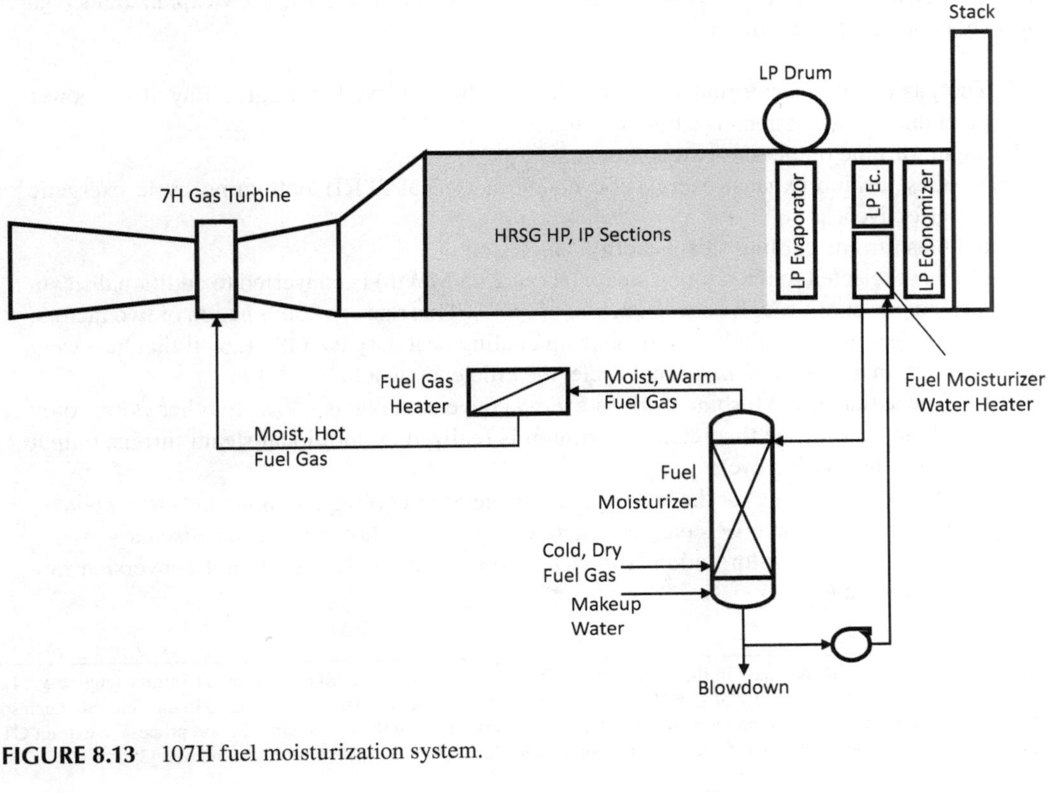

FIGURE 8.13 107H fuel moisturization system.

HRSG effectiveness with and without fuel moisturization is 92.15% and 90.86%, respectively. Corresponding combined cycle efficiencies are 60.5% and 60.2%, respectively, everything else being the same. The efficiency delta of 0.3 percentage points should be considered as an upper limit. With changes in the auxiliary load and steam cycle (note that LP evaporator pressure has to be increased to 110 psia with fuel moisturization), a lower impact can be expected. Estimated NOx reduction is about 4–6 mg/Nm^3.

8.6.4 60% Net (LHV) Bogey

GE designed and advertised the H-System as "60%-capable", which was the original objective of the US DOE's ATS program – to be realized by the end of the 1990s. Alas, the industry had to wait for another decade for the achievement of that benchmark (see Section 8.6.1). The first H-System to reach COD was the 50-Hz 109H in Baglan Bay, Wales, UK, which went through a rigorous "characterization test" program before commercial operation. Neither that unit nor the other five H-System units reached 60%. Why not? The answer is to be found in the bottoming cycle and revolves around the two crucial factors:

- Steam turbine technology (i.e., steam turbine heat rate)
- Site ambient conditions.

The latter impacts the GTCC performance in two ways:

- Economically achievable condenser back pressure (i.e., steam turbine back pressure)
- Auxiliary power consumption.

The impact of these two factors on the H-System performance is clearly illustrated by the data in Table 8.19. Before examining the data in that table and its ramifications, a few explanations regarding several entries therein are in order:

1. The gas turbine performance is very close to that achieved in Baglan Bay 109H power plant during the characterization test in 2003.[15]
2. Steam turbine output has two components:
   a. Gas turbine exhaust exergy conversion at typical 3PRH bottoming cycle exergetic effectiveness
   b. Gas turbine cooling steam exergy conversion.
3. The heat rejected to the cooling steam (about 23.5 MWth) is converted to additional steam turbine power at a rough conversion rate of ~50%. This rate is a combination of two factors:
   a. The exergy-to-enthalpy ratio of steam cooling heat duty is ~60% (i.e., if that heat *were* used in a *hypothetical* Carnot cycle, the efficiency would be ~60%).
   b. The actual steam turbine work-to-exergy conversion rate is ~80%. In other words, only about 80% of the theoretical maximum is realized as additional steam turbine output in the bottoming cycle.
4. A separate accounting for the heat rejected in the rotor cooling air cooler (a *kettle reboiler*-type evaporator where IP steam is generated by cooling hot compressor discharge air) is *not* provided. This information is *buried* in the steam cooling duty and conversion rate used in the table.

[15] To be exact, the performance data in the table is taken from Thermoflow's GTPRO software GT library (engine #114). The performance of the GT PRO's GE 9001H is eerily close to the actual performance measured in the field. So much so that the author used a GT PRO model of 109H as the basis in his paper on the data reconciliation process to extract GT performance from the single-shaft GTCC test measurements during the characterization test in 2003 [35].

In Table 8.19, the columns from left to right primarily differ in the bottoming cycle exergetic efficiency and auxiliary load fraction. All performances are assumed to be at baseload and ISO ambient conditions. In particular,

1. The first column is the *GTW Handbook*-type performance with a bottoming cycle exergetic efficiency corresponding to the "well-designed" bottoming cycle technology curve in the Appendix (Equation C.1). The auxiliary load fraction is 1.6%, which is based on the bare-bones rating-type scope with a once-through open-loop water-cooled condenser. This performance is a *bona fide* 60+% with the given gas turbine performance.
2. The second column reflects the performance that could be achieved in an ideal site where a low back pressure (e.g., 1.2 in. Hg or even lower) can be maintained with a condenser drawing cooling water from a naturally available source such as lake, ocean and river. The other condition is a steam turbine with a large enough exhaust annulus (primarily dictated by the LSB size) to exploit the low back pressure without undue performance penalty in exhaust losses. This performance falls short of 60% but is still a respectable 59+%.
3. The third column reflects the performance that could be achieved in a "real" site where the heat rejection system comprises a mechanical cooling tower. The back pressure is a more realistic 2.0 in. Hg, and the auxiliary load fraction is 2.1%. This performance is very close to what was measured in Baglan Bay 109H in 2003.

What are the contributing factors to low bottoming cycle exergetic efficiency?

TABLE 8.19
109H Performance (Roughly Baglan Bay Conditions)

		GTW Handbook	Ideal	Site
GT output	kW	326,971	326,971	326,971
GT heat rate	Btu/kWh	8,660	8,660	8,660
GT efficiency		39.40%	39.40%	39.40%
Exhaust flow	pps	1,532	1,532	1,532
Fuel flow	pps	36.56	36.56	36.56
Fuel temperature	°F	~400	~400	~400
Exhaust temperature	°F	1,168	1,168	1,168
Cycle PR		23.0	23.0	23.0
TIT	°F	2,732	2,732	2,732
	°C	1,500	1,500	1,500
Firing temperature	°C	1,400	1,400	1,400
SC duty, Q_{SC}	kW	23,500	23,500	23,500
Exhaust exergy	Btu/lb	140.0	140.0	140.0
	kW	226,264	226,264	226,264
Bottoming cycle ε_{ex}		73.5%	70.0%	65.0%
$\dot{Q}_{sc}$ conversion rate		50.0%	50.0%	50.0%
Steam turbine output	kW	180,532	172,547	161,062
Gross CC output	kW	507,504	499,518	488,033
Auxiliary load	kW	8,120	7,992	10,249
Net CC output	kW	499,384	491,526	477,784
Heat consumption	kW	829,898	829,898	829,898
Net combined cycle efficiency		60.2%	59.2%	57.6%

1. H-System steam turbines were supposed to be Toshiba machines. However, time pressure to get Baglan Bay going as soon as possible resulted in equipping that unit with a GE D series machine (specifically, D-10, which represents the early 1990s technology with ~70% bottoming cycle exergetic efficiency at rating conditions).
2. Main cooling steam source for 9H gas turbine is the HP turbine exhaust (cold reheat), which is supplemented by IP steam. Total cooling steam flow for the two stages is more than 125 lb/s (450,000 lb/h). The pressure drop is very high (about 25%–30%), and the steam is returned to the bottoming cycle at the reheater inlet. This imposes a significant reheater pressure drop on the bottoming steam cycle (typical reheater pressure drop for 3PRH steam cycles with D series steam turbines is 10%–12%).
3. The closed-loop heat rejection system with a 10-cell mechanical cooling tower had relatively high condenser pressure (vis-à-vis the rating-type low value of 1.2 in. Hg).

For comparison, the user is encouraged to examine the data in Section 8.6.1 for the world-record holder *Irsching* power plant with Siemens SGT5-8000H gas turbine in Germany. Suffice it to say that if the 109H in Baglan Bay had the Irsching steam turbine along with its steam cycle and condenser pressure with ~75% bottoming cycle exergetic efficiency, it would have easily broken the 60% barrier in 2003.

8.6.5 Epilogue

Due to competition from subsidized solar and wind energy, many German fossil fuel-fired power plants have been running at a fraction of the time needed to be profitable. Even the world-record breaking Irsching GTCC with the H class Siemens gas turbine could not escape this problem. The power plant, owned by the German utility Uniper and renamed "Irsching Unit 4", is currently operated under the so-called grid reserve directive (Netzreserveverordnung). In other words, it is used only when its output is needed to maintain grid stability. Even so, Uniper believes that the utility is insufficiently remunerated for providing this service. In fact, their original intention was to mothball the power plant starting from 1 April 2018.

In June 2019, GE announced that they planned to decommission (and demolish) the IEEC power plant with two steam-cooled 60-Hz H-System single-shaft power blocks. One of the two power blocks was already mothballed in 2017. In their filing with the California Energy Commission, GE stated that the plant is "not designed for the needs of the evolving California market, which requires fast-start capabilities to satisfy peak demand periods."

Earlier, GE had declared the H-System an "orphan technology" and announced that they were going to stop supplying replacement parts for the steam-cooled HGP components. Consequently, although the units had a very successful run at baseload operation in the Tokyo Bay area, TEPCO decided to replace their three 50-Hz H-System gas turbines with the new air-cooled HA class gas turbines. At the time of writing, one H-to-HA replacement had already taken place. This left only the original 109H in Baglan Bay, Wales, as the sole representative of the steam-cooled gas turbine technology.

REFERENCES

1. Gülen, S.C., Jacobs III, J.A., 2003, Optimization of gas turbine combined cycle, *POWER-GEN International 2003*, December 13–15, 2003, Las Vegas, NV.
2. El-Masri, M.A., 1985, On thermodynamics of gas-turbine cycles: Part 1—Second law analysis of combined cycles, *J. Eng. Gas Turbines Power*, Vol. 107, No. 4, pp. 880–889.
3. Horlock, J.H., 1992, The rational efficiency of power plants and their components, *J. Eng. Gas Turbines Power*, Vol. 114, pp. 603–611.

4. Haywood, R.W., 1989, *Equilibrium Thermodynamics ("Single-Axiom" Approach) for Engineers and Scientists/Part 1*, Krieger Publishing Company, Malabar, FL.
5. Kotas, T.J., 1985, *The Exergy Method of Thermal Power Analysis*, Butterworths, London.
6. Sciubba, E., Wall, G., 2007, A brief commented history of exergy from the beginnings to 2004, *Int. J. Thermodyn.*, Vol. 10, No. 1, pp. 1–26.
7. Kano, K., Matsuzaki, H., Aoyama, K., Aoki, S., Mandai, S., Development study of 1500°C class high temperature gas turbine, ASME Paper 91-GT-297, *International Gas Turbine and Aeroengine Congress and Exposition*, June 3–6, 1991, Orlando, FL.
8. Gülen, S.C., 2011, A simple parametric model for analysis of cooled gas turbines, *J. Eng. Gas Turbines Power*, Vol. 133, p. 011801.
9. Stodola, A., 1927, *Steam and Gas Turbines*, Authorized translation from the 6th German Edition by L. C. Löwenstein, McGraw-Hill Book Company Inc., New York.
10. Wicks, F., 2015, Mercury and steam - An early combined cycle using a toxic working fluid set a path for high-efficiency power plants, *Mech. Eng.*, Vol. 137, No. 07, pp. 40–45.
11. Foster-Pegg, R. W., 1966, Utility applications of gas turbines, Chapter 25 in *Gas Turbine Engineering Handbook*, Ed. John W. Sawyer.
12. Seippel, C., Bereuter, R., 1960, The theory of combined steam and gas turbine installations, *The Brown Boveri Rev.*, Vol. 47, No. 12, pp. 783–799.
13. Wunsch, A., 1978, Combined gas/steam turbine power plants – The present state of progress and future developments, *The Brown Boveri Rev.*, Vol. 65, No. 10, pp. 646–655.
14. Foster-Pegg, R.W., 1978, Steam bottoming plants for combined cycles, *ASME J. Eng. Power*, Vol. 100, pp. 203–211.
15. Taylor, W.G., Todd, D.M., 1979, STAG combined cycle power systems, Present and Future, GER-3115, *Electric Utility Gas Turbine State of the Art Engineering Seminar*, September 30 – October 2, 1979, Greenville, SC, General Electric Co.
16. Horner, M.W., Caruvana, A., 1982, High-temperature turbine technology readiness, ASME paper 82-GT-213, *ASME 1982 International Gas Turbine Conference and Exhibit*, April 18–22, 1982, London, England.
17. Rice, I.G., Jenkins, P.E., 1982, Comparison of the HTTT reheat gas turbine combined cycle with the HTTT nonreheat gas turbine combined cycle, *ASME J. Eng. Power*, Vol. 104, pp. 129–142.
18. Patterson, J.R., Walsh, E.T., 1983, A manufacturer's role in heavy-duty gas turbine future technology, ASME 83-GTJ-13, *1983 Tokyo International Gas Turbine Congress*, October 23–29, 1983, Tokyo, Japan.
19. Eckardt, D., 2014, *Gas Turbine Powerhouse*, Oldenbourg Verlag, Munich.
20. Jeffs, E., 1990, Utrect [sic] 225 MW combined cycle certified at 52% efficiency, Gas Turbine World, January-February 1990, pp. 12–16.
21. El-Masri, M.A., 1986, On thermodynamics of gas-turbine cycles: Part 2—A model for expansion in cooled turbines, *J. Eng. Gas Turbines Power*, Vol. 108, No. 1, pp. 151–159.
22. El-Masri, M.A., 1986, On thermodynamics of gas-turbine cycles: Part 3—Thermodynamic potential and limitations of cooled reheat-gas-turbine combined cycles, *J. Eng. Gas Turbines Power*, Vol. 108, No. 1, 160–168.
23. Chin, W.W., El-Masri, M.A., 1987, Exergy analysis of combined cycles: Part 2 – Analysis and optimization of two-pressure steam bottoming cycles, *ASME J. Eng. Power*, Vol. 109, pp. 237–243.
24. Bolland, O., 1990, A comparative evaluation of advanced combined cycle alternatives, ASME Paper 90-GT-335, *Gas Turbine and Aeroengine Congress and Exposition*, June 11–14, 1990, Brussels, Belgium.
25. Elmasri, M.A., 1988, GTPRO: A flexible, interactive computer program for the design and optimization of gas turbine power systems, 88-JPGC/GT-3, *Joint ASME/IEEE Power Generation Conference*, September 25–29, 1988, Philadelphia, PA.
26. Erbes, M.R., Gay, R.R., Cohn, A., 1989, GATE: A simulation code for analysis of gas-turbine power plants, 89-GT-39, *Gas Turbine and Aeroengine Congress and Exposition*, June 4–8, 1989, Toronto, ON.
27. Horlock, J.H., 2002, *Combined Power Plants*, Krieger Publishing Company, Malabar, FL.
28. Kehlhofer, R., 1991, *Combined Cycle Gas and Steam Turbine Power Plants*, Fairmont Press, Lilburn, GA.
29. Farmer, R., 1992, 227-MW Frame 9FA will power EPON's 1.7-GW combined cycle, Gas Turbine World, May-June 1992 issue, pp. 15–24.

30. Prandi, R., 2011, H class Siemens combined cycle plant achieves 60.75% efficiency, Diesel & Gas Turbine Worldwide, July-August 2011, pp. 46–48.
31. Matta, R.K., Mercer, G.D., Tuthill, R.S., 2000, Power Systems for the 21st Century – "H" Gas Turbine Combined-Cycles, GER–3935B.
32. Pritchard, J.E., 2003, H-System™ technology update, ASME Paper GT2003–38711, *ASME Turbo Expo 2003*, June 16–19 2003, Atlanta, GA.
33. Smith, R.W., Johansen, A., Ranasinghe, J., 2005, Fuel moisturization for natural gas fired combined cycles, ASME Paper GT2005–69012, *Turbo-Expo 2005*,June 2005, Reno, NV.
34. Feigl, M., Myers, G., Thomas, S.R., Smith, R.W., 2006, 7H™ combustion system performance with fuel moisturization, GT2006–90281, *ASME Turbo Expo 2006*, May 8–11, 2006, Barcelona, Spain.
35. Gülen, S.C., Smith, R.W., 2009, A simple mathematical approach to data reconciliation in a single-shaft combined cycle system, *J. Eng. Gas Turbines Power*, Vol. 131, p. 021601.

9 Major Equipment

The three major pieces of equipment in the "Power Island" of a gas turbine combined cycle (GTCC) power plant are

1. The gas turbine generator(s) (GTG(s)) (Chapter 4)
2. The steam turbine generator (STG) (Chapter 5)
3. The heat recovery steam generator (HRSG(s)) (Chapter 6).

Steam turbine heat sink is the fourth important "system", which is covered in detail in Chapter 7. Depending on the particular system, the heat sink includes the steam condenser (water or air-cooled), circulating water pump(s), vacuum pumps, the cooling tower and requisite piping, valves and tanks.

From a procurement and construction (and cost estimating) perspective, major GTCC power plant equipment is covered in four separate "packages", i.e.,

- Gas turbine package(s)
- Steam turbine package
- HRSG package(s)
- Synchronous ac generator package(s).

In multi-shaft GTCC power plants, each prime mover has its own generator and thus forms a single unit, i.e., GTG and STG. In single-shaft GTCC power plants, the two prime movers drive a single generator, which is usually placed between the two prime movers. In most modern single-shaft powertrains, the steam turbine is connected to the generator via a SSS[1] clutch (for fast start and/or simple cycle operation with full steam bypass).

9.1 GAS TURBINE PACKAGE

The gas turbine package can be examined in two parts: (i) the "flange-to-flange" gas turbine and (ii) the auxiliary systems. The flange-to-flange gas turbine comprises the axial compressor, combustor, the turbine, the bearings (usually two, one at each end, with one being a combined thrust-journal bearing) along with requisite instrumentation, actuators (e.g., for inlet guide vanes, IGVs, and variable guide vanes, VGVs) and the turning gear. (The exhaust diffuser is usually considered as a separate "system"; it includes a compensator between the turbine casing and the diffuser.)

The combustor section can be one of the three types:

- Can-annular (most common among modern advanced class gas turbines)
- Annular (aeroderivatives and sequential combustion GT24/26)
- Silo (vintage E and F class gas turbines).

Pretty much all modern gas turbine combustors are of the DLN (dry low NOx) type. For combustor technology and combustion calculations, refer to Chapter 12 of **GTFEPG** (Ref. [11] in Chapter 2).

[1] Synchro-Self-Shifting (pronounced "triple S").

Gas turbine auxiliary systems include

- Hydraulic oil for valves and actuators
- Hydraulic oil for hydraulic clearance optimization
- Lube oil with plate-and-frame-type oil-to-water heat exchanger
- Fuel gas system or "skid" (which includes a fuel gas meter)
- Compressor water wash skid (if there are more than one gas turbine, it is shared).

The lubrication system includes oil pumps, coolers, filters, instrumentation (e.g., oil-level transmitters to indicate full, empty, high-level alarm and low-level alarm, thermocouples) and control devices, a mist elimination device and an oil reservoir equipped with oil heaters and drains. The lubrication system comprises redundant pumps to distribute oil from the oil reservoir to the systems that need lubrication. Redundant pumps are used to distribute high-pressure (HP) oil to all hydraulic oil control system components. Typical pumps include

- Dual redundant ac motor-driven main lubrication oil pumps
- A partial flow, dc motor-driven, emergency lubrication oil centrifugal pump as a backup to the main and auxiliary pumps
- Dual redundant ac motor-driven variable displacement hydraulic oil pumps
- An auxiliary generator seal oil pump driven by piggyback ac/dc motors as backup to distribute seal oil to the generator
- A separate lift oil pump to lift bearings prior to turbine roll.

The lube oil is cooled by dual stainless steel plate/frame oil-to-coolant heat exchangers with transfer valve. Dual, full-flow filters clean the oil used for lubrication and hydraulic system. Each filter includes a differential pressure transmitter to signal an alarm through the gas turbine control system when cleaning is required.

Lubrication oil mist particles are entrained in the system vent lines. In order to remove the mist particles, a lube vent demister is used as an air exhaust filtration unit. The demister filters the mist particles and vents the air to the atmosphere while draining any collected oil back to the oil reservoir. The lube vent demister assembly consists of a holding tank with filter elements, motor-driven blowers and relief valve. One assembly is provided for the vent line from the lubrication oil reservoir.

The air-intake system of a gas turbine comprises

- Filter system with pre- and fine filter (multi-stage static filter)
- Inlet air filter house including weather hood, bird screen, internal support structure, instrumentation, lighting, power sockets, access ladders, platforms and doors
- Interconnecting duct work with expansion joint, manhole, damper and silencer
- Evaporative inlet cooler ("evap cooler") – if specified
- Electrical hoist for maintenance
- Dehumidifier for gas turbine standstill
- Nozzle system for compressor cleaning inside air inlet plenum.

In order to prevent ice formation on the filter, in certain sites, a heat exchanger is installed in the filter compartment upstream of the static filtration. A hot liquid (ethylene glycol/water or propylene glycol/water) is utilized to provide the heat.

The gas turbine package includes the control system (e.g., GE's Mark VIe or Siemens SPPA-T3000) with redundant automation processor for closed-loop control functions, I/O modules and turbine protection systems.

Gas turbine electrical accessories include the power control center, the ac power supply (low voltage switchgear, ac motor control centers) and dc power supply (dc voltage distribution, battery and its charger and dc/dc converter).

Pretty much in all cases, the gas turbine is installed in an all-weather protective housing called "enclosure", which provides thermal insulation, acoustical attenuation and fire extinguishing media containment. For safety and optimum performance, the enclosure includes ventilation, heating and cooling features.

Safety-related accessories are the hazardous gas detection system, fire detection system and fire extinguishing system. They are located inside the gas turbine enclosure, which is usually constructed with structural steel and noise abatement panels, both with corrosion protection. The enclosure includes the internal service platforms and ladders, doors with safety windows and internal lighting, including emergency lighting. The enclosure ventilation system comprises air-handling units (equipped with back-draft dampers and fans including mechanical redundancy and silencers) and air outlet openings on top of enclosure (equipped with dampers and silencers).

The largest source of loss in power plants is fire. Consequently, adequate fire protection is indispensable for the protection of major equipment costing millions of dollars to replace (not to mention for insurability). Fixed temperature-sensing fire detectors are provided in the gas turbine and accessory compartments (and in the hot-side bearing tunnel). The detectors provide signals to actuate the automatic multizone fire protection system. The fire detectors are wired as Class A loops, which allow the detection circuit to still function if an abnormal condition occurs, such as a wire break or ground fault. Fire extinguishing is accomplished by injecting carbon dioxide (CO_2) through nozzles in the enclosure and accessory compartments. Per US NFPA 12, fire protection system should be capable of achieving a noncombustible atmosphere in less than 1 min. The CO_2 supply system is composed of a low-pressure (LP) tank with refrigeration system mounted off-base, a manifold and a release mechanism. Initiation of the system will trip the gas turbine, provide an alarm on the annunciator, turn off ventilation fans and close ventilation openings.

Deposits such as dirt, oil mist, industrial or other atmospheric contaminants from the surrounding site environment, reduce airflow, lower compressor efficiency and lower compressor pressure ratio, which reduce thermal efficiency and output of the unit. Compressor water wash is used to remove fouling deposits, which accumulate on compressor blades and to restore unit performance. There are two types of compressor water wash:

- *Online water wash* (i.e., injecting water into the compressor while running at full speed and some percentage of load)
- *Offline water wash* (i.e., injecting cleaning solution into the compressor while it is being turned at turning gear speed).

Online washing is convenient because it does not require a shutdown but it is not as effective as offline washing. It is activated in regular intervals to minimize fouling in between scheduled offline washes, which typically includes injection of a cleaning solution (water and/or detergent) into the compressor while it is being turned at turning gear speed.

The online washing components consist of a piping manifold, spray nozzles and an on/off valve. The turbine control system is equipped with software to perform an automatic online wash. The off-base water wash skid is used for injecting demineralized water for online compressor cleaning and a cleaning solution into the compressor for offline compressor cleaning. The skid contains a water pump, a detergent storage tank, piping and an eductor jet pump capable of delivering solution at the proper flow, pressure and mix ratio. The skid also includes a water storage tank with freeze protection and immersion heaters. The latter is sized to heat the water to an original equipment manufacturer (OEM)-prescribed temperature in a prescribed amount of time, e.g., to 80°C in 12h.

Some gas turbines, e.g., vintage Westinghouse and Siemens F class and ABB/Alstom (now GE and Ansaldo) sequential combustion (reheat) units, utilize *cooling air coolers*. These are typically kettle-type boilers designed to ASME code requirements. Hot air extracted from the compressor flows in the tubes and boils feedwater extracted from the HRSG economizer. Steam generated in the air cooler is returned to the respective HRSG drum. Usually, the gas turbine OEM supplies the kettle boiler and all cooling air piping and supports from the gas turbine to the kettle boilers and back to the gas turbine.

The gas turbine is usually equipped with an *inlet bleed heat* (IBH) system, which comprises a pipe from the compressor discharge to the inlet plenum, a control valve and the associated instrumentation. The IBH system serves as a recirc system (similar to that found in large process centrifugal compressors) during startup to reduce the discharge pressure and prevent stall. The control valve varies the inlet heating air flow as a function of IGV angle. At minimum IGV angles, the inlet bleed flow is controlled to a maximum of 5% of the total compressor discharge flow. During, normal operation, the IBH system protects the compressor from icing at the inlet. It can also be used to lower the minimum emissions-compliant load by extending the DLN combustor's premix mode range. Increased inlet air temperature and reduced flow allows the combustor to operate at a firing temperature high enough to achieve optimal emissions.

For more detailed discussion of gas turbine accessories such as the turning gear and Load Commutated Inverter (LCI), refer to **GTFEPG** (Ref. [11] in Chapter 2).

9.2 STEAM TURBINE PACKAGE

Almost all large combined cycle steam turbines are multi-stage, reheat, straight-condensing turbines and have three major sections, i.e., the HP, intermediate-pressure (IP) and LP sections (see Chapter 5 for different arrangements). The most common configuration is a combined, opposed-flow HP–IP turbine and a separate single- or double-flow LP turbine (with axial-, down- or side-exhaust arrangements). This arrangement is also known as "two-cylinder, tandem compound". Each turbine has three major components: the outer casing (or shell), the inner casing and the rotor. (Older units with a pressure rating of 1,800 psig or lower are single casing/shell designs.) Typically, the rotor consists of a no-bore, one-piece forged shaft with integral coupling flange and inserted rotating blades (buckets).

Three-dimensional cutaway view of a typical combined cycle steam turbine is provided in Figure 9.1. Key features numbered in the figure are described below:

1. Large-diameter, single crossover pipe
2. Down-exhaust from the two-flow LP turbine

FIGURE 9.1 Combined cycle steam turbine. (Courtesy: Siemens.)

3. Spring-back seals in the inner casing (separating IP and HP steam paths – in GE steam turbines, this is referred to as the "N2 packing")
4. Intercept-reheat valve (see below for details) connected to the outer casing's lower half (there are two of them on either side); main steam valve (MSV) on the lower left (there is only one in this case but there may be one on each side)
5. HP reaction turbine steam path
6. Push rod arrangement to allow the LP-inner casing to follow thermal expansion of the shaft for reduced differential expansion
7. Long LP turbine last-stage buckets (see Section 5.1) – also point 9
8. Three-bearing arrangement.

A schematic cross section of the steam turbine in Figure 9.1 is shown in Figure 9.2. (Arrows show the directions of steam flows.) In conjunction with Figure 9.1, key features such as double-casing construction, three-bearing arrangement (combined thrust-journal bearing on the left) and monoblock HP/IP and LP rotors are clearly visible. In single-shaft combined cycle powertrains, the synchronous ac generator is connected to the HP end via a SSS clutch. In multi-shaft combined cycle arrangements, the generator is connected to the LP end.

As long as dimensions and weights do not exceed permissible limits, HP/IP turbine can be shipped complete. Due to the large size and weight involved, however, the LP turbine is erected on site. The LP turbine is shipped in separate parts, which are suitable for transportation and assembly. If the LP rotor, with all blades assembled, exceeds the transportation requirements, the last rotating blade row (containing the longest blades/buckets) may need to be removed for transportation.

Except sites with extreme weather conditions, the STG is located outdoors on an elevated foundation, enclosed in a weather-tight structure provided by the OEM with platforms fully around the operating elevation.

The STG is provided by the OEM with the following:

- Stop and control valve (SCV), which is also known as the HP throttle valve or MSV
- Combined reheat valve (CRV), which is also known as ICV (intercept control valve)
- Non-return valves (NRVs) except the cold reheat (CRH) NRV, which is typically supplied by the engineering, procurement and construction (EPC) contractor
- Hydraulic control oil system with cooling provisions

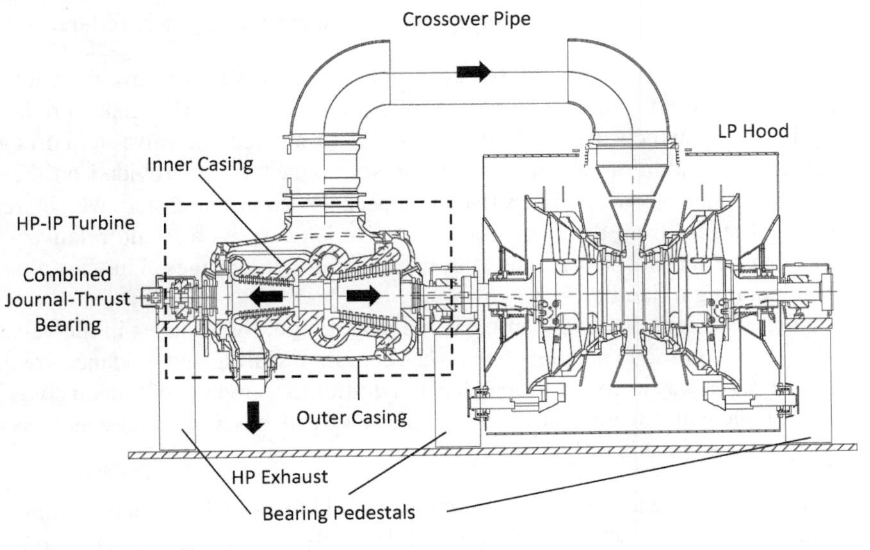

FIGURE 9.2 Combined cycle steam turbine cross section.

- Lube oil systems with cooling provisions
- The crossover pipe
- Exhaust hood spray system
- Gland seal system including condenser with exhausters
- Turning gear
- Water induction prevention system (see below)
- Turbine control system with the distributed control system (DCS) interface.

MSVs, reheat-intercept valves and the crossover pipe between the HP/IP turbine and the LP turbine are the most important components except the "flange-to-flange" steam turbine. The crossover pipe connects the IP turbine exhaust to the LP turbine inlet. It has a flow guide, or rotor deflector, which is located at the LP turbine inlet zone to minimize losses by optimally directing the inlet steam in the axial direction. Steam leaving the last row blades flows through an exhaust diffuser. The exhaust diffuser converts exhaust steam velocity into pressure as the steam flows into a rectangular cross-section at the condenser inlet.

The steam turbine is connected to the condenser via an "expansion joint", which absorbs the differential thermal expansion between the steam turbine and the condenser to minimize the forces and moments acting on the turbine exhaust flange. The expansion joint is typically of a stainless steel bellows type, which is welded to the turbine exhaust and the condenser steam inlet.

9.2.1 Steam Valves

The main (HP) and "hot" reheat (HRH) steam *stop* valves comprise isolation valves mounted in the main steam and reheat steam lines ahead of the turbine. They have protective functions and are tripped by protective devices on demand to interrupt steam flow. For this reason, they are designed to give the shortest possible closing times with maximum possible reliability. The *control* valves adjust steam flow to match load conditions. Stop valve and control valve are combined in a common body. Therefore, main and reheat steam valves are commonly referred to as SCVs. (General Electric jargon is "Combined Stop Valve" or CSV for the main SCV and "Combined Reheat Valve" or CRV for the reheat SCV.) Both valves' stop sections are provided with "strainers" in order to keep foreign particles away from the admission section of the turbine and the blading. The stems and disk bushings of the control and stop valves are specially treated to increase their resistance to corrosion and wear. Both valves have dedicated electro-hydraulic actuators, which move the valves against the force of a spring. The valves are closed by spring force in the event of a power failure.

- *Stop valve.* With the stop valve in closed condition, steam is present above the valve disk. A pilot disk integral with the valve stem is incorporated in the valve disk to reduce the force required to open the valve. Metallic and graphite rings seal the valve stem off against the atmosphere. In addition, a stem leak-off steam connection is provided on the main steam and IP steam stop valves. The valve has a backseat on disk and stem which prevents any leakage of steam from the stem when the valve is open. As a result, relatively large clearances can be set for the seal rings but with only small leakages. This has a positive impact on operating reliability.
- *Control valve.* The control valve is of balanced plug design. Relief holes in the valve disk reduce the force required to position the valve. Valve stem sealing and guidance are implemented in the same way as for the stop valve. In addition, a stem leak-off steam connection is provided on the main steam control valve. This valve disk has a backseat which becomes effective, when the valve is fully open.

Main and reheat SCV packing leak-off connections are routed to a safe location (but *not* to the condenser). Since these lines are normally open, it is not desirable to have vacuum on the downstream side of the connection because it will cause the packing to leak.

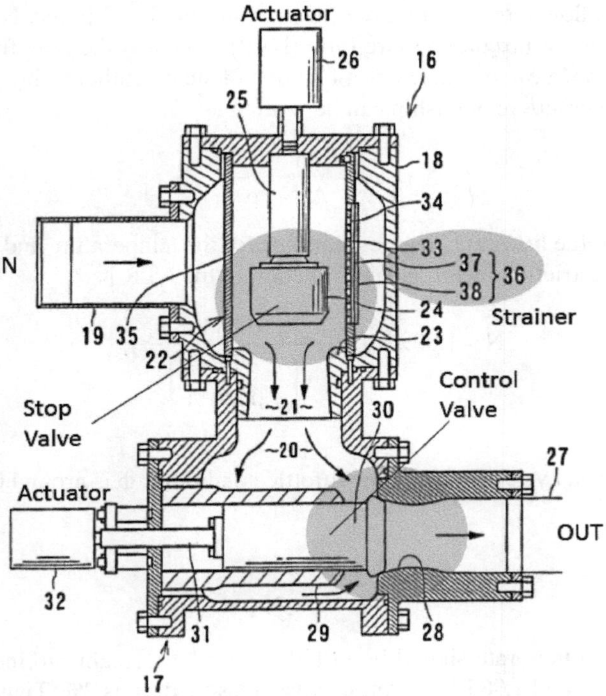

FIGURE 9.3 Main steam SCV (from US Patent 2,284,569 by Takemura).

The drawing of a typical main steam stop-control valve is shown in Figure 9.3. The pressure drop across the main and reheat SCVs is a function of the plant size and valve design. Typical values for the main SCV with denser steam range from 0.8% up to about 2%. Typical values for the reheat SCV with less dense steam range from 1.25% up to about 2.5%.

Some steam turbines are equipped with "overload valves", which can be considered as "relief" valves. What is relieved is excessive main steam pressure at the MSV inlet. Note that any given steam turbine has a fixed "swallowing capacity", which is determined by the "inlet bowl" flow area. This swallowing capacity is quantified by the simple relationship below (from the compressible fluid flow theory):

$$\dot{m} \propto C_q A_e p \sqrt{\frac{1}{T}}, \tag{9.1}$$

where C_q is the flow coefficient, A_e is the effective flow area at the turbine inlet, p is steam pressure and T is steam temperature (at the steam turbine inlet "bowl"). At different operating conditions, the variation in C_q is very small. Choked steam flow through the inlet nozzles can be formulated into a "critical flow parameter", which can be calculated using the *ASME Steam Tables*, to establish the swallowing capacity of a particular steam turbine. Using the following definitions

$$G = \frac{\dot{m}}{A_e N} \tag{9.2}$$

and

$$\phi = G\frac{\sqrt{T}}{p}, \tag{9.3}$$

where G is the steam flow rate per unit area (lb/s per in.^{2} in USCS units), N is a restriction factor (tabulated as a function of nozzle pressure ratio [NPR]) and ϕ is the flow function, which can be looked up from the *ASME Steam Tables* as a function of steam enthalpy, h(p, T) and pressure. The steam turbine flow–pressure relationship can be written as

$$\phi = \frac{\dot{m}}{A_e N}\frac{\sqrt{T}}{p}. \tag{9.4}$$

(Ensure to use lb/s for the flow rate, degrees Rankine for the temperature and psia for the pressure.) The formula for the restriction factor, N, as a function of the NPR is

$$N = 1.009621 - \frac{0.88488}{1 + \left(\frac{NPR}{1.062611}\right)^{8.502739}}. \tag{9.5}$$

For NPR ≥ 1.83, N = 1.

For typical combined cycle steam turbine throttle conditions, ϕ is around 0.4. Another formulation is[2]

$$\frac{\omega}{p} = \frac{\dot{m}}{A_e N}\frac{1}{p}. \tag{9.6}$$

In this equation, steam flow rate should be in lb/h. Consider a steam turbine with design throttle conditions of 2,515 psia and 1,085°F. Throttle valve pressure drop is 2%. Thus, at the steam turbine "bowl", i.e., inlet of the stator nozzles, enthalpy is 1,512.9 Btu/lb, pressure is 2,464.7 psia and ω/p is 38.272. If the steam flow rate at the design conditions is 275 lb/s, the "AeN" of this steam turbine is

$$AeN = 275 \times 3{,}600 / 38.272 / 2{,}464.7 = 10.5 \text{ in.}^2.$$

Equation 9.6, with fixed AeN, allows one to determine steam flow rate-throttle pressure correlation at different operating conditions. For example, a 10% increase in steam flow, at the same steam temperature, would increase the steam inlet pressure by 10% or about 250 psi in the example above. This could push the steam pressure to about 2,700 psia and over the OEM's design limit. This problem could be alleviated by sending the extra steam flow through a second (overload) valve to an admission point downstream of the HP throttle. For a discussion of when and why this capability is needed, refer to the paper by Quinkertz (within the context of ultra-supercritical coal-fired boiler plant steam turbines) [1].

The overload extraction is an optional steam outlet nozzle on the control valve casing part. It provides the overload valve with steam from the MSV.

LP admission valves are hydraulically operated butterfly SCVs. The stop valve has the same functionality as main and reheat steam stop valves. The control valve prevents the LP steam system pressure above a limiting "floor" pressure value so that acceptable steam volumetric flow and velocity is maintained. Typical pressure drop is 3 psi.

9.2.2 Steam Seal Regulator (SSR)

This component is also known as the *steam seal header* (SSH). It is the assembly of pipes and control valves, which combines the steam from the higher-pressure packings and uses it to seal the LP turbine end packings against the in-leakage of air to vacuum. During normal operation, the steam seal regulator (SSR) takes the requisite amount of steam from the header and sends the balance to the

[2] See GER-3642E "Steam Turbine Cycle Optimization, Evaluation, and Performance Testing Considerations," J.S., Wright and Figure 18 therein.

condenser. Normally, flow coming from the higher-pressure packings is sufficient to meet the LP sealing demand. If this is not the case, "pegging" steam is pulled from a suitable location in the steam system. A schematic description of the steam seal system is shown in Figure 9.4 (adopted from Fisher Valves Application Guide D352219X012, August 2013). Before moving on, it should be pointed out that Figure 9.4 depicts "labyrinth" seals. In some newer designs, OEMs use "brush" seals comprising many thousands of flexible pins that can ride on the journal in an effectively "zero" clearance configuration. Consequently, brush seals offer superior leakage control compared to labyrinth seals. (It is noteworthy that sealing accounts for roughly one-third of total stage efficiency loss in a steam turbine.) In addition, brush seals at LP rotor ends reduce sealing steam requirement while brush seals at HP/IP rotor ends reduce supply leakage. Total number of brush seals in the unit and their exact location are determined via a detailed rotordynamic assessment.

Steam supplied to the SSR must be superheated. During startup, rather than specifying an auxiliary boiler that can provide superheated steam, an in-line electric superheater can be installed in the piping feeding the turbine's gland steam system. However, if the steam seal system is the only startup user that requires superheated steam, it can be more economical to have an aux boiler with

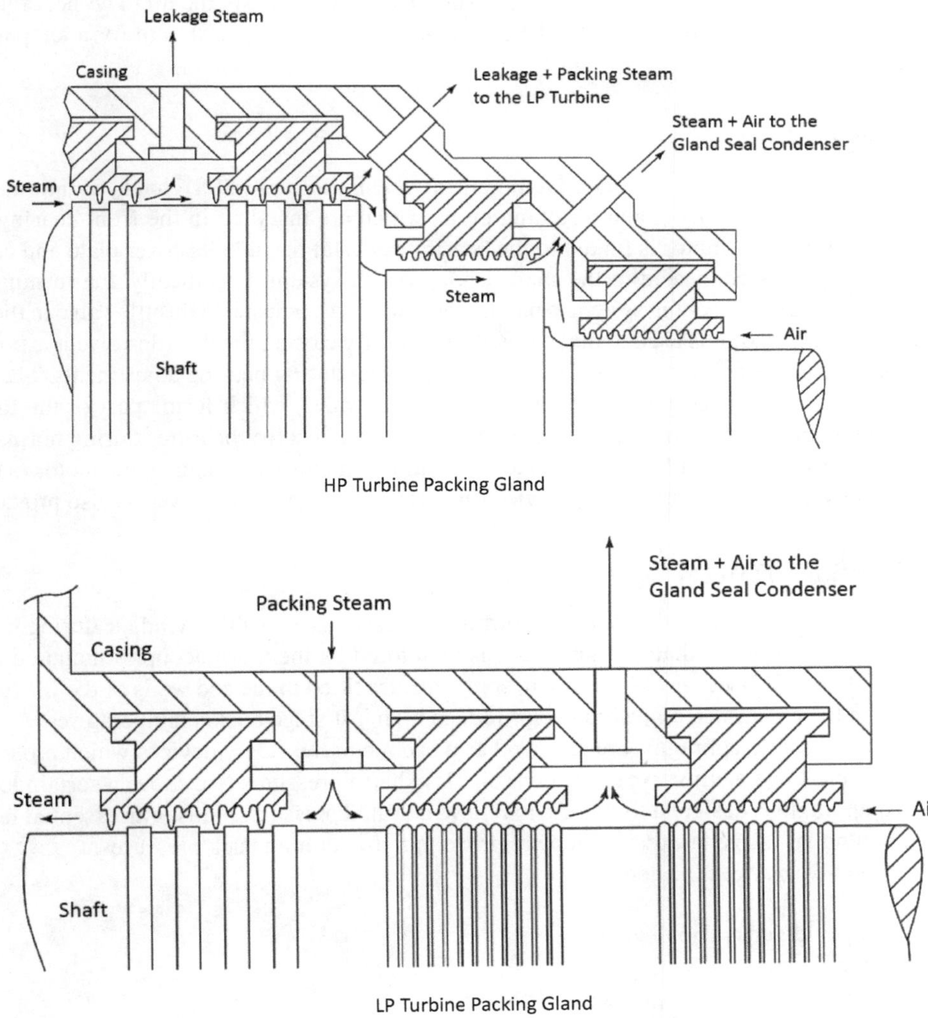

FIGURE 9.4 Steam turbine seal system.

dedicated superheater for the steam seals supply line. An aux boiler can be avoided in a single-shaft GTCC with a SSS clutch, which separates the steam turbine from the rest of the powertrain. Thus, steam turbine roll can be delayed until the HRSG generates requisite steam for the seals.

In theory, in a short-duration event like an overnight shutdown (about eight hours), if the HRSG is bottled up hot via stack dampers, enough steam to maintain seals overnight can be available from a HRSG drum. Note, however, that steam turbine suppliers require seal steam to meet specific superheat or minimum temperature requirements to ensure that the steam-to-rotor metal temperature difference is not too large and to ensure that no moisture is present to damage the seals. This requires a careful evaluation because, even if the HRSG can be bottled up hot, steam from the HRSG loses superheat almost immediately. This is why an electric superheater is requisite in extensively cycled plants.

9.2.3 Gland Seal Condenser (GSC)

This component is also known as the *steam packing exhauster.* It is a condensing heat exchanger operating at slightly lower than atmospheric pressure, which sucks the steam turbine shaft sealing steam (mixed with air for the "air packing"), condenses it and exhausts the air. The heat released by the condensing steam is picked up by the condensate pump discharge stream for a temperature rise of a few degrees.

9.2.4 Turning Gear

The turning gear consists of a hydraulic gear motor, an overrunning (e.g., SSS) clutch, an intermediate shaft and the necessary bearing and fastening parts, which are installed in the front bearing pedestal. The hydraulic gear motor is linked to the bearing pedestal through its cover plate and casing.

The turning gear is connected to the shaft jacking or lift system. Specifically, the turning gear starts moving as the shaft jacking system is put into operation. An adjustable throttle valve in the feed pipe of the hydraulic motor is used to make speed changes. By closing the throttle valve it is possible to isolate the shaft-turning gear from the shaft jacking system during bearing adjustments. The shaft turning gear is also provided with an electrically actuated valve, which forms part of the turbine generator automation system. In order to limit corrosion of the inactive bearings during normal turbine operation, a small flow of lube oil is fed to the hydraulic motor which causes the motor to rotate slowly. Typically, a manual shaft-turning device (in essence, a ratchet and a lever) is also provided.

9.2.5 Protective Features

Temperature in the exhaust section of the LP turbine increases as a result of windage during no load operation. The LP turbine exhaust temperature is monitored by the thermocouples mounted in the LP cylinder. If an OEM-set limit is exceeded, spray nozzles fitted in the end walls of the LP-turbine inject water in the exhaust steam, which helps reduce high last-stage bucket temperatures.

Atmospheric relief diaphragms are provided as a backup to the safety devices which protect the steam turbine back end against excessive pressure. If exhaust pressures rise above a certain level, a breakable diaphragm is designed to rupture to relieve elevated pressures. This is an essential feature for free-standing last stage buckets, which are prone to flutter at high back pressures.

Other protective devices include

- Solenoid emergency trip device
- Overspeed trip device
- Low bearing oil pressure protective device
- Low vacuum protective device
- Thrust bearing protective device.

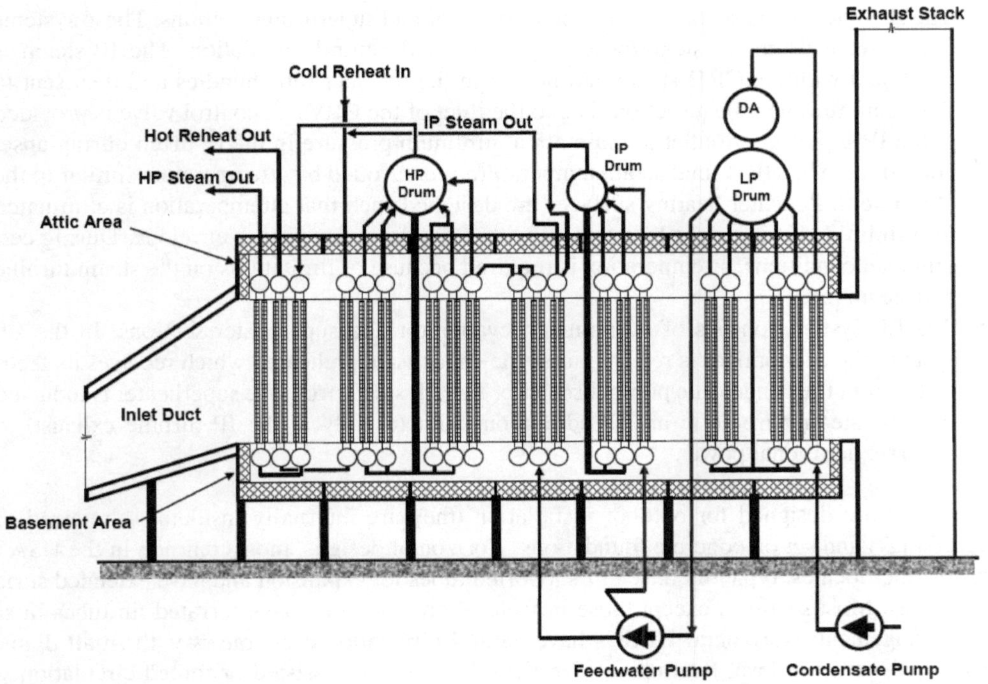

FIGURE 9.5 Typical horizontal HRSG schematic arrangement.

In addition, the main and reheat steam systems have to be designed to prevent steam turbine water damage based on the guidelines of ASME TDP-1-2013, *Prevention of Water Damage to Steam Turbines Used for Electric Power Generation, Fossil Fueled Plants*.

9.3 HEAT RECOVERY STEAM GENERATOR (HRSG)

It is pretty much a design standard that all modern combined cycle power plants with advanced class gas turbines have *three-pressure level, reheat* (3PRH), natural circulation-type HRSGs with horizontal gas turbine exhaust flow through vertical tube heat transfer sections. Nonetheless, in Chapter 13, a case will be made that a *two-pressure reheat* (2PRH) HRSG can make more sense in the USA with cheap natural gas.

The function of the HRSG is to generate steam from the gas turbine exhaust for producing power in the STG (see Chapter 6). Typical schematic arrangement of a 3PRH horizontal HRSG with the LP section dedicated to condensate preheating and deaeration is shown in Figure 9.5.

Typical HRSG system description is as follows:

- The HP system consists of economizer, evaporator and superheater sections. It uses a steam drum for steam/water separation and natural circulation. The HP steam is sent to the steam turbine throttle (i.e., the SCV or MSV inlet). The HP final steam temperature is controlled using a final stage or "terminal" (i.e., downstream of all superheater sections) and an inter-stage (i.e., between superheater sections) spray attemperator (desuperheater). Superheater sections are designed such that HP steam attemperation is eliminated (or minimized if it cannot be avoided) within the design operating envelope. During certain load conditions and startup transients, attemperation is required. (One such example is fast start with full bypass when terminal attemperators control the HP steam temperatures.)

- The IP system consists of economizer, evaporator and superheater sections. These systems use a steam drum for steam/water separation and natural circulation. The IP steam is combined with the CRH steam and heated in the reheater tube bundles and then sent to the steam turbine reheat section (i.e., to the inlet of the CRV). A control valve is provided at the IP superheater outlet to maintain a minimum pressure in the IP drum during upset conditions. The HRH final steam temperature is controlled by attemperators similar to the HP system. Reheater heating surfaces are designed such that attemperation is eliminated (or minimized if it cannot be avoided) within the design operating envelope. During certain load conditions, attemperation is required because of limitations on the steam turbine and steam piping.
- *The* LP system consists of economizer, evaporator and superheater sections. In the LP system, the economizer is referred to as the condensate preheater, which receives its feedwater from the condensate pump discharge. The LP steam from the superheater is inducted into the steam turbine at an LP admission point (usually at the IP turbine exhaust) to generate additional power.

The HRSGs are designed for outdoor installation (they are internally insulated – referred to as "cold casing") and set on concrete foundations. Horizontal designs, most common in the USA and in many other locales, typically have top-supported tubes for expansion and have extended surface, finned tubes. Most sections, except those in front of duct burners, have serrated fin tubes in staggered arrangement. Horizontal HRSGs have natural circulation evaporators with small-diameter tubes to reduce drum level fluctuations. Vertical designs have assisted or forced circulation with 2 × 100% pumps for each pressure level. In most cases, the HRSG is shipped to the construction site in modules with side panels installed and structural steel shipped separately. See Section 6.2.1 for a discussion of different modular HRSG construction options.

From the geometrical point of view, the differences between vertical and horizontal HRSG geometries are as follows:

- Tube length is about 10% shorter in the vertical design vis-à-vis the horizontal
- Total number of tubes is about 15% higher in the vertical design vis-à-vis the horizontal
- Heat transfer surface area is 20%–25% lower in the vertical design vis-à-vis the horizontal.

The difference is mainly due to the geometrical constraints of the existing steel structures.

In Benson-type HRSG, the HP section is of *once-through design* with a separator "bottle" in lieu of a large, thick-walled steam drum. In the Benson design, steam temperature at evaporator outlet is superheated in order to have a sufficient range for HP steam temperature control via feedwater mass flow control only at part load operation and/or at different ambient conditions. During startup or at very low load conditions, the separator splits steam and water in the two-phase flow at the evaporator section exit. Water is recirculated back to the evaporator inlet via the startup recirculation line.

The HRSG consists of heat transfer sections. Each heat transfer section is built in modules. A module consists of a top header, heat absorbing tubes and a bottom header. Heat absorbing tubes usually are provided with fins to increase the area available for heat transfer. (The outside surface area can be increased up to 3,000% by finning.) Tubes are arranged either in-line or staggered (it depends on the manufacturing capabilities of each supplier). Number of tube passes (i.e., how many times feedwater or steam flow crosses the gas flow-path – see Figure 9.6) can be categorized as follows:

- Superheater or reheater – 1 or 2 passes
- Economizer – 2 to 6 passes
- Evaporator – 1 pass.

When the modules are pre-assembled with connecting pipes, inlets, outlets and supports, the assembly is referred to as a *tube bundle*. When the bundles have been assembled, complete with casing and insulation, the assembly is referred to as a *box*.

Since the drum-type, horizontal HRSG is the most common type in the USA and in many locations in the world, description of the key components will be based on that variant. The evaporator has five key components (refer to Figure 9.7): the steam drum (1), the downcomer (2), feeder header and the feeders (3), heat transfer modules (4) and the risers (5). The key design parameter is the *circulation ratio*, which is the ratio of the water flow rate in the downcomer to the steam flow rate out of the drum. Typical value of circulation ratio for modern HRSGs as a function of evaporator drum pressure are:

- 5–6 for 1,600–2,500 psia
- 6–10 for 1,100–1,600 psia
- 8–15 for 500–1,100 psia
- 10–20 for 250–500 psia.

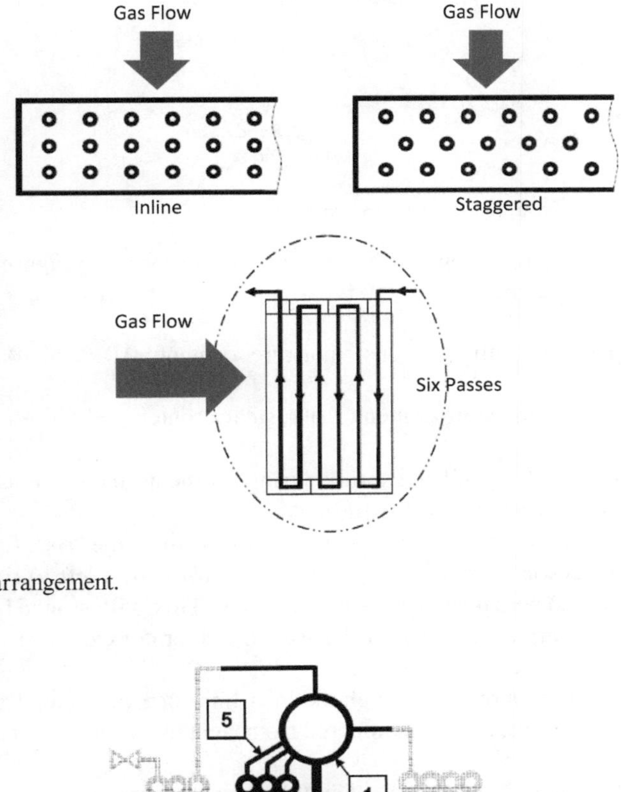

FIGURE 9.6 Tube arrangement.

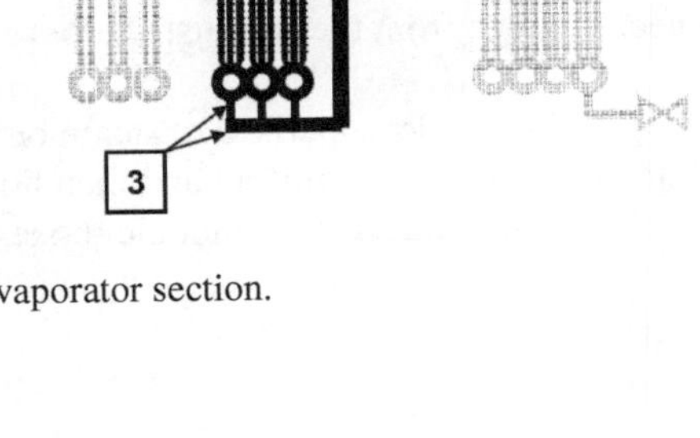

FIGURE 9.7 Typical HRSG evaporator section.

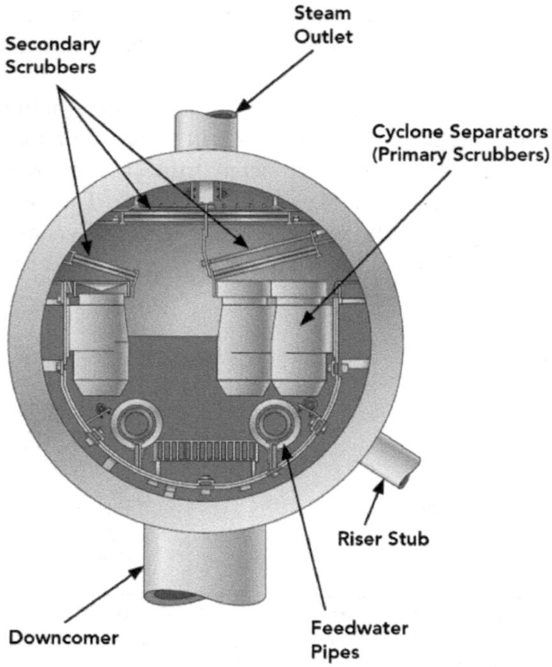

FIGURE 9.8 Typical HRSG evaporator drum internals.

The steam drum internals include separators, baffles and chevrons (see Figure 9.8), which provide required steam quality. In particular,

- Centrifugal separators with chevrons: moisture content: 0.03%–0.3%, steam content: 99.97%–99.7%
- Baffles with chevrons: moisture content: 0.5%, steam content: 99.5%.

The steam drum accommodates swell and shrink during transients and prevents running the HRSG dry for the given holdup time (typically 3–10 min).

The end point of evaporation in a drum-type evaporator is where the "riser tubes" enter the drum from the bottom. In a Benson evaporator, the end point moves from left to right (i.e., in the flow direction of hot exhaust gas) with decreasing steam fraction. This is illustrated by the diagram/chart in Figure 9.9. Note that steam coming out of the second evaporator can be slightly superheated or wet depending on the operating conditions.

HP feedwater enters the evaporator through the inlet lines of Evaporator 1 and is distributed to the tube rows. (Note that evaporators are numbered from right to left in the direction of water/steam flow.)

The distribution of the mass flow to individual tube rows is determined by the tube parameters, the level of heating and the extent of throttling due to orifices at the Evaporator 1 inlet.

Steam of 0.4–0.6 quality enters the downcomer through the outlet lines and is distributed uniformly to the inlet headers of Evaporator 2 through the star distributor and T-pieces which are installed at the end of the connecting lines from the star distributor to increase the number of inlets to Evaporator 2 inlet headers.

Depending on the load and operating mode, superheated steam or wet steam (0.95 quality) exits the second evaporator. It is routed to a common manifold and then flows to the separator bottle.

HRSG casing (including inlet/burner ducts and module boxes) typically consists of (see Figure 9.10):

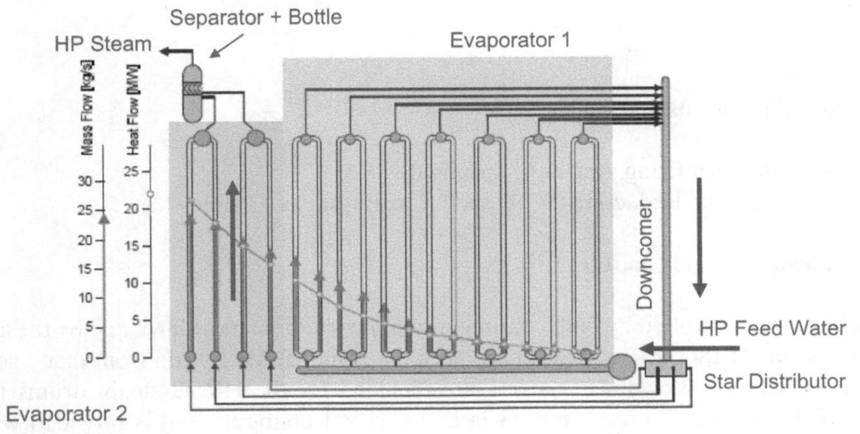

FIGURE 9.9 Benson evaporator configuration.

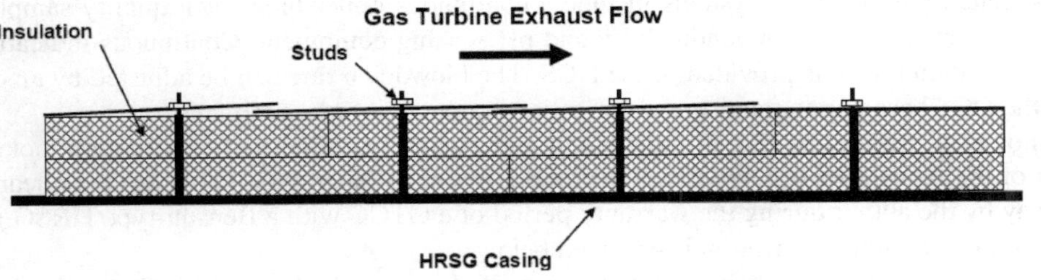

FIGURE 9.10 HRSG casing and insulation.

- Carbon steel outside casing – 0.25 in. thick
- Several layers of insulation – each layer 2 in. thick (overall insulation thickness may reach up to 12–14 in.)
- Stainless or carbon steel liner plates – typically 12-, 14- and 16-gauge thick
- Stainless or carbon steel studs, washers and nuts.

If there is no deaerating condenser, the deaerator is placed on the LP steam drum and becomes an integral part of it (see Figure 9.5 – quite rare in modern GTCC systems). Drum level control valves are downstream of the economizers to prevent steaming in the latter.

With increasingly stringent NOx emissions limits (as low as 2 ppmvd in many sites), HRSGs include a *Selective Catalytic Reduction* (SCR) system for the control of NOx emissions to meet required limits and catalyst for CO and volatile organic compound (VOC) reduction (see Chapter 12).

The self-supporting HRSG stack is equipped with ports for emissions sampling. Platforms and ladders and personnel protection are provided to permit adequate access to the emissions sampling ports. Aircraft warning lights are required. Insulation and/or personnel protection on the inlet duct, outlet duct and stack must be provided.

HRSG components and auxiliaries are enumerated as follows:

- Heating surfaces and tubes (economizers, evaporators and superheaters)
- Steam drums
- Valves (silencers are provided for the safety valves and startup vents)

- Drains
- Casing and ducts
- Expansion joints
- Stairways, platforms and ladders
- Painting
- Supplemental/duct firing system (if specified)
- Selective Catalytic Reduction (SCR) and CO catalyst (see Chapter 12)
- Stack
- Instrumentation and controls.

Water chemistry is controlled by both continuous and intermittent blowdown and by the addition of chemicals into the HRSG drums (e.g., phosphates but some OEMs refrain from that – see Section 16.3 for more details). Blowdown systems are included for the HRSG steam drums to remove dissolved solids. The HRSG (each HRSG in a N × N × 1 configuration) is provided with a continuous blowdown tank venting to atmosphere (located near the respective HRSG). The blowdown water is sent, via the HRSG blowdown sump pumps, to the cooling tower basin for makeup. Refer to Sections 10.1.1 and 10.1.2 (for Benson-type HRSGs) for details.

Associated measurement systems include a continuous boiler blowdown quality sampling and indicating system and cation conductivity and pH sensing equipment. Continuous indication with alarm limit monitoring is provided in the DCS. The blowdown rate can be adjusted by an operator using the DCS based upon the drum water cation conductivity measurement.

Although it may sound like a mundane system, which barely deserves a second look, proper design of the blowdown system is vital to trouble-free plant operation. (This has been learned the hard way by the author during the warranty period of a GTCC with a Benson-type HRSG.) This is why a detailed system description is provided below.

The blowdown tank must be designed in accordance with the ASME Code, Section VIII; Division I; a typical design pressure is 50 psig. It is provided as a full skid with a vent, high liquid-level transmitter and alarm, liquid-level indicator, control valve, pressure gauge and thermometer. Blowdown enters the tank tangentially to take advantage of centrifugal force to separate steam from liquid. A stainless steel wear plate is furnished for erosion protection where the flashing blowdown impinges on the wall of the tank. An anti-swirl baffle is provided at the tank's drain outlet connection to prevent the formation of eddies and blow-through of steam into the water drain system. Adequate steam volume has to be provided to separate condensate without causing swells. The blowdown system is designed to handle two-phase flow and to prevent steam hammer. Continuous and intermittent blowdown valves are specifically designed for blowdown control. These valves have outlets the same size as the inlets. The piping downstream of these valves are designed appropriately for the velocity and service. The tank connection size must not be less than the valve inlet size. Drains from the atmospheric blowdown tanks and other sources, which are to be discharged to the cooling tower basin, or other bodies of water, are ensured not to exceed 140°F. Circulating water is used to quench the blowdown effluent before dumping to the waste system.

Typically, an HRSG includes an A36 structural carbon steel stack with a minimum plate thickness of ¼ in. Height and diameter are specified based on the site-specific and mechanical design criteria (e.g., plume abatement, flue gas flow rate and temperature). A stack silencer and damper are provided. In compliance with Federal Aviation Administration (FAA) regulations, lighting is provided on the stack. Because of the high quality of pipeline natural gas, typically, no stack external insulation is included (based on maximum expected sulfur content) though a metal standoff or equal means of personnel protection is provided at the grade level.

Because under certain operating conditions the flue gas temperature could be below the dew point, the LP economizer (or condensate preheater) of the HRSG is provided with a 1 × 100% condensate recirculation pump to control water temperature at the HRSG inlet such that the stack exit temperatures remain above acid dew point.

In Europe, HRSGs are usually designed according to EuroNorm 12952 (progressively replacing national codes) per the mandatory *Pressure Equipment Directive 97/23/EC* (PED). In the USA and many other places in the world (even in Europe in some cases), Section I of the *ASME Boiler and Pressure Vessel Code* governs the design of the HRSGs. Fontaine and Bonsang (from the HRSG manufacturer CMI Energy) provides a detailed comparison of both codes [2]. Hampson, Robertson and Simandjuntak compare both codes from a design as well as operability perspective [3]. As noted by Robertson, the sole purpose of the ASME Code is to ensure safety—to prevent catastrophic failures and the resultant loss of life, injury and property. It does not give guidance as to the operation of modern HRSGs in cyclic duty behind advanced class gas turbines with exhaust temperatures pushing 1,200°F. The EN code, on the other hand, is a more sophisticated code, which allows further design optimization and greater flexibility in plant operation. The reader is referred to those papers for an in-depth coverage of the subject.

9.4 AC GENERATOR

Alternating current (ac) synchronous generators can be classified into three groups based on the cooling medium used: air, hydrogen or water-cooled. Relative properties and capabilities cooling medium types are summarized in Table 9.1.

Air-cooled generators come in two basic configurations:

1. Open ventilated (OV)
2. Totally enclosed water-to-air cooled (TEWAC).

In the OV design, outside air is drawn directly from outside the unit through filters, passes through the generator and is discharged outside the generator. In the TEWAC design, air is circulated within the generator, passing through frame-mounted air-to-water heat exchangers. As such, TEWAC designs are more effective in keeping the sensitive internal hardware of the generator clean. Air-cooled generators (both stator and rotor) are typically used for ratings up to 350 MVA. A typical synchronous ac generator is shown in Figure 9.11.

In the 300–550 MVA range, hydrogen-cooled machines are the economic choice because they are typically (i) 20% smaller than air-cooled machines at the same rating and (ii) more efficient (e.g., 99% vis-à-vis 98.8% for air-cooled).[3] Beyond that (rare in combined cycle applications), water-cooled stator designs (with hydrogen-cooled rotors) are available.

One OEM offers generators in 370–560 MVA range with the stator winding directly cooled by water and the rotor winding radially cooled by pressurized air (Siemens SGen-2000P). It is claimed to have efficiency comparable to hydrogen-cooled generators with improved safety (no hydrogen) and low operating and maintenance costs.

TABLE 9.1
Relative Comparison of Generator Coolants

Fluid	Specific Heat	Density	Volumetric Flow	Heat Transfer Rate
Air	1.00	1.00	1.00	1.00
H_2 @ 30 psig	16.36	0.21	1.7	5.00
H_2 @ 45 psig	16.36	0.26	1.7	7.50
Water	4.16	1,000	0.03–0.17	140–700

[3] Plus, hydrogen-cooled generators maintain their efficiency at part load better than air-cooled ones.

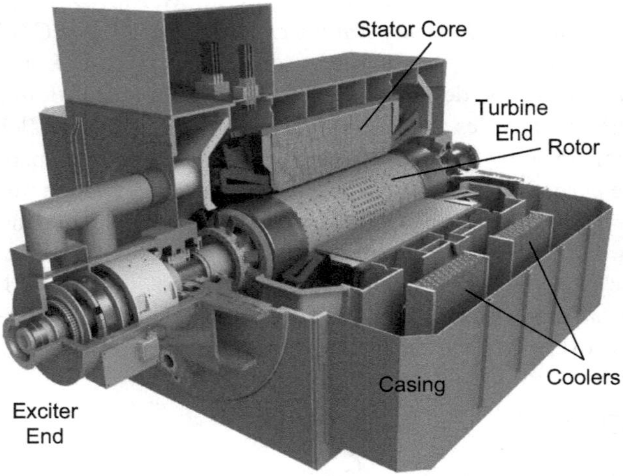

FIGURE 9.11 Combined cycle synchronous ac generator.

Hydrogen's advantage as a coolant comes from its lower density vis-à-vis air (i.e., lower windage losses) and higher heat transfer capability with comparable volumetric flow rate (due to higher thermal conductivity and convective heat transfer coefficients). Its disadvantage is the explosive potential upon mixing with air (between 4% and 75% H_2 in the mixture), which requires special measures to be taken to prevent leakages during operation as well as gassing and degassing (purging).[4] As a side benefit, advanced seals requisite to contain hydrogen also help to keep the generator interior clean. Furthermore, due to the absence of oxygen inside the sealed and pressurized casing, armature insulation deterioration caused by *corona* is less in hydrogen than in air.[5]

OEMs design their generators in accordance with established standards, e.g., IEEE C50.13-2014 (which replaced ANSI C50) in the USA or IEC 60034.[6] IEEE C50.13 governs operational requirements, rating and performance characteristics, insulation systems, temperature limits and test procedures. Familiarity with this standard or IEC 60034 is highly recommended.

The internally recirculating, once-through hydrogen stream dissipates generator heat through gas-to-water heat exchangers. Ventilation fans are mounted at each end of the rotor. The fans facilitate the circulating gas flow for cooling the generator. Cooling of the stator core is accomplished by forcing the coolant through the radial ducts formed by space blocks. The axial length of the core is made up of many individual segments separated by the radial ventilation ducts. The rotor winding, which is typically a directly cooled radial or axial flow design, is largely self-pumping. The rotor is cooled externally by the coolant flowing along the gap over the rotor surface and internally by the coolant that flows through sub-slots under the field coils within the rotor body and passes directly through radial cooling ducts in the copper coils and wedges.

Water is an even better heat transfer medium than hydrogen (plus it is not explosive). In theory, it can be used in any size class, i.e., 300 MVA or smaller. However, due to the complex cooling configuration (i.e., increased manufacturing cost) and additional equipment (auxiliary water cooling and deionization skid, pipes, valves, etc.), water cooling is not cost-effective except for very large

[4] For example, an inert gas, carbon dioxide, is used as an intermediate gas so that air and hydrogen do not mix inside the generator. The purging procedure for the generator has carbon dioxide introduced first to displace air, then hydrogen introduced to displace the carbon dioxide.

[5] Corona refers to an electrical discharge caused by the ionization of the cooling medium (air or hydrogen) surrounding the electrically charged conductor.

[6] International Electrotechnical Commission.

TABLE 9.2
Typical Synchronous AC Generator Design and Performance Data

Rotor Coolant		Air (Radial)	Air (Radial)	H_2 (Radial)	H_2 (Axial)
Stator Coolant		**Air (Indirect)**	**Air (Indirect)**	**H_2 (Indirect)**	**H_2O (Direct)**
	psig			45	45
Rating	MVA	140	300	450	550
Power output	MW	112	240	360	440
	kV	10.5	15.75	22	21
Exciter voltage	V	293	412	426	487
Exciter current	A	1,194	1,359	3,503	4,153
Temperature class		F	F	F	F
Application limit		B	B	B	B
Subtransient reactance		0.18	0.16	0.19	0.19
Transient reactance		0.26	0.24	0.28	0.26
Short circuit ratio		0.48	0.47	0.54	0.48
Heat loss, rotor	kW	316	490	1,410	1,940
Heat loss, stator	kW	335	418	680	1,035
Total losses	kW	1,528	3,470	4,275	5,465
Efficiency	%	98.65	98.58	98.83	98.77
Rotor weight	mt	34	57	54	52
Stator weight	mt	150	260	303	293

ratings. However, in applications to large industrial gas turbines, water *is* used as the stator coolant whereas hydrogen is used to cool the rotor. Cooling water must be of a very high purity and very low conductivity (about 5 μS/cm). This requires a dedicated water treatment system with redundant filters, circulation pumps, coolers, etc. and an ion exchanger, which continuously deionizes a stream of circulating cooling water to ensure low conductivity.

In terms of the "method of cooling", generators can be classified into two groups, i.e., (i) direct and (ii) indirect cooling. In direct cooling, the coolant, flowing through the cooling passages in the copper conductors, comes into direct contact with them, picks up the heat generated by them and disposes it in the external heat exchangers. In indirect cooling, the coolant cools the stator core or the rotor shaft, which is in contact with the exterior of the conductor insulation. In other words, the heat generated by the conductors, before being picked up by the coolant, travels through the insulation layer first.

Representative heavy-duty industrial GTG design (electric and physical) and performance data are summarized in Table 9.2. Note that, according to the IEEE Standard C50-13 (2014), a generator's rating is its apparent output (MVA or kVA), which is available continuously at the terminals at rated frequency, voltage, power factor and primary coolant temperature. For hydrogen-cooled generators, the rating also includes hydrogen pressure and purity. For air-cooled generators, the rating includes altitude (it shouldn't degrade up to 1,000 m above the sea level).

Electrical insulation of motors and generators is rated by standard ANSI/NEMA (National Electrical Manufacturers Association) classifications (per ANSI/NEMA Standard MG 1) according to maximum allowable operating temperature (see Table 9.3). Allowable temperature rises are based upon a reference ambient temperature of 40°C. Maximum allowable operation temperature is the sum total of reference temperature, allowable temperature rise and allowance for "hot spot" winding. Thus, for an F class machine, maximum allowable operating temperature is 40 + 105 + 10 = 155°C. For a more detailed table and other requirements, the reader should consult Table 5 (air-cooled generators) and Table 6 (hydrogen-cooled generators) in IEEE C50.13 (2014).

TABLE 9.3
NEMA Temperature Classifications

Class	Maximum Allowable Operating Temperature	Allowable Temperature Rise	
	°C/°F	°C (SF 1.0)	°C (SF 1.15)
B	130/266	80	90
F	155/311	105	115
H	180/356	125	-

SF, Service Factor.

For example, for hydrogen-cooled generators, no values are prescribed by the IEEE standard; those values are to be reached at by agreement.

As shown in Table 9.2, the armature and field windings of the generators are designed with ANSI/NEMA class F insulation materials. (Stator winding and gas temperatures are monitored and recorded via transducers installed inside the machine.) However, as a safety margin, throughout the allowable operating range, temperature rises are limited to class B specifications.

In some countries, especially those with 50-Hz grids, the practice is to specify generator voltages in proportion to rated power output, i.e., for generators rated at

- Less than 200 MVA, 10.5 kVA
- Between 200 MVA and ~350 MVA, 15.75 kVA
- Between 350 MVA and ~900 MVA, 21 kVA
- More than 900 MVA, 27 kV.

ANSI standards used to prescribe 13.8 kVA for generators rated at less than 150 MVA. Up to about 300 MVA, 16.5 kVA and 18 kVA were popular with the US OEMs (i.e., 60-Hz machines). In IEEE C50.13 (2014), there is no voltage recommendation.

A generator's capability is its highest acceptable continuous output of apparent power (MVA or kVA) at prevailing site ambient and loading conditions. The OEM is required to supply curves of generator capabilities under site conditions over the specified range of generator primary coolant temperature. For a machine with OV air cooling, e.g., this is the same as that of the ambient air surrounding the generator. For a machine with TEWAC air cooling or hydrogen cooling, this is the water inlet temperature of the external cooler.

Gas turbines typically have both a baseload capability and a peaking capability (with a higher firing temperature for a limited duration). Thus, the OEM is required to specify the same capabilities for the generator as well. The generator should be designed to match the capabilities of the gas turbine over the ambient temperature range. This is especially important for performance at low ambient temperatures when the gas turbine output substantially increases due to much higher density of the surrounding air. For OV-type air-cooled generators, the coolant temperature rises in lockstep with the site ambient temperature, which is detrimental to the generator capability. Fortunately, gas turbine output drops with increasing site ambient temperature as well such that this does not create a big handicap.

In addition to rated performance and electromechanical design data, generator OEMs typically supply four types of curves:

1. Reactive capability curve
2. Output versus coolant temperature curve
3. "Vee" curve
4. Saturation curves.

Reactive capability curves show the steady-state operating capability of the generator in terms of real and reactive power as a function of the power factor (leading or lagging). For hydrogen-cooled large machines, there is a capability curve for different hydrogen pressures (e.g., 15, 30 and 45 psig). Load capability of the generator is higher for higher hydrogen pressure.

The "Vee" curve (or the V curve) shows the relation between armature current (y-axis) and field current (x-axis) at constant terminal voltage and constant active power output for different values of the power factor. Consequently, the y-axis also gives the apparent power, S, in "per unit" (pu or p.u.) terms.

Saturation curves show open- and short-circuit characteristics of the generator on an open-circuit voltage (left y-axis) and short-circuit current (right y-axis) versus field current diagram. Since the *short circuit ratio* (SCR) is the inverse of the saturated synchronous impedance, these curves are also known as *impedance curves*. Low value of SCR is not desirable because of low stability but large SCR comes at bigger size and higher cost. Typical value of SCR for combined cycle gas and steam turbine ac generators is around 0.5.

For an in-depth discussion of generator performance and stability along with the underlying electromechanical theory and design principles, the reader should consult Refs. [4–7].

9.5 SCOPE OF SUPPLY

There are *three* types of major equipment supply scope:

1. Equipment Only
2. Power Island/Block
3. Extended Power Island/Block.

In the past (e.g., 1990s and early 2000s), gas turbine OEMs (General Electric (GE), Siemens-Westinghouse,[7] MHI[8] and Alstom[9]) supplied exclusively gas turbines,steam turbines and ac generators.[10] Recently, however, major OEMs acquired HRSG manufacturers and are thus able to offer "complete plant solutions". For example,

- In August 2016, GE acquired the HRSG business of Doosan Engineering and Construction (South Korea) for $250 million. Note that Doosan has been a supplier of HRSGs to both GE and Alstom and has a long history as an Alstom licensee. The acquisition integrated them into GE Power's Gas Power Systems business, which included Alstom's legacy HRSG business as the result of the (now controversial) 2015 acquisition.
- In 2011, Siemens acquired Dutch sister companies NEM B.V. and Nem Energy Services B.V. (NES), which are specialists in HRSGs. NEM designs, engineers and sells HRSGs, while NES is in the associated service business.[11] In 2007, Siemens had also taken over the HRSG business of Balcke Dürr Austria GmbH.

Independent HRSG manufacturers for large combined cycle power plants at the time of writing are Amec Foster Wheeler, Nooter/Eriksen, Vogt Power International and CMI Energy.

Presently (2018–2019), the most common case in large combined cycle power plant projects is the teaming up of the developer/owner exclusively with an OEM to deliver the Power Island (i.e., gas turbine and steam turbine generators and the HRSGs). An EPC contractor is selected through

---

7 Now Siemens.

8 Mitsubishi Heavy Industries – at the time of writing, Mitsubishi Hitachi Power Systems (MHPS).

9 Descendant of BBC and ABB; now owned by General Electric with 50-Hz gas turbine product line divested to Ansaldo (Italy).

10 Interested reader is referred to Chapter 4 of **GTFEPG** (Ref. [11] in Chapter 2) for a comprehensive history of OEM product line development.

11 In 2018, Siemens has rebranded and renamed its full subsidiary NEM Energy into Siemens Heat Transfer Technology (HTT).

a competitive bidding process to engineer, procure and construct the full power plant around the Power Island with *lump sum turnkey* (LTSK) contract (see Chapter 10). In some cases, the OEM and the EPC can and do form a consortium to bid for specific projects.

In the past, there were instances when the EPC contractor (or a utility with its in-house engineering department) would design the combined cycle power plant by mixing and matching major equipment from different OEMs. If done by an experienced EPC contractor or a major utility with a capable and experienced in-house engineering department, this would indeed result in an optimal cost-performance trade-off.

In "equipment only" option, the scope of supply includes gas and steam turbine packages as described above with requisite auxiliary systems, gas turbine air intake and exhaust system, the fuel gas skid, synchronous ac generators with their auxiliary systems, the fire protection system, electrical components, instrumentation and control systems. (Fuel gas performance heater can also be in the supply scope of the OEM.) In some cases, depending on the particular gas turbine model, specialized heat exchangers such as rotor cooling air coolers can also be within the supply scope of the OEM (e.g., sequential combustion GT24/26 by GE/Alstom and Ansaldo).

In "Power Island/Block" option, the scope of supply includes, in addition to the equipment supply scope above, the HRSG, steam turbine condenser including the air removal system (in "extended" power block option), condensate pumps and boiler feedwater pumps (in "extended" power block option), critical valves including steam bypass (especially for "fast start" or "rapid response" options) and plant DCS.

As mentioned earlier, by far the most common GTCC project structure is the Power Island procured by the owner/developer. In this option, the owner/developer is responsible for specifying and procuring the following equipment:

- GTG(s) and accessories (air inlet filters, evaporative coolers and other auxiliary equipment such as the lube oil system)
- HRSG(s) (including SCR, CO catalyst, stack and stack dampers)
- STG and accessories
- Steam bypass valves (for fast start/rapid response systems)
- Terminal steam (HP and HRH) attemperators (for fast start/rapid response systems)
- DCS and CEMS[12].

The owner/developer is also responsible for securing all permits to develop, construct, startup and operate the facility (e.g., emissions, water and wastewater permits). The EPC contractor obtains all necessary local construction permits.

The EPC contractor is responsible for performing the overall plant design, detailed engineering, procurement of non-owner-supplied equipment, project management, construction, startup and commissioning. The EPC contractor is also responsible for all construction activities inside the facility boundaries, e.g., site preparation work, foundations, installation of facility equipment, bulk material and commodities, instrumentation and controls, facility buildings and site finishing work.

Typical list of EPC contractor-procured and installed equipment include (tank sizes for an 800-MW $2 \times 2 \times 1$ fired GTCC)

- Steam turbine condenser
- Condensate system, including $3 \times 50\%$ condensate pumps
- Natural gas system with filters and heating (the latter is usually provided by the gas turbine OEM)
- Number 2 distillate system with pumps, storage tank and either steam or water injection systems

[12] Continuous Emissions Monitoring System

- Auxiliary boiler firing natural gas for steam seals
- Domestic (potable) water system
- Instrument air system
- Service air system
- Closed-loop cooling water fin-fan system for cooling auxiliary equipment, including 2 × 100% closed-loop cooling water pumps
- Service water system for plant services, including 2 × 100% service water pumps
- 500,000 gallon city water storage tank
- Demineralized water system, consisting of 2 × 100% trailers and 2 × 875,000 gallon storage tank
- 500,000 gallon service/fire water storage tank (with 300,000 gallons dedicated for fire)
- Boiler feedwater chemical injection system
- Fire protection system, including one diesel pump, one motor-driven pump and one jockey pump
- Sanitary and wastewater collection system to an off-site discharge
- Administration/control/warehouse/maintenance building, fire water pump enclosure, switchgear and other electrical enclosures
- Heating, ventilating and air-conditioning (HVAC) systems for control room, electrical buildings and offices
- Ventilation for warehouse/maintenance building and other enclosed areas
- 19% aqueous ammonia system for use as SCR reagent with one 20,000 gallon tank
- DCS installation, uninterruptible power supply (UPS) system and CEMS installation
- Miscellaneous hoists
- One 125 ton turbine gantry crane over the GTGs and the STG.
- Heat tracing for process requirements and freeze protection (as required)
- Lighting, grounding and cathodic protection (if required) systems
- Electric power distribution system
- Double-ended medium voltage switchgear
- Generator breakers for each GTG and the STG
- Emergency diesel generator for safe shutdown (1.5 MW)
- In-plant radio communication system
- Facility loop and interior roads, personnel parking and onsite construction access
- Stormwater drainage system with detention ponds
- Fencing for the main power facility, including one motorized gate entrance, one camera and one card reader.

REFERENCES

1. Quinkertz, R., 2008, USC steam turbine technology for maximum efficiency and operational flexibility, *POWER-GEN Asia 2008*, Octobr 21–23, Kuala Lumpur, Malaysia.
2. Fontaine, P., Bonsang, R., 2012, Comparison study between ASME code and EuroNorm 12952 for HRSG design, *POWERGEN EUROPE 2012*, Köln, Germany.
3. Hampson, S., Robertson, D.G., Simandjuntak, S., 2011, Comparing ASME, European boiler codes and their impacts on design, operation, Combined Cycle Journal, 1Q2011 Issue.
4. Fitzgerald, A.E., Kingsley, Jr., C., Kusko, A., 1971, *Electric Machinery*, 3rd Edition, Mc-Graw Hill Book Co., New York.
5. Stevenson, Jr., W.D., 1975, *Elements of Power System Analysis*, 3rd Edition, Mc-Graw Hill Book Co., New York.
6. Kundur, P., 1994, *Power System Stability and Control* (The EPRI Power System Engineering Series), McGraw-Hill Inc., New York.
7. Grigsby, L.L., Editor-in-Chief, 2001, The Electric Power Engineering Handbook, CRC Press, Boca Raton, FL.

10 Balance of Plant

Major gas turbine combined cycle (GTCC) power plant equipment were covered in detail in Chapter 9. In this chapter, the remaining pieces of equipment and systems are covered under the heading of *balance of plant* (BOP), which includes:

1. Electrical equipment
2. Pipes and valves
3. Condensate and feedwater pumps
4. Tanks (water, distillate oil for backup, etc.)
5. Auxiliary boiler
6. Fuel gas booster compressor (if necessary)
7. Fuel gas performance heater
8. Makeup and wastewater treatment systems
9. Miscellaneous pumps, compressors and heat exchangers.

From a conceptual design and performance calculation perspective, most BOP equipment does not play a significant role (except major pumps and the fuel gas-related equipment). They are, however, important from procurement and construction (and cost estimation) perspectives and thus require some consideration. They also require close attention from a total auxiliary power consumption perspective in commercial contracts with precisely spelled-out performance guarantees and liquidated damages.

Before moving on, it should be stated that there are certain systems that need to be provided by the plant owner during construction and operation. Some of these systems and associated equipment are temporary (and portable) and some are permanent (e.g., the switchyard). The engineering, procurement and construction (EPC) contractor is required to provide suitable space for, access to, and necessary interface connections for the following portable equipment and systems furnished by the owner so that their operation is safely integrated with the rest of the facilities:

- Switchyard (substation)
- Gas metering station
- Rental demineralized water trailers
- Chemical feed system
- Water treatment laboratory
- Warehouse equipment, machine shop equipment, and office furnishings/equipment
- Landscaping.

10.1 ELECTRICAL EQUIPMENT

Electrical equipment design is driven by gas and steam turbine generators (STGs) and the bulk electrical system (i.e., the "grid") to which they are connected. The electrical system of the combined cycle power plant can be concisely represented by a *single-line* or *one-line diagram*. A partial one-line diagram for a 2 × 2 × 1 multi-shaft GTCC is shown in Figure 10.1. The output voltage of the three-phase generators (typically 18 kV for large industrial gas turbines and 23.5 kV for large steam turbines) is increased by the *generator step-up transformer* (GSU) to the transmission system voltage (typically 230 or 345 kV). High or medium voltage (MV) (13.8 or 4 kV, respectively) auxiliary power bus is fed by the *unit auxiliary transformer* (UAT). Low-voltage (LV) (480 V) auxiliary

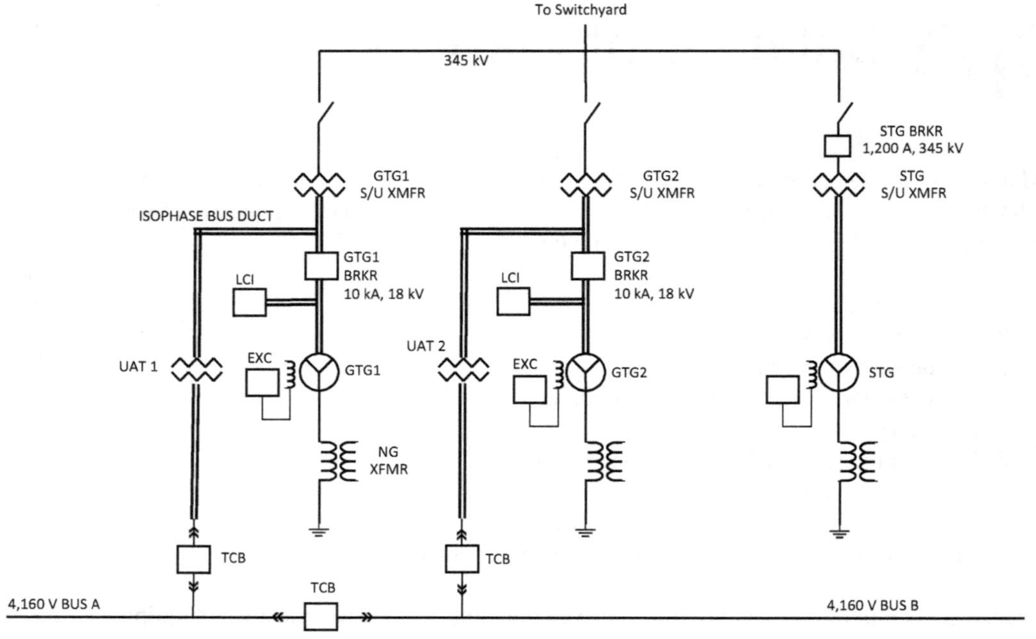

FIGURE 10.1 Electrical one-line diagram (partial) for 2 × 2 × 1 multi-shaft GTCC (TCB, tie circuit breaker; S/U XMFR, step-up transformer; NG XMFR, neutral ground transformer; BRKR, breaker; EXC, exciter).

power bus is fed from the 4 kV bus through the *station service transformer* (SST). Obviously, UAT and SST are "step-down" transformers.

In addition to GSU, UAT and SST, electrical equipment includes a wide variety of permanent plant equipment, e.g.,

- Generator circuit breakers
- Isolated phase bus duct (IPB)
- Generator auxiliaries (excitation transformers and control assembly, neutral grounding cubicle, control, protection and metering panels)
- Packaged electrical and electronic control centers (e.g., MCC[1]/relay panels; DCS[2] cabinets; battery chargers and distribution boards; UPS[3] and distribution boards; lighting panels; heating, ventilating and air-conditioning, HVAC, panels, fire detection panels)
- Instruments (flow, level, pressure, temperature)
- Battery cells, battery racks, intercell links, intercell wiring, electrolyte
- Direct current (dc) systems (includes UPS – see below)
- UPS including inverters, manual and static bypass switches, bypass regulating voltage transformers
- Load center and distribution transformers
- MV switchgear and MCCs
- LV switchgear and MCCs
- Protective relay panels
- CCTV System – cameras, camera control units, monitors, intercoms, VOIP[4] phones

---

[1] Motor Control Center.
[2] Distributed Control System.
[3] Uninterruptible Power Supply.
[4] Voice Over Internet Protocol.

- DCS I/O module and termination unit cabinets
- DCS control room consoles, printers, engineering workstations and servers
- Programmable logic controllers (PLCs)
- Continuous emissions monitoring system (CEMS)
- Steam and water analysis and sampling system.

The IPB – usually made from aluminum – connects the prime mover generators to the step-up transformer LV terminals. Low-side generator circuit breakers are provided for the gas turbine generator(s) (GTGs). Each generator circuit breaker is connected by the IPB to the corresponding generator and GSU.

The dc systems are designed to provide adequate 125 V dc power to safely shut down the plant or protect equipment with minimal risk to people or equipment. Typical dc plant loads include turbine generator emergency lube oil pump, UPS, fire protection, and switchgear control. A careful assessment of the plant dc loads is requisite to determine the appropriate quantities and locations for plant dc systems. (Electrical calculations are performed to verify dc system ratings and interrupting capacities.) Each dc system includes a battery, redundant battery chargers, battery ground detection and dc distribution panels.

The UPS powers all the instrumentation, control and monitoring circuits required for startup, operation, normal/emergency shutdown and offline housekeeping of the facility. In addition to static switch transfers, the UPS provides filtered and regulated power.

Typically, electrical motors are supplied with the driven equipment, e.g., condensate pump, boiler feed pump, circ water pump. Typical nameplate voltages for ac induction motors are

- 4,000 V, three phase
- 460 V, three phase
- 115 V, one phase.

In general, motors 0.5 hp and larger are three phase and motors under 0.5 hp are single phase. Typical voltage for dc motors is 115 V. Direct current motors are capable of starting, accelerating and running without damage within the range of 102–140 V for 115 V motors.

Motors are typically sized to at least a 10% capacity margin above that of the maximum horsepower requirement of the driven equipment. If a service factor >1.0 is provided, the motor's horsepower rating is determined without taking advantage of the service factor even for short duration overloads.

Diligent electrical system calculations are done by the EPC contractor to verify the auxiliary power system performance and motor starting capabilities. Compilation of the electrical load table and the electrical calculations determines the system requirements. (Note that some loads are "continuous", e.g., the boiler feed pump and others are "intermittent", i.e., service water pumps operating in a start/stop cycle.) This is extremely important due to its bearing on accurate auxiliary load assessment and plant net output and heat rate guarantees. The end result can be quite different from the – usually optimistic – estimates obtained from heat and mass balance (HMB) calculations done in the conceptual design phase. In this regard, the reader should also refer to Section 10.3.3; actual pump power consumption obtained once the system sized and procured can be significantly different from front-end conceptual calculations (at least for certain operating conditions).

In general, for switchgear-fed motors, the following generic rules apply:

- Those rated at <300 hp are fed by 480 V (LV) bus
- Those rated up to 5,000 hp are fed by 4 kV (MV) bus
- Those rated at more than 5,000 hp are fed by 6.9 or 13.8 kV bus.

Motors rated at more than 5,000 hp (~3,750 kW) are quite rare in gas turbine simple or combined cycle power plants (except, maybe, very large N × N × 1 GTCCs).

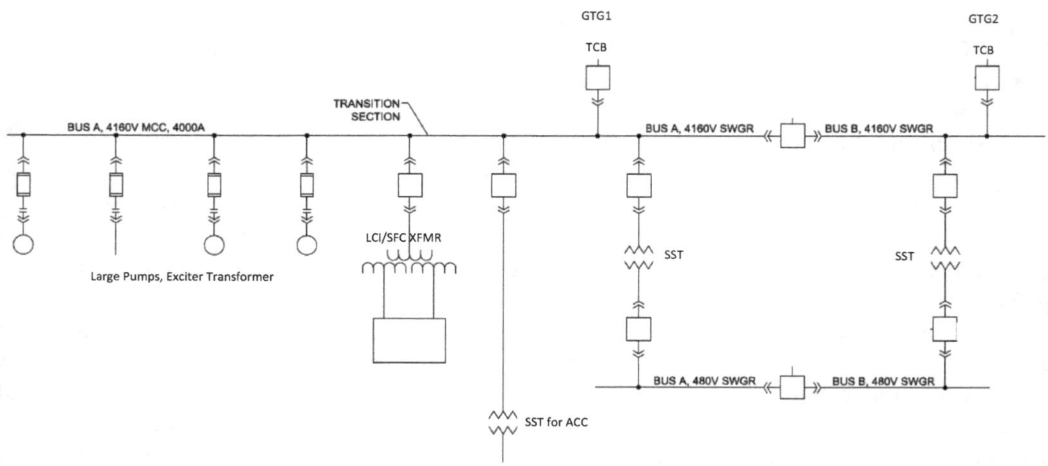

FIGURE 10.2 Electrical one-line diagram (partial) for 2 × 2 × 1 multi-shaft GTCC (SFC, static frequency converter; ACC, air-cooled condenser; SWGR, switchgear).

As shown in Figure 10.1, in a combined cycle power plant with multiple GTGs, auxiliary power is usually extracted from the GTG outputs. This allows a symmetrical arrangement of identical primary and backup source UATs – one from each of the two GTGs as shown in the figure – to be shared by the entire plant. This permits a self-sufficient configuration for simple cycle operation regardless of the status of the STG. In addition, this configuration supports a construction schedule that is sequenced for GTG commissioning first and earlier back-feed of utility power through the GTG UATs for all BOP commissioning. This can be seen in Figure 10.2, which shows 4.16 kV and 480 V bus (partial) for the GTCC of Figure 10.1.

10.1.1 Plant Controls

The prime movers in the GTCC have their own dedicated control systems (e.g., Mark VIe for GE gas turbines). They are integrated into the plant-wide *distributed control system* (DCS). The DCS is typically microprocessor based and consists of redundant process controllers, redundant communications network and operator interface through the operator consoles (*man-machine* or *human-machine interfaces*, MMI/HMIs, in information technology jargon). The DCS interfaces to a "data historian" via ethernet, which records all pertinent process variables and significant status variables.[5]

The following primary systems are also controlled and/or monitored through the plant DCS:

- Heat recovery steam generator (HRSG)
- High-pressure (HP) steam system
- Intermediate pressure (IP) and reheat steam system
- Low-pressure (LP) steam system
- Condenser air extraction system
- Circulating water system
- Auxiliary cooling water system
- Condensate system
- Feedwater system

[5] PI (Process Information), a real-time data historian application with a time-series database structured and developed by OSIsoft, is a prime example.

- Cycle makeup system
- Cycle chemical feed system
- Sampling and analysis system
- Compressed air and nitrogen system
- Potable water system
- Fuel gas system
- Plant electrical system
- Plant load control interface to *remote terminal unit* (RTU)[6]
- Cooling tower blowdown clarification and softening system (if necessary)
- Demineralized water system
- Duct burner systems (if present)
- Power utilities.

The following plant systems are considered *stand-alone* with minimal (if any) operator monitoring from the DCS operator workstations in the control room:

- Fire protection system
- Fire detection system
- Plant HVAC system
- Building/site security system
- Plant communications systems.

The "distributed" nature of the DCS stems from the fact that the prime mover control and I/O cabinets are functionally distributed throughout the facility. The following are the primary locations and functions of the system process control cabinets and remote I/O (input/output) cabinets:

- Steam turbine electrical building – control cabinets for the STG, DCS processor cabinets and BOP area I/O, STG engineering workstation
- Control room – DCS operator workstations, fire detection panel and security system and HRSG evaporator drum level indicator panel
- Control building equipment room – PBX,[7] business LAN hubs/switches, DCS data highway hub/switches, UPS for control room equipment
- Gas turbine(s) PEECC[8] – control cabinets for the gas turbine(s)
- HRSG(s) – DCS I/O in HRSG area.

All BOP systems are controlled and monitored through the plant DCS except the following, which use PLCs and other control systems:

- CEMS
- Air compressor
- Gas turbine inlet evap cooler system (if present)
- Auxiliary boiler
- Cooling tower blowdown clarification and softening system (if present)
- HRSG duct burner systems (if present)
- Demineralized water system.

[6] A remote terminal unit (RTU) is a microprocessor-controlled electronic device that interfaces plant equipment to the plant DCS by transmitting telemetry data to a master system and by using messages from the master system to control connected objects.

[7] Private Branch Exchange.

[8] Packaged Electronic/Electrical Control Compartment.

Nevertheless, remote control and monitoring is provided for these systems from the DCS consoles via hardwired I/O for critical I/O and redundant communications link for alarms and noncritical I/O.

10.1.2 Plant Instrumentation

The starting point of plant instrumentation design is the P&ID (Piping and Instrumentation Diagram). There are PI&Ds for each plant system and subsystem described above showing all required instruments, controls and automation specifics requisite for plant monitoring, control and safety (i.e., alarms and trips). The P&IDs show all instrumentation and DCS tag numbers while clearly identifying all control loops, including all alarm levels. Instrument and I/O tagging is based on the *Instrument Society of America* standard ISA S5.1 *Instrumentation Symbols and Identification*. The reader should consult that standard for a complete list of symbols and instrument designation rules.

P&IDs show required process measurements (e.g., flow, pressure and temperature) and their function (e.g., indication, recording or control). They also show whether the information provided by a particular measuring element is available locally (e.g., a pressure gauge) or in the control room (via the DCS). An example P&ID (for the HP steam system in a GTCC) is shown (in part; the whole diagram would be illegible on the book page) in Figure 10.3.

10.2 PIPES AND VALVES

A typical GTCC is a veritable maze of pipes and valves. The "clean" look of a typical HMB diagram, which shows only the major "arteries" through which the working fluid of the bottoming Rankine steam cycle (i.e., condensate/feedwater and steam) flows during normal operation, is very misleading. There are many auxiliary/BOP systems with myriad fluids in addition to secondary steam and/or feedwater pipes (e.g., bypass or recirculation). Typical above-ground, large bore pipes and pipe materials are as follows:

- Raw water (A106 or A53 Grade B)
- Service water (A106 or A53 Grade B)
- Demineralized water (A312 TP316 and A53 Grade B)
- Potable water (A312 TP304)
- Wastewater (A106 Grade B Lined)
- Feedwater (A106 Grades B and C)
- IP steam (A106 Grade B)
- Main (HP) steam (A335 P22/P91)
- Hot reheat (HRH) (A335 P91)
- Cold reheat (CRH) (A335 P22)
- LP steam (A335 P11)
- Auxiliary steam (A106 Grade B)
- HRSG blowdown (A106 Grade B)
- Condensate (A106 Grade B)
- Condenser vacuum (A106 Grade B)
- Steam turbine drains (A106 Grade B)
- Steam turbine lubrication (A312 TP316)
- Steam seals
- Circulating water
- Closed cooling water (CCW)
- Boiler chemical controls (A312 TP304)
- Compressed air (A312 TP316)
- Fire protection (A53 Grade B/E).

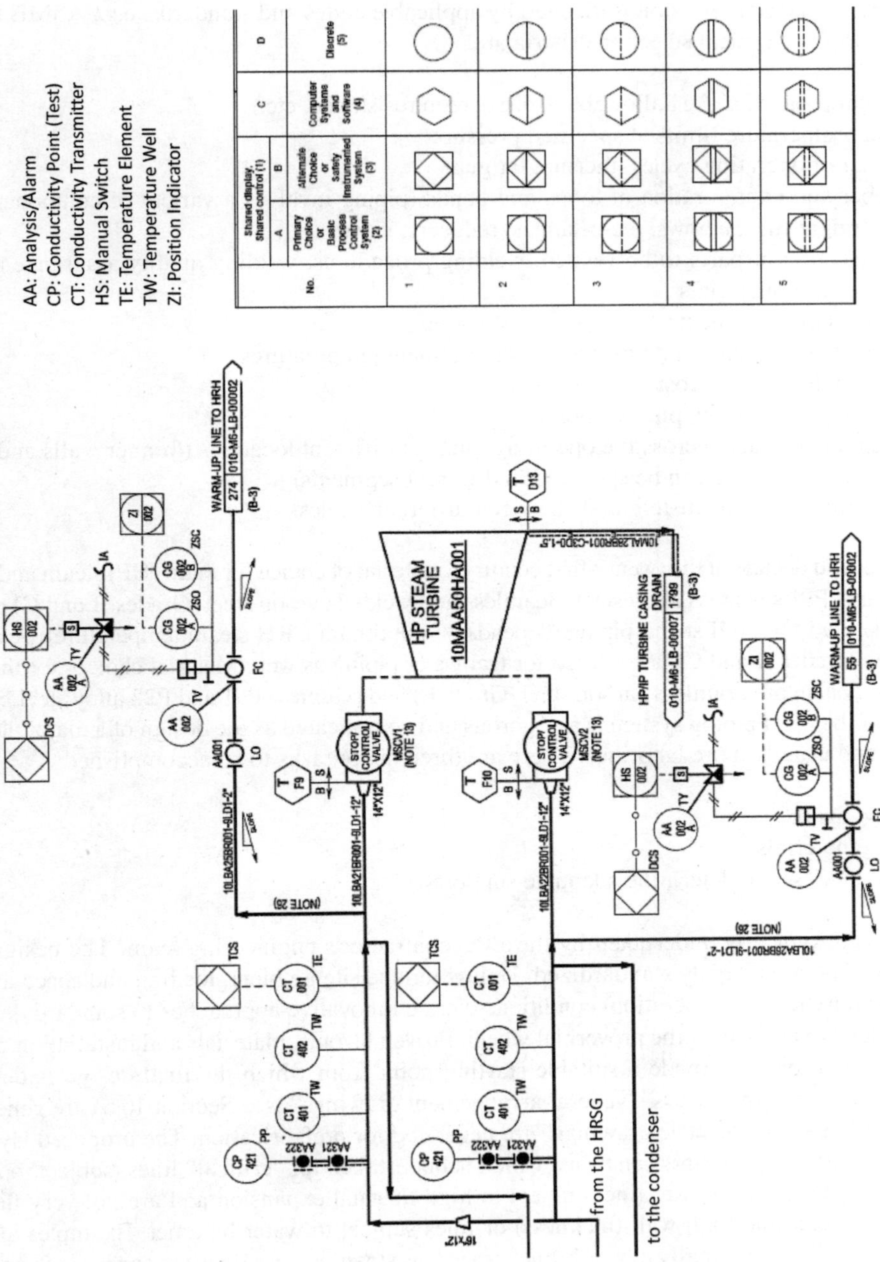

FIGURE 10.3 P&ID example (HP steam system – partial).

ASME SA335 (equivalent designation ASTM A335) designates the family of "seamless ferritic alloy–steel pipe for high-temperature service". In seamless pipe[9] applications, materials in this family (referred to as "P-grade" for pipe applications and "T-grade" for tubes) come in more than 17 grades, e.g., P11, P22 and P91/P92. Such alloy steel pipes are produced in nominal or minimum wall thickness for high temperature and pressure ("high energy") service.

Proper pipe material selection is dictated by applicable codes and standards, e.g., ASME B31.1–2016 Power Piping. Typical selection criteria are:

- Physical properties (i.e., allowable stress, creep resistance, etc.)
- Upper temperature limits at specified pressure
- Number of operating cycles (thermal fatigue)
- Number and configuration of loops and bends (piping involves a variety of components, e.g., straight runs, elbows, tees, flanges, reducers, valves)
- The selection of parts to be welded, welding procedures, welding quality assurance and welding filler materials
- Hardness measurements
- Preheat temperatures and post-weld heat treatment temperatures
- Fabrication time and cost
- Total weight including pipe supports
- Detailed HMB data across the operating range at different locations (thinner walls and/or lower grades of pipe can be specified in different segments)
- Availability and cost differentials between different grades.

Presently (second decade of the twenty-first century), material of choice for main (HP) steam and HRH steam piping is P91 seamless alloy steel. Seamless and welded carbon steel (Grades B and C) or P22 alloy steel is used for CRH steam piping (depends on maximum CRH steam temperature). Seamless carbon steel (Grades B and C) is specified for feedwater piping as well. Material choices for the auxiliary steam system are seamless carbon steel (Grade B) and seamless P91 and P22 alloy steel pipe.

Design of the plant piping system is as rigorous and complicated as the design of a major piece of equipment and usually more laborious. There are three major tasks to be accomplished:

- Piping layout
- Piping materials
- Stress analysis (to determine adequate supports).

These tasks are typically undertaken by the EPC contractor's engineering team. The design of a typical GTCC plant is largely standardized. Unless unique site ambient, loading and space as well as operating duty (e.g., cogeneration) conditions dictate innovative approaches to standard designs, there is no need to rediscover the proverbial wheel. Proven layouts, materials and installation guides used in similar projects provide a suitable starting point from which the final design is derived. At the end of the design process, "general arrangement drawings" (see Section 10.5) are generated for field erection and "isometric drawings" are generated for prefabrication. The proposed layout is verified by detailed piping stress analysis, which is undertaken for "critical" lines (subject to applicable design criteria). These are lines subject to high thermal expansion and are not very flexible (i.e., large diameter and high wall thickness) or lines subject to water hammer. Examples are HP and HRH steam lines. Normally closed branches such as steam bypass lines are sources of acoustic-induced vibrations (they are analogous to Helmholtz resonators) and can lead to serious vibrations

[9] Seamless pipe is made from a solid round steel "billet" which is heated and pushed or pulled over a form until the steel is shaped into a hollow tube. In contrast, welded pipe is manufactured by rolling metal and then welding it longitudinally across its length. Consequently, seamless pipe does not have any joint in its cross section throughout its length.

when the plant is in operation. Stress analysis step is very critical because it determines the loads imposed by the pipe on the supporting structure at the location of its support. Those loads are used by the civil engineers to design the supporting structure.

Field erection of plant piping is by no means to be taken lightly. Advances in gas turbine technology with ever higher exhaust flows and temperatures opened the door to aggressive steam bottoming cycle design parameters. Until recently, typical parameters were 1,800 psig (125 barg) throttle pressure with 1,050°F (565°C) throttle and HRH steam temperatures. With the advanced class H/J class gas turbines, state-of-the-art steam cycle design parameters are 2,600–2,700 psig (180+ barg) and 1,112°F (600°C). (There are even talks of supercritical HP steam systems – see Section 6.4.)

P91 is a modified 9% chromium–1% molybdenum alloy steel (developed in the USA more than 30 years ago), which is preferred over P22 because of its lighter weight (due to significant wall thickness reduction). Per ASME B31.1, pipe wall thickness calculation is

$$t = \frac{P \cdot D}{(2S \cdot E + P \cdot Y)},$$

where P is the internal (i.e., steam) pressure, D is the pipe outside diameter, S is the allowable stress value for the pipe material at design temperature, E is the quality factor (1 for seamless pipe) and Y is a factor between 0.4 and 0.7 depending on the material and temperature (0.7 for ferritic steels above 1,000°F). Thus, for the same operating conditions, higher S and E for P91 leads to lower t and less material (in lb or kg for the same length). As a general rule, for typical main steam piping, an upgrade from the P22 alloy to P91 can reduce wall thickness by nearly two-thirds and component weight by 60%. In addition, it can also

- Raise allowable strength in the 950°F–1,100°F range by up to 150%
- Raise the oxidation limit by 100°F, enabling a lower corrosion allowance
- Increase thermal fatigue life by a factor of 10–12.

All these advantages of P91counterbalance its higher cost (pipe, fittings and valves). However, it has several disadvantages, which makes it a difficult choice from a field constructability and schedule (subject to liquidated damages) perspective, i.e.,

- Highly prone to welding defects (requires highly skilled welders)
- Longer lead times for processing during fabrication and erection and heat treatment both for bends and welds
- Weld filler material more expensive and less available.

10.2.1 Steam System

Bottoming cycle steam system transports HP, LP and reheat steam back and forth between the HRSGs and the steam turbine. This system includes the most important and expensive piping in the entire power plant. The steam system is designed in accordance with ASME standards (see Section 10.2.1) and design practices for the prevention of turbine water induction with deviations as approved by the turbine original equipment manufacturer (OEM).

The steam lines are sized for the appropriate fluid velocities as dictated by the ASME piping standards and the OEM's design practices (see Table 10.1). Typically, pipe sizing is based on 100% flow or *valves wide open* (VWO) operation of the steam turbine. It should be noted that, in almost all cases, HMBs provided by the steam turbine OEM are based on pressure loss "allowances" specified as a fraction of inlet pressure (e.g., 2.5% pressure drop). After rigorous pipe sizing calculations, resulting "actual" pressure drops can be somewhat lower or higher. Note that an HMB does not show/include non-return valves (NRVs), expansion loops, elbows, etc. It is typically the EPC

TABLE 10.1
Pipe Flow Velocities in ft/s – 1 ft = 0.3048 m; for Approximate Conversion to m/s, Divide the Tabulated Velocities by Three

	Small Bore (<2″)		2″–4″		4″–6″		8″ and Larger	
System	**Suction**	**Discharge**	**Suction**	**Discharge**	**Suction**	**Discharge**	**Suction**	**Discharge**
Feedwater	10	15		15–20		20–25		25
Fuel oil	1	2	2	4	2–4	4–6	5	7
Condensate		8		8–12		12–14		15
Service water	4	5.5	5	7	6	8	6.5	8.5
Cooling water	4	5.5	5	7	6	8	6.5	8.5
Compressed gas		28		32		35		35
Saturated steam (<25 psig)		75		88		95		100
Saturated steam (>25 psig)		120		140		150		160
Superheated steam		190		210		250		250

	Suction	**Discharge**
System	**<20″**	**>30″**
Circulating water	6	10

contractor's responsibility to diligently design the main steam piping and account for pressure losses in a cost-effective manner. Not doing so can end up in lost power during the performance test run and liquidated damages. In general,

- Any given pipe "segment" should not have a pressure drop larger than 10% of the inlet pressure to that segment. If so, the segment should be broken up into smaller segments.
- In some cases, required flow rate of steam may need two lines instead of one in order to meet velocity and pressure drop requirements.
- HP lines use seamless pipe, and there is a limit to how large a seamless pipe can be manufactured. For thick-walled main steam pipe (i.e., 2, 3 or 4 in. thick wall), a pipe size of 24–28 in. (diameter) is about the limit. For thin-walled HRH pipe, one can probably go up to 36 in.
- Typical ratio for CRH pressure drop to HRH pressure drop is 2:1 (i.e., two-thirds of the total pressure drop allowed should be on the cold side, one-third on the hot side).

The main steam lines incorporate appropriate drains to ensure proper startup and operation. Materials of construction for each portion of the system are chosen in accordance with the design pressure and temperature conditions imposed on the system. (As discussed above, usually P91 is prescribed for HP and HRH steam pipes.) Main steam system piping drip-pots and low-point drain lines protect the steam turbine against water damage by removing moisture produced by condensation in the main steam piping. Condensate collected in the drip-pots and low-point drains are discharged to the condenser via the drain system. Drip-pot level controls are provided to modulate the air-operated, fail-open drain valves for automatic removal of condensate from the drip-pots.

It is desirable to use carbon steel in CRH piping (downstream of the NRV). This requires that the HP turbine exhaust temperature is below 800°F. Even if this is the case during normal operation, excursions can happen in certain operating regimes, especially during startup. If the expected duration of temperatures above 800°F is short (e.g., when the turbine is at full-speed, no-load before synchronizing), carbon steel can still be adequate for the CRH pipe.

There are three families of advanced class gas turbines with closed-loop steam cooling of the hot gas path (HGP) components:

1. The H-System (a GE trademark), which is not commercially offered anymore
2. G and J class gas turbines by Mitsubishi Hitachi Power Systems (MHPS).

The H-System had steam-cooled turbine stages 1 and 2 (both stationary and rotating components), which utilized CRH steam from the bottoming cycle. In G and J class gas turbines, IP steam is used to cool the combustor transition piece and stage 1 turbine blade ring. Details of these systems can be found in **GTFEPG** (Ref. [11] in Chapter 2). Intricately laid-out and expensive alloy piping used in the H-System to transfer steam from the bottoming cycle to the gas turbine and back was a major contributor to the commercial flop of the technology even though it was successful in all other aspects.[10] (Cooling steam piping is typically within the scope of the EPC contractor.)

The steam system also includes

- Full (100%) steam bypass systems to the condenser including bypass valves, pressure and temperature control stations
- Steam attemperators (desuperheaters)
- Drain connections and accessories for condensate collection and removal
- Startup vents
- Relief valves and silencers
- Sky vents with silencers (as a supplement to the steam bypass system).

Steam flows are measured by means of flow nozzles in the HP, IP and LP steam piping prior to the common header to the STG. The flow nozzles are located upstream of any branch connections on the main steam headers. Note that their accuracy (typically ±2% or higher) is usually not adequate for performance tests associated with commercial guarantees or equipment characterization. Feedwater flow rate measurement and mass balance calculations to determine the steam flow rates is the more accurate option (e.g., with precision-calibrated ASME throat tap nozzles).

Each HRSG in the plant is provided with a steam turbine bypass/HRSG startup system. The HP steam is bypassed from the HRSG HP header to the HRSG CRH header and from the HRSG HRH header to the condenser. This is known as a "cascaded" bypass system (see Section 15.4 for detailed information).

All steam piping is specified for the maximum pressure and temperature design requirements, including transient and upset conditions. Transient and upset conditions are handled as per ASME Standard B31.1.

10.2.2 Valves

In addition to the major steam valves, which are provided by the steam turbine OEM as part of the steam turbine package (see Section 8.2), there are many shutoff, isolation and control valves, whose total number can easily exceed 1,000 in a typical GTCC power plant. They are distributed over the following plant subsystems:

- Fuel gas preheating
- Demineralized water
- Process drainage

---

[10] Complex piping layout also added significantly to the major outage time (up to a month).

- Feedwater piping
- Main steam piping
- HRH piping
- CRH piping
- Auxiliary steam piping
- Main condensate piping
- Steam generator drains
- Sealing and cooling drains
- Condensing system
- Air removal system
- Drain and vent system
- Turbine bypass station, including desuperheating sprays
- Lubricant supply system
- Sealing, heating and cooling steam system
- Circulating water piping and culverts
- Service water piping and culverts
- CCW system

Valves are made from materials suitable for operation at the design pressure and temperature of the piping to which they are connected. Valves in throttling service are selected with design characteristics and materials that resist valve seat erosion when the valves are operated partly closed.

Valves that need to be repositioned during project operation are arranged for convenient operation; if required, they have extension spindles, chain operators or gearing. All valves and in-line equipment are located with emphasis on accessibility for operation and maintenance. A typical accessibility chart is shown in Table 10.2.

Access classifications in Table 10.2 are defined as follows:

- *I: Grade or platform.* Manually operated valves that are normally operated during facility operation and manual block valves that are required for normal plant startup and shut down should be located such that operation can be performed either from grade, platform or by chain wheel or extended stem operator.
- *II: Grade, platform, stairway, permanent ladder, portable ladder or manlift.* Use of ladders or manlift is limited to valves 15 ft or less above grade or platform.
- *III: Portable ladder or manlift.*

TABLE 10.2
Valve Accessibility Requirements

	Access Required		
Valve type	**I**	**II**	**III**
Manually operated valves	X		
Safety/relief valves		X	
Check valves		X	
Hydro vent valves – steam lines			X
High point vent valves			X
Low point drain valves			X
Instrument root valves		X	
Automatic air release valves			X
Fire protection interior header isolation valves			X

Isolation valves are installed per requirements of system redundancy and/or equipment and/or personnel safety (as shown in engineering flow schematics and P&IDs). System isolation capability is critical for normal operation as well as startup and commissioning (e.g., providing, say, steam to one unit, before the other units are complete). Personnel working in units still under construction must be protected from sources of high energy, hazardous chemicals, high temperature steam, flammable mixtures, etc. Consequently, there may be a need for *double isolation* valves, which may not be required for the normal system operation.[11] They should be provided where required to allow for phased turnover or operation.

Steam line low points where condensate can collect are provided with startup and blowdown drains (unless otherwise specified).

Steam lines that are fit with restricting devices such as orifices in the process runs are provided with adequate drainage upstream of the valve to ensure the water does not collect in the line.

Manual valves 8 in. or larger typically have gear operators. Valves are arranged to close when the handwheel is rotated clockwise when looking at the handwheel from the operating position. The direction of rotation to close the valve must be clearly marked on the face of each handwheel. Valves operating at less than atmospheric pressure must have a means to prevent air in-leakage, such as inverted Teflon packing or lantern glands.

LP water valves are either of gated or lugged-body butterfly type and of cast iron construction. Cast iron valves have cast iron bodies, covers, gates (discs) and bridges; the spindles, seats and faces are bronze. (Other materials can also be used provided that the selected material is suitable for the intended service.)

During procurement, butterfly valves with a known C_v should be specified (so that the flow through them can be accurately calculated). This reduces the time spent on system balancing if the valve could be opened or closed to the proper percent on the first try. (The valve vendor should be able to provide C_v curves – they are linear, actually – for their valves.) High-performance butterfly valves must be specified for pump suction and discharge.

Instrument air valves are of the ball, gate, butterfly or globe type. All gate and globe valves 2 in. and larger are of the outside screw and yoke (OS&Y) design[12]. The safety valve on the instrument air receiver must be either piped outdoors or provided with a silencer.

NRVs for steam service are specified and selected in accordance with ANSI standards and are properly drained. Vertical NRVs are equipped with drain lines. Valve bodies must have removable access covers so that internal parts can be examined or renewed without removing the larger valves from the pipeline.

The use of motor actuators for certain valves (i.e., *motor-operated valves*, MOVs) is determined by the EPC contractor or the valve vendor based on each valve's functional requirements, accessibility, size, operating environment and frequency of operation. Motor actuators include torque or limit switches to stop the motor automatically when the valve gate has reached the "full open" or "full closed" position. The motor actuator is positioned relative to the valve to prevent liquid, steam or corrosive gas from leaking from valve joints onto the motor or control equipment. When both are provided, hand and motor actuation mechanisms are interlocked so that the hand mechanism is disconnected before the motor is started. Motor actuator's seating control typically consists of a *slipping clutch* or other torque-limiting device that limits the seating force to an acceptable level.

Pressure vessels, heaters and boilers are required to have safety valves and relief valves as required by the applicable code (e.g., the *ASME Boiler and Pressure Vessel Code*). Safety and relief valves are installed vertically. Piping systems that can be over-pressurized by a higher pressure source are also protected by pressure relief valves (usually equipped with silencers). Equipment or

---

11 Other temporary protection devices and methods are block and bleed arrangements, spectacle blinds (to isolate a pipe section), etc.

12 With an OS&Y valve it is always clear whether the valve is open or closed. That is why they are required by applicable codes for service water lines used for fire protection.

parts of equipment that can be over-pressurized by thermal expansion of the contained liquid are required to have thermal relief valves. No silencers are required for thermal relief valves. Vents must be routed to a safe location (venting to a platform where a plant employee might be walking is rather ill-advised). Safety valves are flanged, except HP and HRH safety valves, which may be welded.

10.3 PUMPS

In addition to the large feedwater, condensate and circ water pumps, there are many pumps of different sizes including

- Demineralized water pump
- Raw water pumps
- Auxiliary boiler feedwater pump (if an aux. boiler is present)
- Auxiliary cooling water pump (closed or open loop)
- Diesel fire pump
- Electric fire pump
- Jockey fire pump (pressure-maintenance pump)[13].

Gas and steam turbine lube oil pumps, generator coolant pumps, etc. are provided by the OEMs as part of their turbine generator packages (usually in compact skids). For the gas turbine, this can include fuel oil forwarding and water injection pumps (for distillate oil backup operation with dual fuel combustor – if needed) and the evaporative cooler water pump (if one is present).

From a performance estimation perspective, one can divide the GTCC power plant pumps (excluding those associated with the heat sink system – see Chapter 7) into two categories:

1. Pumps shown in plant HMB diagram
2. Pumps not shown in plant HMB diagram.

10.3.1 Pumps Shown in Plant Heat and Mass Balance

These are the most important and significant pumps that are shown in plant heat balance diagram, i.e., HRSG feedwater pump(s) and the condensate forwarding pump. Their job is to deliver the steam condensate from the condenser hotwell (at subatmospheric pressure) to the HRSG drums where it evaporates at different pressure levels. To a very good approximation, for modern 3PRH bottoming cycles, they can be estimated as 1.9% of the STG output (this also includes the relatively small (i.e., tens of kilowatts) LP economizer recirculation pump). The power consumption can be estimated simply as

$$\dot{W}_{fp,i} = \frac{1}{\eta_m \eta_i} \dot{V}_{fw,i} \left(1 + \delta_i\right) \frac{P_{HP}}{\pi_i} \tag{10.1}$$

$$\dot{V}_{fw,i} = \frac{\mu_i N_{GT} \dot{m}_{exh}}{62.4}, \tag{10.2}$$

where i = 1, 2 and 3 correspond to HP, IP and LP, respectively, N_{GT} is the number of gas turbines and μ_i is the fraction of the respective admission steam flow rate to the steam turbine exhaust flow rate (see Equation 6.42). Note that flow rates are in lb/s and ft³/s. In Equation 10.1, P_{HP} is the steam turbine throttle pressure (in psia), δ_i is the pressure rise factor between the steam turbine inlet pressure and respective feedwater pump discharge and π_i is the steam turbine pressure drop factor. For most practical purposes,

[13] The jockey pump maintains pressure in the fire protection piping system so the larger fire pump does not need to run all the time. The analogy is to a small jockey sitting on a large horse.

- A good value for δ_i is 0.50
- π_i can be assumed to be 1, 5 and 25 for i = 1, 2 and 3, respectively, (i.e., for HP, IP and LP drums).

For i = 1 (i.e., HP), η_i can be taken to be 75% (for the others 70% is adequate). For the electric motor efficiency η_m, 90% is a reasonable value. For those readers more familiar with standard pump nomenclature and calculations, Equation 10.1 is equivalent to

$$\text{BHP} = \frac{\text{H} \cdot \text{Q} \cdot \text{SG}}{3{,}960\eta}, \tag{10.3}$$

where BHP is the brake horsepower, H is the pump head in ft, Q is the pump flow in gpm and SG is the specific gravity (1.0 for water).

For a state-of-the-art GTCC power plant with 250 lb/s main (HP) steam flow rate and 2,515 psia throttle pressure,

$$\dot{V}_{fw,i} = \frac{250}{62.4} = 4\,\text{ft}^3/\text{s},$$

$$\dot{W}_{fp,i} = \frac{1}{0.90 \cdot 0.75} 4(1+0.5)\frac{2{,}515}{1} 0.19524 = 4{,}365\,\text{kW}.$$

(Note the unit conversion factor 0.19524.) Since 1,000 gpm is equal to 140 lb/s, Q is 1,785 gpm. Thus, from Equation 10.3, with BHP = 4,365/0.7457 = 5,854 hp, pump head H is 8,762 ft or 3,794 psi.

Calculations using Equations 10.1 and 10.2 with actual heat balance data indicate that the total bottoming cycle pump consumption is simply a function of the assumed steam turbine throttle pressure level, i.e., 1.60% and 2.0% of the STG output for P_{HP} = 1,800 psia (125 bar) and 2,200 psia (150 bar), respectively. As a simple approximation, between P_{HP} = 1,400–2,400 psia (95–165 bar), total feed pump power consumption can be estimated as

$$\dot{W}_{fp,tot} = 0.009 \frac{P_{HP}}{1000} \dot{W}_{STG}. \tag{10.4}$$

10.3.2 Pumps Not Shown in Plant Heat and Mass Balance

These are many small (mostly utility) pumps with individual power consumption of the order of tens of kilowatts. As such, they do not merit individual treatment and can be treated via a lump-sum factor or kilowatt number. A recommended allowance is 175 kW per GT-HRSG train (see Table 3.1).

There is, however, one exception; the auxiliary or *closed-loop cooling water* (CCW) pump that services the shell-and-tube heat exchangers that cool turbine bearing lube oil and electric generator coolant. Its power consumption is typically several hundred kilowatts mainly depending on the size of the power plant, which is driven by the size and number of the gas turbines. A good rule of thumb is to assume the closed-loop flow to be equal to 5% of the total condenser cooling water with a 45-m pump head, 80% pump efficiency, and 90% motor efficiency. This would set the auxiliary cooling water pump power to 0.10% of the STG output. Refer to Section 10.8 for more information on the CCW system.

10.3.3 Pump Selection and Design

In contrast to conventional fossil fuel-fired boiler-steam turbine power plants, combined cycle bottoming cycle feedwater pump flows vary significantly in lock-step with the gas turbine exhaust mass flow rate (see Equations 10.1 and 10.2 above). This variation is connected to the changes in ambient temperature (via compressor airflow) and gas turbine loading. Furthermore, the steam turbine in the combined cycle power plant operates in "valves wide open" (VWO) or "sliding pressure" mode, which means that the system pressure varies with the steam flow. Consequently, condensate and feedwater pumps should be sized very carefully in order not to be undersized at different operating conditions. For example, if the pump is sized at the rating point (typically ISO baseload), it will not be able to deliver the requisite discharge pressure when the plant is running on a cold day.

Another factor that impacts the condensate and feedwater pump sizing is the steam turbine bypass operation. Because of the large attemperation (desuperheating) flows required for 100% load bypass operation, pump maximum flows are determined by the flow requirement of this mode of operation. In addition, attemperator (desuperheater) spray pressure requirements should be considered to determine the *total dynamic head* (TDH) of the feed pumps.

In supplementary-fired designs, maximum continuous flow is typically at maximum-fired conditions, on a hot or cold day. There is a big difference in flow-head combinations between unfired and fired operating points. A series of cases are calculated to establish the operating range of the feed pumps. As an example, consider the cases summarized in Table 10.3, which are generated using Thermoflow, Inc.'s GT MASTER® software for a single-shaft, fired GTCC with an advanced class gas turbine. The pump data is for the boiler feed pump with an interstage bleed to supply the IP feedwater. As shown in the table, the minimum flow is at ISO conditions (evap cooler and HRSG duct burners are off) with the gas turbine running at *minimum emissions-compliant load* (MECL), which is around 30%–40% (as a function of the ambient conditions). Maximum flow is when the duct burners are on – (i) on a hot day with evap coolers on and (ii) on a cold day with evap coolers off.

Boiler feed pumps are one of the two types: horizontally split case or radially split segmental with multistage configuration. For the higher HP steam pressures, double-casing barrel-type construction is required (e.g., see Table 10.3). In N × N × 1 power plants with N gas turbine-HRSG trains, each HRSG has its own boiler feed pump. Consequently, the pump flow variation in the emissions-compliant gas turbine load range is less than 2-to-1 (it is about 1.5-to-1 in Table 10.3). In general, boiler feed pumps in GTCC power plants are single-speed designs directly driven by a two-pole motor (3,000/3,600 rpm at 50/60 Hz). A typical single-speed pump characteristic is shown

TABLE 10.3
Boiler Feed Pump Data from Heat and Mass Balances for a Fired, Single-Shaft GTCC (Actual Project Data)

Case		1	2	3	4	5	6	7	8	9	10	11	12	13	14
Ambient temperature	°F	59	59	75	75	75	90	90	90	105	105	105	105	0	0
Relative humidity	%	60	60	45	45	45	50	50	50	14	14	14	14	90	90
Gas turbine load	%	100	MECL	100	100	MECL	100	100	MECL	100	100	100	MECL	100	MECL
Evap cooler		Off	Off	On	On	Off	On	On	Off	On	On	Off	Off	Off	Off
Duct burners		Off	Off	On	Off	Off	On	Off	Off	On	Off	On	Off	On	Off
Suction pressure	psia	79	50	99	80	50	102	82	50	99	78	100	50	99	50
Suction flow	kpph	970	544	1,357	981	559	1,381	998	576	1,348	949	1,354	614	1,381	620
HP discharge pressure	psia	1,618	899	2,765	1,634	940	2,803	1,660	965	2,764	1,580	2,775	1,029	2,766	1,034
IP discharge pressure	psia	448	257	674	453	263	682	460	269	670	437	674	287	670	288
IP flow	kpph	183	87	141	184	93	141	184	96	133	178	104	104	146	108

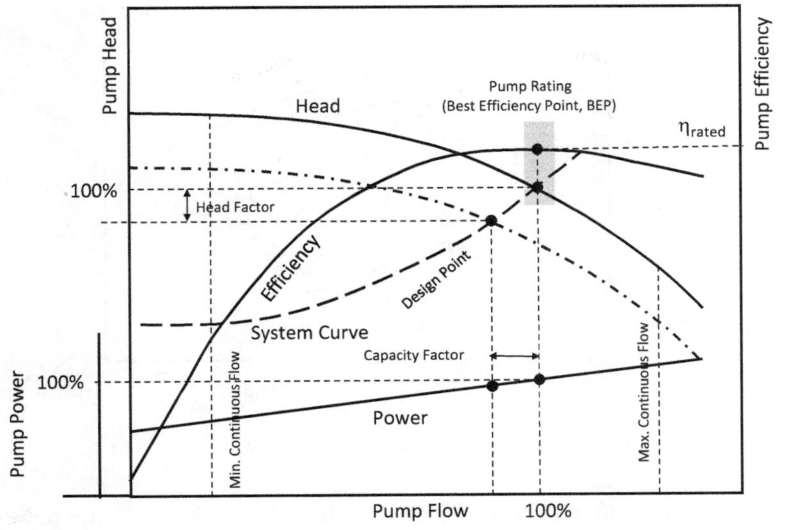

FIGURE 10.4 Typical pump efficiency and head curves (fixed speed).

in Figure 10.4. The design point in the figure is based on pressure rise and flow numbers obtained from the HMB calculations. Usually, when the pump is procured, a safety factor/margin is applied in the pump data sheet provided to the vendors. The safety factor typically comprises head and flow/capacity factors (margins). This accounts for the quite likely case when actual pressure losses and flows in the field are larger than those assumed in HMBs. (In reality, even a head margin itself would provide some capacity margin.)

If the pressure of feedwater at the pump suction is below the vapor pressure, vapor bubbles form in the inlet flow. As the pressure increases above the vapor pressure while passing through the pump impeller, the bubbles collapse at the speed of sound of the water, sending shock waves throughout the water. This situation, known as "cavitation", is damaging to the equipment in the long run. In order to prevent cavitation, *net positive suction head* (NPSH) available at the pump inlet should satisfy the vendor's required value by a typical margin (usually, 1 m of head or 3.3 ft). NPSH is the total (absolute) suction head of water determined at the suction nozzle minus the vapor pressure of water. Note that required NPSH is proportional to the inlet flow; higher pressure head is required at the pump inlet for high flows than low flows. Ideally, most pump operation is expected to be within 70%–120% of *best efficiency point* (BEP) flow. At much lower flows, actual NPSH may fall below required NPSH leading to cavitation. Furthermore, operation at lower than the optimal flow range may require recirculation, which can lead to increased vibration and component damage and, thus, to an unplanned outage.

The "system curve" implied by the data in the Table 10.3 is depicted in Figure 10.5 in normalized form with generic variable speed pump head curves and efficiency contours superimposed. A brief discussion is provided below to clarify the advantages of a variable speed pump from a performance perspective (specifically, parasitic power consumption of the feed pump driver motor).

Obviously, the selected pump must be able to satisfy the highest flow requirement. If one designates Case 1 in Table 10.3 as the design point (100%, 100%), this maximum continuous flow point is Case 6 (142%, 175%). If the selected pump is a single-speed design (N_1 rpm), at lower flow rates (i.e., <142%), its discharge pressure on its head curve will be significantly higher than needed (i.e., >175% head at 100% flow). Thus, the flow-pressure control will be achieved by a throttle valve at the pump discharge with significant throttling (i.e., wasted energy).

The alternative is a *variable speed driver* (VSD), which changes the pump speed. Instead of throttling for flow control, the pump is run slower (i.e., less power consumption). Thus, e.g., at the design

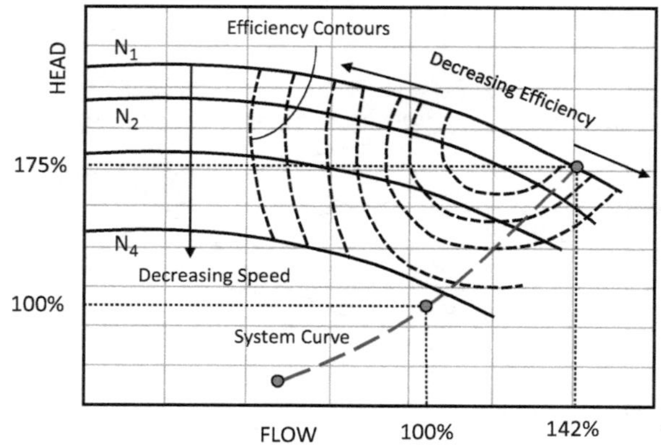

FIGURE 10.5 Typical pump efficiency contours and head curves (variable speed).

point (100%, 100%), the pump will be run at N_4rpm ($N_4 < N_1$) with significant power saving. The actual amount can be easily calculated using the "affinity laws". In addition to power saving, VSD has additional benefits, e.g., reduced wear of flow control valve, pump bearings and seals, improved process control and elimination of throttling noise.

In order to get a feel about the power saving stemming from using VSD, let us consider Cases 1 and 6 in Table 10.3. Using the formula

$$\dot{W}_{bfp} = 0.00122\,\dot{m}[\text{kpph}] \cdot \Delta p[\text{psia}]\text{kW}$$

Case 6 pump power consumption is 4,550 kW (motor power) and Case 1 pump power consumption is 1,820 kW. However, with a single-speed (3,600 rpm) pump rated at Case 6 flow head, Case 1 power consumption would be about 3,800 kW (roughly twice the original estimate).

With a VSD, for 3,600 rpm at the rating point (Case 6), with the lower flow, pump speed in Case 1 would be approximately

$$N = (970/1{,}381)3{,}600 \sim 2{,}500\,\text{rpm}$$

in order to generate the requisite head with the originally estimated power consumption.

Variable speed drives have some disadvantages, too. In addition to being more expensive, they are more susceptible to excitation of a critical speed leading to resonance and high vibrations. Whether a VSD makes sense or not requires a detailed cost-performance trade-off analysis. There are myriad factors to consider including the expected operating envelope of the power plant and the overall hydraulic system including control valves and recirculation/bypass lines rather than the pump and driver alone. A useful reference to consult is *Variable Speed Pumping — A Guide to Successful Applications* developed by a collaboration between the Hydraulic Institute, Europump, and the US Department of Energy's (DOE) Industrial Technologies Program.[14]

Condensate pumps in GTCC plants pump the condensate from the condenser hotwell to the HRSG. They are canned vertical turbine-type pumps. (Usually, there are two of them each with 100% capacity.) For N × N × 1 multi-shaft GTCCs with extensive part load operation, three 50% pumps can be used. Similar to the part load heat rate benefit of multiple GT-HRSG trains, this arrangement leads to power saving by operating the pumps closer to their BEP most of the time at

[14] The full report can be purchased from both the Hydraulic Institute (www.pumps.org, phone: 973-267-9700) and Europump (www.europump.org, phone: +32 2 706 82 30). An executive summary is available free online.

part load. This should be weighed against the higher first cost and more complex controls. The flow variation for a 2 × 2 × 1 GTCC is between 3:1 and 4:1 (one GT-HRSG train at MECL to baseload with both trains on).

During operation with steam bypass, steam redirected to the condenser must be desuperheated via cold water spray, which is supplied from the condensate pump discharge. Especially in GTCC plants with full bypass capability, this increases the condensate pump flow significantly. Therefore, condensate pump specification is made so that the BEP is for a flow rate between normal operation and full bypass to ensure reasonably efficient and smooth operation at both ends. In order to prevent steaming in the LP economizer during startup and low load operation, the drum level control valve of the LP evaporator is located at the LP economizer outlet. This keeps the feedwater in the LP economizer pressurized and prevents steaming. Consequently, those operating cases are requisite for condensate pump sizing calculations.

Boiler feed pump suction flow is usually taken from the LP drum (this ensures NPSH at all operating conditions). As mentioned earlier, IP feedwater is typically supplied from an interstage bleed of the boiler feed pump. The bleed flow is determined from a combination of (i) IP admission steam, (ii) fuel gas performance heater heating water and (iii) LP economizer recirculation flow. The last one is water flow set back to the LP economizer inlet to ensure that the feedwater entering the economizer (combination of cold condensate and warm feedwater) and, hence, the tubes are at a temperature above the dew point of the flue gas. This prevents condensation on the carbon steel economizer tubes. In addition to those three contributors, reheater attemperator spray flows and HP steam bypass desuperheater spray flows (in cascaded bypass systems) are also provided form the boiler feed pump interstage bleed.

Note that the IP drum level control valve is also located at the IP economizer outlet and downstream of the branch to the fuel gas performance heater. That ensures that both flows are controlled independently. Off-design models used for sizing data generation should reflect this configuration. Furthermore, flow increase during full bypass operation is limited only to the pump section upstream of the interstage bleed and is roughly 20%. Thus, the pump rated flow should be based on the full bypass operation mode. In duct-fired systems, this should be compared with the highest, fully fired continuous operation flows. (Obviously, it would be a very misguided operator who would run the plant in full bypass mode while fully fired.)

10.4 TANKS

Tanks in a typical GTCC power plant include

- Fuel oil (backup fuel if the gas turbine is equipped with dual fuel combustors)
- Hydrous ammonia (for the SCR – see Section 12.4)
- Demineralized water (see Section 10.9)
- Raw water
- Neutralized water (see Section 10.9)
- Acid storage
- Caustic storage
- Waste water
- Dedicated fire protection water storage
- Chilled water storage (if there is a gas turbine inlet chiller)
- Process water storage (cogeneration)
- District heating water storage (cogeneration).

Tanks and pressure vessels in the GTCC power plant are designed to have sufficient storage capacity to meet the requirements of the particular system application. They are fabricated from materials suited to the particular service. For demineralized water, raw/fire water and reclaimed water, bolted, field-erected storage tanks from carbon steel are specified. (Atmospheric tanks may be of bolted or welded construction.) Hydrous ammonia tanks are made from stainless steel. The fire–water storage tank should be designed in accordance with NFPA 22. A corrosion allowance should be incorporated into the design such that the tanks have a specified design life (e.g., 25 years or the warranty period). A typical minimum corrosion allowance is 1/16 in.

Unfired pressure vessels are designed, fabricated, tested and installed per ASME Boiler and Pressure Vessel Code Section VIII. They include the following features/accessories:

- Process, vent and drain connections for startup, operation and maintenance
- Materials compatible with the fluid being handled
- Where required for maintenance or cleaning access, a minimum of one manhole and one air ventilation opening (e.g., handhole)
- Relief valves in accordance with the applicable codes.

Tanks are equipped with maintenance drain connections and piping of an adequate size to allow drainage of the tank in a reasonable time period. All tanks have level transmitters and a side glass to monitor level. Outdoor storage tanks, which require freeze protection, are wrapped with at least 2 in. thick insulation (moisture-resistant material). Otherwise, large outdoor storage tanks are typically uninsulated.

Tanks are also equipped with overflow connections and lines of at least one pipe size larger than the inlet connection. Manholes, where provided, are at least 18 in. in diameter and hinged to facilitate removal. Ladders and cleanout doors are provided on storage tanks as required to facilitate access/maintenance. Provisions must be included for proper tank ventilation during internal maintenance.

10.5 AUXILIARY BOILER

The auxiliary boiler is a package boiler that provides low-pressure saturated steam to maintain the steam turbine seals when the power plant is shut down for the night, thereby sustaining the condenser vacuum overnight. Otherwise, during restart, there is a considerably delay while waiting for the HRSG to generate steam to establish the steam seals. Steam seals must be established before condenser vacuum can be established so that the steam turbine can be rolled off of the turning gear. Depending on the shutdown duration, this can take anywhere between one-half to one hour. The auxiliary steam is also used to sparge the condenser hotwell to prevent the condensate from subcooling and picking up dissolved oxygen (DO). DO causes corrosion in both non-ferrous heat exchangers and the carbon steel components of the HRSG, and its concentration in the condensate and feedwater must be limited to 7 parts per billion (ppb). Solubility of DO in water is proportional to its partial pressure in the vapor above the hotwell water. The proportionality factor, the Henry constant, H, is inversely proportional to the water temperature. Consequently, solubility of DO in colder water is much higher. Thus, sparging is crucial to prevent the subcooling of the hotwell water during startup or part load operation.

Typical auxiliary boiler steam conditions are 200 psi and 590°F. In terms of construction, the "aux-boiler" is a bottom-supported, modular-section, shop-fabricated water-tube package with ultra-low NOx combustion system, fuel train, forced-draft fan and economizer. It is designed for outdoor installation. The feedwater system for the aux-boiler includes 2 × 100% feedwater pumps and level control valve. The feedwater is supplied from the steam turbine condensate system.

The aux-boiler is purchased as a factory-made system complete with a burner management system. Remote control and monitoring is provided from the plant DCS. It is typically equipped with emission controls as defined in the GTCC plant's air permit and with its own CEMS.

10.6 FUEL GAS BOOSTER COMPRESSOR

In order to be injected into the gas turbine combustor, the gaseous fuel should arrive at the inlet of the fuel skid at a certain pressure. Normally, for vintage F class heavy-duty industrial machines that are operating at compressor pressure ratios (PRs) of 16:1 to 18:1, the pipeline gas pressure at the point of delivery to the combined cycle power plant is sufficient. For advanced class machines with compressor PRs of 20:1 or higher in some cases but certainly for very high cycle PR units such as ABB/Alstom/GE/Ansaldo "sequential combustion" turbines GT24/26 with PR well above 30:1, the plant should incorporate a fuel gas "booster" compressor to increase the pipeline gas pressure to the required pressure levels. (This is also true for the aeroderivative gas turbines with PRs of as high as 40:1.) Another factor to consider is fluctuations and/or decrease in pipeline gas pressure during peak demand times, especially during unexpected "spikes" caused by extreme weather conditions.[15]

In those applications where the pipeline gas pressure and the required fuel skid inlet pressure dictate a booster compressor, the power consumption of that compressor should be properly accounted for in the plant auxiliary load. The following formula is suggested for that purpose for gas turbines with cycle PR >20:1 (with fuel flow rate in lb/s):

$$\dot{W}_{FCMP}[kW] = \dot{m}_{fuel}\left(80 \cdot \ln PR - 225\right). \tag{10.5}$$

Equation 10.4 is derived from a simple intercooled fuel compressor model using polytropic p-T relationship with 75% efficiency and discharge pressure about 60% higher than compressor discharge pressure. If there are more than one gas turbine in the power plant, either use the total fuel flow rate for all units or multiply the value obtained from Equation 10.4 by the number of units.

All three basic compressor designs have been used for fuel gas pressure boosting, i.e., centrifugal, reciprocating and screw. Each type has different mechanical design characteristics and operating range (see Figure 10.6). In order to get an idea about the flow requirement, consider a large 50-Hz advanced class gas turbine with natural gas flow rate of about 50 lb/s (22.68 kg/s). At 450 psia and 77°F, volume flow rate into the booster compressor is about 3,700 m^3/h (actual). This falls into the applicability range of all three types of compressors.

Reciprocating and screw compressors are "positive displacement" machines; i.e., they draw in and capture a volume of gas in a chamber, then reduce the volume of the chamber to compress the gas.

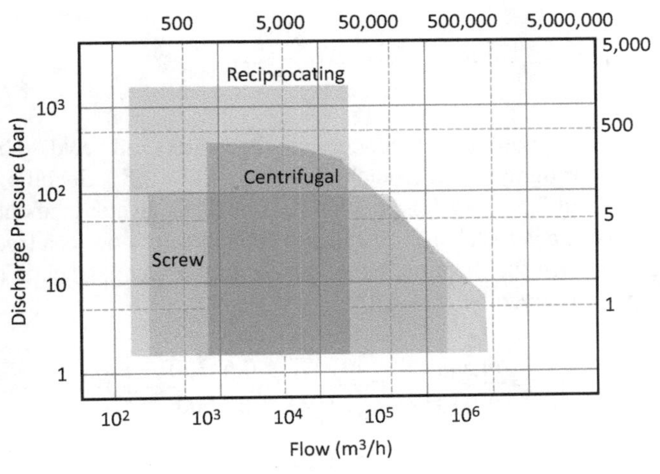

FIGURE 10.6 Compressor design space.

[15] Typically, guaranteed supply pressure to the plant by the gas utility is based on the latter's lowest historical figures.

Centrifugal compressors are "dynamic displacement" machines; i.e., they "speed up" the gas to high velocity (rotor) and then restrict the gas flow (stator/diffuser) so that the reduction in velocity causes pressure to increase. In general, reciprocating and rotary screw compressors are more common in GTCC applications. Due to their initial cost, centrifugal compressors often are considered only in extremely large capacity, high horsepower, multi-turbine installations.

Reciprocating compressors are very efficient, easy to maintain and/or repair (onsite) – although more prone to wear and tear of valves and packings – with the largest experience base, but they are heavier and more expensive than screw compressors and must be evaluated for the "pulsating flow" characteristics.

10.7 FUEL GAS HEATING AND CONDITIONING SYSTEM

Let us start by saying a few words on the thermodynamic principles underlying performance fuel gas heating. Why the moniker "performance"? Liquid-phase hydrocarbons entrained in the fuel gas can cause degradation of gas fuel nozzles and flame flashbacks or re-ignition in the dry low NOx (DLN) combustor. Condensation of moisture must be avoided to prevent the formation of gas hydrates and collection of water in low points of the gas fuel system. Thus, the fuel gas supply to the gas turbine must be free of liquids. Consequently, the fuel gas must be heated above the hydrocarbon and water dew point by a sufficiently high margin to compensate for the temperature reduction as the gas expands across the gas fuel control valves. The superheat requirement is a function of the fuel gas supply pressure. At 500 psia, e.g., it is about 40°F. If this were the only reason for heating the fuel gas, an electric heater would suffice and the fuel gas temperature at the combustor inlet would be around 120°F.

On top of the superheat requirement (which is specified by the OEM based on the site fuel gas characteristics[16]), fuel gas can be heated to a high temperature to increase its energy content via sensible heat and thus reduce its consumption for given temperature rise. This is accomplished in a shell-and-tube or plate-and-frame-type heat exchanger by utilizing the IP feedwater supplied from the discharge of the IP economizer in the HRSG.[17] This is referred to as "performance" heating. In vintage F class units, fuel gas used to be heated to 365°F (185°C) in a shell-and-tube-type heat exchanger using feedwater extracted from the HRSG IP economizer discharge. In advanced class gas turbines, performance heating to 410°F (210°C) is common. From the HMB around the fuel gas heater, it is found that

$$\frac{\dot{m}_{fw}}{\dot{m}_f} = \frac{c_{p,f}}{c_{p,fw}} \frac{\Delta T_f}{\Delta T_{fw}} \approx \frac{c_{p,f}}{c_{p,fw}} = \xi.$$

since temperature change in "hot" (i.e., IP economizer feedwater) and "cold" (i.e., fuel gas) streams in the heat exchanger are roughly equal to each other, i.e., $\Delta T_f \approx \Delta T_{fw}$. Specific heat of natural gas is about 0.6 Btu/lb-R (about 2.5 kJ/kg-K). Specific heat of IP feedwater is about 1 Btu/lb-R (about 4.3 kJ/kg-K), thus $\xi = 0.6$ is a very reliable estimate of IP feedwater flow as a fraction of gas turbine fuel gas flow for performance fuel heating. Heat transfer in the fuel gas heater as a fraction of gas turbine heat consumption is given by

$$q_{fhtr} = \frac{\dot{m}_f\, c_{p,f} \Delta T_f}{\dot{m}_f\, LHV} = \frac{c_{p,f} \Delta T_f}{LHV} \approx \frac{0.6 \cdot 330}{21{,}500} \sim 0.01.$$

---

[16] For example, *Specification for Fuel Gases for Combustion in Heavy-Duty Gas Turbines* (GEI 41040m) by General Electric (January 2011).

[17] For a detailed description of the fuel gas heater system, refer to the General Electric document GER-4189B *Design Considerations for Heated Gas Fuel* by Erickson, Day and Doyle (March 2003).

(This is why ignoring it in simplified, parametric combined cycle efficiency equations introduced earlier does not lead to a significant error.)

Performance fuel gas heating and conditioning system comprises the following pieces of equipment

- Fuel gas meters
- Gas chromatograph
- Fuel gas Wobbe meter
- Separator/filter skid
- Electric startup heater
- Fuel gas performance heater
- Scrubber.

Fuel gas distribution piping from the interface points to all points of consumption (i.e., gas turbines, HRSG duct burners – if present – and the aux boiler) are equipped with flow meters, which are specified with an expected accuracy as listed below, capable of providing input to the DCS:

- Gas turbines, ±0.4%
- HRSG duct burners, ±0.70%
- Aux boiler, ±2.0%.

Wobbe index (WI) is a measure of allowable fuel gas energy content variation, which is a function of the fuel gas heating value (a function of fuel gas composition) and temperature. The WI is defined as the ratio of the lower heating value (LHV) of the fuel (in Btu/scf) to the square root of the specific gravity (SG) of the fuel (relative to air) at reference temperature and pressure (usually 32°F and 1 atm). The "modified" Wobbe index (MWI), which includes the effect of fuel gas temperature (in degrees Rankine), is defined as

$$MWI = \frac{LHV}{\sqrt{SG \cdot T_f}}.$$

A good approximation for SG is the ratio of fuel gas and air molecular weights. Allowable variation in MWI in modern DLN combustors during field operation is typically ±5% due to emissions and combustor dynamics considerations. Recently, OEMs claim as high as ±15% for the DLN combustors in their advanced class heavy-duty gas turbines.

The electric startup heater is required when the gas supply does not meet the minimum superheat requirement and the IP feedwater from the HRSG is not at sufficient temperature and/or quantity. The electric heater is turned off and then bypassed once the performance fuel gas heater is capable of maintaining the required gas temperature on its own.

The filter/separator skid is designed to protect the downstream gas fuel system against the entry of both liquid-phase fuel and particulate contaminants. The scrubber is the final stage of filtration directly upstream of the turbine. The scrubber also removes water droplets from the gas stream following the event of a heater tube leak or rupture. (Note that the IP feedwater pressure in the heater tubes must be higher than the pressure of the fuel gas in the heater shell. In case of a rupture, fuel gas must not be allowed to leak into the feedwater.)

Typical fuel temperature required by the HRSG duct burners is 80°F. Especially, in cold weather, pipeline gas temperature can drop below that due to the throttling between the supply point and the burners. (In some cases, it can go as low as 25°F.) Thus, in some combined cycle power plants, the fuel temperature entering the duct burners is regulated by mixing the heated fuel gas with the unheated gas.

10.8 CLOSED COOLING WATER (CCW) SYSTEM

The CCW system typically comprises a glycol/water mix as the coolant and is designed to provide cooling water up to cooled components in a closed loop at all ambient conditions. Equipment requiring water cooling includes:

- Steam and gas turbine generators
    - Lubricating oil coolers
    - Generator hydrogen coolers
- Feedwater pump coolers
- Sample system coolers
- Air compressors
- Gearbox/hydraulic coupling coolers
- Large pump motors (circ water, condensate, boiler feed)
- Boiler feed pump lube oil coolers.

In a GTCC power plant, typically, 80% of CCW system cooling duty is from the prime movers. The CCW system uses 100% capacity heat exchangers for heat rejection (typically two of them). Makeup water to the CCW system is provided from the demineralized water system. Makeup water is added to the expansion tank to maintain the water level. All piping, valving, instrumentation and controls are designed and provided by the EPC contractor.

For units installed outdoors in cold locations, the coolant is a glycol–water mix (because it lowers the freezing point of the coolant). It may, however, be expensive for indoor units, and heat tracing the outdoor portions may be more cost-effective. Percentage of glycol in the mix and its type (i.e., ethylene or propylene glycol) is specified by the heat exchanger vendor. A typical specification is approximately 36/64 mix of propylene and water. Note that the higher the glycol content, the lower is the coolant-specific heat. Consequently, for the same cooling duty, higher coolant mass flow rate is required (i.e., bigger pumps and pipes).

During normal plant operating modes, cooling water to the CCW heat exchangers is supplied by the circ water pumps in operation (unless seawater is used). To preclude running a circ water pump during extended shutdowns, a single, partial-capacity CCW pump, which ties into the circ pump discharge (downstream of the circ pump discharge valve), is provided. This pump can be operated at plant operator's discretion during shutdowns when an off-site power source is available (i.e., it is not connected to the diesel generator). The materials of construction are specified for the water chemistry of the circulating water and closed-loop cooling water. Plate material should be compatible with high chlorides of the circulating water. A self-cleaning filter is installed to prevent plate heat exchanger fouling.

Cooling water pumps (typically two 100% pumps) are horizontal centrifugal pumps driven by constant-speed motors. Each pump comes complete with case, shaft, impeller, base plate, coupling and the driver. See Section 10.3.2 for estimating their power consumption. For preliminary calculations, assumptions are

- CCW system duty is 5% of the duty of the main steam condenser
- Temperature change is about 10°F–15°F on the hot side of the CCW heat exchanger
- Pressure drop on the hot side is 10 psi.

Typically, CCW heat exchangers are of plate-and-frame type. In some cases, the owner/operator can specifically ask for a shell-and-tube-type heat exchanger. Plate-and-frame-type heat exchangers are less expensive than shell-and-tube type for the same duty, use less space, and are expandable. For one thing, they do not require as much cooling water as the shell-and-tube variants. This has the benefit of reducing plant capital cost by using smaller piping for the circ water. Plate-and-frame heat exchangers have symmetrical flow passages for the hot and cold sides so that the ratio of the flow rates on hot and cold sides is less than 2-to-1 for a well-designed system. Otherwise, the heat exchanger size is controlled by the pressure drop, leading to a larger and costlier heat exchanger.

The CCW system incorporates an expansion tank, which is located at the highest point in the system and connected to the suction side of the circ pump. The expansion is calculated from

$$V \cdot \Delta v,$$

where V is the system volume (pipes and heat exchangers) and the normalized specific volume difference is found from

$$\Delta v = v(T_{max})/v(T_{min}) - 1.$$

A typical number for T_{min} is 50°F (for T_{max} it is 105°F). The "usable" volume of the tank must be able to hold the expansion volume. Overflow should be directed to a safe location if there is glycol in the coolant.

10.9 WATER FACILITIES

The working fluid of the bottoming Rankine steam cycle requires several key plant support systems (also referred to as "facilities") to keep it clean and "usable". Such systems include

- Cooling water systems, including a mechanical draft cooling tower and circulating water pumps, to provide cooling water to the condenser (see Chapter 7)
- Raw water storage system
- Demineralized water system
- Domestic (potable) water system
- Cooling tower blowdown recovery system (if necessary)
- Stormwater collection and discharge system.

Feedwater treatment is an essential part of a fossil fuel-fired power plant with a steam cycle. Therefore, it deserves a closer look at its key attributes.

10.9.1 Why Treatment?

According to the Electric Power Research Institute (EPRI), top major causes of HRSG tube failure, i.e., *flow-accelerated corrosion* (FAC), corrosion fatigue, under-deposit corrosion and pitting, can be traced back to the problems associated with water chemistry. Thus, Rankine steam cycle equipment requires essentially pure water as working fluid. Unfortunately, pure water does not occur in nature. (Bottled water that we drink because it is safe for us is *not safe* for the steam cycle boiler or the HRSG.) To make matters worse, water quality shows a large variability from site to site. Thus, water treatment equipment make up a vital auxiliary system of a combined cycle power plant. Major types of water impurities are summarized in Table 10.4. There are many treatment technologies available to handle one or more of those contaminants (see Table 10.5). Detailed information on these technologies can be found in Ref. [1]. Specific information on power plant water treatment can be found in Ref. [2] as well. The reader can also consult Ref. [3] for water chemistry information specific to steam power plants.

The ultimate goal of water treatment is steam purity, which can have a detrimental impact on performance and availability if not maintained adequately. For example, presence of contaminants can lead to corrosion damage of steam turbine components including pitting, *stress corrosion cracking* (SCC) and corrosion fatigue. The damage is proportional to exposure time and concentration of

TABLE 10.4
Water Impurities and Contaminants [4]

Contaminant Type	Examples
Suspended solids	Clay, dirt, silt, dust, insoluble metal oxides and hydroxides, colloidal materials
Dissolved organic compounds	Synthetic organic compounds, trihalomethanes, humic and fulvic acids
Dissolved ionic compounds	Heavy metals, silica, arsenic, nitrates, chlorides, carbonates
Microorganisms	Bacteria, viruses, protozoan cysts, fungi, algae, molds, yeast
Gases	Hydrogen sulfide, carbon dioxide, methane, radon

TABLE 10.5
Water Treatment Technologies (NA, Not applicable; NR, Not recommended) [4]

Technology	Suspended Solids	Dissolved Organics	Dissolved Ionic Compounds	Microorganisms	Gases[a]
Filters (bed, cartridge or bag)	Very effective	NA	NA	NA	NA
Precoat filtration	Very effective	Partially effective	NA	NA	NA
Activated carbon	NR	Very effective	NA	NA	Partially effective
Reverse osmosis	NR	Very effective	Very effective	Very effective	NA
Distillation	NR	Partially effective	Very effective	Very effective	NA
Electrodialysis	NA	NA	Effective	NA	NA
Electrodeionization	NA	NR	Effective	NR	NA
Ion exchange	NR	NA	Very effective	NA	NA
Ozonation	NA	Partially effective	Partially effective	Very effective	NA
Chlorine	NA	NA	NA	Effective	NA
Ultraviolet radiation	NA	Partially effective	NA	Effective	NA

[a] Gases are removed by deaeration.

contaminants. The objective in establishing proper water and steam cycle chemistry is to minimize deposits and prevent corrosion damage to major equipment (steam turbine, HRSG and condenser) and components (e.g., valves, seals and bearings).

Potential sources and/or causes of contamination are boiler carryover (silica at 900 psig, salts at 1,800 psig), steam attemperation (desuperheating) with contaminated water, silica vaporization and vaporization of organic compounds. Detailed chemical background and quantitative data can be found in Ref. [5].

In drum-type boiler systems, carryover of impurities from the feedwater in vapor or solid phase is the key mechanism of contamination. In once-through systems (e.g., Benson-type boiler), steam contamination is directly tied to the purity of condensate prior to the boiler inlet. Therefore, treatment requirements for once-through boiling systems are more stringent requiring expensive equipment. The differences are summarized in Tables 10.6 and 10.7. For the Rankine steam bottoming cycle of a GTCC, feedwater and steam quality requirements are similar to those in Table 10.7. Slightly more stringent requirements may apply to steam-cooled G and H class systems where cooling steam may leak into the turbine HGP.

In *all-volatile treatment* (AVT), boiler feedwater quality is controlled by adding non-solid chemicals such as ammonia (to control pH to prevent corrosion) and hydrazine (to scavenge oxygen). When the latter is used, the process is referred to as AVT-R. *Oxygenated treatment* (OT) or AVT-O with no hydrazine addition is based on the theory that slightly soluble oxides such as Fe^{+++} iron

TABLE 10.6
Differences in Water Treatment for Subcritical (Drum Type) and Supercritical Boilers

Drum Type Subcritical	Once-through Subcritical
All Volatile Treatment (AVT)	Oxygenated Treatment (OT) Combined Water Treatment (CWT)
Polisher mat not be needed (needed for faster startups)	100% polisher for startup with AVT, OT (CWT) Reduced use during normal operation possible
50% precoat-type demineralizer	Precoat and deep-bed demineralizer

TABLE 10.7
Recommended Boiler Feedwater Limits for Once-Through and High-Pressure Drum-Type Boilers [2]

		Drum	Once-Through	
	Unit	AVT	AVT	OT
pH		9.3–9.6	9.3–9.6	8.0–8.5
TH	ppm $CaCo_3$	0.003	0.003	0.001
Oxygen	ppm	0.007	0.007	0.03–0.15
Iron	ppm	0.01	0.01	0.005
Copper	ppm	0.005	0.002	0.001
Organic	ppm TOC	0.2	0.2	0.2
Cation conductivity	μS/cm	0.2	0.2	0.15

Conductivity is a measure of the concentration of (ionized) dissolved solids in water. Its units are "micro-Siemens per centimeter" or μS/cm. For most public waters, 1.55 μS/cm is about 1 mg/L or 1 ppm (part per million as $CaCO_3$) in total dissolved solids (TDS). TOC stands for total organic carbon.

oxide forming on the boiler tube surface act as a protector against FAC. There are two types of OT, *neutral water treatment* (NWT), in which DO is allowed to coexist in neutral water, and *combined water treatment* (CWT), in which DO is allowed to coexist in weak alkaline water adjusted to a range of pH 8.0–9.3 by ammonia.

In multi-pressure HRSGs, e.g., three-pressure reheat configuration, which is the state of the art with advanced class gas turbines, sodium phosphate is added to the evaporator water in the IP and HP steam drums for pH control. In some cases, however, caustic or acid phosphate corrosion caused by phosphates (due to the tendency of phosphates to precipitate at high temperatures) has been observed [6]. Alternatives to phosphate treatment are caustic treatment or AVT-only treatment. The former can be appropriate for units susceptible to condenser leaks and/or phosphate carryover to superheater and reheater tubes [7].

Based on the temperatures and velocities that favor FAC, it is more of a concern in the LP system. In some cases, neutralizing amines instead of ammonia are used for pH control and minimization of FAC in the LP evaporator. Typical neutralizing amines used in power plant systems are cyclohexylamine (CHA), methoxypropylamine (MPA), monoethanolamine (MEA) and morpholine. The neutralizing amine is typically injected at the condensate pump discharge. Which chemical (or blend of chemicals) to use and in what concentration is highly system dependent and requires expert analysis on a case-by-case basis.

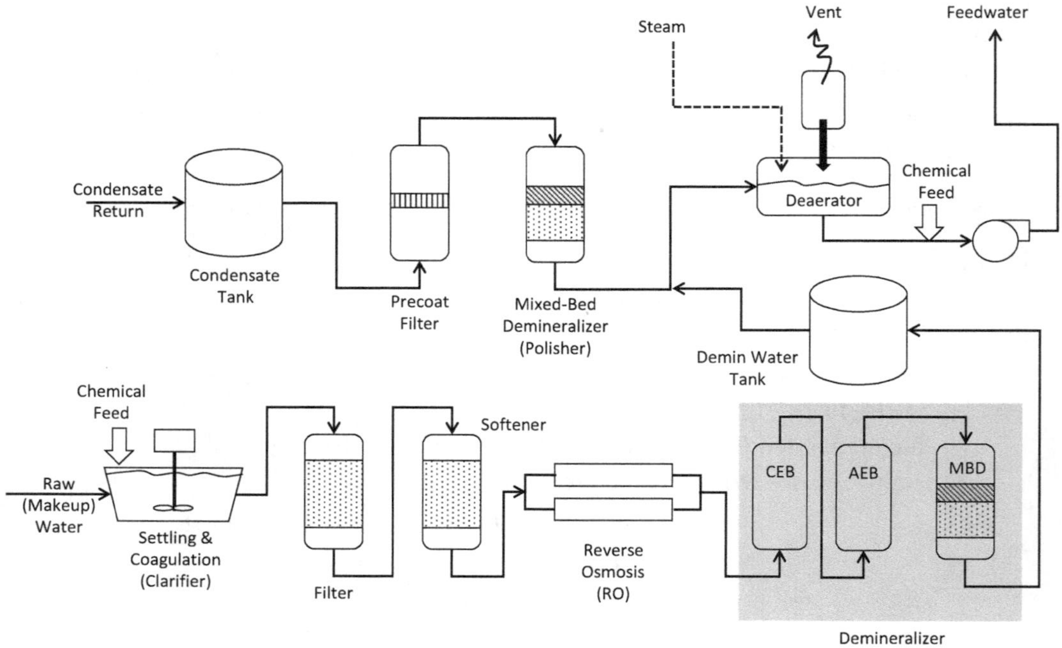

FIGURE 10.7 Conceptual boiler feedwater treatment system for steam power plant (CEB, cation exchange bed; AEB, anion exchange bed).

A typical advanced boiler water treatment system is shown in Figure 10.7. More or less the same type of systems and principles apply to an advanced GTCC with a drum-type or once-through HRSG as well. (Modern GTCC plants usually do not have a separate deaerator or combined LP drum and deaerator. They utilize a *deaerating condenser.*) As shown in Figure 10.7, boiler feedwater is a combination of condensate return and makeup water. The latter is about 1%–2% of the total feedwater flow (unless it is a cogeneration system). Treatment starts with *clarification* of the makeup (raw) water to coagulate and settle suspended matter with the help of chemical additives (alum and iron sulfate). It is followed by *filtering* to remove trace impurities and excess chlorine (which is used to kill microorganisms). Next comes *softening*, which removes calcium and magnesium ions (major hardness ions) by replacing them with sodium salts (i.e., ion exchange, IX). Some other dissolved impurities are also removed during this process. Removal of dissolved solids still present after softening (now including sodium cations as well) is known as *demineralization*. There are several methods of demineralization including IX *reverse osmosis* (RO) and *electrodeionization*. One or more of those (in series) can be used to satisfy the boiler feedwater purity requirements (see Table 10.7). The exact system configuration is a function of the particular site raw water analysis. Complete removal of solids is ideal for once-through boiler systems (e.g., the Benson technology with supercritical HP section). In that case, the process is finished by a *mixed bed demineralizer* (MBD).

In theory, condensate return from the steam turbine condenser should be pure. In reality, it may contain contaminants resulting from corrosion of tubes and pipes, air in-leakage, and in-leakage of condenser cooling water (especially if the circulating coolant is seawater). Thus, *condensate polishing* using MBD is desirable for all modern HP systems with AVT and seawater-cooled condensers. Finally, dissolved O_2 and CO_2 in the feedwater are vented out in the *deaerator.*

Long-term exposure is not necessary for serious damage to equipment. In fact, short duration contamination events associated with accidents and upset conditions (especially during plant commissioning and startup) are sufficient to plant the seeds for future failure. Avoiding such events requires constant vigilance on the part of operators with proper maintenance of the water treatment equipment to prevent leakages and deviations from OEM-specified contaminant levels. Some critical parameters such as *cation conductivity*[18] must be monitored continuously for their target levels (e.g., less than several parts per billion by weight for the former). Others are checked less frequently (e.g., once or twice per week for silica and total organic carbon) or only during commissioning, regular maintenance and troubleshooting periods.

Recommended online analysis/monitoring locations are condensate pump discharge, feedwater pump discharge, deaerator outlet, economizer inlet (HRSG) and boiler (evaporator) drum (blowdown line or downcomer). Some measurements can even lead to a plant shutdown (e.g., if the boiler water pH goes down below 8.0 for an HRSG per EPRI guidelines). For the modern, advanced GTCC plants with fast start features, steam purity achievement is critical. Since the presence of CO_2 in the sample (via air in-leak in the condenser, for example) increases its conductivity (by about 5 μS per ppm of CO_2), *degassed* cation conductivity is recommended in order not to hold up the startup sequence due to artificially high conductivity readings.

10.9.2 Once-Through (Benson) HRSG

Steam drums in conventional HRSGs are designed to concentrate and mechanically remove (via blowdown) impurities in the feedwater. Once-through units, however, do not have the same capability – feedwater evaporates while passing through the tube banks. Consequently, contaminants present in the feedwater are passed on to the steam turbine with the motive steam. Therefore, condensate and feedwater purity must be maintained very precisely and must match the steam turbine OEM's steam purity requirements. *Condensate polishing* is vital to establishing and maintaining this high level of purity in the steam cycle – essentially, only possibilities are OT and AVT.

Under normal operating conditions with leak-tight condensers, only part of the condensate flow has to be polished. However, during startup of the power plant and the first few hours of operation, or in the event of leakage in the condenser, the entire condensate flow should be passed through the condensate demineralizing system for polishing. Due to the criticality of condensate polishing, usually two 100% trains are provided. During regeneration of one polisher train, the other one is kept in service or a backup batch is used to minimize the downtime. For regeneration, requisite diluted chemicals (sulfuric acid and caustic soda) are delivered from the storage tanks of the demineralization system by dosing pumps.

Each condensate polishing train consists of one mixed-bed demineralizer, which removes ammonia, silica and all other cations and anions contained in the condensate by IX. The water quality downstream mixed-bed exchanger is monitored by one specific conductivity measuring device for each unit. Detailed analysis of a sample is done in the water–steam sampling system.

The mixed-bed demineralizers are rated for a maximum of ANSI Class 150 and temperature of 60°C (140°F). Material of components of the system is selected according to the process requirements. Typical, requisite water quality numbers during normal operation are:

- pH value of 8.0–8.9
- Conductivity of 1–2 μS/cm (at 25°C); downstream strong acidic cation exchanger 0.2 μS/cm
- Ammonia 0.2 mg/kg
- Oxygen 0.1 mg/kg
- Silica ≤0.02 mg/kg
- Iron (total) ≤0.02 mg/kg
- Sodium ≤0.01 mg/kg.

[18] A cation is a positively charged ion, e.g., Na^+; they attract oxides and lead to corrosion. Elevated levels of cation conductivity indicate increase in salts such as chlorides and sulfates.

During normal startup, the required quality standards are somewhat more relaxed, i.e.,

- pH value of about 9.2
- conductivity of 5.5 μS/cm; downstream strong acidic cation exchanger about 0.5 μS/cm)
- Ammonia about 0.7 mg/kg
- Oxygen about 0.1 mg/kg
- Silica about 0.05 mg/kg
- Iron (total) about 0.05 mg/kg
- Sodium about 0.02 mg/kg.

10.9.3 Usage Minimization

Quantity and quality of global water resources has been a major concern for some time. Severe drought and resulting problems have even been associated with recent major sociopolitical problems in certain areas of the world. Combined with the uncertainty tied to climate change, water scarcity looms as an urgent problem facing humanity. Thus, conservation of precious freshwater resources is a major initiative of many governmental agencies with a significant impact on electric power generation technology.

Consider that fossil fuel power systems, combined with the nuclear power plants, accounted for nearly 40% of freshwater withdrawals in the USA (mainly as circulating cooling water for steam turbine condensers) while they accounted for 3% of freshwater consumption (via evaporation into the atmosphere or as moisture in disposed solid wastes) [8]. In terms of withdrawal, coal-fired steam power plants (along with nuclear power plants) utilize nearly three times more cooling water than natural gas-fired GTCC (i.e., 25–45 thousand gallons per MWh of generation with once-through open-loop systems or 550–800 gallons per MWh for closed-loop systems with wet cooling tower) [8]. Furthermore, dry or wet FGD systems of coal-fired power plants consume an additional 40–80 gallons per MWh of freshwater (none for GTCC with modern DLN combustion systems).

While the consumption of freshwater by a fossil-fired power plant seems small, consider that a 500 MWe GTCC power plant, with a closed-loop system, consumes annually 7,000 h/years × 500 MWe × 220 gal/MWh ~ 750 million gallons of freshwater. Thus, the incentive for freshwater consumption minimization is quite strong. There are several options:

1. Use water of a quality only as good as required by the power plant processes
2. Use of advanced (dry, hybrid, other) cooling systems
3. Reuse and recycle power plant waste water with requisite treatment
4. Proper selection of water treatment chemicals to reduce hazardous wastes
5. A combination of the above.

One example for the first item is using *membrane distillation* (MD) technology to remove salts and other dissolved solids from brackish groundwater or seawater. Using the waste heat in the flue gas (say, downstream of the economizer prior to discharge to atmosphere through the stack), MD could be used to desalinate water without a penalty in terms of extra parasitic power consumption. There is ongoing research to establish the economic viability of applying MD to coal-fired power plants [9]. Another related concept is carbon nanotube-enabled MD, which can also be applied to cooling tower blowdown treatment [9].

By far, the majority of combined cycle power plants have wet cooling systems, which use water to condense the steam turbine exhaust (see Chapter 7). Open-loop systems utilize cold water from a natural reservoir (e.g., ocean, lake or river) and return the warm condenser discharge (roughly, warmer by 20°F) water back to the same reservoir. Closed-loop or recirculating systems with mechanical or natural draft cooling towers circulate the cooling water, which gives up its heat to air in counter- or cross-flow inside the tower. They began to replace the once-through cooling systems in the 1970s following the restrictions imposed by the Clean Water Act. A high-level comparison of currently available power plant heat rejection technologies, including the dry or air-cooled option, is provided in Table 7.2. (There are also hybrid systems (e.g., Heller system – see Chapter 7), which use a combination of wet and dry cooling methods, e.g., see Ref. [7] in Chapter 7.)

In 2008, about one-third of electricity generated in the US was by plants with once-through wet cooling systems and about half by plants with closed-loop wet cooling systems [10]. Only about 2% was by plants with dry or other cooling systems. However, concerns about water resource scarcity and other environmental regulations have been increasing and expected to grow in future, leading to greater deployment of dry cooling systems (mostly, air-cooled condensers, ACCs, or A-Frame condensers). ACCs essentially reduce cooling water consumption to zero. Unfortunately, this benefit comes at the cost of increased parasitic power consumption by the forced-draft fans pushing the air through the condenser and reduced steam turbine output (higher backpressure). For a typical GTCC, the difference can be as much as 0.3% of gross plant output or even more.

There are several new dry cooling technologies at various stages of development: thermosyphon cooling (TRL of 6), advanced M-cycle dew point cooling (TRL of 4) and adsorption chiller (TRL of 3) [8,9]. A high level assessment of these technologies can be found in Ref. [8,9].

10.9.4 Wastewater Treatment

Boiler or steam generator feed and makeup (raw) water treatment is critical to trouble-free operation of the power plant. Equally important is wastewater treatment prior to discharge to a receiving water body in order to comply with environmental regulations. GTCC power plants do not present as significant a challenge as coal-fired power plants since their wastewater mainly constitutes HRSG and cooling tower blowdown.

There are many different types of wastewater from steam cycle processes, including boiler blowdown, drains and leakages. There are several disposal options for treated wastewater: discharge to surface water (e.g., ocean, lake or river), discharge to sewer, deep-well injection or evaporation pond. The first two are not permitted by regulatory agencies in many places. There are concerns associated with the latter two, e.g., well plugging, insufficient evaporation or lack of space.

Wastewater from the boiler blowdown and certain drains (e.g., feedwater storage tank, boiler, sampling system) are collected in an oil-free basin and either discharged (if permitted) or sent to the *wastewater treatment plant* (WWT). Wastewater from miscellaneous cooler drains, operation leakages, and other "less clean" sources are collected in a drain pit and sent to a central WWT.

Wastewater from maintenance activities (e.g., boiler lay-up drains, boiler cleaning water) are neutralized and collected in a separate pit before being sent to the WWT.

RO is the most widely used technology to purify water. It is a membrane technology using an applied pressure to overcome the osmotic pressure. Thus, water is forced from a region of high concentration through a semipermeable membrane to a region of low concentration. Newer technologies include *forward osmosis* (FO) and *membrane bioreactor* (MBR), which combines biological treatment with conventional filtration. The former is also a membrane technology, which, unlike the RO using a hydraulic pressure as the driving force, uses a "draw" solution of high concentration of ammonium bicarbonate to induce a net flow of water through the semipermeable membrane into the draw solution. Separation of clean water from the diluted draw solution takes places in a heated recovery vessel via evaporation. FO can treat water up to four times as concentrated water as the conventional RO systems.

10.9.5 Zero Liquid Discharge

Power plant wastewater recovery for reuse and recycle is the driving force underlying the high recovery (92–99+%) and *zero liquid discharge* (ZLD) systems. Available technologies can be classified broadly in two groups:

- Thermal evaporation (brine concentrator, BC, and crystallizer)
- Membrane technology (RO)
  - Conventional RO (65%–85% recovery)
  - High-recovery RO (i.e., two-stage RO used in seawater desalination)
  - High Efficiency Reverse Osmosis or HERO™.

These two technologies can be deployed alone or combined resulting in five basic configurations:

- Thermal evaporation
  - BC only (with evaporation pond or deep-well injection)
  - BC and crystallizer (solidification for landfill)
- Combined
  - Softener and RO with BC
  - Softener and RO with BC and crystallizer
- Softener and RO only (with evaporation pond or deep-well injection).

The five configurations listed above are described in detail and compared with each other in terms of cost and performance in Ref. [11]. The most widely deployed system is the thermal evaporation, which is shown in Figure 10.8. The feed 1 is wastewater (e.g., GTCC cooling tower blowdown) heated close to its atmospheric boiling temperature by recovering heat from the distillate 2 (the heat exchanger is not shown). Deaerated feed (to remove corrosive and scale-forming constituents) is introduced into the BC sump where it is mixed with concentrated slurry (brine), which is continuously recirculated to the top of the BC evaporator column. The evaporator is a vertical tube heat exchanger. The brine falls down the evaporator while forming a thin film on the inside of the tubes, from which water evaporates into steam flowing down along with the brine. Upon exiting the tubes at top of the sump, steam flows through mist eliminators and enters the *mechanical vapor*

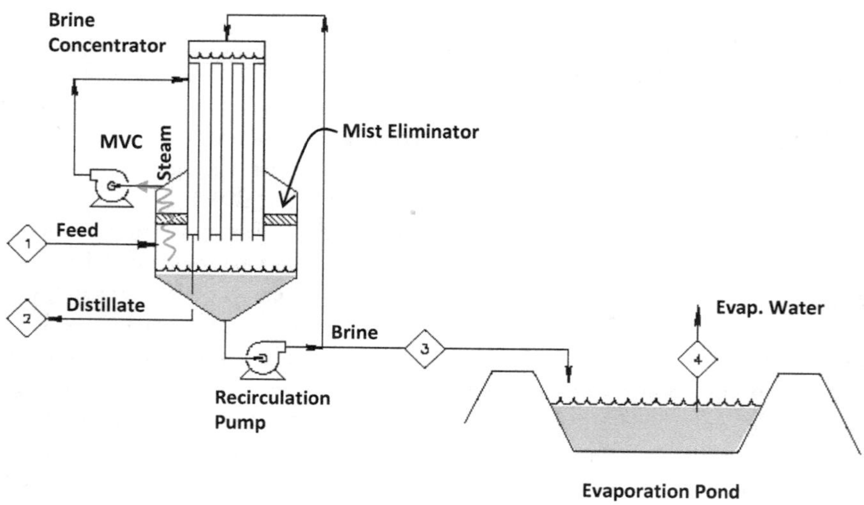

FIGURE 10.8 Thermal evaporation system with evaporation pond [11].

compressor (MVC). The MVC (usually a centrifugal fan) increases the pressure and temperature of steam. It consumes about 70–90 kW per 1,000 gallon of feed per hour. This provides the heat for evaporation of water from the brine, which is supplied by the latent heat of evaporation of compressed stream on the shell side of the evaporator. The condensed water flows down the tubes (outside) and is collected as distillate in the evaporator bottom. The concentrated brine 3 from the BC is sent to an evaporation pond or to a steam-driven *crystallizer* (steam consumption is equivalent to 150–200 kWh per 1,000 gallon of BC feed per hour). The distillate 2 from the BC is ready for reuse in the power plant. The water recovery rate in the BC is in the range of 90%–99%.

Selection of a particular ZLD system configuration requires a careful performance and cost trade-off study on a case-by-case basis. Performance herein is the rate of recovery, i.e., water recovered for re-use in the power plant as a percentage of the ZLD plant feedwater. Main drivers are quantity of feedwater and water quality (i.e., total dissolved solids (TDS), composition of cations and anions, total hardness (TH), silica content, etc.) There are little economies of scale, i.e., multiple trains are required for the larger sizes.

In general, for low TDS feedwater, the best option is a membrane-only system with an evaporation pond. For all others, a combined system comprising pretreatment with a membrane system followed by thermal evaporation is the best option. In such an arrangement with several components or subsystems in series, the overall system reliability is low (product of individual component reliabilities). When one component breaks down, the entire ZLD system breaks down (i.e., forced outage or decreased processing capacity.) Since the power plant cannot run when the ZLD system is down, this has a big impact on plant availability.

The simplest way to address the reliability problem is storage. Storage options are brine and concentrate holding tanks and/or storage ponds. The latter essentially disconnects the ZLD system from the power plant to enable continued power generation while the ZLD system is in forced or planned outage. Factoring in the capital cost of sufficient storage and including design margins to account for system degradation can change the cost picture significantly [12].

Consequently, taking into consideration the probability of a forced outage and capacity degradation when calculating annualized unit cost, one may very well end up with a simple BC (with evaporation pond) or softener plus RO system and evaporation pond (where feasible) as the most optimal solution. It should be pointed out that the latter may not be an option for sites with very high TDS feedwater. Another, guaranteed to be more expensive in terms of upfront capital expenditure, solution is multi-train design with parallel components, i.e., 2 × 50% or 3 × 50%. Depending on the resulting overall system reliability without resorting to large storage capacity, this may also be a feasible path. The ultimate solution can only be determined by a detailed analysis on a case-by-case basis.

General industry experience suggests that ZLD systems are notoriously difficult, costly and time-consuming processes to operate and maintain [12,13]. Investigation of the possibility of other options such as subsurface (deep well) injection and evaporation ponds is critical.

ZLD system selection should be made after a diligent analysis of raw water chemistry (average, best and worst cases), site ambient conditions and environmental regulations and/or permits pertaining to the final disposal of concentrate (liquid and/or solid). Important lessons learned are summarized below:

- System capacity/sizing with ample margin (i.e., the worst-case scenario under *fouled* conditions should be considered – especially BC and crystallizer heat exchangers)
- Redundancy of certain critical components should be carefully evaluated
- Material selection (opting for cheaper materials to save capex may not be a good idea – e.g., use alloy steel for crystallizer feed tank fabrication)
- Availability of ample spares (e.g., smaller pumps, pump motors, MVC impeller, bearings)
- Availability of storage tanks (to ensure continued operation when the system or a subsystem is down)
- Availability of substitute steam source (in case of BC MVC outage).

The impact of ZLD requirement on GTCC parasitic power consumption can be onerous as well. Cooling tower makeup water requirement of a typical GTCC is 220 gallons per MWh. For a 500 MWe power plant, this comes to about 1,850 gpm, which covers the cooling tower water loss via evaporation (ignoring the drift), E, and blowdown, B. For a *cycle of concentration* (CoC) of 5, E and B can be determined from

$$E + B = 1{,}850.$$

$$CoC = 5 = (E/B) + 1$$

as E = 1,480 gpm and B = 370 gpm. At 90 kWh per 1,000 gallons of feed, BC MVC would consume nearly 2 MWe of power. One should also add the lost STG output via steam provided to the crystallizer to that amount (if one is present). The impact on plant performance is significant. Assuming that the GTCC in question is an advanced H or J class unit rated at 60% net efficiency with auxiliary power consumption at 2% of gross, addition of the ZLD reduces the net efficiency to $498/500 \times 60 = 59.76\%$.

There are several new technologies for "volume reduction" (i.e., reducing the feed flow to the BC to reduce its size or completely eliminate it) for high recovery, which are in various stages of development. Respective approaches taken by these technologies and their development status can be found in Ref. [11]. The most common and commercially available approach is HERO™, which involves chemical pretreatment of the feedwater to remove its hardness and raise its pH before undergoing RO. Depending on the particular wastewater characteristics, HERO™ process has 95%–99% recovery effectiveness.

10.9.6 Water Balance

As stated in the beginning of the section, water is a precious commodity and its conservation is of prime importance. In accordance with this principle, water balance is an important design document. A typical combined cycle power plant water balance diagram is shown in Figure 10.9. As shown in the diagram, except for the potable water usage, the plant uses reclaimed water (i.e., treated wastewater). Condensate makeup for the steam cycle is first treated in the demineralized water system, which comprises the RO and IX (EDI) units. As shown in the diagram, gas turbine evap coolers use treated reclaimed water (typically, with a CoC of 3). However, if required by the OEM, treated water from the RO unit can be used as well.

10.9.7 Deaeration

As stated in the beginning of the section, noncondensable gases in the feedwater should be removed before entering the HRSG. Where do those gases come from? For one, air leakage into the condenser operating at subatmospheric pressures is unavoidable. Other gases can also dissolve in the water via other mechanisms such as decomposition of water into O_2 and H_2 by thermal action or by chemical reactions between water and the metal structures. The makeup water is essentially saturated with air, which is soluble in water. Oxygen, CO_2 and other gases dissolved in the water should be eliminated to prevent low pH levels, oxidation and carbonic acid corrosion. A dissolved O_2 level of 7 ppb (parts per billion) or lower is needed to accomplish this in most HP boiling systems such as the HRSG. (A normal condenser would have ~50 ppb DO in the condensate coming out of it.)

The amount of DO in water is inversely proportional to the temperature of water, i.e.,

$$DO[ppb] = \frac{25.07 \cdot 10^6 \, 0.491115 P_{air}[in.\ Hg]}{He(T_{cond})},$$

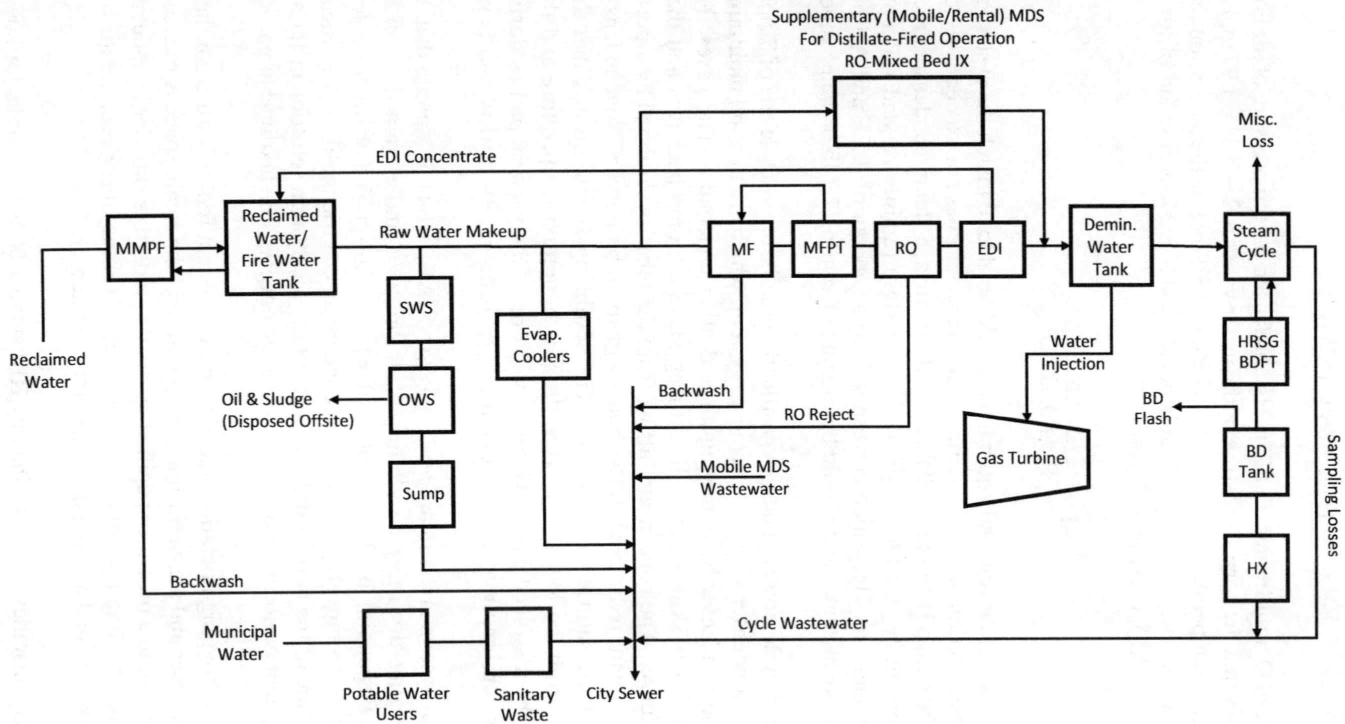

FIGURE 10.9 Water balance diagram (MMPF, multimedia pressure filters; MF, media filters; MFPT, media filter product tank; RO, reverse osmosis; EDI, electro-deionization; BD, blowdown; BDFT, blowdown flash tank; HX, heat exchanger; SWS, service water system; OWS, oil–water separator; MDS, membrane distillation system).

where P_{air} is the partial pressure of air in the gas mixture above the condensate and He is the Henry's constant as a function of condensate temperature. For condensate temperatures typical of combined cycle condensers,

$$He = 2,660 \cdot T_{cond}^{0.645}.$$

Thus, for 7 ppb dissolved O_2 requirement in a 2 in. Hg condenser, partial pressure of air is calculated as 0.03 in. Hg. It follows that the partial pressure of the water vapor is 2 – 0.03 = 1.97 in. Hg with the corresponding saturation temperature of 100.64°F, which is 0.5°F below the saturation temperature at 2 in. Hg. In other words, the water vapor is subcooled by 0.5°F. Mass fraction of air in the gas mixture, i.e., *air fraction* (AF), is estimated from

$$AF = \frac{P_{air}}{P_{air} + 0.6207 P_{vap}}$$

as 0.024, i.e., 2.4%. Thus, the air removal equipment should be able to match the system air ingress to the tune that the mass fraction of air inside the condenser space does not exceed 2.4% of the total gas mass. The Heat Exchange Institute (HEI) standard for surface steam condensers recommends that the venting equipment be sized to handle the gas mixture (mainly air and water vapor) at a temperature of 71.5°F and 1 in. Hg, which corresponds to a subcool of 7.5°F and AF of 0.314. In other words, the air removal equipment should be capable of venting 2.2 lb of water vapor for each 1 lb of air.

In modern combined cycle power plants, degasification to dissolved O_2 levels of 7 ppb or lower is accomplished in the condenser via the air removal system, which creates and maintains vacuum in the shell side of the condenser by removing air and other noncondensable gases. The system typically includes 2 × 100% skid-mounted liquid-ring vacuum pump packages and the connecting piping to the condenser. The condensate falling from the tubes is scrubbed by a portion of the steam turbine exhaust steam redirected to the bottom of the tube bundle. Noncondensable gases have zero solubility at the saturation conditions. The scrubbing steam ensures that the slightly subcooled condensate from the lower tubes is at the full saturation temperature and thus scrubs out and carries away the dissolved gases, which flow to one of the vacuum pumps during normal operation. The power consumption of the vacuum pumps is about 50–100 kW and is part of the plant auxiliary load.

Another air removal system is the *steam jet air ejector* (SJAE), which is a pump that uses steam expanding in a converging–diverging nozzle to pull in the gases and entrain them in the jet via momentum exchange. Depending on the type of condenser, i.e., air-cooled or water-cooled, and the condenser pressure, the air removal system can be based on an electric motor-driven vacuum pump or SJAE or a combination of the two. In order to understand the system selection criteria, consider that the air removal system performs two distinct services: *hogging* and holding. In particular,

- During startup, or hogging operation, the air removal system removes air from the steam space of the condenser and steam turbine. The pressure in the steam space is reduced from atmospheric pressure to a pressure specified by the steam turbine supplier in about 30 min.
- During normal or "holding operation" with the steam turbine in service, the air removal system removes a specified amount of air from the condenser.

Air removal equipment capacities are prescribed by HEI standards. For example, for hogging duty, the recommended capacity is 700 scfm (14.7 psia, 70°F) for condensing steam flow rates of 1 to 2 million lb/h (typical range for most GTCC). The basis for this is evacuation of the air in the condenser and the LP turbine from atmospheric pressure to 10 in. Hg in 30 min (assuming 26 m^3 per 1,000 lb/h of steam for the LP turbine).

A vacuum pump that can handle 700 scfm during hogging (15 scfm during holding) would be rated at 100 hp (79 hp during hogging) at 1 in. Hg suction and 55°F circ water temperature [14]. A two-stage, twin-element SJAE of similar capacity, with 150 psig/400°F motive steam, would consume 6,110 lb/h steam in the hogger [14].

Whether to go with a vacuum pump or SJAE or a hybrid system depends on considerations such as

- The availability of motive steam during startup
- The costs associated with the usage of steam or electricity as the driver
- Reliability, availability and maintainability.

For a thorough discussion of the advantages and disadvantages of different systems, the reader is referred to the paper by Nadig [14]. In combined cycle power plants (no cogeneration), two-stage, twin-element SJAEs are typical for ACCs; liquid-ring vacuum pumps are used with water-cooled surface condensers. Typical air removal system for the ACC comprises

- 2 × 100% two-stage holding ejector trains
- 1 × 200% inter/after condenser
- 1 × 100% hogging ejector
- 1 × 100% deaerator ejector

A deaerating condenser becomes unfeasible if the makeup water exceeds a certain limit, typically a few percent of the total steam flow. Nominally, in the absence of HRSG blowdown and assuming that the system is perfect with zero leaks, makeup water flow would be zero. Accounting for the gas turbine performance fuel gas heater return water and blowdowns (typically about 1%), makeup water requirement is less than 2%. (Typically, makeup water contains high levels of DO, often exceeding 8,000 ppb.) If for some reason the makeup flow exceeds 3% of the total steam flow, a separate deaerator is needed for proper degasification.

The deaerator is a direct-contact feedwater heater that heats the boiler feedwater and removes the dissolved O_2 and CO_2. Heating and scrubbing functions are accomplished by utilizing steam from a steam turbine extraction point or from the HRSG. There are two types of deaerators: spray and tray types. In the spray deaerator, a jet of steam mixes thoroughly with the condensate that is sprayed into the unit. In the tray type, the condensate coming in falls over a series of trays, where it disintegrates into small droplets and mixes with steam. The more common type used in power plants is the tray type.

A separate deaerator has an adverse effect on the combined cycle performance via two mechanisms:

1. Increased HRSG water inlet temperatures adversely impact the HRSG effectiveness due to high stack temperatures. Note that a deaerator operates at slightly higher than ambient pressure and delivers about 230°F feedwater to the HRSG.
2. Extraction of steam to accomplish the heating and scrubbing reduces the steam turbine power output. The steam flow should be high enough to facilitate the heating of the condensate in the deaerator by about 15°F–30°F to the saturation temperature at the deaerator operating pressure. (In some cases, the impact of the lost turbine work can be dampened by lower exhaust flow and exhaust loss.)

Typical natural gas-fired GTCC power plants that are not used in cogeneration applications do not have separate deaerators. In cogeneration applications, a separate deaerator is needed due to the large amount of condensate returning from the process plant that is saturated with oxygen. In combined cycle power plants with high sulfur fuel-fired gas turbines, sulfuric acid condensation and corrosion considerations and high makeup requirements as high as 7% of the steam flow due to water injection in the combustor for NOx control) necessitate the addition of separate deaerators.

REFERENCES

1. Elliott, T.C. (Editor), 1989, *Standard Handbook of Powerplant Engineering*, McGraw-Hill, Book Company Inc., New York.
2. Randtke, S.J., Horsley, M.B. (Editors), 2012, *Water Treatment Plant Design*, 5th Edition, McGraw-Hill, Book Company Inc., New York.
3. 2015, *Steam, its Generation and Use*, 42nd Edition, Babcock & Wilcox Company, ISBN 978-0-9634570-2-8.
4. Cartwright, P., 2006, Process water treatment: Challenges and solutions, *Chem. Eng.*, Vol. 113, No. 3, pp. 50–56.
5. McCoy, J.W., 1981, *The Chemical Treatment of Boiler Water*, Chemical Publishing Co., New York.
6. Tsubakizaki, S., 2015, Advantages and new technologies of high-AVT water treatment in combined cycle plants, *Mitsubishi Heavy Ind. Tech. Rev.*, Vol. 52, No. 2, pp. 105–110.
7. Buecker, B., 2011, Corrosion control in HRSGs, *Power Eng.*, Vol. 115, p. 44.
8. U.S. Government Accountability Office (GAO), Center for Science, Technology, and Engineering, August 2015, Water in the energy sector, Technology Assessment Report, GAO-15-545.
9. Bushart, S., Shi, J., 2014, Advanced cooling and water treatment technology concepts for power plants, *Power*, Vol. 158, pp. 62–67.
10. EPRI, 2008, Water use for electric power generation, EPRI Report 1014026, EPRI, Palo Alto, CA.
11. Mickley, M., 2008, *Survey of High-Recovery and Zero Liquid Discharge Technologies for Water Utilities*, WRF-02–006a, WateReuse Foundation, Alexandria, VA.
12. Sampson, D., 2012, No easy answers: ZLD improvement options for a 720-MW power generation facility, IWC-12-13, *73rd Annual International Water Conference*, November 4–8, 2012, San Antonio, TX.
13. Como, V.A., Cannon, T., Mueller, M., 2013, Operating a ZLD, What does it take? *74th Annual International Water Conference*, November 17–21, 2013, Orlando, FL.
14. Nadig, R., 2016, Evacuation systems for steam surface condensers: Vacuum pumps or steam jet air ejectors? POWER2016–59067, *ASME Power Conference*, June 26–30, 2016, Charlotte, NC.

11 Construction and Commissioning

In Chapters 9 and 10, major combined cycle power plant equipment and *balance of plant* (BOP) equipment are enumerated in some detail. Putting all those pieces together and creating a working power plant requires three major activities: *engineering, procurement and construction* (EPC). This is the subject of this chapter. Let us start with a brief look at the US electric power generation history.

From the emergence of the electric power generation industry in 1880s until the late 1970s, the vertically integrated utility, largely regulated by the states, was the dominant business model. Utilities took care of generating, transmitting and distributing electricity from the power plant to the customer in an environment of natural monopoly. The prevailing power plant construction mode was "design-build-bid" (at that time mostly large coal-fired steam turbine and nuclear power plants). In other words, in-house engineering departments of the utilities designed the power plant, developed the specifications and then solicited bids from the EPC contractors for the work. The lowest bidder was awarded the construction job and managed by the construction group of the utility. This was a costly process (with the "silver lining" of resulting in very well-designed, highly reliable and best-in-class power plants) but the profitability was ensured through the regulatory process (cost-of-service plus a profit).

In 1978, *Public Utility Regulatory Policies Act* (PURPA) created a structure for nonutility generating plants. A provision of PURPA encouraged the development of "cogeneration" plants that produced both electric power and industrial steam, which had better fuel utilization than utility steam-generating plants.

Under PURPA, if a generator was able to supply 5% of its steam output to a nonutility industrial steam host, it would be entitled as a "qualifying facility" to sell its electrical output at a favorable rate, known as the "avoided cost" rate, based on what the electric utility would have had to pay for the same amount of generating capacity in a utility-owned plant. This led to the creation of a new kind of power generator, i.e., the *independent power producer* (IPP).

Unlike the traditional utilities, the IPPs were focused on controlling construction costs and shortening the construction schedule in order to sell the electricity they generate at the maximum profit because their selling price was determined by power supply and demand. Concurrent with this development (and also in part driven by it), a shift from very expensive coal-fired and nuclear power plants with lengthy construction periods (4 years or even longer) to less expensive natural gas-fired gas turbine combined cycle (GTCC) power plants with shorter construction and commissioning schedules took place. On top of that, GTCC power plants were more efficient, cleaner and more flexible than the coal-fired behemoths.

This development rendered utility-style "design-build-bid" model infeasible for the IPPs. They resorted to *cost-plus* (with a *guaranteed maximum price*, GMP) or *Construction Manager as Contractor* (CMC) methods. Under the CMC approach, the functions of contractor and construction manager are merged and assigned to one entity that may or may not give a GMP but which typically assumes control over the construction work by direct contracts with the subcontractors.

At the time of writing, 40% of US power demand is supplied by merchant generators (IPPs) with the remainder supplied by regulated utilities. This is in contrast to the expectation in 1990s, i.e., private generators funded by private risk capital to be the future of electric generation. Three trends can be mentioned as contributors to this:

- Low prices for power and natural gas (at the time of writing, Henry Hub gas prices continue to hover at about \$3/MMBtu higher heating value or HHV) and peak prices for power suppressed by an abundance of wind and solar generating facilities.
- Weak demand in oversupplied markets due to improved energy efficiency, overcapacity (especially due to excessive building of power plants during the so-called "boom" around the turn of the century) and reduced demand caused by the 2008 crisis – all exacerbated by the emergence of renewable generation.
- IPPs facing greater inherent risks vis-à-vis the regulated, monopolistic utilities.

The last one is strongly tied to the fact that IPP revenues generally do not cover the cost of generation, which includes the capital charges, fuel expenditure and fixed/variable costs of operation. (This is measured by the *cost of electricity* (COE), which will be covered in detail in Chapter 13.) In a climate of low natural gas prices and improved GTCC efficiencies with advanced class (i.e., expensive) gas turbines, the COE is dominated by the capital charge, which is incurred and paid whether the power plant operates or not. As a result, a cyclical pattern of prices, profit and investment emerges as explained below:

- When generation capacity is scarce, electricity prices (and capacity payments in certain RTOs[1] such as PJM) must rise to a level that allows recovery of the full COE if new power plants are to be built.
- As capacity is added in response to this price signal, however, the scarcity is alleviated, and competition among generators then drives the price of power down to its variable cost of production, which is a fraction of the COE.

Consequently, developers of IPP/merchant power plants and their investors are very keen on minimizing construction costs and schedules (which reduces the AFUDC,[2] a significant expense item). This led to the emergence of the *lump sum turnkey* (LSTK) as the dominant mode of construction contract. Under the LSTK, an owner agrees to pay a specified contracted amount for completing work to the EPC contractor. Thus, the EPC contractor is responsible for completing the project under his or her financial risk. In addition, the scope of work includes startup of the facility and achievement of the normal operation status under the EPC contractor's responsibility (with performance guarantees) before the proverbial keys are turned over to the owner.

A typical GTCC project development is characterized as follows:

- Project development time is 2–3 years (can be as long as 10 years for a nuclear power plant)
- Project development includes the following phases and/or activities:
  - Siting
  - Permitting
  - Conceptual design
  - Major agreements (e.g., equity, debt, EPC contractor selection (competitive bidding), fuel and water supply, grid interconnection, power purchase agreement)
- Project development investment can be several millions of dollars (venture capital), which is typically recovered as a development fee during the financial closure
- Major project agreements are executed at once during the financial closure, after which the construction is released to start.

[1] Regional Transmission Organization is the designation of an electric power transmission system operator (TSO) that coordinates, controls, and monitors a multi-state electric grid in the USA.

[2] Allowance For Funds Used During Construction (also known as Construction Interest).

Of all the types of large, utility-scale central power generating station projects (e.g., coal-fired boiler-steam turbine power plant, nuclear power plant, gas turbine-based power plants including IGCC), GTCC is the only one that can be developed with private equity and debt. The two prime factors in this distinction are low capital cost (<\$1,000 per kilowatt presently) and short construction time (2 years for a modern GTCC). The latter is crucial in reducing the AFUDC, which can be more than 10% of the total project capital investment for a nuclear power plant with 5-year construction schedule. This is why the subject of GTCC capital investment ("capex") will be covered in detail in Chapter 13.

Presently, in most cases, the owner/developer partners with an OEM as the Power Island supplier. In some cases, the OEM can assume the role of turnkey EPC contractor including construction, commissioning, startup and testing. In other cases, an EPC contractor is selected through a competitive bidding process. In such cases, the RFP (Request for Proposal) is prepared by an "owner's engineer" (OE), which is typically a consulting engineering firm. Large EPC firms typically refrain from OE activities because it precludes them from bidding for the much more lucrative construction job. Yet in other cases, an EPC firm can be named the "preferred" EPC contractor without a competitive bidding process. There are also cases when the OEM, who originally assumed the turnkey plant responsibility as well, transfers that job to an EPC contractor.

A typical OEM/EPC contractor selection process is roughly as follows:

1. The owner solicits firm bids from the several major gas turbine OEMs for a full turnkey GTCC power plant.
2. In conjunction with that, each OEM that decides to bid issues a "non-binding"[3] *Request for Information* (RFI) for EPC services (i.e., engineering, procurement, construction, commissioning, start-up and testing).
3. Using the responses to the RFI, the OEM assesses the EPC contractors' capability by evaluating the responders' power plant construction history, indicative cost alignment, contractual risk alignment, schedule alignment, reference checks, safety record, history with the OEM and the owner and schedule performance.
4. Following its evaluation of the responses to the RFI, the OEM issues a definitive RFP to the selected EPC contractor. This RFP is intended to result in a firm and binding proposal.

The approximate timeline is as follows:

The OEM	The EPC Contractor	Date
Release of RFI		Day 0
	Response to RFI	Day 4
Receive the owner's RFP		Day 10
Release of RFP to the selected contractor		Day 14
	Response to RFP	Day 100
Turnkey bid submittal to the owner		Day 125
Owner's evaluation of turnkey bids		Day 126–215
OEM selection and award		Day 215
Negotiate EPC contract	Negotiate EPC contract	Day 215–230

[3] The owner has the right in its sole discretion to select bidder(s) based on the information supplied in response to the RFI request, or on any other information or method they select; to not continue to negotiate or award any contract to any of the parties submitting information; or to terminate the RFI at any time at no cost to the OEM.

In projects where the developer/owner has not preselected an OEM, an OEM and an EPC contractor can form a *consortium* to bid for the project. When the OEM and the EPC contractor form a consortium, two detailed documents are drafted and agreed upon by the parties including the owner:

1. *The Scope of Work* (SOW) for the project including all activities required to design, engineer, procure, construct, startup, test and place into commercial operation the project, including all ancillary facilities.
2. The *Division of Responsibility* (DOR) to define responsibilities for the SOW to be performed by the OEM, the EPC contractor and the owner of the project. There are typically five such responsibilities, i.e., design criteria, detail design, purchase specification, supply (procurement) and erection.

Design criteria includes the following:

- External influences in the design such as industry standards, federal and state codes, customer contract requirements and OEM engineering practices
- Fundamental basis for performance and capacities in general terms
- Abnormal conditions, if any, to be addressed
- Environmental conditions
- Specific performance requirements, both steady state and transient state, as required
- Overall control philosophy for the system
- Equipment and personnel protection functions such as alarms, trips and safety requirements.

Design criteria are set forth in documents such as:

- Site-specific design criteria
- Civil/structural design criteria
- Power block design basis and functional requirements
- Scope of work and services
- System definitions
- Permitting requirements
- Utility requirements
- Vendor-supplied information.

Detail design includes activities such as preparation of calculations, drawings and lists needed to describe, illustrate and/or define systems, structures or components of the facility. Typically, the following documents are required to fully define the design of the facility:

- Engineering calculations
- Construction drawings
- Engineering lists
- Equipment specifications and material requisitions
- Equipment access concepts
- Safety requirements
- General arrangement drawings
- Building design analysis
- Concrete and reinforcement drawings (equipment foundations)
- Site preparation and grading drawings and specifications
- Structural steel drawings
- Architectural drawings

- Building services drawings (HVAC,[4] plumbing, telephone riser diagrams[5]) to be performed by others
- System descriptions
- Instrument list
- Elementary schematics (BOP scope)
- Instrument calibration sheets (they include setpoints and ranges for BOP instruments and vendor sheets)
- System piping and instrumentation diagrams (P&IDs)
- Piping drawings
- Piping stress analysis
- Yard piping
- HVAC performance specification
- Plant wiring diagrams
- Low, medium and high voltage one-line diagrams
- Raceway[6] arrangement drawings
- Grounding drawings
- Raceway and circuit lists
- Raceway and layout drawings
- Lighting and communication drawings.
- Electrical design studies
- Lightning protection drawings and performance specifications
- Cathodic protection[7] performance specification
- Utility requirements.
- Data acquisition system (DAQ, part of distributed control system, DCS, – supplied by the OEM)
- Man–machine interface (MMI, part of DCS and control room – supplied by the OEM)
- Instrumentation and controls (I&C) design
- Instrument location drawings and installation details
- DCS input/output (I/O) list
- Load list.

Vendor drawings and technical documents are reviewed and approved by the responsible engineering supervisors. Critical vendor information is incorporated into project drawings. Detailed engineering schedule is ensured to be compatible with overall project schedule. Resolution of any non-conformance and field problems are recorded and archived.

11.1 PROCUREMENT

Procurement means acquisition (from the vendor) and supply (to the construction team) of a particular equipment or system. Procurement team responsibilities are well beyond of simply soliciting bids and procurement (where applicable); they include

---

[4] Heating, Ventilating and Air-Conditioning.

[5] A diagram (two-dimensional, in a vertical plane) which shows the major items of pertinent equipment in a building; displays, floor by floor, the feeders and major items of equipment.

[6] A raceway is an enclosed conduit that forms a physical pathway for electrical wiring. Raceways protect wires and cables from heat, humidity, corrosion, water intrusion and general physical threats.

[7] Cathodic protection is used to protect pipelines, water treatment plants, above and underwater storage tanks, and reinforcement bars in concrete structures. Cathodic protection works by connecting the base metal at risk (e.g., steel) to a sacrificial metal that corrodes in lieu of the base metal. Thus, the base metal is preserved by providing a highly active metal that can act as an anode and provide free electrons. By introducing these free electrons, the active metal sacrifices its ions and keeps the less active base metal from corroding.

- Scheduling and ordering the shop fabrication of specialty equipment (where applicable) including preparing shop drawings and other fabrication engineering work
- Providing engineering or other documents (such as studies, manuals or permits)
- Quality control, observing shop tests, as necessary
- Expediting of goods and services to meet the contractual schedules.

The purchase specification document provides sufficient technical detailed information to permit procurement of equipment and materials of appropriate quality to meet the plant design requirements, including the following, as is applicable for the given scope:

- Scope
- Codes and standards
- Required documentation
- Service requirements
- Technical requirements, including control system interface, required operating logic and I/O and instrument requirements
- Design requirements
- Materials
- Quality assurance requirements
- Fabrication requirements
- Shipping, handling and storage requirements
- Inspection and testing
- Spare parts
- Documentation/marking instructions
- Terms and conditions.

Procurement and supply of engineered items including items such as concrete, valves, steel and equipment are ensured to be in accordance with the OEM's or the EPC contractor's specifications. Not doing so can create problems during the warranty period and result in significant monetary loss.

Procurement also includes erection or installation of any temporary works during site construction. This includes any work, services, equipment, material and structure, which are provided to support the implementation of the project on- or off-site during the project implementation and do not become permanent part of the facility. (This is usually the responsibility of the EPC contractor.)

11.2 CONSTRUCTION

Construction of a GTCC power plant comprises three major activities in two engineering disciplines:

- Civil (site work)
- Civil (foundations)
- Mechanical (erection of equipment)

Site work includes surveying, site preparation, excavation, backfill and grading as required to construct the facility and achieve finished site grades.

A topographical site survey is done to establish a permanent baseline grid system for both horizontal and vertical control of construction activities. The EPC contractor maintains the primary control points and provides all detail measurement and layout for the project with qualified survey personnel and certified equipment.

Site preparation involves clearing all trees, shrubs and vegetation to the extent necessary to construct the facility. It is vital to equip the facility with drainage to ensure that no low-lying areas are left that could accumulate water other than those defined on the drawings. All drainage should be directed away from the buildings to a collection system.

Excavation work consists of the removal of earth, sand, gravel, vegetation, organic matter, rock, boulders and debris to the lines and grades necessary for the construction of foundations and underground facilities included in the project. Any blasting (including permits) or other special requirements to achieve this task are the sole responsibility of the site preparation contractor (usually a subcontractor of the EPC contractor). Materials suitable for backfill are stockpiled at designated locations using proper erosion protection, moisture control and safety methods.

Backfilling of excavated areas is done in uniform layers of specified thickness. Soil in each layer is properly moistened or dried to obtain its specified density. All off-site soil materials to be used as fill or backfill are evaluated in accordance with the applicable environmental requirements. To verify compaction, representative field density and moisture-content tests are taken during placement and compaction of each layer.

Grading is the work of ensuring a level base, or one with a specified slope, for equipment foundations, facility buildings and roads. Graded areas are smooth, compacted, free from irregular surface changes and sloped to drain. Final earth grade adjacent to equipment and buildings are at least 6 in. below the top of finished concrete foundations. Furthermore, they are sloped away from foundations as necessary to maintain positive drainage. Minimum slopes are based on surface type. In general, grades are sloped at a minimum pitch of one-half percent (0.5%) to provide drainage to collection points (1% is even better). Maximum slopes are set based on slope stability, maintainability and the site conditions.

Support structures and foundations connect the equipment to the ground and transfers loads from the equipment to the ground. Recommendations for the foundation design can be found in the *Geotechnical Report*, which also includes pile types and design data (if required), excavation, fill, compaction requirements, pavement design and ground water data. The following loads are considered in foundation and supporting structure design:

- Dead load (weights of the structure and all equipment/components of a permanent or semi-permanent nature including, but not limited to, fixed equipment, framing, piping, floors, walls, roofs, partitions, stairs, ductwork, cable tray and any other structures, bins, etc. and the contents of tanks measured at full capacity)
- Live loads (load superimposed by building use and occupancy)
- Dynamic or impact loads (e.g., vibration of rotating equipment)
- Vehicle loads
- Piping loads
- Conduit and cable tray loads
- Earth pressure loads
- Snow loads
- Wind loads
- Seismic loads.

Components of support structure and foundation construction are

- Structural steel
- Grating
- Stairs and ladders
- Structural concrete
- Reinforcing steel.

Requisite materials are ensured to conform to applicable codes and standards (e.g., ACI 301 for structural concrete, ASTM A 615, Grade 60 for reinforcing steel)

In most cases, unless harsh climatic conditions exist, there are two buildings in the GTCC power plant:

- Administrative building (includes the control room, the maintenance shop and the warehouse)
- Water treatment building (it may also enclose the fire water pumps).

Gas turbine has its own enclosure. Steam turbine and HRSG are located outdoors.

Gas and steam turbine support structures and foundations are designed in accordance with the OEM's recommendations and the geotechnical report. In general, the gas turbine structure is a reinforced concrete mat foundation with concrete piers/pedestals provided to match the equipment configuration anchorage requirements. Minimum foundation elevation for the steam turbine allows adequate space to install and maintain the condenser and piping under the pedestal on a mat foundation with a concrete column/beam superstructure provided to match the equipment layout/geometry and anchorage requirements.

The HRSG structures and related equipment are supported by steel frame structures constructed on reinforced concrete foundations. The foundation is a common mat-type foundation sized to accommodate both the HRSG and its stack.

The cooling tower concrete basin (if one is present) is designed and constructed in accordance with OEM's recommendations. All embedments, hardware and other metal components are stainless or galvanized steel including a trash rack with removable screens which is to be provided at the inlet to the circulating water pump basin. Special requirements for water retaining structures are taken into consideration in the design of the concrete. The design of all integral spillways and pump-wells are ensured to be in accordance with the recommendations of the Hydraulic Institute.

Transformer foundations are designed and constructed in accordance with manufacturer's recommendations. Spill containment in the form of reinforced concrete retention pits with a low-point sump to contain 130% of the transformer oil capacity are provided for the prime mover main and unit auxiliary transformers (plus any other oil-filled transformers). Concrete or CMU[8] firewalls are provided between oil-filled transformers and adjacent structures and equipment as required. Transformer containment areas are provided with a lockable indicating type valve (normally closed) with an extension hand-wheel operator draining from the low point of the containment.[9] In some cases, pumping can be required.

A self-supporting steel stack with testing ports, required access and platforms is provided adjacent to the HRSG (see Section 8.3). The stack is supported on a common mat-type foundation sized to accommodate both the HRSG and the stack.

Erection includes the receipt and storage of equipment and material at project site or other designated location (including all support services and materials/equipment required for storage and preservation), installation/construction, cleaning (including cleaning of systems and components during construction and prior to turnover), construction testing and preparing the equipment, system, facility or structure indicated for service.

Erection also includes provision of construction consumables, field engineering and field fabrication, construction equipment and tools and any other material such as grouting and shims that is only necessary for erection purposes. It includes storage and maintenance of equipment/components before and after installation until takeover by the plant owner.

---

[8] Concrete Masonry Unit.

[9] The intention is for plant operators to be able to check accumulated rainwater for oil contamination and if the water is clean, open the valve to drain the water to the closest stormwater retention pond or to the cooling tower basin.

TABLE 11.1
Typical GTCC Project Schedule

	Days	Months
Limited notice to proceed (NTP)	0	0
OEM design inputs received	30	1
Full NTP	61	2
HRSG #1 steel onsite	457	15
HRSG #1 modules onsite	488	16
HRSG #2 steel onsite	519	17
HRSG #2 modules onsite	533	18
Gas turbine #1 onsite	561	19
Gas Turbine #2 Onsite	547	18
Steam turbine generator standards onsite	592	20
Target commercial operation date (COD)	912	30
Guaranteed COD	973	32

Other erection responsibilities are checking the tightness of factory mechanical and electrical connections and re-tightening if necessary. Erection process is ensured to be in accordance with detailed design and construction technical specifications, vendor drawings and manuals.

A typical construction schedule for a 2 × 2 × 1 multi-shaft GTCC is shown in Table 11.1.

11.3 STARTUP AND COMMISSIONING

The next step after completing the construction of the GTCC power plant is to "kick the proverbial tires" and commission it. *Commissioning* is the term for pre-operational testing and setting to work of all systems and equipment such that acceptance tests upon completion can be performed as set out in the contract. For gas turbine commissioning activities, refer to Section 14.5 in of **GTFEPG** (Ref. [11] in Chapter 2). Obviously, in a GTCC power plant, startup and commissioning activities should be planned and coordinated by considering *all* power plant systems as a whole. Such planning and coordination require answers to such questions as

- Which units are started first?
- Which systems are needed early?
- What parts of the power plant are under construction while others are being started?

When a GTCC power plant is designed, plant systems and equipment are sized for full load operation at a design point. For a supplementary-fired system, e.g., this could be the fully-fired operation at specified ambient conditions. This will determine the swallowing capacity of the steam turbine at maximum steam flow rate and throttle pressure, condenser and cooling water to maintain a cost-effective (optimal) back pressure and the HRSG heat transfer surface to generate the requisite steam at prescribed conditions (pressure and temperature). Several other heat and mass balance cases are considered to make requisite design tweaks for stable operation across the intended load range, e.g., from minimum emissions-compliant load (MECL) up to full load. (Refer to Section 10.3.3 for pump selection and design.) Control philosophy and requisite algorithms are also determined with the aforementioned minimum and maximum (continuous) load cases.

The reality in the field is that, before the GTCC plant can get to full load, it must start from standstill and pass through *all* operating points up to the full load. During normal operation, the power plant passes through the operating points *prior to* the lowest design load (usually MECL) rapidly – e.g., it does not linger at them. During initial commissioning, however, it is often necessary to hold at several points along the way for tuning and testing. These points can be below the normal continuous minimum load (e.g., gas turbines at full-speed, no-load or FSNL). Furthermore, it may even be necessary to sit at these points for several days or even for a week at a time. Consequently, it is imperative that the EPC contractor must verify that equipment sizing, design features and control logic are commensurate with those requirements.

Cost-effective design (from the perspectives of initial cost as well as operating and maintenance costs) dictates that utility systems such as potable water, plant air, service water, fire protection, waste water and service gases are provided by a common set of equipment with a service loop distributing the respective fluids to the individual plant users. However, things are quite different during startup and commissioning. Consequently, care must be taken that all such systems are capable of operating to supply one unit at a time and adequate flow-paths are provided for one unit operation while the other is still under construction. Similar considerations apply to pumps and heat exchangers in arrangements such as 2 × 100%, 2 × 50% and 3 × 50%. Such systems should be capable of operating with only one train in service for an extended period of time. For example, during startup and commissioning operation, the first hold point is often at 10% load or less. Plant equipment and systems must be capable of operating at these low loads for extended periods of time. Pumps and compressors should have adequate minimum flow capability to provide such low flows and operate continuously in this region (see Section 10.3.3). As an example, consider a 3 × 50% pump; during the startup, however, the flow can be smaller than the turndown capability of the single 50% unit. How to provide this low flow without adversely affecting the equipment is a problem to be solved (e.g., recirculation at minimum flow point). Thus, the EPC contractor takes care to size the minimum flow lines and the orifices for trouble-free operation. Instrumentation ranges and control valve sizes are also specified to be adequate for functionality at such low loads. (Separate startup control valve, startup flow meter or startup pressure gauges may be needed).

Typically, cooling or service water systems are balanced in the field by use of manual throttling valves provided at individual pieces of equipment. (Detailed pipe network calculations are not provided for those systems.) As designed, such systems are expected to work trouble-free at full load operation. However, they may not be capable to balance flows during startup when only a few users are in service. It may be necessary to have additional flow-paths or balancing valves in the system to allow operation at low load with only a few users in service. Similar problems can be experienced in HVAC systems that serve multiple areas. If all rooms or areas are put into service at the same time, additional features may be required to balance air flows during startup.

11.3.1 HRSG Steam Blow

One of the most important startup activities is steam blow to remove surface/mill scale, weld slog and any other foreign particle remaining in the steam pipes and the HRSG heat transfer sections. If not removed, such solid matter is blown into the steam turbine during plant operation resulting in catastrophic failure of the rotating parts.

In order to be able to perform the steam blow, steam must be produced in the HRSG, i.e., the gas turbine must be running at a low load (the common industry practice) or at FSNL (feasible in some cases) to provide the requisite exhaust gas flow. The alternative to steam blow is air blow, which uses compressed air to remove the debris from the pipes. In this method, temporary air compressors are installed to provide the volume required for air blowing. Even though air blows provide the same flushing action, they do not perform a cleaning action as with steam blows due to the lack of thermal cycling. Thermal cycling is requisite for mill scale (flaky surface of hot rolled steel comprising various iron oxides) removal. Therefore, air blows must be performed in conjunction with mechanical or chemical cleaning.[10] Still, air blows can result in significant fuel and water savings as well as

[10] See GEK 110483b "Cleanliness Requirements for Power Plant Installation, Commissioning, and Maintenance," (March 2004).

schedule savings (e.g., in cases when there are delays in bringing the prime gas turbine fully online, air blow cleaning can proceed without delays).[11]

There are two basic methods to perform steam blow: cyclic and continuous. In the cyclic method, steam pressure is built up behind the temporary steam blow valve (BV). (Steam blow requires installation of temporary valves, piping and silencers.) At a predetermined boiler pressure, the valve is opened to release a burst of steam. Cooldown periods between steam blows provide for thermal cycling. Ideally, this procedure should be done prior to the installation of thermal insulation. Otherwise, longer time is needed between blows (in order to allow the system to cool), which adds to the total construction time. In the continuous method, steam is generated and exhausted for extended periods of time. Continuous steam blowing is enhanced by cyclic operation of temporary and permanent water sprays into the steam flow-path to provide controlled thermal cycling. Sections of steam lines that are not part of the steam line blow path are cleaned by "hydrolazing". In this procedure, a highly pressurized water jet is employed to scour the inside surface of the pipe.

System cleanliness following the steam blow is inferred from the number and type of impingements on a sacrificial piece of polished bar stock inserted into the steam blow path. This piece of metal (typically, carbon steel with a Brinell Hardness of 140–160 polished to a finish of 0.2 μm (rms)) is referred to as "target coupon" (typically, 1-in. bar stock with a length spanning the entire inside diameter of the pipe[12]). Target coupons are located as close to the end of permanent piping as possible in a straight run of horizontal piping at least ten diameters from any elbow, tee, etc.

In essence, debris blown free during steam line blows impacts on the target and leaves a representative sample of dents, scars, etc. After several cycles of steam blow, the expectation is to find a practically unscarred target (i.e., no solid debris left in the system). The acceptance criteria include

- Maximum impact diameter less than 0.5 mm
- Maximum number of impacts per target area 10
- No discoloration (other than heat)
- No embedded material
- Two consecutive passing blows.

A proper steam blow requires careful analysis. This analysis determines the requisite steam conditions and corresponding gas turbine load. An important parameter in this analysis is the *cleaning force ratio* (CFR), which is defined as

$$\mathrm{CFR} = \left(\frac{\dot{m}}{\dot{m}_0}\right)^2 \frac{v}{v_0}, \tag{11.1}$$

where the subscript 0 designates the design (maximum continuous rating or MCR) value of steam flow rate, $\dot{m}$, and specific volume, v. CFR is a measure of the aggressiveness of the cleaning action at the pipe's inner wall during steam blow relative to that experienced at full-load continuous operation. The normal target value of CFR for steam piping is 1.2 or larger. Minimum acceptable value of CFR is 1.0 at every point of the piping being cleaned. Note that different OEMs might have slightly different CFR calculations and/or criteria (e.g., see GEK 110483b cited earlier).

Requisite gas turbine load to achieve an acceptable value of CFR can be found from heat and mass balance calculations. For a 1 × 1 × 1 GTCC with an F class gas turbine (e.g., 1,500 lb/s exhaust flow, 1,125°F exhaust temperature at ISO baseload), a typical steam cycle has 1,815 psia and 1,050°F

[11] See J.M. Jarvis, Babel, P.J., Vieira, A.T., 2017. Advantages of air blow cleaning of steam pipes for cost and schedule savings, PVP2017-65973, *ASME 2017 Pressure Vessels & Piping Conference*, July 16–20, 2017, Waikoloa, HI.

[12] The target coupon is held in the temporary blowout pipe by a target holder or inserter in a vertical position. Ideally, a minimum target area of 15 in.^{2} (100 cm^2) should be achieved.

steam at the main steam valve inlet (specific volume is 0.459 ft³/lb). At baseload, HP steam flow can be estimated as 329 lb/s (see Section 6.2.3). Ensuring that CFR is greater than unity requires that at the steam blow conditions, specific volume is much larger than 0.459 ft³/lb. Thus, one possible combination is 65% of rated steam flow (i.e., 214 lb/s) at 650 psia and 1,050°F (CFR = 1.24). From Figures 15.3–15.6 in the chapter on operability, one needs roughly 50% of gas turbine exhaust energy for these conditions, which corresponds to 20% gas turbine load. This, of course, requires that the grid is available to accept the generated megawatts.

In order to grasp the physics underlying CFR, consider that the main principle behind steam blow cleaning is to create drag forces to break/sheer the mill scale and weld slag away from the pipe wall. These forces, however, are present during normal operation as well. Therefore, drag forces created during the steam blow must be in excess of those created during normal operation. This ensures that the remaining debris (i.e., not removed during the steam blow) will stay put in the steam system during normal operation. The drag forces required for sheer/break the surface scale/weld slag is a product of the dynamic pressure and a drag coefficient, i.e.,

$$F_D = K_D P_{dyn}, \quad (11.2)$$

where K_D is the drag coefficient and P_{dyn} (in psi) is the dynamic pressure given by

$$P_{dyn} = \frac{1}{2}\frac{\dot{m}^2 v}{144 g_c A}. \quad (11.3)$$

In Equation 11.3, g_c is the gravitational constant (32.174 lbm-ft/lbf-s²) and A is the flow area in ft².

A typical steam blow configuration for a 1 × 1 × 1 GTCC is shown in Figure 11.1. Typical steam blow paths are as follows:

- High-pressure (HP) blow path: HP superheater (HPSH) → Main steam header (MSH) → HP temporary pipe (TP) → BV → silencer to the atmosphere
- Cold reheat (CRH) blow path: HPSH → MSH → HP TP → BV → CRH TP → reheat superheater (RHSH) → temporary blowout piping → silencer to atmosphere

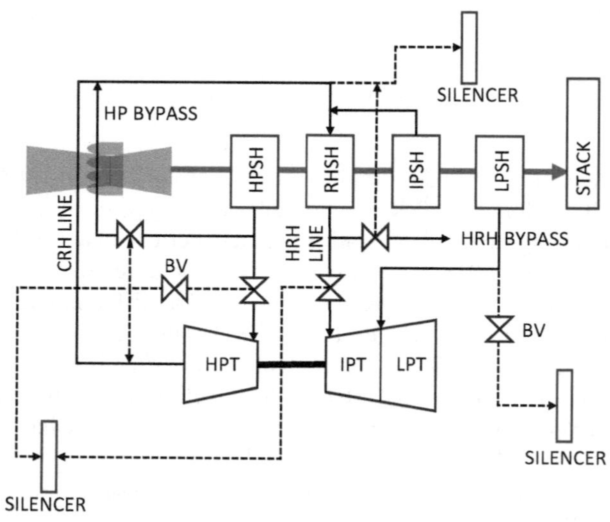

FIGURE 11.1 Steam blow configuration for 1 × 1 × 1 GTCC (dashed lines represent temporary piping).

- Hot reheat (HR) blow path: Same as the CRH blow path but through RHSH → HR line → RH stop valve before the IP turbine inlet → temporary blowout piping → silencer to atmosphere
- Low-pressure (LP) blow path: LP superheater (LPSH) → LP header → LP TP → BV → silencer to the atmosphere.

Common headers in an N × N × 1 GTCC power plants with multiple HRSGs may need mechanical and chemical cleaning in addition to steam blows (due to pressure limits). Further complications stem from cascaded bypass lines in fast start plants or steam export lines in cogeneration power plants. For detailed information, refer to the paper by Jarvis et al.[13]

11.4 ACCEPTANCE TESTS

The final step after erection, startup, commissioning and before turning over the power plant keys to the owner/operator is to run a series of "tests" in order to prove to the owner that the power plant is "acceptable" as stipulated in detail in the EPC contract. The ultimate "proof of the pudding" activity in this endeavor is *performance testing*. Performance testing is preceded by *construction testing* and *functional testing*. Construction testing includes those tests required by the drawings, codes and specifications as well as the requirements and recommendations provided in the applicable vendor documentation or OEM requirements to make the equipment ready to be placed into service. Functional testing includes initial system and system operational tests to verify that the subsystems and systems operate as intended.

Construction tests include

- Megger checks (of insulation resistance)
- Instrument calibration, continuity and I/O checks
- NDE[14] testing per specifications, codes and standards and OEM/vendor requirements
- Pipe hanger surveys and hot/cold settings
- Exercise relays and circuit breakers
- Hydrostatic testing and disposal of effluent
- Pneumatic testing
- Motor bumping and rotation check
- Ventilation/HVAC balancing
- Hot and cold alignment checks (gas turbine, steam turbine, electric motors, pipes)
- Transformer oil filling, conditioning and testing
- Cathodic protection testing
- Circuit breaker calibration and testing
- Hipot testing (to verify electrical insulation), including isolated phase bus and medium voltage systems
- Soil compaction
- Safety and relief valve bench checks
- Concrete field testing
- Grounding testing
- Valve stroking
- Crane load testing
- Chemical cleaning
- Steam blows.

13 Jarvis, J.M., Babel, P.J., Vieira, A.T., 2004. Advances in power plant steam blow cleaning analyses, GT2004-53161, *ASME Turbio Exo 2004*, June 14–17, 2004, Vienna, Austria.

14 Nondestructive examination.

Functional tests include

- Fire suppression system testing
- Testing circuits and interlocks
- Electrical and mechanical testing (e.g., uncoupled and coupled motor run-ins[15])
- Energize and test motor-operated valves, dampers, gates, etc.
- Energize 120 V ac power supply
- Energize 125 and 24 V dc power supplies and battery backup systems
- Energize 6,900 V switchgear
- Energize 480 V switchgear
- Energize auxiliary transformers
- Loop checks[16]
- Synchronization
- Turbine roll
- Verify analog and digital controls.

Once these tests are done and passed, the project achieves "mechanical completion" and the system is then turned over from the construction team to the commissioning team. The commissioning team is responsible for "substantial" and "final" completion of the project. A key part in either endeavor is demonstration of the achievement of performance guarantees.

Plant performance guarantees include net output and heat rate, emissions, noise, availability and reliability. Net output and heat rate guarantees are specified in the contract at base reference conditions. There may be one or more guarantee cases (e.g., with the HRSG duct burners on and off). Conducting the performance tests and correcting the measured results to the base reference conditions are specified in the contract (usually governed by an applicable test code such as ASME PTC 46 and/or OEM's performance test protocols). Key items covered by the contract are:

- Definition of "net" output and heat rate (see Section 3.3.4)
- Test tolerance, which is usually based on post-test uncertainty calculations per ASME PTC 19.1
- "New and clean" conditions, i.e., test must be conducted before accumulating X hours of operation
- Fuel specification (e.g., fuel gas composition)
- Auxiliary load scope (e.g., inclusion or exclusion fuel gas booster compressor).

These requirements are contractually stipulated in minute detail in the contract between the owner and the EPC contractor (or the consortium formed by the EPC contractor and the power island OEM). Thus, there are wide variations on a case-by-case basis. In some cases, the contract may stipulate zero test tolerance. Alternatively, calculated uncertainties can be capped (e.g., ±1.25% for heat rate and ±0.75% for output). Calculation of "equivalent degradation hours" below which the equipment can be assumed to be in "new and clean" condition is determined using rigorous formulae supplied by the gas turbine OEM. If the performance test is conducted after that limit, OEM-supplied degradation curves are applied for correction to base reference conditions (e.g., see Figure 3.6).

[15] Run-in is to "break in" the new motor via an initial period of running, usually under light load, but sometimes under heavy load or normal load.2

[16] Loop check is the operational check of a control loop to find out the deficiencies in control loop, field interface and MMI. It can be done, e.g., between two instruments at different locations simply by checking the continuity of the signals. The same can be done between an instrument and the junction box.

The most intractable problem in performance tests is the discrepancy between site ambient conditions at the time of the test and the guarantee performance ambient conditions. Typically, at least in the USA, plant owners/operators demand performance guarantees at hot ambient conditions (usually, with supplementary-fired HRSG) because such conditions coincide with the highest wholesale power prices (on certain very hot summer days, they can be as high as thousands of dollars per megawatt-hour). This is typically outside the control of the EPC contractor and/or equipment vendors. If the substantial or final completion is in, say, November or December and the guarantee performance is for ambient conditions typical of a hot day in July or August, the only available option is to run the test at existing "cold" ambient conditions and use contractually agreed-upon correction curves to adjust the measure performance to "hot" conditions. The plant equipment most susceptible to variation in site ambient conditions are the gas turbines and the bottoming cycle heat sink, i.e., condenser and/or cooling tower. In some cases, corrected test performance can indicate satisfaction of contract guarantees, but the actual plant performance at actual guarantee conditions can show a shortfall. Normal equipment degradation can usually explain the shortfall. In cases when degradation is not enough to explain the performance shortfall, significant effort on the part of the operators, owner's engineers, EPC contractor and equipment vendors may be required to analyze data and determine the root cause(s) and fix them.

Emission guarantees also include reference conditions and limits of criteria pollutants not to be exceeded (see Chapter 12). Again, there may be more than one reference cases based on gas turbine load and HRSG duct burner status (if one is present). Typical guarantees are

- 2 ppmvd @15% O_2 NOx (this requires SCR in the HRSG)
- 15/10 ppmvd @15% O_2 CO (fired/unfired)
- 10 ppmvd @15% O_2 ammonia (from the SCR slip)
- 4/1 ppmvd @15% O_2 VOC (fired/unfired)
- 15/10 lb/h particulate matter (fired/unfired).

Noise emissions can also change from project to project and is also a function of the power plant site (whether it is in the "boondocks" or within the city limits can make a huge difference). A typical guarantee limit is A-weighted sound pressure level of 75 dBA along the property boundary (see Section 12.3).

Reliability and availability guarantees are determined using specified "period hours" (PH) and forced outage definitions spelled out in the contract in detail. The applicable formulae can be found in Section 16.2. A typical reliability guarantee reads as follows:

> The Contractor guarantees that … the plant shall achieve a reliability factor of at least 95% over a rolling period of 72 hours prior to provisional acceptance (usually after substantial completion).

Period hours should occur with the plant operating at baseload (see Section 3.3) or dispatched load. Forced outage hours are the number of hours during which the plant is not capable of generating megawatts when dispatched by the grid operator.

The availability guarantee is based on the *Equivalent Availability Factor* (EAF), which is contractually defined. A typical definition is

$$\mathrm{EAF} = \frac{\mathrm{kWh}_{\mathrm{Gen}}}{\mathrm{kWh}_{\mathrm{Req}}},$$

where

- $\mathrm{kWh}_{\mathrm{Gen}}$ = Total net kilowatt-hours generated by the plant over the period hours in response to dispatch
- $\mathrm{kWh}_{\mathrm{Req}}$ = Total kilowatt-hours requested by the dispatcher over the period hours.

A typical availability guarantee reads as follows:

> The Contractor guarantees that … the plant shall achieve an equivalent availability factor of at least 97% over a rolling period of 200 hours while operating at dispatched load prior to final acceptance.

Both reliability and availability guarantees are accompanied by several pages' worth of exclusions, caveats, etc.

Performance tests are key to "substantial" and "final" completion of a combined cycle power plant construction project. Successful substantial completion leads to "provisional acceptance". Successful final completion leads to "final acceptance". Provisional acceptance is of prime importance because it designates the time when *schedule liquidated damages* (LDs) stop and the owner releases all (or most) payments withheld to the contractor during construction as part of the retainage on each application for payment. If the provisional acceptance occurs later than the guaranteed date, schedule LDs accumulate at an accelerated pace, e.g.,

- Days 1 through 5 at $125 K/day
- Days 6 through 15 at $175 K/day
- Days 16 and beyond at $200 K/day.

Even though there is usually a maximum cap on schedule LDs, it is not difficult to see that millions of dollars can be at risk. (There are also bonuses for early completion as specified in the contract; usually daily bonus amounts are significantly less than the daily LD amounts.)

After substantial completion, the power plant is sufficiently complete so that the owner can start revenue-making operations (in spite of some odds and ends that might need to be completed or corrected). The list of items that need to be completed and/or corrected by the EPC contractor is known as the "Punch List" (typically, within 1 year). Equipment warranties start at substantial completion, at which the contractor is relieved of responsibilities such as paying for utilities, security, insurance, maintenance and damage to the property (they are passed on to the owner). Once the EPC contractor completes the punch list (plus other items that might have come up after substantial completion) and passes the inspection by the owner, any remaining monies withheld by the owner are released.

In general, substantial completion is achieved when plant net output and heat rate are within 5% of their guarantee values, i.e., 95% for net output and 105% for net heat rate. Final completion requires 100% achievement of net output and heat rate guarantees as well as reliability and availability guarantees. Furthermore, stack emissions (using EPA methods – see Chapter 12) and noise levels should be in compliance.

Clearly, significant financial risk is associated with the performance test for the EPC contractor, major equipment OEMs and the owner. Successful outcome is conditional upon rigorous establishment of performance guarantees and careful test preparation and conduct as dictated by the governing test codes (e.g., ASME PTC 46). This is ensured by a *performance test protocol*, which requires strict adherence to the applicable codes (except for deviations allowed under the contract). A rigorous test procedure must be prepared by the EPC contractor and should include

- Objective of the test
- Parties to the test
- Key personnel and their responsibilities
- Equipment inspection and condition requirements
- Instrumentation requirements

- Frequency of data collection
- Number of test runs to be conducted
- Performance correction equations
- Uncertainty analysis
- Reporting format.

In an LSTK project with an OEM-supplied power island (i.e., gas/steam turbine generators and the HRSGs), the main focus of the EPC contractor is a rigorous calculation of plant auxiliary load based on component design and prior experience with similar projects executed by the contractor. (This is pretty much the standard case at the time of writing.)

A shortfall in performance vis-à-vis guarantee output, heat rate and availability results in performance LDs. Typical values for the output and heat rate LDs can be guesstimated from the value calculations presented in Section 13.4. In particular,

- LDs resulting from a deficiency plant net power output is in the ballpark of the specific plant cost (the owner's cost); thus, it is in the neighborhood of \$1,000/kW.
- LDs resulting from a deficiency plant net heat rate is in the ballpark of the value of 1 Btu/kWh; thus, it is a function of fuel price and project economics.

For example, let us introduce Equation 13.8 here for convenience, i.e.,

$$\mathrm{VHR} \approx 10^{-6} \cdot \frac{\mathrm{Hf}}{\beta} \cdot \mathrm{P_0},$$

where β is the capital charge factor, H is the annual operation hours, f is the fuel price in \$/MMBtu (LHV) and P_0 is the base value of plant output. Let us assume that β = 12%, f = \$4/MMBtu and P_0 = 750,000 kWe. Then,

$$\mathrm{VHR} \approx 10^{-6} \cdot \frac{6{,}000 \cdot 4}{12\%} \cdot 750{,}000 = \$150{,}000.$$

Thus, heat rate LDs in a GTCC built in the USA these days by an IPP can be expected to be in the neighborhood of \$150 K per each missed Btu/kWh in net heat rate.

Deficiencies in plant availability (above the minimum plant availability) result in LDs set at X dollars (say, \$500 K) for each percentage (prorated) above the minimum and below the guaranteed availability.

11.5 GENERAL ARRANGEMENT

A general arrangement drawing that shows the typical layout of a GTCC power plant with two (2) 1 × 1 × 1 single-shaft blocks is shown in Figure 11.2. General arrangement drawings are used to show and locate equipment with respect to the coordinate system within the plant, showing only those dimensions, datum and elevations to refer each equipment to other equipment or structures. Elevation view of one 1 × 1 × 1 block is shown in Figure 11.3.

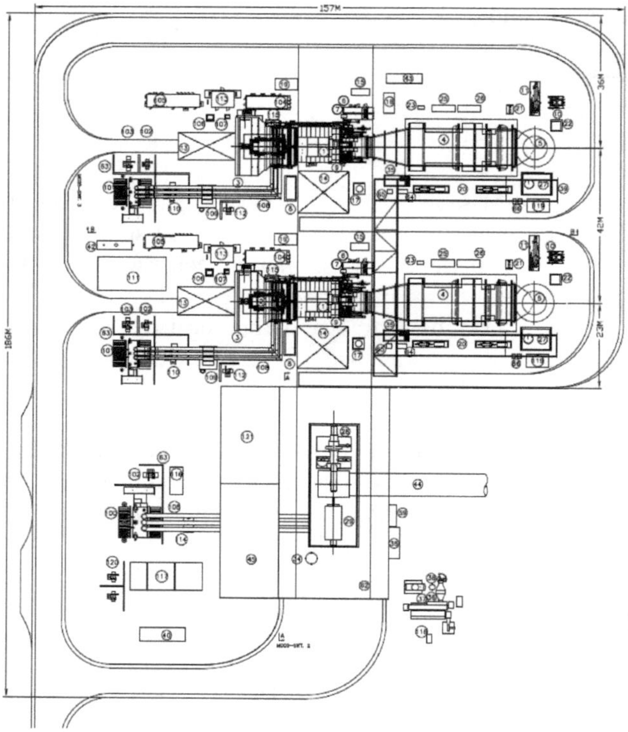

1. Gas Turbine
2. GT Generator
3. Air Inlet Filter
4. HRSG
5. HRSG Stack
6. Lube Oil Module (LOM)
7. LOM Spill Containment
8. GT Fire protection
9. GT Enclosure
10. Fuel Gas Separator
11. Fuel Gas Heater
12. NA
13. Generator Rotor Maintenance Area
14. GT Maintenance Area
15. Liquid Fuel Filtration Skid
16. NA
17. Water Wash Drains Tank
18. Water Wash Skid
19. H2/CO2 Storage
20. Boiler Feedwater Pumps
21. LP Econ Recirc Pump
22. CEMS
23. SCR Control CEMS
24. NA
25. Ammonia Vaporizer Skid
26. Maintenance Area
27. Blowdown Tank
28. Steam Turbine
29. ST Generator
34. Condensate Receiver Drain Tank
35. Stairs
36. Aux Boiler
37. Aux Boiler Deaerator
38. Aux Boiler Blowdown Tank
39. Main Pipe Rack
40. Emergency Diesel Generator
42. Oil Separator
43. Sampling Skid
44. ACC Duct
45. Electrical Building
47. Vacuum Pumps
48. Condensate Pumps
49. Condensate Expansion Tank
50. ST Lube Oil Reservoir
51. Chemical Injection Skid
52. CCW Pumps
53. Booster Pumps
54. CCW Heat Exchangers
55. Air Compressor
56. Air Dryer
57. Air Receiver Tanks
58. ST CO2 Bottle Rack
59. ST H2 Bottle Rack
60. Gland Seal Condenser
61. Seal Steam Superheater
62. Gantry Crane
63. Fire Wall
64. Burner Gas Control Skid
65. Burner Blower Skid
66. IP Feed Pumps
100.ST Main Transformer
101.GT Main Transformer
102.ST/GT Excitation Transformer
103.Isolation Transformer
104.Control Compartment
105.LCI Control Compartment
106. DC Link Reactor
107. AC Reactor
108. Isolated Phase Busbar
109. Generator Circuit Breaker
110. Unit Auxiliary Transformer
111. Power Distribution Center (PDC) – GTGs
112. GT Load Center Transformer
113. Battery Compartment
114. PT Cubicle
115. Generator Aux Skid
116. ST generator Excitation Compartment
117. PDC – BOP/STG
118. Aux Boiler Service MCC
119. PDC – HRSGs
120. MV/LV Transformers
121. Control Building

FIGURE 11.2 General arrangement drawing (plan view) of a GTCC power plant with two (2) single-shaft blocks.

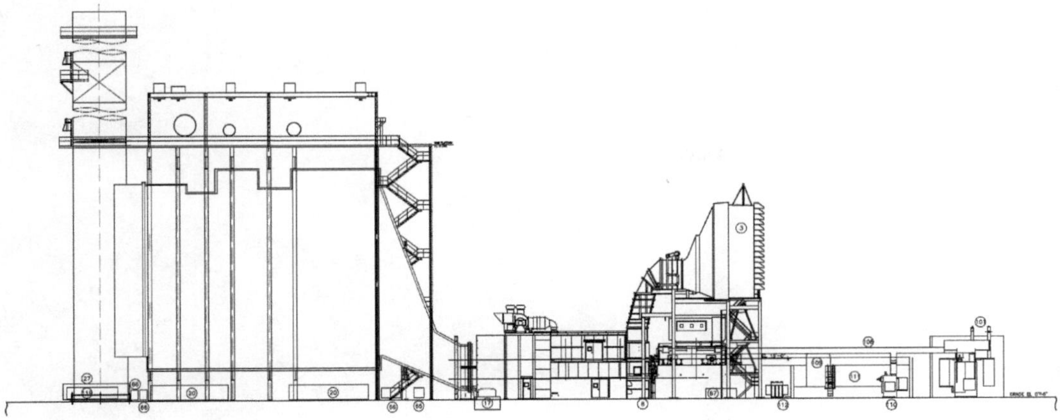

FIGURE 11.3 Single-shaft combined cycle (elevation view) – for legend refer to Figure 11.2.

12 Environmental Considerations

In the second decade of the twenty-first century, probably the biggest challenge facing the humanity is sustainable and "clean" generation of electric power without exacerbating the adverse impact of the climate change already being experienced across the world. Within the context of the subject matter of this book, the relevant considerations are

- Carbon dioxide emissions – primarily from the heat recovery steam generator (HRSG) stack but also from the secondary sources such as the auxiliary boiler or the emergency (black start) diesel generator
- Criteria pollutant[1] emissions (specifically NOx and CO)
- Wastewater treatment (see Section 10.9.4)
- Water conservation (e.g., "zero liquid discharge" – see Section 10.9.5)
- Acoustic (noise) emissions.

Natural gas is undoubtedly the cleanest fossil fuel. In most cases, it is essentially sulfur-free and does not have the particularly nasty emissions problems associated with coal burning power plants (e.g., mercury emissions). In terms of CO_2 emissions, consider the US *Energy Information Administration* (EIA) data for the year 2015 summarized in Table 12.1. Overall, CO_2 emissions from natural gas-fired electric power generation is one-third of the CO_2 generated by coal-fired power plants.

Nevertheless, each new gas turbine combined cycle (GTCC) must go through a rigorous permitting process before it gets financed and construction starts. The driver of this process is *the Clean Air Act* (CAA), which is the comprehensive federal law that regulates air emissions from stationary and mobile sources. This law authorizes the US *Environmental Protection Agency* (EPA) to establish *National Ambient Air Quality Standards* (NAAQS) to protect public health and public welfare and to regulate emissions of hazardous air pollutants (HAPs). In particular, the *New Source Review* (NSR) is the CAA program that requires industrial facilities such as a GTCC power plant to install modern pollution control equipment when they are built or when making a change that increases emissions significantly. The NSR obligates the owners and operators of GTCC power plants to obtain permits limiting air emissions before they begin construction. For that reason, NSR is commonly referred to as the "preconstruction air permitting program".

National Emission Standards for Hazardous Air Pollutants (NESHAP) are stationary source standards for HAPs. HAPs are those pollutants that are known or suspected to cause cancer or other serious health effects, such as reproductive effects or birth defects, or adverse environmental effects. NESHAP 40 CFR 63 Subpart YYYY applies to stationary gas turbines. According to that rule, the EPA identified stationary gas turbines as major sources of HAP emissions such as formaldehyde, toluene, benzene and acetaldehyde. Consequently, the NESHAP implements Section 112(d) of the CAA by requiring all major sources to meet HAP emission standards reflecting the application of the *maximum achievable control technology* (MACT) for combustion.

For gas turbines, the critical issue is the *formaldehyde* limit of 91 ppb at 15% O_2. In general, GTCC power plants equipped with dry-low-NOx (DLN) combustors operating at normal load conditions are not major sources of HAPs. (Mechanical drive units and aeroderivative gas turbines, on the other hand, are.) For example, the EPA conducted risk assessment of nearly 250 gas turbines at

[1] The US EPA sets limits for six principal pollutants, which are commonly called "criteria" pollutants and include ozone, carbon monoxide, particulate matter, sulfur dioxide, lead and nitrogen oxide. The limits are set at levels that protect human health and the environment.

TABLE 12.1
Comparison of Natural Gas and Coal-Fired Power Plant Stack CO_2 Emissions (2015 USA)

	10^6 mt CO_2 per quad Btu	Average HV Btu/cuft, MMBtu/st	Consumption: 10^9 cuft/day, 10^6 st	CO_2, Short tons/year	CO_2 per Heating Value, lb/MWh
Natural gas	52.9	1,035	26.5	584,224,300	398
Coal	96.3	19.149	739.7	1,503,550,840	724
	Average Eff	**CO_2 per MWh Generated, lb/MWh**	**Generation: 10^6 MWh/day**	**10^6 MWh/ year**	**CO_2, short tons/year**
Natural gas	45.5%	875	3.658	1,335	584,224,300
Coal	32.7%	2,218	3.715	1,356	1,503,550,840

major sources of HAPs; only three were found to exceed the limit. Nevertheless, the fact is that the 91 ppb limit started appearing frequently in GTCC air permits. As long as the startup and shutdown times are not unduly extended, this is not expected to be an issue even though the compliance verification standards are quite onerous. In particular, extremely low limit for compliance introduces accurate measurement challenges. If necessary, CO catalyst in the HRSG (see below) is capable of formaldehyde removal as well.

Directive 2010/75/EU of the European Parliament and the Council on industrial emissions (the *Industrial Emissions Directive* or IED) is the main European Union (EU) instrument regulating pollutant emissions from industrial installations. The IED was adopted on 24 November 2010. The limits set by the IED for the gas turbines are 50 mg/Nm^3 (25 ppm) NOx and 100 mg/Nm^3 (80 ppm) CO for both gas and liquid firing. Gas turbines are exempt from particulate matter (PM) limit of 5 mg/Nm^3 for gas-fired units.

However, there is also the Best Available Techniques (BAT) Reference Document (BREF) issued by the European Commission's Joint Research Centre (JRC), which deals with combustion installations with a rated thermal input exceeding 50 MWth.[2] Approved by the EC in April of 2017 and identified as "Reference Document", this document also includes emission limits, which are summarized in Table 12.2.[3] Individual countries in the EU are adopting these standards, which are incorporated into commercial bidding documents.

Another source of emissions guidelines is the World Bank (WB). For gas turbines rated larger than 50 MWth, Table 6(B) in *Environmental, Health, and Safety Guidelines for Thermal Power Plants* (Draft for Second Public Consultation—May/June 2017) has the following values for NOx emissions:

- 50/30 mg/Nm^3 with natural gas in nondegraded/degraded airsheds[4]
- 150/100 mg/Nm^3 with distillate and light fuel-oil in nondegraded/degraded airsheds.

According to the WB, an airshed is considered as degraded if relevant ambient air quality standards (as defined in the WB General Environmental Health and Safety (EHS) Guidelines) are exceeded. Note that these are proposed values because the 2017 draft is not finalized at the time of writing. Current limits do not distinguish between nondegraded and degraded airsheds.

[2] Available on the internet, http://eippcb.jrc.ec.europa.eu/reference/BREF/LCP/JRC_107769_LCPBref_2017.pdf. Last accessed on 21 June 2019.

[3] The table and associated information is extracted from the report prepared by Andrew Dicke, General Electric.

[4] An airshed is a geographical area where local topography and meteorology limit the dispersion of pollutants away from the area.

TABLE 12.2
Emission Levels Set in European Commission's BAT Reference Document

		BAT Achievable Emission Levels				
		Daily Average		Yearly Average		
		NOx (mg/Nm³)	PM (mg/Nm³)	NOx (mg/Nm³)	CO (mg/Nm³)	PM (mg/Nm³)
Natural	SC	25–50		15–35	<5–40[a]	
Gas	CC	15–40		10–30	<5–30[a]	
Distillate	SC		2–10			2–5
Oil	CC		2–10			2–5

[a] Indicative values.

12.1 AIR PERMITS

Air permits are enforceable legal documents that the GTCC power plant owners/operators must comply with. Air permits may place restrictions not only on how the power plant can be operated to meet the stack emission limits but also on the construction activities. In order to assure that a particular GTCC complies with its air permit's emission limits, a permit almost always contains monitoring, record-keeping and reporting requirements (e.g., installation of a *Continuous Emissions Monitoring System*, CEMS).

There are three types of NSR permitting programs, each with a different set of requirements. A facility may have to meet one or more of these sets of permitting requirements, i.e.,

1. *Prevention of Significant Deterioration* (PSD) program, which involves
   a. Installation of the Best Available Control Technology (BACT)
   b. Air quality analysis
   c. Additional impacts analysis
   d. Public involvement.
2. Nonattainment NSR program
   a. Installation of the Lowest Achievable Emission Rate (LAER)
   b. Emission offsets
   c. Public involvement.
3. Minor NSR program.

NSR permits are issued by state or local air pollution control agencies. State and local air pollution control agencies may have developed their own NSR permit programs as part of their *State Implementation Plans* (SIPs) that are approved by the EPA or they may be delegated the authority to issue permits on behalf of the EPA. If a state chooses to develop a SIP and not seek delegation of the federal NSR programs, the EPA would implement the programs and issue the NSR permit.

The starting point of the air permit process is the generation of GTCC heat and mass balances with emissions estimates. Such a set is shown in Table 12.3 (for a 2 × 2 × 1 duct-fired GTCC with an advanced class gas turbine).

Note that the NOx emissions value in Table 12.3 (2 ppmvd at 15% O_2) clearly indicates the presence of an SCR (Selective Catalytic Reduction) system in the HRSG. It is absolutely critical that the assumptions used in emissions calculations are enumerated diligently. For example, in conjunction with the data presented in Table 12.3,

TABLE 12.3

Typical GTCC Heat and Mass Balance Data for Emissions Guarantees and Air Permit Application (Fuel Sulfur Content Is 0.5 Grains/100 SCF @ 60°F, no Fuel-Bound Nitrogen; Gas Turbine Heat Consumption with Permitting Margin – see Note 9)

Operating Point		**1**	**2**	**3**	**4**	**5**	**6**	**7**	**8**	**9**	**10**	**11**	**12**	**13**	**14**	**15**	**16**	**17**
Case description		100%DB Fired	10%DB Fired	100%DB Fired	Unfired	Unfired	Unfired	100%DB Fired	Unfired	Unfired	Unfired	98%DB Fired	Unfired	Unfired	Unfired	Unfired	100%DB Fired	Unfired
SITE Ambient Conditions																		
Temperature	°F	7	7	7	7	7	7	26	26	26	26	68	68	68	68	68	95	95
Pressure	psia	14.7	14.7	14.7	14.7	14.7	14.7	14.7	14.7	14.7	14.7	14.7	14.7	14.7	14.7	14.7	14.7	14.7
Relative humidity	%	80	80	80	80	80	80	63	63	63	63	76	76	76	76	76	45	45
Site Operating Conditions																		
HRSG duct burner		ON	ON	ON	OFF	OFF	OFF	ON	OFF	OFF	OFF	ON	OFF	OFF	OFF	OFF	ON	OFF
SCR		ON	ON	ON	ON	ON	ON	ON	ON	ON	ON	ON	ON	ON	ON	ON	ON	ON
CO catalyst		ON	ON	ON	ON	ON	ON	ON	ON	ON	ON	ON	ON	ON	ON	ON	ON	ON
Evap cooler		OFF	OFF	OFF	OFF	OFF	OFF	OFF	OFF	OFF	OFF	ON	ON	OFF	OFF	OFF	ON	ON
GT load	%	Base	Base	Base	Base	36%	44%	Base	Base	75%	30%	Base	Base	Base	75%	32%	Base	Base
Number of GTs operating		2	2	1	2	2	1	2	2	2	2	2	2	2	2	2	2	2
Fuel																		
Type		NG	NG	NG	NG	NG	NG	NG	NG	NG	NG	NG	NG	NG	NG	NG	NG	NG
GT heat consumption (HHV)		2714.3	2714.3	2717.3	2714.3	1371.0	1528.6	2688.5	2688.5	2125.9	1236.2	2650.8	2650.8	2598.9	2050.8	1219.0	2505.5	2505.5
Duct burner heat consumption (HHV)		498.76	49.876	498.76	0	0	0	498.76	0	0	0	487.67	0	0	0	0	498.76	0
HRSG Stack Gas (Per Unit)																		
Ar	%(v)	0.8832	0.8893	0.8831	0.8900	0.8900	0.8901	0.8832	0.8901	0.8900	0.8900	0.8731	0.8800	0.8800	0.8800	0.8800	0.8528	0.8601
CO_2	%(v)	4.8364	4.1652	4.8460	4.0900	3.7800	3.8104	4.8561	4.1004	4.0700	3.6900	4.8961	4.1400	4.1300	3.9600	3.6200	4.9583	4.1404
H_2O	%(v)	9.4437	8.1454	9.4529	8.0000	7.4000	7.4607	9.6214	8.1608	8.1000	7.3600	11.2491	9.8000	9.6300	9.2900	8.6400	12.4792	10.9211

(Continued)

TABLE 12.3 (*Continued*)
Typical GTCC Heat and Mass Balance Data for Emissions Guarantees and Air Permit Application (Fuel Sulfur Content Is 0.5 Grains/100 SCF @ 60°F, no Fuel-Bound Nitrogen; Gas Turbine Heat Consumption with Permitting Margin – see Note 9)

Operating Point		1	2	3	4	5	6	7	8	9	10	11	12	13	14	15	16	17
N_2	%(v)	74.3109	74.8227	74.3035	74.8800	75.1200	75.0975	74.1921	74.7675	74.7900	75.0800	72.9439	73.5100	73.6400	73.7700	74.0300	72.0422	72.6473
O_2	%(v)	10.5258	11.9774	10.5145	12.1400	12.8100	12.7413	10.4472	12.0812	12.1500	12.9800	10.0378	11.6700	11.7200	12.1000	12.8300	9.6675	11.4311
Molecular weight		28.3680	28.4490	28.3682	28.4581	28.4952	28.4913	28.3503	28.4414	28.4453	28.4916	28.1764	28.2662	28.2836	28.3056	28.3453	28.0462	28.1423
Temperature	°F	175	177	175	180	175	175	175	178	175	175	175	179	178	175	175	175	179
Mass flow rate	lb/h	5,100,900	5,080,100	5,100,900	5,077,800	2,780,400	3,076,500	5,034,800	5,011,700	3,995,300	2,571,500	4,888,300	4,865,700	4,784,900	3,945,900	2,571,500	4,600,100	4,577,000
Emissions (Per Unit)																		
NOx	ppmvd @ 15% O_2	2	2	2	2	2	2	2	2	2	2	2	2	2	2	2	2	2
	lb/h as NO_2	18.8	16.4	18.8	16.1	8.12	9.07	18.7	15.9	12.6	7.33	18.4	15.7	15.4	12.2	7.23	17.6	14.9
CO	ppmvd @ 15% O_2	2	2	2	2	2	2	2	2	2	2	2	2	2	2	2	2	2
	lb/h	14.3	12.2	14.3	12	6.06	6.76	14.2	11.9	9.41	5.47	14	11.7	11.5	9.07	5.39	13.4	11.1
VOC	ppmvd @ 15% O_2	1.8	1.8	1.8	1	1	1	1.8	1	1	1	1.8	1	1	1	1	1.8	1
	lb/h as CH_4	4.75	3.85	4.75	3.43	1.73	1.93	4.73	3.4	2.69	1.56	4.62	3.35	3.29	2.59	1.54	4.57	3.17
NH_3	ppmvd @ 15% O_2	5	5	5	5	5	5	5	5	5	5	5	5	5	5	5	5	5
	lb/h	21.8	18.6	21.8	18.2	9.2	10.3	21.6	18.1	14.3	8.3	21.3	17.8	17.5	13.8	8.19	20.4	16.8
SOx	lb/h as SO_2	5.448	4.656	5.448	4.572	2.304	2.568	5.4	4.524	3.576	2.076	5.316	4.464	4.368	3.456	2.052	5.1	4.212
Particulates	lb/h	15.5	15	15.5	10	8.8	8.9	15.5	10	9.4	8.7	15.4	10	10	9.4	8.6	15.3	9.9
Sulfuric acid mist	lb/h	3.37	2.99	3.37	2.93	1.48	1.65	3.37	2.91	2.3	1.34	3.37	2.87	2.81	2.22	1.32	3.27	2.71

1. Gas turbines and the bottoming steam cycle are in *steady-state* operation.
2. HRSG stack exhaust emissions are reported based on the following conversion rates:
   a. Gas turbine: 95% conversion of sulfur to SO_2 and 5% conversion to SO_3.
   b. Duct burner: 95% conversion of sulfur to SO_2 and 5% conversion to SO_3.
   c. For installations that are equipped with a CO catalyst, it is expected that 10% to about 35% of the SO_2 in the exhaust gas is converted to SO_3. The actual conversion rate used in Table 12.3 calculations is 30%.
   d. For installations with an SCR catalyst for NOx abatement, it is expected that 1%–5% of the SO_2 in the exhaust gas will be converted to SO_3. The actual conversion rate used in Table 12.3 calculations is 5%.
3. HRSG stack NH_3 emissions (from the SCR) are based on assuming no conversion to ammonium salts.
4. Steady-state emissions data in Table 12.3 are estimated values based on the recommendations and methods of the original equipment manufacturer (OEM).
5. Reference conditions for exhaust gas in scf (standard cubic feet or "scuff") are 68°F and 14.6959 psia. Reference conditions for exhaust gas fuel in scf are 60°F and 14.6959 psia.
6. Reference conditions for exhaust gas in Nm^3 (normal cubic meters) are 32°F and 14.6959 psia. Reference conditions for gas fuel in Nm^3 are 60°F and 14.6959 psia.
7. SO_2 emission values have been estimated by assuming that all the sulfur in the fuel is converted to SO_2 and is based on maximum sulfur content in the fuel of 15.862 ppmw for gas. SO_2 values are margined by 20% to account for variation in fuel sulfur content and measurement error.
8. Sulfur mist emission calculations conservatively assume that all SO_3 combines with water to form sulfur mist. In actuality, some SO_3 may form other chemical species. This would include ammonium sulfates in the presence of NH_3. The maximum sulfur mist reported is conservative.
9. The estimated values for heat consumption and the exhaust flows are margined to account for equipment variations, site operating conditions and life-cycle operating parameters.

Without an SCR and no duct firing in the HRSG, the only source of emissions in the GTCC is the gas turbine's DLN combustor. Gas turbine NOx emissions are a function of unit load and ambient temperature. Emissions compliance, say, 25 ppmvd NOx at 15% O_2, can be maintained down to a certain load level, which is referred to as *minimum emissions-compliant load* (MECL). Typical MECL for F class gas turbines with DLN combustors is about 40% at ISO conditions. MECL increases with decreasing ambient temperature; it can be as high as 60% or higher at 0°F. MECL also increases with increasing ambient temperature but at a much lower rate, say, from 40% load to 45% at 100°F. Some OEMs offer "extended" turndown capability via combustor "autotune" or other techniques (e.g., using inlet bleed heat or inlet heating coil – by GE – utilizing HRSG feedwater from the low pressure (LP) section).

Upon review, the permit is issued (or denied) to the developer authorizing the "construction and operation" of the particular GTCC power plant located at the particular locality. A sample permit may read something like this:

1. The GTCC power plant ("the facility") shall comply with applicable requirements of the US EPA regulations in Title 40 Code of Regulation Part 60 (40 CFR Part 60) on Standards Performance for New Stationary Sources (NSPS):
   a. General conditions.
   b. Gas turbines are subject to the applicable requirements of Subpart KKKK, Standards of Performance for Stationary Combustion Turbines.
   c. Auxiliary boilers are subject to the applicable requirements of Subpart D, Standards of Performance for Small Industrial/Commercial/Institutional Steam Generation Units.

   d. Diesel-fired engines are subject to the applicable requirements of Subpart IIII, Standards of Performance for Stationary Compression Ignition Internal Combustion Engines.
2. The facility shall comply with applicable requirements of the US EPA regulations NESHAP in 40 CFR Part 63, promulgated for
   a. General provisions.
   b. Gas turbines are subject to the applicable requirements of Subpart YYYY, Standards for Stationary Combustion Turbines.
   c. The diesel-fired engines are subject to the applicable requirements of Subpart ZZZZ, Standards for Stationary Reciprocating Internal Combustion Engines.
3. Also authorized in the permit is an X kilowatt (or horsepower) emergency generator engine and a Y horsepower (typically, 250 hp) emergency fire protection pump diesel engine. Each is limited to Z hours of operation per year (e.g., 500). The emergency fire protection pump diesel engine is limited to 1 h of non-emergency operation per day.
4. Fuel for the gas turbine, HRSG duct burners and auxiliary boilers authorized by the permit shall be limited to firing pipeline-quality, sweet natural gas containing no more than 5.0 grains total sulfur per 100 *dry standard cubic feet* (dscf) on an hourly basis and 0.5 grain sulfur per 100 dscf on an annual basis. The permit holder shall monitor fuel consumption continuously using a monitoring device that is accurate to ±5% and maintained, calibrated and operated in accordance with the vendor's directions. Monitoring device downtime (including planned maintenance activities as well as unplanned events) shall not exceed 5% of the total annual operating hours. Maximum natural gas usage shall not exceed HC [MMBtu/h] × 8,760 [h/years] = X MMBtu/years where HC is the heat consumption at the guarantee point.
5. Diesel engines are authorized to fire diesel fuel containing no more than 0.05% sulfur by weight.
6. During normal operations, emissions shall not exceed the following (for gas turbines, the numbers are determined based on the HMB data in Table 12.3). There are several caveats such as the definition of a "valid hour" (e.g., consistent with 40 CFR Parts 60.13 and 75). Emissions during initial construction and commissioning are usually excluded. Start of the operations is defined as the date when the facility produces electricity for commercial purposes (excluding the commissioning period).
   a. Typical aux-boiler emission limits are 50 ppmvd CO at 3% O_2, 0.01 lb/MMBtu of NOx and 0.0055 lb/MMBtu VOC (volatile organic compound).
   b. Typical diesel engine emission limits are 2.8 g/hp-h NOx and 7.8 g/hp-h combined NOx and NMHC (non-methane hydrocarbons).
7. Cooling towers shall not exceed a total dissolved solids (TDS) concentration of 2,900 ppmv. (Detailed instructions are provided for measurements requisite for demonstrating compliance.)

Emission sources and maximum allowable emission rates are summarized in a table (*Maximum Allowable Emissions Rates Table* or MAERT). The emission rates are derived from information submitted by the applicant (e.g., see Table 12.3) and are the maximum rates allowed for these facilities. Any proposed increase in emission rates may require an application for a modification of the facilities covered by the particular permit. The structure of MAERT is as follows:

Emission Sources: Maximum Allowable Emission Rates				
Air Contaminants Data				
			Emission Rates	
Emission Point	**Source Name**	**Air Contaminant Name**	**lb/h**	**tons/years**
Gas turbine A	Turbine stack	CO	xx	yy

There are specific requirements pertaining to startup and shutdown events. In general, such events are required to be minimized with the applicable monitoring devices in operation. Definitions of hot, warm and cold starts are spelled out precisely. For example, a hot start can be defined as when the unit in question has not received fuel for 16 h or less or the steam turbine metal temperature is higher than, say, 700°F. Please refer to Section 15.3 for more information on startup definitions.

Startup is usually defined as the period that begins when the gas turbine is "ignited" (i.e., fuel flow starts) and ends when the required mode of combustion is achieved (e.g., "full premix" mode in a DLN combustor at full load), and there are two consecutive CEMS data points in compliance with the aforementioned emission limits.

Startup and/or shutdown events are authorized provided that the NOx, CO and VOC emission rates in lb/h do not exceed those specified in the MAERT and comply with the tons-per-year specified in the MAERT at normal operating conditions. This requirement is extremely important and can be a problem under certain conditions. In other words, the operator of a GTCC must be on top of the accumulated emissions over the course of the year in order not to end up with a permit violation. Consequently, one cannot operate a GTCC in a "willy-nilly" manner, starting and shutting down at any given time.

Single-digit NOx emission limits, which are becoming the norm in the USA, require SCR units in the HRSG. As a result, very strict leak prevention and protection measures are imposed by the air permit. Specifically, all operating practices and procedures are required to conform to the safety regulations for NH_3 by ANSI and the *Compressed Gas Association* guidelines. Installation and operation of a NOx CEMS upstream and downstream of the SCR system is prescribed for the estimation of NH_3 slip (including the formula and measured parameters).

In addition to all these very detailed and specific quantitative limits, the air permit lists requisite measuring, sampling, testing requirements for initial and continuous determination of compliance (in accordance with 40 CFR Part 60 mentioned earlier) as well as requirements for record-keeping and reporting.

Emissions guaranteed by the OEM in compliance with the air permit are based on testing at the HRSG stack in accordance with the following US EPA test methods:

- Demonstration of the NOx guarantee per **US EPA Method 7E** is based on the average of three test runs at each test point. The test points are the minimum and maximum gas turbine loads in the guaranteed load range (e.g., MECL to full load; typical MECL for the advanced class gas turbines with DLN combustors is 30%–40%).
- Demonstration of the NH_3 slip guarantee, per **US EPA Conditional Test Method (CTM) 027 (Modified)** or *Fourier-transform infrared spectroscopy* (FTIR) is based on the average of three 1 h test runs at each test point. The test points are the minimum and maximum gas turbine loads in the guaranteed load range. Modifications to CTM-027 include non-isokinetic sampling[5] and the use of a glass wool plug in lieu of an in-stack filter and nozzle. In addition, analysis by *Ion Selective Electrode* (ISE) may be conducted on site in lieu of, or in addition to, off-site laboratory analysis by *Ion Chromatography* (IC).[6] Each test run should be of sufficient length to collect a minimum sample volume of approximately 30 dscf.
- Demonstration of the CO guarantee per **US EPA Method 10** is based on the average of three test runs at each test point. The test points are the minimum and maximum gas turbine loads in the guaranteed load range.

[5] Stack gas sampling can be either isokinetic or non-isokinetic. Isokinetic sampling is used when the parameter of interest is expected to be present as a particulate or droplet where the actual concentration may various across the stack gas as it flows out the stack. Non-isokinetic sampling is used when measuring gases or vapors whose concentrations are not expected to vary across the flow of the stack gas.

[6] ISE, also known as a specific ion electrode (SIE), is a transducer that converts the activity of a specific ion dissolved in a solution into an electrical potential.

- VOCs are total hydrocarbons (THC) excluding methane and ethane, and are expressed in terms of methane. Demonstration of the VOC guarantee per **US EPA Methods 25A and 18** is based on the average of three test runs at each test point, per US EPA Method 25A, utilizing a THC analyzer calibrated using methane (CH_4) and a not-to-exceed span of 0–10 ppmv (wet). If test results per US EPA Method 25A indicate THC values greater than the VOC guarantees, at least one sample per test run is collected and analyzed per **US EPA Method 18** and the methane and ethane portions are subtracted from the Method 25A results. The test points are the minimum and maximum gas turbine loads in the guaranteed load range.
- Demonstration of the PM guarantee per **US EPA Methods 5 and 202** is based on the average of three test runs at the maximum gas turbine load in the guaranteed load range. The gas turbine should be operating at steady-state conditions at the initial test load for at least 2 h prior to commencement of testing. Each test run is of sufficient length to collect a minimum sample volume of 120 ft^3 (~4 h). The actual fuel flow rate during particulate testing is utilized to determine the exhaust gas flow rate per **US EPA Method 19** when converting from units of concentration to the guaranteed emission rate.

12.2 CONTINUOUS EMISSIONS MONITORING SYSTEM (CEMS)

The CEMS is a completely integrated system capable of monitoring a variety of exhaust gas emissions as well as generating the emissions reports. The system is designed and factory-tested to comply with the US EPA performance specifications. In particular, the CEMS is designed to meet the requirements of 40 CFR Part 75 for measuring, recording and record-keeping of NOx and O_2, and the requirements of 40 CFR Part 60 for measuring, recording and record-keeping of CO, unless extenuating circumstances dictate otherwise. The CEMS generates required and user-defined reports from calculated and recorded CEMS data.

The CEMS is capable of using a variety of analyzers and monitors for the measurement of source emissions. The specific analyzer selection is based upon a thorough review of the application requirements, including expected emission levels, emission limits and certification requirements.

Typically, the CEMS is capable of measuring/recording the following parameters:

- Nitrogen oxides (NOx)
- Oxygen (O_2)
- Carbon monoxide (CO)
- Exhaust stack ammonia (NH_3) slip
- SCR inlet NOx and O_2.

Sulfur dioxide (SO_2) is calculated utilizing the methodology outlined in 40 CFR 75 (Part 75) Appendix D titled "Optional SO_2 Emissions Data Protocol for Gas-Fired and Oil-Fired Units". Exhaust gas flow and carbon dioxide (CO_2) are calculated utilizing the methodology in 40 CFR 60 (Part 60) Appendix A Method 19.

A typical CEMS includes

1. Sample conditioning system
   a. *Sample probe and primary filter.* A sampling probe draws in a representative gas sample from the exhaust stack or duct. The sample probe and primary filter assembly is mounted directly in the exhaust stack or duct. The primary filter assembly, located at the outlet of the sample probe, is electrically heated, if needed, to inhibit condensation and is designed to collect any dust particles that enter through the probe.
   b. A *heated sample line* is used to transport the gas sample from the sample probe to the analyzers. The sample line typically contains tubes for the gas samples and a separate tube for the calibration gas.

   c. The *sample conditioning system* is designed to provide a clean, dry gas sample to the analyzers. Using a refrigerated condenser, the moisture in the gas sample is condensed and continuously removed from the system.
2. The *data acquisition system* (DAS) is designed to collect, process, display and report monitored and calculated air emissions data as well as relevant gas turbine operating parameters in accordance with the application requirements.
   a. *Remote data collection.* Data collection is accomplished via a network of intelligent input and output modules. The I/O Subsystem (*Programmable Logic Controller* (PLC) or *Data Logger*) interfaces between the analyzers and the DAS computer, and performs the basic monitoring system operating functions. Plant signals required for emissions calculations are also input into the I/O Subsystem. An I/O Subsystem is provided as a primary data collection device to simplify the communication between the analyzers and the DAS computer, and increase the reliability and serviceability of the system. Monitored emissions data are sent to the DAS computer and are also hardwired (I/O to I/O) to the control system for display on the operator's console in the central control room.
   b. *DAS computer and software.* Typically, a desktop or laptop PC provides the interface for the entire CEMS and is capable of displaying both real-time and historical emissions data. The computer operating system is a multi-tasking system. It allows the primary task of data collection and computation to operate simultaneously with other tasks, such as writing data to a disk, printing reports or interacting with the operator. The DAS software uses the measured/recorded emissions data to perform calculations and data substitution, and can automatically generate emissions reports including average emissions over specified time intervals.
3. *CEMS Enclosure.* The CEMS is housed in an enclosure. A heating, ventilating and air-conditioning (HVAC) system is provided to maintain a suitable temperature environment for the equipment inside the enclosure. Other features include lighting, bottle racks and a secure lockable door. The DAS computer is typically located in the plant's central control room though it can be located in the CEMS enclosure.

The CEMS vendor provides initial certification testing of the CEMS, including development of a test plan (protocol) and preparation of a test report of the results. The purchaser (i.e., the plant owner) is responsible for completing and submitting all initial certification documentation, as appropriate, to the regulatory agency (e.g., the US EPA, state or local). The vendor initiates the 40 CFR Part 75 (Monitoring and QA/QC Plans) and incorporates the initial vendor- and owner-supplied information. The owner is responsible for completion and submittal of the 40 CFR Part 75 (Monitoring and QA/QC Plans). Certification testing beyond the initial certification testing, including quarterly audits, is the owner's responsibility. The vendor provides the initial set of gases for daily calibration and initial certification testing; after the acceptance, it is the owner's responsibility to supply requisite gases. The owner is also responsible for all required submittals and notifications to the environmental regulatory agency.

12.3 NOISE ABATEMENT

Noise abatement is an important issue especially for those facilities located near residential areas. In those cases, as part of the permitting process, a detailed sound propagation model of the proposed facility is done by consulting firms using specialty software. Plant sound levels are calculated for base load operation, using the closest residences as the model receptor points. The modeling is used to confirm that sound levels attributable to plant operation can remain at or below the standards required by the regulating entity (e.g., state and county).

Depending on the study findings, non-standard noise mitigation treatments can be required to be integrated into the plant design. Typical, non-standard noise mitigation methods are

- Enclose both gas turbine exhaust diffusers (place them indoors)
- High-performance acoustic building walls and roofs
- Silencing of HRSG stack outlets, achieving X dBA at Y feet (per stack outlet)
- Low-noise air-cooled condensers (ACCs), achieving X dBA at Y feet (per condenser)
- Enclose the gas turbine filter house ducting (lag or place inside turbine building)
- Low-noise generator step-up (GSU) transformers, achieving X dBA at Y feet
- Barriers surrounding pipe racks between the HRSGs and turbine building.

Acoustical guarantees are twofold: *near field* and *far field*. Near field refers to the area that surrounds the noise source (e.g., the gas turbine), which is typically defined by a *source envelope contour*. Far field refers to a free environment (e.g., at the plant property boundaries) where the sound field is spreading spherically. A typical near-field sound level guarantee is 85 dB(A), i.e., 85 "A-weighted" decibels. Typical far-field guarantees are around 50–60 dB(A). What does that mean? The threshold of hearing is generally given as 20 micropascals at 1 kHz, at which point the logarithmic decibel scale is set to zero and, strictly speaking, is referred to as "dB SPL[7]". For instance, unsilenced turbine exhaust is 130 dB SPL, whereas the sound level in a rock concert can easily exceed 100 dB SPL. (Note that the threshold of pain is 140 dB SPL.) Since hearing sensitivity varies at different frequencies, A-weighting simulates the frequency response of the human ear.

Note that there are many noise sources in a GTCC power plant. Therefore, even at a point that can be defined as near field for equipment A (say, the gas turbine), contribution of another equipment, B, can result in a sound level higher than the guarantee sound level for equipment A. Consequently, the gas turbine (and its enclosure) must be designed to a level several decibels below the guarantee value (e.g., 80 dB(A) for 85 dB(A) guarantee) in order for sound levels within the turbine building to meet the project requirement at all locations.

Applicable US and international standards are

- ANSI/ASME PTC 36–1985, "Measurement of Industrial Sound".
- ANSI B133.8–1977, "Gas Turbine Installation Sound Emissions".
- ANSI S1.1–1994, "American National Standard Acoustical Terminology".
- ISO 3746, "Acoustics – Determination of sound power levels of noise sources using sound pressure Survey method using an enveloping measurement surface over a reflecting plane".
- ISO 6190, "Acoustics – Measurement of sound pressure levels of gas turbine installations for evaluating environmental noise – Survey method".
- ISO 10494-1993, "Gas turbine and gas turbine sets – Measurement of emitted airborne noise – Engineering/survey method".

For more information, the reader is referred to the General Electric reference documents listed below (available on the internet, www.gepower.com):

- GER-4221, "Power Generation Equipment and Other Factors Concerning the Protection of Power Plant Employees Against Noise in European Union Countries".
- GER-4239, "Power Plant Near Field Noise Considerations".
- GER-4248, "Acoustic Terms, Definitions and General Information".

[7] Sound Pressure Level.

12.4 SELECTIVE CATALYTIC REDUCTION

SCR of gas turbine (and HRSG duct burner if applicable) NOx emissions using ammonia (NH_3) has been widely used in GTCC power plants since early 1990s (the SCR technology was invented in 1957). The process is based on mixing aqueous NH_3 reagent (about 20% ammonia) with the flue gas containing NOx and then passing the gas through a catalyst bed designed to reduce the NOx concentration by reacting it selectively with the NH_3 to form nitrogen and water. The reactions taking place (on the catalyst surface) are

$$4NO + 4NH_3 + O_2 \rightarrow 4N_2 + 6HO_2$$

$$2NO_2 + 4NH_3 + O_2 \rightarrow 3N_2 + 6HO_2.$$

(The catalyst lowers the required activation energy for the reduction reaction and increases the reaction rate.) Complete (100%) reaction between the NOx and NH_3 is not practical; 90% is a reasonably good assumption (i.e., 25 ppm NOx from the gas turbine exhaust is reduced to 2.5 ppm). Catalyst bed exhaust gas thus includes excess NOx and excess NH_3 (slip). The catalyst bed is sized to keep the NH_3 slip below the limit specified in the air permit for the specified number of years. (As discussed above, ammonia is regulated as a toxic air pollutant.)

The SCR system (including the CO catalyst) is installed in the HRSG. The SCR catalyst design and location in the HRSG are based on the required NOx removal and expected flue gas temperature at the SCR location (300°C–350°C, thus, usually between the HP evaporator tube bundles). The catalyst housing interface, supports and piping are designed so that expansion joints are not required. Typically, a trolley and hoist system is provided for loading and removal of the catalyst blocks. The ammonia system components and distribution piping size are determined based on the type and concentration of ammonia solution reactant used.

The two key SCR systems are the *ammonia injection grid* (AIG) and the *catalyst bed*. Important design considerations are pressure drop across the AIG and the catalyst bed (detrimental to gas turbine performance via high backpressure) and catalyst life. Typical values are about 4–5 mbar and 24,000 h, respectively.

The aqueous ammonia system also includes the following elements:

- 2 × 100% aqueous ammonia forwarding pumps
- Shop-fabricated storage tank
- Interconnecting piping and valves
- The vaporization skid.

All tanks, pumps, piping and components in contact with ammonia are manufactured with stainless steel. The ammonia forwarding pumps are located inside the containment wall around the tank. The ammonia pipe is run underground from the discharge of the forwarding pumps to the vaporization skid. All underground piping has double-wall containment. Leak detection is provided at the low point of the containment pipe and in the sump of the tank(s) and pump(s). The system also includes a single minimum-flow recirculation line with a flow control valve and flow element. This line is sized for the minimum flow of one pump.

An oxidation catalyst system is also located within each HRSG to control emissions of CO and VOC. Exhaust gases from the gas turbines will flow through the catalyst bed where excess air will oxidize the CO and VOC to form carbon dioxide (CO_2) and H_2O. The oxidation catalyst system can reduce CO concentrations to 2.0 ppmv (dry, corrected) in the exhaust gas, with or without duct firing at all steady-state load conditions and ambient temperatures. VOC concentrations can be reduced to as low as 1.8 and 0.7 ppmv(dc) on natural gas with and without duct firing, respectively.

12.4.1 NOx Emission Calculations

Emissions of NOx are typically quoted at a reference condition on a volumetric basis, e.g., 25 ppmvd (parts per million by volume dry) at 15% O_2 (for gas turbines) or 5% O_2 (for some gas engines). For the fraction of NOx in the gas turbine exhaust gas, conversion from mass basis to volume basis (dry) can be done as follows:

$$v = \frac{\frac{\dot{m}_{NO_2}}{\dot{m}_{exh}} \cdot \frac{MW_{exh}}{46}}{1 - y_{H_2O}}. \tag{12.1}$$

where the terms in the numerator refer to the mass flow rates of NO_2 (proxy for NOx in the exhaust gas), gas turbine exhaust gas and the molecular weight thereof. The term in the denominator is the fraction of water vapor in the exhaust gas on a volume basis. Correction to x% O_2 results in (assuming air is 21% oxygen)

$$v_c = 10^6 \frac{\left(0.21 - \frac{x}{100}\right) \cdot v}{0.21 - \left(1 - y_{H_2O}\right) \cdot y_{O_2}}. \tag{12.2}$$

For a typical heavy-duty industrial gas turbine, exhaust gas molecular weight is about 29 lb/lbmol; exhaust gas oxygen and water vapor volume fractions are 12% and 9%, respectively (for 100% methane fuel). Thus, for a machine rated at 25 ppmvd NOx (15% O_2), from Equation 12.2, v is found as 0.0000415. If the exhaust gas mass flow rate is 1,175 lb/s, from Equation 12.1, NOx mass flow rate is about 260 lb/h. For a 230 MW unit, this corresponds to about 1.1 lb/MWh NOx emissions. Thus, as a rule of thumb, about 1 lb/MWh is a good NOx estimator for modern units with DLN combustors. Same calculations can be done for CO emissions by using 28 instead of 46 in Equation 12.1.

Another method, provided by EPA, is based on an *emissions index* (EINOx) in units of lb NOx per 1,000 lb fuel, which is proportional to the exhaust NOx emission levels in ppmv by a constant, K [1]:

$$\frac{v_c}{EINOx} = K. \tag{12.3}$$

The equation and K values (for v_c at 15% O_2) were provided by the OEMs for different gas turbines and fuels (e.g., K = 11.6 for methane, K = 12.1 for pipeline-quality natural gas and K = 13.2 for number 2 distillate). The calculation above with NO_2 as a proxy for NOx suggests a K value of 8.9.

EPA's Method 19 allows the use of following estimating factors [2]:

- 1 ppmv NOx (at 15% O_2) = 0.0036 lb/MMBtu (natural gas fuel)
- 1 ppmv NOx (at 15% O_2) = 0.0040 lb/MMBtu (distillate).

Note that Equation 12.3 with the given K values returns a value 0.0040 in lb/MMBtu of NOx emissions for each ppmv of NOx. The corresponding value from the sample calculation above with NO_2 as a proxy for NOx is 0.0052 lb/MMBtu. The difference can be explained by the composition of NOx emissions, which are typically 90%–95% NO with the balance being NO_2. Once the flue gas is out of the stack, most of the NO is eventually oxidized to NO_2 in the atmosphere, which contributes to the so-called "yellow plume". Thus, using a weighted average molecular weight of 31.6 lb/lbmol for NOx in Equation 12.1 (instead of 46 lb/lbmol for NO_2), one obtains a K value of 12.9 and 0.0036 lb/lbmol for each 1 ppmv of NOx from the sample calculation above.

For conversion from mg/Nm^3 to ppm, please use the following as guideline: 50 mg/Nm^3 is

- 26.58 ppm for NO_2 (MW = 46 lb/lbmol) as a proxy for NOx
- 43.66 ppm for CO (MW = 28 lb/lbmol)
- 19.1 ppm for SO_2 (MW = 64 lb/lbmol) as a proxy for SOx.

REFERENCES

1. U.S. Environmental Protection Agency, 1993, Alternative Control Techniques Document: NOx Emissions from Stationary Gas Turbines, EPA-453/R-93-007, January 1993, Office of Air and Radiation, Office of Air Quality Planning and Standards, Research Triangle Park, NC.
2. *Gas Turbine World 2014–2015 Handbook*, Vol. 31, pp. 66, Pequot Publishing Inc., Fairfield, CT.

13 Economics

In this chapter, there will be no presumption of "teaching" the reader how to calculate the "cost" (price?) of a piece of equipment or the entire gas turbine combined cycle (GTCC) power plant itself. The author cannot do it himself and neither can the reader unless he or she is intimately involved in such activities in the organization of an original equipment manufacturer (OEM) or engineering, procurement and construction (EPC) contractor. This is not a reflection on one's technical and other abilities. It is a question of tools, know-how, access to experience-based databases and vendor contacts. Established construction companies have entire departments employing experienced cost estimators, who can draw upon extensive in-house databases (from previous jobs) with equipment prices, craft labor and materials required to erect that equipment in the field and myriad rules of thumbs to do a good job of "estimating" *the lump sum turnkey* (LSTK) or *firm* price of constructing the power plant (and handing the "keys" over to the owner when it is ready to rock and roll) in response to a *Request for Proposal* (RFP) issued by a developer (i.e., "bidding" for the job). This is the most common form of cost estimation when it comes to the construction of GTCC power plants, in the USA as well as elsewhere in the world. In order to get an *inkling* about the intricate details of firm price estimation, the reader is referred to Chapter 3 of the excellent book by Peter Hessler [1]. There are two other forms of construction price estimating, i.e., (i) time and materials (T&M) and (ii) "cost-plus". These are also touched upon briefly in the cited reference. They are, however, rare these days in GTCC power plant construction and more common in chemical and oil-and-gas industries (e.g., construction of LNG stations).

This caveat, of course, brings up the following, quite legitimate question: If so, why is the author wasting paper and ink (and reader's time) on this subject matter? Here is why. The goal of the author in this chapter is to teach the reader how to "think" about cost-performance trade-off inherent in any engineering endeavor in general and GTCC design (and optimization) in particular. Let us face the inconvenient truth first: Nobody is going to ask you, the reader, to cost-optimize a GTCC power plant single-handedly. (No one asks – or ever asked – the author to do that either.) But, you will be able to stay away from "crazy schemes" and "pie-in-the-sky" design assumptions in order to squeeze the last proverbial drop of 0.01 percentage point of efficiency from the metal behemoth. On top of that, once in a blue moon (and do *not* bet on it), you can even be able to dissuade a greedy developer from dumping extra millions of dollars into a project. This, after all, is not a small feat and will be demonstrated by a (quite real) example at the end of the chapter.

13.1 PRICE VERSUS COST

First of all, unless you are in the design group of a gas or steam turbine OEM, you never (ever) try to estimate the "cost" of a prime mover. They are "priced" by the OEM according to the supply-and-demand principles that govern all economic activities from time immemorial. When the demand for gas turbines is high, as it happened during the so-called "boom" around the turn of the present century, a gas turbine, which costs the OEM, say, \$100 to manufacture can be sold for \$200 (note that the increased production numbers will help with economies of scale, and the cost to manufacture each additional unit will go down even further than predicted in the first place). The price of a gas turbine determines the *contribution margin* (CM) charged by the OEM, i.e.,

$$\text{Price} = \frac{\text{Cost}}{1-\text{CM}} \text{ or}$$

$$\text{CM} = 1 - \frac{\text{Cost}}{\text{Price}}.$$

Thus, for the example above, the CM is a hefty 50%. Is this possible? Was it ever recorded? Probably, most likely during the "boom", but this is an extremely sensitive piece of information and difficult to get a handle on. When a new gas turbine is being developed, CMs are determined by diligent market forecasts, extremely detailed bill of materials, supply chain arrangements, etc. in order to recoup the initial investment, which can run into several hundred million dollars or a billion, in X amount of years with Y amount of units sold over that period (with year-by-year orders/sales forecasts). It is a function of the cost of capital for the OEM, and it is reasonable to assume that it is at least 20% or more (you can refer to publicly available information and do the basic financial calculations with debt-equity structure, stock price, annual dividends, so on and so forth).

The foregoing discussion assumed a "new product". On the flip side of the coin, there are the venerable workhorses of the industry, i.e., the E class and vintage F class machines. These machines with their 30- to 40-year-old base architecture and "embellishments" drawing upon new developments in the field under the name of "upgrade packages" (especially in the hot gas path or HGP section) can be very cheap to manufacture and be "cash cows" for the OEMs.

Alas, at the time of writing, the picture for the heavy-duty industrial gas turbines, especially the advanced class behemoths with generator outputs going beyond 500 MW (one gas turbine!), is very bleak. Around 2010, it was foreseen that there would be about 300 large gas turbines sold each year, but in 2013, just 212 were ordered worldwide. In 2017, the number was a puny 122. Major OEMs reported declining sales and, what is even worse, the most advanced turbines were being sold at zero CM or sometimes at a loss (i.e., *negative* CM). At the time of writing (early 2019), it is difficult to gauge whether the decline in large industrial gas turbine sales is just a cyclical problem or whether there is a structural problem in the industry.

Estimated equipment-only budget prices for standard single-fuel gas turbine generator (GTG) (FOB factory) in year-Y US dollars are published in *Gas Turbine World*'s annual handbooks. The numbers published in *GTW* are "budgetary" prices for a well-defined scope of supply, flange-to-flange gas turbine (gas-only dry-low-NOx, DLN, combustor – dual fuel capability is optional at added cost), synchronous ac generator and the auxiliary systems. The latter include the inlet air filter, ducting and silencer, the exhaust ducting and the stack (short) with silencer and the controller (e.g., Mark VIe for GE gas turbines) and electrical systems (e.g., batteries, motor control center, voltage regulator and surge protection.) The equipment-only budgetary price does not include engineering, construction or owner project costs. As an example, an excerpt from the *2013 GTW Handbook* simple cycle price table is shown in Table 13.1.

As can be seen in Table 13.1, the specific price of a heavy-duty industrial gas turbine in 2013 was $230/kW. In fact, for units rated at 150–200 MW or more, this number does not vary a lot. In essence, the higher cost of materials and manufacturing for the larger, advanced class gas turbines

TABLE 13.1
2013 (Selected) Simple Cycle Gas Turbine Budgetary Prices (*GTW 2013 Handbook*)

	MW	Btu/kWh	%	MM$	$/kW
SGT6–8000H	274	8,530	40.0	64.98	237
M501GAC	276	8,574	39.8	63.4	230
SGT5–4000F	292	8,567	39.8	68.16	233
9F 5-Series	298.174	8,855	38.5	68.49	230
GT26	326	8,467	40.3	74.89	230
M501J	327	8,325	41.0	75.12	230

with higher firing temperatures pretty much negates any economies of scale that might otherwise have been realized. Note that

- \$3 million should be added for inlet "evap cooler" and accessories
- \$22–\$24 million should be added for the dual fuel capability.

The numbers cited above are in 2014 dollars [2]. The higher number for the dual fuel capability is for simple cycle power plant due to the cost of the demineralized water package (which is already present in the combined cycle for the bottoming steam cycle).

In the academic literature, there are many *cost estimation relationships* (CERs), which purport to estimate the cost of a gas turbine as a function of design parameters (e.g., firing temperature, airflow, and cycle pressure ratio). In most cases, there is a separate CER for the compressor, combustor and turbine. Due to the wide variance in design approaches of different OEMs in terms of HGP materials, casting and coating technologies, film cooling techniques and other enablers (e.g., rotor air cooling), which is reflected in the data presented in Table 13.1, those gas turbine cost estimation methods should be ignored. A third party researcher's best bet is the most recent *GTW Handbook* equipment-only budgetary price data.

A similar source of published price data for steam turbine generators (STGs) does not exist. In general, STG prices do not vary significantly within a class, e.g., 100–150 MWe single-flow axial low-pressure (LP) turbine unit with (nominal) steam pressures up to 1,815 psia and steam temperatures up to 1,050°F. Such units (turbine *and* generator) come at \$25–\$30 million (in 2015 dollars). This is why it is not easy to come up with a specific price (note, *not* cost) for an STG as a function of generator output. (Those readers who work for a steam turbine OEM can certainly put their hands on that information and develop CERs.[1]) The caveat stated above for the gas turbine CERs is also valid for STG prices. One's best bet is to get an actual budgetary price quote by the OEM for the particular problem at hand. A very rough rule of thumb is \$150–\$175/kW.

In order to estimate the price of the remaining "power island" (including the heat recovery steam generator (HRSG), the duct firing system (if any), the SCR (if any), distributed control system (DCS) and continuous emissions monitoring system (CEMS)), one can add 30%–40% to the STG price. In other words, if the STG price is \$30 million, the HRSG price (including the SCR and duct firing) is \$39–\$42 million. (Note that this is equipment-only price and does not include erection labor and materials.)

For conceptual studies, it is easier to use a bottoming cycle price, which can be extracted from the *GTW Handbook* simple and combined cycle budgetary price data. For example, from *GTW 2014–2015 Handbook*

$$\text{BC\_Price} = 1{,}238 \cdot \exp(33.4 / \text{STG\_MW})\$/\text{kW}. \tag{13.1}$$

The data used to derive the curve-fit formula, Equation 13.1, is summarized in Figure 13.1. For a 2 × 2 × 1 GTCC power plant with F class gas turbines, STG output is about 200 MW. From Equation 13.1, the bottoming cycle price is \$1,463/kW but it can be anywhere between \$1,350 and \$1,550/kW. Note that the GTW equipment scope *excludes* the SCR (and the CO catalyst) and assumes an *unfired* HRSG. Included are standard balance of plant (BOP) equipment, i.e.,

- Water-cooled condenser and mechanical draft cooling tower
- Condensate and feed pumps and piping

[1] Even then it may not be as easy as it sounds unless one is actually involved in costing. From the author's own experience, such information is jealously protected and shared on a "need to know" basis.

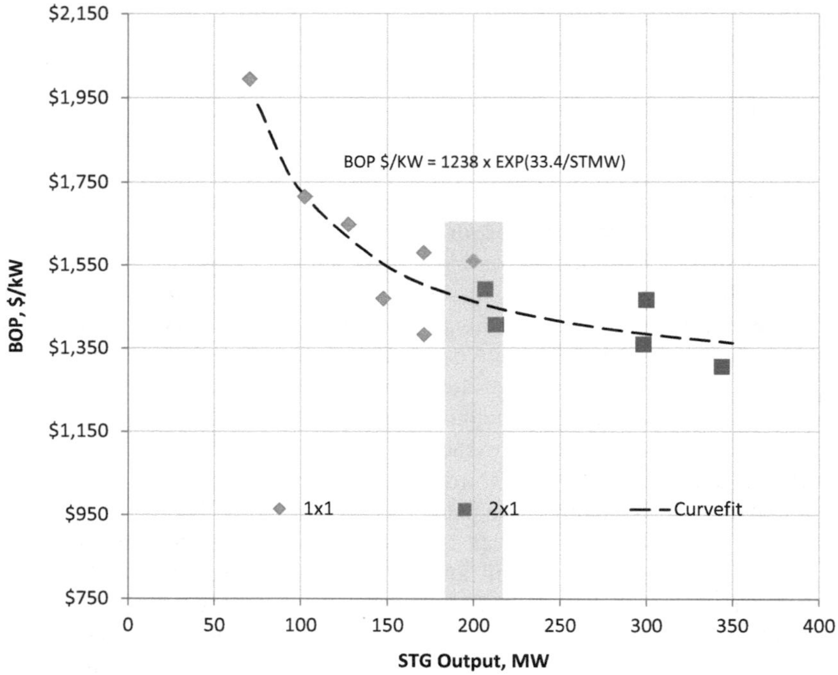

FIGURE 13.1 Bottoming cycle price (*GTW 2014–2015 Handbook*).

- Unit auxiliary transformers, switchgear, breakers and isolated phase bus duct (but step-up transformers are excluded)
- Installation/erection including the foundations (non-union labor, US Gulf Coast).

The price in Figure 13.1 is the "contractor's price", which would be charged by the EPC contractor on a turnkey basis. It includes the detailed design and engineering plus procurement of all plant equipment, materials and labor, construction costs and initial startup. It is an "overnight" price that *excludes* escalation and interest during construction.[2] Also excluded are the owner's costs such as land, legal fees, development fees and royalties.

13.2 COST ESTIMATION

There are different "classes" of cost estimation based on the level of project definition completeness, methods used in estimating and expected range of accuracy, i.e., –X% to +Y%. The accuracy range is an indication of how far off the actual cost will be from the estimate, i.e., it can be less than the estimate by X% or higher than the estimate by Y%. AACE International Recommended Practice No. 18R-97, *Cost Estimate Classification System – As Applied in EPC for the Process Industries*, identifies five cost estimate classes summarized in Table 13.2, which are based on the "maturity level" of project definition. As outlined in the AACE 18R-97, there are other factors that can impact the accuracy of the cost estimate significantly, i.e., estimate accuracy is also driven by other systemic risks such as the complexity of the project, quality of the reference data and assumptions used in preparing the estimate, experience and skill level of the estimator, etc.

A typical, state-of-the-art GTCC power plant with off-the-shelf (i.e., from the respective OEMs' product line catalogs) gas and steam turbine generators, three-pressure reheat (3PRH) HRSG and

[2] Allowance For Funds Used During Construction (AFUDC).

TABLE 13.2
Cost Estimate Classification (AACE 18R-97, March 1, 2016)

	Project Definition % Complete	Purpose	Methodology	Expected Accuracy
Class 5	0%–2%	Concept screening	Capacity-factored, parametric models, judgment or analogy	Low: −20% to −50% High: +30% to +100%
Class 4	1%–15%	Study or feasibility	Equipment-factored or parametric models	Low: −15% to −30% High: +20% to +50%
Class 3	10%–40%	Budget authorization or control	Semi-detailed unit costs with assembly-level line items	Low: −10% to −20% High: +10% to +30%
Class 2	30%–75%	Control or bid/tender	Detailed unit cost with forced detailed take-off	Low: −5% to −15% High: +5% to +20%
Class 1	65%–100%	Check estimate or bid/tender	Detailed unit cost with detailed take-off	Low: −3% to −10% High: +3% to +15%

standard BOP represent a commercially proven technology that the "maturity level" of the project definition is quite high from the get-go. This is why, even in the concept phase, say, 2 × 2 × 1 GTCC with OEM A's Frame F gas turbines and STG (with an air-cooled condenser because of site limitations on water usage), fired 3PRH HRSG in County C in State S, the developer's architect engineer can put together a Class 3 or even Class 2 level cost estimate.

Based on the preceding paragraph, one might argue that there is little need for simplified cost estimation techniques in a book on GTCCs. *Prima facie*, this may be an appropriate argument. As will be discussed in more detail below, however, prior field experience-based cost databases, cost estimating skills and resources (references, time and tools) requisite for Class 3 or lower accuracy are only available in large organizations with dedicated staff. Even if one is employed in one of those organizations, mobilizing those resources for small R&D type studies, "what if" exercises and other front end, conceptual endeavors may not be a feasible proposition. Therefore, one has to have a basic tool set for high-level combined cycle cost estimation to facilitate basic calculations and studies.

13.2.1 Simplified

As discussed in Section 13.1, it is fairly straightforward to estimate the "price" of a GTCC power plant. This is of course not a big help to the design engineer interested in doing a cost-performance trade-off study to answer "what if" questions during the conceptual design phase. Such questions mostly pertain to the bottoming cycle design, e.g.,

- Conventional versus advanced steam cycle parameters (pressure and temperature)
- Steam turbine back pressure (i.e., heat rejection system sizing)
- Performance fuel gas heating (fuel gas temperature)
- HRSG sizing (approach and pinch temperature deltas).

Unfortunately, even when one is armed with reasonably accurate equipment cost estimates, rolling them up to a final power plant cost is not easy. A typical GTCC power plant cost roll-up is as follows:

1. Installed equipment cost (E + M + L)
   a. Unit cost (E)
   b. Installation materials (M)
   c. Installation labor (L)

2. Total installed (direct) costs
   a. Installed equipment cost
   b. Buildings
   c. Service facilities
   d. Site work
3. Indirect costs (% of total direct costs)
   a. Engineering (including supervision)
   b. Contingency
   c. EPC contractor's mark-up
4. Owner's costs
5. Total capex (direct + indirect + owner's costs).

Installation materials and labor cover the following mechanical, civil and electrical components of the process of erecting a given piece of equipment (i.e., pump, compressor, heat exchanger, prime mover, etc.) in its appropriate location in the plant area ready to be connected to the rest of the plant (see Figure 13.2):

- Foundation (concrete and rebar)
- Piping
- Electrical
- Wiring
- Insulation
- Instrumentation
- Paint.

Typical combined cycle power plant equipment can be grouped into *four* categories:

- Engineered equipment – mechanical
  - Gas turbine package (see Section 9.1)
  - Steam turbine package (see Section 9.2)
  - HRSG (see Section 9.3)
  - Water-cooled condenser (Section 7.1) or
  - Air-cooled condenser (Section 7.4)
  - Mechanical draft cooling tower (Section 7.2)

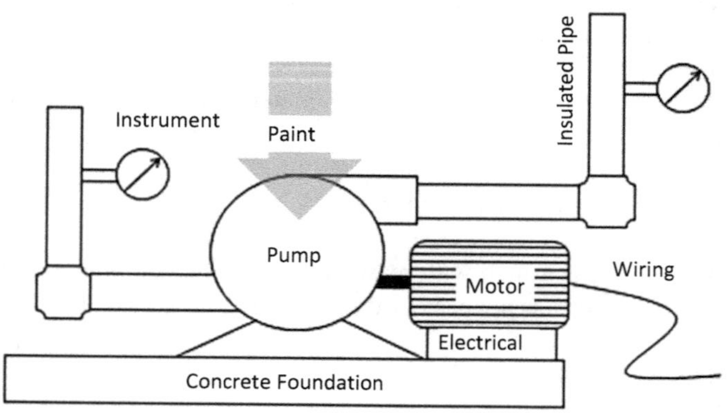

FIGURE 13.2 Illustration of installed equipment cost concept.

- Engineered equipment – electrical (Section 10.1)
    - Transmission voltage equipment (transformers, circuit breakers)
    - Generating voltage equipment (isophase bus duct, circuit breakers)
    - Medium voltage equipment
    - Low voltage equipment
- Control system (Section 10.1.1)
    - DCS
    - CEMS
    - Instrumentation (Section 10.1.2)
- BOP
    - Feedwater and condensate pumps (Section 10.3.1)
    - Circ water pumps (Section 7.3)
    - Other pumps (Section 10.3.2)
    - Tanks (Section 10.4)
    - Aux boiler (Section 10.5)
    - Fuel gas compressor (if necessary, Section 10.6)
    - Performance fuel gas heater (Section 10.7)
    - Closed cooling water (CCW) heat exchangers (Section 10.8)
    - Water treatment system (Section 10.9).

If steam conditions (pressure and temperature) do not change and the cost of a particular piece of equipment (e.g., pump, condenser) is known from a previous job or project (or some other source), it can be used to make an order of magnitude estimate via *capacity scaling*:

$$C = \left(\frac{Q}{Q_0}\right)^{\alpha} C_0, \tag{13.2}$$

where C_0 is the known cost of a particular piece of equipment or plant with a representative capacity of Q_0 (e.g., flow rate of a pump, heat transfer surface area of a heat exchanger) and Q is the capacity of the same piece of equipment under study. The cost of the latter, C, can thus be referenced to its *scaled* capacity via the *scaling exponent* α. The most commonly used exponent is 0.6 from the well-known "six-tenth rule of thumb". Exponents for different equipment and plant types are available in process industry references [3]. Unfortunately, no such information is available for major combined cycle equipment (to the best knowledge of the author). The recommendation here is to obtain budgetary quotes from the vendors and OEMs for cases under study. If an in-house cost database from previous projects is available, it can be used for developing a cost-capacity relationship à la Equation 13.2. In developing such CERs, one has to pay attention to the two confounding factors:

1. Cost escalation between the time frames of the cost data and the present project
2. Material impact.

The first one is relatively straightforward to take care of. If the available cost data is from, say, 2014 and one is interested in estimating the equipment cost in 2018, the best approach to adjust the 2014 number using a cost index. One such example is the *Chemical Engineering Plant Cost Index* (CEPCI). First introduced in 1963, the CEPCI consists of a composite index assembled from a set of four sub-indexes: equipment, construction labor, buildings and engineering/supervision. The equipment index is further divided into categories such as heat exchangers and tanks, process machinery, pumps and compressors. A brief summary of CEPCI data is published each month on the last page of the *Chemical Engineering* magazine. The online database contains much detailed information (at additional subscription cost). Annual CEPCI values between 2010 and 2018 (April – final value) are summarized in Table 13.3.

TABLE 13.3
Chemical Engineering Plant Cost Index (CEPCI)

Years	2010	2011	2012	2013	2014	2015	2016	2017	2018
CEPCI	550.8	585.7	584.6	567.3	576.1	556.8	541.7	567.5	595.0

For example, if there is a vendor price quote for a feed pump (including the driver motor) from, say, 2014 for $500 K, a quick estimate of the 2018 price is

$$\$500\text{K}(595/576.1) = \$516.4\text{K}.$$

Note that, going back more than 10 years is not recommended. If the numbers in the particular database are, say, from mid-1990s, it is best to obtain new price quotes.

In using Equation 13.2, one should also be cognizant of the fluid conditions (i.e., steam or water pressure and temperature). A particular equipment made from carbon steel for the then-prevailing specs may not be appropriate for the new specs even if the size parameter (e.g., flow in gpm or heat transfer surface area) does not change appreciably. If the new specs require, say, 316 stainless steel instead of carbon steel, the base cost has to be multiplied by the appropriate material adjustment factor. In the case of tanks, columns and other type of pressure vessels, one might have to account for the vessel pressure as well. For certain process equipment, this information is available in publications such as Ref. [3].

Installation materials and labor can be accounted for using the *installed cost factor*, CFINST, which is defined as

$$\text{CFINST} = \frac{\text{E}+\text{M}+\text{L}}{\text{E}} = 1 + \frac{\text{M}+\text{L}}{\text{E}}. \tag{13.3}$$

One of the most comprehensive and earliest methods to estimate CFINST was by Guthrie (see pages 9–15 in Ref. [4]). Examples of Guthrie method factors are provided in Table 13.4.

Thus, for the feed pump example above, total installed cost is

$$\$516.4\text{K}(1+2.42) = \$1.766\text{M}.$$

Total installed cost is found by adding up the (E + M + L) estimates for individual pieces of equipment. Indirect costs are usually accounted via factors applied to the total installed equipment cost. The sum of total direct and indirect costs gives the *total plant cost* (TPC). It is also known as the *fixed capital investment* (FCI). Some rough guidelines are as follows [3]:

TABLE 13.4
Guthrie Method Cost Factors

	Shell–Tube Heat Exchanger	Vertical Pressure Vessel	Pump (Including Motor)
E	1.00	1.00	1.00
M	1.71	2.05	1.72
L	0.63	0.95	0.70
M+L	2.34	3.00	2.42

- Engineering and supervision is approximately 30% of purchased equipment cost (E) or 8% of total direct and indirect costs
- Contingency can range from 5% to 15% of total direct and indirect costs
- EPC contractor's fee can be assumed to be 2%–8% of total direct costs.

There are three basic routes one can take for simplified combined cycle cost estimation:

1. *Gas Turbine World Handbook*'s combined cycle plant price data
2. Cost roll-up from a basic equipment list (obtained from a heat and mass balance diagram) with material (M) and labor (L) factors
3. Use a software package (e.g., Thermoflow, Inc.'s PEACE®).

The first one is the easiest, but one must (i) have a copy of the handbook available (it is not cheap) and (ii) be cognizant of the pricing scope. In 2017 dollars, the best-fit correlation for combined cycle power plants rated at or over 600 MW is

$$k[\$/kW] = 610 + 5 \cdot 10^{10} \cdot \dot{W}[kW]^{-1.55}. \tag{13.4}$$

Thus, for a 750-MWe GTCC, k ~ \$650/kW so that the total price is \$487 million. The accuracy ascribed to this price (note: *not* cost) by the *GTW Handbook* editors is ±15%. The scope reflected by the *GTW* price is on an "overnight construction" basis (i.e., AFUDC is not included). Also excluded are owner's costs such as project development and financing fees, permits and insurance. In addition to the power block equipment (including the steam turbine water-cooled condenser) and the DCS, GTW price scope includes limited mechanical and electrical BOP as well as EPC engineering and construction costs (based on "open shop" or non-union labor). Significant items such as cooling tower (almost a *sine qua non* in the USA with a water-cooled condenser; more expensive air-cooled condenser is even more likely), generator step-up transformers, water treatment and wastewater systems are excluded. As will be shown below, these can add significantly to the plant price.

A simple heat balance diagram or flowsheet can identify power island equipment and major mechanical BOP items (e.g., condensate and boiler feed pumps and cooling tower). Major steam piping, electrical BOP, tanks, small pumps, etc. have to be added separately and require a basic knowledge about those systems (e.g., see Chapter 10). It turns out that this "simple" cost roll-up is not so simple after all and requires quite a bit effort. The advantage is that, once a basic structure is set up (e.g., in an Excel spreadsheet) and a CER for each identified piece of equipment is available, repeated calculations for optimization and "what if" type studies can be automated. For conventional plant equipment (e.g., pumps, tanks, heat exchangers), CERs can be obtained from the chemical process industry (CPI) literature, e.g., Ref. [3], which is probably the best reference book out there. Alas, no such reference is available for the power island equipment. The best one can do is to obtain reliable cost information for a GTCC of comparable vintage/size and use the key design parameters with Equation 13.2 for adjustment to the problem at hand. One such source of information is Table 20 (page 27) of the report prepared for PJM by a consulting firm and their A&E (Architecture & Engineering) partner [2]. The GTCC is multi-shaft 2 × 2 × 1 with GE's 7FA.05 gas turbines with inlet evap coolers. The heat sink incorporates a mechanical cooling tower. Unfired net summer capacity is 587 MWe (in counties along and just west of I-95 in Virginia). A heat and mass balance analysis by the author using GT PRO® came up with the following details (assuming 93.7°F – 47.2% site ambient at 390 ft elevation per Table 13.5 of Ref. [2]):

- 395 MWe from two GTGs (90% effective inlet evap cooler)
- Gas turbine exhaust 2,200 lb/s (two gas turbines total), 1,126°F
- 204 MWe from the STG (3 in. Hg condenser pressure)

- 1,800 psia/1,065°F/1,065°F 3PRH steam cycle
- Condenser duty 350,000 Btu/s
- HRSG stack temperature 199°F
- Heat transfer from exhaust gas in the HRSG (assuming average c_p of 0.27 Btu/lb-F)

$$\dot{Q} = 2{,}200 \cdot 0.27(1{,}126 - 199) \approx 550{,}000\,\text{Btu/s}.$$

Cost roll-up from Ref. [2] is summarized in Table 13.5. In *GTW 2018 Handbook*, the price for the same GTCC (independent system operator or ISO baseload) is $500 million ($661/kW at ISO baseload rating of 756 MW). For a similar rating (i.e., 581 MW with two GT13E2-2 gas turbines – formerly Alstom now offered by GE), the price in *GTW 2018 Handbook* is $375 million ($645/kW).

The cost scope in Table 13.5 is as follows:

- Prime mover generators and the HRSGs are furnished by the owner
- "Other equipment" includes inside-the-fence equipment required for interconnection and other miscellaneous equipment and associated freight costs (includes mechanical draft cooling tower)
- Reclaimed water is used in the cooling systems
- Gas turbine inlet evaporative coolers and associated equipment add $3 million per gas turbine
- The CO catalyst system increases the cost of emissions control equipment by $2.4 million (in 2014 dollars)
- Dual fuel capability added $22 million to the total capital cost (gas turbine price plus fuel oil testing, commissioning, inventory and the capital carrying charges on the additional capital costs)
- The 345 kV switchyard is assumed to be within the plant boundary and is counted as an EPC cost under "Other Equipment," including generator circuit breakers, main power and auxiliary generator step-up transformers and switchgear

TABLE 13.5
Overnight Cost Roll-Up of 2 × 2 × 1 GTCC (GE 7FA.5) – 2018 Dollars [2]

Gas Turbines (2 7FA.5)	**$97.2**
HRSGs (2 with SCR)	$43.5
Steam turbine	$35.5
Condenser	$4.2
Other equipment	$60.3
Total equipment	$240.7
Construction labor	$146.9
L/E	0.61
Materials	$37.8
M/E	0.16
Total E+M+L	$425.4
Other labor (indirect)	$39.1
Indirect/(E+M+L)	0.09
Total	$464.5
$/kW	$791

- "Other labor" includes engineering, procurement, project services, construction management, and field engineering, start-up, and commissioning services
- "Materials" include all construction material associated with the EPC scope of work, material freight costs and consumables during construction.

Note that the cost estimate was done in 2014 and escalated to 2018 dollars by using an escalation rate of 2.65% for equipment and materials and 3.75% for labor. In Table 13.5, the following are not included

- Sales tax ($17.4 million total, $8.8 million for owner-furnished equipment)
- EPC contractor's fee (12% of EPC costs, $57.8 million)
- EPC contingency (10% of EPC costs, $54 million)
- Owner's costs (total of $114.2 million, ~16% of total overnight capital cost of $708 million or $1,073/kW).

All equipment and material costs were estimated by the A&E consultant using proprietary data, vendor catalogs or publications. Both labor rates and materials costs were estimated for the specific locality (mix of union and non-union rates). Estimates for the number of labor hours and quantities of material and equipment needed to construct the GTCC power plant were based on the A&E consultant's experience on similarly sized and configured facilities.

The data presented here can be used as a rough guide to estimate the GTCC plant cost for similar vintage/technology plants, e.g.,

- GTG price ~$250/kW (incl. evap coolers and dual fuel combustor)
- STG price ~$175/kW
- HRSG (incl. SCR, unfired) ~$80 per Btu/s heat transfer from flue gas
- As a very crude check,
  - HRSG cost is roughly 50% of gas turbine cost
  - Steam turbine cost is roughly 75% of HRSG cost
- Water-cooled condenser ~$12 per Btu/s of heat rejection
- Other equipment (including the mechanical draft cooling tower[3]) is 25% of total equipment cost
- Labor factor, L/E = 0.15–0.20 (see below for more on this)
- Material factor, M/E = 0.15–0.20
- Indirect costs 8%–10% of total E + M + L.

The better alternative is to use a software package that can do that for you. There are two commercially available products:

- SOAPP (State-of-the-Art Power Plant) software, currently offered by EPRI[4];
- Thermoflow, Inc.'s PEACE® (an add-on to GT PRO® and other software in Thermoflow Suite).

---

[3] Cooling tower is usually subcontracted to the cooling tower OEM (equipment plus installation materials and labor).

[4] SOAPP was first conceived by Sargent & Lundy in 1991; in 1994 SEPRIL Services, LLC, was formed with ownership split equally between EPRI and Sargent & Lundy to further commercialize the SOAPP product line. In 1999, EPRI acquired 100% interest in SEPRIL Services, LLC and changed the name to SOAPP, LLC, which became a wholly owned subsidiary of EPRI, managed by EPRIsolutions. EPRI continues to develop, distribute and support the SOAPP family of software products.

TABLE 13.6
Overnight Cost Roll-Up of 2 × 2 × 1 GTCC (GE 7FA.5)

	Ref. [2]	PEACE	Recommended
Gas turbines (2 7FA.5)	\$97.2	\$70.2	\$90.0
HRSGs (2 with SCR)	\$43.5	\$46.0	\$45.0
Steam turbine	\$35.5	\$28.9	\$30.0
Condenser	\$4.2	\$3.2	\$3.5
Other equipment	\$60.3	\$44.8	\$75.0
Total equipment	\$240.7	\$193.1	\$243.5
Construction labor	\$146.9	\$35.0	\$40.0
L/E	0.61	0.18	0.16
Materials	\$37.8	\$34.1	\$35.0
M/E	0.16	0.18	0.14
Total direct (E+M+L)	\$425.4	\$262.1	\$318.5
\$/kW	\$725	\$447	\$543
Other labor (indirect)	\$39.1	\$19.2	\$25.0
Indirect/(E+M+L)	0.09	0.07	0.08
Total	\$464.5	\$281.4	\$343.5
\$/kW	\$791	\$479	\$585

The author used PEACE® for the cost roll-up of the example GTCC in Ref. [2]. A comparison of cost estimates from Ref. [2] and PEACE® is presented in Table 13.6. (Note that PEACE® estimates are in August 2017 dollars.)

As shown in Table 13.6, the biggest disconnect between the two estimates is in three buckets, i.e.,

- GTGs (incidentally, 7FA.05 simple cycle price in *GTW 2018 Handbook* is \$53 million or \$220/kW including BOP but excluding inlet evap cooler and dual fuel option)
- Other equipment
- Construction labor.

Of those three, by far the most difficult to reconcile is the construction labor. In PEACE, construction labor hours were estimated as 850,000 h (including buildings) for an average (composite) labor cost of \$40/h. This is a pretty realistic estimate (see the next section). Details of the construction labor estimate in Ref. [2] are not known. However, at \$40/h, implicated construction labor hours are 3,672,500 – unrealistically high. One possibility is inclusion of fringe benefits and burdens in labor cost estimate. This would roughly double the dollars per hours value and imply about 1.8 million man-hours, which is still high for this type of facility.

13.2.2 Detailed

Detailed cost estimate for an LSTK project is based to a large extent upon the EPC contractor's prior experience in similar projects. Key quantity data for the current project estimate is based on that experience "database" to which adjustments are applied to account for site-specific requirements of the project location. Such adjustments are based on information from the scope document, geotechnical report, water availability and applicable state/local regulations.

As a very crude reference/guide, let us assume a hypothetical 750-MWe fired GTCC in 2 × 2 × 1 multi-shaft configuration with F class gas turbines, which costs about \$750–\$800/kW (in 2015 dollars) or about \$600 million. It requires about 1,000,000 craft labor hours, which is roughly \$40 million (open shop or non-union with a productivity rate of 1.1). (This does not include the subcontractor's crew and labor costs.) In locations where union labor must be used, this cost can be much higher. Productivity is also a strong function of location and must be evaluated carefully by considering all pertinent factors (e.g., the particular site might require considerable travel between the power block and construction facilities and services). In some places, a premium has to be paid for certain key skilled trades.

A typical manpower distribution for the sample project is assumed as follows:

- 50 site construction management staff
- 500 peak craft manpower
- 30 months of construction.

Estimation of temporary facilities such as change trailers, office trailers, safety equipment, testing and training is made on the basis of the requisite manpower. Other indirect costs are

- Construction equipment
- Temporary utilities
- Small tools and consumables
- Testing and startup expenses
- Permits.

As a rule of thumb, total indirect costs are 15% of the direct construction costs.

A key document for the EPC contractor's cost estimation is the mechanical equipment list. The power island is typically procured separately by the owner developer. For the sample project herein, representative numbers are

- $90 million for two F class GTGs (about 235 MWe each)
- $75 million for two duct-fired HRSGs
- $40 million for the STG (about 280 MWe)
- $3 million for the power island auxiliary equipment.

For this plant, BOP equipment to be procured by the EPC contractor are estimated at

- Mechanical equipment (pumps, etc.) about $50 million
- Electrical equipment $35 million.

Total equipment cost (E) is thus $293 million.

Construction materials' cost estimate is built up from actual quantity takeoffs from the 3D model, which is key to the whole plant design and cost estimation effort. It shows the major equipment, foundations, piping 75 mm (3 in.) and over, conduits, foundations and structural steel. The model provides detailed quantity takeoffs, which are used by estimating, scheduling and construction to cost, plan and execute the job. The model can be quickly revised to allow studies to be made and results quantified for various design options such as changes in pipe sizes, redundancy and motor versus steam turbine-driven feedwater pumps. The model also detects interferences and creates plan and elevation drawings. The following quantities are developed:

- Piping lengths by diameter, system and class, both aboveground and underground
- Valve quantities
- Instrument and DCS input/output quantities by system
- Foundation quantities by cubic meter
- Foundation reinforcing quantities
- Structural and miscellaneous steel

- Electrical circuits by system
- Electrical raceway and duct bank quantities
- Yard quantities, including excavation and paving.

For the example herein, construction material cost is estimated at $52 million so that M/E ~ 0.18, which is in good agreement with the estimates in Table 13.6. Labor cost at $40 million results in L/E ~ 0.14.

Mechanical equipment procured and installed by the EPC contractor does not include the mechanical draft cooling tower, which is assumed to be subcontracted at $15 million. Total subcontracted scope is assumed to be $35 million. Altogether, direct cost (E + M + L) is $420 million or $560/kW (cf. $585/kW in Table 13.6).

For the generic scope of the mechanical and electrical equipment list by the EPC contractor, refer to Section 9.5 (specifically the bullet list at the end of the section). A partial list for the sample project is provided in Table 13.7.

A typical (partial) rental construction equipment list is shown in Table 13.8.

TABLE 13.7
Mechanical Equipment List Example (Partial)

19% Aqueous Ammonia Storage Tank	12,000 gallons, Carbon Steel (CS)	1 × 100%
Ammonia forwarding pumps	Positive displacement, 316 SS	
Common (facility) fuel gas flow meter		
Common fuel gas chromotograph		
Common fuel gas pressure reduction station		
Common fuel gas filters and knockout drum		
Main fuel gas filter separator		1 × 100% per GT
Pilot fuel gas filter separator		1 × 100% per GT
Fuel gas performance heater (FGPH)		2 × 50% per GT
FGPH knockout drum		1 × 100% per GT
GTG fuel gas flow meter		1 × 100% per GT
HRSG final pressure reduction station		1 × 100% per HRSG
STG lube oil package		1 × 100%
Boiler feedwater pump hoist and trolley	Electric chain hoist/trolley	1 × 100% per HRSG
Boiler feed water pump jib crane		1 × 100% per HRSG
Maintenance building hoist	Manual hoist	1 × 100%
Cooling tower hoist and trolley	Electric chain hoist/trolley	1 × 100%
Plant air compressor	Air-cooled, 400 scfm @ 125 psig min.	2 × 100%
Compressed air driver	Dual-tower, heatless, desiccant type with dew point of –40°F or lower with pre-filter, after-filter and bypass filter	2 × 100%
Compressed air receiver	Vertical 750-gallon capacity, CS	1 × 100%
Electric fire pump	Horizontal centrifugal, 2,000 gpm @ 300 ft TDH, 316 SS internals	1 × 100%
Diesel-driven fire pump (DDFP)	Horizontal centrifugal, 2,000 gpm @ 300 ft TDH, EPA Tier 3 emissions engine, incl. a totalizing hour meter with minimum 9,999 h display, 316 SS internals	1 × 100%
DDFP fuel storage tank	400 gallon	1 × 100%
Jockey pump	CS	1 × 100%
Fire hydrants		1 × 100%

TABLE 13.8
Construction (Rental) Equipment List Example (Partial)

Equipment	Quantity	Months
Hydraulic excavator (3 CY)	1	12
Tractor backhoe (1.5 CY)	3	12
Bulldozer (100 hp)	1	12
Articulating 4WD loader (100 hp, 2 CY)	1	18
Crane crawler 300 ton (for the HRSG)	2	12
Welding machine (8 Arc, 250 Amp)	12	20
Air compressor (1,500 cfm)	6	18
…	…	…

Detailed labor cost estimation is obtained from "work breakdown sheets" (WBS). A typical WBS is shown in Table 13.9. Labor costs are rolled from this type of take-off into total man-hours multiplied by wages, fringes and burdens (if union labor is used). Fringes include items like welfare and pension funds. Burdens include FICA, Medicare, etc. A unionized "journeyman" boilermaker earning X dollars per hour in wages could cost the employer 2X dollars per hour with fringes and burdens. This is why construction costs between "open shop" (non-union) and union locations can make a big difference in total installed cost for the same technology/vintage/capacity power plant. As mentioned earlier, for this sample project, total construction labor man-hours were estimated at about 1,000,000 at an average cost of $40/h (non-union). For a specific "real" project it can be much higher.

Other cost adders are as follows:

- Design engineering (8% of direct costs)
- Project management (1% of direct costs)
- Warranty reserve ($1 million)
- Contingency (6.5% of direct plus indirect costs)
- EPC fee (5%–10% of direct plus indirect costs).

Total overnight cost (TOC) thus comes to about $600 million or $800/kW.

TABLE 13.9
Man-Hours Estimate for a Generic Construction Work [1]

Activity	Craft Type	#	Normal Time	Overtime	Shifts	Man-Hours
A	Boilermakers	2	8	0	4	64
	Trade helper	3	8	0	4	96
B	Welders	2	8	0	2	32
	Trade helper	2	8	0	2	32
…	…					…
	…					…
Total						986
Duration (weeks)						2
Hours/week						40
Average craftsmen						12

13.3 COST OF ELECTRICITY

In evaluating electric power generation technologies using dollars and cents, the most widely used metric is the *levelized cost of electricity* (LCOE), which combines plant output, efficiency and operating and maintenance (O&M) with capital investment and fuel expenditure in a simple formula. This metric is useful when comparing power generation alternatives that use *similar technologies*. The standard formulation of the *first-year* COE is the sum of capital, fuel and O&M costs of plant ownership, i.e., [6]

$$\mathrm{COE} = \frac{\beta \cdot C}{P \cdot H} + \frac{f}{\eta} + \left\{ \frac{OM_f}{P \cdot H} + \mu \cdot OM_{v,b} \right\}, \tag{13.5}$$

where

β = Capital charge factor (i.e., cost of money)
C = Total capital investment ($)
H = Annual operating hours
P = Net rated output (kW)
f = Fuel cost in $ per kWh (in lower heating value LHV)
η = Net rated LHV efficiency
OM_f = Fixed O&M costs ($ or $/kW-year)
$OM_{v,b}$ = Variable O&M costs for baseload operation ($/kWh)
μ = maintenance cost escalation factor (1.0 for baseload).

The cost of generation provided by this COE formula can be interpreted as the price at which electricity must be sold in order to cover all fixed and variable generating expenses and to match the return on company's equity implicit in the assumed *capital charge factor* (β). In other words, it is the price of electricity that would make the *net present value* (NPV) of the power project in question zero. The COE is limited to a single operating condition, typically new and clean rated performance, and is usually calculated at ISO baseload.

The fuel cost used in Equation 13.5 should be in $/kWh (LHV) and is given by

$$f = \frac{f_0 \cdot h}{293.071}, \tag{13.6}$$

where

h = Ratio of fuel higher heating value (HHV) to LHV (1.109 for natural gas, NG, as 100% CH_4)
f_0 = Base fuel cost in $ per MMBtu (HHV)
The levelized COE can be calculated as
LCOE = LF × COE
where LF is the *levelization factor.*

Total capital investment, C, can be determined per Department of Energy National Energy Technology Laboratory (DOE NETL)'s quality guidelines for energy systems [7]. It is the *total* "as-spent" *capital* (TASC), which includes interest during construction on top of the *total* "overnight" *capital* (TOC), which is a sum total of

- Bare erected cost (BEC), which is the sum total of direct and indirect costs
- EPC fee plus contingency
- Owner's costs.

BEC for each case can be obtained from Thermoflow's PEACE® software (Contractor's Internal Cost) or from detailed cost estimation. EPC cost and contingency together can be assumed to be 20% of BEC (see Section 13.2). Owner's cost can be 15%–30% of the subtotal (BEC plus EPC cost plus contingency).

Typical values of TASC/TOC factor, ϕ, are shown in Table 13.10 for *investor-owned utilities* (IOUs) and *independent power producers* (IPPs) [7] for three and five capital expenditure years.

Typical values for the financial parameters that can be used in the LCOE equation are [7]:

- For IOUs, β is 10%–13%
- For IPPs, β is 15%–20+% (see Table 13.11)
- For IOUs, LF is 1.268, for IPPs 1.169.

Financial parameters are subject to change from year to year (prevailing economic conditions) and from developer to developer. At the time of writing, for the IPPs, a more appropriate value of β is in the range of 11%–15%. The reader should always strive to use the most reliable, up-to-date (and pertinent) information in his or her studies.

A good rule of thumb for fixed O&M cost for GTCC is \$10–\$15/kW-year (higher values for advanced H and J class machines). For the variable O&M, one can use \$0.50 to 1.50/MWh. The lower value is appropriate for base-loaded plants; the higher value for cyclic operation. Somewhat higher values may be adopted for H and J class machines comprising advanced single-crystal alloys and coatings. The reader is encouraged to refer to Ref. [5] for more information on generic O&M costs for gas turbine and other fossil fuel technologies. In a GTCC power plant, gas turbine is the largest O&M cost contributor. Gas turbine maintenance comprises three major activities: combustor inspection, HGP inspection and major inspection [8]. Periodicity of these maintenance activities is a function of fired operating hours and start–stop cycles. For the state-of-the-art advanced machines comprising high-technology parts and operating at high firing temperatures, the prudent but costly approach is to enter into *long-term service agreements* (LTSAs) with the OEM.[5] For the more mature technologies such as the older F or E class machines, independent (third party) service providers can be the less costly choice with minimal risk [9].

Prevailing and historic US NG spot prices can be easily found on the internet. In the past, before the shale gas glut hit the market, it was as high as \$13 per MMBtu (HHV). Depending on unexpected weather conditions (pushing up the demand for home heating), it can still show significant spikes (e.g., winter of 2014). For the time being, \$3 NG seems to be a reality in the USA, which makes widespread acceptance of advanced but proven technologies such as IGCC (especially with

TABLE 13.10
TASC/TOC Factors

	Three Years	Five Years
High-risk IOU	1.078	1.140
Low-risk IOU	1.075	1.134
High-risk IPP	1.114	1.211
Low-risk IPP	1.107	1.196

TABLE 13.11
Capital Charge Factors

	Three Years	Five Years
High-risk IPP (%)	17.7	21.4
Low-risk IPP (%)	14.9	17.6

[5] Also known as contractual service agreements (CSA).

their cost and schedule creep problems) extremely difficult. Overseas, especially in countries dependent on imported NG or LNG, it is around \$7–\$8 at the time of writing (2016–2017) but it went up as high as \$15 in the recent past (*landed* price, i.e., as received at the terminal). This is one reason why coal-burning technologies such as advanced USC and carbon capture are expected to make significant inroads in Europe and Japan well before the USA.

Strictly speaking, the basic LCOE formula has a very limited applicability; it is only appropriate for comparing *similar technologies of the same vintage*, i.e., an F class, 3PRH GTCC power plant with another GTCC of the same configuration with the same features. The technologies must have similar rated performance, part load and ambient condition efficiency lapse, degradation, *RAM (Reliability, Availability and Maintenance)*, emissions, etc. In other words, for a given operating scenario, they must deliver approximately the same annual megawatt-hour generation. Unfortunately, it has been used for comparing different technologies (e.g., IGCC with GTCC) with total disregard of the aforementioned technology characteristics. The pitfalls and corrective remedies are discussed in detail by Gülen and Mazumder [6]. In simple terms, for a better handle on the LCOE estimate, it is better to use mean-effective values of output and efficiency in Equation 13.5 (instead of, say, ISO baseload rating values).

Mean-effective values of net power output and efficiency, $\overline{P}$ and $\overline{\eta}$, respectively, take into account variability in combined cycle plant ambient site and loading conditions over the course of a year. This is accomplished by the *load factor* (LF). Consider the cyclic operation of an advanced GTCC power plant over the course of a given time period as shown in Figure 13.3.[6]

The LF is the equivalent combined cycle part load, which would generate the *same* megawatt-hours of electricity over the same time as the actual generation *if the plant ran at that constant load.* For various modes of GTCC cyclic operation, LF can vary between 0.70 and 0.80. The corresponding mean-effective combined cycle efficiency can be found from the particular cycle part load curve. For a modern GTCC with F class gas turbines, a typical part load table (normalized) is provided in Figure 13.3.

As far as the average site ambient is concerned, it can be shown that, in most climates, the annual load-weighted average ambient temperature is somewhere between 60°F and 65°F. (This is so *unless* the plant has a peculiar operating regime, e.g., running in summer and waiting in winter or something along those lines.)

Consequently, mean-effective GTCC performance parameters, $\overline{P}$ and $\overline{\eta}$, as functions of their ISO baseload values can be found as follows:

$$\overline{P} = LF \cdot K1 \cdot P_0 \tag{13.7}$$

$$\overline{\eta} = K2 \cdot K3 \cdot \eta_0, \tag{13.8}$$

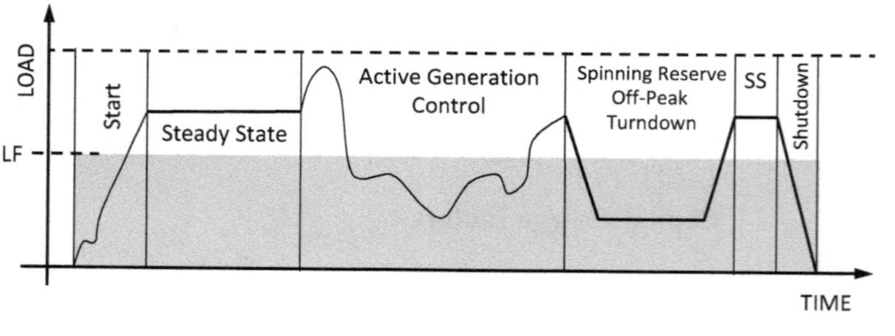

FIGURE 13.3 Typical combined cycle operation profile (qualitative).

[6] Adopted from a presentation by General Electric to EPRI (2011).

where the subscript 0 denotes the ISO baseload values and

LF = Load factor
K1 = Ambient correction for output
K2 = Efficiency correction for load
K3 = Ambient correction for efficiency.

Using the mean-effective values of output and efficiency, the simple LCOE metric obtained from a single formula represents a better approximation of the "true" LCOE value, which one would obtain from a full-blown, *pro forma* calculation with detailed day-to-day generation and fuel consumption data over the economic life of the power plant.

Equation 13.5 can be rewritten as follows (where $k = TOC/P_0$ is the *specific capital cost*):

$$LCOE = LF\left\{\frac{\beta\cdot\phi\cdot k}{LF\cdot K1\cdot H}+\frac{f}{K2\cdot K3\cdot\eta_0}+\frac{OM_f}{LF\cdot K1\cdot P_0\cdot H}+\mu\cdot OM_{v,b}\right\}. \quad (13.9)$$

Capacity factor (CF) is the ratio of a power plant's actual generation to its maximum potential generation. The relationship between CF and H is given below

$$CF = \frac{LF\cdot K1\cdot H}{8,760}. \quad (13.10)$$

Using the symbol om_f for fixed O&M cost per rated kW and Equation 13.10, Equation 13.9 is recast into the following form, which clearly illustrates the "tug-of-war" between performance and operational/economic parameters (see Section 13.5.1 below).

$$LCOE = LF\left\{\underbrace{\frac{\beta\cdot\phi\cdot k+om_f}{CF\cdot 8,760}}_{\text{Fixed}}+\underbrace{\frac{f}{\eta}+\mu\cdot OM_{v,b}}_{\text{Variable}}\right\}. \quad (13.11)$$

Even with the expanded version proposed in Ref. [6] and represented by Equation 13.9 herein, the fact still remains that the LCOE concept is a "relic" from the days when the electricity markets were dominated by regulated utilities (e.g., see the brief history in the beginning of Chapter 11). When it first came out, LCOE represented the average lifetime cost for providing a kilowatt-hour of electricity for given full-load hours. In other words, as mentioned earlier, it gave the utility planners the constant price of electricity for which the NPV of the investment they were planning (say, a 750-MWe coal-fired power plant for baseload generation) would be equal to zero. This of course helped the utility planners to assess the level of electricity tariffs.

Since 1990s when the electricity markets were liberalized in the developed world, the LCOE metric is not a reliable gauge to assess the competitiveness of a given power generation investment. This is so because competitive markets establish prices that reflect the *marginal costs* rather than the *average costs* in the LCOE formula, which are independent of system requirements. In the USA, three major market developments can be cited in that regard:

- Passed in 1996, *FERC Order 888* forced utilities to provide non-discriminatory market access to merchant generators. As a result, IPPs were able to sell power into different markets for the first time. In conjunction with FERC Order 2000, Order 888 also established the framework for ISOs/Regional Transmission Organizations opening up investment access to many new investors.

- Around 2000, ISOs introduced *locational marginal pricing* (LMP), which allows more certain pricing data to be known across many different points in a marketplace, as opposed to a single zonal price. These better price signals help developers site new generation in locations that are most in need and offer the highest returns.
- First introduced in 2006, *capacity markets* provide an incentive for new plant investment by providing a more predictable revenue stream. Capacity markets are functioning in ISOs across the USA such as New York ISO and PJM.

Once electricity prices are input into investors' profitability calculations (rather than being outputs of them), LCOE loses its predictive value. Investors use detailed *pro forma* spreadsheets to calculate the NPV to assess whether the cash flow of a new power project is sufficient to reimburse their investment to finance it. Those calculations are based on expected external electricity prices as well as their variation and uncertainty over time.

In pro forma calculations, usually two types of revenues are considered: capacity and energy. The former is a simple product of the power generation asset's capacity and the capacity price. Energy revenue is a product of three factors: capacity, dispatch hours and the electricity price during dispatch hours. The level of dispatch depends on the bid made by the generating asset operator. In a competitive market, the bid price reflects the variable costs of the generation asset, i.e., fuel price, O&M costs, any environmental allowance costs plus any *uplift* payments[7] required by the operator. (In a competitive market, the hourly dispatch of a plant will be based on simple arithmetic. If the plant's variable costs are lower than the hourly market price, the plant will be dispatched, otherwise not.)

Competitive wholesale or spot electric energy prices are determined on an hourly basis by matching supply (i.e., available generating resources) with demand. In each hour, the prevailing spot price of electric energy will be approximated by the short-run marginal cost of production of the most expensive unit operating in that hour. Reasonably accurate projections of spot and capacity prices and combining them with detailed performance data of the generating asset to come up with a reliable cash flow stream and NPV are requisite for making investment decisions that can run into a billion dollars or even more. For a detailed look into such modeling endeavors, the reader is referred to a 2010 report issued by the US DOE's NETL, *Investment Decisions for Baseload Power Plants*, (NETL Report 402/012910, January 29, 2010), which is available online free of charge.

13.4 VALUE OF 1 BTU/KWH OF HEAT RATE

For quick evaluation of technology improvements for a given system, a useful tool is *maximum acceptable increase in capital cost* (MACC), which is the capital cost equivalent of a given plant performance improvement (output and/or heat rate or HR) with no change in COE. Using the basic COE formula and ignoring the change in O&M costs, MACC is calculated by equating before and after values COE for a given plant improvement, i.e.,

$$\mathrm{MACC} = \frac{\mathrm{H} \cdot \mathrm{f}}{\beta}\left(\frac{\Delta \mathrm{P}}{\eta_0} - \Delta \mathrm{HC}\right) + \mathrm{k}_0 \cdot \Delta \mathrm{P}. \tag{13.12}$$

where f is the levelized fuel cost (in $ per kWh in LHV), ΔP is the change in plant net output (kW), ΔHC is the change in plant heat consumption (HC) (in kW), η_0 is the base plant net efficiency and k_0 is the base plant-specific capital cost ($/kW).

[7] Uplift is the gap between the revenues collected in market settlements, and the compensation that generators require (based on their bids). The exact magnitude of the uplift depends on the methodology used by the ISO to calculate market prices.

The two terms on the right-hand side of the MACC formula in Equation 13.12 give the value of plant HR (or, its equivalent, efficiency, η) and output, P. In general, an improvement in combined cycle HR (efficiency) comes with a change in output and heat (i.e., fuel) consumption, HC. If the improvement is limited to the bottoming cycle or gas turbine HGP section, however, there is no change in gas turbine HC, i.e., $\Delta HC = 0$. Using the well-known definition of efficiency and HR, i.e., $\eta = P/HC$ and $HR = 3{,}412/\eta$ Btu/kWh in US customary system, it can be shown that the value of 1 Btu/kWh reduction in HR is given by

$$VHR = 10^{-6}\frac{H\cdot f}{\beta}\cdot P_0\left(1+\frac{\Delta P}{P_0}\right)\approx 10^{-6}\frac{H\cdot f}{\beta}\cdot P_0. \tag{13.13}$$

where P is in kW with f in \$/MMBtu (LHV) and P_0 is the base value of plant output.

The value of each kilowatt of output is simply k_0. These two value numbers form the theoretical basis of metrics used for evaluation of bids in commercial projects as well for the payments to be made to the owner for missed performance guarantees (i.e., liquidated damages or LDs).

Note that, for the special cases where ΔHC is zero, Equation 13.13, which is directly derived from the COE equation is incorrect and overstates the value of HR by a very large margin. This was shown via detailed mathematical analysis in Ref. [10]. In particular, if the HR of the GTCC is reduced by a design improvement in the bottoming cycle, the change in the fuel bill of the plant operator is zero.

An "intrinsic" value of HR, which is fully independent of any change in power output, can be evaluated by applying a *realization factor*, RF, which is rigorously evaluated in Ref. [10] and can be estimated as

$$RF = 0.2355\cdot \ln(E) + 1.6634. \tag{13.14}$$

The independent variable, E, in Equation 13.14 is the *efficiency improvement factor*, which is defined as percent change in η per percent change in P. Thus, the intrinsic value of 1 Btu/kWh of HR is

$$VHR = 10^{-6} RF\frac{H\cdot f}{\beta}\cdot P_0. \tag{13.15}$$

As an example, consider a combined cycle power plant with 500 MWe net output and 58% net LHV efficiency. The gas turbine compressor is upgraded with new technology to give 0.7% higher combined cycle output and 0.3% higher combined cycle efficiency (i.e., the new combined cycle performance is 503.5 MWe and 58.17%). Therefore, assuming H = 4,250 h/year, β = 12% and LF = 1.169

- The efficiency improvement factor E is 0.3/0.7/100 ~ 0.0043 (or 0.43% for each 1% in P).
- Using this value, from Equation 13.14, RF is calculated as 0.38.
- At \$3 fuel, using Equation 13.15, the intrinsic value of 1 Btu/kWh in HR is \$26,180.

In addition to the significant difficulty and uncertainty involved in capital investment estimation, the uncertainty in predicting future price of fuel, load or demand growth; inflation and other economic and/or financial fluctuations (e.g., interest rates) resulting from social and political turmoil makes LCOE a very tricky tool. As such, to the extent possible, probabilistic methods such as *Monte Carlo simulation* should be preferred over deterministic comparisons. By assigning probabilities to the key parameters in the LCOE formula (by no means an easy task), the end result from any LCOE comparison should be the *probability* of "option A being lower/higher than option B" and *not whether* "option A *is* lower/higher than option B".

Application of *real options theory* to the valuation of power generation assets is a powerful technique that should be superior to *deterministic* methods such as LCOE evaluation. This is especially true for the currently (and most likely in future as well) prevalent deployment mode of fossil generation assets – especially the simple and combined cycle gas turbines. While the real options theory provides a powerful *stochastic* tool for economic analysis, its application requires sophisticated mathematical modeling and computer programming expertise. Nevertheless, the reader is strongly encouraged to consult the introductory book by Mun [11] to obtain an idea on the key principles underlying the real options theory, which should also help with applying probabilistic techniques such as Monte Carlo simulation to the LCOE modeling.

13.5 BOTTOMING CYCLE "OPTIMIZATION"

The best way to illustrate the cost (or price?) versus benefit analysis using the metrics developed above is via a realistic problem. The problem in question is defined as follows: What is the optimal bottoming cycle choice for a combined cycle power plant with an advanced class gas turbine?

As discussed earlier in the book, current state of the art in GTCC design is 3PRH steam bottoming cycle with steam generation at three different pressure levels. The goal is to maximize total steam generation and STG power output for a given gas turbine exhaust energy and, thus, to maximize combined cycle efficiency. There are three "knobs" available to the designer to dial in that maximum – all dictated by fundamental thermodynamic considerations:

- Steam mass flow rate
- Steam availability (exergy)
- Heat rejection temperature (steam condenser pressure).

All three knobs have a significant impact on bottoming cycle equipment size (footprint and weight) and cost via following mechanisms:

- HRSG heat transfer surface area
- Condenser and cooling tower heat transfer surface area
- Pipe, tube and steam turbine/valve casing/shell materials (high grade stainless and/or alloy steels)
- Steam turbine exhaust annulus area (last stage bucket (LSB) length).

Thermodynamic drivers of steam generation at multiple pressure levels were explored in detail in Chapter 6 (to minimize the heat transfer irreversibility). As mentioned earlier, *three* pressure levels (high, intermediate and low, HP, IP and LP, respectively) are the industry standard. See Section 8.2.2 for typical design parameters.

The term "maximizing" used in reference to the STG output has two connotations: "make as much as possible" and "make the best use of [what?]". The [what?] in question is, of course, capital cost of the resulting system. Otherwise, simply building equipment as large as possible using the most exotic materials with no regard to cost, footprint and ease of construction would lead to higher and higher performances (up to a certain limit, of course, set by the second law of thermodynamics – more on this later). This, in fact, is pretty much the approach taken by the OEMs in advertising "world record" combined cycle efficiency ratings as well as achieving such "world record" performances in showcase power plants with highly advantageous site characteristics (e.g., proximity to a year-round available cooling water source such as a river or ocean). Unfortunately, this is not exactly a widely reproducible and/or sensible business approach to the problem at hand.

In fact, each bottoming cycle is a tailor-made system specific to the particular project strongly dependent on owner/developer's financial criteria, site conditions and prevailing (or projected) economic climate. There are some discontinuities (or break points) introduced mainly by steam turbine OEMs' product line portfolios (essentially casing/shell configuration and LSB size) and, to some extent, other equipment vendors' in-house design practices (e.g., HRSG "box" sizes, cooling tower cell/fan sizes) but, otherwise, this is a fairly continuous design spectrum.

As in any rational business endeavor, selection of the proper bottoming cycle is a process of trade-off between performance (net electric power output and thermal efficiency) and total capital investment and operating costs to maximize a suitable financial criterion, e.g., NPV or internal rate of return (IRR). This process of optimization requires quite a long list of inputs and methods, either of which can be classified into two categories:

- Deterministic (known – past, present and future)
- Probabilistic or stochastic (past and present known, future subject to uncertainty).

With the exception of guaranteed performance of the particular power plant (power output and thermal efficiency plus emissions and maybe some other criteria) at specified boundary conditions (site ambient and loading), almost all financial, economic and thermodynamic parameters, which are inputs to the said optimization problem, are essentially probabilistic with varying degrees of uncertainty and randomness.

LCOE is the most convenient metric combining a power plant's ownership costs (capital and operating) and thermal performance (output and efficiency). The deterministic approach to the optimization problem cited above is *minimization* of life cycle LCOE.

For a capital investment with 20–30 years of economic life, it is not realistic to expect that a deterministic solution found at time $t = 0$ is going to be a robust answer. While there are highly complex methods to optimize systems with stochastic boundary conditions (e.g., "real options" method), it is highly unlikely that such studies are undertaken in actual power plant project evaluation. Nevertheless, there are less rigorous "robustness tests" such as Monte Carlo simulations to assess the "resistance" of deterministic answers to random fluctuations (i.e., future uncertainty) in boundary conditions (e.g., fuel price). In the investigation presented below, the deterministic path is taken.

13.5.1 LCOE Dissected

There are *four* important parameters in the LCOE formula, Equation 13.11, and the "tug-of-war" between them constitutes the key to optimization (i.e., LCOE *minimization*):

- The tug-of-war between specific capital cost, k in $/kW, and (i) plant LF and (ii) annual operating hours, H
- The tug-of-war between fuel price, f, and thermal efficiency, $\overline{\eta}$.

In particular, investing a lot of capital into a power plant (i.e., high k) in order to buy as much efficiency as possible (i.e., high $\overline{\eta}$) can only be justified if

- The expected/projected electric energy generation (kWh or MWh) is commensurately large, i.e.,
    - High plant LF (i.e., more kilowatts or megawatts)
    - high annual operating hours (i.e., high CF)
- The fuel price f is high.

Each parameter is looked at separately below.

What amount of extra capital investment into the bottoming cycle is justified by the improved combined cycle efficiency via increased STG output? This is a fundamental question, whose answer is dictated by the basic principles of GTCC power plant thermodynamics and economics. This can be easily verified by data extracted from the *budgetary price* numbers listed in *Gas Turbine World 2014–2015 Handbook* for simple and combined cycle gas turbine power plants. For a large heavy-duty GTG of a 300+ MWe size, say, each kilowatt from the bottoming cycle costs more than *six times* that from the topping cycle (see Figure 13.1 – roughly \$1,500/kW vs. \$250/kW).

In order to put the disparity in topping and bottoming cycle prices into a plant-level quantitative perspective, assume a 500 MWe GTCC power plant at 60% net efficiency (5,687 Btu/kWh). Additional 5,000 kW bottoming cycle output costs about \$7.5 million in budgetary price and "buys" 0.6 percentage points of efficiency or 57 Btu/kWh HR. In other words, assuming \$675/kW for the combined cycle budgetary price, each one Btu/kWh reduction in net HR comes at a cost of \$75,000.[8] Is this a good trade-off? In order to answer this question, we need to point out several key factors based on available industry data.

Annual operating hours are typically expressed in terms of the CF. US Energy Information Administration, Electric Power Monthly, Table 6.7a provides monthly CFs for 16 different fossil and non-fossil fuel and technology combinations. The data for NG-fired combined cycle power plants is summarized in Figure 13.4. Prior to 2010, these plants were run at a very low CF but the situation changed quite dramatically in recent years. Clearly, shale gas "boom" and ensuing low NG prices played a significant role in this. Even so, it is hard to envision that the annual average CF for NG-fired combined cycle power plants will be much higher than 55%–60% in the foreseeable future (especially with increasing renewable resource penetration). Translation from CF to annual hours, H, is subject to uncertainty since the annual average LF is not known and there is significant HRSG supplementary firing and gas turbine inlet conditioning to boost output, especially during summer. A wide variation from plant to plant is to be expected (more on this later). For example, a daily-cycled combined cycle power plant with weekend shutdown and 2 weeks of scheduled maintenance

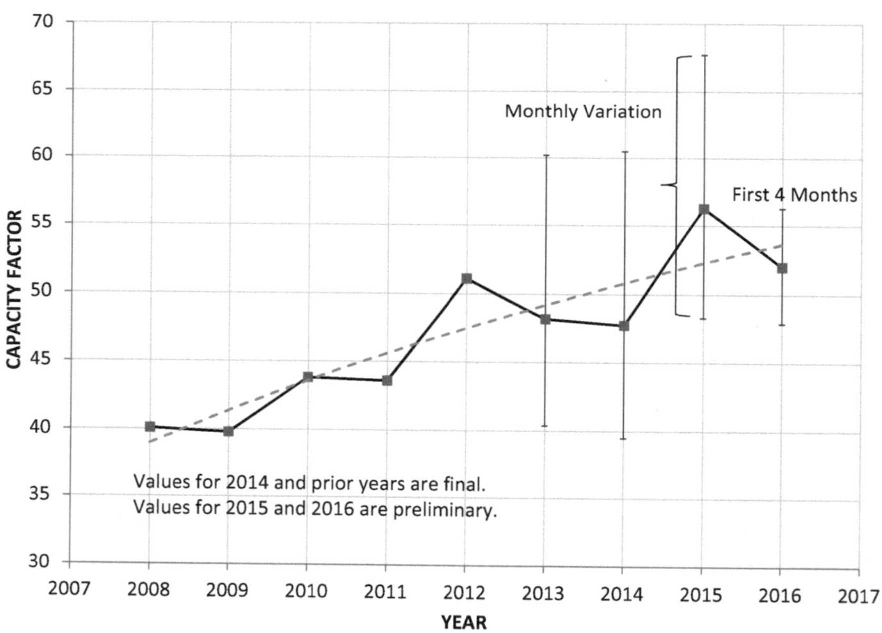

FIGURE 13.4 US EIA (Electric Power Monthly, Table 6.7a) CF data for NG-fired combined cycles.

[8] $(7,500,000 - 675 \times 5,000)/57 = 73,660 \sim \$75\text{K}$.

will run only 50 × 5 × 16 = 4,000 h/year, which corresponds to a CF of 0.75 × 4,000/8,760 = 35% (LF of 0.75, no supplementary firing or gas turbine inlet conditioning). However, CFs in Figure 13.4 are significantly higher, which is an indication of significantly higher LF (e.g., more hours at full load), power augmentation (via supplementary firing and/or gas turbine inlet conditioning) or a combination thereof.

Long-term NG price forecasts are difficult to make as illustrated by the chart in Figure 13.5,[9] which superimposes outlooks by the US DOE (consistently predicting increasing scarcity and rising prices) and The National Petroleum Council (NPC), with the latter comprising *pessimistic* (reactive path) and *optimistic* (balanced future) scenarios.[10]

Except for the 2003–2008 period, when prices spiked above historical levels due to a tight market caused by several factors, i.e., weak supply and growth in demand for peaking power in particular, long-term expectation of annual ~2% growth in NG prices pretty much held (going back to the Carter administration era and the *Alaska Natural Gas Transportation System* (ANGTS) project). Right after that peak price period, development of new sources of shale gas, driven by hydraulic fracturing and horizontal drilling technologies, has more than compensated for the decline in conventional supply and has led to major increases in reserves of US NG. The so-called shale gas boom, although not a guarantee by any means, is expected to prevent non-seasonal, year-long price spikes and exorbitant long-term growth rates in the USA. The situation is somewhat different in Europe and Japan (see Figure 13.6).

What about the thermal efficiency? Historical GTCC efficiencies (rating, i.e., "advertising", numbers as well as selected "field-clocked" values) are depicted in Figure 13.7 (this is an updated version of Figure 2 in a paper on combined cycle evolution by Gülen [12]). Also included in the same graph are the average efficiency of *top twenty* (in terms of HR) gas fired GTCC plants in the US in 2004–2015 – including duct-fired units – which squeaked past 55% (LHV basis) only in the last couple of years.[11] (This was probably driven by the commissioning of the more advanced FA/H class units and increasing LF – it is difficult to glean from the data, which includes only generation and fuel consumption numbers.) As illustrated by the min–max range, a select few were as high as 57% whereas most plants (remember: these are among the 20 best in terms of performance – for a reality check, see the average 2015 efficiency of NG-fired generation in Table 12.1) were clocked at only about 53%! Recent data from the most advanced units in the field is summarized in Table 13.12.

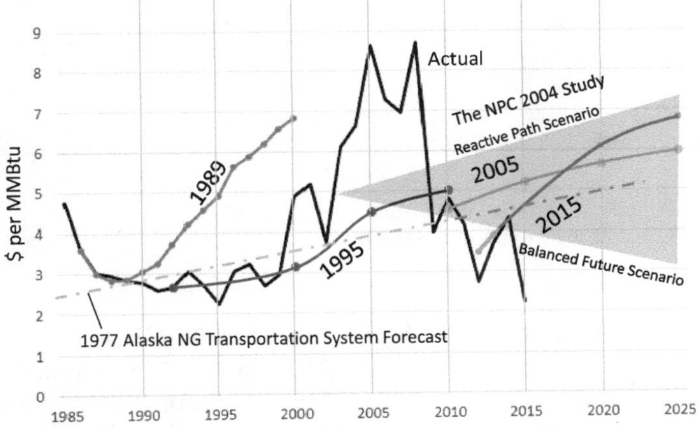

FIGURE 13.5 Evolution of US DOE NG price forecasts and actual prices.

9 Lynch, M., 2016, "The Confusion About Natural Gas Prices," Forbes Blog, 5/23/2016.
10 The two scenarios are classified in terms of public policy conflict between supply and demand.
11 U.S. Energy Information Administration, Form EIA-923 Data, www.eia.gov/electricity/data/eia923/.

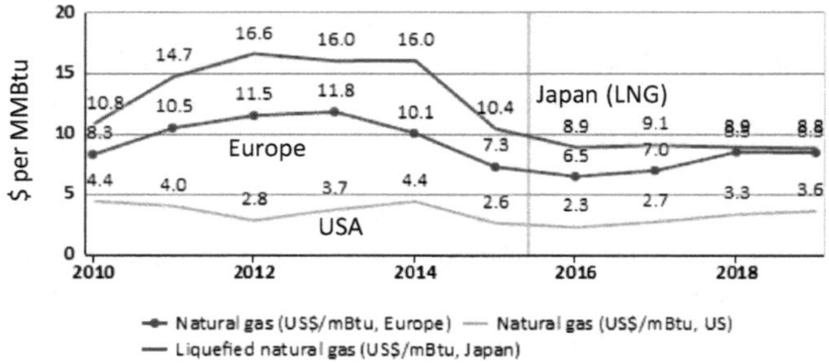

FIGURE 13.6 Comparison of gaseous fuel prices (past and short-term outlook) in USA, Europe and Japan (EIU Economic and Commodity Forecast, December 2015).

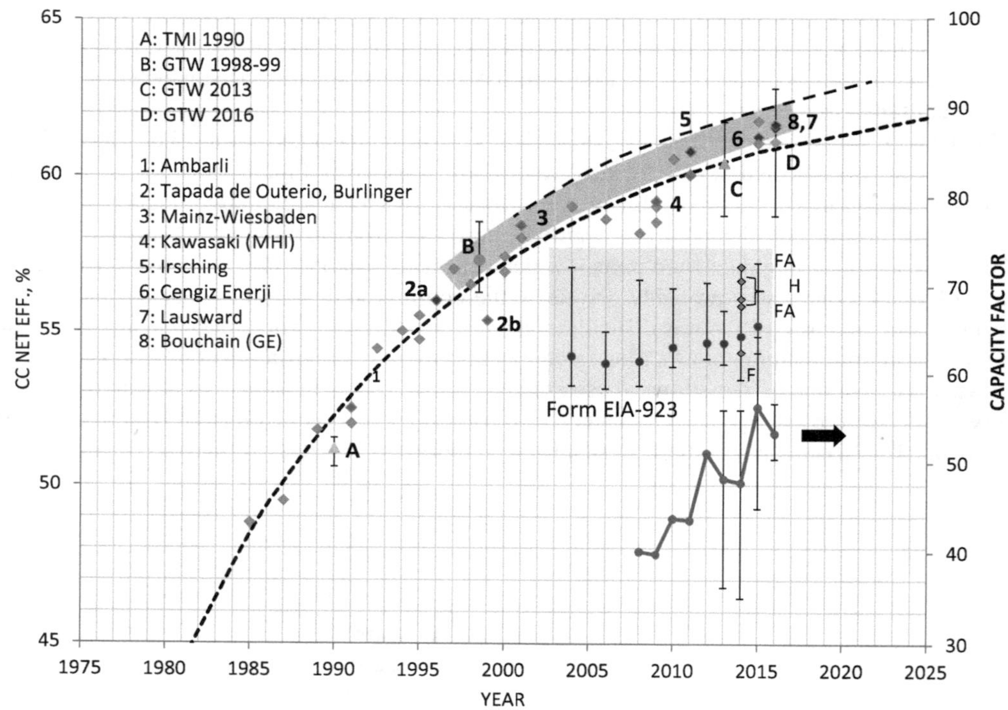

FIGURE 13.7 GTCC efficiency evolution 1985–2015.

Coming back to the question posed at the beginning (i.e., additional 5 MW bottoming cycle output at $7.5 million extra cost – a good trade-off or not?), using the LCOE formula and assumptions outlined in Tables 13.13 and 13.14, the answer is "it depends". For a GTCC plant with cyclic duty, the value of the proposed improvement of extra 5 MW output is about $6 million for a fuel price of $4/MMBtu (HHV) or about $50,000 per each one Btu/kWh reduction in net HR. The fuel price to make it worthwhile at $7.5 million cost adder is about $6.50/MMBtu (HHV). Alternatively, at $4 fuel, the plant should run around 5,800 h/year at baseload duty (LF of 0.9) to justify the $7.5 million cost adder.

TABLE 13.12
Selected Form EIA-923 Data (January–September 2018)

			Total Electricity Generation	Total Fuel Consumption	Mean-Effective HR	Mean-Effective Efficiency	
OEM	Class	Configuration	MWh	MMBtu HHV	Btu/kWh HHV	% HHV	% LHV
A	F	2 (2 × 2 × 1)	2,315,620	16,954,109	7,322	46.60	51.68
A	F+	2 × 2 × 1	2,159,244	15,275,314	7,074	48.23	53.49
A	H	3 × 3 × 1	5,482,523	36,275,531	6,617	51.57	57.19
A	H	3 × 3 × 1	5,413,559	36,020,550	6,654	51.28	56.87
B	H+	2 × 2 × 1	2,315,067	16,682,609	7,206	47.35	52.51
B	H+	2 × 2 × 1	1,983,351	13,060,540	6,585	51.82	57.46

TABLE 13.13
LCOE Parameters

β	16%
HHV/LHV	1.109
O&M fixed, million $ per year	$7.50
O&M variable (baseload), $/MWh	$2.60
Levelization factor, LF	1.169
Total "as-spent" capital factor, ϕ	1.107

TABLE 13.14
Plant Operating Regimes

	Baseload	Cyclic Load
Annual operating hours, H	6,500	4,000
Average site ambient conditions	65°F/60%RH	65°F/60%RH
Variable O&M multiplier, μ	1.0	2.0
Load factor, LF	0.90	0.75
Load correction – ambient, K1	1.0233	1.0247
Efficiency correction – load, K2	0.9983	0.9860
Efficiency correction – ambient, K3	1.0033	1.0028

13.5.2 Two-Pressure or Three-Pressure?

In Chapter 6, an in-depth look into the thermodynamic principles governing the HRSG design illustrated the "diminishing marginal return" in adding successive pressure levels. This trend is accelerated by the advanced class gas turbines with high turbine inlet and exhaust temperatures (the latter getting close to or even exceeding 1,200°F). The qualitative explanation for this can be gleaned from the two figures in Chapter 6, i.e.,

- Figure 6.4, which illustrates the relatively small contribution of the third pressure section to total steam production
- Figure 6.6, which shows the "depressing" effect of increasing gas turbine exhaust temperature on the HRSG stack temperature.

(In the limit, one can envision that the HRSG will be analogous to a conventional boiler with single-pressure steam production. This is in fact what happens in supplementary-fired HRSGs – see Section 6.3.) In this section, we will take an even deeper look into this question, i.e., whether a third pressure level is cost-effective or not, using the LCOE approach.

Two cases are evaluated in order to estimate capital investment savings in a 1 × 1 × 1 single-shaft GTCC similar to that proposed for an actual combined cycle project. The base case is set as follows:

- 1,600°C turbine inlet temperature (TIT) class gas turbine (NG-fired, with inlet evaporative cooler)
- Design ambient conditions 90°F, 40% relative humidity
- 3PRH unfired steam cycle: 2,415 psia HP throttle with 1,050°F for HP and hot reheat steam admission
- Air-cooled condenser at 3.5 in. of mercury
- HRSG evaporator pinch deltas 15°F.

The second "cheap" case is based on a 2PRH steam cycle with 1,815 psia cycle and 25°F HP evaporator pinch delta. LP admission pressure is the same as in the base 3PRH case. In this case, gas turbine fuel gas performance heating (to 410°F, same as in the base case) utilizes hot feedwater from a dedicated economizer section. The gas turbine is fired 8°F higher to maintain the same GTCC net output as the base case. (The implicit assumption here is that the gas turbine in question is one of the latest H/J class machines with the most advanced technology – superalloys, coatings, cooling schemes, etc. – allowing the OEM limited "wiggle room" about the nominal TIT of 1,600°C. For a gas turbine quoted by the OEM at its extreme capability, of course, this is not a feasible option.) Cycle performances are calculated using Thermoflow's GTPRO® software. TOC is calculated using the PEACE® add-in with calibration per above. LCOE assumptions are listed in Tables 13.13 and 13.14. The results are summarized in Tables 13.10–13.13.

Gas turbine and generator prices are taken from *GTW 2014–2015 Handbook* budgetary price data. Steam turbine price is from PEACE® with calibration based on in-house data.

Three HRSG OEMs are provided heat and mass balance data roughly corresponding to two variants with (i) same NG-fired gas turbine exhaust gas conditions (~1,500 lb/s and 1,175°F) at ISO baseload and (ii) same HP throttle conditions (nominal 1,800 psig and 1,050°F):

1. Base (conventional) case: 3PRH with normal HRSG pinch deltas (15°F)
2. "Cheap" case A: 3PRH with large HRSG pinch deltas (25°F)
3. "Cheap" case B: 2PRH with normal HRSG pinch deltas.

In addition to the price quotes from the OEMs, quantities and man-hour savings resulting from the elimination of the IP section are estimated as well. The latter is based on a similar HRSG unit by one of the three OEMs (in terms of size, configuration and steam production), which was erected at a recent US combined cycle project. Deleted commodities and associated labor included large bore pipe, valves, supports, and welds; small bore piping, IP steam drum, pressure relief valves (flanged), IP relief valve silencers and their support steel and platforms; instruments, HRSG hydro testing, chemical cleaning and pipe installation and IP two-row wide box deletion. The resulting saving was equivalent to slightly above 10,000 man-hours.

It was found that nearly $1 million saving in equipment price is achievable via a cheaper HRSG (i.e., by the removal of the IP section) whereas nearly half a million dollars can be saved by going to a smaller HP section. Including the savings in erection materials and labor, one ends up with $1.7 million total saving per HRSG (i.e., ~$1.5 million in price plus ~$750 K for construction).

Furthermore, per OEM feedback, HRSG price delta between 2,400 and 1,800 psig cycle is around $500 K based on tube, drum and pipe thicknesses with consideration for valve classification. Thus, the HRSG equipment *price* difference in Table 13.15 can be broken down as follows:

- $500 K for 2,400–1,800 psig steam cycle
- $525 K for HP evaporator pinch increase by 10°F
- $1 million for IP section elimination.

The "Mechanical" cost bucket in Table 13.16 includes onsite transportation, rigging, equipment erection assembly plus piping (materials plus labor). The difference of about $2.4 million between the "Base" and "Cheap" versions can be broken down as follows:

- $750 K for IP section elimination
- $1 million for 2,400–1,800 psig steam cycle
- $600 K for HP evaporator pinch increase by 10°.

The latter two are estimated by the PEACE® program and mainly driven by smaller and lighter HRSG. They have not been verified by detailed construction material take-off and labor estimates.

TABLE 13.15
Engineered Equipment Cost Breakdown

	Base	Cheap
Total engineered equipment	161,900,000	160,195,000
Gas turbine + generator	82,500,000	82,500,000
Steam turbine	14,400,000	14,150,000
Heat recovery boiler	14,350,000	12,325,000
Air-cooled condenser	28,000,000	28,570,000
CEMS	400,000	400,000
DCS	1,500,000	1,500,000
Transmission voltage equipment	12,750,000	12,750,000
Generating voltage equipment	8,000,000	8,000,000

TABLE 13.16
TOC Roll-Up

	Base	Cheap
Engineered equipment	161,900,000	160,195,000
BOP equipment	7,000,000	6,345,000
Civil	20,700,000	18,315,000
Mechanical	27,486,000	25,111,000
Electrical assembly and wiring	8,650,000	8,000,000
Buildings and structures	3,450,000	3,450,000
Engineering and plant startup (8%)	18,335,000	17,713,280
BEC	247,520,000	239,130,000
EPC fee plus contingency (20%)	49,500,000	47,825,000
EPC contractor's price	297,025,000	286,955,000
Owner's costs (10%)	29,700,000	28,695,500
TOC	326,727,500	315,650,650

The "Civil" cost bucket in Table 13.16 includes site work, excavation and backfill and concrete foundations (including rebar). The difference of about $2.4 million between the "Base" and "Cheap" versions can be broken down as follows:

- $900 K for 2,400–1,800 psig steam cycle
- $400 K for HP evaporator pinch increase by 10°
- $900 K for IP section elimination.

All three are estimated by the PEACE® program and mainly driven by reinforced concrete foundation material and labor for the smaller and lighter HRSG. They have not been verified by detailed construction material take-off and labor estimates as described in Section 13.2.2.

Detailed performance comparison of base 3PRH and "cheap" 2PRH bottoming cycles is presented in Table 13.17. The assumptions used in LCOE calculation are listed in Tables 13.13 and 13.14. The reason for 7.9°F higher firing temperature in the "cheap" 2PRH variant is to equalize

TABLE 13.17
Performance and Cost Comparison ($5/MMBtu Fuel)

Case	Base	Cheap
CC configuration	1 × 1 × 1 single shaft	
BC configuration	3P, reheat	2P, reheat
HP/HRH steam temperatures, °F	1,050	
HP/IP/LP steam pressures, psia	2415/450/74	1815/342/65
Heat rejection	ACC/3.5″ Hg	ACC/3.5″ Hg
Gas turbine	1,600°C TIT class	
Fuel	100% CH_4	100% CH_4
Fuel temperature, °F	410	410
Fuel heat source	IP economizer	Dedicated economizer
Firing temperature	BASE	+7.9°F
Evaporative cooler	ON	
GT exhaust loss, in. H_2O	17.2	16.1
GT shaft, kW	325,313	328,062
GT HC, kWth	788,773	793,783
GT fuel, lb/s	34.75	34.97
GT exhaust temperature, °F	1,191	1,197
GT exhaust flow, lb/s	1,466	1,466
ST shaft, kW	153,755	150,082
CC generator efficiency, %	99.10	99.10
CC gross output	474,736	473,827
CC gross efficiency, %	60.19	59.69
Plant auxiliary load, %	2.41	2.22
CC net output	463,290	463,290
Δ output		0
CC net efficiency, %	58.74	58.36
CC net heat rate, Btu/kWh	5,809	5,846
Δ heat rate		+37
CC TOC	$326.7 million	$315.7 million
CC TOC delta		–$11.1 million
LCOE – cyclic, $/MWh	$97.82	$96.58
LCOE – baseload, $/MWh	$68.19	$67.67

the net output of the two cases. Otherwise, the cheaper 2PRH design would have 3.7 MWe lower output (with about the same cost delta and slightly higher LCOE, i.e., \$96.92 and \$67.87/MWh for cyclic and baseload duties, respectively, but still lower than those for the more expensive 3PRH variant). To reflect the impact of the higher firing temperature on the parts life (LTSA cost), \$160 K is added to the annual fixed O&M cost (see below for a more in-depth discussion of this aspect).

Clearly, even at \$5 NG, which is on the expensive side for the US market in the foreseeable future, investing into the bottoming cycle for a few Btus of HR clearly does not pay off. At \$5 fuel, for LCOE parity between base and "cheap" cycles

- TOC saving of ~\$2.6 million is sufficient for cyclic operation whereas
- TOC saving of ~\$4.1 million is required for baseload operation.

At ~\$11 million TOC saving, for LCOE parity between base and "cheap" cycles

- Fuel price must exceed \$30/MMBtu for cyclic operation whereas
- Fuel price must be nearly ~\$16/MMBtu for baseload operation.

It is amply clear that, unless NG prices are exorbitantly high and/or the power plant in question is intended to assume a truly baseload duty, there is no case to be made for an expensive bottoming cycle. (Note that even if PEACE "Mechanical" and "Civil" estimates are off by 50%, the TOC saving is \$8.7 million and contains enough margin to support this conclusion.)

One may justifiably object to the comparison in Table 13.17 by pointing out the nearly 8°F higher firing temperature for the "cheap" case. The chain of reasoning for this tweak goes as follows:

- By going from the cheaper bottoming cycle to the expensive one, extra 3.7 MWe output is "bought" by paying \$11 million.
- This is equivalent to 45 Btu/kWh better HR – at exact same fuel consumption!
- The question to ask is this: Which one is cheaper?
  - Buying 3.7 MWe output for extra \$11 million
  - Buying 3.7 MWe output by extra fuel consumption.[12]

The answer, via LCOE analysis, turns out to be the latter. (Note that HR improvement more than compensates for marginally higher fuel burning and the HR delta improves to 37 Btu/kWh.[13])

Another reasonable objection would be "what about the expensive bottoming cycle and the 8°F higher firing temperature?" In order to answer that, consider the performance and LCOE comparison of four possible design permutations in Table 13.18.

TABLE 13.18
Performance and Cost Comparison of Four Design Permutations

	Base	Base +7.9°F	Cheap	Cheap +7.9°F
CC net output	463,290	467,031	459,579	463,290
CC net efficiency, %	58.74	58.81	58.29	58.36
CC net heat rate, Btu/kWh	5,809	5,802	5,854	5,846
LCOE – cyclic, \$/MWh	\$97.82	\$97.34	\$96.92	\$96.58
LCOE – baseload, \$/MWh	\$68.19	\$67.92	\$67.87	\$67.67

12 While extra capital investment is negligible, a higher LTSA price should be expected (herein annual \$160 K is added to the fixed O&M).

13 The impact of 8°F higher firing temperature on emissions and component life for a 1,600°C (2,912°F) TIT technology is minuscule (i.e., extra ~5 lb/MWh extra CO_2, ~1.3 ppm NOx).

On a truly "apples-to-apples" basis, obviously, the performance delta between the "expensive" and "cheap" bottoming cycles is about 3.7 MWe and 45 Btu/kWh of HR. At $5 fuel, the LCOE comparison favors the latter.

The next logical question is "what if the impact of the 7.9°F higher firing temperature on parts life (i.e., LTSA costs) is much higher?" From Figure 10 in GER-3620N, the parts life effect corresponding to increases in firing temperature for F class gas turbines is given by

$$A_p = \exp(0.023 \cdot \Delta TF).$$

where A_p is the peak-fire severity factor and ΔTF is the firing temperature adder in degrees Fahrenheit. For 7.9°F, A_p is 1.2. If the power plant operates at baseload for 3,000 h out of a total of 6,000 h, the maintenance factor, i.e., factored fired hours (FFHs) divided by actual fired hours, is

$$(3{,}000 \cdot 1.2 + 3{,}000) / 6{,}000 = 1.1.$$

Thus, the adjusted inspection interval, AI, is reduced by 9%, i.e.

$$AI = I / 1.1 = 0.91 \cdot I.$$

For example, 25,000-h inspection interval becomes 22,750 h. The impact of 10% higher FFH on the LTSA price is difficult to estimate because it is a complex, commercial calculation (see Section 16.1). Here is the upper limit: in order to eliminate the LCOE advantage of the "cheap" version in Table 13.17,

- For cyclic duty, each additional degrees Fahrenheit of firing temperature should add $210 K to the fixed O&M cost ($7.5 million per year) in Table 13.13
- For baseload duty, each additional degrees Fahrenheit of firing temperature should add $180 K to the fixed O&M cost.

Thus, the "bogey" in annual fixed O&M cost is about $1.6 million (~$200 × 8°F). At $275/FFH of LTSA (see Section 16.1), this corresponds to almost 6,000 extra FFH. At double the cost, nearly 3,000 extra FFH is necessary to erase the advantage of the "cheap" version.

On a "gross margin" basis (the difference between the market price of energy and the variable generation costs), it is true that the "base" variant has a slight advantage. However, it is not significant enough to severely impact the eventual place of the particular GTCC configuration selection (from owner/developer perspective) in the "merit/economic dispatch order" in a large ISO. Thus, the difference between the "levelized revenue requirement" quantified by the LCOE and the forecasted gross margin is the determinant in the selection of one variant over the other. In this case, the "cheap" variant with the smaller difference should be the preferred configuration.

In general, as long as there is marginal improvements within a small band, i.e., ±1% or less on net output and HR, erring on the side of less capex is probably a good bet. Beyond that, however, commercial considerations, which cannot be encapsulated in a simple metric like LCOE, can take precedence. In addition, the uncertainty aspect may become more critical to the extent that erring on the side of better performance (a more robust number than fuel prices over the next 20 years) may be the more prudent course of action.

REFERENCES

1. Hessler, P., 2015, *Power Plant Construction Management*, 2nd Edition, PennWell, Tulsa, OK.
2. Newell, S.A., Hagerty, J.M., Ungate, C.D., Wrobe, J., 2014, Cost of New Entry Estimates for Combustion Turbine and Combined Cycle Plants in PJM, The Brattle Group, Inc., Cambridge, MA.
3. Peters, M.S., Timmerhaus, K.D., West, R.E., 2004, *Plant Design and Economics for Chemical Engineers*, 5th Edition, McGraw-Hill International Edition, New York.
4. Couper, J.R., Hertz, D.W., Smith, F.L., 2014, *Perry's Chemical Engineers' Handbook*, 8th Edition, Section 9, Process Economics, McGraw-Hill, New York.
5. U.S. Energy Information Administration, April 2013, Updated Capital Cost Estimates for Utility Scale Electricity Generating Plants, Washington, DC.
6. Gülen, S.C., Mazumder, I., 2013, An expanded cost of electricity model for highly flexible power plants, *J. Eng. Gas Turbines Power*, Vol. 135, p. 011801.
7. U.S. Department of Energy Office of Scientific and Technical Information, April 2011, Cost Estimation Methodology for NETL Assessments of Power Plant Performance, U.S. DOE Report DOE/NETL-2011/1455.
8. Balevic, D., Burger, R., Forry, D., 2004, Heavy-Duty Gas Turbine Operating and Maintenance Considerations, GER-3620K, GE Energy.
9. Nagy, D., Savic, S., 2015, Alternative Gas Turbine Maintenance Concepts by Independent Service Providers, *Powergen International 2015*, December 8–10, Las Vegas, NV.
10. Gülen, S.C., 2015, What is the Worth of 1 Btu/kWh of Heat Rate, *Power*, June 2013, pp. 60–63, ssswww.powermag.com.
11. Mun, J., 2005, *Real Options Analysis: Tools and Techniques for Valuing Strategic Investment and Decisions*, 2nd Edition, John Wiley & Sons, Inc., Hoboken, NJ.
12. Gülen, S.C., 2013, Étude on gas turbine combined cycle power plant: Next 20 years, *J. Eng. Gas Turbines Power*, Vol. 138, p. 051701.

14 Cogeneration

Cogeneration describes fossil-fired power plants that generate multiple product streams, usually thermal energy and electricity. Outside the USA, the term commonly used for cogeneration is *combined heat and power* (CHP). In most cogeneration ("cogen" for short in the industry jargon) applications, the thermal energy product comprises one or more streams of steam at different pressures and temperatures that are used by an end-use customer for district heating and/or cooling, manufacturing process needs, or similar industrial and/or residential uses.

Before moving on, let us clarify one persistent deficiency in the terminology. By definition, *all* heat engines, including the gas turbine, are cogeneration (i.e., *combined* heat and power generation) devices. This can be easily deduced from the simple gas turbine heat balance given by Equation 4.3:

$$\dot{Q}_{exh} + \dot{W}_{shft} \approx HC$$

where heat consumption (HC) quantifies the fuel energy consumption rate by the unit (i.e., the product of fuel mass flow rate and its heating value). In short, a gas turbine burns fuel and generates two products: shaft output (work) and exhaust energy (heat). In advanced class heavy-duty machines, the former is roughly equal to 40% of HC (thermal efficiency) and the latter to about 60% (i.e., the remainder). Consequently, what is really meant by the terms of cogeneration and CHP is that the second product, i.e., exhaust gas energy/heat, is utilized for some "useful" purpose (instead of being dumped into the atmosphere through the exhaust stack).

In a gas turbine combined cycle power (GTCC) plant, exhaust gas energy of the gas turbine is used for steam generation in the heat recovery steam generator (HRSG) – at up to three pressure levels. In a pure power application, 100% of generated steam is used in the steam turbine generator. In essence, the GTCC power plant is the *ultimate* or *perfect* cogeneration system. In this chapter, alas, the focus is on *less-than-perfect* cogeneration systems.

In a cogen application, some of the steam is sent to a "steam host", e.g., a factory, university campus, hospital or commercial building, to satisfy its thermal energy needs. It can be shown that a gas turbine cogen plant requires less fuel than equivalent separate heat and power systems to produce the same amount of energy services (e.g., a fired boiler to provide steam, i.e., "heat", and average fossil-fueled electricity, i.e., "power", with transmission and distribution losses). Benefits of cogeneration are especially pronounced in "distributed generation" with smaller-size power plants located near the steam host. This approach can relieve the grid congestion, increase the energy security of the heat and power customers, and eliminate the losses that normally occur in the transmission and distribution of electricity from a central, utility-scale power plant to the end user (can be as high as 7%).

A 2008 report by the Oak Ridge National Laboratory (ORNL), titled *Combined Heat and Power: Effective Energy Solutions for a Sustainable Future*, enumerated four key benefits of cogeneration:

- Significant reduction in CO_2 emissions
- Reduced business costs
- Ability to use locally sourced, renewable fuels (depending on the generation technology chosen)
- Relieving grid congestion and improving energy security.

Although there is no question about the benefit of cogeneration's effectiveness in fuel utilization and reduced emissions, what has been missing so far is a standardized and consistent definition of cogeneration efficiency that would enable industry participants to measure and rank a wide variety of cogeneration plants on a thermodynamically consistent basis.

The most commonly used concept for rating of cogeneration power plants is "heat charged to power" or "fuel charged to power" (FCP), which is essentially a modified or *incremental* heat rate. Let us start with the generic definition of the heat rate (HR), i.e.,

$$HR = FC/W_{net}. \tag{14.1}$$

In an unfired GTCC power plant, FC is the fuel consumed by the gas turbine combustors. In a fired GTCC power plant, FC also includes the fuel consumed by the HRSG duct burners. The denominator is the net plant output, i.e., prime mover generator output minus the auxiliary power consumption (pumps, fans, lighting, etc.).

In a cogen application, the numerator is corrected by subtracting the amount of fuel that would be required to generate an equivalent amount of steam, Q_s, in a conventionally fired boiler with lower heating value (LHV) efficiency of η_b. (A commonly adopted value for η_b is 90%.) Thus,

$$FCP = \frac{FC - Q_s/\eta_b}{W_{net} + W_{aux,b}}, \tag{14.2}$$

where $W_{aux,b}$ is the auxiliary power consumption of the conventionally fired boiler plant (pumps, fans, etc.). Note that

- The numerator of Equation 14.2 is *additional* fuel burned in the cogen power plant to supply the original thermal kilowatts *plus* additional electric kilowatts.
- The denominator of Equation 14.2 is the additional electric kilowatts supplied by the cogen power plant. (Since the conventionally fired boiler plant *consumes* power, its supply is a *negative* value, thus the plus sign in front of it in Equation 14.2.)

This is the explanation of the logic behind the concept of FCP, which is the marginal efficiency of the additional kilowatts provided by the cogen power plant.

Another measure commonly used for cogeneration effectiveness is the "fuel utilization" effectiveness (FUE), which is defined as

$$FUE = \frac{W_{net} + Q_s}{FC}. \tag{14.3}$$

In Equation 14.3, the numerator is the total "product kilowatts" mixing proverbial apples (i.e., *thermal* kilowatts) with oranges (i.e., *electric* kilowatts). It can be a very high number, e.g., 80% or 90% or even higher. Nevertheless, even though some practitioners in the industry, deliberately or not, refer to FUE as "efficiency", it is not a *bona fide* efficiency. This is so because exergetically *more* valuable electric kilowatts are treated on an equal footing with exergetically *less* valuable thermal kilowatts.

In small, distributed generation applications, the cogeneration plant can simply be an aeroderivative or small gas turbine with a single-pressure HRSG supplying steam to a steam host (i.e., no steam turbine). Another similar variant has a noncondensing steam turbine (i.e., without a condenser) with the exhaust steam sent to the steam host.

One example for the former is as follows:

- GE's Frame 6 gas turbine (43,802 kWe, 33.32% LHV efficiency).
- Single-pressure HRSG supplying 100 psia 62.67 lb/s saturated steam to the steam host (68,709 kWth).
- Gas turbine fuel consumption is 131,467 kWth (LHV).
- Net plant output is 43,078 kWe (net electric efficiency is 32.77%).
- HRSG stack temperature is 243.2°F.

Using Equation 14.2, FCP is calculated as (assuming 100 kW for $W_{aux,b}$)

$$FCP = \frac{131,467 - 68,709/0.9}{43,078 + 100} 3412.14 = 4,356 \text{ Btu/kWh}.$$

In efficiency terms, FCP is 3,412/4,356 = 0.783, i.e., more than 78% LHV. In other words, the marginal efficiency of 43 MWe electricity produced by the cogen plant is nearly 20 percentage points better than that of an advanced F class GTCC.

Similarly, using Equation 14.3, FUE is calculated as

$$FUE = \frac{43,078 + 68,709}{131,467} = 0.85 \text{ or } 85\%.$$

The second example, with the same gas turbine, is as follows:

- Single-pressure HRSG supplying 1,200 psia and 970°F steam to a noncondensing steam turbine.
- Saturated 100 psia steam from the steam turbine exhaust, 40.33 lb/s, is supplied to the steam host (44,217 kWth).
- Steam turbine output is 9,131 kWe.
- Net plant output is 51,856 kWe.
- HRSG stack temperature is 403.7°F.

Using Equation 14.2, FCP is calculated as (assuming 65 kW for $W_{aux,b}$)

$$FCP = \frac{131,467 - 44,217/0.9}{51,856 + 65} 3412.14 = 5,411 \text{ Btu/kWh}.$$

In efficiency terms, FCP is 3,412/5,411 = 0.631, i.e., about 63% LHV. Similarly, using Equation 14.3, FUE is calculated as

$$FUE = \frac{51,856 + 44,217}{131,467} = 0.73 \text{ or } 73\%.$$

Note that the net electric efficiency of this cogen plant is 39.44%.

Even though the net electric efficiency of the cogen plant with the noncondensing steam turbine is 7% points better than the cogen plant with no steam turbine, it is a poorer choice from a fuel utilization perspective. This can be traced back to the much higher HRSG stack temperature for the former (by about 200°F). The reason for that is the much higher pinch point in the HRSG with the steam turbine (1,200 psia steam generation) vis-à-vis the HRSG without the steam turbine (100 psia steam generation). Since the exhaust gas temperature is the same (because the gas turbine is the same), stack temperature is lower in the HRSG with the lower pinch point. Since lower HRSG stack temperature is a measure of much better utilization of the "waste heat source", it is a better design from a CHP perspective. The reader is encouraged to revisit Section 6.1 in order to have a better appreciation of this nuance.

On the other end of the spectrum, a recent example is the Lausward GTCC power plant in Düsseldorf, Germany. This is a single-shaft GTCC with Siemens' SGT5-8000H gas turbine. In full power generation mode, it is rated at 603.8 MW and 61.5% net LHV efficiency. The bottoming cycle has a Benson-type HRSG (once-through high-pressure, HP, section) generating 170 bar-600°C HP steam. In cogeneration mode, steam extracted from the steam turbine is sent through three pipes to the heat exchangers in an adjacent building (see Figure 14.1). From there, hot water at more than 95°C flows through an insulated

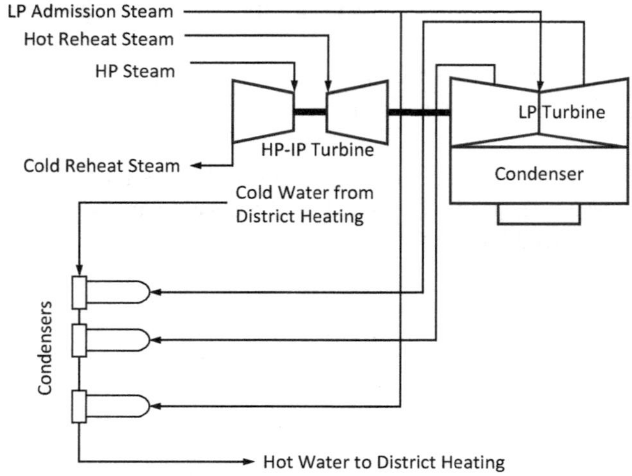

FIGURE 14.1 Three-stage steam extraction at Lausward GTCC.

pipe underneath the river Rhine to the Düsseldorf city center. Thus, when the power plant is priced out of the generation markets, it can operate in a heat-optimized mode to provide district heating. To maximize this potential, a 40-m-high tank to hold 35,000 m³ of hot water is provided (several days' worth of storage at of 1,480 MWh and mass flow of 5,000 m³/h), enabling the plant owner to decide what mix of electricity and heat the power plant should supply. Full district heating extraction is about 300 MWth (1,025 MMBtu/h); at that point, net electric output goes down to 520 MWe.

Cogeneration applicability space of a GTCC is illustrated in Figure 14.2. The diagram in the figure is for a natural gas-fired 85-MWe GE Frame 7EA gas turbine. Performances used to develop the applicability space in Figure 14.2 are presented in Table 14.1. The data points in the figure are described as follows:

1. Point A is for an unfired low-pressure (LP) HRSG with LP process steam supply.
2. Point C is for a two-pressure, unfired HRSG (HP at 1,250 psig, 900°F (87.2 bars, 482°C), LP at 150 psig (11.4 bars) saturated),with a noncondensing steam turbine exhausting to a 150 psig/10.3 bar process steam header.

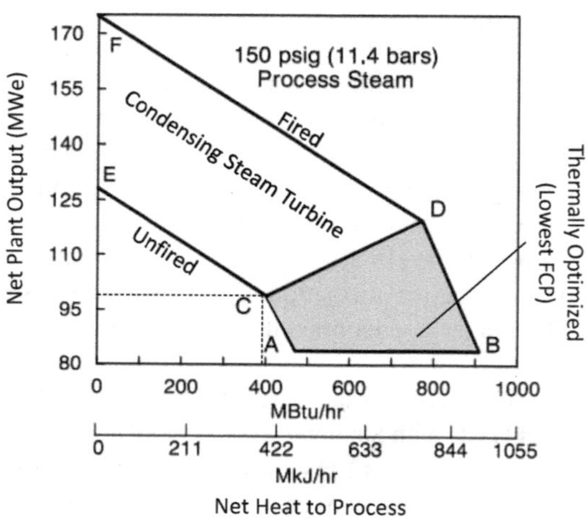

FIGURE 14.2 Performance envelope for a GTCC cogeneration system.

TABLE 14.1
Performance Data for the Envelope Points in Figure 14.2 (84% Process Boiler Efficiency Is Assumed for FCP Calculation)

Point	A	B	C	D	E	F
Net output, MWe	84.4	84.3	98.6	119.7	128.1	174.4
Net heat to process, MMBtu/h	473	906	402	769	0	0
Net heat to process, MWth	139	266	118	225	0	0
W_{aux}, kWe	202	386	171	328	0	0
FCP, Btu/kWh HHV	5,010	4,590	5,150	4,590	7,700	8,410
FC, MWth HHV	289	430	289	430	289	430

3. Points B (LP process steam supply) and D are for an HRSG with supplementary firing to a 1,600°F/871°C.
4. Point D steam generation is at 1,250 psig, 900°F (87.2 bars, 482°C) with noncondensing steam turbine generator.
5. Points E and F are with condensing steam turbines and pure power generation.
6. Process returns and makeup enter the integral deaerating heater at 44 psia (3 bars) at a mixed temperature of 150°F (66°C).

The area {A-B-D-C-A} in Figure 14.2 represents the most thermally optimized use of a gas turbine in a cogeneration application (i.e., it provides the lowest FCP). Operation along the line CE, DF or any intermediate point to the left of the line CD represents the use of a condensing steam turbine. Note that

- Points E and F represent combined cycle operation without any heat supplied to process.
- Points A and B represent gas turbine-only output (84.4 MWe) because there is no steam turbine.

Consequently, the chart in Figure 14.2 illustrates the following sequence of design modes:

- From point E to point C, one can see the reduction in steam turbine output with increasing heat supply to an external user.
- Further increase of heat supply renders a condensing steam turbine infeasible; the only option is to use a noncondensing or back pressure steam turbine for marginal power generation (about 12 MWe at point C).
- At point A, all steam generated by the gas turbine exhaust is exported to the external user – there is no steam turbine.
- For even higher steam needs, supplementary firing in the HRSG is requisite.
- Maximum heat supply (i.e., maximum supplementary firing) is represented by point B.

Let us look at point C in Figure 14.2 and Table 14.1 in some detail, i.e.,

- Delta output from pure power (point E) is 128.1 – 98.6 = 29.5 MWth
- Net heat to process (NHP) is 117.8 MWth
- Thus, the marginal efficiency of NHP is 29.5/117.8 = 0.25 or 25%.

Repeating the same calculation for point A, the marginal efficiency is (128.1 – 84.4)/138.6 = 0.315 or 31.5%. For the Lausward cogen power plant discussed earlier, the marginal efficiency is (603.8 – 520)/300 = 0.279 or 27.9%.

Numerical examples above indicate that the useful work production potential of the LP steam extracted from the bottoming cycle is equivalent to a "mini" Rankine steam cycle with 25%–30% efficiency. The exact value is a function of the pressure and temperature of the extracted steam. (HP steam extraction to process is quite rare in cogen plants.) This will be investigated in more detail below using the second law concept of exergy.

For the electric power product of the GTCC power plant, by definition, the exergy is exactly equal to the power generated. As dictated by one of the two fundamental corollaries of the second law, the Kelvin–Planck statement, the maximum useful work generating potential of any material stream is exactly equal to its thermodynamic property exergy.

The most common material stream that is a product of a cogen power plant is steam or hot water at a known pressure and temperature. Using steam tables, the exergy of steam or hot water can be exactly calculated. The numerical value that is obtained is exactly equal to the power that can be generated in a *Carnot engine*, which utilizes that particular product stream as its heat source. According to the second law of thermodynamics, the Carnot engine is the most efficient energy conversion device possible when operating under a given set of temperature conditions.

A simple numerical example illustrates the calculation process using data taken from the steam tables. Suppose that a GTCC power plant supplies an industrial customer with 25,000 pph of saturated steam at 125 psia. The enthalpy of the steam is 1,191.1 Btu/lb for a total thermal energy supply of 29.8 million Btu/h (or 8,725 kWth). The exergy of 125 psia (saturated) steam is 369.9 Btu/lb for a total thermal exergy supply of 9.25 million Btu/h or 2,710 kW. In other words, a hypothetical Carnot engine utilizing 29.8 million Btu/h of saturated steam at 125 psia as its sole heat source and rejecting heat to the ambient at 59°F would generate 2,710 kW of power, which implies a maximum theoretical thermal efficiency of 31.1%. Note that these two numbers – 2,710 kW and 31.1% – are not subject to any misinterpretation because they are directly derived from a fundamental law of physics. Figure 14.3 shows the calculated results for a range of steam pressures and temperatures in terms of β, which is the ratio of steam exergy to steam energy (i.e., enthalpy).

Obviously, it is practically impossible to design a Carnot engine. What is practically possible is, say, a steam turbine with 85% isentropic efficiency discharging to a condenser at a pressure of 2.5 in. of mercury. Using 29.8 million Btu/h of saturated steam at 125 psia, this steam turbine would generate 1,840 kW of shaft power for a thermal efficiency of only 21.1%. Though this seems to be a paltry

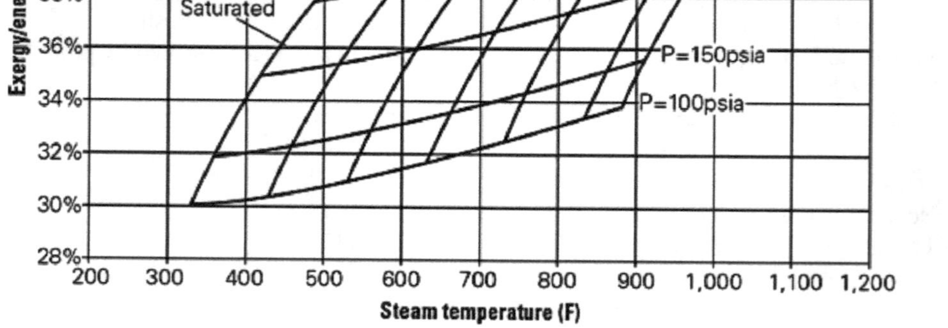

FIGURE 14.3 Exergy-to-energy ratio for steam.

number, consider that, when compared to the theoretically possible maximum of 2,710 kW (only from a hypothetical Carnot engine), the second law or "rational" efficiency of the example steam turbine is 21.1%/31.1% = 67.9%. The term "rational" conveys the underlying concept of using a reference point that is theoretically possible instead of using a reference point that is (even theoretically) impossible.

For 150 psig (164.7 psia) saturated steam, from Figure 14.3, β is about 32.3%. For point C in Figure 14.2, the incremental efficiency calculated earlier, 25%, suggests a rational efficiency of 25/32.3 = 0.775 or about 78%. For point A, the implied rational efficiency is 31.5/32.3 = 0.976 or nearly 98%, which is highly improbable for saturated 150 psig steam by itself (the high value reflects the two-pressure HRSG design for the pure power option).

At this point, it should be obvious that the rational efficiency of a cogen power plant with $\dot{W}$ MWe electric power output and $\dot{Q}$ MWth process heat supply is exactly equal to the efficiency of the same power plant operating in "pure power" mode. In that mode, it is easy to show that the electric power output of the power plant is

$$\dot{W}_{max} = \dot{W} + \eta_{rat}\beta \cdot \dot{Q} \tag{14.4}$$

or

$$\dot{W}_{max} = \dot{W} + \eta_{mar} \cdot \dot{Q}, \tag{14.5}$$

where η_{rat} is the "rational" efficiency, and η_{mar} is the "marginal" efficiency of $\dot{Q}$. Furthermore,

$$\dot{W}_{Carnot} = \beta \cdot \dot{Q} \tag{14.6}$$

with

$$\beta = \frac{a}{h} \tag{14.7}$$

and

$$a = (h - h_0) - T_0(s - s_0). \tag{14.8}$$

In Equations 14.7 and 14.8, the symbol for exergy is *a* (from the term "availability" common in US textbooks). The subscript 0 denotes the zero exergy reference (typically 59°F and 14.7 psia). Enthalpy, h, and entropy, s, can be evaluated from the *ASME Steam Tables* as a function of pressure and temperature.

At the risk of being too pedantic, it should be pointed out that Equations 14.7 and 14.8 implicitly assume that steam delivered to the process (i.e., mass flow out of the bottoming cycle control volume) is compensated by "makeup" water at p_0 and T_0. This is illustrated schematically in Figure 14.4a. In most cogen applications, however, there usually is a "process return" stream (e.g., "warm" water) at p_r and T_r, which is illustrated schematically in Figure 14.4b. The exact location of process return entry into the bottoming cycle is a function of p_r and T_r. (It cannot be simply added to the condenser because of deaeration requirements. The same goes for the makeup water as well.)

It should also be pointed out that there is a slight error in Equation 14.7, which should have been

$$\beta = \frac{a}{h - h_0}. \tag{14.9}$$

This is so because the zero enthalpy reference for the *ASME Steam Tables* is the triple point of the water.[1] Consequently, at the zero exergy reference of 59°F and 14.7 psia, h_0 is not zero; it is 27.1 Btu/lb. Obviously, choosing the triple point as the zero exergy reference would have prevented

[1] The triple point of pure water is at 0.01°C (273.16 K, 32.01°F) and 4.58 mm of mercury (611.2 Pa, 0.089 psia).

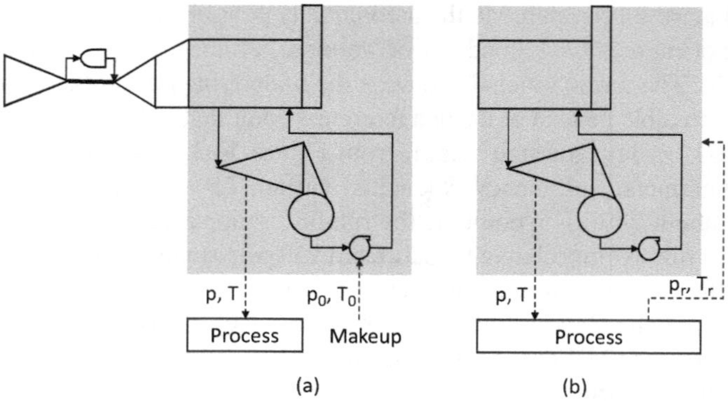

FIGURE 14.4 Process return and makeup water options.

the introduction of this error. But then, it would be silly to assume makeup water at 32°F and <0.1 psia. (This, however, is really not an issue if one goes the thermodynamically rigorous route; please read on.) As far as the error introduced, the enthalpy of 150 psig saturated steam is 1,195.5 Btu/lb, and its exergy is 385.7 Btu/lb. From Equation 14.7, β is calculated as 0.3226 vis-à-vis 0.3301 from Equation 14.9. As long as one is fully aware of the underlying assumption or simplification, this is an error that one can live with.

For the more generic case of process return at p_r and T_r, the applicable equations are

$$\beta = \frac{\Delta a}{h - h_r} \tag{14.10}$$

and

$$\Delta a = (h - h_r) - T_0 (s - s_r). \tag{14.11}$$

Combining Equations 14.10 and 14.11,

$$\beta = 1 - T_0 \left(\frac{s - s_r}{h - h_r} \right). \tag{14.12}$$

From the discussion of "mean-effective" Brayton cycle heat addition and rejection temperatures in Chapter 4, it is hoped that the reader will easily recognize that

$$\overline{T}_p = \frac{h - h_r}{s - s_r}, \tag{14.13}$$

where $\overline{T}_p$ is the "mean-effective" temperature of process supply and return streams. (Note that, from Equation 14.13, $\overline{T}_p$ is calculated in degrees *Rankine*; subtract 459.67 to convert it to degrees *Fahrenheit*. The same caveat applies to the SI units as well with degrees *Kelvin* and *Celsius*; subtract 273.15 to find the temperature in the latter units.) Consequently,

$$\beta = 1 - \frac{T_0}{\overline{T}_p}, \tag{14.14}$$

where β is the efficiency of a Carnot cycle operating between the two temperature "reservoirs" at $\overline{T}_p$ and T_0.

Using the same example as above, i.e., process steam supply at 150 psig (saturated), let us assume that the process return is at 40 psia and 90°F. In this case,

- From Equation 14.13, $\overline{T}_p$ is 324.6°F (note that saturated steam temperature at 150 psig is 365.9°F).
- From Equation 14.14, β is found as 0.3383 or 33.8%.

Note that this rigorous approach can be applied to the makeup water case in Figure 14.4a as well. One simply assumes 14.7 psia and 59°F for p_r and T_r, respectively. In that case,

- From Equation 14.13, $\overline{T}_p$ is 315.1°F.
- From Equation 14.14, β is found as 0.3301 or 33% (which, of course, is what we calculated above).

Note that this would also be the answer one would get if one chose the zero exergy reference point as the triple point. All this effort to pinpoint a number, which, depending on one's assumptions (or the boundary conditions of the problem), varies between 0.3226 and 0.3383, can be deemed to be a waste of resources. This is indeed a fair assessment if one is only interested in "back of the envelope" type estimates. Nevertheless, for a thermodynamically rigorous analysis, it is best to dot the proverbial i's and cross the proverbial t's – especially since the days of pen and paper with a $10 drugstore calculator are long gone.

If one is designing the cogen system using a heat and mass balance simulation software, these types of calculations are not really necessary. Simply by turning the process export stream "on" and "off", one can calculate the marginal efficiency of the NHP. As demonstrated earlier, the *bona fide* "efficiency" of the cogen power plant, with or without the process steam export, is exactly equal to the net electric efficiency of the facility in pure power mode. (As stated in the beginning of the chapter, the GTCC is the ultimate or perfect cogeneration system.) Trying to "inflate" the cogen's importance (it is, by the way, very important) by adding exergetic "apples" to "oranges" and quoting "efficiencies" of 80%, 90%, etc. is thermodynamic cheating. Fuel utilization "effectiveness" is acceptable but not really informative – one can even get a number greater than 100% if use is made of the lowest grade energy is system (e.g., utilizing the condenser heat rejection for water heating to take showers with it).

Fuel or heat chargeable to power is indeed a meaningful metric as far as it goes, but it also lacks from information deficiency. Consider the unfired points E, C and A in Figure 14.2. From Table 14.1, FCP from E to C to A improves from 7,700 to 5,150 to 5,010 Btu/kWh in HHV. Translated to efficiency terms, the improvement from E to C to A is 44.3% to 66.3% to 68.1%, respectively. In LHV terms, the efficiencies are 49.1%, 73.5% and 75.5%, respectively. However, of all these numbers only 44.3% (HHV) and 49.1% (LHV) are the concrete metrics providing the full information on the "goodness" of the gas turbine in question (GE's Frame 7EA in this example) and its bottoming cycle. The others are meaningful only in the presence of an "apples-to-apples" benchmark to compare against in a rational, capex and opex cost-based manner.

The usual benchmark for comparison with cogen FCP is the average grid efficiency. In other words, if one plugs into the grid to obtain $\dot{W}$ MWe needed for the operations of the industrial host facility, the average efficiency of all the power plants (including fossil fuel-fired ones as well as nuclear, hydro or other renewable generation assets on the particular grid) is at best at low 40s (in percent LHV terms – please refer to Table 12.1). This, however, is not exactly true. Here is why.

Let us look at point C in Table 14.1 and round its electric output to 100 MWe for convenience. This cogeneration power plant generates 100 MWe at a marginal efficiency of 73.5% (LHV). Let us assume that there is a particular advanced class GTCC on the grid, which is rated at 500 MWe and 60% net LHV efficiency. This GTCC typically runs at 80% load. Consequently, the 100 MWe needed by this particular host facility can be supplied by increasing the loading of this particular GTCC from 80% to 100%. What is the marginal efficiency of the additional 100 MWe generated by this GTCC? Let us calculate it.

- At 100% load, the fuel/heat consumption of this GTCC is 500/0.6 = 833 MWth.
- At this point, its heat rate is 5,687 Btu/kWh.
- At 80%, from Equation 3.5, normalized heat rate is 1.0315, which translates into 5,866 Btu/kWh and fuel/heat consumption of 688 MWth.
- Thus, the additional fuel burn for the additional 100 MWe is 833 − 688 = 146 MWth.
- The marginal efficiency of generating the additional 100 MWe by increasing the loading of this GTCC from 80% to 100% is 100/146 = 68.6%.

This calculation can be done for the entire grid, which would, of course, be very tedious albeit straightforward. At the other end of the spectrum, for example, one can hypothesize that the additional 100 MWe can be supplied by the grid by increasing the capacity of a big hydro or nuclear power plant with zero additional fuel/heat consumption. In that case, the marginal heat rate would be zero (i.e., infinitely large marginal thermal efficiency). Since the generating asset portfolio of a large grid is a mixture of fossil-fired and renewable technologies, it is easy to recognize that the true grid average of the marginal heat rate will be somewhere in between of the best and worst assets (from a part load heat rate lapse perspective). In any event, though, it is guaranteed that the marginal efficiency thus obtained will be much higher than the average grid efficiency, which is at best around 40% (LHV) or so in the USA.

While the cogen power plant with 73.5% marginal efficiency can still be the better option, as this sample calculation demonstrates, the "real" improvement in heat rate is rather modest. As such, a careful cost-performance trade-off is requisite before declaring cogeneration as the winner "hands down".

It is difficult to propose a one-size-fits-all economic analysis for cogeneration projects. They differ enormously in size, intent, industry, location, local laws and regulations, etc. One should also consider that cogeneration is widely used not only in industrial and manufacturing facilities but also at large institutional sites such as universities, hospitals and military bases. According to the American Council for an Energy Efficient Economy (ACEEE), as of 2016, over 82 GW of cogen capacity existed at more than 4,400 industrial, institutional and commercial facilities across the USA. This number includes smaller installations in commercial buildings with continuous thermal loads such as casinos and multifamily buildings. Thus, evaluation of cogen project requires close examination of a broad range of technical, financial, political and regulatory issues.

In the most basic example, an industrial customer may want to produce its own electric power instead of buying it from the local utility while at the same time satisfying its steam and/or hot water needs. In that case, the first thing to do is to make a comparison of making one's own power vis-à-vis purchasing it from the utility. This requires calculation of FCP and conversion of that FCP to fuel expenditure. If it is less than half of the power purchase expenditure, looking further into cogeneration is advisable.

Even if the simple comparison described above is quite appealing, further study accounting for costs associated with environmental and other regulations including permits (air and noise), codes, etc. as well as insurance requirements can easily turn that initial appeal into a mirage. This can be through a combination of the particular fees and payments and additional (and expensive) equipment such as SCR and advanced, high-efficiency turbines or engines.

There is more than simple "spark spread" comparison to evaluating a cogen project for facilities in locations with very poor utility service. If the local grid is highly unreliable, i.e., subject to wild frequency excursions and/or frequent blackouts, having a cogeneration facility capable of operating in an islanded mode may be the biggest economic driver. This situation can be exacerbated by financial consideration such as standby rates.

Standby rates are charges levied by utilities when a cogeneration system experiences a scheduled or emergency outage, which requires power purchase from the grid. There are two components of standby charges: standard or conventional energy charges, in dollars per kilowatt hours, and demand charges, in dollars per kilowatt, which are intended to recover the costs to the utility of providing capacity to meet the peak demand of the cogen facility. The second one is not always rational and can be punitive because utilities look down on cogen facilities as loss-of-revenue sources.

15 Operability

The best way to introduce the reader to the subject of operability is using a modern sports car as an example. (Note that a subcompact would do just as well but the author suspects that a sports car would be more fun to read.)

Chevrolet Corvette Stingray (body type C7), introduced in 2014, is powered by its fifth-generation (Gen V) small-block LT1 engine. This engine is rated at 455 hp at 6,000 rpm and 460 lb-ft of torque at 4,500 rpm (base/stock car, not the turbocharged Z06 or the ultimate ZR1 version).[1] The original equipment manufacturer (OEM)-rated fuel consumption of this car is 17 mpg (city driving – speed not specified) and a very respectable 29 mpg (highway driving – speed not specified). Electronically limited top speed is 186 mph. The combination of numbers cited above is analogous to the rating performance of a gas turbine (simple cycle) that one finds in *Gas Turbine World Handbook*, e.g., 500 MWe and 60% lower heating value (LHV) efficiency (net). What does that really mean?

LT1 is a pushrod, two-valve, all-aluminum, direct-injection V8 (90°) with 6.2-L displacement and 11.5:1 compression ratio. The power rating number cited above is strictly for the engine, i.e., "brake horsepower" (bhp); it is measured at the *flywheel*. The power output measured at the wheels is the "wheel horsepower" (whp), which can be determined from a *dyno run*. Due to the losses in the driveline (including clutch and transmission), whp is typically about 10%–15% lower than bhp. Thus, a stock Corvette Stingray is expected to deliver $0.90 \times 460 = 414$ whp at the rear wheels. In fact, this is confirmed by the dyno test conducted by Edmunds.com in 2014 (www.edmunds.com/car-reviews/track-tests/2014-chevrolet-corvette-stingray-z51-dyno-test.html) as shown in Figure 15.1.

Let us now look at some "real", field-measured numbers. The 2015 Corvette Stingray (curb weight of about 3,300 lb) has a seven-speed manual or automatic transmission. On a typical US highway (65 or 70 mph speed limit), the car comfortably runs at about 1,500 rpm in the sixth gear at 75 mph. From Figure 15.1, the power at the wheels is about 110 whp. When tested at GM's Milford Proving Ground with 93 octane premium gasoline, the car clocked a trap speed of 117.3 mph (at the end of a quarter mile) – no information on elapsed time (ET) was provided. Note that quarter-mile trap speed depends on the car's initial acceleration from standstill. A novice driver fumbling with the gear shifter at the start is destined to fall way below the *ideal* 123 mph at 11.7 s ET. Stock 2015 C7

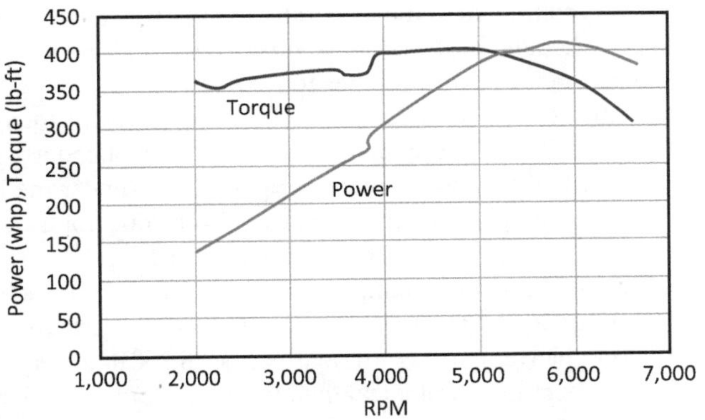

FIGURE 15.1 Dyno-measured power and torque numbers (Chevy Corvette Stingray 2014) – from Edmunds.com.

1 The performance exhaust option adds 5 hp and 5 lb-ft of torque.

performance clocked and reported in dragtimes.com was 115.873 mph in 11.894 s. At that point, the LT1 engine was reported to churn out slightly above 400 whp.

Clearly, one does not pull a stint like that day in and day out. For one thing, it would severely reduce the tire life and a set of C7 tires does not come cheap. For another, it is absolutely no fun to get clocked by "the Smokey" – in real life, being "the Bandit" is quite different from what one sees in that classic movie. While the Stingray is a perfectly good daily driver, it is highly unlikely that one's commute is a straight 75 mph, low-traffic, four-lane highway cruise. So, what is the "normal"?

Well, everybody has a different "normal", which is a combination of location (i.e., climate, roads, traffic conditions, etc.), driving skills and commute characteristics (distance, frequency, time of day, etc.). For the author, whose commute is about 50 miles each way through back roads and a stretch of highway with not-too-bad traffic, the average speed over a period of a month (including incidental trips to here or there) is 41–42 mph, and average fuel consumption is about 22–23 mpg. For some luckier souls, somewhat better numbers are possible. Since nobody in his or her right mind uses a sports car for city driving, this should be considered as a reasonable "normal" (except for those who transport their cars to a track to have some weekend fun of course).

In conclusion, what one reads about high-end sports cars in car magazines are the OEM's "chest thumping" performances, which can only be obtained in ideal conditions, on controlled tracks, with professional drivers and all the bells and whistles that can be piled on the stock car. When the proverbial rubber hits the road, the "mean-effective" performance that is clocked in the stock car over an extended period is as ho-hum as one would get from a low-end economy sedan.

And this is exactly what one should expect from a top-of-the-line advanced class gas turbine in a combined cycle configuration with similar/analogous caveats. ISO baseload ratings advertised in *Gas Turbine World* or *Turbomachinery International* handbooks annually are "on paper" performances with very aggressive bottoming cycle design assumptions, bare minimum auxiliary loads (e.g., step-up transformer losses are not included) and very low condenser pressures (typically, 1.2 in. of mercury) with an open-loop, water-cooled condenser. The same gas turbine combined cycle (GTCC), however, when built in a certain geographic location would perform quite differently depending on the particular climatic conditions and the availability of cooling water. Other deviations could and most likely would be caused by site-specific boundary conditions (e.g., the owner and/or developer's ability to finance the cost of the facility from conception to receiving the proverbial keys from the EPC contractor, environmental regulations and permits, rules and regulations imposed by the grid operator, etc.). During its operation throughout the year, the plant's performance would vary with seasonal ambient variations, becoming worse on hot days and somewhat better on colder days. Furthermore, a GTCC power plant does not operate in a monotonic manner. It has to start from standstill, synchronize to the grid, ramp up to its full load at a certain rate, stay at that load or at a lower load as dictated by the system demand and, finally, shut down when its electric power generation is no longer required (or becomes too expensive).

Typical operating rhythm of a modern combined cycle power plant (qualitatively, of course) was illustrated in Figure 13.3. Over a given time period (say, one calendar year), total megawatt-hours (MWH_TOT) generated by this power plant is equal to area under the load curve. Dividing MWH_TOT by the time period in question (in hours, e.g., T = 8,760 h for a calendar year) gives us the *mean-effective load* (MEL) (in megawatts) of the power plant.

Furthermore, by looking at the ambient temperature variation over the selected time period T, one can calculate a straight or load-weighted average ambient temperature. Such an example is shown in Figure 15.2. A straight average over a 1-year period (8,760 h) gives 57.6°F ± 18.1°F. Using a straight average of the pertinent meteorological data (if available) is sufficient if the plant operation is evenly distributed across the entire year. Otherwise, one has to do the more "exact" (but pretty straightforward) load-weighted averaging, which would return a different answer by several degrees. For example, if the plant were more frequently dispatched for longer durations during the hot summer period, the load-weighted average would be higher, the exact number dependent on the particular plant duty. Let us call the average ambient temperature TAMB_AVG.

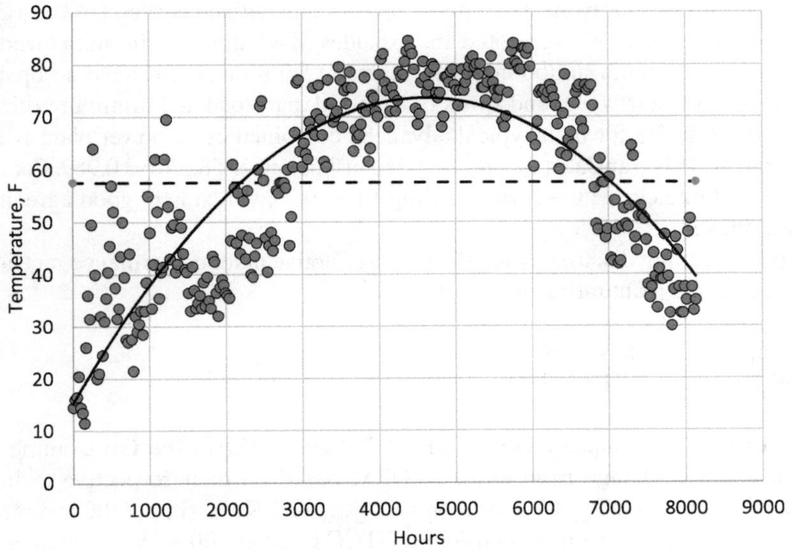

FIGURE 15.2 Annual ambient temperature variation (Washington, DC, in 2018).

Combined cycle full load (i.e., IGVs fully open with the machine running per the control schedule or model-based algorithm) at TAMB_AVG is a fraction of the ISO baseload, i.e.,

$$\text{MW\_TAVG} = \text{K1} \cdot \text{MW\_ISO}.$$

(You can read K1 from a "correction curve" similar to that shown in Figure 3.4.) Thus, the *load factor* (LF) of this power plant is

$$\text{LF} = \frac{\text{MWH\_TOT}}{\text{T} \cdot \text{MW\_TAVG}}.$$

For a typical cyclic operation duty, LF is somewhere between 0.7 and 0.8. *Mean-effective efficiency*, MEE, of the power plant can be calculated by dividing the total heat (fuel) consumption, HC_TOT, by MWH_TOT.

For a representative selection of US combined cycle power plants, the results for the time period of January–September 2018 were summarized in Table 13.12.[2] Note that all six power plants in the table are equipped with supplementary-fired heat recovery steam generators (HRSGs) (heavily deployed during the summer time). Unfortunately, the Energy Information Administration (EIA) data files do not include the design performance data so that back-calculating LF is not possible.

How can this "field reality" be translated into a meaningful rating parameter? The answer was already provided by General Electric (GE) in 2011: *FlexEfficiency*, which is defined to account for both the profitability of power production and the annual fuel consumption for cyclic operation.[3] FlexEfficiency (FE), definition per GE is

$$\text{FE} = \frac{\text{Annual MWh}}{\text{Annual HC}}.$$

---

2 Actual US plant data can be obtained from the statistics published by the US EIA – see Form EIA-923 data available online at www.eia.gov/electricity/data/eia923/.

3 FlexEfficiency* 50 Combined Cycle Power Plant, July 2011, Guy DeLeonardo, Marcus Scholz, Chuck Jones, GEA19089 (07/2011).

In other words, FE is essentially the MEE of the power plant defined earlier. (In GE's definition, the numerator is the *profitable* MWh generated and excludes MWh during minimum turndown, and the denominator includes fuel consumption during startup.) With this metric and an operating profile that includes about 200 starts/year and a mix of baseload, part load and minimum turndown hours, GE calculated FE ~ 58.5% for their typical advanced combined cycle power plant with a baseload efficiency over 61%. This translates to a normalized efficiency of 58.5/61 = 0.959. For a typical part load efficiency/heat rate curve, the corresponding LF is 0.75, which is in good agreement with the assertion made above.

Based on the preceding discussion and the mean-effective plant performance metric, an easy-to-implement and realistic plant rating definition is

- Rated output: 0.75 × MW_ISO
- Rated efficiency: K2 × EFF_ISO

where K2 is the efficiency lapse factor at LF = 0.75 (i.e., 0.959 in the GE example above). The exact value of K2 will change from OEM to OEM based on their respective technologies (see Figure 3.3 for a typical GTCC – it is based on heat rate, and K2 is simply the inverse of the value read from the curve). As an example, consider a GTCC rated at 500 MW and 60% net LHV (ISO base load). This is analogous to the 455 hp at 6,000 rpm rating of the Stingray with 29 mpg (highway). With K1 = 0.75 and K2 = 0.959, average performance of this GTCC in typical cyclic operation is 0.75 × 500 = 375 MW and 0.959 × 60% = 57.5%. This is analogous to the average performance of the Stingray at 42 mph and 22 mpg under "normal" driving conditions (see above). The average engine output is a paltry 70 hp.

15.1 STEADY-STATE OPERATION

Steady state refers to the operating state of the system when key system parameters do not change with time. In mathematical terms, the time derivative of a given parameter, X, is equal to zero, i.e., $\partial X/\partial t = 0$. This is the basis of all thermodynamic performance calculations. In the brick-and-mortar power plant, the counterpart of steady state is "stability". According to the ASME PTC-22, stability is achieved when continuous monitoring of plant operation via the distributed control system (DCS) indicates that the instrument readings have been within the maximum permissible variation established by the OEM. This reflects the practical reality that achieving, say, perfectly constant X MW output corresponding to a perfectly flat line on an X-versus-time plot, is impossible. The best representation of X in actual plant operation monitoring is $X_{AVG} \pm X_{VAR}$ over a time period of T, where

- X_{AVE} is the average of X over the time period T.
- X_{VAR} is the maximum allowable variation from that average.

According to the ASME PTC-22 (Table 3-3.5 on page 10), maximum permissible variation in electrical power output is ±1.3% (population standard deviation of the data sample). For example, if the "true" steady-state value of X at the prevailing operating conditions is 300 MW and if the standard deviation of the data sample taken over, say, 15 min, is ±1.3% × 300 = ±3.9 MW or smaller, stable operation assumption is acceptable.

In any event, majority of performance calculations in this book and elsewhere are steady-state calculations. There are two types of steady-state performance: design and off-design. The most common design calculation is the ISO baseload rating performance. Rating performance data is published in trade publications and can be readily analyzed using fundamental thermodynamic principles. This has been covered in detail in Chapters 4–7. The off-design performance calculations of most practical interest are done to find "part load" performance of the power plant at different site ambient conditions, i.e., temperature and humidity.

A combined cycle power plant is designed for operation across a wide range of site ambient and loading conditions. Unless one considers the particular requirements of a specific job, the design point is fixed at standard reference conditions. Usually, that standard is ISO ambient conditions with the plant components running at 100% load. A common term for the 100% load condition is "base" or sometimes "full" load. There is a certain ambiguity about the definition of the base or full load. Please see Section 3.3.2, at the end of which the difference between the two definitions is explained in sufficient detail. In essence, the gas turbine is said to be running at baseload when it is operating at a specified, design ambient (in most cases ISO) and at full load. At any given time, a gas turbine is said to be running at "part load" when via controlling inlet guide vanes (IGVs), variable stator vanes (VSVs), which are present in pretty much all advanced class machines, and/or firing temperature, the gas turbine is operating at less than full load capacity.

Note that the part load can be defined in reference to the base load (at the design ambient) or the full load at the given ambient. Consequently, one should clearly understand the reference load, which a particular part load is referring to.

In a combined cycle power plant, the steam turbine operates in a "sliding pressure" or "valves wide open" (VWO) mode (until the steam control valve "floor" pressures are reached; thereafter, high-pressure (HP) and intermediate-pressure (IP) inlet pressures at valve inlet are constant). As such, the steam turbine generates the power that it can naturally generate with the following inputs:

1. Steam generated in the heat recovery boiler that is commensurate with the exhaust energy provided by the gas turbine (assuming an unfired system)
2. Condenser pressure that is commensurate with the temperature of circulating cooling water at the particular ambient.

In other words, the steam turbine is a "slave" to the gas turbine, which determines the loading of the combined cycle power plant. Consequently, two major (primary) factors determine the "off-design" performance:

1. Ambient conditions, particularly the temperature and the humidity of the air
2. Gas turbine loading (i.e., full or part load).

Secondary factors such as generator power factor, fuel composition (i.e., LHV) and component degradation also impact the off-design performance and can (and should) be accounted by using appropriate corrections. The latter are covered in great detail in the applicable performance test codes listed in Section 2.3.

Computer simulation using detailed, high fidelity and, therefore, quite complicated software is a must for plant off-design performance calculations due to two key reasons.

1. First of all, calculation of the performance of "fixed" hardware operating at off-design is much more complicated than the simple enthalpy and mass balances that are sufficient for the design calculations.
2. Second, simulation of the control loops that govern the performance of the plant components is complicated and highly iterative.

Examples include but are not limited to attemperation (desuperheating) sprays, low-pressure (LP) economizer bypass and/or recirculation, pump flow-head and efficiency curves. Note that these are in addition to myriad control loops that govern the operation of the gas turbine.

The aforementioned fact that sophisticated computer models are requisite for reliable combined cycle plant off-design performance calculations does not preclude application of proper engineering judgment to the problem for developing relationships suitable to reasonable performance estimation.

The gas turbine performance is readily amenable to a simplified treatment via correction curves. For a particular unit in the field, curves provided by the OEM are (usually) available. OEM engineers run highly accurate computer models over the entire gas turbine operating envelope and generate these curves. For quick estimates and conceptual studies, generic normalized curves are adequate for most heavy-duty industrial gas turbines.

Until recently, performance variation with ambient conditions (pressure, temperature and humidity) and inlet/exhaust losses for air-cooled machines would not display a significant deviation between different OEMs. This statement may, however, not be entirely true anymore – especially for those machines with model-based (adaptive) controls in lieu of linear control schedules with multiple segments. For a detailed discussion of gas turbine controls, the reader is referred to the **GTFEPG** (Ref. [11] in Chapter 2), particularly Chapter 18 therein.

In any event, for advanced machines such as steam-cooled H-System (GE) or sequential combustion (reheat) gas turbines (former ABB/Alstom, now GE and Ansaldo), product-specific curves should be preferred.

Gas turbine part load performance is dependent on the particular control philosophy adopted by the OEM. Nevertheless, two basic approaches can be identified:

1. Constant turbine inlet temperature (TIT) with inlet flow modulation via IGVs and VSVs
2. Constant turbine exhaust temperature with TIT and inlet flow modulation.

In theory, a third method, i.e., TIT variation at constant airflow, is possible, but it is detrimental to combined cycle performance due to significant reduction in exhaust gas temperature and ensuing deterioration in bottoming cycle performance. For an in-depth discussion of how each gas turbine control philosophy impacts the combined cycle performance, please refer to Chapter 18 in **GTFEPG** and the paper by Gülen and Joseph (Ref. [4] in Chapter 3).

For the aeroderivative gas turbines or the reheat gas turbines, complex control schedules for the former and the existence of a second burner for the latter result in part load characteristics different from the standard heavy-duty industrial gas turbines (especially the exhaust characteristics).

As stated earlier, OEM-provided correction curves for simple and combined cycle performance across the site ambient and load envelope are adequate for most practical purposes. In fact, for modern machines with model-based controls, they are the only reliable and accurate means available to the non-OEM parties. The drawback of using correction curves is the "black box" nature of the process, which hides the "physics" behind the system response to changes in boundary conditions and operator commands. This will be investigated below in some detail by using an F class gas turbine in 1 × 1 × 1 combined cycle configuration.

Figure 15.3 shows the standard gas turbine part load control path of GE "frame" machines (in terms of exhaust gas flow and temperature) and the resulting steam turbine and combined cycle output (both normalized). In order to maintain the highest combined cycle part load efficiency, gas turbine is brought to a lower load by closing the IGVs (i.e., reducing the airflow). The gas turbine firing temperature is kept at the base load value until the exhaust temperature reaches a maximum value, which is referred to as the "exhaust isotherm". For the vintage F class units, the exhaust isotherm is set to 1,200°F. The isotherm is dictated by the last turbine stage (third or fourth) material limits as well as the limits set by the HRSG parts life concerns (especially HP superheater headers and tubes). State-of-the-art H and J class gas turbines have ISO baseload exhaust temperatures significantly above 1,200°F. In fact, one major OEM (680°C (1,256°F) ISO baseload exhaust temperature) has a cooled fourth stage blade in its next-generation advanced class product (Siemens HL class). Heat recovery boiler OEMs (some of them have been acquired by the major gas turbine OEMs) are expected to specify their equipment in accordance. Furthermore, some OEMs (notably GE) already use model-based (adaptive) control in their machines. Consequently, 1,200°F exhaust isotherm can be considered obsolete for the latest product offerings, for which pertinent control philosophy information must be obtained directly from the OEM. Nevertheless, basic governing

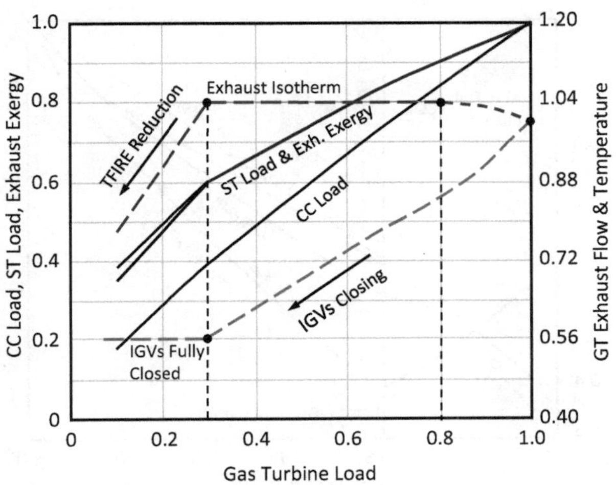

FIGURE 15.3 Part load characteristics of a typical F class GTCC.

principles do not change; that is, exhaust temperature must be kept as high as possible for optimum bottoming cycle performance, but it cannot be increased indefinitely beyond a limit set by gas turbine and HRSG materials and parts life concerns.

Once the exhaust isotherm is hit (80% load in Figure 15.3), gas turbine and plant load is lowered by simultaneously reducing the firing temperature and closing the IGVs to maintain the exhaust isotherm. When the IGVs reach their fully closed position (i.e., the gas turbine airflow reaches its minimum value – 30% load in Figure 15.3), further reduction in load is only possible via firing temperature reduction. At constant (minimum) airflow, this lowers the exhaust temperature as well. Curves in Figures 15.3–15.6 are generated using a GateCycle model of a Frame 9FB GTCC power plant (single shaft with three-pressure reheat or 3PRH bottoming cycle and water-cooled condenser).

The impact of this control strategy on the bottoming cycle is via gas turbine exhaust energy and exergy. Let us recap the significance of the latter once more. Gas turbine exhaust gas exergy, just like the exhaust gas energy (i.e., enthalpy), is a fluid property (a function of gas composition, pressure and temperature via the equation of state). It quantifies the theoretical maximum of bottoming cycle output set by the Kelvin–Planck statement of the second law of thermodynamics. In a real bottoming cycle, only a fraction of that maximum is achieved (about 0.75). Consequently, one should expect that the steam turbine output is proportional to the gas turbine exhaust gas exergy. This is indeed what is observed in Figure 15.3.

Figure 15.4 shows the HRSG steam production and the steam turbine output as a function of gas turbine exhaust gas energy. All parameters are normalized with respect to their full load values. Until the gas turbine reaches 30% load, at which point the exhaust gas energy is at about 60% of its full load value, steam turbine output is directly proportional to the exhaust gas energy. From Figure 15.3, this is the point when the IGVs become fully closed. Furthermore, HRSG steam production is also proportional to the exhaust gas energy following a power law with varying exponents.

In particular, for gas turbine loads higher than or equal to 30% (at which point combined cycle load is about 40% from Figure 15.3), when IGVs are modulated,

$$\frac{\dot{W}_{ST}}{\dot{W}_{ST,0}} \propto \frac{\dot{Q}_{exh}}{\dot{Q}_{exh,0}}, \tag{15.1}$$

$$\frac{\dot{m}_{LP}}{\dot{m}_{LP,0}} \propto \left(\frac{\dot{Q}_{exh}}{\dot{Q}_{exh,0}}\right)^{1.82}, \tag{15.2}$$

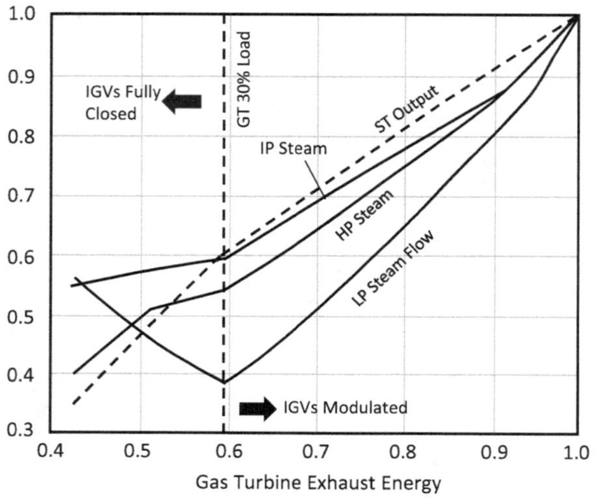

FIGURE 15.4 Three-pressure HRSG steam production (flow) in each section.

$$\frac{\dot{m}_{HP}}{\dot{m}_{HP,0}} \propto \left(\frac{\dot{Q}_{exh}}{\dot{Q}_{exh,0}}\right)^{1.16}, \tag{15.3}$$

$$\frac{\dot{m}_{IP}}{\dot{m}_{IP,0}} \propto \left(\frac{\dot{Q}_{exh}}{\dot{Q}_{exh,0}}\right)^{0.96}. \tag{15.4}$$

For gas turbine loads lower than 30% (exhaust gas energy at 60%), when IGVs are fully closed,

$$\frac{\dot{W}_{ST}}{\dot{W}_{ST,0}} \propto \left(\frac{\dot{Q}_{exh}}{\dot{Q}_{exh,0}}\right)^{1.60}. \tag{15.5}$$

However, LP, IP and HP steam flow/production trends diverge significantly from their simple power-law correlation with the exhaust energy. In particular,

- The rate of reduction in HP and IP steam flow/production becomes slower.
- The trend in LP steam flow/production, however, is completely reversed.

The explanation for this can be found in the combined effect of (i) steam attemperation (desuperheating), (ii) steam pressures, and (iii) exhaust gas temperature before and after the fully closed position of the IGVs. In reading the discussion below, to have a copy of the HRSG schematic in Figure 6.3 might come in handy.

Until the turn of the twenty-first century, allowable HP and hot reheat (HRH) steam temperatures as dictated by pipe and steam turbine material limits used to be 565°C (1,050°F). Advanced class gas turbines with high exhaust temperatures necessitated introduction of alloy steam pipes and turbine rotor materials, which pushed this limit to 600°C (1,112°F). During plant operation, superheater tube wall temperatures are essentially at the same temperature as the steam flowing inside. Thus, steam temperatures should not be allowed to exceed the design value above a design margin (about 25°F) for prolonged times. A similar limit is also imposed by the steam turbine design margins. For example, over any given year, average HP or HRH steam admission temperature is not allowed to exceed the rated temperature (i.e., 1,050°F or 1,112°F). Limited duration excursions are allowed within that period for specified durations per event and aggregate duration over the year.

Typically, operation at a steam temperature, say, 50°F above the rated is allowed only for 15 min at any given time and not more than 80 h total.[4]

If nothing is done, at part load operation with lower steam flow rates at lower exhaust gas energy (but 1,200°F or higher gas temperature), steam temperatures from the (now effectively *oversized*) superheaters would go above their design or rated values. Consequently, "cold" feedwater injection between the HP and reheat superheaters (RHSHs) (typically from the feed pump discharge or interstage bleed) is required to maintain allowable steam temperatures.[5] Note that these "interstage" spray attemperators (desuperheaters) are different from the "terminal" attemperators placed downstream of the HRSG. The latter are used for precise steam temperature control during startup.

Below 30% gas turbine load, however, exhaust gas temperatures drop rapidly so that attemperation is not needed. Exhaust gas temperatures (normalized with respect to the full load value) as a function of exhaust gas energy (also normalized) at the inlet of HP evaporator (HPEV), IP evaporator (IPEV) and LP evaporator (LPEV) are shown in Figure 15.5.

Lower steam flow rates through the steam turbine result in lower pressures due to the fixed "swallowing capacity" of the turbine. This was explained in detail in Section 9.2.1 in conjunction with the overload valve (e.g., see Equation 9.1). Beyond a certain limit, commonly known as the "floor pressure", further drop in steam pressures cannot be allowed and steam admission valves are modulated to maintain the floor pressure. The reason for that is to prevent excessive steam velocities (via lower density and higher volume flow) in HRSG tubes and steam pipes. Typically, floor pressure is set at 40% of the design value for the main or HP steam (e.g., 720 psig if the design value of HP steam pressure is 1,800 psig). It is controlled by modulating the main steam valve (MSV) opening (see Section 9.2.1). For the HRH steam, the floor pressure is controlled by modulating the intercept control valve (ICV) opening at 50%–60% of its design value. For the LP steam, the floor pressure is 50 psia (typically). Exact values should be obtained from the OEMs. Heat recovery boiler HP, IP and LP drum pressures (normalized with respect to the full load value) as a function of exhaust gas energy (also normalized) are shown in Figure 15.6.

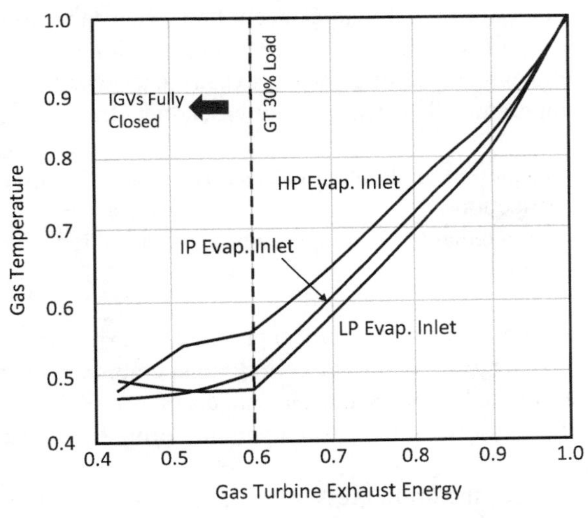

FIGURE 15.5 Gas temperatures at the inlet of each HRSG evaporator (normalized).

[4] GEI 67510F *Allowable Pressure and Temperature Variations – Reheat Units with Overpressure*, General Electric (1991).

[5] In some newer designs, saturated steam from the HP drum is used for attemperation. The advantage of this method is elimination of the risk of thermal shock damage to the piping or steam turbine due to incomplete evaporation of spray water. The disadvantage is reduced steam flow through the superheater sections and resulting higher tube metal temperatures.

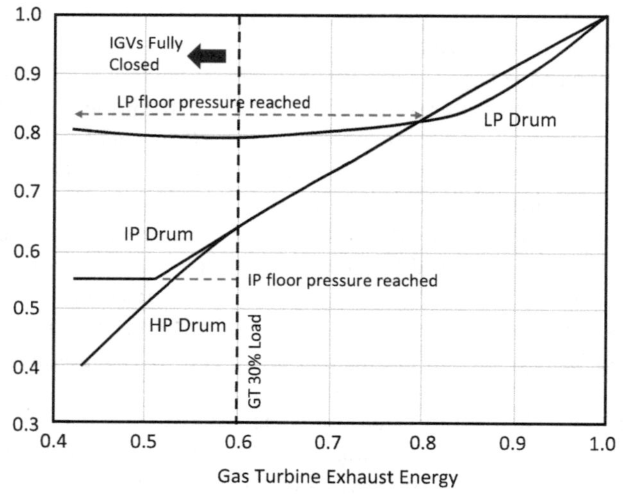

FIGURE 15.6 HP, IP and LP drum pressures in the HRSG (normalized).

Examination of Figures 15.5 and 15.6 provides the clues to the change in steam production trends observed in Figure 15.4. Reduction in gas temperature in the LP section, which is farthest from the HRSG inlet (see Figure 6.3), is more pronounced at a given load. Thus, LP steam flow drops faster, and the LP admission valve starts controlling the LP floor pressure (typically at 50 psia) well before the other floor pressures are reached.

Until 30% gas turbine load (40% combined cycle load and 60% exhaust energy), everything "slides" in proportion to the exhaust gas flow (controlled by the IGVs) at constant exhaust gas (isotherm) temperature. All steam valves are fully open (except the LP admission valve but LP steam contribution is very small) with the steam turbine in VWO mode so that there are no "hiccups" to disturb the monotonous downward trend in total steam flow and steam turbine output.

Once the IGVs reach their fully closed position, further reduction in gas turbine and plant load is achieved by reducing the firing temperature and the exhaust energy change is driven mainly by the change in exhaust gas temperature. This leads to two cascading effects:

- Reduction in and elimination of attemperation spray reduces the total steam flow through the front superheater sections.
- This reduces the drop in exhaust gas temperature and puts a "brake" on the drop in steam production in the evaporators.

The second effect is amplified by the drop in drum pressure (as explained above), which is accompanied by the drop in saturation (i.e., feedwater boiling) temperature. From the shape of the vapor–liquid equilibrium "dome" of the H_2O substance, latent heat of evaporation is inversely proportional to the boiling pressure. Thus, for the same gas temperature drop, less steam can be generated at a lower evaporator pressure.

The proportionality between the drum pressures and the exhaust energy (before floor pressures are reached) is given by the following equations:

$$\frac{p_{LPd}}{p_{LPd,0}} \propto \left(\frac{\dot{Q}_{exh}}{\dot{Q}_{exh,0}}\right)^{1.08}, \tag{15.6}$$

$$\frac{p_{HPd}}{p_{HPd,0}} \propto \left(\frac{\dot{Q}_{exh}}{\dot{Q}_{exh,0}}\right)^{0.91}, \tag{15.7}$$

$$\frac{p_{IPd}}{p_{IPd,0}} \propto \left(\frac{\dot{Q}_{exh}}{\dot{Q}_{exh,0}}\right)^{0.94}. \tag{15.8}$$

For quick estimates, the power-law exponents in Equations 15.6–15.8 suggest that assuming direct proportionality is sufficiently accurate.

Normalized curves in Figure 15.3 suggest that one can estimate the combined cycle part load performance by making a few assumptions and using the second law approach outlined in Chapter 4. Before tackling this subject in more detail, a few clarifying remarks are in order. As mentioned earlier, the particular control scheme in Figure 15.3 is characteristic of vintage "frame" machines of GE (i.e., Frame 7 for 60 Hz and Frame 9 for 50 Hz applications). Especially, the later variants of those machines with three-stage turbines were characterized by high exhaust temperatures. As a result, closing IGVs to reduce combined cycle load at constant TIT would bring the exhaust temperature quickly to its limit. Further reduction of load by closing IGVs required TIT reduction at constant exhaust temperature (the "isotherm"). Note that when IGVs are closed and airflow is reduced, turbine pressure ratio goes down as well because of the reduction in turbine inlet pressure. (This is a direct result of the turbine inlet with constant cross-sectional flow area acting as a choked nozzle.) This is the main reason why not reducing the TIT would result in higher exhaust temperature.

In comparison, competitor gas turbines of the same vintage as GE frame machines had four-stage turbines and lower exhaust temperatures. For example, a Siemens "V class" gas turbine comparable to a GE F class had an exhaust temperature less than 1,100°F (about 1,080°F) vis-à-vis about 1,150°F for the latter. Clearly, the basic physics of a gas turbine does not change from OEM to OEM. Furthermore, maintaining the highest possible combined cycle efficiency at any given load is a common goal. However, differences in turbine design parameters and control philosophies can lead to different exhaust gas characteristics. For example, a machine with lower exhaust temperature can go lower in airflow at constant TIT before reaching its isotherm. Even so, however, particular axial compressor characteristics can preclude a wide range of airflow modulation because of getting close to the surge limit (see Chapter 11 in **GTFEPG** for an in-depth explanation of compressor off-design behavior). This would require load reduction via TIT reduction earlier than suggested in Figure 15.3 so that a particular gas turbine's part load exhaust temperature curve might not display an isotherm at all. The bottom line is that for a particular gas turbine, one should strive to obtain the OEM's information for accurate estimates. As an example, consider the part load exhaust temperature and flow schedule of a Siemens SGT6–5000F gas turbine in Figure 15.7.[6] For comparison, typical GE F class part load exhaust parameters from Figure 15.3 are included as well.

Combined cycle output is simple algebraic sum of gas and steam turbine generator outputs minus plant auxiliary load. Thus,

$$\dot{W}_{CC,Net} = \left(N_{GTG}\dot{W}_{GTG} + \dot{W}_{STG}\right)(1-\alpha), \tag{15.9}$$

where N_{GTG} is the number of gas turbines in the GTCC, and α is the plant auxiliary load as a fraction of the plant gross (generator) output. For simplicity, i.e., reduction of clutter in the formulae, the discussion is continued for a 1 × 1 × 1 GTCC with $N_{GTG} = 1$. Steam turbine generator output is proportional to the gas turbine exhaust exergy, i.e.,

$$\dot{W}_{CC,Net} = \left(\dot{W}_{GTG} + \varepsilon A_{exh}\right)(1-\alpha) \tag{15.10}$$

$$A_{exh} = \dot{m}_{exh} \cdot a_{exh}\left(T_{exh}\right). \tag{15.11}$$

[6] SGT6-PAC 5000F Gas TurbinePower Plant Application Handbook, A96001-W90-A414-X-4A00, Siemens (2005).

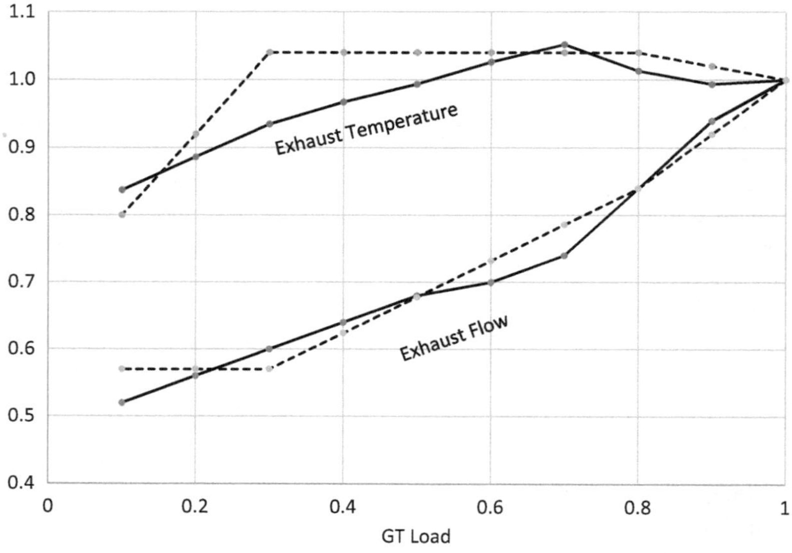

FIGURE 15.7 Siemens SGT6–5000F part load exhaust gas parameters (dashed lines are for a typical GE "frame" machine from Figure 15.3).

Our first assumption is that α does not change much between design and off-design operation. Thus, designating the design conditions by the subscript 0 and dropping the designation "Net" for clutter reduction, we have

$$\frac{\dot{W}_{CC}}{\dot{W}_{CC,0}} = \frac{\dot{W}_{GTG} + \varepsilon A_{exh}}{\dot{W}_{GTG,0} + \varepsilon_0 A_{exh,0}}. \tag{15.12}$$

Equation 15.12 is a bit unwieldy for quick pen-and-paper or simple spreadsheet calculations. As demonstrated earlier (e.g., see Section 3.3.1), to a very good approximation

$$\frac{\varepsilon_0 A_{exh,0}}{\dot{W}_{GTG,0}} \approx 0.5$$

so that Equation 15.12 can be reformulated in a simpler manner as

$$\frac{\dot{W}_{CC}}{\dot{W}_{CC,0}} = \frac{1}{3}\left(2\frac{\dot{W}_{GTG}}{\dot{W}_{GTG,0}} + \frac{\varepsilon}{\varepsilon_0}\frac{A_{exh}}{A_{exh,0}}\right). \tag{15.13}$$

Since the exhaust gas exergy is a linear function of the exhaust gas temperature (e.g., see Equation 4.4), our third assumption is that

$$\frac{A_{exh}}{A_{exh,0}} \sim \frac{\dot{m}_{exh}}{\dot{m}_{exh,0}} \cdot \frac{T_{exh}}{T_{exh,0}}. \tag{15.14}$$

Thus, Equation 15.13 becomes

$$\frac{\dot{W}_{CC}}{\dot{W}_{CC,0}} = \frac{1}{3}\left(2\frac{\dot{W}_{GTG}}{\dot{W}_{GTG,0}} + \frac{\varepsilon}{\varepsilon_0}\frac{\dot{m}_{exh}}{\dot{m}_{exh,0}}\frac{T_{exh}}{T_{exh,0}}\right). \tag{15.15}$$

Using the designation κ for the non-dimensional parameters in Equation 15.15, which can be looked up from available curves, tables, etc. (e.g., Figure 15.3), we end up with

$$\frac{\dot{W}_{CC}}{\dot{W}_{CC,0}} = \frac{1}{3}\left(2\kappa_W + \frac{\varepsilon}{\varepsilon_0}\kappa_m\kappa_T\right). \quad (15.16)$$

The only remaining task is to come up with a relationship for $\varepsilon/\varepsilon_0$ term in Equation 15.16. From the part load schedule in Figure 15.3, this is obtained as shown in Figure 15.8. The "exact" curve-fit for gas turbine load of 30% or higher is

$$\frac{\varepsilon}{\varepsilon_0} = 1 + 0.13 \cdot \ln(X), \quad (15.17)$$

where X is the gas turbine load as a fraction. A simple linear estimate is adequate for most practical purposes:

$$\frac{\varepsilon}{\varepsilon_0} = 0.8 + 0.15 \cdot X. \quad (15.18)$$

For an unfired GTCC, plant and gas turbine heat consumption (HC) are the same. Consequently,

$$\frac{HR_{CC}}{HR_{CC,0}} = \frac{HC_{GT}}{HC_{GT,0}}\frac{\dot{W}_{CC,0}}{\dot{W}_{CC}}. \quad (15.19)$$

For the frame machine in Figure 15.3, complete part load behavior is summarized in Table 15.1. For a typical Siemens F class machine (discussed above – see Figure 15.7), combined cycle part load performance from simple cycle part load data is estimated using Equations 15.16 and 15.17. The results are summarized in Table 15.2.

Ambient correction data for three different steam turbine heat rejection systems are shown in Tables 15.4 and 15.6 (for Siemens SGT6–5000F[7]). No duct firing or gas turbine inlet conditioning (e.g., inlet evaporative cooler or chiller) is present. Clearly, the heat rejection system has a significant impact on combined cycle output and heat rate. The primary driver is the condenser pressure (i.e., steam turbine back pressure). Due to the difficulty (size/cost and parasitic power consumption)

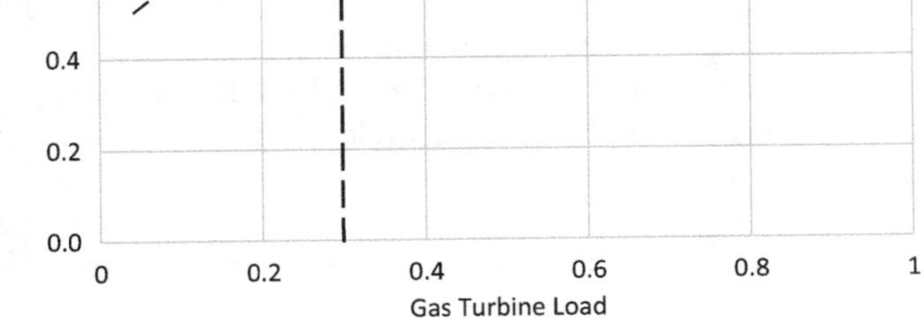

FIGURE 15.8 Bottoming cycle (gross) exergetic efficiency (normalized) as a function of gas turbine load.

[7] From the charts in *Siemens Gas Turbine SGT6-5000F Application Overview*, E50001-W210-A104-V2-4A00, 2008.

TABLE 15.1
Part Load Performance Parameters of a GE F Class Machine (Typical)

GT Load	Exhaust Temperature	Exhaust Flow	GT HC	GT Heat Rate	CC Load	CC HC
1.00	1.00	1.00	1.000	1.000	1.00	1.000
0.90	1.02	0.92	0.925	1.028	0.91	1.012
0.80	1.04	0.84	0.850	1.062	0.83	1.026
0.70	1.04	0.79	0.775	1.108	0.74	1.043
0.60	1.04	0.73	0.700	1.166	0.66	1.065
0.50	1.04	0.68	0.624	1.249	0.57	1.093
0.40	1.04	0.62	0.549	1.371	0.49	1.129
0.30	1.04	0.57	0.472	1.574	0.40	1.181
0.20	0.92	0.57	0.373	1.867	0.29	1.288
0.10	0.80	0.57	0.272	2.724	0.18	1.513

TABLE 15.2
Part Load Performance Parameters of a Siemens F Class Machine (Typical)

GT Load	TEXH	MEXH	GT HC	GT HR	CC Load	CC HR
1.00	1.00	1.00	1.000	1.000	1.00	1.000
0.90	0.99	0.94	0.918	1.020	0.91	1.009
0.80	1.01	0.84	0.837	1.046	0.81	1.031
0.70	1.05	0.74	0.756	1.080	0.72	1.053
0.60	1.03	0.70	0.677	1.129	0.63	1.072
0.50	0.99	0.68	0.600	1.200	0.55	1.086
0.40	0.97	0.64	0.525	1.312	0.47	1.124
0.30	0.94	0.60	0.452	1.507	0.38	1.190
0.20	0.89	0.56	0.380	1.900	0.29	1.318
0.10	0.84	0.52	0.290	2.900	0.19	1.506

of achieving low condensation pressure and temperature with the tower-based systems (wet or "dry", especially the latter), output and heat rate penalty at high ambient temperatures is more severe. For a detailed discussion of the underlying principles, the reader is referred to Chapter 7.

A simplified treatment of the combined cycle performance as a function of the ambient temperature similar to that for the part load performance described above is more problematic. In principle, Equation 15.6 is still valid. The problem is in obtaining the simple cycle correction factors κ_1, κ_2 and κ_3. The main reason is the TIT-IGV schedule at high ambient temperatures to compensate for the drop in airflow (via lower density) and provide the bottoming cycle with the highest possible exhaust gas exergy. In many cases, especially in the USA, supplementary firing in the HRSG along with evaporative cooling of the gas turbine inlet is quite common.

In theory, one can use a base, unfired case from Tables 15.3–15.5 as appropriate and then try to estimate the impact of duct firing on the bottoming cycle using, say, Figure 6.18. The difficulty in that approach is that one also has to consider the impact of the increased steam flow on the condenser pressure. Thus, even simplified calculations, while not impossible, can become cumbersome to the point that one's best bet is to use either OEM-provided corrections or a computer software.

15.1.1 Hot Day Power Augmentation

The dramatic drop in gas turbine power output accompanying the increase in ambient temperature is clearly visible in the tables presented in the preceding section. This is a significant problem

TABLE 15.3
Combined Cycle Ambient Temperature Performance (Once-through Condenser)

Ambient, °F	Ambient, °C	CC Output	CC Heat Rate	CC/GT HC
0	−17.8	1.120	1.010	1.132
10	−12.2	1.100	1.009	1.110
20	−6.7	1.080	1.007	1.087
28	−2.2	1.063	1.005	1.069
40	4.4	1.039	1.003	1.042
50	10.0	1.018	1.001	1.020
59	15.0	1.000	1.000	1.000
70	21.1	0.968	0.999	0.967
80	26.7	0.939	1.000	0.938
90	32.2	0.909	1.001	0.910
100	37.8	0.880	1.004	0.884
110	43.3	0.851	1.009	0.858
120	48.9	0.821	1.015	0.834

TABLE 15.4
Combined Cycle Ambient Temperature Performance (Mechanical-Draft Cooling Tower)

Ambient, °F	Ambient, °C	CC Output	CC Heat Rate	CC/GT HC
0	−17.8	1.120	1.007	1.128
10	−12.2	1.100	1.007	1.108
20	−6.7	1.080	1.006	1.086
28	−2.2	1.063	1.004	1.067
40	4.4	1.039	1.001	1.040
50	10.0	1.018	1.000	1.018
59	15.0	1.000	1.000	1.000
70	21.1	0.963	1.003	0.966
80	26.7	0.930	1.008	0.937
90	32.2	0.896	1.018	0.912
100	37.8	0.863	1.033	0.891
110	43.3	0.829	1.053	0.873
120	48.9	0.795	1.080	0.859

because this performance loss takes place during the time of the year when the demand for electric power is at its highest. One obvious remedy is supplementary firing in the HRSG to boost steam production and steam turbine power output. This method is described in detail in Section 6.3. In addition, there are several gas turbine inlet air cooling techniques to overcome this problem so that the power plant can generate enough power to keep the residential and industrial air conditioners humming when people are escaping the blistering summer heat.

There are five major compressor inlet air cooling technologies:

1. Evaporative cooling
2. Inlet fogging (no overspray)
3. Wet compression (fogging with overspray)
4. Mechanical (electric) chilling
5. Absorption chilling.

TABLE 15.5
Combined Cycle Ambient Temperature Performance (Air-Cooled Condenser)

Ambient, °F	Ambient, °C	CC Output	CC Heat Rate	CC/GT HC
0	−17.8	1.120	1.009	1.130
10	−12.2	1.100	1.008	1.109
20	−6.7	1.080	1.005	1.085
28	−2.2	1.063	1.003	1.067
40	4.4	1.039	1.000	1.039
50	10.0	1.018	0.999	1.018
59	15.0	1.000	1.000	1.000
70	21.1	0.960	1.005	0.965
80	26.7	0.923	1.013	0.936
90	32.2	0.887	1.027	0.910
100	37.8	0.850	1.046	0.889
110	43.3	0.813	1.071	0.872
120	48.9	0.777	1.105	0.858

Evaporative cooling or "evap cooling" systems are widely used in warm and dry climates. The basic psychrometric process comprises a wetted media through which the gas turbine inlet air flows and absorbs moisture via evaporation before entering the compressor. Sensible heat transfer from the air to supply the latent heat of evaporation is the basic cooling mechanism. The minimum achievable temperature is equal to the ambient wet-bulb temperature. How close one can get to that ideal is quantified by the evap cooler effectiveness. Typical design practice is 0.90 (ratio of achieved temperature drop to the maximum possible). The process in the evap cooler is amenable to calculation with pen and paper (or spreadsheet) and *ASME Steam Tables*. The formulae and examples can be found in any thermodynamics textbook (e.g., Section 12.8.1, page 586, in Moran and Shapiro, Ref. [2] in Chapter 2). For a thorough discussion of evap cooling with wetted media, the reader is referred to the paper by Chaker and Meher-Homji [1].

Inlet fogging with no overspray is similar to evap cooling in the underlying psychrometric principle. Therefore, the minimum achievable compressor inlet air temperature is the ambient wet-bulb temperature. The system comprises a multiple nozzle distribution manifold at the turbine entrance through which demineralized water is injected into the air stream. Large contact area between the air and microscopic droplets (<50 μm in diameter) is beneficial in two respects:

- Very high effectiveness (i.e., practically 1.0 or inlet air at ambient wet-bulb temperature)
- Practically no inlet pressure loss (because there is no wetted media).

The drawback is the impact of surviving water droplets on stage 1 compressor blades causing leading edge erosion. Some OEMs (e.g., GE), at least until recently, did not permit using inlet foggers with their equipment (otherwise, the warranty is voided). Please consult Bhargava et al. [2] and the references cited in there for more details.

Mechanical chilling systems are essentially very large refrigerators that provide cold water to a cooling coil in front of the gas turbine. Inlet air stream flowing across the inlet coil can be cooled below the ambient wet-bulb temperature. This major advantage of the mechanical chilling systems over the evaporative cooling systems is not unlimited. A typical low temperature limit is 45°F–50°F dictated by ice forming at the compressor inlet.

Absorption chillers are also very large refrigerators that can cool the inlet air stream below the ambient wet-bulb temperature. Their advantage over the mechanical chilling systems is the fact that the driving energy source is thermal without the need for a gas compressor that consumes a large amount of electric power and inflates the plant auxiliary load. Absorption chillers are especially advantageous in the presence of a waste heat source, e.g., the last economizer section (between the LP evaporator and the stack) in the HRSG. For example, hot water extracted from the LP economizer can be used in the absorption chiller's generator. Basic thermodynamic principles underlying the absorption chiller operation can be found in any thermodynamics textbook (e.g., see Section 10.5, page 462, in Moran and Shapiro).

Each inlet cooling system described above has pros and cons that will dictate their applicability to a given situation. Key characteristics that should be considered and carefully evaluated are

- Cost of purchase, installation and operation (e.g., cost of demineralized water)
- Impact on net combined cycle output and heat rate
- Impact on gas turbine inlet pressure loss
- Impact on heat sink size (e.g., increased steam flow and condenser circ water flow)
- Impact on plant design (especially for the absorption chiller, e.g., steam or hot water extraction source)
- Impact on component life (e.g., compressor blade erosion due to microscopic water droplet impact)
- Parasitic power consumption (e.g., electric motor-driven compressors for mechanical chillers).

Excellent studies are available for more information about the comparative features of the inlet air cooling systems discussed above [3,4]. In the case of mechanical and absorption chillers, considerable additional installed cost is a prime factor. For the mechanical chiller, excessive parasitic power consumption is another hit. Absorption chillers are quite complex systems with a lot of piping and valves crisscrossing the plant layout and suffer significant loss in efficiency when the plant is operating at low loads. Mechanical chillers are mostly used in GTCC power plants in hot and humid climates (e.g., Southeastern USA). In some cases, the inlet chiller is combined with a *thermal energy storage* (TES) system. In this arrangement, chilled water is produced by electricity supplied from the grid at night-time hours (at low, off-peak pricing) and stored for use the next day during the peak demand for gas turbine compressor inlet air cooling. The economic justification for this application is strongly dependent on the price arbitrage between off-peak and peak times. For an in-depth discussion of how the inlet chiller plus TES system works, the reader is referred to the paper by Ebeling et al. [5]. The author is not aware of a commercial absorption chiller installation in a utility-size power plant.

Wet compression is different from inlet fogging and similar inlet water spraying techniques in that its main benefit is derived from the intercooling effect achieved within the frontal compressor stages. This is achieved by overspraying demineralized water either in front of the compressor inlet or in between the first few stages. The water sprayed, in excess of the amount required for 100% relative humidity at the inlet, will evaporate during the compression as the temperature of the air increases and cool the front stages of the compressor via extraction of the latent heat of evaporation from the compressed air. For a comprehensive overview and extensive bibliography, the reader should consult the paper by Obermüller et al. [6]. For a simplified calculation procedure, please refer to Wettstein [7].

15.1.2 Drum-Type versus Once-Through (Benson) Control

In a drum-type boiler, whether with natural or forced circulation, feedwater flow rate from the economizer to the evaporator is controlled by the drum water level. This is accomplished by a *three-element system* as shown in Figure 15.9. Level and steam flow elements correct for unmeasured disturbances within the system such as boiler blowdown and boiler/superheater tube leaks. The feedwater flow element responds rapidly to variations in feedwater demand, from either the steam flow rate *feed-forward* signal (i.e., a pound of feedwater change is made for every pound of steam flow change) or feedwater pressure or flow fluctuations.

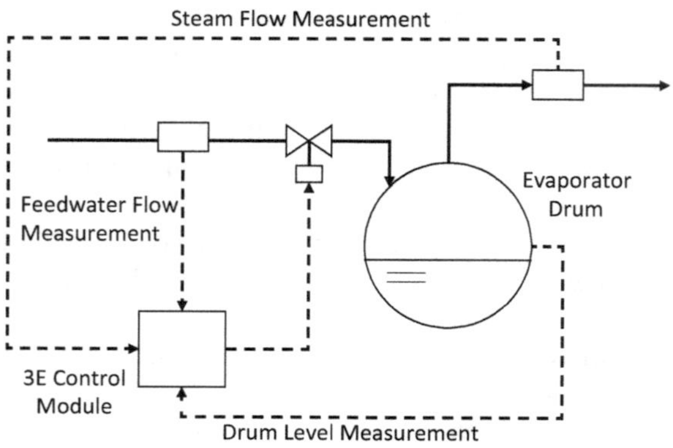

FIGURE 15.9 Three-element control logic for drum-type evaporator.

The output signal of the feedwater flow transmitter is the process variable to the feedwater flow controller and is compared to the output of the feedwater flow computer (set point). The feedwater flow controller produces the necessary corrective signal to maintain feedwater flow at its set point by the adjustment of the feedwater control valve. Compensation for load change is accomplished largely by the feed-forward system. The drum level portion of the control scheme is used only in a complementary capacity.

In the control scheme described above, feedwater flow does not affect the steam temperature. Steam from the drum is always saturated and "dry", i.e., with a quality of $x = 1.0$. Steam from the Benson evaporator, on the other hand, can be superheated or wet (i.e., $x < 1.0$) as will be discussed below.

In a fixed design, the more energy input into the evaporator (i.e., higher gas turbine exhaust gas mass flow rate and/or temperature), the more steam is produced. This increases the steam pressure in the sliding operation mode (i.e., fixed steam turbine "swallowing capacity") and, consequently, steam boiling point (saturation) temperature in the evaporator. Thus, feedwater flow is influenced indirectly via the water level inside the drum.

In a once-through Benson evaporator, the evaporation end point is not fixed (see the discussion in Section 9.3). Thus, feedwater flow control must be adjusted precisely with changing energy input because it influences not only the evaporation end points but also steam pressure, temperature and mass flow rate.

In a Benson once-through system, HP feedwater control covers *three* regimes, one pre-operating (filling) and two operating:

- Control the feedwater flow rate while *filling* the evaporators prior to HRSG startup.
- Control the evaporator *minimum flow* after the heat input to the boiler has started.
- Calculate the *feed-forward control* flow set point for two different operation modes:
  - HP Separator Level Control (no Benson operation)
  - HP Enthalpy Control (Benson operation).

The two operating regimes (a total of three modes, i.e., minimum flow, level control and enthalpy control, labeled I, II and III, respectively, in Figure 15.10) are dictated by energy input (i.e., gas turbine exhaust gas flow and temperature). This is illustrated in Figure 15.10, which shows steam enthalpy, feedwater flow and exhaust gas flow (from bottom to the top).

The two operation modes of the third operating regime require calculation of energy input from the gas turbine exhaust gas into the HP feedwater (as a function of the gas turbine exhaust gas mass flow rate and the enthalpy difference between evaporator inlet and outlet).

- The enthalpy at the evaporator inlet is calculated via the directly measured exhaust gas temperature.

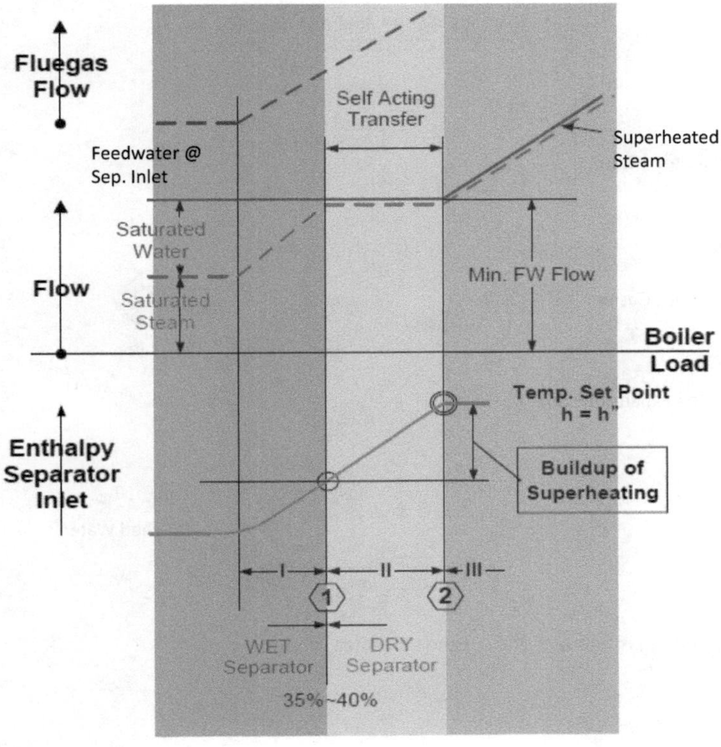

FIGURE 15.10 Benson once-through evaporator operating modes.

- The enthalpy at the evaporator outlet is detected by the saturated steam temperature of the feedwater and the estimated temperature difference between the feedwater and the exhaust gases as a function of the feedwater mass flow rate (see Figures 15.11 and 15.12), i.e.,

$$\dot{m}_{fw} \propto \frac{\dot{m}_g \left(h_{g,1} - h_{g,2}\right)}{\Delta h_{w/s}}. \tag{15.20}$$

$$T_{g,2} = T_{fw} + \Delta T_p + \Delta T_{sc} \tag{15.21}$$

$$T_{g,2} = T_{sat}\left(P_{HP}\right) + \Delta T_p \tag{15.22}$$

$$\Delta h_{w/s} = h\left(T_{sat}, x\right) - h_{fw}\left(T_{sat} - \Delta T_{sc}\right), \tag{15.23}$$

where

$\dot{m}_{fw}$ = Feedwater mass flow rate from the HPEV into the separator
$\dot{m}_g$ = Gas turbine exhaust gas mass flow rate
$h_{g,1/2} = f\left(T_{g,1/2}\right)$ = Exhaust gas enthalpy at the HPEV inlet and exit
$T_{g,1/2}$ = Exhaust gas temperature at the HPEV inlet and exit
$\Delta h_{w/s}$ = Enthalpy change of water/steam in the HPEV.
x = Quality of steam (ratio of vapor mass to total steam mass)

In theory, the actual (measured) flue gas temperature at the evaporator outlet could be used for flue gas enthalpy change in the calculation. However, feedwater temperature measurement is more accurate and preferred over the less accurate gas temperature measurement.

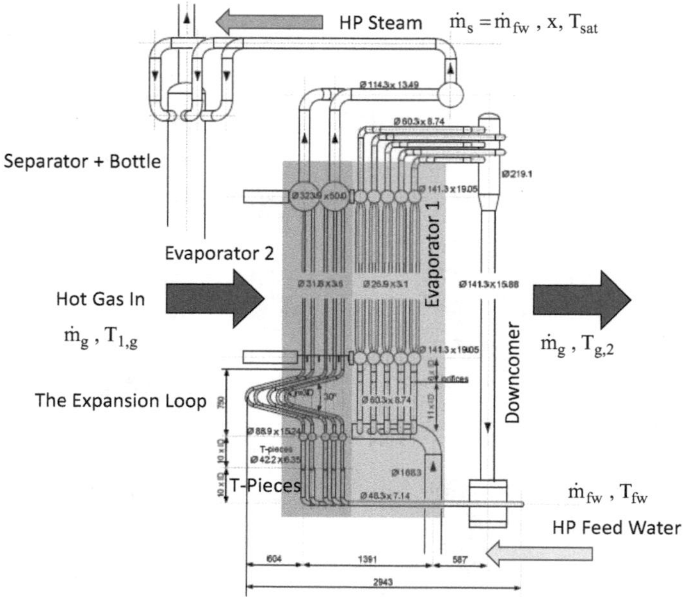

FIGURE 15.11 Heat and mass balance of the Benson boiler section.

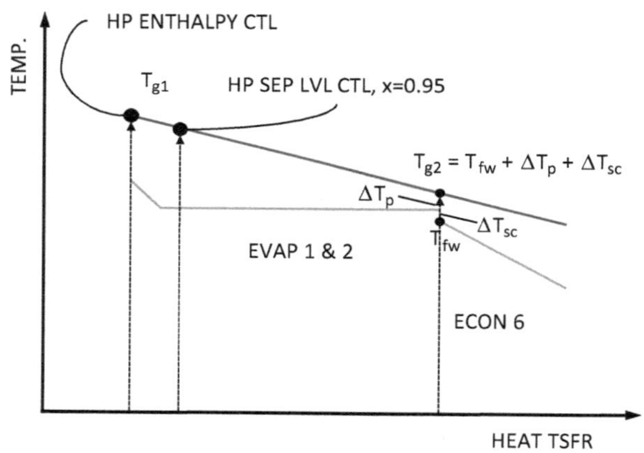

FIGURE 15.12 Temperature-duty diagram of the HP evaporator (conceptual). The straight line at the top, from left to right, is the flue gas temperature transferring heat to steam/water. The lower line, from right to left, is water/steam receiving heat from the flue gas.

In the control algorithm, the total difference between the feedwater temperature at the economizer exit and the exhaust gas temperature at the HP evaporator exit is estimated from a simple ratio of exhaust gas energy as follows:

$$\Delta T = \Delta T_p + \Delta T_{sc} \tag{15.24}$$

$$\Delta T_{est} = \frac{\dot{m}_g}{\dot{m}_{g,des}} \cdot \frac{\left(T_{g1} - T_{fw}\right)}{\left(T_{g1} - T_{fw}\right)_{des}} \cdot \Delta T_{des} \tag{15.25}$$

$$T_{g2,calc} = T_{fw} + \Delta T_{est}, \tag{15.26}$$

where the subscripts "est" and "des" refer to "estimated" and "design", respectively. Thus, the exhaust gas temperature at the HP evaporator exit, which is used in the control algorithm, is ΔT degrees above the measured economizer exit feedwater temperature.

It is thus possible to calculate the feedwater mass flow rate set point (feed-forward control flow set point), which is required for

- Maintaining wet steam with a steam quality of x = 0.95 in the case of *HP Separator Level Control* operation.
- Maintaining superheated steam with a superheating of ΔT (18°F–126°F or 10°C–70°C in SI units) in the case of *HP Enthalpy Control* operation.

This calculation is performed by generating an enthalpy difference between the feedwater enthalpy at evaporator inlet and the steam enthalpy in the separator. Here, the different enthalpy set points (for wet and superheated steam) are used, depending on the operation mode (with/without Benson) which has been selected by the operator.

- In the case of *HP Separator Level Control* mode, the related feedwater feed-forward control flow set point is fine-adjusted by the HP separator level correction controller.
- If *HP Enthalpy Control* mode is selected, the related feedwater feed-forward control flow set point is corrected by the HP enthalpy correction controller in such a way that the steam temperature at HP separator outlet ensures a sufficient superheating and simultaneously, the HP steam temperature controls are utilized in an optimal way (i.e., to minimize the periods of injection and the amount of the injected water).

15.2 TRANSIENT OPERATION

The opposite of "steady state" is, strictly speaking, "unsteady state", which means that system parameters can be expressed as functions of time. In mathematical terms, the time derivative of a given parameter, X, is nonzero, i.e., $\partial X/\partial t \neq 0$. While the term "unsteady state" is used in thermodynamics textbooks, the term common in the industry jargon is "transient".

The calculation and optimization of the transient (unsteady-state) performance of combined cycle power plants require sophisticated simulation tools, which are also utilized for the design and development of active plant control systems (see Chapter 2). Due to the inherent complexity of the governing conservation equations (differential instead of arithmetic) and the requisite solution techniques, the transient simulation tools are extremely complex computer software with steep learning curves, long model development and CPU run times. In addition to the aforementioned software platforms, there are numerous published studies describing the differential equations governing the unsteady heat, mass and momentum conservation processes in major combined cycle subsystems, i.e., the gas and steam turbines, the HRSG and the balance of plant (BOP). The reader is referred to the paper by Gülen and Kim for a selected list [8].

The gas turbine is most amenable to the block diagram approach to simulate its governor controls, which can then be implemented in Simulink for simulation studies. This subject is covered in detail in **GTFEPG** (Ref. [11] in Chapter 2). The reader should consult Chapter 19 of that book for a detailed discussion and pertinent references. In fact, due to its large thermal mass and associated time lags in addition to the complex heat exchanger arrangement, the bulk of the published studies are devoted to the simulation of HRSGs and, especially, the evaporator drum controls. Even with the simplest approaches, full solution of the resulting system of equations presents a considerable challenge for 3PRH steam bottoming cycles, which represents the current state of the art with the advanced heavy-duty gas turbines. Thus, whether using a commercially obtained tool or setting up a system of equations for numerical solution, the analysis of GTCC transients requires in-depth knowledge of several engineering disciplines, differential calculus and numerical techniques (which, in turn, requires knowledge of at least one programming language such as C++ or Fortran). This renders transient performance estimation and optimization the privileged domain of a select few.

There are several types of combined cycle transient operation modes, e.g., plant startup from standstill to full or specified load, controlled shutdown or trip, load ramps, and primary and secondary responses to a drop in grid frequency. All of them are important, but the discussion herein will focus on the combined cycle startup, especially the "fast" variant, which includes all salient aspects of the bottoming cycle dynamic response. The reason for that is simple. At the time of writing and in the foreseeable future, it is fairly certain that simple and combined cycle gas turbine power plants will play a dual role in worldwide electric power generation portfolio:

- High-efficiency baseload power generation in locations where other resources are not sufficient and/or cost-effective to meet the demand
- Load following and backup generation in locations where renewable assets have an increasingly larger share in generation portfolio.

Countries in the Pacific region and Southeast Asia fall into the first category. In those countries, very expensive gas (mostly imported LNG) and dearth of other sustainable resources (i.e., nuclear, hydro and solar/wind) make 60+% efficient (nominal rating) large GTCC blocks with the latest super-heavy H and J class gas turbines very attractive from initial (i.e., low dollars per kilowatt) as well as life-cycle cost perspectives.

In the developed world, especially in Europe, GTCC power plants are mostly relegated to a supporting role, i.e., the second category above. The reader is referred to Sections 8.6.1 and 8.6.2 for prime examples of this dynamic and how it impacts GTCC operation.

A modern GTCC with advanced F, G, H and J class gas turbines is expected to be a "jack of all trades". Realistically or not, this is the market expectation and major OEMs are designing their equipment to meet it. This means that a state-of-the-art GTCC is expected to have the following features – *simultaneously*:

- High thermal efficiency, i.e., 60+% net LHV (ISO baseload)
- Large gas turbines, i.e., 400 MW in 60Hz and 500 MW in 50Hz, for power density and low dollars per kilowatts
- Emissions-compliant turndown to low loads (e.g., 30% or even lower)
- Limited heat rate degradation at part load and/or high ambient temperatures
- Cyclic operation (e.g., nightly and weekend shutdown) with minimal or no impact on RAM
- Fast start (e.g., 30min or faster from turning gear to full load after overnight shutdown)
- Fast load ramps (up or down)
- Fuel flexibility (e.g., ±15% Wobbe index).

The inherent conflict in the aforementioned characteristics should be obvious to the reader already familiar with the preceding chapters. High thermal efficiency is driven by high firing temperatures and pressure ratios, which put the onus on turbine hot gas path materials, coatings and cooling technologies. At the same time, combustor designs should be able to handle high temperatures while simultaneously meeting increasingly stringent emissions regulations (as low as 2 ppmvd NOx, e.g., in California) without diluent injection (detrimental to efficiency and wastes water).

High thermal efficiency also demands high steam temperatures and pressures in the bottoming cycle. This, of course, requires HRSG, steam turbine and BOP components (e.g., large steam pipes and valves) with large metal mass and thick walls – not particularly suitable to fast operational swings with resultant thermal stresses and their adverse impact on parts life.

Key principles underlying GTCC performance have been explored in detail in Chapters 4–8. For an in-depth coverage of gas turbine-specific items (e.g., combustors, emission characteristics and design options), the reader is referred to the relevant chapters in **GTFEPG**. Model-based (adaptive) control is the key technology in meeting emissions-compliant low load operation (in addition to dry low NOx or DLN combustors with axial fuel staging) as well as off-design operation with minimal efficiency lapse (i.e., part load and/or high ambient temperature). Those aspects do not require further elaboration in this chapter.

15.3 GTCC STARTUP: BASICS

A generic gas turbine startup sequence is shown in Figure 15.13. For more details, the reader is referred to Section 19.3.1 in **GTFEPG** (e.g., Figure 19.9 therein and the accompanying text). The particular sequence in Figure 15.13 is for a "hot" startup, which typically takes place after an overnight shutdown (8–12 h)[8].

A modern, heavy-duty industrial gas turbine is started by the *static starter system*, which is also known as the LCI (*Load Commutated Inverter*). The LCI runs the synchronous ac generator as a "motor" to crank the powertrain to the firing speed and help it thereafter until the gas turbine becomes self-sustaining. In the past, when gas turbines were much smaller, a small diesel engine or electric motor was used for cranking the machine. (This is not practical with modern machines rated at several hundred megawatts; one would need a small industrial gas turbine for that job.)

Fresh air "purge" of gas turbine and HRSG (including the transition duct and the bypass stack if present) of combustible vapors accumulated therein during shutdown is a requirement by National Fire Protection Association (NFPA) 85, *Boiler and Combustion Systems Hazards Code* (2011). In accordance with that, the LCI first accelerates the gas turbine to the HRSG *purge speed* (typically, 25% of synchronous speed) and keeps it there for up to 15 min. (This is not shown in Figure 15.13.) Thereafter, the gas turbine coasts down to the ignition or light-off speed (about 15%), at which point the combustors are ignited. The LCI disengages at about 90% (self-sustaining) speed. The gas turbine is then accelerated to FSNL (full speed, no load), and the controller synchronizes it to the grid.

Up to the synchronization, the gas turbine start sequence does not change between different plant start categories, which are differentiated by the gas turbine loading to the FSFL (full speed, full load). Generally speaking, this period can be divided into two distinct regimes:

1. The gas turbine is brought up to a part load, which is about 40% in Figure 15.13, and held there for a certain period of time, which is a function of the shutdown time. This particular load level typically coincides with the *minimum emissions-compliant load* (MECL) of the particular unit. The longer the shutdown period prior to the unit start, the longer the gas turbine load hold time.

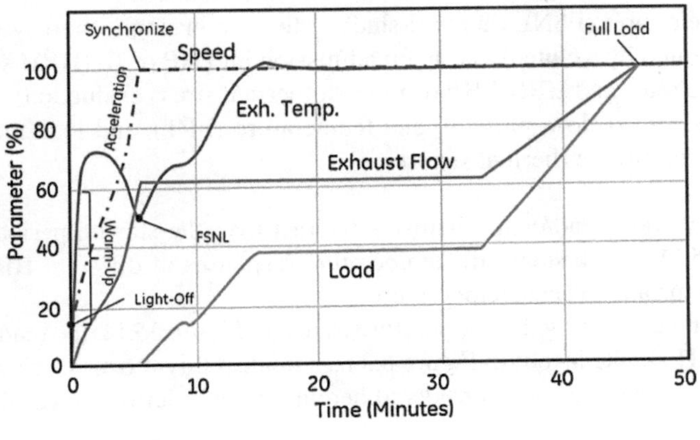

FIGURE 15.13 Typical gas turbine start characteristics (combined cycle hot start).

[8] Other startup types are commonly referred to as "warm" and "cold" starts. (The classification is usually made on the basis of plant shutdown duration; however, the driving factor is the steam turbine rotor and casing metal temperatures.)

2. Once the so-called *steam temperature matching* period, during which the IGVs are at their closed position, is over, the gas turbine is loaded to FSFL in a controlled manner by opening the IGVs to their open position.

The underlying logic here is to ensure a gradual increase of the steam turbine metal temperatures (and metal temperatures of other thick-walled plant components such as the HP drum of the HRSG) in order to prevent excessive thermal stresses.

It should be mentioned that this particular method of holding the gas turbine at its MECL requires that the HRSG is equipped with *terminal* attemperators for precise control of steam temperatures. The reason for that is, as shown in Figure 15.3, at 40% load, the exhaust gas is at its maximum (isotherm) value. Depending on the model and vintage of the gas turbine in question, this can be 1,200°F or higher. Standard interstage deaerators, which are used for steam control during normal operation, cannot provide the requisite precision. In the absence of terminal attemperators (which add to plant cost and complexity), steam temperature is controlled directly by modulating the gas turbine firing and exhaust gas temperature. This, of course, means that the gas turbine runs for an extended period at very low load (about 20%) with increased NOx emissions. This obsolete method is not possible anymore due to the increasingly strict environmental regulations putting rather onerous limits on criteria pollutant emissions.

Recently, in accordance with the changing landscape of electric power generation, gas turbine OEMs developed technologies to further accelerate the conventional start sequence described above. These technologies, marketed under names such as "rapid response" or "fast start", aim to eliminate the low load hold period altogether and load the gas turbine from FSNL to FSFL in one continuous step. In order to achieve this, such fast/rapid startup technologies rely upon advanced control and design modifications such as HRSG *purge credit*, model-based control and cascaded steam bypass with terminal attemperation. This will be discussed in more detail in Section 15.4.

Bottoming cycle dynamic response to the gas turbine exhaust flow and temperature (i.e., energy input to the system as a function of time) is a combination of the following items taking place in a chronological order:

- Time-dependent warm-up of HRSG heat exchange section metal (tubes, headers, drums, etc.)
- Time-dependent steam generation in the HPEV, IPEV and LPEV;
- Steam pressure and temperature at steam turbine admission points (HP, IP, or HRH, and LP) as a function of time
- Achievement of steam turbine steam admission start permissives
- Steam turbine ramp to FSNL (in multi-shaft systems or in single-shaft systems with prime movers; separated by a clutch) via steam admission into HP or IP (HRH steam) section
- Steam turbine rotor and HRSG HP drum wall thermal stress evaluation
- Steam turbine steam flow, pressure and temperature (FPT), and load ramps dictated by allowable steam turbine thermal stresses.

How these different events and/or mechanisms interact over the startup period is illustrated succinctly in Figure 15.14, which shows the temperature responses of different HRSG sections to the variation in the gas turbine exhaust temperature.

In interpreting and explaining the temperature trends in Figure 15.14, the reader is encouraged to use the typical 3PRH HRSG layout in Figure 6.3 as a guide. Only the most relevant heat exchanger sections (tube banks or bundles) are considered herein. In the order of the gas flow direction (from left to right in Figure 6.3),

- First HP superheater (HPSH1 in Figure 6.3)
- HP evaporator (HPEV) drum
- IP superheater (IPSH)

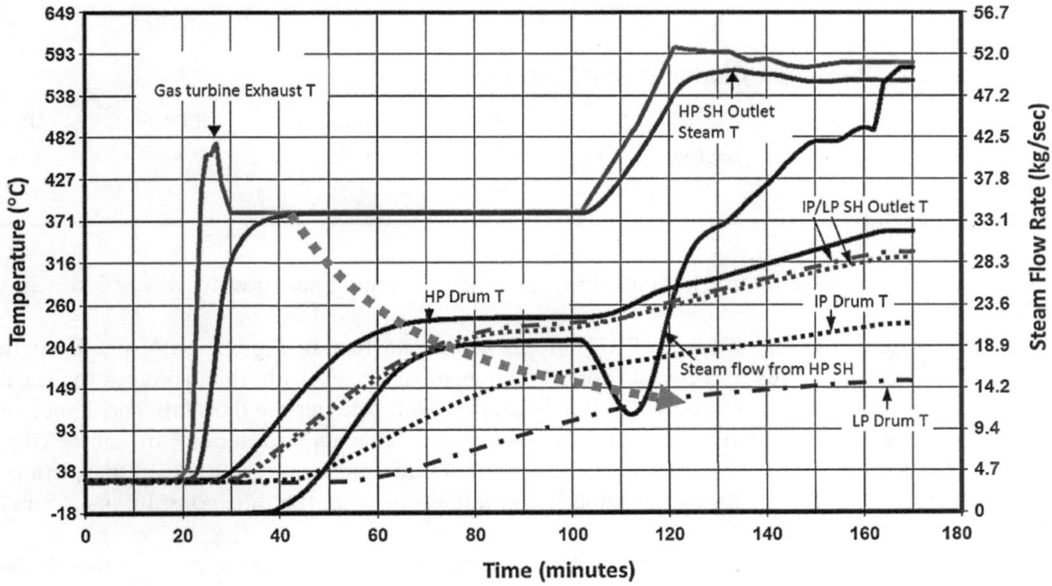

FIGURE 15.14 Dynamic response of selected HRSG heat exchanger sections during a cold start (drawn after the startup simulation results in Briggs [9]).

- LP superheater (LPSH; the LPSH and IPSH are either adjacent or very close to each other and their steam temperature profiles pretty much coincide.)
- IP evaporator (IPEV) drum
- LP evaporator (LPEV) drum.

It is clearly seen that the more a section is removed from the HRSG inlet, the slower it is to respond. In this case, the slowest responder is the LP evaporator. Furthermore, as indicated by the dashed arrow in the figure, the response time constant of each component increases in proportion to its proximity to the "driving force". The driving force of the transient event is the gas turbine exhaust energy (i.e., flow and temperature – and composition, to be precise) or, rather, the change therein.

The particular start event in Figure 15.14 is for a vintage F class combined cycle; that is, there are no terminal attemperators. Thus, steam temperature is controlled directly by the gas turbine load and exhaust temperature. This is clearly seen by the coinciding exhaust gas and HP steam temperature lines in Figure 15.14. Due to the heat transfer characteristics of the gas and the steam, superheater tube metal and steam are essentially at the same temperature as the hot exhaust gas.

Only the first HP superheater absorbs the full brunt of the driving force of the hot exhaust gas. Subsequent components experience only a fraction thereof; the farther away the component, the smaller the fraction. That fraction can be represented reasonably well by a very simple relationship, i.e.,

$$T_{gas} \approx \kappa T_{exh}, \tag{15.27}$$

where T_{gas} is exhaust gas temperature at the inlet to a particular HRSG section. The value of the proportionality factor κ is 1.0 for the first HP superheater. For the others, its value diminishes in the direction of the gas flow.

The other parameter of interest is the characteristic response time τ of each HRSG section, which is the driver behind the similarity between the temperature profiles of the HRSG sections in

TABLE 15.6
HRSG Thermal Response: Characteristic Parameters

	RHTR	HPEV	IPSH	LPSH	IPEV	LPEV
κ	0.92	0.70	0.50	0.50	0.40	0.30
τ, min	1.35	1.50	1.00	2.75	0.75	0.75

Figure 15.14. That response time is a characteristic of the governing heat transfer process, which is described in Appendix D.

Typical values of κ and τ for a 3PRH HRSG similar to that in Figure 15.14 are listed in Table 15.6. There is, however, a caveat. For a component separated from the driving force (heat input) by other components, not only its own thermal capacitance but the thermal capacitance of all the components in between must be considered. A reasonable approximation is to set the time constant of a particular HRSG section to the sum total of individual time constants of the sections upstream of it. For example, for the LP superheater and evaporator, the time constant is 12.5 and 20 min, respectively.

This basic thermal response mechanism characterized by hardware-dependent lag times is the key principle underlying the transient analysis of GTCC bottoming cycle. In mathematical terms, everything revolves around the solution of a modified version of the well-known *exponential decay function*, $\exp(-t/\tau)$, for heating (or cooling for shutdown problems) of major metal components (e.g., HRSG tube banks, evaporator drums, steam turbine casing and rotor) under the assumption of "lumped capacitance". For a thorough discussion of the actual mathematic formulations and their solution, please refer to

- Chapter 19 of **GTFEPG** (Ref. [11] in Chapter 2) for gas turbines
- The paper by Gülen and Kim for the combined cycle and bottoming cycle [8].

A very brief introduction to thermal response basics is provided in Appendix D.

15.3.1 Steam Turbine Roll

Steam turbines with cascaded steam bypass are typically started by admitting steam from the reheat superheater into the IP section. Admission steam conditions *should* be sufficient to overcome the rotational inertia (in lb-ft^2) of the entire steam turbine and its generator, I_{rot}, and accelerate it from the turning gear speed (a few rpm) to FSNL (3,000 or 3,600 rpm). Based on available steam conditions and initial IP rotor temperature, using the relationship between steam turbine power generation (expansion from the IP turbine inlet to the condenser), rotor torque and rate of change in angular speed, ω, the roll time can be estimated as 2–15 min. The key formula is the simple relationship between power, torque and rate of change in rotational speed (i.e., acceleration):

$$\dot{W} = \tau \frac{2\pi N}{60}, \tag{15.28}$$

where N is the rotational speed in rpm, τ is the torque in lbf-ft and $\dot{W}$ is the shaft power in kilowatts (conversion factors are not shown but must be included in the calculations). The shaft power requisite to overcome the steam turbine drivetrain inertia and accelerate it from the turning gear speed (about 10 rpm) to generation speed (i.e., 3,000 or 3,600 rpm depending on the grid frequency) is generated by the steam flow, i.e.,

$$\dot{W} = \dot{m}_{stm}\left(h_{in} - h_{ex}\right). \tag{15.29}$$

The enthalpy drop can be rewritten using the isentropic efficiency:

$$\dot{W} = \dot{m}_{stm} h_{in} \eta \left(1 - \frac{h_s}{h_{in}}\right) \quad \text{with} \quad h_s = h(p_{ex}, s_{in}). \tag{15.30}$$

Turbine exit enthalpy and pressure are set by the condenser conditions. The inlet enthalpy, pressure and entropy are evaluated at the IP turbine *inlet bowl.* The efficiency in Equation 15.30 is the ratio of the actual enthalpy drop across the IP and LP turbines to the isentropic enthalpy drop for the same inlet and exit pressures, Δh_s. For most off-design conditions, η does not vary appreciably from its design value. However, during the turbine acceleration, steam flow changes from zero to a fraction of its design value along with changing inlet conditions. Thus, the variation in η should be expected to be significant at least during the initial part of the process. Following an empirical relationship developed for off-design performance of fossil power plant steam turbines to be used in dynamic simulation models and control system design, a simple scaling relationship is proposed for using in Equation 15.30:

$$\eta = \eta_0 \alpha \left(\frac{\Delta h_s}{\Delta h_{s,0}}\right)^{\beta}, \tag{15.31}$$

where η_0 is the design point efficiency. Constants α and β are determined as 0.93 and 0.69 using actual operating data.

The torque is the product of the total rotational inertia of the steam turbine drive-train, I in lb-ft^2, and the rotational acceleration, i.e.,

$$\tau = I\frac{d\omega}{dt} = I\left(\frac{2\pi N}{60}\right)\frac{dN}{dt}. \tag{15.32}$$

Combining Equations 15.28 through 15.32 and noting that the shaft power during acceleration is a function of time yields the following formula:

$$\dot{W}(t) = \dot{m}_{stm} h_{in} \eta \left(1 - \frac{h_s}{h_{in}}\right) = I\omega(t)\frac{d\omega}{dt}. \tag{15.33}$$

Integrating Equation 15.33 and noting that $\omega = 2\pi N/60$, the result is

$$\frac{1}{I}\int_{t_0}^{t} \dot{m}_{stm} h_{in} \eta \left(1 - \frac{h_s}{h_{in}}\right) dt = \frac{1}{2}\left(\omega^2 - \omega_0^2\right),$$

$$N(t) = \frac{30}{\pi}\sqrt{\left(\frac{2\pi N_0}{60}\right)^2 + \frac{2}{I}\int_{t_0}^{t} \dot{m}_{stm} h_{in} \eta \left(1 - \frac{h_s}{h_{in}}\right) dt}. \tag{15.34}$$

Setting t_0 to 0 and noting that the turning gear speed N_0 (a few rpms as stated earlier) can be ignored, we obtain

$$N(t) \approx \frac{3{,}000}{\pi}\sqrt{\frac{5}{I}\int_{0}^{t} \dot{m}_{stm} \eta \Delta h_s \, dt}, \tag{15.35}$$

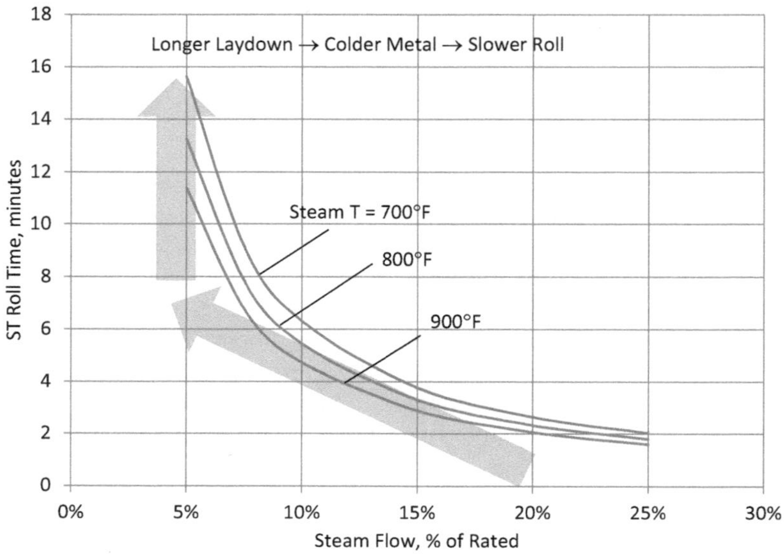

FIGURE 15.15 Steam turbine roll times for varying steam flows and temperatures (steam pressure 120 psia, rotational inertia 700 kp-ft^2, rated IP turbine inlet flow 681.5 kpph).

where the argument of the integral on the right-hand side of Equation 15.35 is the power (in Btu/s) generated by steam expanding between IP turbine inlet and condenser. Representative speed profiles obtained from Equation 15.35 are plotted in Figure 15.15.

15.3.2 Steam Turbine Stress Control

Steam temperature matching mentioned in the beginning of the section refers to bringing the HRSG-generated steam temperature to the allowable value set by the turbine stress controller prior to admission into the turbine. The allowable values are determined such that axial differential expansion of the rotor and thermal stresses acting on it do not lead to component failure. There are three drivers of rotor expansion and thermal stress:

1. Steam-to-metal temperature mismatch
   a. At the start of steam admission
   b. At the start of steam temperature ramping.
2. Steam temperature ramp rate.

Initial mismatch is controlled automatically by the permissives to initiate the turbine roll and transfer to *forward flow* (if the steam turbine is rolled by steam admission into the IP turbine). The permissive to initiate temperature ramping requires that both the reheat (IP and LP) and HP sections of the steam turbine have been heated sufficiently to prevent rotor overstress and long rotor differential expansion during ramping.

The goal of any steam turbine startup is to raise turbine rotor and shell metal temperatures from initial levels to rated conditions in the shortest possible time without damaging the unit and reducing its life via exceeding material limits. The underlying principles are covered in Appendix E. Everything hinges on careful evaluation of allowable steam temperatures as a function of initial metal temperatures. Salient aspects of this endeavor are illustrated below using an example.

The chart in Figure 15.16 shows the first 2 h of steam turbine roll, warm-up and loading phases for an initial metal temperature of 180°F (about 5–6 days of downtime per Figure E.1). Steam is

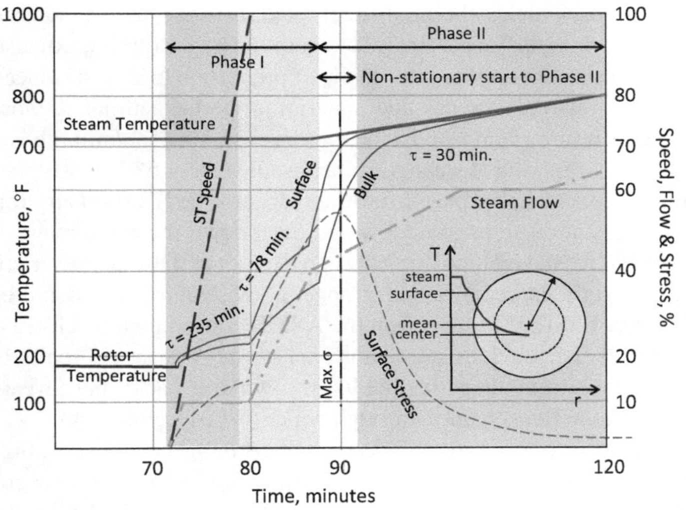

FIGURE 15.16 Typical steam turbine roll via IP steam admission and the ensuing warm-up period [8]. Representative of a single-shaft GTCC cold start (total 3 h). Note how the quasi-stationary Phase II is preceded by a short non-stationary period.

admitted into the IP turbine at 710°F and 120 psia at a flow rate of 10% of its rated value at full load. This is sufficient for acceleration from turning gear to synchronization in 8 min (see Figure 15.15). Initial steam–metal temperature mismatch is more than 500°F, and *prima facie*, this may seem excessive (see Table E.3 for maximum allowable steam admission temperatures). However, this large temperature delta is acceptable due to the low value of the heat transfer coefficient (HTC) (<30 Btu/h-ft^2-F per Table E.2) and the ensuing low thermal stress as predicted by Equation E.7 (also very high time constant, i.e., $\tau > 200$ min).

The thermal response of the steam turbine rotor metal to the hot steam flow is characterized by different time constants, which leads to the rotor surface and bulk temperature profiles shown in Figure 15.16. The basic mathematical principles are briefly presented in Appendix D. Detailed discussion can be found in Gülen and Kim [8]. In Phase I in Figure 15.16 with constant steam temperature, there are two distinct thermal "speeds" (i.e., time constants of 235 and 78 min), which are dictated by steam admission flow rate (and, hence, HTC). The first one covers the turbine roll from turning gear to FSNL with IP steam flow at 10% of its rated value. After synchronization, steam flow is ramped to 40% of its rated value, while steam temperature is kept constant at 710°F.

Following synchronization, IP steam flow is ramped steadily to 40% to accelerate the warm-up process via increased HTC. Once the steam–metal temperature mismatch (based on rotor surface temperature inferred via IP inner bowl thermocouple) reaches about 250°F, steam temperature is ramped (via terminal attemperator control) at a rate defined by the cyclic life expenditure (CLE) curve (about 3–4°F/min for an acceptable life of 4–5,000 cycles from Figure E.2).

15.3.3 HRSG Stress Control

The other component subject to low-cycle fatigue (LCF) damage due to cycling is the cylindrical HP drum of the HRSG (4–5 in. of wall thickness). The limiting thermal stress is at the inner drum wall controlled by saturated steam pressure and temperature inside the drum. During startup, mechanical stress due to internal drum pressure and thermal stress due to thermal expansion are in opposite directions, while they are in the same direction during shutdown. For a thorough discussion of the impact of cycling and fast starts on the HRSG life and associated damage mechanisms, please refer to the article by Taylor and Pasha [10].

Unlike the steam turbine, which is thermally decoupled from the gas turbine via terminal attemperators, HRSG sections are directly "under fire". They respond to gas turbine exhaust temperature transients much faster than the steam turbine rotor in direct proportion to their distance from the inlet (see Figure 15.14). Thermal stress calculations and material properties similar to those described above limit the pressure–temperature ramp rate inside the drum to 10–15°F/min (about 50 psi/min max.) for units designed up to ~1,800 psig at steam turbine throttle (about 6%–10% higher at the HP drum).

Advanced steam cycles with 2,400 psig throttle and drum-type HRSGs (very thick walls, e.g., 6 in. or 150 mm wall thickness at those pressures) would push down the ramp rate to a few degrees per minute (see Equation E.8 for the relationship between dT_{stm}/dt and the characteristic length, L_c). This can be alleviated to a certain degree by using stronger alloy steel (obviously more expensive) and/or designing the HRSG per EN-12952 rather than the ASME code, which results in thinner walls (e.g., see Ref. [2] in Chapter 9). One obvious solution is *once-through* design of the HP evaporator, which eliminates the thick-walled drum altogether but has its own drawbacks and caveats (e.g., see Ref. [3] in Chapter 6 for more on the Benson once-through boiler (OTB) technology).

A recent design approach proposes to replace the HP drum by a cylindrical, thin-walled knockout vessel with external separator bottles and thus avoid the thermal stress problem in cold starts (Siemens-NEM's DrumPlus™) [11]. A similar concept is adopted by Babcock & Wilcox, which replaced the thick-walled HP drum with two vertical steam separators [12]. Alstom Power (now a part of GE) developed OCC™ (Optimized for Cycling and Constructability) design for enhanced cycling performance [13]. In a nutshell, the OCC design uses smaller diameter tubes and headers in top-supported single-row "harps" (i.e., tube banks) vis-à-vis conventional designs with multiple tubes connecting to bigger headers with larger diameters. Note that the HRSG design in Ref. [12] also uses the single-row harps in superheaters. Design approaches by another HRSG supplier can be found in Refs. [14–15].

In "hot" starts, HP and reheat superheaters subjected to very steep gas temperature ramps are critical in terms of HRSG life consumption. Still, according to HRSG suppliers, cold starts ($T_{drum} < \sim400°F$) are 20 times more damaging than warm starts ($T_{drum} < \sim500°F$), whereas hot starts ($T_{drum} > 500°F$) do not impact LCF life. Consequently, HRSG design life is a strong function of the operating characteristics of the combined cycle power plant. In fact, HRSG designers use CLE curves similar to those used by steam turbine designers (see Figure E.3 in the Appendix and the accompanying text). These curves describe the relationship between the temperature ramp rate in the component (e.g., the HP drum) and total temperature change during the transient event (e.g., startup) and the fraction of total life consumed per each cycle. A typical comparison of HRSG life under different scenarios is presented in Table 15.7 [13]. The implication of the data in the table is that each cold start reduces the HRSG life by about 600 EOH (Equivalent Operating Hours). Corresponding life expenditure for each warm and hot start is about 30 and 10 EOH, respectively.

Natural pressure–temperature decay of the HP drum can be described by Equation E.1 with τ_c of 60–80 h. It takes about 2–3 days for the pressure to decay to the atmospheric conditions. As a rule of thumb, HP drum steam pressure is typically 60–200 psig, after a shutdown of 10–20 h and above 500 psig after an overnight shutdown.

Bottling up the HRSG via stack dampers with insulation up to the damper, steam sparging (requires auxiliary boiler) or running the SCR ammonia vaporizer heaters helps keep the HRSG warm and pressurized over limited duration shutdowns to enable gas turbine starts with no low load hold.

TABLE 15.7
HRSG Life in Equivalent Operating Hours (EOH) [13]

Cold Starts	200	200	125
Warm starts	1,000	900	200
Hot starts	4,000	1,000	375
HRSG life, EOH	100,000	130,000	200,000

Beyond about 3 days, however, this is increasingly impractical, and even in plants designed for fast starts, limited duration gas turbine holds are needed to accomplish HP drum warm-up in two steps (somewhat similar to that shown in Figure E.5).

HRSG purge during startup, in addition to increasing the startup time and consuming electric power (LCI running the generator as a motor), has another drawback. Relatively cold purge air pumped into the HRSG by the gas turbine cools the tube banks leading to metal and water/steam temperatures (and, thus, steam pressures) even lower than they were to start with. This effect can be prominent to the point that the front-end superheaters can act as "supercondensers" for steam trapped in the tubes. The condensate accumulated in the tubes and lower headers must be drained in order to prevent water ingestion in the steam turbine and blockage of steam flow through the HRSG tubes as well as thermal distortion and shock. Unless draining is accomplished in a speedy manner through an adequately sized drain system (pipes, valves, etc.), it will further reduce steam pressure and temperature in the HP section and add to the start time. For a thorough discussion, please refer to the article by Moelling and Jackson [16].

15.4 GTCC STARTUP: PRACTICAL CONSIDERATIONS

There is a long list of operability and maintenance considerations in a successful GTCC start from standstill, which are discussed in detail elsewhere [17–19]. Correct steam chemistry, establishment of steam seals, vibration, overspeed and thrust controls are all vital for acceptable component life and RAM (Reliability, Availability and Maintenance). These features are summarized in the diagram in Figure 15.17. They are briefly explained below (list numbers correspond to the labels in the figure):

1. "Trickle" steam flows are extended to (closed) major steam valves during steam bypass operation, e.g., MSV and the intercept valve (IV), so that they are already warm when they are opened to admit steam flow into the steam turbine.
2. In a single-shaft arrangement, the steam turbine is decoupled from the gas turbine and the generator via a SSS clutch. Thus, during steam bypass operation with the gas turbine running and the HRSG making steam, the steam turbine is still on turning gear.
3. When steam is admitted into the IP and HP turbines to roll the steam turbine first to full speed at no load (FSNL) and then start loading, steam–metal temperature difference must be controlled. The underlying principles and how it is done are explained in detail in Appendix E.

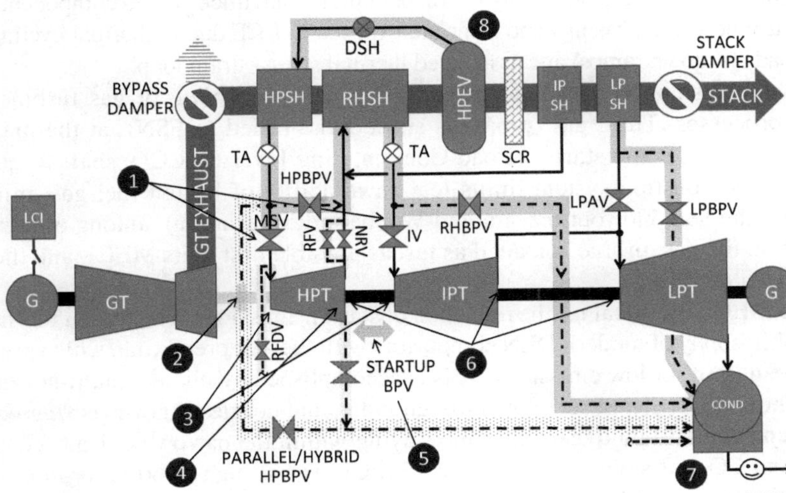

FIGURE 15.17 Operability considerations in conjunction with GTCC startup.

4. Controlling HP back pressure to prevent overheating of the turbine exit stages (via high cold reheat pressure and limited expansion) is crucial during steam turbine bypass operation and loading. The non-return valve (NRV) at the HP turbine exhaust ensures that bypassed steam does not pressurize the back end of the HP turbine.
5. In opposed-flow HP-IP configurations, IP and HP steam flows should be controlled to maintain the thrust balance.
6. Prior to startup, steam seals must be established. This typically requires the inclusion of an auxiliary boiler in the system because the HRSG does not produce steam at the early stages (additional cost and complexity). This can be avoided by the presence of the SSS clutch, which ensures that the steam turbine is idle until it starts to roll (typically via admission of steam into the IP turbine in GE steam turbines). This ensures that there is steam available from the HRSG for the steam seals (see Section 9.2.2 for more on this system).
7. Establishment of the water chemistry (primarily, dissolved oxygen in the condensate and feedwater and pH control) is another crucial startup requirement. Detailed coverage of steam cycle feedwater treatment can be found in Section 10.9.
8. In theory, complete decoupling of the gas turbine and the bottoming cycle is possible by the inclusion of a bypass stack in the duct between the gas turbine in the HRSG. This is rarely the case in modern GTCC power plants. Thus, once the gas turbine starts, HRSG is active and starts making steam. Controlled rise of steam pressure and temperature in the HP evaporator (with a thick-walled, cylindrical steam drum) to prevent large thermal stresses in the drum walls is important. Other control subjects are drum swell and Selective Catalytic Reduction (SCR) temperature. Achieving the operating gas temperature of the SCR (usually between the evaporator tube banks) as soon as possible is vital from emissions control perspective.

When all is said and done, the single most important issue from a fast start perspective is steam turbine thermal stress management. Furthermore, if the HRSG is drum-type, HP drum thermal stress management becomes an integral part of the problem.

In a nutshell, GTCC startup optimization problem can be formulated as to minimize the time required to reach the dispatch power (e.g., full load or a specific part load) without "breaking anything" in the process – literally. The failure mode to avoid is crack initiation and propagation. Failure to control thermal stresses results in cracks via LCF and brittle fracture. In fact, LCF (typically, failures occurring at less than 10,000 cycles) is found to account for roughly two-thirds of steam turbine rotor life with the remainder attributable mainly to creep. In particular, thick-walled components such as HP drum, steam turbine valves, casings and rotor are exposed to LCF due to thermal cycling (start–stop sequence or load up–down ramps) and associated thermal stress–strain loop.

In principle, the solution is simple enough: thermal *decoupling* of gas turbine and steam turbine start processes. Thus, gas turbine is started and rolled to FSNL at the maximum rate dictated by the size of static starter (Load Commutating Inverter, LCI), shaft torque limit and particular DLN combustion system limits (e.g., availability of heated fuel gas, minimum fuel requirement by the lean blowout margin and Wobbe index variation), among others. Following synchronization, the gas turbine is loaded as fast as possible first to its MECL and then to its full load at full speed (FSFL).

The rush to MECL is critical for the reduction of startup emissions. The reason for that lies in the basic design philosophy of modern DLN combustors with fuel-air premixing, which are designed to run near the lean limit for low emissions. This is accomplished by piloted, multi-nozzle fuel injectors via sequential activation of fuel flow through individual nozzles (known as *staging*) to prevent lean blowout and combustion dynamics while staying within the narrow equivalence ratio band to control NOx and CO emissions. For older units, MECL was as high as 60%; for modern units, the low load limit is around 40% (30% for most advanced systems or even lower). The exception to the rule is *sequential combustion* (reheat) gas turbines, which can turn off their second combustors to operate at 20% or lower load while emissions-compliant.

Two steps are instrumental in reducing gas turbine start time; i.e., elimination of

- HRSG purge sequence (by performing it right after shutdown in compliance with NFPA 85)
- Hold time at low load with reduced exhaust energy (flow and temperature) to control HRSG steam production rate and steam temperatures (at the HP drum and HP superheater exit).

Elimination of direct HRSG steam temperature control via gas turbine load and exhaust energy is the "thermal decoupling", which is the key enabler of fast start. There are several ways to accomplish this, e.g.,

- A bypass stack and modulated damper in the GT-HRSG transition duct controlling the exhaust flow to the HRSG
- Venting steam to the atmosphere ("sky venting")
- "Air attemperation" of the gas turbine exhaust gas flow via air injection into the transition duct.

Sky venting is an obsolete technique that wastes expensive demineralized water and does not constitute a viable option anymore. Bypass stacks are problematic due to the difficulty associated with perfect sealing to prevent leakage of hot exhaust gas through the gap between the damper and the duct. Air attemperation is a recently proposed technique, and the author is not aware of its commercial deployment in a utility-scale GTCC. Once we eliminate the "obvious" solutions of making "just enough" steam in the HRSG via exhaust gas bypass (the Goldilocks solution if you will) and/or venting excess steam into the atmosphere, there is only one solution available to us: bypassing the steam turbine and dumping the excess steam into the condenser. There are two types of steam bypass operation: parallel ("dry" reheater) and cascaded ("wet" reheater) bypass.

15.4.1 Steam Bypass Systems

In parallel bypass operation, steam generated in the HP and IP sections is sent to the condenser separately through their own bypass piping. Consequently, there is no steam flow through the reheat superheaters; that is, they operate "dry". Bypassing LP steam is typically avoided by suppressing steam generation in the LP evaporator by bypassing the LP economizer and supplying the LP evaporator with cold condensate. Any excess steam generated (a relatively minor amount) is vented to the atmosphere. This method has several drawbacks (e.g., expensive reheater tube material and increased alloy pipe length) and is pretty much obsolete.

The currently accepted method is the "cascaded" steam bypass with *terminal attemperators*. Steam generation and temperature–pressure ramp rates in the HP drum are dictated by gas turbine exhaust energy, whereas final steam temperature control is accomplished by terminal attemperators. Until steam temperatures reach acceptable levels for admission into the steam turbine, steam is bypassed via a route including the reheat superheater so that the latter is pressurized and "wet" (i.e., cooled by steam flow obviating the need for expensive alloys).

As shown in Figure 15.17, cascaded steam bypass is accomplished as follows:

1. Steam admission valves, MSV, IV and LPAV (LP steam admission valve) are closed.
2. HP steam bypass valve (HPBPV) and LP steam bypass valve (LPBPV) are opened.
3. HP steam from the HP superheater (HPSH) flows through the HPBPV into the reheat superheater (RHSH).
4. From the RHSH, steam flows through the *reheat bypass valve* (RHBPV) into the condenser.
5. Not shown in the diagram in Figure 15.17 are the HP and HRH bypass steam pressure control and attemperation systems. A schematic diagram of the latter system is shown in Figure 15.18. The arrangement for the HP and LP bypass systems is similar.

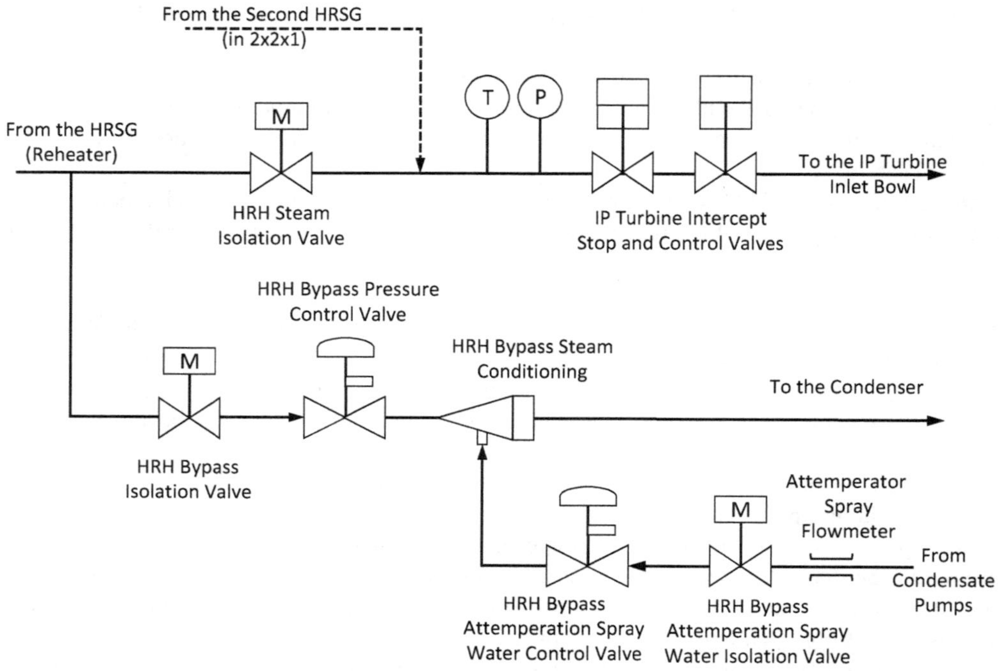

FIGURE 15.18 HRH steam cascade bypass pressure control and attemperation.

In a conventional bottoming steam cycle design with cascaded bypass, steam temperature coming out of the superheaters is controlled by the gas turbine exhaust temperature. Since the gas turbine exhaust temperature is a function of the gas turbine load, this method requires that the gas turbine is run at a sufficiently low load to ensure that the exhaust gas (and, hence, steam) temperature is low (e.g., see Figure 15.13). This method has several drawbacks, i.e.,

- Long start-up time and excessive fuel burn
- High NOx and CO emissions because the gas turbine load is below the MECL.

The obvious remedy is to start the gas turbine as fast as possible and go to MECL, or even better, to FSFL and control the steam temperatures at the HRSG exit by the terminal attemperators (TA in Figure 15.17). For the fastest possible start, the latter option is obviously the best because a gas turbine can get from turning gear to FSFL in 30 min or even less. This option, however, has its own drawback: excessive steam production in the HRSG because of high exhaust gas energy, which cannot be handled by the conventional cascaded bypass system described above without excessively large bypass valves and steam pipes. The solution of this problem is the "hybrid" bypass system.

In a hybrid cascaded bypass system, part of the HP steam is directly sent to the condenser through a second HPBPV (see Figure 15.17) just like in the parallel bypass method. This ensures that *all* HP steam generated in the HRSG can bypass the steam turbine. This, of course, adds significantly to the plant cost through extra alloy piping, valves and bypass attemperation stations. The cheaper alternative is to let the gas turbine wait at MECL for a period until steam temperature matching is achieved and then ramp up to FSFL. This ensures that the GTCC is in emissions compliance during the startup, which, of course, would take longer than the fastest possible variant with the hybrid bypass.

15.4.2 HP Turbine Exhaust Temperature Control

As mentioned above in conjunction with item 4 in Figure 15.17, keeping the HP turbine back end "cool" during startup is of prime importance. When the HP steam undergoes a truncated expansion with lower-than-design efficiency during startup, its temperature at the HP turbine exit is higher than its normal operating value, further exacerbated by windage heating (see Figure 15.19). This detrimental effect can be alleviated via different means depending on the steam turbine rolling method. There are two options, i.e., via steam admission:

- Into the IP turbine
- Into the HP turbine.

The second method is common for multi-shaft systems or single-shaft systems with a SSS clutch. Either method is used by different OEMs.

In the first variant, *reverse flow discharge valve* (RFDV in Figure 15.17; it is also known as the "ventilator" valve) is opened to hold the HP turbine at nearly condenser vacuum until steam admission into the HP turbine, which is referred to as "transfer to forward flow". At hot or warm starts (see Figure E.1 in the Appendix for how "hot" or "warm" starts are classified), when a certain speed is reached (i.e., increased windage), the *reverse flow valve* (RFV in Figure 15.17) is opened to create a cooling flow through the HP turbine. Transfer to forward flow is initiated by closing RFV and RFDV and opening the MSV.

An alternative solution is the provision of a "startup line", which is a pipe extending from the HP turbine exhaust to the condenser. This is a possible solution for systems with steam turbine rolling by admitting steam into the HP turbine. In this case, the NRV in the cold reheat line isolates the HP turbine from the high pressure in the reheat section. The HP turbine exhaust pressure is kept low through its connection to the condenser via the startup line.

15.4.3 Fast versus "Slow"

Conventional and fast gas turbine start speed/load profiles are compared in Figure 15.20. State points in the diagram denote the following events:

0. Start command
1. LCI engages – ramp to purge speed
2. HRSG purge starts

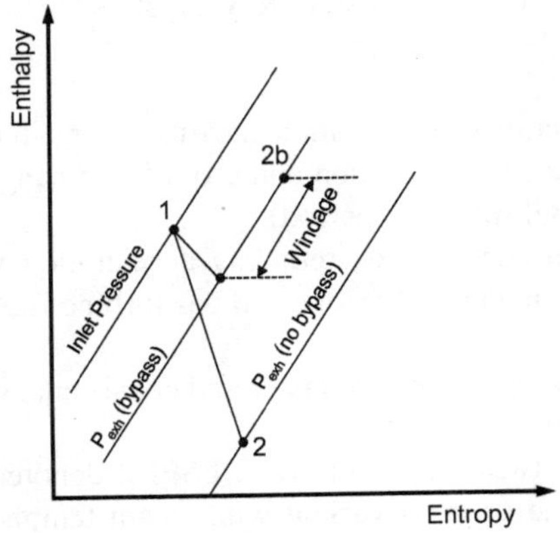

FIGURE 15.19 HP turbine expansion with (1 to 2b) and without bypass (1 to 2).

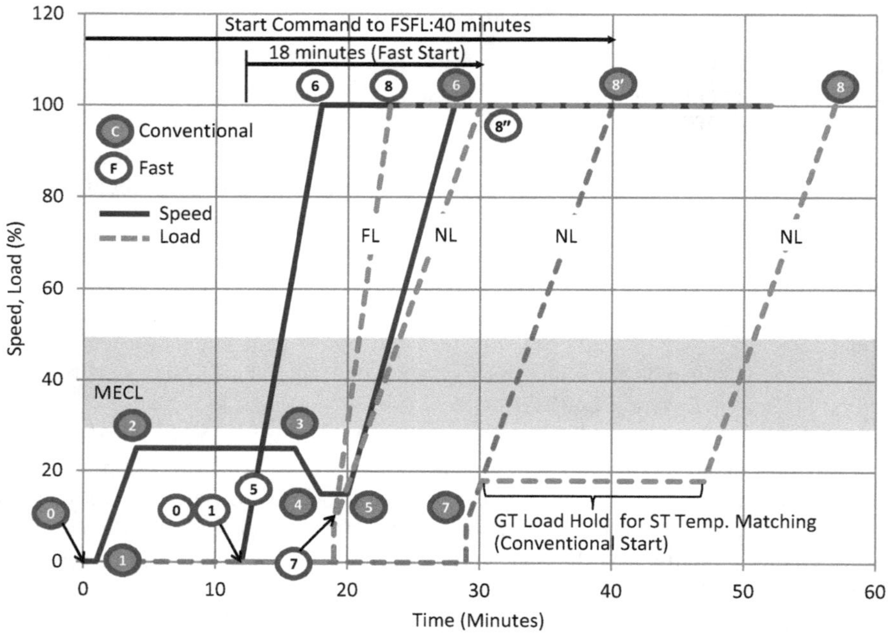

FIGURE 15.20 Gas turbine speed and load profiles during startup.

3. HRSG purge ends
4. Coast to ignition speed
5. Gas turbine fire and warm-up
6. Full speed no load (FSNL)
7. Synchronization
8. Full speed full load (FSFL).

There are two load ramp rates in the figure:

- Fast loading (FL) at about 20% per minute (i.e., 80 MW/min for a 400 MW unit)
- Normal loading (NL) at about 8% per minute (i.e., 25 MW/min for a 300 MW unit).

In the fast start option, steps 2–4 are eliminated. Key enablers for this modification are

- HRSG purge credit
- LCI preconnect (i.e., establish and maintain starting means in engaged state)
- "Fire on the fly" (i.e., successful ignition and cross-fire during acceleration ramp without the need for the traditional warm-up period)
- Closed-loop acceleration control (i.e., reducing acceleration variability due to seasonal ambient variation by balancing LCI torque and gas turbine fuel schedule from the turning gear to FSNL)
- Advanced control system optimized for faster and consistent synchronization to the grid.

For the conventional version, two end points (i.e., FSFL), denoted by 8 and 8′, are shown in Figure 15.20. The former is the original variant with steam temperature control via gas turbine exhaust temperature. The latter is the modern variant with steam temperature control via terminal attemperators. As mentioned earlier in Section 15.3, this method is rendered obsolete by strict air permit requirements. (There is a third variant, not shown in Figure 15.20 to minimize clutter in the chart, where the gas turbine load hold is at MECL.)

For the fast start version, too, two end points, denoted by 8 and 8″, are shown in the figure. The former is the *ultimate* fast start where the gas turbine reaches FSFL in less than 25 min with 20% per minute load ramp ("hot" start with purge credit). Even with a normal ramp rate, however, the gas turbine can be brought to FSFL in 30 min.

This brings up the question of how to define "fast". The 4-min mile of fast start capability is roughly 30 min from a standstill to combined cycle full load for a "hot" start (e.g., following an overnight shutdown). It can be as short as 18 min with purge credit as shown in Figure 15.20. This is generally compared to a conventional hot start, which can take as long as 1 h with the now-obsolete conventional method. Note that cited start times can change whether one starts the chronometer from the time when the operator pushes the START button or from the time of ignition of the gas turbine combustor.

Combining the elements discussed in Section 15.3 and illustrated by the steam turbine roll example in Figure 15.16, a representative steam turbine start curve can be established as a function of the key controlling parameter, namely, steam turbine metal temperature at the startup initiation (Figure 15.21). Appropriate gas turbine start time (from start command to the point when steam turbine roll begins) should be added to that for total GTCC start time.

The underlying physics discussed in Section 15.3 briefly and summarized in Figure 15.21 makes it clear that the commonly accepted monikers such as "hot", "fast" and "warm" are just discrete points in a continuum of start scenarios driven mainly by the downtime preceding the pushing of the start button.

As shown in Figure 15.21, the shorter the downtime, the higher the turbine metal temperature (as represented by the HP bowl thermocouple) and

- The higher the rate of turbine roll from turning gear to FSNL.
- The higher the ramp rate of steam temperature during loading.
- The shorter the overall start time.

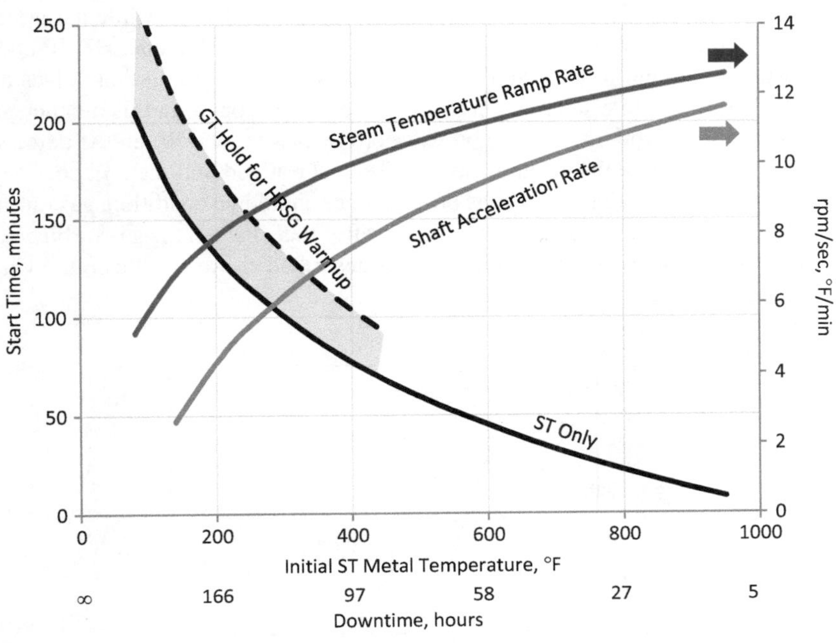

FIGURE 15.21 Startup time of a typical steam turbine (ferritic steel used in its construction) in a modern GTCC with drum-type HRSG (from the start of steam turbine roll to the point when all bypass valves are closed and all admission valves are fully open).

For long downturns of several days duration, start time is controlled not only by steam turbine stress control but also by HRSG warm-up as dictated by HP drum stress control.

Nevertheless, OEMs usually classify the expected start times for their combined cycle products into three groups. Although there may be some slight difference between individual OEMs, to a very good approximation, those three groups are as follows:

- Hot start (after an overnight shutdown of about 8 h), around 30 min from turning gear to full load
- Warm start (after a shutdown period not to exceed 2 days), around 1.5–2 h from turning gear to full load
- Cold start (after a shutdown period of more than 2 days), around 2.5–3 h from turning gear to full load.

In the above definitions, "full load" should be understood as

- Gas turbine at full load with IGVs at their fully open position
- All steam bypass valves fully closed
- All steam admission valves fully open.

In other words, due to the exponential decay nature of bottoming cycle thermal response, it would take the combined cycle power plant significantly longer to achieve its "true" steady-state, state-flow performance predicted by heat and mass balance models.

15.5 GTCC SHUTDOWN

Normal plant shutdown consists of gradual unloading and complete shutdown of the gas turbine and steam turbine generators from part or baseload operation. The sequence starts by unloading of the gas turbine to MECL. This reduces steam generation in the HRSG while maintaining steam temperatures. Since the steam turbine operates in a "sliding pressure" mode, HP, IP and LP steam admission pressures go down in lock-step with steam flows. Steam admission valves are ramped closed at a controlled rate, while steam bypass valves are ramped open. In this manner, steam pressure control is transferred from the admission valves to bypass valves. When the admission valves reach their minimum position, the steam turbine is desynchronized and the unit coasts down to its turning gear speed. After the steam turbine is brought to its shutdown condition, gas turbine unloading from the MECL starts in a fired shutdown mode. After fuel shutoff, the gas turbine continues to coast down to its turning gear speed. Individual steps described above are illustrated via speed and load profiles in Figure 15.22 (for a single-shaft GTCC).

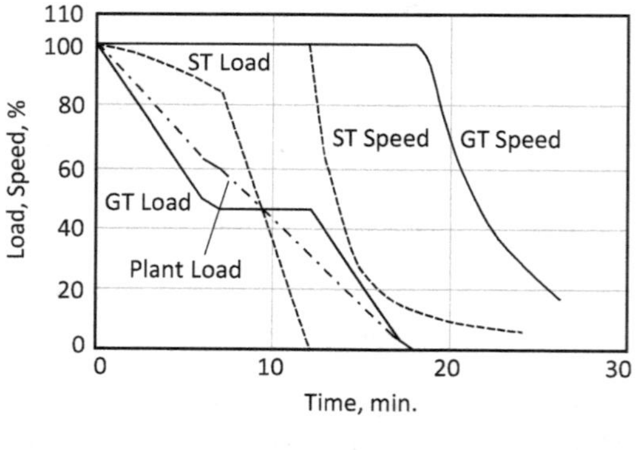

FIGURE 15.22 Combined cycle normal shutdown (single-shaft 1 × 1 × 1 GTCC).

Shutdown with purge credit is similar to the sequence described above except a second hold at 700°F exhaust temperature to cool the HP sections before starting the purge. Purge credit has two parts: gas turbine and HRSG. The former involves purging the gas turbine fuel system with nitrogen from any fuel traces for ~5 min. The HRSG purge is performed after gas turbine shutdown is complete as described earlier.

Trips are unexpected, sudden shutdowns initiated by the DCS as a result of a critical alarm such as high rotor vibration. In the case of a steam turbine trip, the gas turbine can remain in operation in steam bypass mode. In such an event, the HP and HRH steam bypass valves open sufficiently fast to control HRSG steam pressures for the sudden changes in steam turbine flow (steam admission valves are closed fast). This type of operation is of limited duration until the steam turbine is quickly brought back on line (if possible). Especially at high ambient temperatures, the gas turbine is run at part load.

HRSG-related protective trip actions include

- HP/IP/LP drum level too high
- HP/IP/LP drum level too low
- HRSG stack damper "Not Full Open" above a prescribed speed (e.g., 150 rpm).

In case of an HRSG trip, the gas turbine is also tripped (if there is no bypass stack).

In multi-shaft combined cycle power plants, if one gas turbine trips, the plant can remain operation at about 50% load as long as needed.

In a scheduled (normal) shutdown, if the plant is expected to restart the next day or after the weekend, the goal is to minimize the cool-down of the steam turbine and the HRSG. This will reduce the time required for the subsequent startup and resultant cyclic stresses.

If the plant is scheduled for a major outage to perform maintenance work, the goal is to expedite the cool-down process so that the maintenance work can start as soon as possible and the plant can be back online faster.

These two diametrically opposed requirements lead to two methods of steam turbine cooling: natural and forced. In a natural cooling process, following the steam turbine shutdown, while on the turning gear (<10 rpm), the temperature differential between the turbine metal (casing, rotor and nozzles/buckets) and the ambient air reach an equilibrium state via natural convection. In general, natural cooling in an unexpected shutdown event (e.g., turbine trip at operating temperatures) takes longer than that in a planned shutdown. As shown in Figure 15.22, in a planned shutdown, steam turbine load and steam temperatures are reduced (the latter to their minimum allowable values). This shortens the overall cool-down time due to the lower starting temperature of the natural cooling process.

Forced cooling process speeds up the heat transfer between the turbine metal and the air by creating a forced airflow (i.e., forced convection). A method developed by a major OEM involves using the existing condenser air removal system (i.e., the vacuum pumps) to draw air from the turbine hall through the blade path of the turbine and into the condenser, where it is ultimately exhausted from the system.[9] The air is drawn in via the control valves through the nozzles provided for connecting dehumidifiers at the admission valves.

Another major OEM's approach to forced cooling in "maintenance shutdowns" is to run the gas turbine at constant load for 2 h with fully open IGVs to reduce the exhaust temperature to 700°F. The IGV opening rate is determined by the steam turbine controller based on clearances. One option to enhance the forced cooling process is to use the terminal attemperators (if they are available) and reduce the steam temperatures. (Current practice is to lock the terminal attemperators at closed position after the startup due to water induction concerns.)

When the GTCC power plant is shutdown, the HRSG is "bottled up" by closing all the valves and, if available, the stack dampers. This preserves the most advantageous steam conditions for overnight

[9] *Forced Cooling of Steam Turbines*, Siemens Energy Services Division, Brochure E50001-G520-A170-X-4A00.

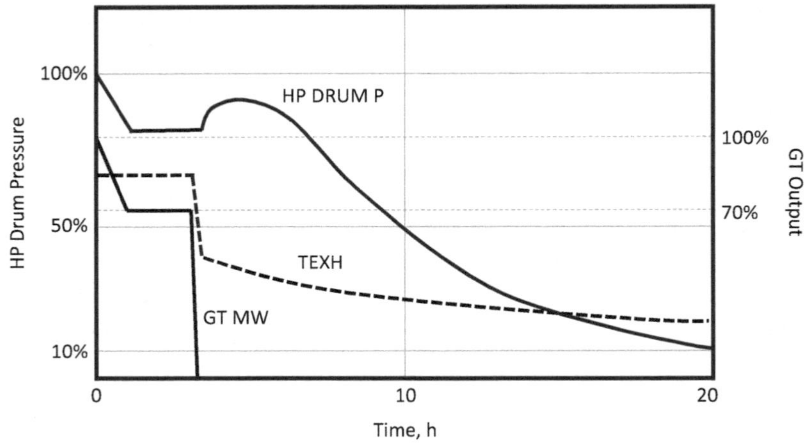

FIGURE 15.23 Typical HP drum pressure decay during shutdown [10].

and after-weekend startups. During those temporary shutdowns, condenser vacuum and steam turbine seals are maintained via auxiliary steam to the gland seal system (supplied by the auxiliary boiler). This ensures that the steam purity is maintained by preventing O_2 and CO_2 dissolving in the feedwater. (Otherwise, waiting for the required steam purity adds to the next startup time.) For long maintenance outages, the goal is to cool the system as fast as possible so that the maintenance crew can start the work on the components. Thus, the HRSG stack dampers are kept open.

Immediately after shutdown, steam and water trapped in the heat exchanger tubes and headers and the gas trapped in the HRSG box start cooling via heat loss to the surroundings. As the steam cools and condenses (evacuated via steam drains), steam pressure in the system becomes progressively lower. Specifically, steam pressure in the evaporator drums is a function of the shutdown period. A typical HP drum pressure decay curve during shutdown is shown in Figure 15.23. Thus, for warm or hot shutdowns, HP drum pressure can end up at <10% or 50% of its rated operating value, respectively. Cold starts with more than 72 h of downtime would probably lead to drum pressures close to the ambient.

15.6 EMERGENCIES

There are two major emergencies in a combined cycle power plant: load rejection and trip. Both emergencies include the prime mover generators (the *turbogenerators*) in the plant, i.e., the gas and steam turbine generators. In either case, what happens in a nutshell is that the turbogenerator is disconnected from its *load*. (The load in question is the electric load.) In the case of load rejection, the culprit is on the load side (e.g., a disturbance or fault event in the grid). In the case of trip, the culprit is on the turbogenerator or plant side. Both emergencies require proper response from the turbine control and protection systems to ensure the safety of plant personnel, avoid immediate damage and minimize cumulative damage to the equipment.

In the case of load rejection, the turbogenerator is disconnected from the grid by opening the generator and/or line breakers. If nothing is done, the overspeed caused by the enormous, unbalanced kinetic energy of the turbogenerator and resulting centrifugal forces can destroy the unit and its surroundings within a few seconds leading to major loss of property and, in worst cases, to injuries and death. Typically, the turbine protection system has two settings for overspeed protections. When the unit is synchronized, the trip value is set at 110% (i.e., 3,300 rpm for a 50-Hz unit). When the unit is not synchronized, the trip value is set at 105% (i.e., 3,150 rpm for a 50-Hz unit). These are typical numbers for large steam turbines. For a particular turbogenerator, OEM's manual should be consulted.

In the case of the gas turbine generator, the protection system shuts down fuel flow to the combustor immediately in a trip or load rejection event. This is sufficient to prevent a catastrophic overspeed loss because the energy stored in gas turbine's relatively small internal volume (with very large airflow) is small vis-à-vis the steam turbine. In addition, the axial compressor of the gas turbine acts as a very effective *brake* (load). Thus, a quick closure of the fuel valves produces a quick reduction of turbine power such that control of overspeed after sudden load loss is not a major issue.

A steam turbine does not have a compressor to act as a brake/load. Upon load rejection, immediate shutdown of the stop and control valves is imperative. The two-valve-in-series admission valve structure (stop and control valves) ensures that if one fails the other will operate and stop the steam flow. Even so, "entrained steam/energy" trapped in the large pipes downstream of the admission valves can be significant to prolong the overspeed for a few seconds even after the valves are slammed shut. This applies in particular to the LP steam admission because main and reheat steam valves are provided by the steam turbine OEM (normally) and they are practically adjacent to the turbine casing. Thus, the OEM provides the EPC contractor designing the LP steam piping with guidance to ensure that entrained energy limits for safe operation are not exceeded.

Even if there is nothing wrong on the load side (i.e., the grid), certain events and conditions in the prime movers, their auxiliary systems or the BOP equipment may force a trip (essentially an *uncontrolled* shutdown). For a steam turbine, such conditions are

- Thrust failure (shaft position protection)
- Excessive vibrations
- Low main (HP) steam temperature
- High HP turbine exhaust steam temperature (to prevent windage heating of the turbine blades)
- High LP exhaust steam temperature (to prevent deformation of LP turbine casing)
- High exhaust pressure (to prevent last-stage bucket, LSB, flutter)
- Low lube oil supply pressure
- Low grid frequency (to prevent LSB resonance).

(Obviously, most of them, e.g., vibrations and lube oil problems, equally apply to the gas turbines.) Conditions listed above are usually monitored by the control/protection system with triple-redundant sensors. For example, an "alarm" is issued when the average reading of the three sensors is over the alarm value (or the *HI* value). A trip signal is made by the two-out-of-three "voting" logic in the system (or the *HI-HI* value).

In a protective trip event, the protection system trips the steam turbine but does not trip the generator (i.e., the breakers are not opened). Consequently, generator power decays to zero as the energy entrained in the steam turbine is dissipated. The generator breakers are finally opened by the reverse power element of the generator protection system (to prevent the synchronous ac generator to become a synchronous ac *motor*). In this manner, the torsional impact on the turbogenerator shaft is minimized. Note that operating in reverse power condition is damaging to the steam turbine because there is no steam flow to cool the blades heating up via windage. Therefore, reverse power or motoring protection cannot come before overspeed protection. In other words, the breakers are not to be opened until it is known that the generator output is negative (i.e., the unit is motoring).

In a load rejection event, the gas turbine can continue to operate at a reduced output, i.e., at FSNL or at a low enough output to supply the plant's "house load", which is the amount of power required to keep the plant equipment running until the plant is reloaded or shut down. Depending on the size of the power plant and the heat rejection system, the house load can be anywhere between 1% and 2% of the gross power output. As required by the applicable grid code of the country or the supranational entity (e.g., the European Union), load rejection to house load could happen (and must be accommodated by the power plant) from each permissible operating point, i.e., MECL to baseload.

In multi-shaft GTCC power plants, each prime mover has its own generator. In a $2 \times 2 \times 1$ power plant, if one of the gas turbines trips, the plant can operate at a part load with one GT-HRSG train supplying about half the rated steam flow to the steam turbine. In a 1 × 1 × 1 multi-shaft power plant, if the steam turbine trips, the unit can operate in an "island" mode with the gas turbine at a low load, e.g., 15% or lower. (In island mode, the power plant is disconnected from the grid and generates enough power to run its own house load and other local power user such as an industrial facility.[10]) Steam generated by the HRSG is bypassed to the condenser unless the gas turbine is equipped by a bypass stack. In the latter case, the bottoming cycle is fully disconnected from the gas turbine and can be shut down completely. The gas turbine can run at full load, i.e., essentially 67% plant load, and continue to supply power to the grid. This would not be possible without a bypass stack because it is detrimental to the plant BOP to run at full steam bypass mode for an extended period of time.

The scenario described in the preceding paragraph is possible in a 1 × 1 × 1 single-shaft GTCC if the steam turbine is coupled to the generator via a SSS clutch. Upon trip, the steam turbine is decoupled from the generator and coasts down to the turning gear speed.

15.7 GRID CODE COMPLIANCE

The *raison d'etre* of a power plant is continuous, reliable generation of electricity to sustain modern life. Electric power generated by the prime mover generators of a combined cycle power plant is stepped up to a higher voltage, at which it connects to the transmission network, which carries it to the wholesale customers (usually the company that owns the local distribution network), often across the state and international boundaries. Upon arrival at the substation, electric power is stepped down from the transmission-level voltage to the distribution-level voltage. Upon exiting the substation, electric power is transmitted through the distribution wiring to the service locations. At that point, electric power is stepped down from the distribution voltage to the required service voltage(s) at the demand centers (factories, offices, residencies, etc.). All these components form an interconnected network, the *electric grid*, for delivering electricity from suppliers to consumers.

A key problem in simple and combined cycle gas turbine operability is system response to grid events such as large swings in grid frequency due to loss of load or generation. In particular, fast and adequate frequency response is a critical power plant system requirement to ensure safe and stable operation of the electric grid, and it is enforced by local codes and regulations (referred to as *grid codes*). A proper discussion of this subject matter is impossible within the scope of this book (not to mention that it is beyond the abilities of the author). It requires a solid background in electrical equipment design and operability. For an in-depth discussion of the electromechanical theory and design principles governing synchronous machines, the reader should consult Refs. [4–7] in Chapter 9. The key piece of combined cycle equipment from a grid code compliance perspective is the gas turbine and its control. For fundamental gas turbine control principles, please refer to Chapter 19 in **GTFEPG** (Ref. [11] in Chapter 2). Herein, salient aspects of grid codes and their impact on GTCC operability will be discussed in relatively broad strokes.

The grid code is a technical document containing the rules governing the operation, maintenance and development of the transmission system. Its purpose is to safeguard the electric power system and ensure reliable delivery of electric power to the end users. Technical power plant design and operational aspects specified by a grid code include

- Quality of electricity supply (e.g., voltage and frequency variation)
- Equipment protection (e.g., loss of excitation and pole slips)
- Generator specification (e.g., power factor, short circuit ratio and automatic voltage regulation)
- Metering and monitoring of electric power.

[10] In an island mode, the power plant and its turbine generators are isolated. For stable operation, the control system should be capable of maintaining the frequency of the connected system within reasonable limits. In fact, in island mode, the gas turbine will operate with an isochronous governor, which will maintain a constant frequency regardless of the load.

In different countries and supranational organizations such as the European Union, different grid codes have been developed and are enforced by respective *Transmission System Operators* (e.g., the National Grid in the UK and Verband der Netzbetreiber in Germany).

There is no national power grid in the USA, which is divided into three main power grids:

1. The Eastern Interconnected System or The Eastern Interconnect
2. The Western Interconnected System or The Western Interconnect
3. The Texas Interconnected System or The Texas Interconnect.

There are ten regional reliability councils of the *North American Energy Reliability Council* (NERC), whose members come from all segments of the electric industry, i.e., investor-owned utilities; federal power agencies; rural electric cooperatives; state, municipal and provincial utilities; independent power producers (IPP); power marketers; and end-use customers. Examples are Electric Reliability Council of Texas (ERCOT) and Western Systems Coordinating Council (WSCC).

Frequency response is one of the most important grid code requirements. Transmission system frequency is a continuously changing variable that is determined and controlled by striking a careful balance between system demand and total generation. If demand is greater than generation, the frequency falls (generators connected to the grid slow down). If generation is greater than demand, the frequency rises (generators connected to the grid speed up). When generation and demand are in balance, the frequency stabilizes at its rated value (50 or 60 Hz).

The best mechanical analogy to this is driving a car on the highway. When the power generated by the internal combustion engine (ICE) is exactly balanced by the resisting forces (friction, air resistance, etc.), the car runs at a constant speed, v, and the acceleration $a = dv/dt = 0$ (generation is equal to demand). If the car comes to an uphill section of the road, the gravity is added to the resistance and the car slows down, i.e., $a = dv/dt < 0$ (demand is greater than generation). To counter that, the driver has to press the gas pedal further down, i.e., increase the fuel flow to the ICE. Alternatively, if the driver does this on the level road, the automobile will accelerate, i.e., $a = dv/dt > 0$ (generation is greater than demand).

In fact, for a hypothetical grid with only one generator (e.g., a gas turbine), this is exactly how the system frequency is controlled. This situation is referred to as *isochronous control*, and it is indeed the logical choice for a small grid (such as a small island with only one generator) because the generator can set its speed and, thus, the grid frequency.

For a generator connected to a large grid of tens of gigawatts, it is the grid that determines the frequency and speed of the particular electric generator. In other words, the grid disturbance and frequency variation are too big to be influenced by a single generator. The mechanical analogy to this situation is a very large road vehicle driven by multiple ICEs of varying sizes, each controlled by its own driver. When that vehicle comes to an uphill section, each driver trying to react to the increased load demand individually (i.e., each driver pressing on the gas pedal on his or her own) would lead to utter chaos.

The remedy is *droop governor*, which is a straight proportional controller in which the output is proportional to the *speed error*. The droop of a generator is a dimensionless ratio (expressed as a percentage):

$$d = \frac{\Delta f}{f_0} \frac{\dot{W}_0}{\Delta \dot{W}},$$

where the subscript 0 denotes the rated value (e.g., $f_0 = 60\,\text{Hz}$ or $\dot{W}_0 = 300\,\text{MWe}$). A typical droop value is 4% (0.04). Thus, if $\Delta f = 0.6\,\text{Hz}$, using the example values above, change in power with 4% droop is

$$\Delta \dot{W} = \frac{\Delta f}{f_0} \frac{\dot{W}_0}{d} = \frac{0.6}{60} \frac{300}{0.04} = 75\,\text{MWe}.$$

In other words, if the system frequency goes up from 60 to 60.6 Hz (e.g., loss of load), generator output should go from 300 MWe down to 225 MWe (75% load). Similarly, if the system frequency goes up from 60 to 59.4 Hz (i.e., increased load), generator output should go from 300 MWe up to 375 MWe (125% load). The former is within the realm of possibility; the latter is most likely not.

In a droop governor, the key principle is to compare the actual speed to a reference or setpoint value. This setpoint value, determined by the operator, is known as the *speed changer*. The difference between the actual and setpoint value is referred to as the *speed error*. Gas turbine fuel flow (and load change) is a function of this error. This is illustrated in Figure 15.24, which shows droop governor lines for 50%, 75% and 100% outputs at 100% speed with 4% droop.

In a combined cycle power plant connected to the grid, gas and steam turbines run at constant speed. In other words, the actual speed does not change (e.g., 3,600 rpm at 60 Hz). Fuel flow to the gas turbine combustor(s), and therefore the plant output, is held proportional to the speed error by the droop governor. Thus, the plant output can be controlled by the operator by changing the speed error via the setpoint value. In Figure 15.24, it is illustrated how this is done. Output is set to 50%, 75% and 100% by changing the setpoint value to 2%, 3% and 4%, respectively, with 4% droop.

The speed error can also be affected by the actual speed, i.e., the system frequency. Consider the 3% speed error setpoint value case in Figure 15.24. The unit is running at 75% load at 100% speed. Assume that the system frequency drops by 1%, e.g., from 50 to 49.5 Hz. This will cause the unit speed to decay also by 1% such that the total speed error becomes 4% even though the setpoint value is still 3%. Thus, the droop governor will increase the unit load from 75% to 100% for 25% load increase.

In practice, the range of adjustment is between highest allowable and lowest allowable shaft speeds. Typical values are 107% and 95%, respectively. The correspondence between the droop governor and the actual fuel flow is explored in Chapters 18 and 19 of **GTFEPG**.

In general, a change in grid frequency is caused by loss of generation (supply) or load (demand). It should be realized that the grid frequency is never exactly 50.000 or 60.000 Hz. It fluctuates continuously due to normal variations in load and generator output. For stable operation of the transmission system, however, the grid frequency should be held within narrowly defined limits. Minor deviations from the frequency reference value (say, 200 mHz) indicate that there is a balance of power generation and demand. Sample frequency distributions for selected countries are summarized Table 15.8.

FIGURE 15.24 Four percent droop governor operating lines.

TABLE 15.8
Frequency Distributions Summary for Selected Countries

Country	Database	Standard Deviation, mHz	Outliers (beyond 6σ)	Largest, Hz	Smallest, Hz
Turkey	5 months	64	28	50.55	49.28
UK	5 months	59	2	50.36	49.44
Germany	5 months	40	11	50.49	49.65
Italy	10 days	20.6	1	NA	48.6435
Spain	5 months	21.5	0	NA	NA

Faults in the system resulting from the loss of power plants, shutdown of loads, short circuits and so on result in deviations and gradients of varying magnitudes. These faults can lead to grid instability or outage. There are two types of faults:

- Faults within the anticipated range, controlled by provision of reserve power.
- Serious system faults that are counteracted by disconnection from the interconnected system and measures such as *load shedding.*

Controllable faults result in frequency fluctuations that remain within a control band defined by the grid operator, e.g., ±200 mHz. The goal is to ensure that the loss of the largest generator in the system should be ridden out with the frequency staying within the control dead band. This operation is referred to as "frequency control".

Fast decreases in frequency in the case of serious system faults cannot be counteracted solely by measures on the generating side. Protection devices are implemented to switch off loads (load shedding) in case of a specific under-frequency. In the case of fast decreases in frequency where the frequency remains above a disconnection limit, it is the task of the generators to remain in a stable load operation mode.

As an example, consider the *National Grid Code* in the UK, which stipulates that each generating unit must satisfy the following minimum requirements for frequency control:

- Fast-acting proportional speed governor to provide continuous, automatic and stable response across its entire operating range
- Speed governor capable of being set to a droop of 3%–5%
- Minimum speed governor dead band no greater than ±15 mHz
- Load control capability with target frequency setting of 50 ± 0.1 Hz either continuously or in 0.05 Hz (50 mHz) steps
- Capability to control frequency to below 52 Hz in island operation
- If operating at full load, capability to maintain power output if frequency falls to 49.5 Hz, thereafter a reduction in power output no more than pro rata with frequency down to 47 Hz.

These requirements are summarized in Figure 15.25.

The under-frequency power demand characteristic in Figure 15.25 is of the "ramp" type. It is adopted in Mexico (60 Hz), the UK, Norway and Thailand. Two other grid code power demand types are "step" (adopted in Germany) and "intrinsic" (adopted in Brazil (60 Hz), Argentina, Australia, Greece, Italy, Spain and Turkey, among others). Clearly, designing a gas turbine product to meet all possible grid code requirements anywhere in the world is a monumental task. One logical path towards that goal is to gather actual frequency trends and statistical data along with probability distributions and to design a product that can meet the requirements for the most likely events from a representative data set covering the major gas turbine markets. This requires extensive studies with contributions from all component and subsystem teams. Key considerations are

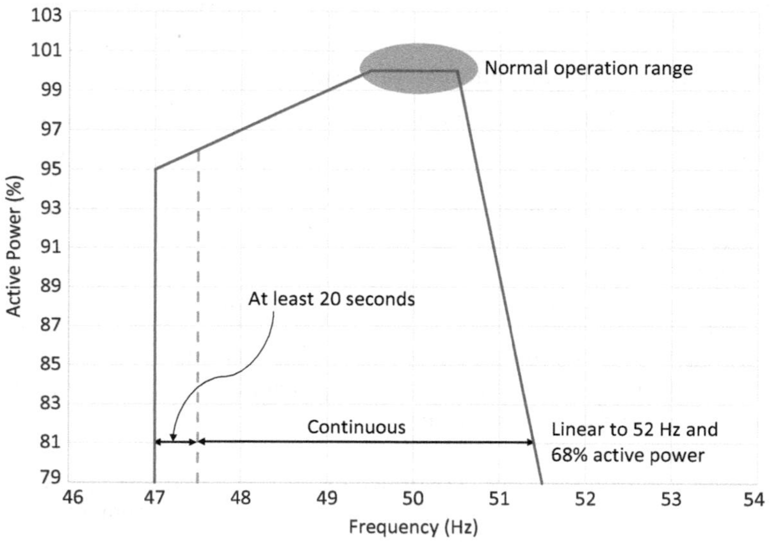

FIGURE 15.25 Active power output for frequency control per UK National Grid Code.

- Equipment scope
  - Simple cycle gas turbine
  - Power island (i.e., gas and steam turbines plus the HRSG).
- Product design
  - Gas turbine firing temperature
  - Water wash
  - IGV control
  - Steam attemperation
  - Steam turbine throttle valve control
  - Steam turbine overload valve (see Section 9.2.1).
- Steady-state and dynamic simulation (baseload and part load)
- Component limits
  - DLN combustor dynamics and blowout
  - Turbine load/torque, corrected speed and exhaust temperature limits
  - Hot gas path parts life
  - Control system response time lag and sensitivity.

For a detailed discussion of GTCC frequency response to grid events, the reader should consult Section 19.5 in **GTFEPG**. In essence, while the specific requirements vary from code to code, the basic grid code compliance framework for a GTCC power plants has a two-step structure of plant *capability*:

- The primary response (P)
- The secondary response (S).

P capability is the minimum increase in active power output between 10 and 30 s after the start of the ramp *injection*. This increase in active power output should be released increasingly with time over a period of 0–10 s from the time of the start of the ramp injection.

S capability is the minimum increase in active power output between 30 s and 30 min after the start of the ramp injection.

Due to its large thermal inertia, the bottoming steam cycle does not have P capability. Thus, for the primary response, the onus is on the gas turbines in the power plant. In an under-frequency event, the drop in the rotational speed and airflow dictates that the gas turbine must make up for that loss first and then contribute even more to help the grid stop the downward ramp in frequency and restore it. There are three control parameters to accomplish this:

- Firing temperature (increase by 70°F–100°F)
- IGVs (open to the maximum position)
- Online water wash (turn it on to compensate for lost airflow).

These control actions increase gas turbine output and exhaust energy. The latter is bound to increase the steam turbine output and contribute to the overall GTCC power output. This, however, takes several minutes due to the thermal inertia of the bottoming cycle and the associated time lag (see Section 15.2). Steam turbine response can be enhanced in several ways, e.g., reducing steam attemperation for steam temperature excursion to higher-than-rated value for a limited time (e.g., from 1,050°F to 1,085°F).

Another option is to run the GTCC in a "grid control" mode by partial closing of steam turbine main steam admission valve. Typically, this feature is combined by running the GTCC at part load (say, 90% load) with the gas turbine IGVs partly closed. This provides a readily available reserve capacity to be deployed when the grid frequency drops. Running the power plant in this mode at lower-than-rated capacity comes with a plant heat rate penalty. This loss should be balanced against the bonus payments obtained for the ancillary service provided (for grid frequency control) and hot gas path parts-life saving due to reduced overfiring.

REFERENCES

1. Chaker, M., Meher-Homji, C.B., 2013, Evaporative cooling of gas turbine engines, *J. Eng. Gas Turbines Power*, Vol. 135, p. 081901.
2. Bhargava, R.K., Meher-Homji, C.B., Chaker, M., Bianchi, M., Melino, F., Peretto, A., Ingistov, S., 2007, Gas turbine fogging technology: A state-of-the-art review: Part I: Inlet evaporative fogging: Analytical and experimental aspects, *J. Eng. Gas Turbines Power*, Vol. 129, pp. 443–453.
3. Cortes, C.R., Willems, D., 2003, Gas turbine inlet air cooling techniques: An overview of current technologies, *In Proceedings of the Power-Gen International*, December 9–11, 2003, Las Vegas, NV.
4. Jones, C., Jacobs III, J.A., 2000, Economic and Technical Considerations for Combined Cycle Performance Enhancement Options, GE Power Systems, GER-4200.
5. Ebeling, J., Balsbaugh, R., Blanchard, S., Beaty, L., 1994, Thermal Energy Storage and Inlet Air Cooling for Combined Cycle, 94-GT-310, *International Gas Turbine and Aeroengine Congress and Exposition*, June 13–16, 1994, The Hague, Netherlands.
6. Obermüller, M., Schmidt, K.-J., Schulte, H., Peitsch, D., 2012, Some aspects on wet compression: Physical effects and modeling strategies used in engine performance tools, *Deutscher Luft- und Raumfahrtkongress 2012*, Berlin, Document ID: 281210.
7. Wettstein, H.E., 2015, Polytropic change of state calculations with condensation or evaporation and its use for thermal power cycles, GT2015–42075, *ASME Turbo Expo 2015*, June 15–19, 2015, Montreal, Quebec, Canada.
8. Gülen, S.C., Kim, K., 2014, Gas turbine combined cycle dynamic simulation: A physics based simple approach, *J. Eng. Gas Turbines Power*, Vol. 136, p. 011601.
9. Briggs, R.J., 1999, Repowering a 2400 psig steam cycle with natural circulation HRSG, *Powergen International*, November 30–December 2, 1999, New Orleans, LA.
10. Taylor, D., Pasha, A., 2010, Economic Operation of Fast-Starting HRSGs, *Power*, June 2010.
11. Allen, B., 2011, Using Today's Design for Tomorrow's Plant, *Energy-Tech*, December 2011, pp. 12–14.
12. Albrecht, M.J., Arnold, W.A., Jain, R., DeVitto, J.G., 2013, Rapid startup analysis of a natural circulation HRSG boiler with a vertical steam separator design, *Powergen International*, November 12–14, 2013, Orlando, FL.
13. Bauver, W., Perrin, I., Selby, G., Singh, H., 2011, Alstom's HRSG: Optimized for today's flexibility requirements, *Powergen International*, December 13–15, 2011, Las Vegas, NV.

14. Fontaine, P., Galopin, J.-F., 2007, HRSG optimization for cycling duty, *Power Eng. J.*, Vol. 111, No. 11, pp. 170–184.
15. Rengarajan, G., Taylor, D., Pankow, B., 2010, Study and Design Considerations of HRSG Evaporators in Fast Start Combined Cycle Plants, Power Engineering White Paper (CMI Energy).
16. Moelling, D.S., Jackson, P.S., 2012, Startup Purge Credit Benefits Combined Cycle Operations, *Power*, June 2012.
17. Chrusciel, A., Zachary, J., Keith, S., 2001, Challenges in the Design of High Load Cycling Operation for Combined Cycle Power Plants, *Power-Gen International 2001*, Las Vegas, NV.
18. Akhtar, Z., 2006, Design Features for Minimizing Start-Up Time in Combined Cycle Plants, *Power-Gen Europe 2006*, Köln, Germany.
19. Ugolini, D.J., Bauerschmidt, J.R., 2006, Optimization of Start-Up Times for Combined Cycle Power Plants, *Electric Power Conference 2006*, Atlanta, GA.

16 Maintenance

The purpose of maintenance is to maximize power plant reliability and availability. Using passenger automobiles as an analogy,

- *Availability* means that your car is ready to go when you need it (i.e., it is not in your mechanic's shop undergoing repairs) whereas
- *Reliability* means that it starts and runs when you turn the ignition key and takes you to your destination without breaking down.

In order to ensure that your car is available and reliable for as long as possible, you must pay attention to

- Routine inspection (i.e., check the oil, check the tire pressure, check other important fluids)
- Scheduled maintenance as recommended by the manufacturer.

Once in a while, even if you diligently take care of your responsibilities as an owner/driver, unexpected problems are bound to happen. These problems can pop up suddenly and leave you stranded in a remote location before you can limp to a service station.

For a combined cycle power plant, the exact same considerations apply. Modern power plants are designed for trouble-free operation with minimal operator attendance (just like a modern passenger car). There are routine daily checks of mechanical and electrical equipment and other systems (e.g., water analysis), which are taken care of by the plant personnel. There is an operator in the control room monitoring the plant systems via human–machine interface (HMI) stations. Severe malfunction of a key system typically leads to an automatic trip without operator intervention.

Combined cycle power plant maintenance activities can be grouped into two major categories:

1. Operating (i.e., while the equipment is running)
2. Non-operating (i.e., the equipment is shut down)

In modern power plants, the most significant operating maintenance activity is periodic online wash of gas turbine compressor for performance recovery due to fouling. The reader is referred to Section 16.4 in **GTFEPG** (Ref. [11] in Chapter 2) for more details and references.

Non-operating maintenance can be done during scheduled (or planned) or unscheduled (or unplanned) outages. Unscheduled outages can be "forced" (i.e., due to equipment malfunction and/or trip) or "unforced" (i.e., the power plant is waiting to be dispatched by the grid operator). Minor inspection and/or repairs can be done during unscheduled, unforced outages based on the information obtained from the plant distributed control system (DCS; e.g., an "alarm" from a critical sensor) or from routine visual inspection during plant walkdowns.

Scheduled maintenance is taken care of during "planned outages". The determinant of a combined cycle planned outage is the gas turbine because it is the only major piece of equipment operating under extreme conditions (e.g., hot gas at temperatures pushing 3,000°F flowing through parts rotating at 3,000 or 3,600 rpm). The frequency of planned gas turbine outage is a function of the type of primary fuel and the operating duty cycle, i.e., number of start/stops, time at low loads.

Gas turbine maintenance schedules are determined by each original equipment manufacturer (OEM) for their product line. There are some differences between major OEMs in calculating the "intervals" between inspections but, in general, they follow a similar logic. A highly recommended resource is General Electric document GER-3620N "Heavy-Duty Gas Turbine Operating And Maintenance Considerations" (October 2017 version). The reader is encouraged to consult that document (it is available online – just google it) for an in-depth learning of gas turbine maintenance issues.

In general, there are *three* inspection intervals: combustor, hot gas path and major. Hot gas path inspection includes combustor and turbine inspections. Major inspection includes hot gas path and compressor inspections. Typically, a major inspection can cover the rotor as well because its removal may be requisite for turbine inspection.

Advanced class gas turbines with DLN combustors rarely burn liquid fuel (except as a backup fuel in emergency situations) and do not have diluent injection (steam or water) for NOx control. Thus, their standard inspection intervals are impacted primarily by number of starts, trips, load rejections and peak-fired running time. General Electric accounts for the impact of these events by the calculation of "factored hours" (FH) and "factored starts" (FS) (see GER-3620N for a detailed description of FH and FS calculation for different frames). A similar concept adopted by Siemens uses "equivalent hours" and "equivalent starts".

Nevertheless, in dual fuel gas turbines, inspection of the fuel oil system is necessary. This involves removal, cleaning and inspection of the fuel nozzle assemblies and inspection of the interior surfaces of the combustors and transition pieces through the nozzle openings to verify that the nozzles are clean and free of debris and leaks and the combustor baskets are clean.

For the vintage E and F class gas turbines, typical combustor, hot gas path (i.e., the turbine) and major inspection intervals used to be 8,000, 24,000 and 48,000 FH, respectively. Modern, advanced class gas turbines have longer intervals to increase machine availability. For example, GE's HA class eliminates separate combustion inspection and combines it with hot gas path inspection at 25,000 FH or 900/720 FS ("dot-oh-one" and "dot-oh-two", respectively[1]), whichever comes first. Major inspection interval is 50,000 FH or 1,800/1,440 FS.

Another major OEM, Alstom, also used the *equivalent operating hours* (EOH) concept with operator-adjustable service intervals. (After having been acquired by GE in 2015, Alstom product line is divided between GE and Ansaldo.) There are two such intervals based on operation modes, i.e.,

1. Maximum continuous load (MCL) with 24,000 EOH major inspection interval
2. Lifetime-optimized or extended load (MXL) with 36,000 EOH major inspection interval.

These intervals were later increased to 32,000 and 48,000 EOH for MCL and MXL modes, respectively.

Reduced downtime and increased turbine availability to maximize the owner's revenue stream are the drivers behind the OEM efforts to increase the inspection intervals and combine the combustion and hot gas path inspection. For a base-loaded or intermediate duty combined cycle power plant, combustor inspection with 8,000 FH or EOH can take place every 6 months to a year. Inspected components include fuel nozzles, combustor cans, cross-fire tubes, transition pieces, the "U" ducts between the compressor discharge and the combustor inlet (for reverse-flow cannular combustors). For an F class machine, this inspection can take 600 man-hours or more (e.g., 12 2-hour shifts of a 6-man crew).

Hot gas path inspection with 24,000 FH or EOH is less frequent (1–2 years for intermediate duty machines, 2–3 years for base-loaded ones) but takes more man-hours to complete, i.e., 2,500 man-hours or more. In addition to the combustion section, stationary and rotating components of the turbine are inspected. They include stator nozzle vane assemblies, vane segments, blade tip shrouds, blades themselves, seals, packings and bearings.

[1] Each HA class gas turbine, 60-Hz 7HA and 50-Hz 9HA, has two variants, e.g., 7HA.01 and the larger 7HA.02 or 9HA.01 and the larger 9HA.02. As an example, 7HA.01 is "seven HA dot oh one".

A major inspection requires close to 5,000 man-hours. Depending on the crew size and number of daily shifts, it can easily eat up 2–3 weeks. It involves the complete disassembly of the gas turbine including the compressor, combustors and the turbine along with the removal of the rotor.

In the past, big utilities had their own in-house engineering and maintenance departments employing seasoned technical personnel, who took care of routine maintenance as well as scheduled inspection/overhaul of the plant equipment (including repairs). This is not the case anymore due to two main developments:

1. Merchant power plants owned and operated by the independent power producers (IPPs) do not have the personnel to take care of anything other than minor inspection and repairs.
2. Advanced technology in major plant equipment, especially in gas turbines, with complex digital controls requires OEM-trained expert personnel for maintenance.

As an analogy to the second item above, consider a modern car engine with computer-controlled fuel injectors and valves vis-à-vis a 1970s vintage engine with carburetors and analog/mechanical controls. In the past, many car owners did their own routine maintenance and most repairs, whereas today, one has to take his or her car to a fully equipped and trained mechanic (usually the dealer's service shop).

The objectives of any gas turbine combined cycle (GTCC) power plant maintenance program are obvious: (i) maximize revenue, (ii) minimize operating costs and (iii) minimize risk of catastrophic failure. In order to achieve these objectives, one has to

- Maximize planned outage intervals
- Minimize unplanned outages
- Minimize outage durations.

There is a range of options available to combined cycle power plant owners/operators to take care of their maintenance needs. There are still many older facilities out there with vintage gas turbines. Those can be handled by in-house personnel with the help of *independent service providers* (ISPs) for major overhauls and after-market parts. Next level of maintenance agreement would require original parts obtained directly from the OEMs. For most advanced machines, the more common but costlier practice is *long-term service agreements* (LTSAs), which are also known as *contractual service agreements* (CSAs). Ultimately, the services of the power island OEM can be retained to operate and maintain the power plant under an operations and maintenance (O&M) contract for the first few years of operation or longer. The hierarchy of these options from technical and contractual risk as well as cost perspectives is illustrated in Figure 16.1 (adopted from Nagy and Savic [1]).

In order to have a rough idea about the magnitude of the LTSA costs, consider that, for an advanced class gas turbine, a 15- to 20-year LTSA/CSA contract can cost as much as the gas turbine generator itself (for base-loaded operation). For cyclic operation, the contract cost can be higher by as much as 30%.

16.1 MAINTENANCE COSTS

In a report documenting the "cost of new entry" for simple and combined cycle gas turbine power plants in the PJM area, the largest component of the fixed operating expenses was found to be the staff labor costs (see Ref. [2] in Chapter 13). They accounted for about half of the total fixed O&M costs depending on plant type and location. On average, for an F class (GE's 7FA.05) 2 × 2 × 1 GTCC, in 2015 dollars, first year annual fixed O&M expense distribution was as follows:

- Facility staff labor costs were $3.5 million
- Consumables were $0.30 million

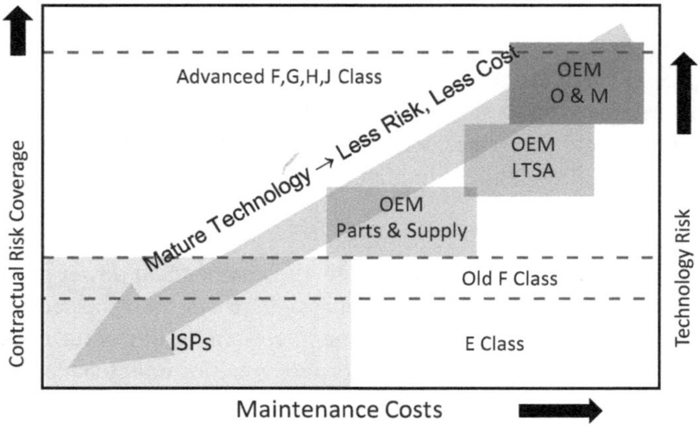

FIGURE 16.1 Maintenance service options.

- Office costs were \$0.21 million
- Maintenance and minor repairs were \$0.92 million
- Corporate and administrative charges were \$0.43 million.

Thus, the total came to \$5.36 million dollars for a 627-MWe GTCC with an operating duty profile of 5,000 h/year and 150 starts or \$1.71/MWh.

For the variable O&M expenses, the same report had on average (for the first year in 2015 dollars)

- \$0.85/MWh variable "facilities" O&M (mainly consumables)
- LTSA cost of about \$9,500 per factored fired start
- \$275 per factored fired hour (FFH).

The assumptions used in coming up with the LTSA cost were

- Seventeen-year contract
- Escalation with 2.5% inflation rate after the first year
- If the plant operation is >27 FFH/FFS, the maintenance intervals are hours based
- Otherwise, they are starts based.

FH and FS were calculated per General Electric GER3620. (In a recent 2015 project with an F class turbine, LTSA prices were about \$150/FFH for a baseload profile (8,400 FFH and 50 FFS) and about \$375/FFH for a cyclic duty profile (4,200 FFH and 250 FFS).)

Assuming a plant online date of June 1, 2015, the LTSA inspection schedule was assumed to be as shown in Table 16.1.

In US EIA's 2013 report "Updated Capital Cost Estimates for Utility Scale Electricity Generating Plants," for the natural gas-fired GTCC with advanced gas turbine (400 MWE),

- Fixed O&M was listed as \$15.37/kW-year
- Variable O&M was listed as \$3.27/MWh.

The assumptions underlying the fixed O&M component in the EIA estimate were

- Staffing and monthly fees under pertinent operating agreements
- Typical bonuses paid to the plant operator
- Plant support equipment which consists of equipment rentals and temporary labor

TABLE 16.1
Combined Cycle Gas Turbine Inspection Schedule

Date	Days from Online	Years from Online	Unit	Inspection Type	Inspection Duration (Days)
1/26/2017	751	2.06	GT02	CI	6
11/9/2017	1,038	2.84	GT01	CI	6
1/26/2020	1,846	5.06	GT02	HGP	11
11/9/2020	2,134	5.85	GT01	HGP	11
1/26/2023	2,942	8.06	GT02	CI	6
11/9/2023	3,229	8.85	GT01	CI	6
1/26/2026	4,038	11.06	GT02	HGP	11
11/9/2026	4,325	11.85	GT01	HGP	11
1/26/2029	5,134	14.07	GT02	CI	6
11/9/2029	5,421	14.85	GT01	CI	6
1/26/2032	6,229	17.07	GT02	MI	27
11/9/2032	6,517	17.85	GT01	MI	27

CI, combustor inspection; HGP, hot gas path inspection; and MI, major inspection.

- Plant-related general and administrative expenses (postage, telephone, etc.)
- Routine preventive and predictive maintenance performed during operations, e.g.,
  - Maintenance of equipment such as water circuits, feed pumps, main steam piping and demineralizer systems
  - Maintenance of electric plant equipment, which includes service water, DCS, condensate system, air filters and plant electrical systems
  - Maintenance of miscellaneous plant equipment such as communication equipment, instrument and service air and water supply system
  - Plant support equipment which consists of tools, shop supplies and equipment rental and safety supplies
- Maintenance of structures and grounds
- Other fees required for a project to participate in the relevant *North American Electric Reliability Corporation* (NERC) region and be in good standing with the regulatory bodies.

EIA assumptions underlying variable O&M included

- Major maintenance
- Production-related costs which vary with electrical generation, e.g.,
  - Raw water
  - Waste and wastewater disposal expenses
  - Purchase power (which is incurred inversely to operating hours), demand charges and related utilities
  - Chemicals, catalysts and gases
  - Ammonia for selective catalytic reduction (SCR) (if present)
  - Lubricants
  - Consumable materials and supplies.

The major maintenance buckets corresponded to the gas turbine outage activities typically covered under an LTSA, e.g., overhaul service cost, labor and spare parts, as well as major balance of plant (BOP) equipment maintenance.

In the PJM report cited at the beginning of the section, fixed O&M cost was $5.36 million for a 627-MWe GTCC with an operating duty profile of 5,000 h/year. This comes to 1.71 mils per kWh. The US EIA number is very close to that, i.e.,

$$15.37 / 8,760 = 1.75 \text{ mils per kWh}.$$

Comparing the variable O&M numbers is somewhat more difficult. At $275/FFH and 8,400 FFH per annum, we find that the LTSA cost is

$$\$275 \cdot 8,400 / 5,000 = 0.74/\text{MWh}.$$

Thus, the total variable O&M comes to $0.85 + $0.74 = $1.59/MWh, which is about half of that in the US EIA report.

16.2 IMPORTANT METRICS

Definition of key RAM (reliability, availability and maintainability) metrics can be found in two industry standards:

1. **IEEE 762** "IEEE Standard Definitions for Use in Reporting Electric Generating Unit Reliability, Availability, and Productivity"
2. **ISO 3977–9:1999** "Gas turbines – Procurement – Part 9: Reliability, availability, maintainability and safety"

Several key RAM metrics defined in those standards are covered below.

The *reliability factor*, RF, is the probability that a unit, major equipment or component will not be in a forced outage condition at a point in time when it is needed. It is given by the following formula:

$$\text{RF} = 1 - \frac{\text{FOH}}{\text{PH}}, \tag{16.1}$$

where FOH is *forced outage hours* and PH is *period hours*. The fraction on the right-hand side of the equation is the *forced outage factor*, FOF. In other words, the RF is the complement of the FOF.

The *availability factor*, AF, is the probability that a unit, major equipment or component will be usable at a point in time, whether it is needed or not, based on the past experience with that specific gas turbine. It is given by the following formula (ISO 3997):

$$\text{AF} = 1 - \frac{\text{FOH} + \text{POH}}{\text{PH}} = 1 - \frac{\text{UH}}{\text{PH}}, \tag{16.2}$$

where POH is *planned outage hours* and UH is the total unavailable hours.

Equivalent reliability factor is the probability of a multi-shaft combined cycle power plant not being totally forced out of service when the unit is required. It combines the contributions of gas and steam turbine outputs to the total plant output in the following formula:

$$\text{ERF} = 1 - \frac{\text{FOH}_{\text{GT}}}{\text{PH}} - \text{B}\left(\frac{\text{FOH}_{\text{HRB}}}{\text{PH}} + \frac{\text{FOH}_{\text{ST}}}{\text{PH}}\right), \tag{16.3}$$

where B is the contribution of the steam turbine generator output to the total combined cycle power plant output. (The subscript HRB denotes the *heat recovery boiler*, i.e., the heat recovery steam generator, HRSG.) As demonstrated in Chapter 3, a typical value for B is 0.33 (e.g., see Section 3.3.2 and Equation 3.6). Each component is assumed to have the same period hours, PH.

Equivalent availability factor is the probability of a multi-shaft combined cycle power plant being available for power generation whether the unit is needed or not. It includes all unavailable hours and combines the contributions of gas and steam turbine outputs to the total plant output in the following formula:

$$\text{EAF} = 1 - \frac{\text{UH}_{\text{GT}}}{\text{PH}} - \text{B}\left(\frac{\text{UH}_{\text{HRB}}}{\text{PH}} + \frac{\text{UH}_{\text{ST}}}{\text{PH}}\right). \qquad (16.4)$$

Note that Equations 16.3 and 16.4 are valid only for a 1 × 1 × 1 GTCC. For a typical N × N × 1 GTCC with N GT-HRB trains, the generic formulae become quite cumbersome. The reader is referred to the paper by Lambert et al. for details [2]. For the most common version of multi-shaft GTCC, i.e., 2 × 2 × 1, the equivalent availability formula is

$$\begin{aligned} \text{EAF} = 1 &- \left(\frac{\text{UH}_{\text{GT,1}} + \text{UH}_{\text{GT,2}}}{3\text{PH}}\right) - \left(\frac{\text{UH}_{\text{HRB,1}} + \text{UH}_{\text{GT,1}}}{6\text{PH}}\right) \\ &- \left(\frac{\text{UH}_{\text{HRB,2}} + \text{UH}_{\text{GT,2}}}{6\text{PH}}\right) - \left(\frac{\text{UH}_{\text{ST}}}{3\text{PH}}\right). \end{aligned} \qquad (16.5)$$

Equation 16.5 is based on the following assumptions:

1. Both gas turbines are identical
2. Each gas turbine and the steam turbine contribute one-third of the total plant output
3. There is a bypass stack in each GT + HRB train
4. There is full steam bypass to the condenser
5. There are no steam turbine unavailable hours coinciding with either GT+HRB train's unavailable hours.

The significance of assumption #3 is the ability to run GT1 (or GT2) even if HRB1 (or HRB2) is down. (This may not be the case in newer combined cycle power plants due to the problems associated with imperfect sealing with the bypass stack dampers.) The significance of assumption #4 is the ability to run GT1 or GT2 or both even if the steam turbine is down. If a bypass stack is not included in the design, when HRB1 (or HRB2) is down, GT1 (or GT2) has to be shut down as well (even if there is nothing wrong with it). In that case, Equation 16.5 becomes

$$\text{EAF} = 1 - \frac{\left(\text{UH}_{\text{GT,1}} + \text{UH}_{\text{HRB,1}}\right) + \left(\text{UH}_{\text{GT,2}} + \text{UH}_{\text{HRB,2}}\right)}{3\text{PH}} - \ldots \qquad (16.6)$$

The similar logic is behind the third and fourth terms of Equation 16.5. When GT1 (or GT2) is unavailable, so is HRB1 (or HRB2) even if there is nothing wrong with it.

If some of the steam turbine unavailable hours coincide with either GT + HRB train's unavailable hours, Equation 16.6 is modified as

$$\begin{aligned} \text{EAF} = 1 - \ldots &- \left(\frac{\text{UH}_{\text{HRB,1}} + \text{UH}_{\text{GT,1}} - \text{UH}'_{\text{ST}}}{6\text{PH}}\right) \\ &- \left(\frac{\text{UH}_{\text{HRB,2}} + \text{UH}_{\text{GT,2}} - \text{UH}''_{\text{ST}}}{6\text{PH}}\right) - \left(\frac{\text{UH}_{\text{ST}}}{3\text{PH}}\right), \end{aligned} \qquad (16.7)$$

where UH'_{ST} represents the steam turbine outage hours coinciding with GT1 + HRB1 train and UH''_{ST} represents the steam turbine outage hours coinciding with GT2 + HRB2 train. This prevents "double-dipping" in accounting for the contribution of a steam production source (i.e., a GT + HRB train) when there is no possibility of making use of produced steam in the first place.

In Equations 16.5–16.7, unavailability hours of each component (out of a total of five components, i.e., GT1, GT2, HRB1, HRB2 and ST) as a fraction of period hours are weighted by the contribution of that component to the total combined cycle power plant output. Thus, the factor of ⅓ for GT1, GT2 and ST is clear. Each HRB contributes 50% (i.e., ½) to the ST output, which is one-third of the total plant output; thus, each HRB contributes ½ × ⅓ = 1/6 to the total plant output.

Reliability, availability and other key RAM metrics are used by the NERC, which maintains the *Generating Availability Data System* (GADS) and by the *ORAP®* (operational reliability analysis program) system offered by Strategic Power Systems, Inc. (SPS).

GADS was initiated by the electric utility industry in 1982; it expands and extends the data collection procedures begun by the industry in 1963. GADS maintains complete operating histories on more than 7,700 generating units, representing over 90% of the installed generating capacity of the United States and Canada. As of January 1, 2013, GADS became a mandatory industry program for conventional generating units 20 MW and larger.

ORAP is a RAM-reporting system with a specific focus on gas turbines in simple and combined cycle arrangements, various applications and duty cycles and across various manufacturers. ORAP was initiated by General Electric in 1976 in response to major design issues encountered in Frame 7 gas turbines. Field data was required to characterize issues experienced by gas turbine operators and measure the effectiveness of engineering fixes developed to solve those issues. At the time, of course, gas turbines were mostly relegated to peaking service with service factors (SFs) of about 10%. In 1978, ORAP was extended to cover simple and combined cycle power plants on an equipment basis, which required addition of many more measurements to the system. In 1980, ORAP transitioned from being a customer support system to a *bona fide* engineering design support system. At present, ORAP is a commercial product supported by all major OEMs.

SPS monitors a large number of legacy (e.g., GE's old Frame 5), F, G, H and J class gas turbines manufactured by major OEMs. The list includes units operating under very different regimes (peaking, cyclic and base loaded). The mission profile of a gas turbine is represented by the following parameters:

- Annual service hours (SH) and starts
- Service hours per start
- Service factor (SF), which is the ratio of SH to PH (usually 8,760, for a calendar year)
- Capacity factor (CF)
- Availability factor (AF)
- Reliability factor (RF)

Combined cycle power plants are rarely (if ever) used for peaking duty (except very old ones). By the same token, simple cycle gas turbines are not used for baseload duty. Thus, SPS ORAP data for base-loaded gas turbines is a good measure of availability and reliability of gas turbines in combined cycle applications. Numbers from the period 2007 through 2012 are summarized in Table 16.2.

Large industrial gas turbines for electric power generation are mostly deployed in combined cycle configuration. This brings the RAM of the other two major pieces of equipment, i.e., the HRSG and the steam turbine generator, into the picture. Average F class GTCC outage factors in Table 16.3 provide a good measure of technology reliability (they are also from the ORAP database) [3]. As shown in the table, gas turbine-only unexpected outage rate is 2.4%, whereas for the entire plant, i.e., gas turbine, steam turbine and HRSG combined, the unexpected outage rate increases to 3.9% (2005–2009 time frame). What this means is that, on average, one should expect annually ~300 h of lost power generation opportunity due to unforeseen events.

Gas turbine forced outage data in Table 16.3 also illustrates the "learning" or "maturity" effect. As a technology matures, OEMs and operators become more adept in its design, operation and maintenance. Nevertheless, as shown in Figure 16.2, effects of equipment and changes in the duty cycle of a plant can be detrimental to reliability (and availability). This was indeed supported by data

TABLE 16.2
ORAP RAM Metrics for E and F Class Gas Turbines (SPS, Inc.)

	2007–2011		2012	
	E Class	**F Class**	**E Class**	**F Class**
Annual service hours	6,860	6,027	6,284	5,900
Annual starts	32	62	40	59
Service hours/start	213.8	97.0	157.6	99.2
Service factor, %	78.3	68.8	71.7	67.3
CF, %	69.2	56.5	64.5	55.1
Availability, %	93.0	92.5	92.5	89.4
Reliability, %	98.1	98.0	98.2	97.9

TABLE 16.3
Average Outage Factors for F Class GTs in Combined Cycle Configuration and Other Major Equipment (Based on Units with Minimum 6,500 Annual Operating Hours)

	1995–1999	2000–2004	2005–2009
	Gas Turbine Subsystem		
Forced outage, %	2.67	2.09	1.59
Unscheduled maintenance, %	1.78	0.55	0.79
Service factor, %	78.7	62.4	60.9
	HRSG Subsystem		
Forced outage, %	0.11	0.28	0.21
Unscheduled maintenance, %	0.93	0.26	0.26
	Steam Turbine Subsystem		
Forced outage, %	0.36	0.39	0.81
Unscheduled maintenance, %	0.59	0.24	0.26

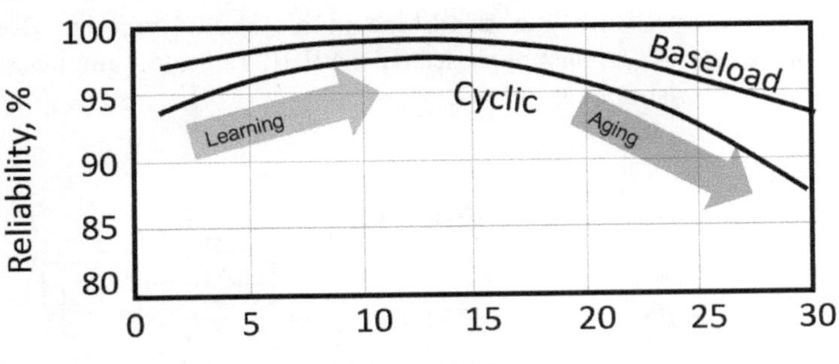

FIGURE 16.2 Impact of cyclic operation on GTCC reliability. (Adopted from Walsh [5].)

presented in a paper by Steele et al., which showed that the *mean time between failures* (MTBFs) for base-loaded F class units decreased by 7% between 1996–2000 and 2001–2005 periods when their service hours per start dropped from 137 to 60 [4].

16.2.1 Availability Calculation Example

Let us consider a 600-MWe 2 × 2 × 1 multi-shaft GTCC power plant with two 200-MWe gas turbines and a 200-MWe steam turbine. For simplicity, let us assume that each GT + HRB train is equipped with a bypass stack and the plant has 100% steam bypass capacity. Let us assume that

- Gas turbine FOF is 5% and POH are 876 (both gas turbines simultaneously)
- HRB FOF is 2% and POH are scheduled during gas turbine outages
- Steam turbine FOF is 1% and POH are scheduled during gas turbine outages.

The period hours are 8,760 (one calendar year). During the planned outage, the entire plant is shut down. In other words, the maximum plant availability is 99% by default. The equivalent availability factor from Equation 16.5 (with a slight modification to account for the planned outage) is

$$\text{EAF} = (1\% - 1\%) - \left(\frac{5\% + 5\%}{3}\right) - \left(\frac{2\% + 5\%}{6}\right) - \left(\frac{2\% + 5\%}{6}\right) - \left(\frac{1\%}{3}\right)$$

$$= 99\% - 3.33\% - 2.33\% - 0.33\% = 93\%.$$

If the plant is not equipped with bypass stacks, we have to use Equation 16.6, i.e.,

$$\text{EAF} = (1\% - 1\%) - \left(\frac{5\% + 2\% + 5\% + 2\%}{3}\right) - \left(\frac{2\% + 5\%}{6}\right) - \left(\frac{2\% + 5\%}{6}\right) - \left(\frac{1\%}{3}\right)$$

$$= 99\% - 4.67\% - 2.33\% - 0.33\% = 91.67\%.$$

Thus, the benefit of the bypass stack is 1.33% in EAF.

For a closer look at combined cycle EAF, consider the simple functional block diagram in Figure 16.3 (for a 2 × 2 × 1 multi-shaft GTCC). In the diagram, bypass stacks are labeled as EB1 and EB2, steam bypass as SB and HRBs as B1 and B2. Common plant equipment is represented by CE.

The outage characteristics of the GTCC in Figure 16.3 are graphically illustrated in Figure 16.4 in a logic tree. Each number node represents an operating combination of turbines and HRBs. There are six such combinations summarized in Table 16.4.

In order to demonstrate the logic tree, let us consider node 8 as an example. This node represents a particular operation mode with only one gas turbine, in this case G1, running. Both HRBs and the steam turbine are down and G1 exhaust bypasses B1 via EB1. Table 16.4 implies that (ignoring the CE outage case (node 17) for simplicity).

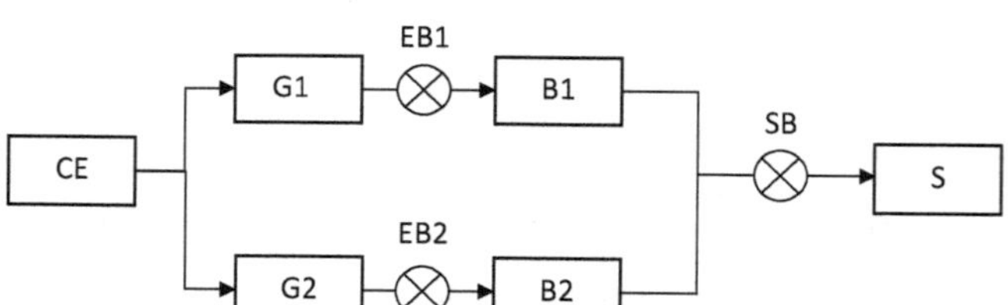

FIGURE 16.3 Functional block diagram of a 2 × 2 × 1 multi-shaft GTCC.

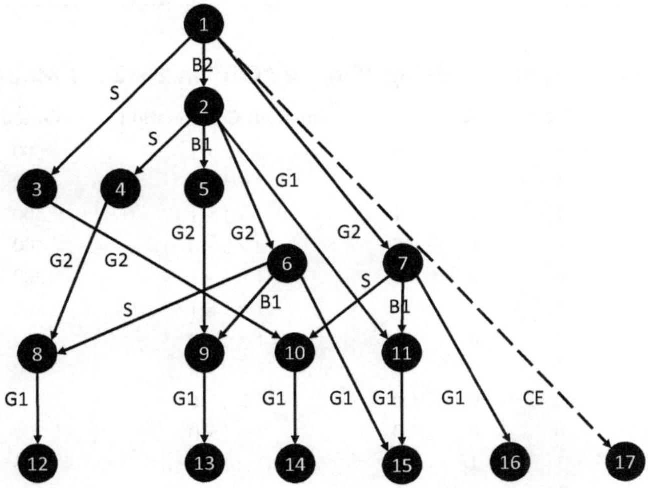

FIGURE 16.4 Outage characteristics logic tree for the GTCC in Figure 16.3.

TABLE 16.4
Possible 2 × 2 × 1 Multi-Shaft GTCC Running Combinations

Node in Figure 16.4	Combination	Output, MW
1	2 × 2 × 1	600
2	2 × 1 × 1	500
3–5	2 × 0 × 0	400
6,7	1 × 1 × 1	300
8–11	1 × 0 × 0	200
12–17	0 × 0 × 0	0

1. There are six operating combinations of two gas turbines, two HRBs and one steam turbine in a 2 × 2 × 1 multi-shaft GTCC equipped with bypass stacks and full steam bypass
2. There are 16 different paths to those six operating combinations.

In reality, however, the logic tree with 16 nodes Figure 16.3 (i.e., 16 different paths to six operating combinations of five components) is only one-half of 32 possible paths. In the second logic tree with the additional 16 paths to 1 × 0 × 0 configuration, nodes 8, 9, 10 and 11 would have G2 running in.

There are 32 possible permutations of five components with each of them either running (1) or down (0); they are listed in Table 16.5. Only nine of those permutations represent feasible operating conditions "as is". For example, G1+B1+S (permutation 5) is a feasible 1 × 1 × 1 operation with 300 MW output whereas G1+B2+S (permutation 27) is not. In the latter case, when G2 is down, B2 is out of action as well; this would lead to S shutdown and the only possible operation is 1 × 0 × 0 with G1 running (permutation 7). Going through all those cases, the six combinations and their permutations are as follows:

- One 2 × 2 × 1 (600 MWe)
- Two 2 × 1 × 1 (500 MWe)
- Five 2 × 0 × 0 (400 MWe)

TABLE 16.5
Permutations of Five Components Being "On" or "Off" in 2 × 2 × 1 Multi-Shaft GTCC

	G1	B1	G2	B2	S	Apparent Combination	Output	True Combination
1	1	1	1	1	1	2 × 2 × 1	600	Same
2	1	1	1	0	1	2 × 1 × 1	500	Same
3	1	0	1	1	1	2 × 1 × 1	500	Same
4	1	0	1	0	0	2 × 0 × 0	400	Same
5	1	1	0	0	1	1 × 1 × 1	300	Same
6	0	0	1	1	1	1 × 1 × 1	300	Same
7	1	0	0	0	0	1 × 0 × 0	200	Same
8	0	0	1	0	0	1 × 0 × 0	200	Same
9	0	0	0	0	0	0 × 0 × 0	0	Same
10	1	1	0	0	0	1 × 1 × 0	200	1 × 0 × 0
11	0	0	1	1	0	1 × 1 × 0	200	1 × 0 × 0
12	0	1	1	0	0	1 × 1 × 0	200	1 × 0 × 0
13	1	0	0	1	0	1 × 1 × 0	200	1 × 0 × 0
14	0	0	0	1	1	0 × 1 × 1	0	0 × 0 × 0
15	0	1	0	0	1	0 × 1 × 1	0	0 × 0 × 0
16	1	0	0	0	1	1 × 0 × 1	200	1 × 0 × 0
17	0	0	1	0	1	1 × 0 × 1	200	1 × 0 × 0
18	1	1	0	1	0	1 × 2 × 0	200	1 × 0 × 0
19	0	1	1	1	0	1 × 2 × 0	200	1 × 0 × 0
20	0	1	0	0	0	0 × 1 × 0	0	0 × 0 × 0
21	0	0	0	1	0	0 × 1 × 0	0	0 × 0 × 0
22	1	1	1	0	0	2 × 1 × 0	400	2 × 0 × 0
23	1	0	1	1	0	2 × 1 × 0	400	2 × 0 × 0
24	1	1	0	1	1	1 × 2 × 1	300	1 × 1 × 1
25	0	1	1	1	1	1 × 2 × 1	300	1 × 1 × 1
26	0	1	1	0	1	1 × 1 × 1	200	1 × 0 × 0
27	1	0	0	1	1	1 × 1 × 1	200	1 × 0 × 0
28	1	0	1	0	1	2 × 0 × 1	400	2 × 0 × 0
29	1	1	1	1	0	2 × 2 × 0	400	2 × 0 × 0
30	0	1	0	1	1	0 × 2 × 1	0	0 × 0 × 0
31	0	1	0	1	0	0 × 2 × 0	0	0 × 0 × 0
32	0	0	0	0	1	0 × 0 × 1	0	0 × 0 × 0

- Four 1 × 1 × 1 (300 MWe)
- Twelve 1 × 0 × 0 (200 MWe)
- Eight 0 × 0 × 0 (0 MWe).

Thus, the 32 permutations from Table 16.5 are reordered in the order of decreasing plant output and the probability of each combination is calculated. In particular,

- Probability of a gas turbine, HRB or steam turbine to be running is 0.95, 0.98 and 0.99, respectively
- Probability of a gas turbine, HRB or steam turbine to be not running is 0.05, 0.02 and 0.01, respectively.

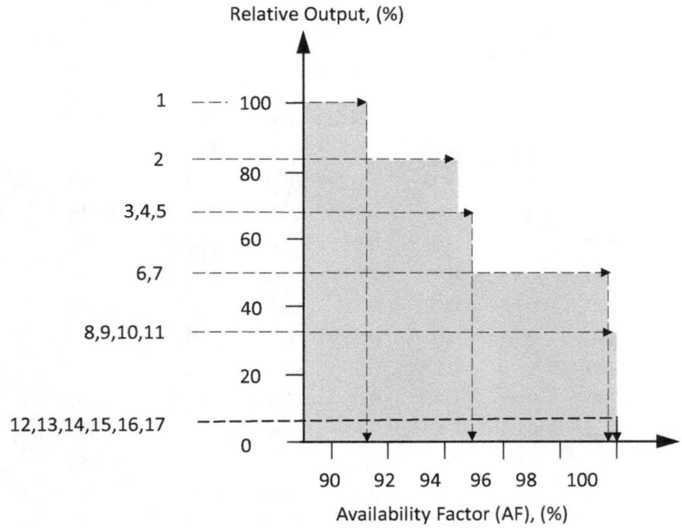

FIGURE 16.5 Availability output chart.

Combined probability of each permutation is the product of individual probabilities. Multiplication of the combined probability of each permutation with the period hours (8,760 for one calendar year in this example) gives the annual hours. Multiplication of hours with plant output gives the megawatt-hours for each permutation. This is shown in Table 16.6.

From Table 16.6, total megawatt-hours is 4,943,601 MWh. Annual average output is 4,943,601/8,760 = 564 MW. Using the information in the table, available time for running at X MWe or higher can be calculated as shown in Table 16.7. For example, the time availability, TA, of generating 400 MWe or more is

$$\text{TA} = (7{,}517 + 307 + 82) / 8{,}760 = 0.903.$$

Time availability and output data in Table 16.7 are used to construct the availability diagram in Figure 16.5. The shaded area gives the EAF, which is 0.941. Subtracting 0.01 for the planned outage, we find 0.931, which is the result obtained earlier using Equation 16.5.

16.3 FAILURE MECHANISMS

Based on more than 150 insurance claims between 2005 and 2013, a major insurance and risk management company prepared a report on causes of large losses in global power industry.[2] In terms of value of loss in dollars and number of events, machine breakdown was by far the largest cause. Failure of turbine blades, transformers and generators comprised 77 of the 108 machinery breakdown losses and they also ranked high in value of loss in dollars.

In combined cycle power plants, the gas turbine ranks number one in failure frequency followed by the HRSG and the steam turbine. As a whole group, however, BOP equipment ranks higher than the steam turbine.

In addition to normal wear and tear, dominant failure mechanisms in the gas turbines are creep, *high-cycle fatigue* (HCF) and corrosion. In the HRSG, *flow-assisted corrosion* (FAC) ranks as the

[2] *Common Causes of Large Losses in Global Power Industry*, Marsh Risk Management Research, September 2013.

TABLE 16.6
Combined Probabilities of Combinations in Table 16.5

	Combination	Output	G1	B1	G2	B2	S	Probability	Hours	MWh
1	2 × 2 × 1	600	0.95	0.98	0.95	0.98	0.99	0.858	7,517	4,510,139
2	2 × 1 × 1	500	0.95	0.98	0.95	0.02	0.99	0.018	153	76,703
3	2 × 1 × 1	500	0.95	0.02	0.95	0.98	0.99	0.018	153	76,703
4	2 × 0 × 0	400	0.95	0.02	0.95	0.02	0.01	0.000	0	13
5	2 × 0 × 0	400	0.95	0.98	0.95	0.02	0.01	0.000	2	620
6	2 × 0 × 0	400	0.95	0.02	0.95	0.98	0.01	0.000	2	620
7	2 × 0 × 0	400	0.95	0.02	0.95	0.02	0.99	0.000	3	1,252
8	2 × 0 × 0	400	0.95	0.98	0.95	0.98	0.01	0.009	76	30,371
9	1 × 1 × 1	300	0.95	0.98	0.05	0.02	0.99	0.001	8	2,422
10	1 × 1 × 1	300	0.05	0.02	0.95	0.98	0.99	0.001	8	2,422
11	1 × 1 × 1	300	0.95	0.98	0.05	0.98	0.99	0.045	396	118,688
12	1 × 1 × 1	300	0.05	0.98	0.95	0.98	0.99	0.045	396	118,688
13	1 × 0 × 0	200	0.95	0.02	0.05	0.02	0.01	0.000	0	0
14	1 × 0 × 0	200	0.05	0.02	0.95	0.02	0.01	0.000	0	0
15	1 × 0 × 0	200	0.95	0.98	0.05	0.02	0.01	0.000	0	16
16	1 × 0 × 0	200	0.05	0.02	0.95	0.98	0.01	0.000	0	16
17	1 × 0 × 0	200	0.05	0.98	0.95	0.02	0.01	0.000	0	16
18	1 × 0 × 0	200	0.95	0.02	0.05	0.98	0.01	0.000	0	16
19	1 × 0 × 0	200	0.95	0.02	0.05	0.02	0.99	0.000	0	33
20	1 × 0 × 0	200	0.05	0.02	0.95	0.02	0.99	0.000	0	33
21	1 × 0 × 0	200	0.95	0.98	0.05	0.98	0.01	0.000	4	799
22	1 × 0 × 0	200	0.05	0.98	0.95	0.98	0.01	0.000	4	799
23	1 × 0 × 0	200	0.05	0.98	0.95	0.02	0.99	0.001	8	1,615
24	1 × 0 × 0	200	0.95	0.02	0.05	0.98	0.99	0.001	8	1,615
25	0 × 0 × 0	0	0.05	0.02	0.05	0.02	0.01	0.000	0	0
26	0 × 1 × 1	0	0.05	0.02	0.05	0.98	0.99	0.000	0	0
27	0 × 1 × 1	0	0.05	0.98	0.05	0.02	0.99	0.000	0	0
28	0 × 1 × 0	0	0.05	0.98	0.05	0.02	0.01	0.000	0	0
29	0 × 1 × 0	0	0.05	0.02	0.05	0.98	0.01	0.000	0	0
30	0 × 2 × 1	0	0.05	0.98	0.05	0.98	0.99	0.002	21	0
31	0 × 2 × 0	0	0.05	0.98	0.05	0.98	0.01	0.000	0	0
32	0 × 0 × 1	0	0.05	0.02	0.05	0.02	0.99	0.000	0	0

TABLE 16.7
Cumulative Probabilities of Generating X MW

Load (%)	MWe	Hours	TA
100	600	7,517	0.858
83	500	307	0.893
67	400	82	0.903
50	300	807	0.995
33	200	25	0.998
0	0	0	1.000

number one failure mechanism. The primary culprit in FAC is the failure to maintain requisite feedwater chemistry, but it is compounded by cyclic operation. The LP section of the HRSG, the evaporator and the economizer are hardest hit by FAC. Cyclic operation also impacts adversely HP superheater sections and the HP evaporator drum via *low-cycle fatigue* (LCF) and creep. Other common HRSG failure mechanisms are

- Thermal shock (e.g., condensation in superheater tubes)
- Corrosion in tubes (inside or outside)
- Differential expansion of adjacent tubes in tube banks at different temperatures.

Proper design and maintenance of seemingly mundane BOP equipment is crucial to trouble-free operation of the major plant equipment such as the HRSG. For example, if the condensate formed in the cold reheat steam pipe during outage is not properly drained, it is carried into the reheater during startup with the bypass steam and can cause temperature differences in excess of 500°F between the tubes. Leaking desuperheater spray valves can cause similar temperature differences and lead to tube failures.

Water chemistry problems are also at the top of the list for steam turbine failures via deposits, erosion and corrosion of the turbine buckets. Last-stage blade failure due to flutter and *stress corrosion cracking* (SCC) is another significant (and very costly) steam turbine failure mechanism. SCC refers to the (usually unexpected and sudden) failure of a normally ductile metal under tensile stress in a corrosive environment. Three factors are at play in SCC:

- Material susceptibility to SCC
- Exposure to a corrosive environment
- Tensile stresses above a threshold.

In general, though, as long as the water chemistry is maintained properly and the OEM guidance regarding low load and/or high condenser pressure operation is followed, the steam turbine is the most robust component in the GTCC. Steam purity requirements during operation and commissioning are specified in detail by the OEMs, e.g., see GEK 72281C "Steam Purity Recommendations for Utility Steam Turbines" by General Electric (2004). For a general discussion of the steam purity subject, refer to Section 10.9 in this book. Briefly, steam "impurity" is caused by

- Suspended solids (e.g., metallic particles)
- Dissolved solids (e.g., Na, Cl, CO_4)
- Gases such as CO_2 and O_2
- Vaporized silica
- Organics.

Ideal steam turbine steam should be dry and pure H_2O with a pH value of about 9.0 (i.e., basic or alkaline). Potential sources and/or causes of steam contamination are

- Boiler carryover (typically salts for systems at high pressures, e.g., 1,800 psig or higher)
- Attemperation (desuperheating) with contaminated feedwater
- Silica vaporization
- Vaporization of organic compounds.

During HRSG commissioning, steam is generated from impure water in piping contaminated during manufacturing and field storage due to exposure to the elements and/or chemicals. It is imperative that HRSG-generated steam is pure enough for admission to a steam turbine. Even a short-time exposure to impurities during startup (as short as 8 h) can be damaging enough to cause catastrophic failure during normal operation within a few years even if the water chemistry is maintained diligently in accordance with OEM requirements during the entire period.

HCF observed in gas turbine compressor and hot gas path and in steam turbine buckets is most likely a design issue. The failure is usually due to excitation of a natural frequency by a harmonic (also known as the "engine order") of the machine speed (e.g., integer multiples of 50 or 60 Hz) leading to resonance. This is unavoidable for a very short period during startup and shutdown (e.g., see point C in Figure 16.6). It is, however, quite rare to happen during normal operation unless there is a design and/or manufacturing defect. In addition, if the machine is not kept clean (this is where gas turbine inlet filter maintenance and proper feedwater chemistry play significant roles), deposits on the blades can be significant enough to change the component natural frequency and lead to HCF over time.

Flutter of the long last-stage blades is caused by self-excitation due to the dynamic interaction of structural and aerodynamic forces acting on the component. If not prevented, it can rapidly lead to catastrophic blade failure or to HCF in the long term. Flutter can be identified as the cause of blade vibration by examining the vibration data. If the vibration frequency is very close to a natural frequency (i.e., a blade "mode") and is not near a harmonic, the vibration is likely to be symptomatic of blade flutter (e.g., see points B and C in Figure 16.6).

Compressor surge and stall can destroy the machine within seconds. Axial compressors have rather narrow bands of stable operation. Gas turbine controller ensures that the machine stays within that band. (See Chapter 11 of **GTFEPG** for more details.) Nevertheless, it is critical that the compressor is kept clean by frequent online washing (and offline washing during planned outages), inlet filter maintenance and monitoring the operating parameters. Erosion of compressor blades or fouling of the turbine (changing its flow capacity) can lead to pressure rises and create surge conditions.

Synchronous ac generators are quite reliable. Since they are highly vulnerable to contamination, their tightly sealed enclosures ensure that they do not experience significant performance deterioration (e.g., unlike the gas turbine compressor). Their major overhaul interval is typically several years. However, in-situ inspection (especially with robotic devices) can prevent freak accidents and mishaps due to hard-to-detect conditions such as copper dusting or coil distortion in addition to reducing the major inspection downtime. Most common failure modes involve stator/rotor winding and the rotor itself. Typical causes of failure are short circuits in stator core and rotor winding, broken windings and deteriorated insulation of windings. Generators can also suffer from cyclic operation (e.g., elongation and breaking of rotor winding ends). For an in-depth discussion of large

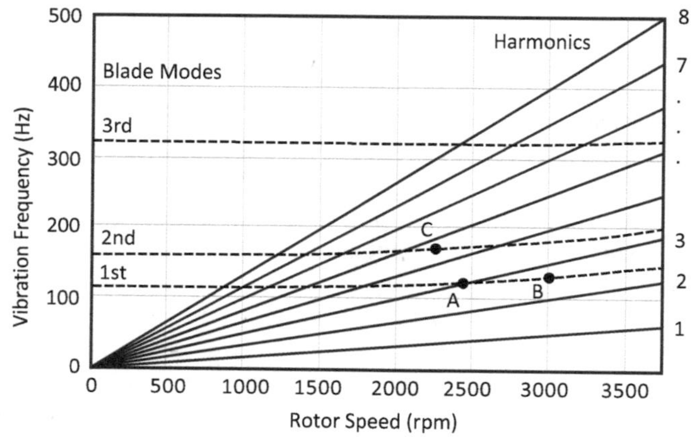

FIGURE 16.6 Campbell diagram (for a 60-Hz unit).

ac generator maintenance and inspection techniques, the reader is referred to the General Electric white paper GER-3954C.[3]

Steam turbine components are not subject to very high temperatures experienced by gas turbine hot gas path components. But they operate at pressures higher by an order of magnitude and their construction comprises bulky casings/shells and solid rotors. These components are vulnerable to thermal stresses caused by rapid temperature changes during plant transients such as startup and load ramps. Therefore, thermal stress and associated failure mechanisms are at the top of the list. In addition to the usual suspects, i.e., LCF and creep rupture, one should also be cognizant of the *brittle fracture* in solid rotors with a central bore.

Many steam turbines in service have hollow bores with diameters of a fraction of an inch. The reason for that is that they have been machined at the manufacturing stage to remove material of inferior mechanical and metallurgical properties from the central portion of the original forging. The thermal stress at the rotor bore, if uncontrolled, can lead to crack formation and propagation. In particular, at temperatures below the *fracture appearance transition temperature* (FATT), the rotor material loses its ductility in the presence of cracks. Therefore, especially in cold starts, the turbine warm-up and roll should proceed at a closely controlled slow pace until the metal temperature at the rotor bore is above FATT (typically, 200°F–250°F[4]). If this is not done and the bore is overstressed, the otherwise acceptable bore defect can result in a rotor burst.

Like any other type of rotating equipment, uncontrolled/excessive vibration is the biggest source of mechanical failure in gas and steam turbines [6]. Detailed discussion of the subject within a gas turbine context can be found in **GTFEPG**. The underlying principles are the same for the steam turbines as well. Major causes of vibration in steam turbines are rotor unbalance, rubbing between rotating and stationary parts and oil whipping in the bearings. There are other less common sources of vibration such as mechanical runout in the couplings, water or oil in the rotor bore and temporary misalignment due to temperature changes.

Vibration monitoring is vital for safe operation. Vibration due to unbalance is synchronous with the shaft speed and can be corrected by balance weight adjustments. Sudden change in synchronous vibration levels is usually symptomatic of rotating part failure (e.g., loss of a bucket). Shaft misalignment or a bowed shaft can also be the source of synchronous vibrations. They can also result from resonance in the system (i.e., the frequency of the excitation coinciding with a natural frequency of the system – e.g., point C in Figure 16.6). Synchronous vibrations can occur at whole number multiples of the synchronous speed (e.g., at 120, 180 or 240 Hz for a 60-Hz system).

Note that vibration, even if below or at alarm levels (typically, 0.5–0.7 in./s), is damaging to other plant equipment while the turbomachinery itself is safe. Examples of secondary vibration effects, which can lead to trips and/or costly repairs, are electrical wiring becoming loose or broken (especially where the wiring makes a turn around a sharp metal edge), cracks in the enclosure walls or foundations, problems with fixators (used to align the ac generator instead of shims), and other problems of equipment detachment and looseness (e.g., fire protection system water spray fixtures).

Nonsynchronous vibrations have frequencies that are not whole number multiples of the synchronous speed. (Note that blade flutter is a special case of nonsynchronous vibration.) Nonsynchronous vibrations at frequencies lower than the synchronous speed are called *sub-synchronous* vibrations. Nonsynchronous vibrations at frequencies higher than the synchronous speed are called *super-synchronous* vibrations. Super-synchronous vibrations can be caused by bearing defects, cavitation (in feed pumps) and other fluid-dynamic instabilities (vortex shedding, rotating stall, etc.) or electrical faults. Blade flutter is an example of nonsynchronous vibration

---

[3] Generator In-Situ Inspections – A Critical Part of Generator Maintenance Cost Reduction, GER-3954C, C. Markman and R.J. Zawoyski, March 2012. Last downloaded from the internet in June 2019.

[4] It should be noted that, after years of service at high temperatures, the rotor material will gradually embrittle and FATT value will increase.

(e.g., points B and C in Figure 16.6). Sub-synchronous vibrations can be caused by looseness in the system (i.e., the foundation) or rubbing between rotating and stationary parts.

Prime movers in the GTCC power plant are equipped with vibration monitoring using proximity probes mounted on radial/journal and thrust bearings. Typically, two proximity probes are mounted in an orthogonal configuration at each radial bearing (usually on the bearing housing) to record the relative motion between the bearing liner and the journal (i.e., the shaft). Similarly, two proximity probes are used to monitor the axial position of the shaft within the thrust bearing clearance. Readings obtained by the probes are connected to the DCS for continuous monitoring. There are alarms and trip levels associated with each reading to prevent catastrophic failure of the unit (e.g., see Table 16.9 in **GTFEPG**) Usually, if the vibration levels reach about 1 in./s, the controller initiates a turbine runback (i.e., bringing the power output down to, say, 60%) or trip. The exact values are specified by the equipment manufacturer.

Vibration signals picked up by transducers can be composites of all three types of vibrations taking place simultaneously. Proper analysis requires mathematical conversion of the composite waveform into spectral data via application of *Fast Fourier Transform* (FFT) that identifies individual, independent frequencies. In order to do this analysis while the unit is running and to facilitate improved asset management, diagnostic and condition monitoring software should be a part of the DCS.

In addition to the vibration issues in the prime movers and other rotating equipment in the GTCC, close attention should be paid to vibration in high-energy pipelines, which can cause very costly (to equipment as well as life and limb) steam pipe failures. By far, the most dominant mechanism is flow-induced vibration (turbulence, valve leaks, Helmholtz resonance). The other significant cause of pipe vibrations is mechanical excitation from turbomachinery unbalanced forces and moments. For a detailed coverage of the causes, analysis and mitigation of piping vibration, the reader is referred to the paper by Wachel et al. [7].

There is no active monitoring of piping vibrations in the GTCC power plant. Typically, they are detected by the plant personnel as early as during commissioning or after a major uprate (or equipment replacement). Diagnosis of the problem's root cause and its mitigation requires installation of transducers for vibration testing, data collection and analysis by rotordynamics specialists. In order to determine whether the piping vibration amplitudes recorded during the testing are acceptable or not, the resultant dynamic stresses caused by the vibrations must be compared to the allowable endurance stress limit. As a first-order attempt to gauge the "severity" of the recorded vibrations, one can refer to experience-based vibration amplitude (peak-to-peak mils) versus frequency charts (e.g., the chart in Figure 11 in Ref. [7]).

In ASME OM-S/G-20072015 Standard, a screening value of 0.5 in./s (zero to peak) is given for pipe vibrations (Appendix D). The standard does not include any frequency dependency as the pipe stress level is a function of pipe distortion only. The 0.5 in./s (12.5 mm/s) is a conservative value based on worst-case correction factors for concentrated weights, pipe diameter, pipe configuration and pipe frequency factor. Any vibration velocity measurement below 0.5 in./s indicates a safe level for any type of piping configuration from dynamic stress standpoint and no further analysis is required.

A similar guideline is provided in "Field Criteria for Pipe Vibration" by S. Maten (Hydrocarbon Processing, July 1984 issue) as 0.6 in./s. In *Guidelines for the Avoidance of Vibration Induced Fatigue Failure in Process Pipework* (published by Energy Institute, London, 2008), a diagram of acceptable, concern and problem vibrations versus frequency up to 300 Hz is provided. An almost identical diagram is included in VDI 3842-2004, *Vibrations in Piping Systems*, which is reproduced in Figure 16.7.

Obviously, it would be better to evaluate the severity of piping vibration deflection via comparison of the maximum resonant vibration-induced dynamic stresses to an allowable endurance limit stress. Such information is available in the ASME Codes. For example, ANSII ASME Code OM3–1987 uses stress versus cycles-to-failure (S–N) curves as a basis for specifying criteria for

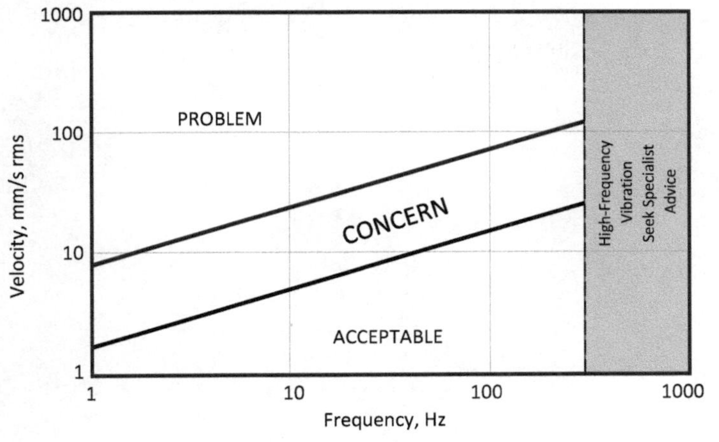

FIGURE 16.7 Acceptable, concern and problem vibrations (VDI 3842-2004).

evaluating the vibration-induced stresses in nuclear power plant piping for preoperational and startup testing. API Standard 618 also uses the same data to specify the allowable dynamic stress level for steel pipes as a design requirement.

REFERENCES

1. Nagy, D., Savic, S., 2015, Alternative Gas Turbine Maintenance Concepts by Independent Service Providers, *POWERGEN 2015*, December 8–10, 2015, Las Vegas, NV.
2. Lambert, D., Steele, R.F., Rosenfelder, M., DellaVilla, S., Equivalent Availability Measurement for Combined Cycle Power Plants: A New Approach, 2001-GT-0579, *ASME Turbo Expo 2001*, June 4–7, 2001, New Orleans, LA.
3. Grace, D., Christiansen, T., 2012, Risk Based Assessment of Unplanned Outage Events and Costs for Combined Cycle Plants, GT2012–68435, *ASME Turbo Expo 2012*, Copenhagen, Denmark.
4. Steele Jr., R.F., Paul, D.C., Rui, T., 2007, Expectations and Recent Experience for Gas Turbine Reliability, Availability, and Maintainability (RAM), GT2007–27655, *ASME Turbo Expo 2007*, May 14–17, 2007, Montreal, Canada.
5. Walsh, D., 2015, Impact of Cyclic/Two Shift Operation on the Reliability, Economics, and Emissions of Fossil Generation (Coal and Combined Cycle Gas Turbine), *Powergen International 2015*, December 8–10, Las Vegas, NV.
6. Meher-Homji, C.B., 1995, Blading Vibration and Failures in Gas Turbines, Part A: Blading Dynamics & the Operating Environment, 95-GT-418, *International Gas Turbine and Aeroengine Congress and Exposition*, June 5–8, 1995, Houston, TX.
7. Wachel, J.C., Morton, S.J., Atkins, K.E., 1990, Piping Vibration Analysis, *Proceedings of 19th Turbomachinery Symposium*, Texas A&M University, College Station, TX.

17 Repowering

Repowering involves the addition to or replacement of aging power generation equipment with obsolete technology with newer equipment comprising up-to-date technology while retaining still-usable components to more effectively use an existing site, which involves one or more of the following [1]:

- Improve generation economics
- Extend plant life
- Improve environmental performance
- Enhance operability and maintainability.

Generic definitions can be made more specific as follows:

1. The *aging power generation equipment with obsolete technology* in question herein is a fossil-fired boiler plant with a steam turbine generator (STG).
2. The *newer equipment comprising up-to-date technology* in question herein is natural gas-fired gas turbine in simple or combined cycle.
3. The *reason* for repowering is twofold:
   a. Increase performance (power output and thermal efficiency)
   b. Reduce and/or eliminate harmful stack emissions.
4. The *hurdle* for repowering is the **life cycle cost** of generation: i.e., the new power plant must have better LCOE[1] (capital expenditure, fuel, O&M[2]) than the old one.

Typically, the hurdle for repowering is a very tall one. The aging steam power plants (at the time of writing, i.e., 2010, the average age of *ripe-for-retirement* coal power plants is nearly 50 years!) are fully paid-off generating assets, whose generating costs are limited to the variable expenses for fuel (typically coal) and O&M. Unless the equipment is in such a terrible shape that O&M costs to maintain a reasonable efficiency (in order to keep the fuel costs in check) become intolerable, these plants could be operational for quite a long time (albeit at a reduced capacity factor).

The hurdle is significantly lower nowadays primarily due to much more stringent emissions control requirements enforced by government agencies that make the aging fossil power plants impossible to keep running. The investment into new equipment to reduce harmful stack emissions and improve the efficiency is simply exorbitant. The situation is made even worse by low natural gas prices driven by increased production of shale gas using new technology such as fracking. Thus, natural gas-fired gas turbine technology with efficiencies pushing 60% becomes economically a much more attractive alternative.

There are *four* major gas turbine-based repowering options:

- Site repowering (a completely new gas turbine combined cycle (GTCC) power plant)
- Heat recovery repowering, i.e., boiler replacement with a gas turbine plus HRSG train
- Hot windbox repowering (utilizing a gas turbine as "forced draft fan")
- Feedwater heater (FWHR) repowering (utilizing gas turbine exhaust gas for feedwater heating).

[1] Levelized Cost of Electricity – see Chapter 13.
[2] Operating and Maintenance (Expenses).

The options listed above are described in detail in Ref. [2]. *Heat recovery repowering* is the most common type of repowering being implemented in the USA, where the existing boiler is replaced by a gas turbine and an HRSG (see Figure 17.1). The existing plant's net generating capacity can be increased by up to 200% accompanied by an improvement in efficiency (up to 40%) and reduction in NOx emissions. This approach is ideal for older units rated less than 250 MW and with steam pressures up to 1,800 psia.

Table 17.1 shows typical power output and net heat rate changes for each repowering option (excluding the site replacement). The performance change is application specific and depends on the match of the new gas turbine with the existing power plant [1]. Since heat recovery repowering delivers the biggest "bang for the buck", most of the existing repowering experience involves this system type.

There are myriad factors that enter the decision-making process in repowering of an existing coal-fired Rankine cycle steam turbine power plant. An incomplete list is given below:

- Capacity factor
- Installed capacity per fuel in the service area
- Cost of fuel

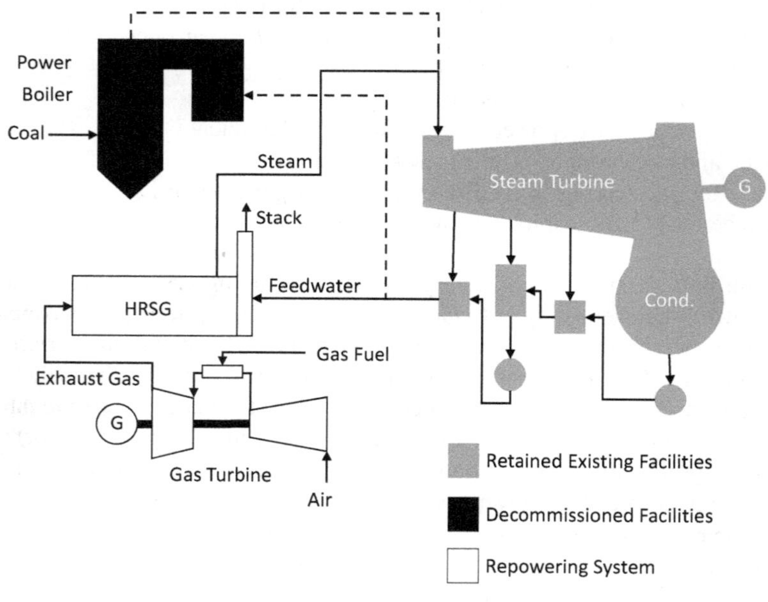

FIGURE 17.1 Schematic diagram of heat recovery repowering.

TABLE 17.1
Typical Repowering Performance Changes

Repowering Type	% MW Increase	Heat Rate Decrease
Feedwater heating	40%	10%
Boiler windbox	40%	5%–10%
Heat recovery	200%	30%

- Net plant efficiency (heat rate)
- Cycling requirements (integration with renewable generation)
- Size of the power plant
- Other operating costs
- Anticipated life
- Outstanding or anticipated *air quality control system* (AQCS) equipment upgrades
- Location of the power plant
- Community support for or against the plant
- Other.

Similarly, there are myriad factors involved when considering natural gas-fired gas turbine-based repowering options. (Note that one way to implement FWHR repowering is to utilize *solar thermal energy* instead of a gas turbine.) Some of those factors are listed below:

- Age of the steam turbine and anticipated life (for GT + HRSG replacing the boiler)
- Steam conditions (efficiency) of the steam turbine
- Size/capacity of the steam turbine
- Power demand vs. installed capacity
- Cycling requirements
- Other operating costs
- Other.

17.1 WHICH REPOWERING?

The *raison d'etre* for repowering of interest in present day (i.e., 2010s) is the expected retirement of a significant number of coal-fired units by 2020 in compliance with new Environmental Protection Agency (EPA) regulations, especially MATS[3] [3]. The number of *ripe-for-retirement* units is estimated to be anywhere between ~150 and ~350 [4]. Since the coal-fired boiler is eliminated, only two of the four possible types of GT-based repowering options are applicable:

1. Site replacement (a new *brownfield* GTCC project)
2. Replacement of original boiler with a GT + HRSG train (heat recovery repowering).

The second option is extremely appealing in terms of reduced engineering complexity and construction risk, minimum life cycle cost as measured by *levelized cost of electricity* (LCOE) and maintaining baseload generation capability with minimal additional capital investment. The assumptions leading to this assertion are as follows:

1. The steam turbine of the retired coal-fired plant is in good shape (age, performance, etc.)
2. The heat rejection system (condenser and cooling tower) is in good shape
3. The switchyard is in good shape and can be expanded to accommodate the second generator's output.

The definition of the term "good shape" above is as follows:

> The equipment in question can be continued to be used as is (1) with minimum repair and/or upgrades and (2) without major redesign and/or retrofitting to a capacity above or below their original rating.

[3] Mercury and Air Toxics Standards.

While the first item in the definition above is obvious, the second item needs some elaboration. What is meant by that clause in terms of specific equipment is as follows:

1. The steam turbine will operate in a recuperative (i.e., with FWHRs) cycle with the same steam conditions (e.g., 1,450 psia throttle pressure and 1,000°F for main and hot reheat steam) and same steam flows with all FWHRs operating as designed. In other words, there will be no *plugging* of existing bleed holes and/or drilling of new ones for additional steam admission. The unit will generate the same output at its rated conditions.
2. The condenser will operate with the same cooling water flow and temperature rise to condense the same amount of steam as designed at its rated conditions.

The assertion of minimal additional capital investment to maintain the baseload generation capability can be justified by simple arithmetic. Let us assume the following

TRCC = Total coal-fired generation capacity to be retired
NGUC = New generation unit capacity to replace TRCC
RPF = Fraction of TRCC amenable to the proposed repowering scheme.

Thus, we have

$$NGUC = TRCC - (1 + \alpha) \times RPF \times TRCC$$

where α is the increased generation capacity multiplier applied to the steam turbine capacity of the repowered coal-fired plant to account for the gas turbine. In other words, if the coal-fired boiler of a fossil plant with 100 MW net rated output is replaced by an F class gas turbine (say, General Electric Frame 7) and *mildly fired* HRSG, the net rated output of the new (repowered) combined cycle power plant becomes $(1 + \alpha) \times 100$ MW. The expected (average) value of α is 1.75 (the origin of this number will be clear in the example below).

Consequently,

- With no repowering at all, i.e., RPF = 0 → NGUC = TRCC
- If only 10% of TRCC is amenable to repowering, i.e., RPF = 0.1 → NGUC = 0.725·TRCC.

In other words, if TRCC estimate is taken at its high value of 60 GW, only 10% repowering saves

$$0.275 \times 60 \text{ GW} = 16.5 \text{ GW} = 16{,}500 \text{ MW of NGUC.}$$

Assuming all of that, NGUC is new GTCC at $1,000/kW and, as will be shown below with an example, extra

$$1.75 \times 0.1 \times 60 \text{ GW} = 10.5 \text{ GW} = 10{,}500 \text{ MW}$$

(from GT + HRSG + BOP for repowering) comes at $500/kW, the net saving in capital expenditure is

$$16{,}500 \times 1{,}000 \times \$1{,}000/\text{kW} - 10{,}500 \times 1{,}000 \times \$500/\text{kW} = \mathbf{\$11.25\ billion}.$$

17.2 COST OF REPOWERING

According to a 1997 EPRI Report [5], the single most significant driver prompting generating companies to evaluate repowering has been the "desire to minimize the cost of production and thereby increase capacity factor". Since reduced capital investment is a key factor in that respect, it is important to have an idea about the cost of repowering. Table 17.2, using data from Ref. [6], gives a summary of repowering projects from about two decades ago.

TABLE 17.2
Sample List of Repowering Projects [6]

			Heat Rate			Capacity			$/kW	
Plant	**Type**	**Completed**	**Old**	**New**	**Δ**	**Old**	**New**	**Δ**	**Net New**	**Incremental**
Lauderdale 4,5	GT + HRSG	1992	10,000	8,400	−16.0%	313	820	162.0%	$570	$920
Clark 5,6,10	HRSG	1993	13,500	8,500	−37.0%	296	486	64.2%	$288	$737
Comanche	HRSG	1986	8,900	8,630	−3.0%	240	260	8.3%	$78	$1,019
Burlington 10	GT + HRSG	1993	14,000	8,800	−37.1%	190	255	34.2%	$545	$2,138
Butler Warner	HRSG	1988	12,350	10,400	−15.8%	180	250	38.9%	$198	$706
Harbor	GT + HRSG	1995	14,000	7,900	−43.6%	90	230	155.6%	$1,196	$1,964
Grayson	GT + HRSG	1975	14,500	9,000	−37.9%	40	102	155.0%	$165	$272
El Centro 2	GT + HRSG	1992	12,500	8,426	−32.6%	30	115	283.3%	$574	$776
Larsen	GT + HRSG	1993	13,780	8,700	−36.9%	28	116	314.3%	$457	$602
Vero Beach	GT + HRSG	1992	13,910	8,700	−37.5%	16.5	56.5	242.4%	$523	$739
Animas	GT + HRSG	1994	14,500	8,000	−44.8%	7.2	25	247.2%	$752	$1,056

For the heat recovery repowering projects in Table 17.2, ignoring the two outliers (one from 1975 and less than $200/kW and the other from 1995 at nearly $1,200/kW), the specific cost per new capacity is between $400 and $800/kW (covering a period between 1986 and 1994). Note the cost numbers are reported "as is" with no revision to ensure that they are on a completely "apples-to-apples" basis.

A more recent cost estimate from 2013 in Table 17.3 is more or less in agreement [7]. This is somewhat surprising when the time period separating the two tables (about two decades) is considered. Then again, no information is provided on the cost scope in either reference so a direct comparison is not possible.

As a "real" example, consider the $519 million Fort Myers repowering, which added 960 MW to the 1960s-vintage oil-fired steam plants it replaced. (The 6 × 2 repowered GTCC was rated 1,500 MW vis-à-vis 540 MW of the old units.) The plant was formally commissioned on May 30, 2002. In today's dollars, this is roughly equivalent to $750–$800 million or about $800 per additional kilowatt of capacity. It is a unique example of heat recovery repowering, which contains useful information and "lessons learned". The full story appeared in *POWER* magazine's August 2003 issue [8].

TABLE 17.3
Repowering Cost Estimates [7]

Type	**Performance**	**Benefit**	**Cost**
Site	Matches "greenfield" in performance	Savings in land, permitting, transmission and socioeconomic	$700–$1,000/kW (savings of $100–$200/kW over "greenfield")
Steam turbine	Increases capacity by 150%–200%; improves heat rate by 30%–40%	Reuse steam turbine and cooling system	$450–$700/kW based on total capacity
FWHR	Improves plant heat rate by 6%	Adds peaking capacity and improves GT efficiency	$75–$110/kW based on total capacity
Hot windbox	Adds up to 20% capacity; improves heat rate by 10%–20%	Most reuse of existing equipment; flat heat rate curve	$150–$250/kW based on total capacity

17.3 AN EXAMPLE CALCULATION

Let us assume that the coal-fired power plant slated for retirement is a subcritical pulverized coal (PC) unit rated at 120 MW (Figure 17.2). This is a *fictitious* plant. In terms of size and performance, it is *likely* to fall in the average rating class of "ripe for retirement" power plants.

The reheat steam turbine (similar to General Electric's A series fossil steam turbines with separate high-pressure (HP) casing and combined intermediate-/low-pressure, IP/LP, casing) has five (5) FWHRs and operates at relatively *modest* steam conditions (i.e., 1,450 psia/1,000°F/1,000°F). The closed-loop heat rejection system comprises a water-cooled condenser (about 1.8 in. Hg pressure at ISO baseload) and wet cooling tower. There is no IP or LP steam addition.

The goal is to utilize the STG along with its recuperative feedwater heat exchanger train (including the boiler feed pumps) and the plant's heat rejection system *with no changes*, i.e.,

- Same steam cycle conditions
- Same HP steam flow
- Same extractions for feedwater heating
- Same condenser back pressure
- Same STG output at rated conditions.

The fossil boiler is removed and replaced by a single-pressure HRSG, which utilizes the exhaust gas of an advanced F class gas turbine (e.g., GE's 7FA.05) and a duct burner (DB) in front of the HP evaporator section for steam generation, superheating and reheating.

The final, repowered plant arrangement is shown in Figure 17.3. Note that the HRSG contains an economizer section for water heating. The hot water (450°F) at 650 psia is utilized to heat the gas turbine fuel gas (assumed to be 100% methane) to 410°F for increased efficiency (*performance* fuel gas heating).

The placement of the DB upstream of the evaporator aims to *avoid* the added cost and complexity of oversized *attemperation* (desuperheating) equipment (piping, valves, spray nozzles, etc.). The DB fuel gas flow (same as the gas turbine fuel) is controlled to achieve the desired HP steam generation at various site ambient conditions. Another benefit of the DB placement and minimal

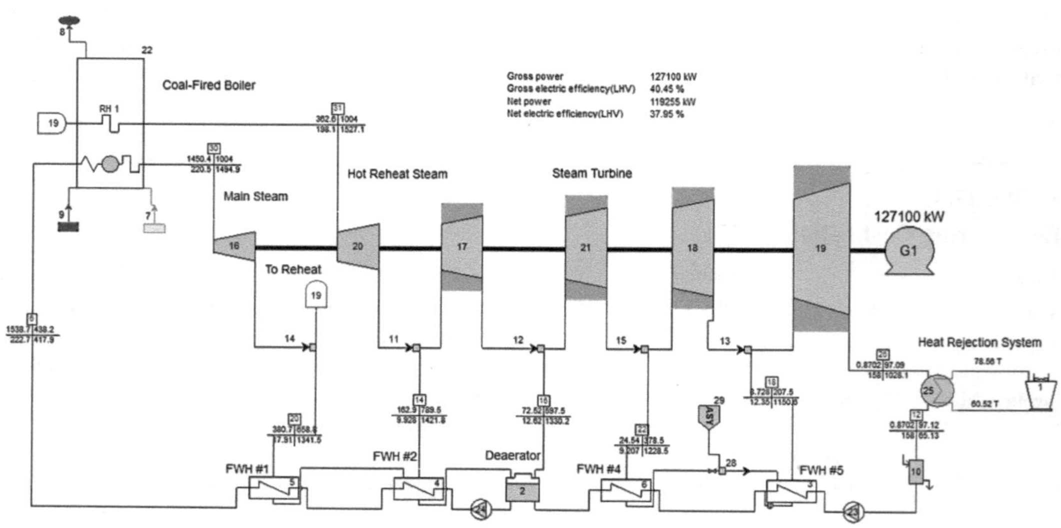

FIGURE 17.2 Arrangement of the example coal-fired power plant to be retired (THERMOFLEX® model).

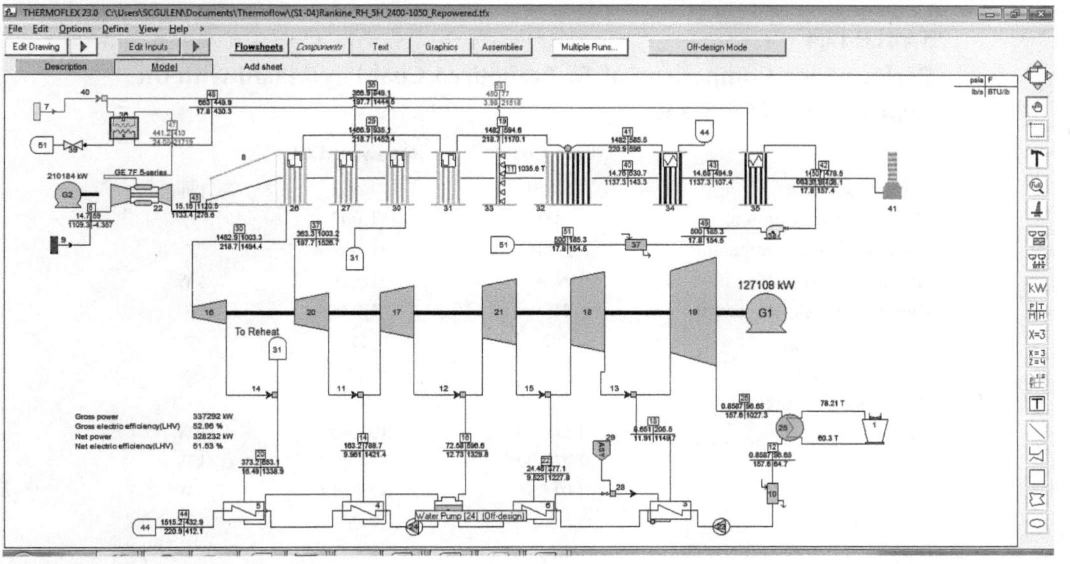

FIGURE 17.3 Repowered plant arrangement (THERMOFLEX® model).

firing (less than 1,100°F at the DB exit) is to limit the size and cost of the selective catalytic reduction (SCR) for NOx and CO abatement. The SCR is not included in the calculations below. It can add around $2 million to the equipment cost.

The performance of the original Rankine cycle steam turbine plant and the GTCC plant after repowering (at ISO conditions) is summarized in Table 17.4.

The key takeaways from Table 17.4 are as follows:

- 175% increase in net plant output (~209 MW)
- 25% reduction in net plant heat rate (~13.5 percentage point increase in net plant efficiency)
- 50+% reduction in *specific* CO_2 emissions (in lb/MWh)
- Elimination of all toxic stack emissions such as mercury, arsenic and acid gases (*not* obvious from the table but a fact given the difference between the two fuels, i.e., coal and natural gas and combustion systems)
- The marginal (incremental) efficiency is ~65%!

The last item is the ratio of incremental plant net output (about 209 MW) to incremental plant heat consumption (about 323 kW in LHV). In other words, *the repowered plant generates the additional 209 MW of electricity (net) at an efficiency significantly above that for an advanced H class GTCC.*

The key question is the cost of the significant benefits provided by the repowered GTCC power plant. A rough cost estimate is given in Table 17.5 (from Thermoflow, Inc.'s PEACE® software). The equipment cost (total $78 million) is about $375 per additional kilowatt. The total installed cost (TIC) is estimated at about $645 per additional kilowatt (about $135 million) *including* the demolition and removal of the existing boiler. Dismantling of the existing coal-fired boiler, removal of debris (minus the value of saleable scrap materials), is estimated at $40/kWth boiler rating.

In order to put the cost and performance of the proposed repowering in perspective, consider a new GTCC power plant. According to *Gas Turbine World 2013 GTW Handbook*, the closest candidate

TABLE 17.4
Performance Comparison of To-Be-Retired Coal-Fired Plant with the Repowered GTCC Plant

	Base Coal	Repowered CC	
Throttle pressure	1,450	1,453	psia
Main steam temperature	1,004	1,003	°F
Hot reheat steam temperature	1,004	1,003	°F
Gas turbine (7FA.05)	N/A	210,184	kW
Duct burner exit temperature	N/A	1,035	°F
Stack temperature	280	479	°F
Throttle steam flow	220.5	218.7	pps
Condenser pressure	1.77	1.75	in. Hg
Steam turbine	127,100	127,108	kW
Gross output	127,100	337,292	kW
Net output	119,255	328,232	kW
Auxiliary power	7,845	9,060	kW
	6.6%	2.8%	
CO_2 emissions	64.52	77.48	lb/s
	1,948	850	lb/MWh
Efficiency	37.95	51.53	%
Heat rate	8,991	6,622	Btu/kWh
GT heat consumption	N/A	546,673	kW
Duct burner HC	N/A	90,355	kW
Total heat consumption	314,232	637,028	kW
Δ output		208,977	kW
Δ heat consumption		322,796	kW
Marginal efficiency		64.74%	
Marginal heat rate		5,271	Btu/kWh

(on an *incremental performance* basis) is a 1 × 1 plant (MPCP1) with Mitsubishi's M701DA gas turbine [9]:

- 212.5 MW net output (ISO baseload – 70.4 MW from the STG)
- 51.4% net plant efficiency
- Budget plant price $156.4 million ($736/kW).

According to Ref. [9], [they] *estimate the* ***budget price*** *as the EPC contractor's* "out-of-pocket" *cost, plus a reasonable allowance for profit, to supply and build a combined cycle plant on an owner's site.*

It is quite difficult to establish an "apples-to-apples" cost comparison basis with such scant information. As a rough first cut, excluding 15% from the repowered plant cost estimate for owner's costs and Allowance For Funds Used During Construction (AFUDC), which are excluded from *GTW Handbook's* budget price, the budget price number for the repowered plant is about $123 million.

Similarly, the closest candidate *on a total performance basis* is a 1 × 1 plant with General Electric's 7FA.05 gas turbine [9]:

- 323.0 MW net output (ISO baseload – 112.0 MW from the STG)
- 58.2% net plant efficiency
- Budget plant price $224.8 million ($696/kW).

TABLE 17.5
Cost Estimate for the Repowered GTCC Power Plant

Steam turbine	$0	Existing
Condenser	$0	Existing
Cooling tower	$0	Existing
Switchyard	$0	Existing
Feed pumps	$0	Existing
FWHRs	$0	Existing
HRSG	$23,184,000	PEACE
GT plus generator	$51,770,000	GTW 2013 Hbk
Fuel heater circ pump	$150,000	PEACE
Fuel heater	$500,000	Guesstimate
Piping and valves	$2,500,000	Guesstimate
Total equipment	**$78,104,000**	
STG refurbishment	$0	Guesstimate
Dismantling of boiler	$12,569,281	
HRSG installation	$10,488,000	PEACE
GTG installation	$10,000,000	Guesstimate
Indirects	$23,431,200	30% added to equipment
TIC	**$134,592,481**	

Source: From PEACE®.

Note that, based on the same scope and indirect cost assumptions, budget plant price calculated with PEACE® was $255.2 million.

Thus, the cost-performance comparison between the proposed repowered GTCC plant and a new GTCC is summarized in Tables 17.6–17.9 below on incremental and total performance basis. The assumptions used in LCOE calculations are as follows:

- $4 natural gas fuel (per million Btus in higher heating value or HHV)
- 6,132 h/year at full load (70% capacity factor)
- Capital charge factor of 12%
- Levelization factor of 1.25
- O&M costs 10% of LCOE for the new GTCC, 15% of LCOE for the repowered GTCC.

Note that the *GTW Handbook* performance numbers for the new GTCC plants are ISO baseload values with 1.2 in. Hg back pressure and once-through (open loop) water-cooled condenser. The auxiliary power consumption for GTW ratings is about 1.6% of the gross output (see Section 3.3.4), whereas the auxiliary power consumption of the repowered GTCC of the example case herein is 2.8% (see Table 17.4). The difference is mainly driven by the heat rejection system's parasitic power consumption. Accounting for this discrepancy impacts the comparison between the options (on a total performance basis) as shown in Table 17.8. The impact of not demolishing the existing plant (simply removing the old equipment) on the repowering option is demonstrated in Table 17.9.

According to a recent article, available plant data identified several steam turbine units nationwide rated between 100 and 200 MW and installed between the mid-1960s and mid 1970s [4]. One such station is equipped with two units rated at 110 MW with gas-fired boilers and reheat steam turbines (208.3 lb/s throttle steam flow and 1,900 psig/1,000°F cycle) at a gross turbine heat rate of 9,800 Btu/kWh (about 35% efficiency).

TABLE 17.6
Comparison on *Incremental* Performance Basis

	Base	MPCP1	Repowered CC
Net output, MW	119.3	212.5	209.0
Net efficiency	37.95%	51.40%	64.74%
Budget price, MM$	N/A	$177.5	$122.9
$/kW	N/A	$835.4	$588.0
LCOE, $/MWh	N/A	$59.06	$47.92

TABLE 17.7
Comparison on *Total* Performance Basis

	Base	7F.05	Repowered CC
Net output, MW	119.3	323.0	328.2
Net efficiency	37.95%	58.20%	51.53%
Budget price, MM$	N/A	$224.8	$122.9
$/kW	N/A	$696.0	$374.4
LCOE, $/MWh	N/A	$51.25	$51.82

TABLE 17.8
Comparison on *Total* Performance Basis (New GTCC Performance Adjusted for Auxiliary Power)

	Base	7F.05	Repowered CC
Net output, MW	119.3	319.2	328.2
Net efficiency	37.95%	57.51%	51.53%
Budget price, MM$	N/A	$224.8	$122.9
$/kW	N/A	$704.3	$374.4
LCOE, $/MWh	N/A	$51.87	$51.82

TABLE 17.9
Comparison on *Total* Performance Basis (New GTCC Performance Adjusted for Auxiliary Power) with No Demolition (i.e., the Old Boiler Building Is Left Standing – with Equipment Removed)

	Base	7F.05	Repowered CC
Net output, MW	119.3	319.2	328.2
Net efficiency	37.95%	57.51%	51.53%
Budget price, MM$	N/A	$224.8	$110.3
$/kW	N/A	$704.3	$336.1
LCOE, $/MWh	N/A	$51.87	$50.93

No credit is given for equipment and scrap material salvage.

This might be a more representative heat rate for a retirement candidate unit than that in the example above, i.e., 8,436 Btu/kWh (~40.5%). Applying 10% heat rate degradation to the example steam turbine and its condenser above, the repowered GTCC case is recalculated. The HRSG and the DB are sized to achieve 127.1 MW STG output with the degraded ST power island. The nominal baseload case is evaluated at 115.8 MW STG output. The cost-performance comparison is shown in Tables 17.10–17.12.

TABLE 17.10
Comparison on *Incremental* Performance Basis

	Base	MPCP1	Repowered CC
Net output, MW	108.1	212.5	219.5
Net efficiency	34.80%	51.40%	63.26%
Budget price, MM$	N/A	$177.5	$138.9
$/kW	N/A	$835.4	$632.8
LCOE, $/MWh	N/A	$59.06	$49.76

TABLE 17.11
Comparison on *Total* Performance Basis

	Base	7F.05	Repowered CC
Net output, MW	108.1	319.2	327.6
Net efficiency	34.80%	57.51%	49.81%
Budget price, MM$	N/A	$224.8	$138.9
$/kW	N/A	$704.3	$424.0
LCOE, $/MWh	N/A	$51.87	$54.45

TABLE 17.12
Comparison on *Total* Performance; New GTCC Budget Price from PEACE, No Demolition of Existing Plant Buildings

	Base	7F.05	Repowered CC
Net output, MW	108.1	319.2	327.6
Net efficiency	34.80%	57.51%	49.81%
Budget price, MM$	N/A	$255.2	$126.3
$/kW	N/A	$799.4	$385.7
LCOE, $/MWh	N/A	$53.93	$53.57

17.4 TAKEAWAYS

In this chapter, the feasibility of a simple heat recovery repowering concept has been used to illustrate the key aspects of applying gas turbines to "revitalize" old steam turbine power plants. The heart of the concept is to use the steam turbine, feedwater heating train and heat rejection system of the existing Rankine cycle power plant with *minimal or no changes*. The fossil-fired boiler (it can be an oil, gas or coal-fired unit) is decommissioned and replaced by a natural gas-fired GT/HRSG train.

The gas turbine is preferably an advanced F or H class unit with high efficiency.

The HRSG is a single-pressure unit with DBs upstream of the evaporator section. A separate economizer is included to facilitate performance fuel gas heating to boost the gas turbine efficiency.

The heat balance calculations *unequivocally* show that the performance boost is similar to what is expected from a full-blown heat recovery repowering (i.e., three-pressure reheat HRSG, IP and LP steam admission with steam turbine – and possibly heat rejection system – modifications and upgrades:

- 175% increase in net output
- 25% reduction in net heat rate
- 50+% reduction in CO_2 emissions (lbs/MWh basis)

The feasibility of the concept, just like any repowering concept, hinges around the reduced capital investment and project technology risk. The latter is a known *Achilles' heel* for repowering projects. Thus, the first and foremost requirement is a diligent evaluation of the equipment (age, degradation, requisite repairs, modifications and/or upgrades to ensure continued service with no excessive O&M expenditures) in the fossil-fired plant slated for retirement (and thus a candidate for repowering). This evaluation should be done on a case-by-case basis to ensure that the plant in question is suitable to any repowering concept.

Capital cost estimate, especially at such an early stage and with scant information, is subject to high uncertainty – especially so for "brownfield" projects such as repowering. The reduced scope of the simple repowering concept somewhat alleviates this concern but does not eliminate it. The estimated capital investment for the proposed concept is between $120 and $140 million TIC (depending on whether the existing boiler building is demolished and removed). On a TIC per incremental kilowatt basis, this corresponds to a capital investment of $550–$625/kW. Note that this is based on an incremental output of about 210 MW. (As a comparison, consider that the aforementioned FP&L's Fort Myers project with 960 MW incremental output in 2002 came with a bill of nearly $800/kW in today's dollars – probably more [8].)

A *bona fide* economic evaluation of repowering should be based on a power generation system planning study. Such a study requires detailed simulation of how the power system generating units operate to meet the load demand over a period of time (typically 20 years). Capacity addition decisions are made to meet the required reserve margin or a generation system reliability target. If a repowering decision is implemented, the future generation addition schedule is impacted and may result in future capacity needs. Capacity savings along with the resulting fuel and O&M savings from more efficient operation comprise the repowering benefits and savings [1]. The reader is referred to the book by Marsh for a detailed description of system planning studies [10].

REFERENCES

1. Stoll, H.G., Smith, R.W., Tomlinson, L.O., 1994, Performance and Economic Considerations of Repowering Steam Power Plants, GER-3644D.
2. Stenzel, W.C., Repowering Existing Fossil Steam Plants, SEPRIL, LLC, 180 North Wacker Dr., Chicago, IL, http://soapp.epri.com/papers/Repowering_Fossil_Plants.pdf.
3. www.power-eng.com/articles/2014/02/eia-90-of-coal-capacity-to-retire-by-2016-due-to-mats.html.
4. Cleetus, R., Clemmer, S., Davis, E., Deyette, J., Downing, J., Frenkel, S., November 2012, Ripe for Retirement: The Case for Closing America's Costliest Coal Plants, Union of Concerned Scientists.
5. EPRI Report TR-106908, 1997, Strategic Assessment of Repowering.
6. Kalany, F., 2013, Review of combustion turbine based repowering options of rankine cycle power plants, *Energy Generation Conference*, January 29–31, 2013, Bismarck, ND.
7. Developments to Watch: Combined Cycle with regeneration—Tweaking the Brayton, Rankine Cycles to Take Advantage of New Market Opportunities, Combined Cycle Journal, www.ccj-online.com/3q-2013/developments-to-watch-combined-cycle-with-regeneration/.
8. https://online.platts.com/PPS/P=m&s=1029337384756.1478827&e=1096495472414.2240023308785804128/?artnum=20040qtK7B2I6u15460P2i_1.
9. Gas Turbine World, 2013, *GTW Handbook*, Vol. 30, Pequot Publishing, Inc., ·Guilford, CT.
10. Marsh, W.D., 1980, *Economics of Electric Power Utility Power Generation*, Oxford University Press, New York.

18 Integrated Gasification Combined Cycle

The commercially available technology for syngas production from all forms of hydrocarbon feedstock is *gasification*. Coal gasification goes back to the 1800s when it was used to make "town gas" for local cooking, heating and lighting. Gasification is the substoichiometric reaction of coal with oxygen and steam under high pressure (HP) and temperature to form a gaseous product consisting primarily of carbon monoxide and hydrogen. The gasification process typically takes place at temperatures of 1,200°C–1,400°C (2,200°F–2,600°F) and pressures of 30–60 bar (435–870 psi). The resulting gas contains CO, H_2, CH_4 and other fuel constituents, which can be referred to as the desirable products (i.e., suitable to be used as gas turbine fuels) in addition to the undesirable products such as CO_2, H_2S, NH_3, particulate matter (PM) and neutral products such as N_2 and H_2O vapor.

When the destination of the gaseous product of gasification is a gas turbine combustor, the entire system is referred to as *Integrated Gasification Combined Cycle* or IGCC (because the power block is generally a gas turbine combined cycle, GTCC). For a concise but well-rounded review of IGCC technology with different gasifiers, refer to Khartchenko [1]. For detailed coverage of the fundamental physical and chemical principles and gasification technologies, see Probstein and Hicks [2], Higman and van der Burgt [3] and Rezaiyan and Cheremisinoff [4]. A simplified method for IGCC performance estimation can be found in the study by Gülen and Driscoll [5].

The IGCC power plant has been the most promising clean coal technology (CCT) for the last three decades. Several large-scale commercial IGCC power plants have been in operation for many years to the point that the technology can safely be considered field-proven [3]. Nevertheless, the large capital cost driven by the immense size, complexity and construction time of IGCC power plants prevented the technology from a more widespread acceptance. Even so, IGCC's favorable emission characteristics (vis-à-vis the conventional coal-fired thermal power plants) and especially its amenability to carbon capture and sequestration (CCS) have so far ensured that research and development of the technology has never ceased completely.

Environmental regulations and gas turbine hardware design and life requirements preclude the direct utilization of the raw syngas as turbine fuel. It must be cleaned of particulate and sulfur compounds. Thus, it is first cooled by making steam at a high pressure or via quenching and "scrubbed" with water to remove particulates. Scrubbing also removes some NH_3 and further cools the syngas to about 200°C. (This is the so-called *cold gas cleanup* process. Alternative *hot gas cleanup* processes have been investigated in the past for their potential benefits (lower capital cost and higher efficiency), but they did not overcome their technical difficulties. For the most common variant of gasification technology for IGCC applications, slagging entrained-bed systems, the efficiency benefit was about 1% point.)

If the IGCC power plant is designed for CCS, the next step is the so-called "sour shift" process, where most of the CO in the raw syngas reacts with water vapor in a catalyst-filled reactor to produce H_2 and CO_2 (which is then removed from the predominantly H_2 stream prior to combustion in the gas turbine). This reaction is exothermic and typically generates intermediate pressure (IP) steam, which is exported to the steam turbine for power generation. It can also be used to increase the syngas H_2/CO ratio from less than 1.0 (for most gasifiers) to 2.0 (for subsequent conversion to liquid fuels such as diesel or gasoline) or 3.0 (suitable for methane production which is also referred to as *substitute natural gas* or SNG).

In plants without CCS, a hydrolysis reactor is required to react the carbonyl sulfide (COS) in the syngas to H_2S, which is easier to remove. In plants with CCS, this conversion takes place simultaneously with the shift reaction and no separate hydrolysis reactor is needed.

Either way, the syngas is still too hot so that it is cooled down to near ambient temperature suitable for the cleanup process, which is referred to as *acid gas removal* (AGR). A significant part of the heat is recovered during the cooling process to generate low pressure (LP) or IP steam, which is sent to the power block (parts might be utilized in the process block).

There are a variety of processes for AGR, either commercially available or under development. The most common technique involves absorption into a liquid with chemical (e.g., methyl diethanolamine (MDEA)) or physical (e.g., Selexol™, Rectisol®) solvents. The selection of the optimal AGR technology in the IGCC is controlled by factors such as the required degree of H_2S control and the need for removal of CO_2 (i.e., a plant to be built with or retrofitted to CCS). Detailed cost/performance trade-off is required for the final selection. However, the most likely choice for the high-pressure systems, especially if with CCS, is a physical solvent process. Whatever the chosen technology, the process involves *absorption* (removal of acid gases from the syngas into the liquid absorbent) and *solvent regeneration* (removal of the acid gases from the liquid absorbent). The latter process requires significant energy, which is typically supplied by the LP steam extracted from the power block's bottoming cycle. In the plants with CCS, staged flashing is the technique for regeneration of the CO_2-rich, physical solvent from the CO_2 absorber, which is the second stage of the AGR. A promising but not yet commercially proven alternative technology is gas separation membranes [6]. Carbon capture via amine-based chemical absorption is covered in more detail in Chapter 19.

Syngas is a medium heating value fuel (*medium-Btu* or MBTU are commonly used terms). While it is eminently suitable for gas turbine combustion, it requires modification to the existing gas turbine hardware designed for natural gas, especially in the combustor and either turbine or compressor design. Alternatively, the IGCC plant can then convert the syngas into SNG or methane through a process called *methanation* [2]. The SNG is higher in energy content than syngas and, in fact, is the fuel for which modern gas turbines are designed for. Thus, SNG can be burned in a heavy-duty industrial gas turbine without any modification. The negative impact on efficiency, however, would make conversion to SNG a consideration only if the gasification and gas turbines were long distances from one another.

From a practical gas turbine-centric point of view, the IGCC plant is simply a gas-fired GTCC power plant with a "giant fuel gas skid." (The GTCC power plant and the "fuel skid" are commonly referred to as power and process blocks, respectively; e.g., see Figure 18.1.)

From a pure performance perspective (i.e., generation of net electric power and the associated thermal efficiency or, equivalently, fossil fuel consumption), the process block of an IGCC gas turbine has *four* key attributes:

- Consumption of solid or liquid feedstock (e.g., coal, petcoke, oil shale);
- Consumption of electric power, which is debited to the gross (generator) output of plant prime movers
- Net consumption and/or generation of useful thermal energy, which, in the form of steam, is utilized by the steam turbine of the power plant for electric power generation
- Generation of clean, gaseous fuel (which is referred to as synthesis gas or, in short, syngas).

The process block has three major components, which are bona fide plants in their own right:

- Gasification plant (O_2 or air-blown; includes syngas cooler)
- Gas cleanup plant (GCP)
- Oxygen plant (if the gasifier is O_2-blown) or air separation unit (ASU)

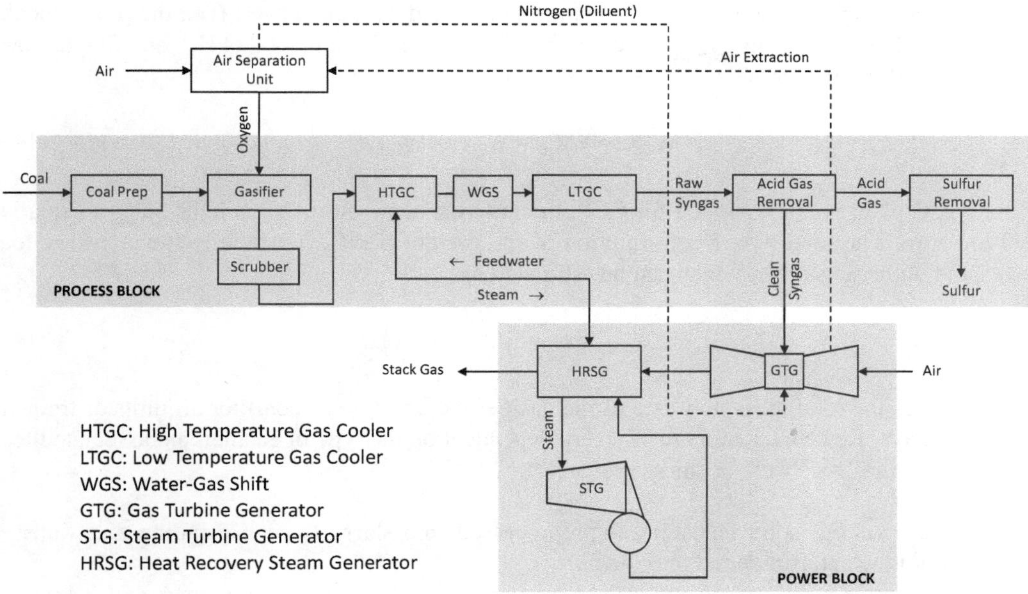

FIGURE 18.1 Block diagram of a typical IGCC power plant.

The gasification plant, as shown in Figure 18.1, has two subsystems:

1. Gasifier
2. High-temperature gas cooling (HTGC)
   a. Radiant syngas cooler (RSC)
   b. Quench

Strictly speaking, the next three subsystems are parts of the GCP:

1. Scrubber
2. CO shift reactor
3. Low-temperature gas cooling (LTGC)

From a conceptual perspective, however, it is more convenient to consider them an integral part of an extended gasification *island*.

The IGCC plant performance is expressed in terms of net plant output and efficiency. Plant net output is found by rolling up the total output of the prime movers and syngas expander (if present) and subtracting from that the total auxiliary power consumption:

$$\dot{W}_{IGCC} = \dot{W}_{GTG} + \dot{W}_{STG} - \dot{W}_{AUX}. \tag{18.1}$$

If there are more than gas turbine generators in the power block, $\dot{W}_{GTG}$ represents their total output. (If present, the syngas expander output can be added to the mix.) Total plant auxiliary power consumption is found by rolling up the individual items, i.e.,

$$\dot{W}_{AUX} = \dot{W}_{GSF} + \dot{W}_{GCP} + \dot{W}_{ASU} + \dot{W}_{CC}. \tag{18.2}$$

Subscripts ASU and CC refer to the air separation unit and combined cycle (i.e., the power block), respectively. The total power consumption of the GCP $\dot{W}_{GCP}$ is the sum total of H_2S and CO_2 removal (capture) auxiliary power and CO_2 compressor power, i.e.,

$$\dot{W}_{GCP} = \dot{W}_{H_2S} + \dot{W}_{CO_2} + \dot{W}_{CO_2C}. \tag{18.3}$$

If the IGCC plant is not equipped for CCS, the last two items on the right-hand side of Equation 18.3 are zero. The total power consumption of the gasifier plant is essentially the auxiliary load associated with the gasifier, which can be estimated as

$$\dot{W}_{GSF} = \alpha_{GSF}\left(43.2 \cdot \dot{M}_{feed}\right), \tag{18.4}$$

where $\dot{M}_{feed}$ is the total mass flow rate of the gasifier feedstock (e.g., coal) for all units or trains in the plant in lb/s. Key contributors to $\dot{W}_{GSF}$ are dependent on the type of gasifier and associated feed system (e.g., "dry" vs. "wet" or slurry systems), i.e.,

- In *wet systems*, solid feedstock is preprocessed into slurry by fine grinding and water addition, which is pumped into the burner.
- In *dry systems*, pulverized and dried feedstock is pressurized in lock hoppers and fed into the gasifier with a transport gas by dense-phase conveying.

Depending on the particular system, individual power consumers include coal grinding and slurry preparation, slurry feed pump, slag crusher and slag handling. A reasonably good value for α_{GSF} is 2.0 kW/STPD,[1] which is adequate for auxiliary power consumption bookkeeping purposes.

IGCC net efficiency is found by dividing the net plant power output by total feedstock consumption (in terms of higher heating value or HHV):

$$\eta_{IGCC} = \frac{\dot{W}_{IGCC}}{\dot{M}_{feed} HHV_{feed}}. \tag{18.5}$$

Contribution of natural gas or another fuel burned in the Claus unit furnace or the coal dryer (if present) should be included in the denominator of Equation 18.5.

The IGCC efficiency defined by Equation 18.5 can vary depending upon the HHV of feedstock (e.g., coal) used in the calculation. One option is to use HHV on *as received* (ar) basis, which includes the moisture and ash content of the feedstock. In almost all practical applications, pure electric power generation or combined heat and power (CHP), no use can be made of the latent heat contained in the water vapor in combustion products. Thus, one can argue that using lower heating value (LHV) (ar) in the denominator of Equation 18.5 presents a more realistic efficiency number (even though the feedstock is purchased on an HHV basis).

Another source of confusion stems from the widely used solid feedstock heating value reporting on a *moisture and ash free* (maf) or *dry and ash free* (daf) basis, which excludes the moisture and ash content of the fuel in per unit mass heating value definition. For good coals with low moisture and ash content, this does not introduce a significant error (i.e., the denominator of the fraction in Equation 18.5 is approximately constant). However, for poor coals such as lignite with very high moisture content, especially for the dry-fed gasification systems, this can be a source of large error and/or confusion. In fact, with certain coal drying systems making use of the vapor released during the coal drying process, LHV-based efficiency definition can be misleading.

---

[1] Standard Ton Per Day.

It is easy to estimate the upper limit of the IGCC efficiency defined by Equation 18.5. If the denominator of the fraction on the right-hand side of Equation 18.5 is total syngas consumption (LHV), it simply represents conventional GTCC efficiency, which is around 60% net LHV. The ratio of syngas to gasifier feedstock (e.g., coal) consumption is the *cold gas efficiency* (CGE, see Section 18.3.2). The ultimate value of CGE that one can expect at the present stage of development is about 80% [3]. Thus, maximum IGCC efficiency, before accounting for process block auxiliary power and net steam consumption, is $0.6 \times 0.8 = 0.48$ or 48% net LHV. In other words, net IGCC efficiency in LHV is highly unlikely to be more than low 40s.

18.1 SYNGAS-FIRED GAS TURBINE

Gas turbines burning syngas in general and hydrogen in particular are not designed from the proverbial "blank sheet" specifically for these particular gaseous fuels. There is simply not a big enough market to justify the huge cost and lengthy development time requisite for such an endeavor. Thus, major OEMs either modify their existing machines for specific applications or, as is the case recently, make their new models "fuel flexible" to the extent possible.

As far as modification of existing products is concerned, there are two main issues:

- Burning syngas with high hydrogen content (30% or more by volume) in premix dry-low-NOx (DLN) combustors
- Compressor–turbine rematch.

Presently, the first issue is resolved by using a conventional *diffusion combustor* with nitrogen (from the ASU) or steam (from the HRSG) injection for NOx control. In lieu of steam injection, a combination of syngas moisturization and nitrogen injection is a better option from overall plant heat integration and efficiency optimization perspective. In near future, DLN combustors capable of burning high-hydrogen gas fuels will be offered by the OEMs. (See the relevant section in **GTFEPG**.) The second issue is the focus of this section and will be explored below in depth.

Let us start with the basics. The pressure ratio (PR) across the axial compressor of a gas turbine is dictated by the *swallowing capacity* of the turbine downstream; specifically, the throat area of the nozzles formed by stage 1 stator vanes (S1N). From basic gas dynamics, the correlation between gas flow parameters (mass flow rate, total temperature and total pressure) at the turbine inlet in the "choked nozzle" condition is

$$\mu_3 = \frac{\dot{m}_3\sqrt{\dfrac{T_{3,t}}{MW_{gas}}}}{p_{3,t}} \approx \text{const.} \tag{18.6}$$

(Subscript t denotes *total* or *stagnation* values.) Thus, according to Equation 18.6, when the gas flow rate at the turbine inlet (state point 3) increases, at the same temperature (i.e., turbine inlet temperature, TIT, controlled by combustor fuel flow rate), total pressure at the turbine inlet will also increase. This pressure "signal" will propagate upstream, amplified by combustor pressure loss (about 4%–6%) and set the compressor discharge pressure (state point 2).

Typically, LHV of coal syngas from a GE-Texaco gasifier is about 20% of that of typical pipeline natural gas (per unit mass basis). Thus, for the same TIT, requisite syngas supply to the combustor is nearly seven times the natural gas supply. Consequently, turbine inlet hot gas flow is about 15% higher. It follows that, via Equation 18.6, one should expect a similar rise in total inlet pressure and a corresponding rise in compressor discharge pressure and PR. The question becomes: can the compressor handle this rise? This is best answered by examining qualitative or generic compressor and turbine maps (see Figure 18.2).

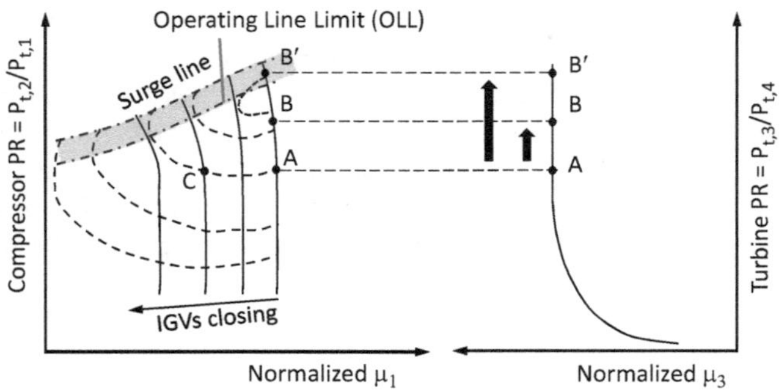

FIGURE 18.2 Generic axial compressor and turbine maps.

The horizontal axis of the compressor map on the left is the compressor flow parameter evaluated at the inlet conditions, i.e.,

$$\mu_1 = \frac{\dot{m}_1\sqrt{\dfrac{T_{1,t}}{MW_{air}}}}{p_{1,t}}$$

normalized via division by the design point conditions. The horizontal axis of the turbine map on the right is the turbine flow parameter in Equation 18.6, also normalized by its design point conditions.

Let us assume that point A on both component maps represents ISO baseload operation with natural gas. When switched to syngas, due to higher fuel and turbine inlet gas flows, the operating point will slide upwards to B. If point B is sufficiently distant from the *operating line limit* (OLL), this represents a new, stable operating condition. Otherwise (see point B′ in Figure 18.2), changes have to be made to turbine controls and/or hardware. From a *controls* perspective, there are three common approaches in this regard:

- Reduce TIT (i.e., $T_{3,t}$) so that, at constant $\dot{m}_3$, $p_{3,t}$ will also go down in proportion and B′ will slide back to A (or close to it safely below the OLL)
- Close the compressor inlet guide vanes (IGV) so that increase in $\dot{m}_3$ via fuel flow is compensated by decrease in $\dot{m}_1$ (point C in Figure 18.2)
- A combination of the two.

From a *hardware* perspective as well, there are three basic approaches:

- "open up" S1N throat area via staggering the stator vanes;
- Extraction of air from the compressor discharge (sent to the ASU or to the gas turbine exhaust)
- Modify the compressor
    - By "tip cut", i.e., shortening the blade heights (e.g., Mitsubishi [7])
    - By an additional stage (e.g., Siemens [8]).

In doing these changes, two mechanical design limits must be considered:

- Turbine exit Mach number limit
- Shaft torque limit.

As a rule of thumb, turbine exit Mach number should not exceed 0.8 (to minimize exit losses) and should not be lower than 0.35. Initial design of a new product usually sets this value around 0.5 to leave room for cold day performance (maximum airflow) and future growth. Turbine flow-path design process aims to set a balance between AN^2 (see Section 4.4.3) between exit Mach number via mass flow rate, exhaust temperature and last-stage blade length.

Shaft torque limit is another result of mechanical design optimization. Obviously, by designing a solid rotor/shaft with a large diameter, one can achieve an arbitrarily large torque capability. From the basic design considerations outlined in Section 4.4.3, one can easily see that this is not a feasible approach. It would result in large pitch-line radii and supersonic blade tip speeds leading to low aerodynamic stage efficiency (i.e., high ψ, shock losses, etc.) The exact value of the shaft torque limit is a question of particular OEM design practice. As an example, for General Electric's 1990s-vintage 7FA, rated at about 162 MWe ISO baseload with natural gas fuel, syngas rating was set to 197 MWe limited by the torque-carrying capability of the shaft. This corresponds to an increase of about 20%, which cannot be generalized as a rule of thumb but should give a rough idea about what to expect.

One other consideration is the increase in heat flux from the hot gas to the turbine vanes and blades via the change in gas composition, specific heat and gas velocity (i.e., higher mass flow rate). Estimating the impact of this increase on hot gas path parts life requires complex CFD modeling and heat transfer calculations. A conservative approach is to reduce the firing temperature by X degrees for each percentage point increase in water vapor content of the combustion gas. (This actually helps with the compressor PR rise but has negative impact on thermal efficiency.) Reduction in stage 1 rotor blade life as a function of moisture in combustion gas is illustrated by the curve in Figure 18.3 [9]. How much firing temperature reduction is requisite can be estimated using the *peak fire severity factor* formula (see Section 13.5.2 for its application), e.g., see Figure 10 in GER-3620N [10]. In other words,

- Read the life fraction ratio for syngas and natural gas (for which the hot gas path parts are designed) operation
- Calculate the maintenance factor (i.e., the ratio of the two)
- Find the firing temperature delta that gives the same factor in the formula in the reference cited above.

Note that heat flux issue can be a significant factor in the presence of steam injection or syngas moisturization for NOx control in the diffusion combustor of the syngas-fired gas turbine (i.e., very high moisture in the combustion gas). A typical syngas moisturization loop, using hot feedwater from the HRSG economizer in a packed column, can increase the H_2O content of the syngas to about 15% by volume. Steam injection at a steam/syngas mass flow ratio of 0.5 can push the exhaust moisture content above 20%(v). Since nitrogen is less effective than steam as a diluent, typical N_2/syngas mass flow ratio is about 1.3. In this case, the increase in plant aux load via N_2 compression should

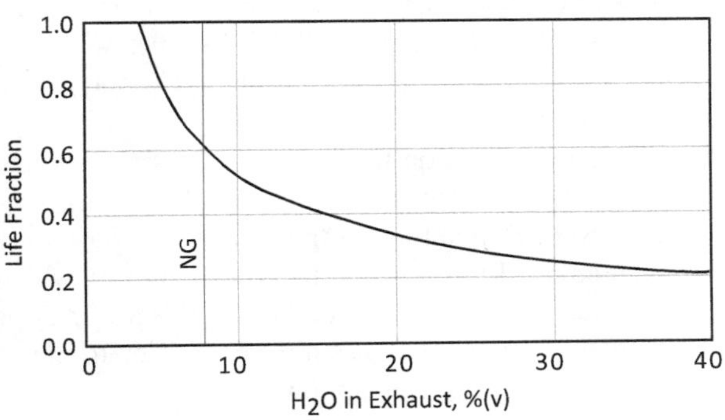

FIGURE 18.3 Impact of combustion gas moisture on hot gas path parts life [9].

be balanced against reduced exhaust moisture (<10%(v)). One interesting option is moisturizing nitrogen as well to reduce the parasitic load of nitrogen compression (at the cost of extra hardware).

"Full or partial integration" with the ASU via air extraction from the compressor discharge and nitrogen injection in the combustor can allow the OEM to use the standard gas turbine without hardware modification. This was indeed the case in two commercial IGCC power plants in Europe, Buggenum and Puertollano, where 100% of ASU air demand was met by air extracted from the gas turbine (roughly 25% of the compressor airflow [11]). This is the "full" integration option. The other extreme is no integration, i.e., no ASU air supply from the gas turbine compressor (this was the case in the IGCC demonstration plant TECO Polk in Florida). Any particular IGCC system can be optimized, on a case-by-case basis, between zero and full integration. For a discussion of the advantages and disadvantages of full GT-ASU integration, the reader is referred to Section 3.2 of Ref. [11]. Apart from the question of IGCC efficiency versus output, i.e., thermal design problem, field experience from the two European plants strongly suggest that, from a reliability, availability and maintenance (RAM) perspective, GT-ASU integration is probably not a feasible path to future IGCC gas turbine designs.

From the foregoing discussion, it should be clear that there is a large number of permutations available to the designer for gas turbine cycle tweaks towards IGCC performance optimization, i.e.,

- Integration with the ASU (none, partial or full)
- Syngas moisturization and heating
- Hardware modification (opening up S1N throat area, compressor modification)
- Maximizing TIT (with better alloys and/or coatings in hot gas path components)
- Combustor selection (diffusion vs. DLN)
- Diluent selection (with diffusion combustors), i.e., nitrogen (from the ASU) or steam (from the bottoming cycle)
- SCR in the HRSG (for NOx and CO emissions reduction).

A General Electric study performed for the US Department of Energy (DOE)/National Energy Technology Laboratory (NETL) in 2005 looked at 18 cases comprising combinations of the options enumerated above. The findings were not surprising; highest IGCC efficiency required an advanced gas turbine with highest possible cycle PR and firing temperature with DLN combustors (SCR in the HRSG) and partial (50%) ASU integration. The most interesting case was the 2 ppmvd NOx case, which was obtained with diffusion combustors with moisturized syngas and nitrogen diluent (no SCR in the HRSG) and lowest firing temperature. The efficiency penalty vis-à-vis the best-efficiency cases was about 2 percentage points.

18.2 BOTTOMING CYCLE

Bottoming cycle performance of the IGCC can be estimated using the same method (based on the second law analysis) outlined at the end of Section 8.3.1. In other words, total exhaust exergy of the syngas-fired gas turbine is converted to net bottoming cycle output using a suitable value of exergetic efficiency.

There are two considerations in translation from a natural gas or methane-fired application to a syngas-fired application. The exhaust exergy formula in the Appendix, i.e., Equation A.4, can be used for syngas application as well. Depending on the syngas composition, the error is –1.5% (no carbon capture in the gas cleanup block) to +2% (with carbon capture), which is not worth developing a more complex transfer function calculation and, if desired, a property package such as *JANAF Tables* can be used with the known composition. A simple correction to Equation A.4 for known gas turbine exhaust gas moisture fraction, y_{H_2O}, is as follows:

$$a_{exh} = \left(0.9536 + 0.526 y_{H_2O}\right) 0.001628 \cdot T_{exh}^{1.60877}. \quad (18.7)$$

Typical IGCC bottoming cycles for gasifiers with RSGs are two-pressure reheat (2PRH) systems with HRSG stack temperatures of about 250°F. The reason for only two pressure levels is the large HP economizing and superheating duties in the HRSG associated with the RSG. This leaves relatively little exhaust gas energy for IP steam production (maybe for a few thousand pounds per hour). Economically, it makes sense not to build an IP section (superheater and evaporator) for that meager IP steam production opportunity. The high exhaust gas stack temperature is a direct result of the high-temperature condensate (~230°F) coming from the plant deaerator (to prevent sulfuric acid condensation on economizer tubes). Therefore, application of a reduction of 3–4 percentage points to the exergetic efficiency (Equation C.1 in the Appendix) is proposed to establish a suitable basis for IGCC applications. This should return a value of ~68% at the gas turbine exhaust temperatures typical for syngas-fired units.

In an IGCC power plant, the bottoming cycle has significant (net) exergy input in addition to the gas turbine exhaust flow. In general, this additional exergy source is in the form of net steam import from the process block (i.e., the gasifier radiant and/or convective syngas coolers). By far, the largest contributor is the HP steam generated in the RSG of the gasifier, followed by IP/LP steam generation in the LTGC section. The additional steam turbine power output derived from the energy imported from the process block via LP, IP and/or HP steam produced in various syngas coolers (designated by the subscript i in the formulas below) can be estimated as follows:

$$\dot{W}_{ST,i} = \varepsilon'\dot{Q}_i\left(1 - \frac{T_0}{\overline{T}_{stm,i}}\right). \tag{18.8}$$

(For the value of ε′, see Table 18.1.) Net energy imported from the process block is given by

$$\dot{Q}_i = \dot{m}_{stm,i}\left(h_{stm,i} - h_{fw,i}\right), \tag{18.9}$$

where h_{stm} is the enthalpy of the imported (usually saturated) steam, h_{fw} is the enthalpy of the boiler feedwater exported to the process block (from the HRSG economizer) and $\dot{m}_{stm}$ is the amount of steam generated in the syngas coolers. The mean-effective steam temperature is evaluated from the HRSG feedwater supply and steam return (import) conditions as

$$\overline{T}_{stm,i} = \frac{h_{stm,i} - h_{fw,i}}{s_{stm,i} - s_{fw,i}}. \tag{18.10}$$

However, for most practical purposes, it is sufficiently accurate to assume that $\overline{T}_{stm,i}$ is equal to the saturated steam temperature at the HP evaporator pressure.

Thus, the total steam turbine power output can be estimated as

$$\dot{W}_{STG} = \frac{\varepsilon_{BC}\dot{m}_{exh}a_{exh}}{1 - \lambda_{fp}} \pm \sum_i \varepsilon'\dot{Q}_i\left(1 - \frac{T_0}{\overline{T}_{stm,i}}\right). \tag{18.11}$$

TABLE 18.1
Gasifier Heat Recovery Parameters

		θ	ε′	Pressure (psia)
GSF	HR	10%–13%	~90%	1,500–2,000
	QU	5%–7%		
GC	HR	1%–3%	~70%	100–300
	QU	3%–5%		

HR, heat recovery in radiant and/or convective syngas coolers; QU, quench-cooled gasifier.

This generic formulation assumes that there are several different net steam import (usually only two) streams from various syngas coolers at different pressures. Note that the "minus" sign is for the net steam export from the power block bottoming cycle. One example of a significant exergy export from the bottoming cycle is gas turbine diluent steam injection. Typically, hot reheat or IP steam at 23 bar (absolute) and 345°C is used as a diluent in the gas turbine combustor for NOx abatement. As can be seen from the above formula, this is detrimental to the IGCC performance due to the lost steam turbine output. While less steam is needed than N_2 for the same level of NOx reduction (about one-third of N_2 on a mass flow basis), in general, utilizing N_2 as a diluent is more advantageous in IGCC systems.

18.3 GASIFICATION

18.3.1 Gasifier Types

Gasifiers are classified based on their reactor types (i.e., moving or fluidized bed, entrained flow), gasifying medium (e.g., steam and air, steam and oxygen) and reaction temperatures (i.e., high-temperature "slagging" or low-temperature "dry ash"). Key characteristics of existing gasification technologies are given in Refs. [2–4]. For a comparison of oxygen and air-blown gasification systems with a gas turbine perspective, the paper by Parulekar [12] is recommended. The simplified model developed herein is exclusively for electric power generation. As such, due to its dominance in the existing plants, only the entrained flow technology is considered.

From a purely performance impact perspective, differentiating features of major entrained flow gasification systems are type of coal feed, syngas cooling and reactant, which determine the following key model parameters:

- CGE
- Gasification sensible heat recovery
- ASU power consumption, which is a strong function of the degree of integration with the gas turbine.

The composition and heating value of the clean syngas are of prime importance to the gas turbine combustor and turbine design. Typical values are cited in the literature [2,4]. The H_2/CO ratio is normally around 0.5 for dry-fed gasifiers and around 1.0 for the slurry-fed gasifiers (about 0.3 for the air-blown gasifier). In terms of the raw syngas at the scrubber inlet, syngas from the dry-fed gasifier has very low moisture, e.g., less than 5% (vol) vis-à-vis slurry-fed gasifier raw syngas with 20+%(v). This is a disadvantage for the dry-fed systems with CCS due to the large steam need (about 2 for minimum steam/carbon ratio) for the *water-gas shift* (WGS) reaction, which is diverted from the power block at the expense of steam turbine power output.

18.3.2 Cold Gas Efficiency

The key gasifier metric of interest is the CGE. This is the ratio of the heating value of the clean syngas consumed by the gas turbine to the heating value of the feedstock consumed by the gasifier (i.e., IGCC plant heat consumption). In terms of absolute values, it is difficult to make a reliable assessment. Gasifier CGE is a strong function of the coal type (HHV, sulfur content, etc.), feedstock delivery system, gasifier pressure and other design features such as syngas or CO_2 recycle. For the entrained flow gasifiers with high gasification temperatures (~1,100°C or higher), the energy supplied to the gasifier (mostly feedstock enthalpy, i.e., heating value) is roughly distributed as follows:

- Raw syngas heating value (~85%)
- Sensible heat in the raw syngas (12%–13%)
- Heat losses and slag (2%–3%).

In gasifiers with RSCs, most of the sensible heat in the raw syngas is recovered during the gas cooling process by means of steam production. (Quench-cooled gasifiers allow only partial recovery.) Before being burned in the gas turbine combustor, raw syngas is cleaned in a separate "chemical plant" (AGR), where sulfur is removed. The clean syngas at the end of the process retains approximately 75% of the HHV of the feedstock (approximately 72% on LHV basis). The basic energy balance described above leads to the definition of CGE, i.e.,

$$\text{CGE} = \vartheta \frac{(\dot{M} \cdot \text{LHV})_{\text{SG,Clean}}}{(\dot{M} \cdot \text{HHV})_{\text{feed}}}, \tag{18.12}$$

where ϑ is the HHV/LHV ratio of the feedstock. Thus, for a known gas turbine and gasifier (specified by its CGE), gasifier feedstock mass flow rate can be calculated as

$$\dot{M}_{\text{feed}} = \vartheta \frac{(\dot{M} \cdot \text{LHV})_{\text{SG,Clean}}}{\text{CGE} \cdot \text{HHV}_{\text{feed}}}. \tag{18.13}$$

Note that there is no universal agreement on the definition of CGE. Ideally, CGE should be defined with two key assumptions:

- The energy basis for syngas and feedstock is LHV
- The numerator of the efficiency formula in Equation 18.12 is clean syngas to the gas turbine.

The rationale for these assumptions is simple: Clean syngas flow to the gas turbine combustor and its LHV are readily available parameters, which are critical to gas turbine performance calculation. (Note that the gas turbine performance is the anchor for the IGCC performance calculation.)

In the literature, different CGE definitions are available (e.g., based on syngas and feedstock HHV). In one commonly used definition, the numerator of the formula is the raw syngas at the exit of the scrubber or syngas coolers, i.e.,

$$\text{CGE}' = \frac{(\dot{M} \cdot \text{LHV})_{\text{SG,Raw}}}{(\dot{M} \cdot \text{HHV})_{\text{feed}}}. \tag{18.14}$$

Comparing the two CGE definitions, Equations 18.12 and 18.14, one can write

$$\frac{\text{CGE}'}{\text{CGE}} = \nu \frac{\vartheta'}{\vartheta}, \tag{18.15}$$

where ϑ' is the HHV-to-LHV ratio of the clean syngas and ν is the ratio of the total HHV energy content of the raw syngas stream to the clean syngas stream. Typically, ν is 1.07 with no carbon capture (in the gas cleanup block) and 1.11 with 90% carbon capture. The energy content ratio ν depends on different factors such as

- Gasifier feedstock's sulfur content (i.e., raw syngas H_2S content)
- Gas cleanup process and effectiveness (i.e., with or without carbon capture)
- Whether the clean syngas is moisturized or not.

These are the factors that impact the gas flow rate and heat content (note that H_2S has an LHV of 15,235 kJ/kg). In general, a good value for ϑ' in non-capture and capture systems is 1.02 and 1.10, respectively. Thus, e.g., if a gasifier is quoted with 80% CGE′ with a bituminous coal feedstock (ϑ is 1.05), the corresponding CGE to be used in the current model is:

- $80\% \times 1.05/(1.02 \times 1.07) = 77\%$ for a system with no carbon capture
- $80\% \times 1.05/(1.10 \times 1.11) = 69\%$ for a system with 90% carbon capture.

Thus, a reduction in CGE from its base value is to be expected when CCS is introduced. The actual value is dependent on the specific gasification and gas cleanup system (typically, 5–9 percentage points).

18.3.3 Gasifier Heat Recovery

In non-quench-type gasifiers, the bulk of the heat recovery via HP steam production takes place in the RSC, which can be expressed as a fraction of the gasifier feedstock energy (LHV) as

$$\dot{Q}_{GSF} = \theta_{GSF}(\dot{M} \cdot LHV)_{feed}. \tag{18.16}$$

This steam is utilized in the bottoming cycle of the combined cycle power plant to generate electric power. The power contribution can be estimated via Equation 18.8 with $\dot{Q}_{GSF}$ and $\overline{T}_{stm}$, which is evaluated from the HRSG feedwater supply and RSC steam return (usually saturated at 115–140 barg) conditions using Equation 18.10. Feedwater conditions (if not known) can be estimated using a 15 bar adder (to saturated return steam pressure) and 20° subcool.

Depending on the gasification and gas cooling technology, a second (but much smaller) contribution to the steam turbine power output can be estimated from the LP (or IP) steam produced in LTGC heat exchangers. The net heat import and the steam turbine power contribution can be evaluated in a manner similar to that for the gasifier via Equations 18.16 and 18.8 by replacing θ_{GSF} with θ_{GC}. Typical values for non-capture systems are provided in Table 18.1.

The exact amount and quality of steam production via syngas cooling in multiple locations (e.g., gasifier, LTGC, gas shift reactor) are subject to system-level optimization (e.g., cost-performance trade-off), heat exchanger design considerations such as fouling and available power block hardware to make use of generated steam. (Note that heat import/export via condensate or HRSG feedwater heating, heat rejection to plant cooling water, etc. are ignored due to their low exergy content.)

Ideally, the values of θ_{GSF} with θ_{GC} should be determined from the available information on the particular gasification technology (e.g., technical papers, reports, published articles or a detailed system model). Note that there are two other uses for the raw gas sensible heat: clean syngas fuel heating and moisturization. (A third option is diluent nitrogen moisturization.) For suitably high gasifier pressures (e.g., 60–70 bar or higher), a syngas expander can be considered in lieu of heat recovery. The numbers in Table 18.1 are commensurate with HTGC and LTGC heat recovery for dry syngas fuel at 175°C–200°C at the gas turbine fuel skid inlet with no syngas expander.

Fuel or diluent nitrogen (if available) moisturization is an effective use of low-grade waste heat for improved system efficiency and reduced combustor flame temperature to control NOx emissions [13]. Adjustments to the heat recovery parameters listed in Table 18.1 can be made to account for the syngas expander, fuel moisturization and/or fuel heating to higher temperatures.

Steam (usually HP) is supplied to the WGS reactor in carbon capture-equipped systems. Especially dry-fed gasifiers require significant steam import from the power block because the raw syngas has insufficient H_2O vis-à-vis slurry-fed gasifiers. The exothermic shift reaction transfers the fuel heating value from CO to H_2 and converts the carbon from CO to CO_2. Heat generated during the exothermic shift reaction is recovered for IP or LP steam generation to be utilized in the steam turbine. The net steam import resulting from these two mechanisms should be debited to the steam turbine power generation in a manner similar to that for gas turbine diluent steam injection.

18.3.4 Air Separation Unit

Cryogenic air separation is the commonly used commercial technology for the oxygen plant of the O_2-blown gasification systems. The technology has high capacity and reliability in producing oxygen at purities exceeding 99.5%. For IGCC performance calculations, the key ASU characteristic is the parasitic power consumption associated with the cryogenic distillation process, which is primarily gas compression work and can be estimated as follows:

$$\dot{W}_{ASU} = \alpha_{ASU}\left(3{,}600\xi\dot{M}_{feed}\right) \tag{18.17}$$

$$\xi = \frac{\dot{M}_{O_2}}{\dot{M}_{feed}}, \tag{18.18}$$

where ξ is the ratio of the O_2 flow rate to the gasifier feedstock flow rate (total of all trains) and α_{ASU} is the ASU power consumption per unit O_2 flow rate.

The value of ξ is a feature of the gasifier technology and the particular feedstock. In general, for the same type of coal, slurry-fed gasifiers require more O_2 than the dry-fed ones since more heat is needed to vaporize all the water in the slurry. Typical values of ξ range between 0.7 and 1.2; for quick estimations, a good default value is 0.8 for dry-fed and 1.0 for slurry-fed systems. (This parameter is subject to confusion and errors due to the definition of the oxygen and feedstock flows. Herein, the assumption is that

- O_2 from the ASU is 95% pure
- Feedstock is on as received basis.

When calibrating the model to published data, care must be given to the definitions used by respective authors.)

ASU power consumption is the sum total of the power consumption of three compressors:

1. Main air compressor (MAC)
2. Oxygen compressor
3. Diluent nitrogen compressor.

In general, ASU power consumption is a function of the ASU cold box pressure, gasifier pressure (to which the O_2 is compressed), gas turbine PR, the diluent N_2 consumption of the gas turbine (for the diluent N_2 compressor) and the gas turbine air extraction (which will reduce the power consumption of the MAC).

The ASU is the only process block system that has significant interaction with the gas turbine via air extraction and diluent nitrogen streams and can significantly impact the gas turbine and plant performance. Adjustments to the ASU power can be made for specified air extraction and diluent injection by simple ratios. A reasonable value for α_{ASU} is 250 kW/kpph of O_2 (~2 MJ/kg) with no gas turbine air extraction (i.e., all ASU air is supplied by the MAC) and N_2 sent to the gas turbine as a diluent. This assumes a reasonably state-of-the-art compressor with 88% polytropic efficiency, 95% mechanical/electric efficiency and an F class gas turbine with nominal compressor PR of 17:1–18:1. The other important consideration that goes into this number is related to the individual ASU compressor PRs, which are functions of the ASU pressure, N_2 and O_2 pressures at the inlet to their respective compressors, gasifier pressure and gas turbine compressor PR. The number cited above (i.e., 250 kW/kpph = 2 MJ/kg) is based on the following assumptions:

- ASU pressure of 5 bar (75 psia)
- Gasifier pressure of 45 bar (650 psia)
- N_2 and O_2 pressures equal to 90% of ASU pressure
- 95% O_2 purity
- All N_2 injected to the gas turbine combustor.

Deviations from this basis will strongly impact α_{ASU} and a range of 150–300 is expected. Oxygen purity is a strong driver, especially above 98% (e.g., see Rubin et al. [14] for a chart, which can be used for adjustment). Adjustments for gas turbine air extraction (replacing air supplied by MAC) and diluent injection can be made via basic rules of thumb resulting in the following formula:

$$\alpha_{ASU} = [250 - 450AX - 35(2.9 - DNO)]\left(\frac{p_{GSF}}{650}\right)^{\frac{1}{3}}, \tag{18.19}$$

where AX is air extraction as a fraction of gas turbine inlet airflow and DNO is diluent nitrogen-to-oxygen ratio.

Equation 18.19 contains a correction for the gasifier pressures different than the base value 45 bar (650 psia). For example, for a 1,200 psi gasifier (most likely a quench system), this would add ~20% to the base estimate. Obviously, diluent N_2 injection is limited by the amount of N_2 available from the ASU, which is approximately

$$\dot{M}_{N_2} \approx 3.3\xi\dot{M}_{feed}. \tag{18.20}$$

Typically, for systems without capture, 100% of nitrogen from the ASU is about 1.2–1.4 in terms of diluent-to-fuel ratio, which corresponds to 2.6–3.1 in DNO. Air extraction cannot exceed about 15%–20% of the gas turbine compressor airflow. (For O_2-blown gasification systems, 10% air extraction would replace about 40% of the MAC airflow.) Previous studies and past experience have shown that the best GT-ASU integration strategy is a combination of air extraction and injection of all N_2 from the ASU [15]. This ensures the lowest NOx emissions with maximum possible turbine output (to the extent hardware limits are not exceeded).

IGCC plants with air-blown gasifiers also include a small ASU to supply pressurized nitrogen, which is used for coal transport. Oxygen generated by the ASU is mixed with air extracted from the gas turbine (100% of the oxidant) to the gasifier [12]. Power consumption is about 15% of the ASU for the same size of oxygen-blown gasifier per Equation 18.19.

18.3.5 Syngas Cleanup

From an overall IGCC plant performance perspective, the gas cleanup plant (GCP) is a consumer of electric power (i.e., pumps and compressors) and heat energy (i.e., steam from the power block and/or various syngas coolers in the process block). In that sense, the following subsystems are of importance:

1. Acid gas removal (AGR)
   a. H_2S removal
   b. CO_2 removal or capture (with CCS)
2. CO_2 compression (with CCS)
3. Claus plant (sulfur recovery)
4. Tail gas treatment unit (TGTU).

As far as H_2S removal is concerned, the key performance metric is pumping power consumed per unit flow of raw or clean syngas (or acid gas removed). The heat energy is usually small (68 MJ/kmol of acid gas removed for estimating purposes [2]) and of low exergy, i.e., the equivalent of a few megawatts of steam turbine power output. While it is relatively easy to keep track of this steam turbine power debit due to AGR regenerator steam load, ignoring it does not introduce a big error into the final roll-up. (The values for θ_{GC} listed in Table 18.1 are *net* of the AGR steam load.) However, heat energy (i.e., steam) consumption can be significant for the CCS systems and should be tracked separately.

Data from studies indicate that H_2S removal auxiliary power consumption can be estimated as

$$\dot{W}_{H_2S} = \alpha_{H_2S}\left(3.6\dot{M}_{SG,Clean}\right) \tag{18.21}$$

$$\alpha_{H_2S}[kW/kpph] = 4.5 + 1.5(S\text{–}3\%). \tag{18.22}$$

A reasonably good value for α_{H_2S} (Selexol-based system) is 4.0 kW/kpph (~32 kJ/kg). This is for a typical US bituminous coal from Illinois or Appalachia with sulfur content of about S = 3%(w).

18.3.6 Syngas Expander

Gasifier pressures are typically high (30–60 bar) and thus present an opportunity to generate additional power via expansion of syngas to pressures commensurate with the gas turbine fuel skid inlet. The latter is typically about 50% higher than the combustor inlet, e.g., around 25 bar for most existing F class gas turbines. Whether a syngas expander is present or not is subject to economic optimization due to the design problems associated with gland seals, leakage at valves and flanges and gas path component materials commensurate with high pressures and temperatures. In general, pump-based slurry-fed systems are more amenable to higher gasification pressures requisite for feasible expander designs than dry-fed systems with lock hoppers and transport gas conveyor systems (i.e., about 62 bar or higher).

For a reasonable design with inlet temperature of 538°C (1,000°F) (uncooled parts), the following relationship can be used:

$$\dot{W}_{SGX} = 100\ln(\pi_{SGX})\dot{M}_{SG,Clean}, \tag{18.23}$$

where π_{SGX} is the syngas expander PR (about 2:1–3:1). For a reasonable estimate, using the gas turbine compressor PR, the following formula can be used:

$$\pi_{SGX} = \left(\frac{p_{GSF}}{p_{Amb}}\right)\frac{1}{2\pi_{GT}}. \tag{18.24}$$

Note that the syngas expander work from Equation 18.23 is no longer available for HTGC heat recovery and θ_{SGX} should be subtracted from θ_{GSF}:

$$\theta_{SGX} = \frac{\dot{W}_{SGX}}{(\dot{M}\cdot LHV)_{Feed}}. \tag{18.25}$$

18.3.7 Syngas Heating and Moisturization

Utilizing the low-grade waste heat to heat the gas turbine fuel is a standard combined cycle feature to improve efficiency. Advanced F class units in natural gas-fired applications typically utilize IP economizer feedwater from the HRSG to heat the fuel gas to about 200°C (about 400°F). In IGCC plants, there is enough waste heat in the HTGC and LTGC sections to accomplish this level of fuel heating. Higher temperatures are possible, both in natural gas-fired combined cycle and IGCC power plants, up to ~540°C (1,000°F) or even higher in the latter. However, studies have shown that a point of diminishing returns is reached at around 315°C, beyond which no significant performance improvement is achieved to warrant the cost and complications associated with materials and design of piping, performance heater and gas turbine fuel skid systems [16].

Fuel gas moisturization is a direct contact heat and mass exchange process (typically in a packed column referred to as fuel gas saturator) utilizing low-level waste heat (HRSG LP economizer feedwater in natural gas-fired combined cycle) to simultaneously heat the gas turbine fuel gas and increase its mass (see Section 8.6.3.3). The added benefit of moisturization is the reduction of combustor flame temperature for lower NOx emissions. The same benefit is equally available to the cold (~38°C) and dry syngas fuel from the AGR in the IGCC power plant. Utilizing the waste heat in the LTGC, moisturizing the clean syngas up to about 20%(v) of H_2O content at about 150°C is possible. This can be accounted for using the following relationship:

$$\theta_{SAT} = 0.18\% + 0.162\mu_{SG}, \quad (18.26)$$

where μ_{SG} is the clean syngas moisture content by volume at the exit of the fuel gas saturator and θ_{SAT} is the net heat required to moisturize the fuel gas as a fraction of the gasifier feedstock heat input (LHV). In a study involving fuel gas moisturization, θ_{SAT} from Equation 18.26 should be subtracted from the θ_{GC} listed in Table 18.1 (unless air extracted from the gas turbine can be utilized for the same purpose). Note that part of the moisturization water can be extracted from the HRSG downstream of the LP evaporator, making use of waste heat, which otherwise would be lost through the stack (i.e., no exergy penalty). This is limited by the allowable HRSG stack temperature to prevent sulfuric acid condensation on economizer tubes.

For 260°C moisturized syngas fuel at the gas turbine fuel skid inlet, θ_{SAT} is given by the formula

$$\theta_{SAT} = 1.39\% + 0.183\mu_{SG}. \quad (18.27)$$

This level of fuel heating would require HP feedwater from the HRSG or HP steam from the gasifier syngas coolers (e.g., θ_{SAT} is about 5% for 20% fuel moisture). Moisturization heat duty given by Equation 18.27 should be debited to θ_{GC}. The remainder (if any) and the difference in θ_{SAT} between Equations 18.26 and 18.27 should be debited to θ_{GSF}.

Air extracted from the gas turbine (at more than 370°C) is a high grade heat source. In IGCC plants with ASU-GT integration, it is almost always utilized to heat the diluent N_2 from the ASU. Remaining heat can be utilized to heat the fuel gas or nitrogen saturator circulating water, other feedwater or for partial heating of the fuel gas. Possibilities are myriad and highly dependent on the gas turbine and ASU nitrogen compressor designs, which determine extraction air, diluent N_2 and moisturized syngas flows and temperatures. Based on the gas turbine simulation output, requisite adjustments can be easily made to the heat recovery parameters.

18.3.8 Carbon Capture and Sequestration

Standard IGCC configuration comprises an AGR for H_2S removal. If designed for carbon capture, the AGR system should be designed for selective removal of H_2S and CO_2. (One example is a two-stage Selexol unit.) Prior to the AGR, the raw syngas has to go through a WGS reactor to convert CO in the syngas to CO_2, i.e.,

$$CO + H_2O \rightleftarrows H_2 + CO_2.$$

(The WGS reaction can also be present in non-capture systems to readjust the syngas H_2/CO ratio.) The WGS catalyst in the reactor can also catalyze the COS hydrolysis reaction

$$COS + H_2O \rightleftarrows H_2S + CO_2.$$

Note that some AGR units can remove COS along with H_2S (e.g., Rectisol) but others, especially amine-based systems, require upstream COS hydrolysis for most efficient sulfur removal.

Auxiliary power consumption of the carbon capture block in the gas cleanup section (if one is present) can be estimated as

$$\dot{W}_{CO_2} = \alpha_{CO_2} \dot{M}_{CO_2}. \quad (18.28)$$

A reasonable default value for α_{CO_2} is 80 kW per lb/s of CO_2 flow (175 kJ/kg) for the solvent-based capture unit via chemical absorption (e.g., see Hoffmann et al. [17]). Rigorous models of the actual system, based on chemical or physical solvents, can provide a more accurate number, which can be substituted (if available) but is unlikely to change the final answer by much. Thermal energy consumption by the stripper reboiler is about 1,500 Btu/lb of CO_2 (usually LP steam). No separate accounting is made for the thermal energy. In the absence of specific information, using the values listed in Table 18.1, the following rules of thumb are recommended:

- **Dry-fed gasifier**: Set θ_{GSF} to zero and multiply non-CCS value of θ_{GC} by 3.5.
- **Slurry-fed gasifier**: Set θ_{GSF} to two-thirds of the non-CCS value and multiply non-CCS value of θ_{GC} by 2.5.

Estimation of $\dot{M}_{CO_2}$ (in lb/s) for generic IGCC systems with CCS is based on captured CO_2 flow rate expressed as a mass fraction of the total raw syngas flow rate χ_{CO_2}:

$$\chi_{CO_2} = \frac{\dot{M}_{CO_2}}{\dot{M}_{SG,Raw}}, \tag{18.29}$$

$$\dot{M}_{CO_2} = \chi_{CO_2}\dot{M}_{SG,Raw}, \tag{18.30}$$

$$\sigma_{SG} = \frac{\dot{M}_{SG,Clean}}{\dot{M}_{SG,Raw}}, \tag{18.31}$$

$$\dot{M}_{CO_2} = \frac{\chi_{CO_2}}{\sigma_{SG}}\dot{M}_{SG,Clean}. \tag{18.32}$$

In Equations 18.31 and 18.32, σ_{SG} is clean-to-raw syngas ratio. For quick estimations or generic IGCC system evaluations, a reasonable value for χ_{CO_2} is 0.75. Typical values for σ_{SG} are 0.2 and 0.95 for systems with 90% carbon capture and without (coal sulfur 3% by weight), respectively. Following adjustments are proposed for carbon capture effectiveness, η_{Cap}, (with CCS) and coal sulfur by weight S (no CCS):

$$\sigma_{SG} = 0.2 - 0.8\left(\eta_{Cap} - 90\%\right), \tag{18.33}$$

$$\sigma_{SG} = 0.95 - 1.5(S - 3\%) \tag{18.34}$$

Carbon dioxide compressor power consumption can be estimated as

$$\dot{W}_{CO_2C} = \alpha_{CO_2C}\dot{M}_{CO_2}. \tag{18.35}$$

The value of CO_2 compressor power consumption and consequently α_{CO_2C} is a function of storage pressure (typically ~150 bar), the distance of storage location to the plant (which determines the pipeline and associated pressure loss allowance) and the compression system. For a typical application with intercooled and integrally geared gas compressors, a good value for α_{CO_2C}, representing the state of the art is 105 kW/pps of CO_2 flow (~230 kJ/kg). This assumes 85% stage efficiency for the compressor with three flashing pressure levels. This is adequate for auxiliary power consumption bookkeeping purposes, which is essentially insensitive to gas turbine optimization.

For a single flash level, a good value of α_{CO_2C} is 135 kW/pps (~300 kJ/kg). There are studies showing 20% reduction CO_2 compression power if the CO_2 is liquefied during the compression process and then pumped to the final pressure as a liquid [18]. Even further reduction is possible via refrigeration during liquefaction, but the benefit in parasitic loss saving is offset by the lost steam turbine power output as a result of steam extraction required to drive the refrigeration cycle (Table 18.2).

TABLE 18.2
IGCC Performance Estimation Parameters

Parameter		Range	Default
CGE (Equation 18.15)	CGE	70%–80%	70%
Oxygen to feedstock (e.g., coal)	ξ	0.8–1.0	0.8 (dry-fed) 1.0 (slurry-fed)
Feedstock HHV/LHV	ϑ	1.02–1.06	1.04
Clean syngas HHV/LHV	ϑ'		From composition
Raw to clean syngas HHV energy content	ν		1.07 (No CCS) 1.11 (90% CCS)
Sulfur in coal in %(w)	S	0.3%–5%	3%
Gasifier aux load per feedstock [kW/STPD]	α_{GSF}	1.0–3.0	2.0
H_2S removal aux load per clean syngas flow [kW/kpph]	α_{H_2S}	1.0–7.0	4.0
ASU aux load per O_2 flow [kW/kpph]	α_{ASU}	150–300	250
Clean-to-raw syngas flow ratio	σ_{SG}	0.2–1.0	0.95 (No CCS) 0.2 (90% CCS)
Gasifier RSC duty as a fraction of feedstock input	θ_{GSF}	See Table 18.1	12%/6% (HR/QU)
LTGC duty as a fraction of feedstock input	θ_{GC}	See Table 18.1	2%/4% (HR/QU)
Air extraction (fraction of compressor airflow)	AX	0%–20%	Gas turbine optimization
Diluent N_2-to-O_2 ratio	DNO	0–3.1	Gas turbine optimization
CO_2 capture aux load per CO_2 flow [kW/pps]	α_{CO_2C}	50–100	80 (90% CCS)
CO_2 compression power per CO_2 flow [kW/pps]	α_{CO_2C}	100–135	105 (90% CCS)
CO_2 flow as a fraction of raw syngas flow	χ_{CO_2C}	0.5–0.8	0.75

18.4 EXAMPLE

The example is a 1990s-vintage 60-Hz F class gas turbine (similar to GE's 7FA) with 2,360°F (1,357°F) TIT and cycle PR of 14.6:1. The model used for gas turbine calculations is from THERMOFLEX® with stage-by-stage turbine modeling using GASCAN (model S2–22). The IGCC plant case is similar to the *Tampa Electric Polk Power Station* (TECO Polk) as described in the paper by Dennis et al. [19]. Before a brief description of the plant itself, it should be mentioned that the cited paper is an excellent source of pertinent information on commercially deployed IGCC technology. At 33 pages of length, it is more of a "report" than a conference paper. It presents detailed information on five coal-based IGCC power plants, three from the USA (Wabash, Tampa and Ashtabula) and two from Europe (Buggenum and Puertollano). Data presented in the report are culled from public literature and system studies undertaken by the US DOE's NETL. Process conditions and parameters related to the gas turbine (simple and combined cycle) and gasification are summarized in detailed tables. GT PRO® models of the power island for each IGCC are included as well.

TECO Polk IGCC was built as part of the US DOE's CCT demonstration program. It is a nominal 250-MWe with a GE (Texaco) single-stage, slurry-fed and oxygen-blown entrained flow-type gasifier. The plant became operational in 1996 and underwent a 5-year demonstration campaign. It has been in commercial operation since 2001.

The gas turbine in TECO Polk is GE's 7FA gas turbine (designated 7221FA at the time, it is the forebear of the present 7FA.04). This is a nominal 162 MWe (natural gas) gas turbine with an 18-stage axial compressor and three-stage turbine. It was modified for syngas operation as follows:

- Lower inlet airflow
- Lower firing temperature
- Higher turbine flow

- Nitrogen from the ASU as diluent in the combustor for NOx control
- Simple cycle output of 192 MWe.

The bottoming steam cycle is three-pressure reheat with 1,465 psia and 1,000°F HP steam and 1,000°F hot reheat steam. HP steam generated in the gasifier radiant and convective syngas coolers is imported from the process block. Exports from the bottoming cycle include feedwater to the syngas coolers and LP steam to the cold gas cleanup section (for the stripper reboilers).

TECO Polk IGCC could not achieve its design efficiency of 38.6% (HHV) due to lower-than-expected component performance and design issues. The actual field-measured efficiency was 35.4% (HHV).

From Table 1 in Ref. [19],

- Gasifier feedstock is Kentucky No. 9 subbituminous coal with 11,510 Btu/lb HHV (as received).
- Clean syngas composition includes 43.1%(v) CO and 13.8%(v) H_2 with 4,509 Btu/lb LHV.
- Diluent N_2 flow to the gas turbine combustor is 117 lb/s.
- Performance roll-up is
    - 192.4 MWe gas turbine generator
    - 122.4 MWe steam turbine generator
    - 61.5 MWe plant auxiliary load (~20% of gross output).
- Net IGCC performance is 253.3 MWe and 38.5% HHV (40.3% LHV).

Base gas turbine model with 100% methane fuel (cycle PR = 15:1, TIT = 2,475°F) is presented in Figure 18.4.

Syngas-fired version of the gas turbine is presented in Figure 18.5. Key modifications are:

- IGVs halfway closed, 10% lower compressor airflow
- 105°F lower TIT (2,370°F)
- S1N inlet area 13% larger
- 117.2 lb/s nitrogen diluent flow (N_2-to-syngas ratio is 1.17).

A comparison of methane-fired base and syngas-fired IGCC gas turbines is provided in Table 18.3. Key IGCC performance model parameters and data are summarized in Tables 18.4–18.6.

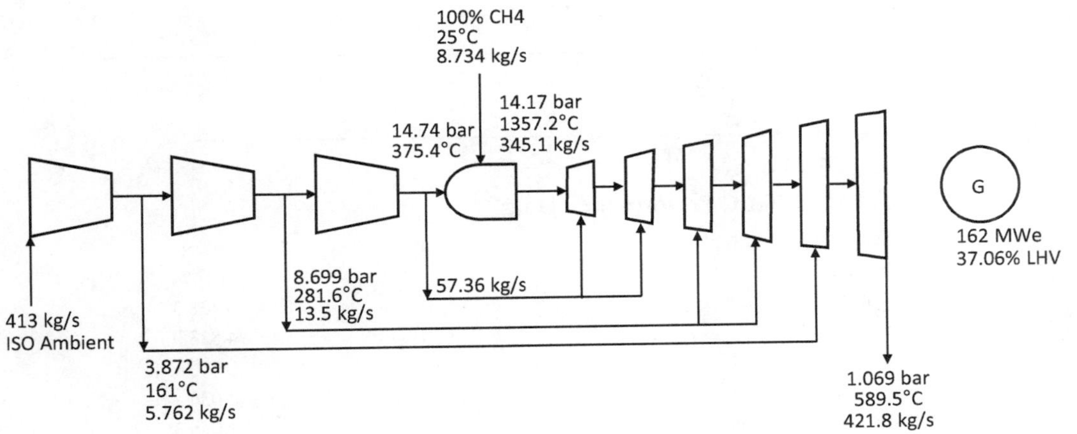

FIGURE 18.4 Base gas turbine (7221FA with 100% methane fuel).

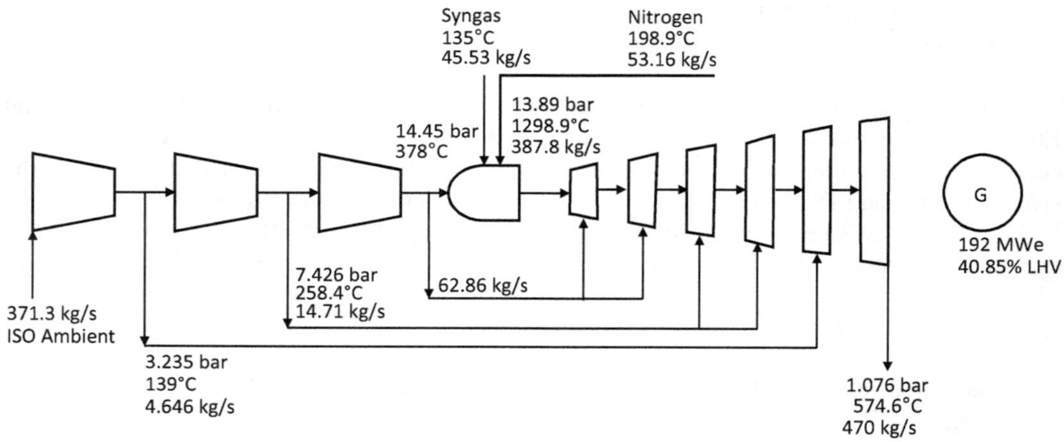

FIGURE 18.5 IGCC gas turbine (7221FA with low-Btu syngas fuel and diluent N_2 injection).

TABLE 18.3
Gas Turbine Performance Comparison

	Base	IGCC	Delta
Airflow, kg/s	413	371	−10.1%
Compressor PR	14.6	14.3	−2.0%
TIT, °C	1,357	1,299	−58
Fuel flow, kg/s	8.74	45.53	421.2%
Diluent flow, kg/s		53.2	
Turbine inlet flow, kg/s	345.1	387.8	12.4%
H_2O(v) in gas, %	9.9	7.7	−2.1
Fuel–air ratio	0.021	0.123	
Fuel LHV, kJ/kg	51,574	10,330	−80.0%
Exhaust flow, kg/s	422	470	11.4%
Exhaust temperature, °C	590	575	−15
Exit Mach number	0.53	0.59	
Output, MW	162	192	18.5%
Efficiency, %	37.06	40.9	3.8

TABLE 18.4
Selected IGCC Performance Data

Syngas flow	lb/s	100.4
N_2 Flow	lb/s	117.2
Syngas LHV	Btu/lb	4,438.3
Syngas HHV	Btu/lb	4,823.0
ϑ'		1.09
Exhaust flow	lb/s	1,036.2
Exhaust temperature	F	1,066.3
BC exergetic efficiency (Equation C.1)		72.2%

(Continued)

TABLE 18.4 (*Continued*)
Selected IGCC Performance Data

BC exergetic efficiency (corrected)		68.2%
Exhaust moisture		0.06549
Exhaust exergy (Equation A.3)	Btu/lb	121.0
Exhaust exergy (corrected)	Btu/lb	119.6
GTG output	kW	191,988
STG output (Equation 18.11)	kW	122,998
Aux load (Equation 18.2)	kW	51,442

TABLE 18.5
IGCC Performance Model Parameters

Feedstock HHV	Btu/lb	11,510
ϑ		1.05
CGE		0.75
Feedstock flow rate (Equation 18.16)	lb/s	54.2
S		2.94%
$(\dot{m} \times LHV)$ feed	Btu/s	593,999
θ_{GSF}		11%
θ_{GC}		2%
$\dot{Q}$ (GSF)	Btu/s	62,370
$\dot{Q}$ (GC)	Btu/s	11,880
Import steam flow	lb/s	135
ξ		0.8
$\dot{m}$ O_2 (Equation 18.21)	kpph	156.1
Available N_2 flow (Equation 18.23)	lb/s	143.1
DNO		0.75
α_{ASU}	kW/kpph	250
$\dot{W}$ (ASU) (Equation 18.20)	kW	39,015
α_{H_2S}	kW/kpph	4.0
$\dot{W}(H_2S)$ (Equation 18.24)	kW	1,445
α_{GSF}	kW/STPD	2.0
$\dot{W}$ (GSF) (Equation 18.4)	kW	4,682

TABLE 18.6
Steam Import Contribution to STG Output

HP steam pressure	psia	1,547
HP steam temperature (sat.)	F	600
Average steam temp. (Equation 18.10)	F	600.3
ε'		0.80
$\dot{W}$ (Equation 18.8)	kW	31,995

REFERENCES

1. Khartchenko, N.V., 1998, *Advanced Energy Systems*, Taylor & Francis Group, Washington, DC.
2. Probstein, R.F., Hicks, R.E., 2006, *Synthetic Fuels*, Dover Publications, Inc., Mineola, NY.
3. Higman, C., van der Burgt, M., 2008, *Gasification*, 2nd Edition, Gulf Professional Publishing, Imprint of Elsevier, Oxford.
4. Rezaiyan, J., Cheremisinoff, N.P., 2005, *Gasification Technologies–A Primer for Engineers and Scientists*, CRC Press, Taylor & Francis Group, Boca Raton, FL.
5. Gülen, S.C., Driscoll, A.V., 2013, Simple parametric model for quick assessment of IGCC performance, *J. Eng. Gas Turbines Power*, Vol. 135, p. 010802.
6. Kaldis, S.P., Skodras, G., Sakellaropoulos, G.P., 2004, Energy and capital cost analysis of CO_2 capture in coal IGCC processes via gas separation membranes, *Fuel Process. Technol.*, Vol. 85, pp. 337–346.
7. Komori, T., Hara, H., Arimura, H., Kitauchi, Y., 2003, Design for F Class Blast Furnace Gas Firing 300 MW Gas Turbine Combined Cycle Plant, International Gas Turbine Congress, November 2–7, 2003, Tokyo, Japan.
8. Becker, B., Schetter, B., 1992, Gas turbines above 150 MW for integrated coal gasification combined cycles (IGCC), *J. Eng. Gas Turbines Power*, Vol. 114, pp. 660–664.
9. Brdar, D.B., Jones, R.M., 2000, GE IGCC Technology and Experience with Advanced Gas Turbines, GER-4207.
10. Eggart, J., Thompson, C.E., Sasser, J., Merine, M., 2017, Heavy-Duty Gas Turbine Operating and Maintenance Considerations, GER-3620N.
11. Maurstad, O., 2005, An Overview of Coal-Based IGCC Technology, MIT LFEE 2005-002 WP, Massachusetts Institute of Technology, Cambridge, MA.
12. Parulekar, P.S., 2011, Comparison Between Oxygen-Blown and Air-Blown IGCC Power Plants: A Gas Turbine Perspective, GT2011-45154, ASME Turbo Expo 2011, June 6–10, 2011, Vancouver, Canada.
13. Jones, R.M., Shilling, N.Z., 2003, IGCC Gas Turbines for Refinery Applications, GE Energy, Schenectady, NY, Paper No. GER-4219.
14. Rubin, E.S., Berkenpass, M.B., Frey, H.C., Chen, C., McCoy, S.T., Zaremsky, C.J., 2007, Technical Documentation: IGCC Systems With Carbon Capture and Storage, Final Report, U.S. DOE Contract DE-AC21-92MC29094.
15. Frey, H.C., Zhu, Y., 2006, Improved system integration for integrated gasification combined cycle (IGCC) systems, *Environ. Sci. Technol.*, Vol. 40, No. 5, pp. 1693–1699.
16. Electric Power Research Institute (EPRI), 1980, Gasification Combined Cycle Plant Configuration Studies, EPRI, Palo Alto, CA, EPRI Report AP-1393.
17. Hoffmann, S., Bartlett, M., Finkenrath, M., Evulet, A., Ursin, T., 2009, Performance and cost analysis of advanced gas turbine cycles with precombustion CO_2 capture, *J. Eng. Gas Turbines Power*, Vol. 131, p. 021701.
18. Botero, C., Finkenrath, M., Belloni, C., Bertolo, S., D'Ercole, M., Gori, E., Tacconelli, R., 2009, Thermoeconomic Evaluation of CO_2 Compression Strategies for CO_2 Capture Applications, ASME Turbo Expo 2009, Orlando, FL, June 8–12, ASME Paper No. GT2009-60217, pp. 517–526.
19. Dennis, R.A., Shelton, W.W., Le, P., 2007, Development of Baseline Performance Values for Turbines in Existing IGCC Applications, ASME Turbo Expo 2007, Montreal, Canada, May 14–17, ASME Paper No. GT2007-28096, pp. 1017–1049.

19 Carbon Capture

Over the next decades, the planet's energy challenge will be resolving how hundreds of millions of people can gain access to electricity while meeting climate targets. To do so means cutting the carbon emission intensity of power plants - that is, reducing the amount of CO_2 that is spewed out for each megawatt generated.

International Energy Agency (IEA) press release (September 23, 2016)

Carbon dioxide (CO_2) constitutes the largest fraction of greenhouse gases (GHG), which are widely believed to be a major contributor to climate change. As such, significant research and development effort has been dedicated to reduce and/or eliminate emissions of CO_2 into the atmosphere. Combustion of fossil fuels, especially coal, in electricity generating power plants is a significant source of CO_2. To date, post-combustion CO_2 removal from the stack gases via the deployment of aqueous amine-based absorber-stripper technology is the only commercially available option, which is applicable to new units as well as to retrofitting the existing plants [1].

In a fossil fuel-fired power plant with post-combustion capture, a continuous scrubbing system is used to separate the CO_2 from the flue gas stream by chemical absorption [2]. The system consists of two major components:

1. An **absorber** in which the CO_2 is removed.
2. A **regenerator (stripper)** in which the CO_2 is released in a concentrated form and the solvent is recovered.

Prior to the CO_2 removal, the flue gas (at around 90°C (180°F–190°F) at the stack for gas turbine combined cycle power plants) is typically cooled to about 45°C (about 110°F) and then treated to reduce particulates that cause operational problems and other impurities, which would otherwise cause costly loss of the solvent (e.g., in a direct-contact cooler or "quench tower"). The solvent absorbs the CO_2 (together with traces of NOx) by chemical reaction to form a loosely bound compound. A booster fan (blower) is requisite to overcome the pressure loss in the capture plant and is a significant (parasitic) power consumer.

The largest power consumption by the CO_2 capture system is due to the large amount of heat required to regenerate the solvent. The temperature level for regeneration is normally around 120°C. This heat is typically supplied by steam extracted from the bottoming cycle and reduces steam turbine power output and, consequently, net efficiency of the power plant significantly.

As for all other carbon capture technologies, electrical power is consumed to compress the captured CO_2 for transportation to the storage site and injection into storage. Consequently, two of the most taxing post-combustion CO_2 capture plant (CCP) design challenges have always been to minimize regeneration energy by selecting a solvent with a relatively low reaction energy and to use the lowest possible exergy steam extraction to provide the requisite energy.

Before continuing with the post-combustion subject, two other alternative technologies must be mentioned for the sake of completeness. The first one is "pre-combustion" carbon capture, which is predicated upon the "water-gas shift (WGS) reaction", which converts the CO-H_2 syngas (mostly) from the gasifier of an IGCC power plant into a gaseous mixture of CO_2 and H_2 (typically, utilizing steam from the bottoming cycle of the gas turbine combined cycle part of the IGCC). Carbon dioxide can be separated from this mixture utilizing a standard acid gas removal process with a solvent like Rectisol (see Section 18.3.8 for more on this subject). None of the IGCC power plants in operation at the time of writing has this feature. (Duke Edwardsport IGCC has 60% capture option as a future retrofit.)

The other pre-combustion carbon capture option (so far only on paper) is steam-methane reforming followed by WGS and CO_2 scrubbing. In either case, the gas turbine fuel is (nearly) pure hydrogen. The difficulties associated with combustion of 100% H_2 fuel have been discussed in detail in Chapter 12 of **GTFEPG**.

Oxyfuel combustion is second technology intended for carbon capture. In this case, (nearly) pure O_2 generated in an air separation unit (ASU) similar to one utilized in IGCC is the oxidant. Consequently, with methane as the fuel, combustion products are CO_2 and H_2O, which leads to the possibility of CO_2 separation simply via condensation of H_2O from the mixture in a condenser.

In pre-combustion carbon capture technologies, the gas turbine in question is an existing product from the major OEMs with requisite modifications to the combustion system to handle the hydrogen fuel. In oxyfuel combustion, a unique turbine system has to be designed from the blank sheet. There are several approaches to this problem (pretty much all of them – except one – are so far on paper and expected to remain so for the foreseeable future). These are discussed in the review chapter written by Gülen [1]. Another source containing a good overview of all capture technologies is the chapter written by Soothill et al. [3]. The most promising oxyfuel combustion technology (at least a pilot plant is being constructed) is the Allam cycle [1].

19.1 POST-COMBUSTION CARBON CAPTURE BASICS

19.1.1 State of the Art

A typical solvent-based CO_2 capture system is shown in Figure 19.1. Flue gas stream from the stack of the power plant goes first to the quench tower, which is a direct-contact heat exchanger utilizing water to cool the flue gas (usually after SO_2 removal). The circulating water is cooled and filtered, providing a means of removing particulates that may be present in the flue gas stream. After leaving the quench tower, the cooled flue gas stream is passed into a booster fan (blower) to maintain the required pressure in the inlet flue gas ducts.

On exiting the blower, the flue gas stream enters the absorber (a vertical, packed column with counterflow arrangement) where it is contacted with a typically amine-based solvent. Carbon dioxide is absorbed from the flue gas stream into the lean solvent as it goes up the column. Treated flue gas stream passes out through a wash section at the top section of the absorber.

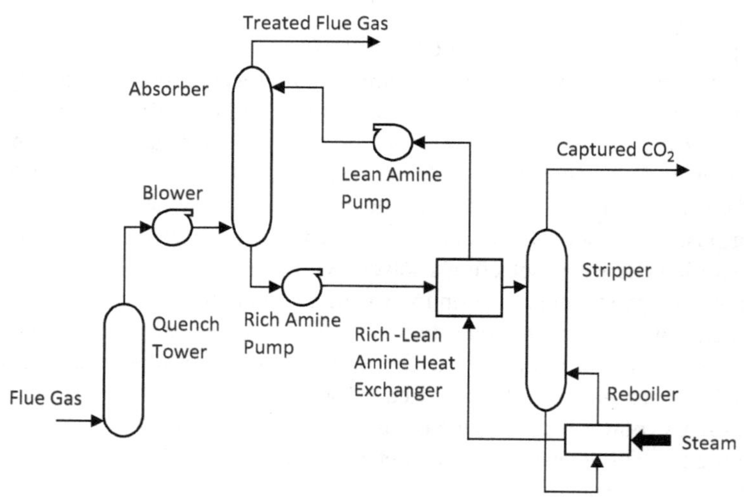

FIGURE 19.1 Carbon dioxide capture from flue gas via aqueous amine-based absorption.

The CO_2 rich solvent stream leaves the absorber and is sent for regeneration in the stripping system, where heat is provided to reversibly release the CO_2 from solution. Before entering the stripper, rich solvent is preheated by exchanging heat with the hot, lean solvent exiting the stripper in a heat exchanger.

A wash section at the top of the stripper ensures that a minimal amount of entrained and vaporized solvent leaves the column in the product CO_2 stream.

Prior to the CO_2 removal, the flue gas (at around 180°F–190°F at the heat recovery steam generator (HRSG) stack for the most efficient GTCC power plants) is typically cooled to about 110–120°F (e.g., in a direct-contact cooler or "quench tower") and then treated to reduce particulates and other impurities, which would otherwise cause operational problems and costly loss of the solvent. The solvent absorbs the CO_2 (together with traces of NOx) by chemical reaction to form a loosely bound compound. A booster fan (blower) is requisite to overcome the pressure loss in the capture plant and is a significant (parasitic) power consumer.

The largest power reduction caused by the CO_2 capture system is due to the large amount of heat required to regenerate the solvent. The temperature level for regeneration is normally around 250°F. This heat is typically supplied by steam extracted from the bottoming cycle and reduces steam turbine power output and, consequently, net efficiency of the GTCC significantly. In addition, as for all other carbon capture technologies, electrical power is consumed to compress the captured CO_2 for transportation to the storage site and injection into the storage cavern.

Technologies for gas sweetening and syngas purification using alkanolamines and other absorbents have been extensively utilized in chemical process industry over the past century. Nevertheless, large-scale recovery of CO_2 from flue gas poses several serious challenges. Most important of these challenges in a GTCC context are low CO_2 partial pressure (about 4%(v) in the flue gas) and high regeneration energy (about 1,500 Btu/lb of CO_2). In addition, oxygen in the flue gas (about 12% by volume at the HRSG stack) can cause corrosion and solvent degradation. Due to the absence of many impurities, which are amply present in coal-fired power plant flue gases, e.g., SOx (negligible), soot, fly ash and mercury, the only significant degrading agent to worry about in GTCC flue gas is oxygen. While inhibitors have been reasonably effective in mitigating these effects, the need for continuous removal of unavoidable solution contaminants adds to the operating costs.

For a typical F class GTCC with post-combustion, amine-based carbon capture,

- Net efficiency drops from 56%–58% (no capture) to 47%–49%
- Specific capex (per kilowatt of net output) increases from about \$500 to \$600 to anywhere between \$1,100 and \$1,600 [4].

To summarize, in a natural gas-fired GTCC framework, post-combustion CCP design challenges are as follows:

- To minimize regeneration energy by selecting a solvent with a relatively low reaction energy
- To use the lowest possible exergy steam extraction to provide the requisite energy
- To cool the gas turbine exhaust gas to the lowest possible temperature in the HRSG
- To maximize the CO_2 content of the HRSG stack gas.

There are several ways to attack the problem. Two of them, one involving a new system and another involving a new operating philosophy are discussed below. Furthermore, the reader is referred to papers published by selected authors, which approach the problem from different angles [5–10].

19.1.2 A Novel Method

A recently proposed solution [11] meets *three* key design challenges of post-combustion CO_2 capture from the stack gas of a GTCC power plant using aqueous amine-based scrubbing method by offering the following:

- Low HRSG stack gas temperature
- Increased HRSG stack gas CO_2 content
- Decreased HRSG stack gas O_2 content.

This is achieved by combining *two* bottoming cycle modifications in an inventive manner:

- High supplementary (duct) firing in the HRSG
- Recirculation of the HRSG stack gas.

A detailed system diagram is shown in Figure 19.2. This diagram is used to explain the key features of the GTCC with carbon capture (henceforth *GTC4*) per the earlier discussion and how they are implemented to result in a final coherent system. Furthermore, a detailed sample calculation with all the pertinent numbers will be presented to demonstrate the significance of GTC4 vis-à-vis current state of the art (SOA).

The gas turbine combined cycle system of GTC4 comprises the following major components:

1. Main gas turbine generator (Main GTG)
2. Single-pressure HRSG with reheat and supplementary firing
3. Steam turbine generator (STG)
4. Recirculation gas turbine generator (recirculation GTG).

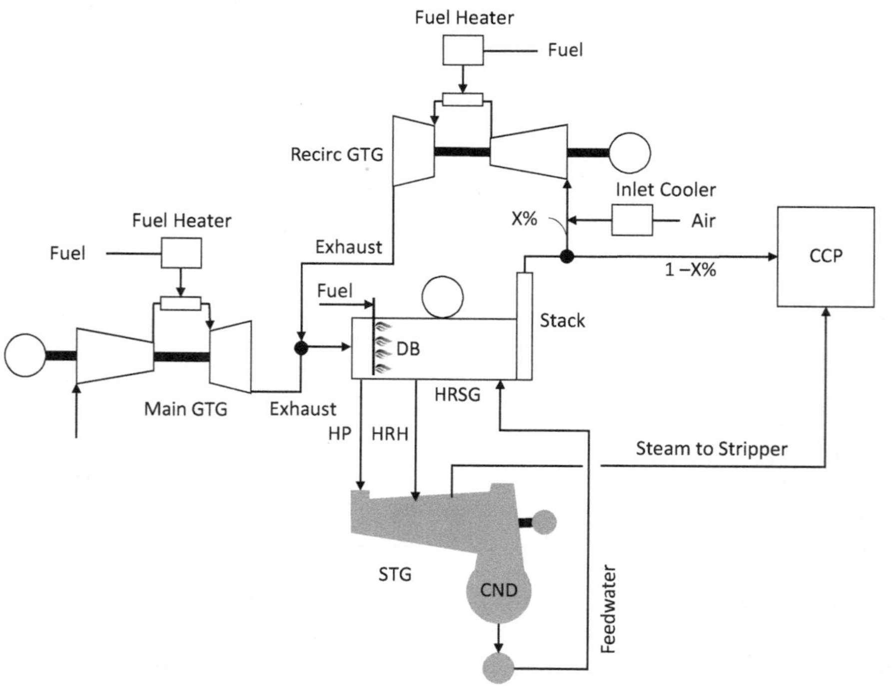

FIGURE 19.2 System diagram of the GTCC optimized for carbon capture.

Main GTG, HRSG and STG comprise the current SOA in terms of gas turbine combined cycle plant arrangement. HRSG stack gas is forwarded to a post-combustion CCP.

The proposed system, GTC4, is based on the diversion of a portion of HRSG stack gas from the CCP. Diverted gas is mixed with ambient air cooled in a heat exchanger (i.e., evaporative cooler, electric chiller, etc.). The remaining gas is forwarded to the CCP. The combined air-gas flow is the motive air of the recirculation GTG, which generates further electric power. The exhaust gas from the recirculation GTG is mixed with the exhaust gas from the main GTG. The combined exhaust gas enters the HRSG, and its energy is increased by the duct burner (DB). The rest of the steam cycle is similar to the current SOA.

The carbon capture plant can be based on any post-combustion capture technology and has the following features:

- It is inclusive of CO_2 compression and conditioning for pipeline transportation to the final storage or usage location (e.g., sequestration cavern or oil field for enhanced oil recovery (EOR)).
- It includes electric motor-driven equipment such as compressors and pumps, whose power consumption is debited to the gross power generation of the GTCC power plant.
- It utilizes steam at specified pressures and temperatures to provide energy requisite for capture processes (e.g., the reboiler of the stripper/regenerator column of the aqueous amine-based capture plant in Figure 19.1).

Steam requirements of the CCP are met by steam extracted from suitable locations in the bottoming cycle of the GTCC, e.g., the HRSG and/or the STG. One example is low-pressure (LP) steam extraction from the STG, which is shown in Figure 19.2. Another option is to supply the low-pressure reboiler steam from an auxiliary boiler. Final selection is subject to a cost-performance trade-off and operability impact study, which can be done on a case-by-case basis.

The recirculation GTG supplementary air flow requires cooling for optimal gas turbine performance. This is especially important for plant operation on hot days. The inlet cooler in Figure 19.2 can be an evaporative cooler, which is projected to be the most cost-effective option in most cases. However, it can also be one of myriad possibilities including electric chiller or absorption chiller (utilizing steam or hot water extracted from the HRSG or the STG), among others. The final selection should be determined on a case-by-case basis via diligent cost-performance trade-off.

Recirculation GTG can be identical to the main GTG (most likely to be the ideal configuration) or can be of a different type and size. The final selection should be determined on a case-by-case basis via diligent cost-performance trade-off. Gas turbine fuels can be of the same type (e.g., both natural gas) or different (i.e., one natural gas and the other distillate). Similarly, the HRSG DB can use the same fuel as the GTGs or a different one.

Other important design parameters and decisions subject to optimization are the DB exit gas temperature, the location of the DB and the HRSG stack gas recirculation rate (commonly referred to as *exhaust gas recirculation*, EGR). Higher EGR, although beneficial from a stack gas CO_2 and O_2 content perspective, results in warmer motive air for the recirculation GTG (plus with reduced O_2 for the combustor). The discussion herein is based on calculations with 30% EGR. This is believed to be roughly the optimal rate. Nevertheless, a diligent optimization study is requisite to pin down the best EGR rate on a case-by-case basis.

EGR is being considered by one original equipment manufacturer (OEM) for their next-generation gas turbines with 1,700°C turbine inlet temperature (TIT) to reduce NOx emissions. In the system adopted by that OEM, recirculated HRSG stack gas, after being cooled and mixed with ambient air, is admitted into the gas turbine compressor inlet. Tests have been conducted in full-scale combustors at medium and high pressures to demonstrate operability and NOx reduction capability with up to nearly 30% EGR. Another OEM has also demonstrated the effect of EGR on operability, efficiency and emission performance under conditions of up to 40% EGR. Recirculation gas turbine compressor and turbine operability considerations due to changing gas composition and molecular weight (MW) should be evaluated by the OEM at the detailed design phase.

The significance of the GTC4 is its immense capital cost benefit of about $200 million (nominal 750 MWe net), which makes it quite attractive even at expensive fuel gas (at the same net MWe output). The advantage of the system is demonstrated by detailed heat and mass balance simulation of the power block (using Thermoflow Inc.'s THERMOFLEX® and GT PRO® software) and the amine-based post-combustion capture system (using ProMax®v3.2 with a hypothetical amine of 50%(wt) MDEA and 5%(wt) Piperazine). Details can be found in Ref. [11]. In particular, GTC4 has the following advantages:

- $182 million lower installed cost
- 25+% lower specific cost ($/kW)
- 15+% lower capture penalty (relative basis).

Note that there is a variant of GTC4 (not discussed herein but included in Ref. [11]), which can reduce the capture penalty by 65% (relative basis) with the same specific cost advantage.

19.1.3 A Novel Operating Strategy

As discussed above, two of the most taxing post-combustion CCP design challenges have always been to minimize regeneration energy by selecting a solvent with a relatively low reaction energy and to use the lowest possible exergy steam extraction to provide the requisite energy. The assumption has always been that the power plant would be base loaded. The realities of grid operations in today's environment require these assumptions to be reexamined.

There is no doubt that minimizing regeneration energy is still a worthy goal. (Using the steam source with the lowest possible exergy is a *sine qua non* design objective in any case.) However, recognizing the ultimate limit of this "design knob", an alternative approach is to minimize the "financial damage" it does to the power plant's bottom line. In fact, in a recently published white paper, Electric Power Research Institute (EPRI) stated that "[t]he challenges with carbon capture and storage are partly technical and mainly, it seems, linked to the lack of viable business models and regulatory frameworks necessary to attract investment".[1] Thus, tackling the problem with a system and operation philosophy based on the exploitation of significant differences in electricity price over the course of a calendar year to maximize the revenue from power sales is an attractive proposition.

In order to illustrate the idea, consider a sample operational scenario with the following assumptions (please refer to Figure 19.3, which is a very simplified version of the more detailed system description in Figure 19.1) [12]:

- Generic fossil fuel-fired power plant with amine-based post-combustion capture plant.
- There are 4 h of peak operation in any 24 h period, separated into two periods with maximum three hours in each period and minimum 7 h in between.
- The stripper system includes the rich-lean heat exchanger (in order to store amine at ambient temperature) and all downstream CO_2 handling system.
- Although not shown explicitly in Figure 19.3, CO_2 separated from the rich solvent in the stripper is sent to a CO_2 compressor, which compresses it to the requisite pipeline pressure for transmission to the storage location. Consequently, the CO_2 compressor is ON, when the stripper is ON.
- Lean amine from the stripper column must be cooled before storage. This is best done by using the heat exchanger against stripper feed. (Hot amine storage would need to be in

[1] From "*The Integrated Energy Network*" by EPRI published in February 2017 (document number 3002009917).

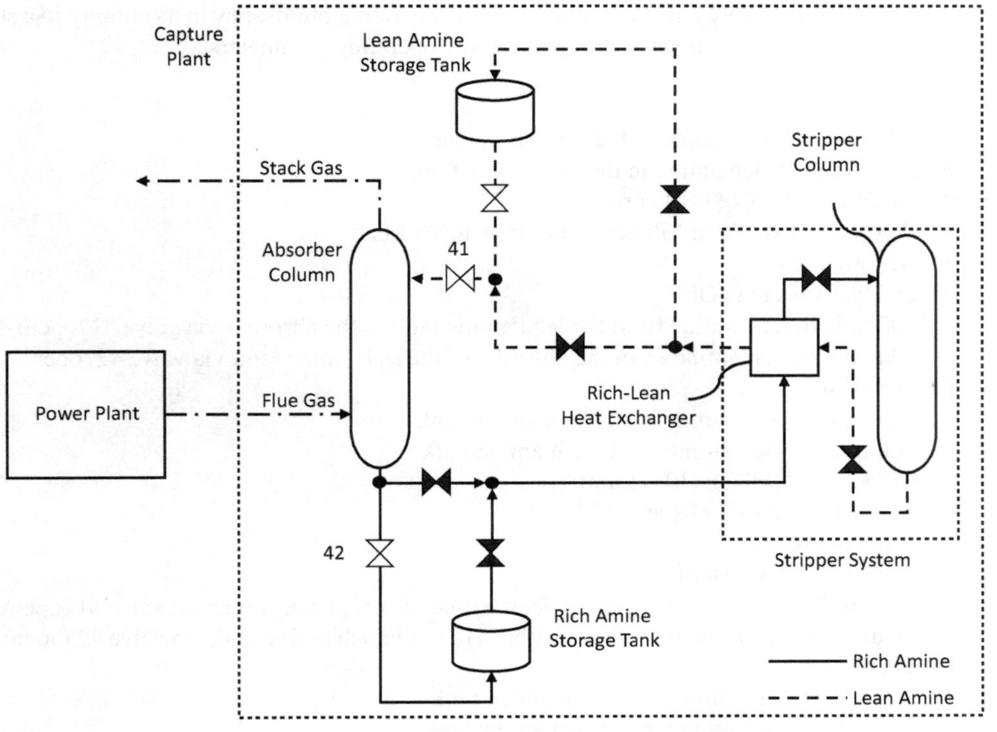

FIGURE 19.3 PEAK OP (7–10 am) and STRIPPER STARTUP (10–11 am).

pressurized vessels which would be prohibitively expensive for this volume. Degradation at temperature would also be an issue. Rich amine may need either some additional cooling or some dilution with lean amine to be stored safely in atmospheric tanks.)

- Rich amine can be stored directly from the absorber, as it is already cool.
- Total volume of amine whether lean or rich stored in their respective tanks remains constant (volume increase for CO_2 absorption ignored).

The absorber requires 100 units/h of lean amine. Morning peak is 7–10 am, and the evening peak is 5–6 pm. The stripper system is assumed to be *not* running during peak operation. (Note that 1 h is required to restart stripper system.) Thus, the total daily operation time for stripper system is

$$24 - 4 - 2 = 18 \text{ h}.$$

In contrast, the absorber is always in operation. Same quantity of CO_2 is in flue gas when power plant is operating in full power mode or steam extraction mode. Consequently, the absorber size is 100%. (Percentages refer to equivalent percentage for a system which would operate with all parts in operation 24 h a day.)

The stripper system, including the rich-lean heat exchanger, is sized (100+X)%, where X is 33% because 24 h worth of rich amine has to be processed in 18 h when the power plant runs at 100% load throughout the entire period. This is the maximum possible operating and processing capacity of the power plant and the stripper system, respectively.

A sample chronology of 1-day (24 h) system operation is given below to demonstrate the operating philosophy of the process proposed herein. Note that this particular operating regime (i.e., 24 h, non-stop, with 4 h of 100% plant load and 20 h of 25% plant load) is not intended to reflect a realistic dispatch scenario. It is merely intended to illustrate the operating philosophy in its entirety in a step-by-step manner with numerical values assigned to key operating parameters.

- BEGIN (7 am)
  - 400 units of lean amine in the lean amine tank
  - Zero units of rich amine in the rich amine tank.
- PEAK OPERATION (7–10 am)
  - Power Plant is ON in full power mode at 100% Load
  - Absorber is ON
  - Stripper system is OFF
  - 100 units/h lean amine from the lean amine tank to the absorber via valve 41 (open)
  - 100 units/h rich amine from the absorber to the rich amine tank via valve 42 (open).
- Tanks (10 am)
  - 100 units of lean amine in the lean amine tank
  - 300 units of rich amine in the rich amine tank.
- STRIPPER START-UP (10–11 am)
  - Power Plant 200 is ON at 25% Load
  - Absorber is ON
  - Stripper STARTING UP
  - 25 units/h of lean amine from the lean amine tank to the absorber via valve 41 (open)
  - 25 units/h rich amine from the absorber 110 to the rich amine tank via valve 42 (open).
- Tanks (11 am)
  - 75 units of lean amine in the lean amine tank
  - 325 units of rich amine in the rich amine tank.
- OFF-PEAK OPERATION (11 am to 5 pm)
  - Power Plant is ON at 25% Load
  - Absorber is ON
  - Stripper is ON running at 50% Load
  - 25 units/h of rich amine from the absorber to the stripper system
  - 25 units/h of rich amine from the rich amine tank to the stripper system
  - 25 units/h of lean amine from the stripper system to the absorber
  - 25 units/h of lean amine from the stripper system to the lean amine tank.
- Tanks (5 pm)
  - 225 units of lean amine in the lean amine tank
  - 175 units of rich amine in the rich amine tank.
- PEAK OPERATION (5–6 pm)
  - Same as before.
- Tanks (6 pm)
  - 125 units of lean amine in the lean amine tank
  - 275 units of rich amine in the rich amine tank.
- STRIPPER STARTUP (6–7 pm)
  - Same as before.
- Tanks (7 pm)
  - 100 units of lean amine in the lean amine tank
  - 300 units of rich amine in the rich amine tank.
- OFF-PEAK OPERATION (7 pm to 7 am)
  - Same as before.
- Tanks (7 am)
  - 400 units of lean amine in the lean amine tank
  - Zero units of rich amine in the rich amine tank.

Although the basic philosophy of operation is described using a generic amine, the described operational sequence is applicable to most solvent-based capture systems with any type of fossil fuel-burning power plant. The heart of the described operation is "time-shifting" of the solvent regeneration process and CO_2 compression. This method, in turn, is enabled by the two-tank modification to the basic, amine-based post-combustion capture plant, which results in a new system.

The exact amount of additional revenue is highly dependent on site and time-specific "boundary conditions". A sample calculation, using ERCOT as an example, showed that it can be significant [12]. A typical fossil power plant would be "out-of-the-money" for nearly 7,000 h in 2016 at average \$19/MWh. Using a 670-MW fossil plant (before capture), if one assumed that it would run as proposed at those times of the year when the wholesale price is \$100/MWh or higher, the extra income generated would be \$3.8 million [12]. In addition to this extra revenue scenario, there are other possibilities to generate ancillary revenue with this operating method. One example is spinning reserve revenue for 240 MWe when CO_2 capture is ongoing during normal operation or for a portion of it during off-peak operation.

19.2 SIMPLE CALCULATIONS

In order to calculate the performance penalty imposed by a solvent-based, post-combustion capture system similar to that shown in Figure 19.1, one needs to model it in a chemical process simulation software such as Aspen HYSYS or ProMax. Such software packages are extremely expensive with hefty maintenance fees and very steep learning curves. In short, they are only available to those who are employed by private or public R&D organizations and OEMs.

Nevertheless, it is still possible to make very good estimates starting from the best principles using an Excel spreadsheet. By far the best method is based on the "minimum separation work" (MSW) principle, which is described in minute detail with step-by-step examples in the superb book by Wilcox [2]. For anyone who is interested in the subject matter, this book is *the* must-have reference.

The starting point is Equation 1.11 in Section 1.5 of Ref. [2], which gives the minimum thermodynamic work for CO_2 separation. It requires information on three streams described in Figure 19.4.

- Stream A is fully defined by the stack gas of simple or combined cycle gas turbine (e.g., from a simulation software such as Thermoflex).
- Stream B is defined by the capture rate (e.g., 90%) and the purity of captured CO_2 (e.g., 99.9%).
- Stream C is simply the difference between streams A and B.

Requisite information is quite simple and limited:

- Flue gas mass flow rate, molecular weight (MW) and composition
- Flue gas temperature
- Capture rate and CO_2 purity.

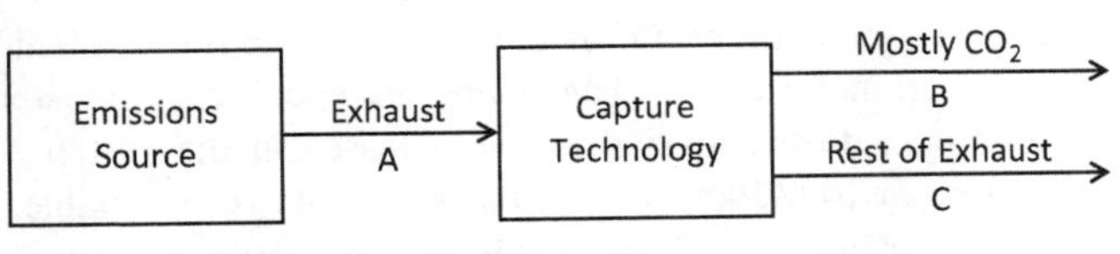

FIGURE 19.4 Carbon capture process inlet/outlet material streams.

TABLE 19.1
Interpolation Table for Second Law Efficiency

	CO_2	Second Law Efficiency
Pulverized coal	0.120	27.38
Gas turbine CC	0.038	21.54
Interpolate	0.041	21.78

An Excel VBA function calculating the MSW as described in Ref. [2] is provided in the Appendix F (`Min _ Sep _ Work`). MSW is the equivalent of gas turbine exhaust exergy, which is extensively used in this book. In order to translate it into "actual" separation work, we need a conversion factor, which is analogous to the bottoming cycle exergetic efficiency. This conversion factor is provided by the chart in Figure 1.12 in Wilcox [2], which gives the relationship between the exergetic (second law) efficiency and CO_2 concentration of initial gas mixture. Two data points from the chart can be used for a simple interpolation to provide an estimate for the problem at hand (see Table 19.1).

Example:
For an F class gas turbine combined cycle, rated at 400 MWe net, the following data is available:

- Flue gas flow rate 1,264.1 lb/s (HRSG Stack)
- Flue gas temperature 192.3°F
- Flue gas MW 28.33 lb/lbmol (kg/kmol)
- Flue gas molar composition
  - Nitrogen 74.13%
  - Oxygen 11.711%
  - Water vapor 9.148%
  - Carbon dioxide 4.118%
  - Argon 0.893%.

For 90% capture and 99.9% CO_2 purity, using `Min _ Sep _ Work`, we obtain 8.8 MW for MSW. From Table 19.1, second law efficiency is ~22%. Thus, actual separation work is 8.8/0.22 = 40.5 MW. Auxiliary power consumption for CO_2 compression and conditioning is 15–16 MJ/kmol (caution: *not* lbmol) of captured CO_2.

Capture CO_2 molar flow rate is

$$(1264.1/28.33)\times 0.04118\times 0.90/2.2046 = 0.75 \text{ kmol/s}$$

so that total compression power requirement is

$$0.75\times 16 = 12 \text{ MW}.$$

Another 2 MJ/kmol should be allocated to amine recirculation and other capture unit pumps.

Typically, the post-combustion capture unit is connected to the power plant via a flue gas duct, whose total length, cross-sectional area and layout are site-specific. A booster fan is requisite to push the flue gas and compensate for myriad pressure losses in the system. Without a detailed design at hand, the allocation of 10 MJ/kmol for booster fan duty is reasonable.

Thus, total BOP and booster fan power requirement of the capture plant is

$$0.75\times(10+2) = 9 \text{ MW}.$$

Finally, total capture penalty is estimated as

$$40.5 + 12 + 9 \approx 62 \text{ MW},$$

which amounts to 62/400 = 15.5% of the plant rated output with no capture.

A key number to be used in amine-based carbon capture calculations is the stripper reboiler energy requirement, which determines the extraction steam flow from the bottoming cycle and, as a result, "lost" steam turbine power output. Based on the numbers used for this example, CO_2 flow rate is

$$0.75 \times 44 = 33 \text{ kg/s}.$$

Actual separation work is calculated as 40.5 MW, which is the lost steam turbine work. Typical conversion efficiency is 35%–36%. This means that the net energy input to the reboiler is

$$40.5/0.35 = 115.7 \text{ MWth}.$$

Before moving on, it behooves to spend a paragraph on this subject.

Firstly, setting the stripper reboiler pressure is an important optimization problem. In the stripper column, absorbed CO_2 (from the flue gas) in the "rich" amine is released as a concentrated stream by adding heat to reverse the chemical reaction of capture. From an absorption perspective, the colder the amine the better. Thus, typical "rich" amine temperature (solvent plus absorbed CO_2) is about 45°C. For the desorption process, higher temperature is desirable. Heat exchange between the rich and "lean" amines ensures that warmer stream goes into the stripper and colder one goes into the absorber. Ultimately, optimal heat exchanger design sets the stripper bottom temperature, i.e.,

- Higher values would be advantageous from CO_2 compression perspective.
- Lower values would be advantageous from steam extraction (from the bottoming cycle of the GTCC) and "lost" STG output perspective.

Literature is full of optimization studies (e.g., see Warudkar et al. [13]). For the most common and proven solvent, 30% MEA, the ideal was found to be around 120°C (~250°F). This corresponds to a pressure of about 1.9 bara in the stripper bottom and sets the steam pressure at 3–3.5 bara (about 50 psia).

Using the saturated vapor/steam and liquid enthalpies at 3.5 bara, steam flow requisite is found as

$$115{,}700/(2{,}731.5 - 582) = 53.8 \text{ kg/s}.$$

Furthermore, heat of desorption can be found as

$$115{,}700/33 = 3{,}500 \text{ kJ/kg of captured } CO_2 \text{(about 1,500 Btu/lb)}.$$

This is indeed about right for 30% (by weight) MEA solvent.

REFERENCES

1. Gülen, S.C., 2017, Advanced fossil fuel power systems, Chapter 13 in *Energy Conversion, 2nd Edition*, Eds. D.Y. Goswami and F. Kreight, CRC Press, Boca Raton, FL.
2. Wilcox, J., 2012, *Carbon Capture*, Springer, New York.
3. Soothill, C.D., Bialkowski, M.T., Guidati, G.L., Zagorskiy, A., 2013, Carbon Dioxide Capture and Storage for Gas Turbine Systems, Chapter 15 in Ref. [11] in Chapter 2.

4. Gülen, S.C., Hall, C., June 2017, Optimizing post-combustion carbon capture, *Power Eng.*, Vol. 121, pp. 14–19.
5. Gülen, S.C., Hall, C., 2017, Gas Turbine Combined Cycle Optimized for Carbon Capture, ASME Paper GT2017-65261, ASME Turbo Expo 2017, June 26–30, 2017, Charlotte, NC.
6. Elliott, W.R., Gülen, S.C., 2017, Cost-effective post-combustion carbon capture from coal-fired power plants, *International Pittsburgh Coal Conference*, September 5–8, 2017, Pittsburgh, PA.
7. Botero, C., Finkenrath, M., Bartlett, M., Chu, R., Choi, G., Chinn, D., 2009, Redesign, optimization, and economic evaluation of a natural gas combined cycle with the best integrated technology CO_2 capture, *Energy Procedia*, Vol. 1, pp. 3835–3842.
8. Botero, C., Finkenrath, M., Belloni, C., Bertolo, S., 2009, Thermoeconomic Evaluation of CO_2 Compression Strategies for CO_2 Capture Applications, ASME Paper GT2009-60217, ASME Turbo Expo 2009, June 8–12, 2009, Orlando, FL.
9. Baldwin, P., Williams, J., 2009, Capturing CO_2: Gas compression vs. liquefaction, *Power*, June 2009, pp. 68–71.
10. Jonshagen, K., Sipöcz, N., Genrup, M., 2010, A novel approach of retrofitting a combined cycle with post combustion CO_2 capture, *J. Eng. Gas Turbines Power*, Vol. 133, No. 1, p. 011703.
11. Sipöcz, N., Jonshagen, K., Assadi, M., Genrup, M., 2010, Novel high-performing single-pressure combined cycle with CO_2 capture, *J. Eng. Gas Turbines Power*, Vol. 133, No. 4, p. 041701.
12. Jonshagen, K., Sammak, M., Genrup, M., 2011, Postcombustion CO_2 capture for combined cycles utilizing hot-water absorbent regeneration, *J. Eng. Gas Turbines Power*, Vol. 134, No. 1, p. 011702.
13. Warudkar, S.S., Cox, K.R., Wong, M.S., Hirasaki, G.J., 2013, Influence of stripper operating parameters on the performance of amine absorption systems for post-combustion carbon capture: Part I–High pressure strippers, *Int. J. Greenhouse Gas Control*, Vol. 16, pp. 342–350.

20 What Next?

Basic thermodynamics tells us the theoretical upper limit of combined cycle efficiency for given gas turbine inlet temperature. It is the "enhanced" Brayton cycle efficiency developed in Chapter 8, Equation 8.2, which is repeated below for convenience

$$\eta_{CC} = 1 - \frac{\ln\left(\frac{\tau_3}{PR^k}\right)}{\tau_3 - PR^k} \tag{20.1}$$

with $k = 1 - 1/\gamma$. In Equation 20.1,

- PR is the cycle pressure ratio of the gas turbine (p_2/p_1 in standard cycle notation)
- $\tau_3 = T_3/T_1$ (T_3 is the turbine inlet temperature, TIT).

State-of-the-art technology can only achieve a fraction of that upper limit. That fraction is known as the "technology factor" (TF). In a *2016 TMI Handbook* article[1], TF for H and J class gas turbine combined cycles, based on published 2015 ISO baseload ratings, was around 0.82. This is a resounding testament to the engineers designing these power plants and the equipment therein. The high value of TF and its asymptotic trend are also indicative of the cost and difficulty involved in achieving further gains.

Combined cycle efficiency estimates (ISO baseload ratings) projected for three different TITs are summarized in Table 20.1. Cycle PRs are approximate optimum values (to maximize the gas turbine specific output). Estimates are provided for two TF predictions. Today's state of the art, based on published field-measured performances (after discounting the significant marketing hyperbole), supports the lower value. If or when the more aggressive technology push for the higher TF will bear fruit remains to be seen.

In order to investigate where the opportunities for improvement lie, a sample calculation is done. A typical advanced class gas turbine with 1,600°C (2,912°F) TIT and cycle PR of 22:1 (efficiency 41.4%) is taken as the starting point. For 22:1 PR, from Equation 4.12, the theoretical

TABLE 20.1
Projected Combined Cycle Efficiencies

TIT, °C	1,500	1,600	1,700
TIT, °F	2,732	2,912	3,092
PR	18	22	25
T_3/T_1	6.15	6.50	6.84
Carnot efficiency	83.7%	84.6%	85.4%
Brayton efficiency	56.2%	58.7%	60.1%
Brayton "enhanced" efficiency	74.4%	75.8%	76.8%
Combined cycle efficiency, TF = 0.825	61.4%	62.5%	63.4%
Combined cycle efficiency, TF = 0.85	63.2%	64.4%	65.3%

[1] See Figure 5 on page 26 in "Combined cycle trends: Past, present and future," S.C. Gülen, *Turbomachinery International Handbook* 2016, pp. 24–26.

efficiency entitlement for this gas turbine is 58.7% (with zero losses, isentropic components and no turbine cooling). Going through the elimination of loss mechanisms in an orderly fashion, one ends up with the "stairstep" chart shown in Figure 20.1. Not surprisingly, the "hot gas path", i.e., the turbine, provides the largest opportunity for improvement.

Before moving on to what can be done to reduce the turbine losses, the reader should note that the improvements in Figure 20.1 add up to about 12 percentage points. Thus, the maximum possible efficiency is **53.4%**. The difference between 53.44% and 58.7% is attributable to the *exergy destruction* (irreversibility) in the combustor. (Combustor losses in Figure 20.1 comprise pressure and heat losses, which are relatively minor.) There is nothing that can be done to alleviate that fundamental loss mechanism.

The loss mechanisms active during expansion of hot combustion gas through the turbine are quite difficult to categorize in a concise manner due to the heat transfer between the hot gas and the coolant, irreversible mixing of the two (i.e., entropy generation) and the resulting aerodynamic imperfections. In general, the problem can be reduced to achieving the following three objectives:

- Reduction of cooling flows via improvements in and/or introduction of
  - Component materials (i.e., superalloys, ceramic matrix composites)
  - Casting techniques (directionally solidified, single crystal)
  - Film/effusion cooling
  - Thermal barrier coating.
- Brush/honeycomb seals and reduced clearances
- Three-dimensional (3D) CFD for airfoil profiles and vane/blade geometries.

Technologies that can help in this endeavor are

- Additive manufacturing (e.g., blades with complex internal coolant flow channels and architecture to facilitate better film cooling)
- Advanced computer hardware and software to facilitate faster 3D-Aero design/optimization analysis

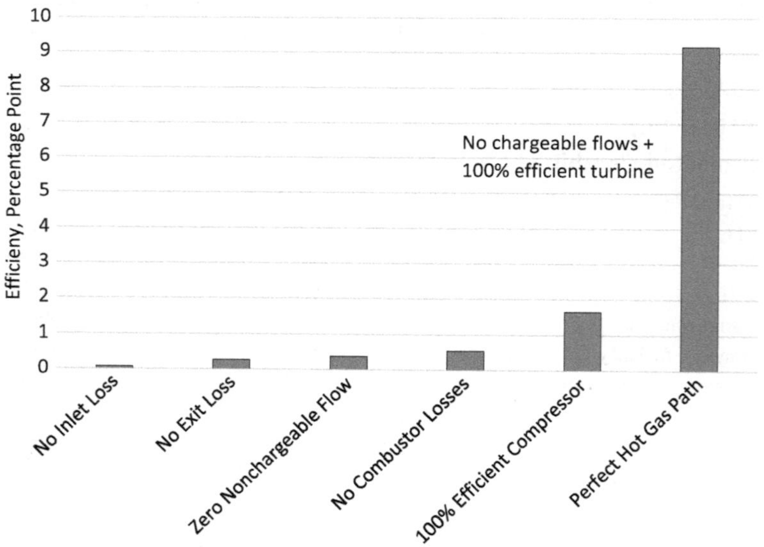

FIGURE 20.1 Gas turbine Brayton cycle entitlement stairsteps.

- Data analytics (to facilitate condition-based maintenance and push the design margins to lowest possible levels)
- Model-based (adaptive) controls.

As stated earlier, from a thermal efficiency perspective, there is not much to be done in the combustor. As far as this component is concerned, the design goal is stable, premixed combustion for low NOx at 1,700°C. The enabling technology in this respect is "axial fuel staging". The concept can be thought of as reheat or sequential combustion without a turbine stage in between the combustors. The technology also enables the turndown of the gas turbine to low loads while staying emissions-compliant.

Modern axial compressors are also quite close to entitlement with polytropic efficiencies of 93% or a bit higher. Similar to the combustor, the focus of the designers is stable operation without surge or stall across the entire operating regime (ambient and load conditions). The enabling technologies are 3D-Aero design codes and multiple stages of variable stator vanes with model-based (adaptive) controls. As the gas turbines become bigger (i.e., higher airflow) and hotter (i.e., increasing PRs), it is becoming imperative to optimize the component size (i.e., lower stage number and flow-path profile). One original equipment manufacturer (OEM)'s next-generation machine has a 12-stage axial compressor with a PR of 24:1 (Siemens HL class gas turbine). Successful design of advanced transonic compressors with highly loaded stages makes, in addition to the aforementioned 3D design tools, diligent, full-scale factory and/or field testing imperative.

In addition to the generic technologies enumerated above, OEMs have their own unique technologies and accessories outside the flange-to-flange gas turbine. (Performance fuel gas heating in combined cycle is a universally accepted technology.) Among those are

- Sequential (reheat) combustion
- Rotor and/or hot gas path (or combustor) cooling air cooler
- Steam cooling of hot gas path components
- Welded rotor (vis-à-vis bolted-disk construction)
- Hydraulic clearance optimization (see Section 8.6.1)
- Serrated (Hirth) or CURVIC coupling of disks
- Fuel moisturization (see Section 8.6.3.3).

In-depth discussion of these technologies from a historical background as well as thermal performance and operability perspectives can be found in **GTFEPG** (Ref. [11] in Chapter 2).

From a bottoming cycle perspective, it is safe to say that no "pure" technology drivers are left. As turbine inlet and exhaust gas temperatures steadily creep up, the key thermodynamic design principle is to match "source" (i.e., exhaust gas) and "sink" (i.e., steam) temperatures to minimize exergy destruction. This is primarily a material and cost issue. The existing state of the -art is 600°C (1,112°F) in main and hot reheat steam temperatures. At existing steam cycle pressures (as high as 180 bar or 2,600 psi for high-pressure steam), alloy tubes (in the heat recovery boiler) and pipes are required. This adds significantly to the construction cost of the power plant (see Section 10.2). Even higher steam temperatures are requisite for an optimal match with exhaust temperatures pushing 650°C or higher.

Rating performances are typically quoted with minimal auxiliary equipment loads and once-through water-cooled condenser (lowest possible steam turbine back pressure with minimum parasitic loss). In many places in the world, definitely in the USA, water scarcity and environmental regulations would make this an impractical choice. Another site-specific limit is the stack gas temperature from a plume abatement perspective. On paper, it is possible to design a heat recovery boiler and steam cycle to bring the stack temperature to very low levels, which would be built (even when the cost is completely ignored) only in a few places in the world.

There is no question that natural gas-fired gas turbine combined cycle is by far the most efficient fossil fuel-fired electric power generation technology. This is true at the time of writing and will be true no matter where the technology ends up in the next few decades. Superior thermal efficiency comes with the added benefit of reduced stack CO_2 emissions. This is already apparent from the recent field performance of natural gas and coal-fired power plants in the USA. As illustrated by the data in Table 12.1, at practically the same level of electricity generation, natural gas-fired units emit only one-third of the CO_2 emitted by coal-fired units. In other words, simple coal-to-gas conversion would reduce US CO_2 emissions from electric power plants by almost 60%.

Carbon dioxide is a potent greenhouse gas, and anthropomorphic CO_2 emissions since the dawn of the industrial age in early nineteenth century have been identified as the primary driver of global warming. As a direct result of this realization, renewable energy sources have been playing an ever-increasing role in the worldwide generation portfolio. Capture and storage of fossil power plant stack CO_2 and combustion of alternative fuels such as H_2 (no CO_2 generation) and biomass (net zero CO_2 impact) are among the technologies considered to stop and eventually turn the tide. One interesting (and appealing) possibility is production of hydrogen through electrolysis utilizing energy generated by renewable sources such as wind or solar at times of low demand. (Recall that wind and solar are not "dispatchable", i.e., they generate power when the wind is blowing or the sun is shining irrespective of whether there is a power demand or not.) Gas turbines can burn H_2 for efficient, zero-carbon power generation provided that there is an infrastructure for its storage and transportation.

Gas turbine combined cycle power plants can also be retrofitted with post-combustion capture units for very low-carbon footprint, which cannot be matched by any other heat engine technology with comparable efficiency and cost-effectiveness. (See Chapter 19 for a more detailed discussion.)

High-efficiency, low-cost, low-carbon footprint (even without capture) and fuel flexibility features of gas turbines are further reinforced by their agility (i.e., fast start and shutdown, fast load ramps) in simple and combined cycle configuration and ensure that the technology is going to be a major factor in electric power generation deep into the twenty-first century.

Appendix A
Property Calculations

The heart of power plant performance analysis is calculation of thermodynamic properties as a function of state-point pressure, temperature and composition (primarily for combustion products). The most needed properties are enthalpy and entropy. They are calculated from pressure and temperature (and composition, if necessary) using a proper equation of state (EOS).

For water and steam properties, there is a globally accepted industry standard: *The ASME Steam Tables*. There are two versions of ASME steam properties:

- **IFC-67**: This is the now-obsolete industry standard for the calculation of steam properties first published in 1967.
- **IAPWS-IF97**: This is the current standard developed and published by the *International Association for the Properties of Water and Steam*. This formulation was adopted by ASME in 1999 to replace the IFC-67 formulation.

Difference in calculations made with one or the other version is quite small, i.e., a few hundred kilowatts in generator output out of 230+ MWe. Note that IFC-67 is still in use for compatibility with legacy models and data. New calculations should always use IAPWS-IF97.

For gas properties, the standard package to use is the *NIST-JANAF Thermochemical Tables*. ASME and NIST-JANAF property "tables" are available as printed hard copy and preprogrammed subroutines. Many textbooks contain limited tables and/or polynomial curve-fits (e.g., the Appendix of Moran and Shapiro cited in Chapter 2). In any event, unless one has ready access to such resources through his or her organization, it is a veritable chore to program them into one's own code (e.g., as VBA function in Excel spreadsheets).

Luckily, these property packages are built into all commercially available heat and mass balance (HMB) simulation software. Thus, one does not have to worry about them too much in a modern treatise such as the one you are reading. Nevertheless, for many practical analyses, e.g., to extract "real" numbers from published rating data and similar information, one needs good estimates of gas turbine exhaust gas enthalpy and exergy (see Section 4.4). Building rigorous models and running them for such quick estimates and other front-end information purposes is not an efficient use of one's time. Therefore, reasonably accurate but very simple "transfer functions" are indispensable tools to have.

Whether one uses a rigorous EOS or simplified transfer functions introduced below, the most important item that requires attention is the "zero enthalpy reference temperature". In case of entropy calculations, required by exergy analysis, one also has to be cognizant of the "zero entropy reference pressure". Finally, "dead state" for exergy (i.e., when the system is at full equilibrium with its surroundings and cannot generate any work) is the third reference datum requiring precise specification. For details the reader is referred to **GTFEPG** (Ref. [11] in Chapter 2) or the following performance test codes and standards:

- ASME PTC 22–2005 (or the latest version) "Gas Turbines"
- ASME PTC 4.4 2008 (or the latest version) "Gas Turbine Heat Recovery Steam Generators"
- ISO-2314:2009 (or the latest version) "Gas turbines – Acceptance tests".

The HMB calculations presented in Section 4.4.1 use the methodology outlined in ASME PTC 4.4 for air and gas enthalpy calculations. The constituent enthalpy equations used in PTC 4.4 are derived from NASA correlations (Ref. NASA Reference Publication 2002–211556, September 2002), which essentially boils down to an up-to-date version of JANAF tables. The zero enthalpy reference temperature for the PTC 4.4 calculations is 60°F. This is why the air enthalpy at 59°F in Table 4.7, e.g., is a negative number.

As long as one is interested in enthalpy *differences*, which is, after all, what really matters in HMB calculations, the reference temperature selection is immaterial. If the knowledge of enthalpy with a specific reference temperature is needed, one has to apply a correction. For example, in Table 4.7, gas turbine exhaust gas enthalpy at 1,164°F is 294.92 Btu/lb (T_{ref} = 60°F). At 77°F, gas enthalpy is 4.28 Btu/lb. Thus, for T_{ref} = 77°F, the exhaust gas enthalpy at 1,164°F is 294.92 – 4.28 = 290.64 Btu/lb.

As noted earlier, the zero entropy reference datum includes the pressure 14.7 psia (since ideal gas entropy is a function of temperature *and* pressure). The proverbial wrench in the wheels is the H_2O in the combustion products forming the exhaust gas of the gas turbine. At typical reference conditions, i.e., 14.696 psia (one atmosphere) and 59°F (15°C) or 77°F (25°C), H_2O in the exhaust gas condenses out of the gas mixture. Typical H_2O content of gas turbine exhausts gas between 7% and 11% by volume or mole. Dew point T_{dp} at 14.696 psia as a function of H_2O mole fraction y_{H_2O} (as a percentage) is given by

$$T_{dp} = 55.761 \cdot y_{H_2O}^{0.3143}. \tag{A.1}$$

According to Equation A.1, the dew point range at 14.696 psia is between 104°F and 116°F (40°C and 47°C). Consequently at 59/77°F (15/25°C), instead of "dry" gas, one has a mixture of dry gas (including some H_2O vapor in the composition) and liquid H_2O. In addition to the added complexity of property calculation (which one should not worry about when using a software package), there is the practical aspect of setting a "dead state" for exergy evaluation. Going with the exact thermodynamic calculations results in a much higher exergy for the exhaust gas with the implication that the latent heat of condensation of the H_2O in the gas has value for useful work production.

Strictly speaking, one can use a condensing heat exchanger downstream of the last economizer section and make use of the latent heat of condensation of knocked-out water (e.g., by using an *Organic Rankine Cycle* (ORC) engine). However, this is cost-prohibitive and not a practical option. Therefore, in analogy to using the lower heating value (LHV) of natural gas fuel in efficiency calculations instead of higher heating value (HHV), it is more reflective of the engineering reality to ignore the latent heat of condensation of H_2O in exergy calculation.

This assertion can be demonstrated by an example calculation. Let us consider exhaust gas properties below

- 15.35 psia, 1,200°F with molar composition of N_2 = 73.841%, O_2 = 10.91%, CO_2 = 4.481%, H_2O = 9.869% and Ar = 0.899%
- Corresponding mass fraction of H_2O can be calculated as 5.3% (T_{dp} of about 114°F from Equation A.1 with $y_{H_2O} = 9.869$).

At 59°F, mole fraction of H_2O in the gas mixture is only about 1.7%. The remainder, i.e., the difference between 9.9%(v) and 1.7%(v), will condense out of the gas mixture, carrying out the latent heat of condensation of 1,060 Btu/lb (from the *ASME Steam Tables* at 59°F) with it.

Typical stack gas temperature of a modern combined cycle is 180°F (enthalpy 26.14 Btu/lb). For an advanced 50-Hz gas turbine with 2,000 lb/s exhaust flow, one can cool the gas from 180°F to 59°F (impossible in practice, of course, but consider this as a *Gedankenexperiment*) for total heat transfer of

$$2{,}000(26.14 - 60.79) = 173{,}860 \text{ Btu/s}.$$

Mean-effective temperature is 117.4°F (e.g., see Equation 4.10 for a sample mean-effective temperature calculation). Thus, if one could build a Carnot engine to make use of this heat, its efficiency would be

$$1-519/(117.4+460)\sim 10\%.$$

The futility of building a heat engine whose thermodynamic potential is only 10% is obvious. There may very well be an ORC to squeeze maybe a few hundred kilowatts out of this available heat source but its cost would be significantly above the acceptable value of $1,500/kW (roughly the capex of a state-of-the-art bottoming steam cycle – see Chapter 13). (As a reference magnitude, consider that the steam turbine generator output of the sample gas turbine combined cycle is well above 250,000 kWe.)

Using THERMOFLEX®, at 15.35 psi and 1,200°F, enthalpy and entropy, h and s, are found as 301.81 Btu/lb and 0.3581 Btu/lb-R, respectively. At 59°F and 14.7 psia with water as liquid in the dry gas-plus-liquid mixture, h and s are −60.79 Btu/lb and −0.0488 Btu/lb-R, respectively. Note that gas properties in THERMOFLEX® are based on specific heats determined from fifth-order polynomials, derived as curve-fits to data in the JANAF tables. Zero enthalpy/entropy reference is 77°F. This is the reason why h and s at 59°F have negative values. The equation for the exhaust gas exergy is (e.g., see Moran and Shapiro)

$$a_{exh}=\left(h_{exh}-h_0\right)-T_0\left(s_{exh}-s_0\right)$$

with respect to a "dead state" denoted by the subscript 0. If the dead state is chosen as 59°F and 14.696 psia, the substitution yields

$$a_{exh}=(301.81-60.79)-(59+459.67)(0.3581-0.0488)=151.55\,\text{Btu/lb}$$

(*a* for *availability* – the name for exergy in US textbooks). In this example, dead-state properties are calculated rigorously with water as liquid in the dry gas-plus-liquid mixture. If the minuscule and practically worthless contribution of H_2O condensation is to be ignored, extrapolation of enthalpy and entropy when the gas is "dry" to the dead-state temperature is one way to estimate a *pseudo* exergy reference. Assuming 59°F for the dead state, with the extrapolated enthalpy and entropy values, –7.86 Btu/lb and 0.0491 Btu/lb-R, respectively, the exhaust gas exergy is found as

$$a_{exh}=(301.81-7.86)-(59+459.67)(0.3581-0.0491)=149.40\,\text{Btu/lb},$$

which is within 2 Btu/lb of the "true" thermodynamic value obtained above (–1.4% error). Furthermore, this value is also within 2 Btu/lb of what one would obtain if the dead state had been set exactly at the dew point of H_2O in the exhaust gas at 14.7 psia, which is 114°F. Note that zero enthalpy and zero entropy reference selection is immaterial for exergy calculation because it is based on enthalpy and entropy *differences* between the actual fluid state and the dead state. What *does* make a difference is the dead-state conditions, especially the temperature.

From a practical perspective, the difference between 59°F (15°C) or 77°F (25°C) as the dead-state temperature is immaterial. As discussed in detail in Chapter 6, heat recovery steam generator (HRSG) stack gas temperature is unlikely to be below about 150°F (about 65°C) due to plume dispersion and economic considerations (see Section 6.2.4). The key point of attention is (i) consistency in usage of a given reference and (ii) awareness of assumptions used in other references for proper evaluation and/or comparison of findings. With that in mind, the following linear transfer functions are developed for use in simplified calculations. The independent variable is the exhaust gas temperature. Pressure has a very small effect through the entropy in the exergy definition. A series of combustion runs have been made in THERMOFLEX® to arrive at exhaust temperatures commensurate with firing temperatures from older E class to the most advanced H/J class gas turbines (using sample machines from the built-in "engine library"). Thus, the composition effect is implicit in the formulas.

The exergy of gas turbine exhaust gas burning natural gas fuel, to a good approximation, is given by

$$a_{exh} = 0.1961T_{exh} - 86.918\,\text{Btu/lb} \quad (A.2)$$

with a dead-state temperature of 59°F (T_{exh} is in °F). With a dead-state temperature of 77°F, the transfer function becomes

$$a_{exh} = 0.1909T_{exh} - 87.0\,\text{Btu/lb}. \quad (A.3)$$

A similar formula used by the author in his earlier publications is (e.g., Ref. [8] in Chapter 8)

$$a_{exh} = 0.001628T_{exh}^{1.60877}. \quad (A.4)$$

Equation A.4 is valid between 900°F and 1,200°F with a zero enthalpy reference of 59°F for exhaust gas composition typical of 100% methane gas turbine fuel. The difference between the exergy predictions of Equations A.2 and A.4 is summarized in Table A.1.

Similarly, the enthalpy of the exhaust gas is given by

$$h_{exh} = 0.3003T_{exh} - 55.576\,\text{Btu/lb} \quad (A.5)$$

with a zero Btu/lb enthalpy reference of 59°F and H_2O in the mixture in gaseous form. With a zero Btu/lb enthalpy reference of 77°F, the transfer function becomes

$$h_{exh} = 0.3003T_{exh} - 59.897\,\text{Btu/lb}. \quad (A.6)$$

The difference in enthalpy calculation between the two reference point assumptions is 4.3 Btu/lb.

Equations A.2 through A.6 can be used for the temperature range 900–1,200°F with reasonable accuracy. The error resulting from using these simple equations in lieu of *bona fide* property calculations should not be more than ±1% to 2%. The information implicit in Equations A.5 and A.6 is the specific heat of the gas turbine exhaust gas, i.e., $c_p = 0.3003$ Btu/lb-R (1.2573 kJ/kg-K), which is approximately, but not exactly, constant in the range of its applicability.

TABLE A.1
Comparison of Different Exergy Predictions

Exhaust Temperature	Equation A.2	Equation A.4	Δ	
°F	Btu/lb		Btu/lb	%
900	89.6	92.1	2.5	2.8
950	99.4	100.5	1.1	1.1
1,000	109.2	109.1	0.0	0.0
1,050	119.0	118.0	−0.9	−0.8
1,100	128.8	127.2	−1.6	−1.2
1,150	138.6	136.7	−1.9	−1.4
1,200	148.4	146.3	−2.1	−1.4

In a similar fashion, HRSG stack gas enthalpy can be estimated by

$$h_{stck} = 0.2443T_{stck} - 13.571\,\text{Btu/lb} \tag{A.7}$$

with a zero Btu/lb enthalpy reference of 59°F and H_2O in the mixture in gaseous form. With a zero Btu/lb enthalpy reference of 77°F, the transfer function becomes

$$h_{stck} = 0.2443T_{stck} - 17.892\,\text{Btu/lb} \tag{A.8}$$

The information implicit in Equations A.7 and A.8 is the specific heat of the HRSG stack gas, i.e., $c_p = 0.2443$ Btu/lb-R (1.0228 kJ/kg-K). For the exergy of the HRSG stack gas, use

$$a_{stck} = 0.0519T_{stck} - 6.1539\,\text{Btu/lb} \tag{A.9}$$

with a zero Btu/lb enthalpy reference of 59°F and H_2O in the mixture in gaseous form. With a zero Btu/lb enthalpy reference of 77°F, the transfer function becomes

$$a_{stck} = 0.045T_{stck} - 5.7748\,\text{Btu/lb.} \tag{A.10}$$

Appendix B
Standard Conditions for Temperature and Pressure

This book heavily uses property calculations for design and off-design performance estimates of combined cycle power plants. The two most important thermodynamic properties in those calculations are enthalpy and *exergy*, which is a combination of enthalpy and entropy. For enthalpy, entropy and exergy, numerical values are meaningless unless the "zero value" reference conditions are specified. This has been touched upon in Appendix A in conjunction with the enthalpy and exergy transfer functions.

Zero enthalpy or zero entropy reference conditions are usually set to one of the widely used standard conditions for temperature and pressure (STP). These are standard sets of pressure and temperature defined by certain organizations to allow comparisons to be made between different sets of data.

In a perfect world, there would be one and only one STP set defined by a supranational organization. Alas, this is not the case. The most widely used standards are those of the *International Union of Pure and Applied Chemistry* (IUPAC) and the *National Institute of Standards and Technology* (NIST). However, there are other organizations that have established a variety of alternative definitions for their standard reference conditions.

For the gas turbines, the universally accepted set of STP is defined by the *International Organization for Standardization* (ISO) as 59°F (15°C) and 14.696 psi (101,325 Pa or 760 mm Hg), which also specifies the relative humidity as 60%.

The STP definition by the United States Environmental Protection Agency (EPA) differs from that by the ISO in temperature, i.e., 77°F (25°C). There is no relative humidity definition.

There are many other definitions, which use 32°F, 60°F or 70°F among others with slight variation in pressure (usually one atmosphere). For example, conditions for normal cubic meters (Nm^3) are 0°C (32°F) and 760 mm (29.92126 in.) of mercury (Hg). Conditions for standard cubic feet (SCF) are 60°F and 14.696 psia. Unfortunately, many technical publications including equipment specs and brochures fail to specifically spell out the STP set used in their data sets and calculations. Merely stating "x cubic feet or cubic meters per minute at standard conditions" or "y Btu or kilowatt exhaust energy" does not provide enough information and can lead to gross errors and/or misconceptions.

For this reason, utmost importance is placed upon zero enthalpy and entropy references in property calculations involving the combustion products (i.e., the exhaust stream of the gas turbine), which are vital in accurate assessment of steam (bottoming) Rankine cycle of the gas turbine combined cycle.

Appendix C
Exergetic Efficiency

Based on the premise that there is a well-designed (or optimal) bottoming cycle (Rankine cycle with steam/water as the working fluid) for a given gas turbine exhaust flow and temperature, different technology curves in the form of $f(T_{exh})$ can be generated. Such a curve is given in the paper by Gülen and Smith (Ref. [14] in Chapter 5). This optimal curve goes through the data points representing the bottoming cycle (steam Rankine) exergetic efficiencies extracted from the trade publication data and is adequately described by the following formula:

$$\varepsilon_{BC} = 0.2441 + 0.0746\left(\frac{T_{exh}}{100}\right) - 0.00279\left(\frac{T_{exh}}{100}\right)^2. \tag{C.1}$$

(Note: exhaust temperatures are in degrees Fahrenheit.) Equation C.1 is very well suited to combined cycle analysis of air-cooled gas turbines with the implicit assumption of a well-designed Rankine bottoming cycle based on 2005–2010 vintage most optimistic state-of-the-art technology and aggressive equipment design specifications. Specifically, it is representative of three-pressure reheat (3PRH) systems based on advanced F class gas turbines with exhaust temperatures between 590°C (1,094°F) and 670°C (1,238°F) and heat recovery steam generator (HRSG) stack temperatures of about 82°C (~180°F). For advanced H and J class gas turbines (around 2020) with aggressive bottoming cycle designs, predictions from Equation C.1 can be bumped up by a few percentage points.

When Equation C.1 is plotted, it is obvious that beyond exhaust temperatures of ~700°C (~1,300°F), there is essentially no room for improvement in ε_{BC} (see Figure C.1). The key reason for that is that it is not yet possible or cost-effective to increase steam temperatures in lock-step with the

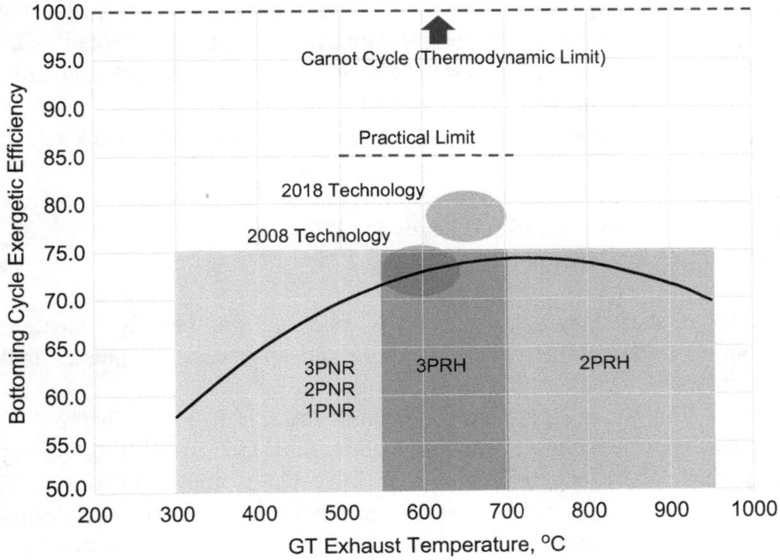

FIGURE C.1 GTCC bottoming cycle exergetic efficiency.

gas temperature due to material limitations (current state of the art is 600°C or 1,112°F main and/or hot reheat steam). Secondary reasons are (i) elimination of HRSG IP and LP sections progressively and (ii) limits on steam pressure increase.

C.1 BOTTOMING CYCLE EXERGY BALANCE

Bottoming cycle exergetic efficiency (or, rather, effectiveness) of 75%–80% implies that 20%–25% of maximum work producing of the gas turbine exhaust gas, quantified by the thermodynamic property exergy, is somehow "lost". For a given system control volume, there are two basic loss mechanisms:

- Exergy "destruction" (i.e., irreversibility)
- Exergy transfer *out* of the control volume.

In order to identify the individual exergy loss buckets, let us look at a sample problem. For starters, let us specify an advanced class gas turbine with the following data:

- Exhaust flow is 2,250 lb/s
- Exhaust temperature is 1,172°F
- 225°C (437°F) fuel gas performance heating.

Thermoflow's GTPRO® heat and mass simulation software is used for combined cycle performance calculation. Very aggressive (i.e., expensive to procure and construct) bottoming cycle design assumptions are used, i.e.,

- 10°F pinches in high, intermediate and low pressure (HP, IP and LP) evaporators
- 2,600 psig/1,112°F/1,112°F steam cycle
- Once-through HP evaporator (i.e., effectively economizer approach subcool is zero)
- Open-loop water-cooled condenser at 1.2 in. Hg
- 8.5% reheater pressure loss.

The steam turbine output from the GTPRO model is 259.3 MWe (245.8 Btu/s). The bottoming cycle exergy balance output of the GTPRO model is summarized in Table C.1. Note that GTPRO exergy calculation basis (also known as the "dead state") is 77°F (25°C) and 14.696 psia with H_2O as vapor in the gas mixture (i.e., a *pseudo* state – see Appendix A).

In Table C.1, bottoming cycle losses can be combined into three major buckets:

- Exergy destruction (heat transfer irreversibility) in the HRSG about 9%
- Total steam turbine generator (STG) losses about 7%
- Heat rejection irreversibility about 3.5%.

The remaining 4% is shared by stack gas exergy and fuel gas heating water exergy (transfer out of the bottoming cycle control volume) equally and miscellaneous plus a small amount for unaccounted-for losses.

A comparison of 2008 (from Figure 5 in Gülen and Smith, Ref. [14] in Chapter 5) and 2018 state of the art in bottoming cycle technology is provided in Table C.2. In order to facilitate an "apples-to-apples" comparison, the exergy breakdown in Table C.1 is converted to a 59°F "dead-state" basis. The major difference is in exergy transfer (loss) with heat rejection to cooling water in the condenser, which is zero in Table C.1. The reason for that is that the mean-effective cooling water temperature, about 68°F, is lower than the "dead-state" temperature, 77°F.

TABLE C.1
Bottoming Cycle Exergy Balance

Exergy Loss Source	Btu/s	%
Gas turbine exhaust exergy	308,112	100.00
Duct loss	433	0.14
HRSG inlet gas exergy	307,679	
Heat transfer exergy loss	26,516	8.61
Blowdown and feed pump losses	298	0.10
Other HRSG loss	4,999	1.62
Feedwater to fuel gas heater	3,242	1.05
Steam turbine expansion exergy loss	16,195	5.26
Mechanical/electrical/gear loss	3,483	1.13
Other Steam turbine loss	1,941	0.63
Condenser irreversibility	11,000	3.57
Stack gas exergy	2,958	0.96
STG electric output	245,800	
Feed pump power	4,756	−1.54
Bottoming cycle net output	241,044	78.23

TABLE C.2
Bottoming Cycle Technology Progress 2010–2018

	2018 State of the Art (%)	2008 State of the Art (%)
HRSG (including duct)	8.53	10.10
Feedwater to fuel gas heater	1.03	0.80
Steam turbine loss	6.84	7.70
Condenser irreversibility	3.37	3.30
Condenser heat rejection	1.89	3.60
Stack loss	1.43	2.60
Miscellaneous	1.68	1.50
Bottoming cycle net output	76.26	70.50

Higher steam temperatures and tighter heat exchange approach temperature deltas clearly resulted in higher steam production, lower stack gas temperature and reduced HRSG heat transfer irreversibility and stack loss. More efficient steam turbine and lower condenser pressure (1.2 vis-à-vis 1.5 in. Hg) resulted in lower steam turbine expansion irreversibility and lower heat rejection in the condenser. Overall, nearly 6 percentage points of improvement in bottoming cycle exergetic efficiency is achieved. (Each percentage point in bottoming cycle exergetic efficiency is worth about 0.25 percentage points in combined cycle efficiency.)

Appendix D
Thermal Response Basics

In mathematical treatment of transient (time-dependent) phenomena, the most important parameter is the time constant of the process under investigation. The time constant is the characteristic time scale of a physical process where thermodynamic (pressure, temperature and density, for example) and hydrodynamic properties (mass flow rate or velocity) of the working fluid change with time, in response to a perturbation, and reach equilibrium. In *steady-state, steady-flow* (SSSF) processes, the time constant is of no consequence because the properties do not change with time. In *uniform-state, uniform-flow* (USUF) processes, the time scale and, more specifically, its magnitude are of prime importance.

If the time constant is too small, e.g., of the order of milliseconds, and the change in properties are not too big, they can be ignored and the process can be treated as a *quasi*-SSSF process. In other words, at each time instant, t_i, the fluid is assumed to be at an equilibrium state with the new values of properties, which may or may not be different from those at the previous time instant, t_{i-1}.

If the time constant is too small but the change in properties is too big, the process can be treated as a discontinuity in an otherwise SSSF process. The most familiar example is the normal shock wave in a compressible fluid flow. Upstream and downstream of the shock wave, the fluid is at equilibrium with much higher pressure and temperature (and subsonic flow, $Ma < 1$) downstream vis-à-vis upstream (supersonic flow, $Ma > 1$).

If several processes are taking place simultaneously and they have different time scales, a hybrid model is appropriate. A well-known example is the *detonation* wave, which is the combination of a normal (hydrodynamic) shock wave, which is treated as a discontinuity across which temperature and pressure of an upstream combustible fluid mixture increase within a few microseconds, and a downstream relaxation zone, across which the chemical reactions triggered by the sudden rise in temperature and pressure are completed in a few milliseconds (i.e., three orders of magnitude slower). Those chemical reactions are assumed to be *frozen* during the forerunner shock; i.e., there is not enough time for them to start and reach completion.

In mathematical representation of combined cycle transient performance, the gas turbine operation is the "driving event". In other words, it is the *perturbation* to which the bottoming cycle responds dynamically with a certain *time lag*. Thus, the key endeavor in combined cycle transient analysis is the estimation of the bottoming cycle thermal response to the changes in gas turbine exhaust flow and temperature.

A common combined cycle transient problem is the startup from standstill and the thermal behavior of the bottoming cycle components with large metal mass during that particular transient. The physical process taking place in the heat recovery steam generator (HRSG) is the heat transfer from the hot flue gas to the heat exchangers in the HRSG (i.e., superheaters, evaporators and economizers) and, consequently, to the working fluid therein (water, steam or a mixture of the two).

The process in a particular HRSG heat exchanger (in essence, vertically arranged rows of tubes in cross-glow arrangement) can be described in *four* identifiable subprocesses, which take place simultaneously:

1. Heat transfer from the hot flue gas to the cold heat exchanger tubes (metal)
   a. Cooling of flue gas
   b. Heating of the heat exchanger metal mass (tubes and evaporator drums)

2. Heat transfer from the heat exchanger tubes (metal) to the water/steam flowing therein
   a. Cooling of the heat exchanger metal mass
   b. Heating (including evaporation) of water and/or steam.

For these four subprocesses, four time constants can be identified:[1]

1. For the external heat exchange
   a. For the flue gas, which is of the order of milliseconds (0.1 s)
   b. For the metal, which is of the order of minutes (100 s)
2. For the internal heat exchange
   a. For water/steam, which is of the order of seconds (1 s)
   b. For the metal, which is of the order of minutes (100 s).

Of the four time constants listed above, the first and third one can be ignored for practical engineering purposes. They take place sufficiently fast so that the enthalpy exchange between the gas and water/steam can be handled using quasi-equilibrium or quasi-steady assumption, i.e.,

$$\Delta H_{gas} = (\hbar A)_{ext}\left(\overline{T}_{gas} - T_{met}\right) \tag{D.1}$$

$$\Delta H_{w/s} = (\hbar A)_{int}\left(T_{met} - \overline{T}_{w/s}\right) \tag{D.2}$$

where

ΔH_{gas} = Total enthalpy change of flue gas, Btu/h
$\Delta H_{w/s}$ = Total enthalpy change of steam or feedwater, Btu/h
$\hbar A_{ext}$ = External overall heat transfer coefficient, Btu/h-F
$\hbar A_{int}$ = Internal overall heat transfer coefficient, Btu/h-F
$\overline{T}_{gas}$ = Average flue gas temperature, °F
$\overline{T}_{w/s}$ = Average steam or feedwater temperature, °F
T_{met} = Bulk heat exchanger metal temperature, °F

The *lumped capacitance method* (LCM) can be used to obtain the transient (bulk) heat exchanger metal temperature. The entire metal mass of the heat exchangers in the HRSG gas path comprises bare tubes and (solid or serrated) fins on those tubes. (Note that evaporator drums are outside the HRSG gas path. They are situated on top of the HRSG box and exposed to the ambient air.) In a horizontal HRSG (the most prevalent type in the USA), the finned tubes are positioned in rows hanging from the headers in the flow-path of the gas turbine exhaust gas. Fin density and proximity of the tubes are such that the approximation of gas flow around a "lumped" metal object with the same mass (i.e., volume and density) and heat transfer surface area of the finned-tube, cross-flow HRSG heat exchanger is reasonably good.

In graphical terms, the real HRSG heat exchanger section, which is three dimensional with rows of vertical finned tubes and flue gas in cross flow, can be replaced by a simple, one-dimensional (hypothetical) counterflow heat exchanger. This hypothetical heat exchanger is a cylinder with a wall thickness of L_c (see below for its definition) and its geometry (basically its length and diameter) is such that to have the same overall heat transfer rate and metal mass as the actual heat exchanger.

A particular HRSG tube bank follows the changes in the temperature of the hot flue gas (and settles into a new steady state) with a characteristic time delay. In mathematical terms, this is the time at which the system completes approximately two-thirds (to be exact, $1 - e^{-1}$) of its total

[1] From Dechamps, P.J., 1995, Modelling the transient behavior of heat recovery steam generators, *Proc. Inst. Mech. Engrs.*, Vol. 209, pp. 265–273

change before reaching the new equilibrium. The parameter that determines the "slowness" of the thermal response of an HRSG section (and thus the temperature of steam) is the time constant in the fundamental heat transfer equation governing the "bulk" tube metal temperature,

$$T(t) = T_0 + (T_\infty - T_0)\left(1 - \exp\left(-\frac{t}{\tau}\right)\right) \tag{D.3}$$

where T(t) is the tube metal temperature at time t, T_0 is its initial (cold) temperature, T_∞ is the temperature of the hot gas flowing around the tubes and τ is the time constant, which is given by

$$\tau = \frac{Mc}{\hbar A} = \frac{\rho c L_c}{\hbar}. \tag{D.4}$$

Note that Equation D.3 introduces a mathematical expression, which is of utmost importance in power plant transient problems, i.e., the *exponential decay function* (EDF) $\exp(-t/\tau)$. In particular,

- Time-dependent cooling or slowing processes are characterized by EDF;
- Time-dependent heating and acceleration processes are characterized by (1 − EDF).

In Equation D.4, M is the total tube metal mass, c is the specific heat of the tube metal, $\hbar$ is the convective heat transfer coefficient of the hot gas flow, ρ is the tube metal density and A is the total heat transfer surface area. The characteristic length L_c used to determine the time constant is the length scale corresponding to the maximum spatial temperature difference for the particular heat exchanger. As such, a sensible choice is the tube wall thickness.

The external (convection) heat transfer coefficient governing the heat exchange between the hot flue gas and the cold tube metal, $\hbar_{ext}$, is of the order of 10 Btu/h-ft^2-F. The internal heat transfer coefficient governing the heat exchange between the hot tube metal and cold water or steam, $\hbar_{int}$, is of the order of 100 Btu/h-ft^2-F. In particular, for a three-pressure, reheat HRSG with an advanced F class gas turbine with 1,150°F–1,200°F exhaust gas temperature at ISO baseload, typical values for the internal heat transfer coefficient are

- ~500 Btu/h-ft^2-F for high-pressure (HP) superheaters
- ~125 to 150 Btu/h-ft^2-F for reheat and intermediate-pressure (IP) superheaters
- ~1,000 Btu/h-ft^2-F for HP evaporator
- ~1,500 Btu/h-ft^2-F for economizers
- ~400 Btu/h-ft^2-F for IP and low-pressure (LP) evaporators.

Thus, clearly, the heat transfer is governed by the external gas-to-metal heat transfer. Using the typical values of 475 lb/ft^3 for the tube metal density and 0.1 Btu/lb-F for the tube metal heat capacity, the time constant for a heat exchanger with 0.2 in. tube wall thickness can be found from Equation D.4 as

$$\tau = \frac{475 \cdot 0.1\left(\frac{0.2}{12}\right)}{10} 60 \cong 5 \text{ min}.$$

Differences from this value occur as a result of variation in metal properties (driven by the material used in heat exchanger construction, e.g., T91, T22, carbon steel, and, to a lesser extent, gas temperature variation) and tube geometry, i.e., tube wall thickness. A range of 4–10 min should be expected. Note that this value is higher than the characteristic time constant of 100 s quoted above, albeit at the same order of magnitude.

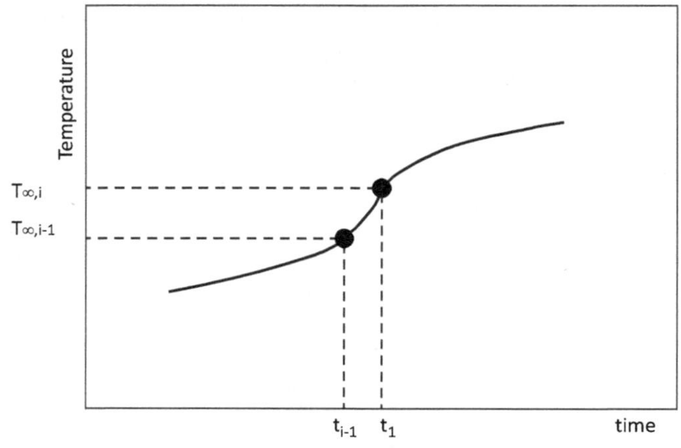

FIGURE D.1 Approximate solution setup for an arbitrary gas temperature profile.

The simple LCM formula, Equation D.3, is valid for a constant free stream temperature. During startup, gas turbine exhaust temperature profile is highly nonlinear (e.g., see Figure 15.13). The next level of complexity is a linear free stream temperature profile, i.e.,

$$T_\infty(t) = a + bt. \tag{D.5}$$

For this temperature profile, after some tedious but straightforward calculus, it is found that Equation D.3 becomes

$$\frac{T(t) - T_{\infty,0}}{T_0 - T_{\infty,0}} = b\tau \frac{\left\{\frac{t}{\tau} + \left[\exp\left(-\frac{t}{\tau}\right) - 1\right]\right\}}{T_0 - T_{\infty,0}} + \exp\left(-\frac{t}{\tau}\right) \tag{D.6}$$

with

$$T_\infty(t) = T_{\infty,0} + \left(\frac{T_{\infty,f} - T_{\infty,0}}{T}\right)t \tag{D.7}$$

where $T_{\infty,0}$ is the initial free stream temperature ($t = 0$) and $T_{\infty,f}$ is the final free stream temperature. For more complicated free stream temperature profiles, a closed-form solution is not possible. However, an arbitrary temperature profile can be discretized into linear segments with a small-enough time step as shown in Figure D.1. Specifically, for a transient process taking t minutes, one can divide it into N sections of duration t/N, during which the temperature change is, to a very good approximation, linear. In that case,

$$a = T_{\infty,i-1}, \tag{D.8}$$

$$b = \frac{T_{\infty,i} - T_{\infty,i-1}}{t_i - t_{i-1}}. \tag{D.9}$$

In this manner, Equation D.6 can be applied successively to each linear segment.

Appendix E

Steam Turbine Stress Basics

Steam conditions (flow, pressure and temperature) acceptable for admission into the steam turbine are dictated by metal temperatures (primarily valves, casings or shells and the rotor). The critical component is the rotor, whose temperature cannot be measured directly and inferred by proxies (e.g., high- and low-pressure, HP and IP, inner bowl). Steam turbine metal temperature, T_m, is a direct function of unit downtime and ambient temperature as shown in Figure E.1 (unless *forced* cooling is applied to start maintenance as soon as possible to minimize the downtime). The *natural* cooling time depicted in Figure E.1 is represented by the exponential decay function (EDF)

$$\frac{T_m - T_{amb}}{T_{m,0} - T_{amb}} = \exp\left(-\frac{t}{\tau}\right) \tag{E.1}$$

with a characteristic *cooling time constant*, τ, as a function of the ambient temperature, T_{amb}, and the starting value (denoted by subscript 0). This temperature is the main gas turbine combined cycle (GTCC) startup classification gauge instead of widely used but fuzzy terms such as "hot" or "warm", whose definitions vary from one source to another.

Component metal temperature and, more precisely, its variation in a metal structure across a characteristic dimension, L_c, along a characteristic dimension, x, is the key determinant of thermal stress via the following formula:

$$\sigma = E'\alpha\Delta T_m \tag{E.2}$$

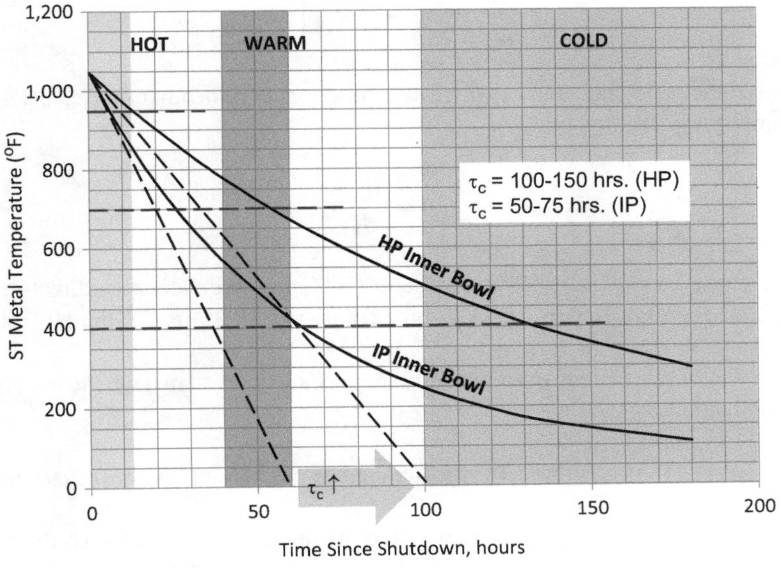

FIGURE E.1 Typical steam turbine cool-down profiles (as measured at HP and IP inner bowls). Shaded regions indicate typical time windows for "hot", "warm" and "cold" start classifications. Dashed lines indicate average metal HP bowl temperature corresponding to the same.

TABLE E.1
Definition of Key Material Parameters and Their Typical Values

Modulus of Elasticity	E	26,000 (180)	ksi (GPa)
Linear coefficient of thermal expansion	α	$6–7 \times 10^{-6}$ $(1.08–1.026) \times 10^{-5}$	1/R (1/K)
Poisson's ratio	ν	0.30	
Thermal conductivity	k	18.0 (31.1)	Btu/h-ft-F (W/m-K)
Density	ρ	490 (7,850)	lb/ft³ (kg/m³)
Heat capacity	c	0.125–0.175 (0.523–0.733)	Btu/lb-R (kJ/kg-K)
Thermal diffusivity	δ	$0.20–0.25\ (18.6–23.2) \times 10^{-3}$	ft²/h (m²/h)

where E' = E/(1 − ν). Key material parameters appearing in thermal stress formulae and their typical values are summarized in Table E.1. Note that there are two classes of materials (stainless steels) used in steam turbine construction: ferritic and austenitic. In general, for temperatures above 580°C, austenitic steels are used in turbine component construction. Modulus of elasticity, thermal conductivity and heat capacity are strong functions of temperature. *Thermal diffusivity*, $\delta = k/\rho c$, quantifies the *speed* with which the temperature of a heated or cooled body changes. It is the controlling parameter in the Fourier's law of heat conduction (Equation E.6 below), which is used to calculate the variation of heat flow with time and the resulting temperature distribution in a heated or cooled body. Charts for k, c and δ for ferritic and austenitic steels can be found in Ref. [1]. Due to their difference in thermal conductivity/diffusivity, i.e., much lower for austenitic steels, thermal response time of components constructed from ferritic steels is much faster.

For the steam turbine rotor, ΔT_m in Equation E.2 is the difference between rotor surface or bore and mean body (bulk) temperatures for surface and bore stresses, respectively. For a given steam temperature, T_{stm}, bulk rotor body metal temperature varies according to 1 − EDF, i.e.,

$$\frac{T_m - T_{m,0}}{T_{stm} - T_{m,0}} = 1 - \exp\left(-\frac{t}{\tau}\right) \tag{E.3}$$

with a characteristic time constant, τ, which is a function of rotor material (e.g., 1% CrMoV) and size *cum* geometry represented by L_c,

$$\tau = \frac{\rho L_c c}{\hbar} \tag{E.4}$$

where $\hbar$ is the *convective heat transfer coefficient* (HTC) between steam and metal. A reasonable selection for L_c is the diameter of steam turbine rotor – 20 to 25 in. for the HP rotor of modern GTCC units.

Equations E.1–E.4 tell the entire steam turbine thermal stress management story in the concise language of mathematics:

1. Thermal stress is determined by the temperature gradient in the rotor (essentially a cylinder) via Equation E.2
2. The latter is determined by the initial steam–metal ΔT (denominator of the ratio on the left-hand side of Equation E.3) with a time lag, which itself is dictated by the HTC in Equation E.4
3. Everything hinges on the initial value of the metal temperature, $T_{m,0}$, which is a function of the cooling period given by Equation E.1.

In physical terms, this translates into a mechanism to control steam conditions into the steam turbine at initial values sufficient

i. To roll the unit from turning gear (TG) speed to *full-speed, no-load* (FSNL)
ii. To warm the steam turbine rotor until steam–metal temperature mismatch decreases to an acceptable level
iii. To ramp them up at acceptable rates to their rated levels while ensuring that thermal stresses do not exceed prescribed limits.

Steam flow enters the picture via HTC in Equation E.4, which controls the rate of heat transfer between steam and the rotor surface as described by the *heat flux* balance at the steam–metal boundary (x = 0)

$$\dot{q} = \hbar(T_{stm} - T_m) = k\frac{dT_m}{dx}\bigg|_{x=0}. \tag{E.5}$$

This equation introduces the dimensionless *Biot* number, $Bi = \hbar \cdot L_c/k$, which is a relative measure of the uniformity of temperature gradients inside a heated or cooled body. Determination of HTC is one of the most uncertainty-prone undertakings in transient heat transfer problem in a complex geometry such as steam turbine/valve flow channels. Its dependence on steam flow is based on the well-known *Nusselt* number correlation for heat transfer in internal flows, i.e., $\hbar \propto \dot{m}_{stm}^{0.8}$. The heat transferred from steam to the rotor at the surface increases the rotor's bulk temperature according to Fourier's law of heat conduction, i.e.,

$$\frac{dT_m}{dt} = \delta\frac{d^2T_m}{dx^2} \tag{E.6}$$

in Cartesian coordinates and one dimension. (The formula is much more complicated in three dimensions and in cylindrical coordinates but does not provide additional insight.) Typical values for the key parameters governing steam turbine rotor thermal transients are given in Table E.2.

TABLE E.2
Representative Values of Major Parameters Characterizing the Transient Heat Transfer during Steam Turbine Warm-Up for Typical Steam Flow, Pressure and Temperatures

m/m_o	P	T	$\hbar$	Bi	δ	τ
[-]	psia	F	Btu/h-ft²-F	[-]	ft²/h	min
1.0	120	700	116	7	0.26	37
1.0	120	1,050	100	6	0.21	54
1.0	1,200	700	958	56		5
1.0	1,200	1,050	701	41		8
0.2	120	700	32	2		135
0.2	120	1,050	28	2		196
0.2	1,200	700	264	15		16
0.2	1,200	1,050	193	11		28

For ferritic steels used in modern GTCC units, k and ρ do not show significant variation with pressure. Thus, δ is primarily a function of temperature and changes by about 25% between 700°F and 1,050°F; i.e., rate of change of metal temperature is 25% faster at the *lower* temperature. The data in Table E.2 can be summarized as follows:

> higher steam flow and/or pressure result in higher rates of heat transfer between steam and metal, which is quantified by higher Biot numbers and shorter time constants (i.e., faster heating or cooling).

In conjunction with the data in Table E.2, Equations E.5 and E.6 identify the two distinct phases in steam turbine start with thermal stress control:

i. Low flow and high steam–metal temperature difference with low HTC until temperature gradients settle down (*non-stationary* phase or **Phase I**)
ii. Increasing steam flow, pressure and temperature to load the unit with high HTC and nearly constant, low steam–metal temperature difference (*quasi-stationary* phase or **Phase II**).

Equation E.5 describes Phase I via its simplified solution for a cylindrical geometry given by [1]

$$\sigma_{max} = E'\alpha\phi_Z K_T \Delta T \tag{E.7}$$

with

$$\phi_Z = \frac{Bi}{2.8 + Bi + \sqrt{Bi}}, \tag{E.8}$$

which gives the maximum thermal stress implied by a given step rise in steam temperature at time $t = 0$ (with a time lag characterized by the Biot number). Equation E.8 for typical geometries in a graphical form is presented in Ref. [1] (Figure 28 in the cited work). For quick estimates in typical hot/warm startup problems, 0.6–0.7 can be assumed for ϕ_Z.

Note that the base stress formula of Equation E.2 is amplified by a *stress concentration factor* K_T, which accounts for the presence of geometric discontinuities on the rotor (which is not a perfect cylinder after all). Similarly, Equation E.6 describes Phase II via its simplified form given by

$$\frac{dT_{stm}}{dt} = \frac{\delta}{\phi_F E'\alpha L_c^2}\sigma_{max} \tag{E.9}$$

where ϕ_F is the *form factor* (0.125 for a cylinder [1]). Material-related parameters in Equation E.9 can be combined into a material factor ϕ_M defined as

$$\phi_M = \frac{E'\alpha}{\delta} \tag{E.10}$$

so that

$$\frac{dT_{stm}}{dt} = \frac{\sigma_{max}}{\phi_F \phi_M L_c^2} \tag{E.11}$$

or

$$\sigma_{max} = \phi_F \phi_M L_c^2 \frac{dT_{stm}}{dt}. \tag{E.12}$$

Typical values for ϕ_M at 400°F (~200°C) are [1]

- 0.83 ksi-h/ft^2-R (111 MPa-h/m^2-K) for 1% Cr steel
- 1.14 ksi-h/ft^2-R (153 MPa-h/m^2-K) for 12% Cr steel
- 1.46 ksi-h/ft^2-R (194 MPa-h/m^2-K) for austenitic steel.

In passing, combining Equations E.7 and E.10 yields a simpler form of the former, i.e.,

$$\sigma_{max} = \delta\phi_M\phi_Z K_T \Delta T \quad \text{(E.13)}$$

or

$$\Delta T = \frac{\sigma_{max}}{\delta\phi_M\phi_Z K_T}. \quad \text{(E.14)}$$

Equation E.11 gives the allowable steam temperature ramp rate for a given maximum *allowable* stress, σ_{max}, which is dependent on rotor material and typically lies in a range of 50–80 ksi.

For the cited range, with the data in Table E.2, Equation E.14 suggests that

- For low HTC (~100 Btu/h-ft^2-F or less, $\phi_Z \sim 0.6$) steam–metal ΔT can range from 200°F–300°F (high K_T) to 500°F and higher (low K_T)
- For high HTC (~650 Btu/h-ft^2-F, $\phi_Z \sim 0.8$), steam–metal ΔT can range from 100°F–200°F (high K_T) to about 400°F (low K_T).

Similarly, using Equation E.11 with typical values of ϕ_M, it can be seen that allowable values for dT_{stm}/dt range from 3 to 10°F per minute depending on the maximum allowable stress and component size (e.g., rotor radius).

The allowable stress is *not* a precisely defined material property. (For ferritic steels used in steam turbine rotor construction, 0.2% tensile yield strength lies between 70 and 90 ksi for temperatures 600°F –1,000°F.) It is derived from the *S–N curves* relating total strain to cycles to failure, which gives the fatigue life of the material in question (for low-cycle fatigue or LCF life of CrMoV alloy see Figure E.2). Based on the relationship between stress and strain, ε, via the modulus of elasticity,

$$\sigma = E' \cdot \varepsilon. \quad \text{(E.15)}$$

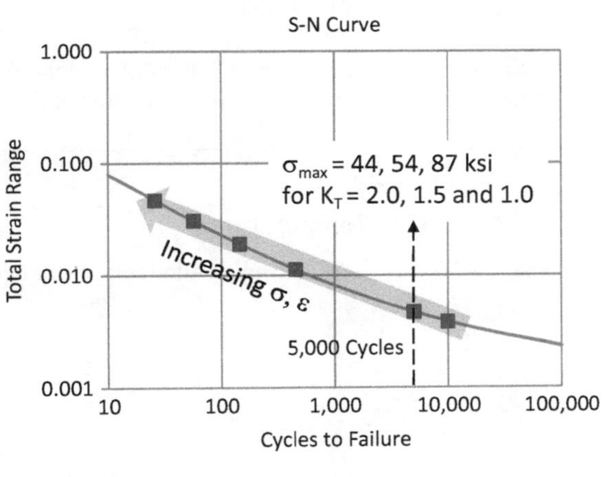

FIGURE E.2 Typical S–N curve for steam turbine rotor LCF (CrMoV) – calculated using the method of universal slopes, Equations 4–11 in Viswanathan [2] with E = 26,000 ksi, ultimate strength, σ_u = 60 ksi (415 MPa).

This curve is used to determine σ_{max} for a defined fatigue life. Note that the y-axis of the S–N curve can be strain/stress range or amplitude. If it is the former, the amplitude is found by dividing the range by two. For the mean value of stress/strain, one has to calculate the average of the minimum and maximum values. One cycle is the time between two consecutive minimum or maximum stress/strain values.

In practice, the relationship between σ and ΔT allows the translation of the S–N curve into *Cyclic Life Expenditure* (CLE) curves, which determine the allowable steam temperature ramp rates (Figure E.3). In other words,

- Read total strain range from the S–N for desired cyclic life (N cycles)
- Translate it into maximum allowable stress (Equation E.15)
- Calculate allowable steam–metal temperature delta from Equation E.14
- Read dT_{stm}/dt from the CLE curves (or equation E.11).

Depending on the rotor material, size and geometry and its temperature at start initiation, the range is limited to about 5–10°F per minute except for very hot "restarts" after a few hours of downtime.

The other limiting parameter is steam-to-metal temperature mismatch at the start of steam admission into the turbine. Typical values are tabulated for the IP turbine in Table E.3 (metal temperatures refer to the IP turbine bowl upper thermocouple reading). Minimum allowable steam temperature is the largest of 450°F or the initial metal temperature.

Minimum allowable steam temperatures are set to ensure sufficient superheat and prevent wet steam and/or water induction into the steam turbine. Maximum allowable steam temperatures are set to limit the rate of rotor/shell heating such that warming in shortest possible time is achieved without exceeding limits for thermal stress and differential expansion.

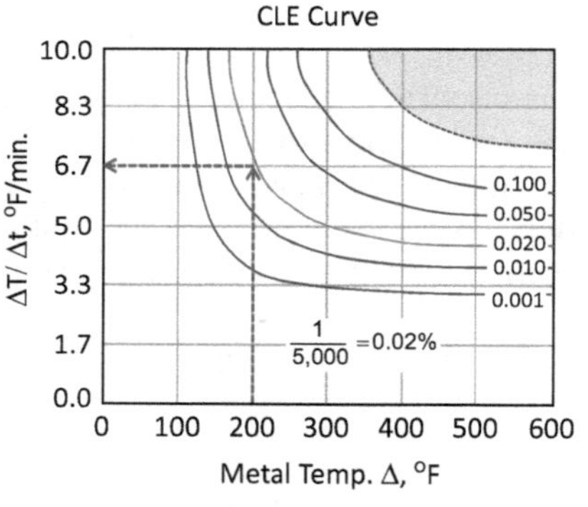

FIGURE E.3 Typical CLE curves for steam turbine rotor LCF (CrMoV). Metal ΔT represents the *total* temperature change between initial and final states (beyond 600°F, curves are flat).

TABLE E.3
Allowable Steam Admission Temperatures

Initial Metal Temperature, °F	Maximum Allowable Steam Temperature, °F
200	710
400	750
450	800
700	1,050
800	1,050
950	1,050
1,050	1,050

In hot starts, steam temperature must be greater than the turbine inlet bowl metal temperature to prevent forced cooling or quenching of the parts. Forced cooling is detrimental because of

- Increased probability of radial rubbing
- "Rotor short" condition
- Reverse stress cycles imposed on the rotor/shell
- Extra time (longer start) to rewarm cooled metal.

For "hot" starts (i.e., metal temperature 700°F or higher), admission at rated temperature (e.g., 1,050°F) is acceptable because of low flow requisite for rolling (and lower HTC). However, this can result in elevated temperatures in the LP turbine exit stages if synchronization and loading is delayed.

E.1 DIFFERENTIAL EXPANSION

Differential axial expansion of the rotor is a result of the faster expansion rate of the turbine rotor vis-à-vis the turbine casing (another term used for casing is "shell") due to its lower mass and different metallurgy. This is referred to as the "rotor long" condition, which, if not controlled, can lead to clashing of rotating and stationary parts (i.e., the buckets and nozzle vanes). Rotor long condition is prevalent during startup and requires strict control of steam flow and temperatures to allow heating and expansion of the turbine casing in lock-step with the turbine rotor.

The other differential expansion problem is the "rotor short" condition, which can occur during startup as well as normal operation. This condition is caused by "quenching" of the rotor by a large drop in steam temperature and/or flow.

Thermal growth of the rotor from its "cold", stationary state to the "warm", rated operation state is unavoidable. If the rotor is clamped or restricted from such growth at both ends, it will create enormous tensile or compressive stresses in the component. This can be illustrated by a quick example. The magnitude of the thermal stress resulting from a temperature change from T_i to T_f can be estimated from Equation E.2., i.e.

$$\sigma = E'\alpha(T_i - T_f).$$

Assume that the low-alloy steel rotor of a steam turbine is at 700°F and subjected to a steam flow at 900°F. Ignoring the rotor construction details and assuming that it is basically a solid cylinder, if the rotor were constrained from thermal growth at both ends, the stress would be

$$\sigma = 200\,\text{GPa} \cdot 12 \cdot 10^{-6} \frac{1}{\text{K}} \frac{(700 - 900)}{1.8} \text{K} = -270\,\text{MPa}.$$

Since the final temperature is higher than the initial temperature, the negative value indicates *compressive* stress. Otherwise, one would obtain a positive value, indicating *tensile* stress. At the given temperature range, the average tensile strength of the rotor material is about 550 MPa. Thus, the calculated compressive stress magnitude is about 50% of the tensile strength of the rotor material.

Consequently, the rotordynamic design of the turbine must allow for unrestricted axial growth of the rotor inside the casing in one direction. That direction is typically fixed by the location of the thrust bearing (TB) as shown in Figure E.4, which shows the bearing arrangement of a two-casing steam turbine for multi-shaft GTCC applications [3]. As shown in the figure, the opposed-flow HP–IP casing and the double-flow, down-exhaust LP casing are fixed to the foundation (known as "key" to the foundation). The rotor is fixed by the TB in the *front standard*. Thus, when the machine warms up, the rotor grows axially towards the generator. At the same time, however, the HP–IP casing, which is keyed to the foundation at the front standard, also expands in the same direction when heated. The net result is maintenance of the cold-state clearances during normal operation – unless, of course, the machine is started too fast to allow the slower expanding casing to catch up with faster-growing rotor.

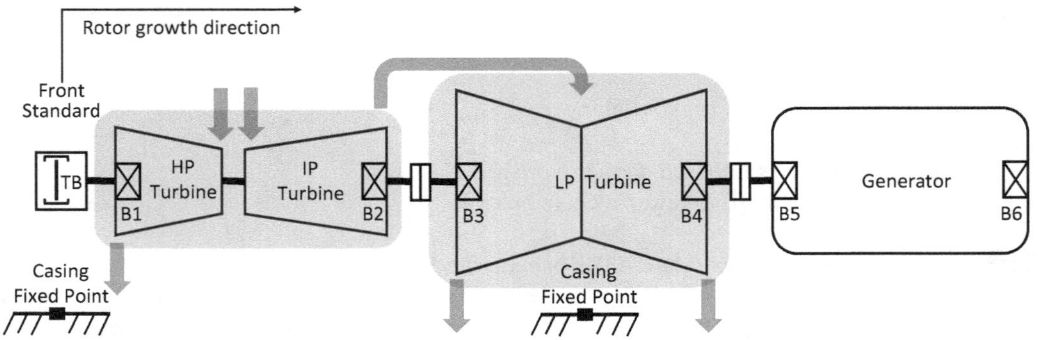

FIGURE E.4 Combined cycle reheat steam turbine arrangement (two-casing, double-flow LP with down-exhaust).

Note that, in the arrangement shown in Figure E.4, HP–IP turbine casing or shell is keyed to the foundation at the front standard. The HP–IP turbine can be of single- or double-shell (usually for steam pressures above 1,800 psig) construction. In the latter case, the outer shell is attached to the front pedestal. The inner shell is keyed to the outer shell via brackets in the horizontal plane and via pins on upper and lower vertical center lines to fix it transversally to the turbine axis. In the case of the LP turbine, it is the LP "hood" that is keyed to the foundation. The inner casing is keyed to the hood by supporting pads for axial and transverse location (usually four).

The arrangement in Figure E.4 is representative of only one original equipment manufacturer (OEM) (i.e., General Electric) and one product line (i.e., old D series). It is used to illustrate the basic principles of differential rotor expansion. Other configurations are also possible for different casing configurations by different OEMs. One such example is illustrated in Figure 9.2 (Siemens steam turbine SST6-5000). In that arrangement, as the rotor grows axially towards the generator, the HP–IP outer shell and the LP inner casing are thread together via the thrust bolts and grow axially in the same direction to minimize the differential growth between the rotating and stationary components.

E.2 ROTOR THERMAL STRESS

In order to analyze heating or cooling of components with large metal mass and/or thick walls, one has to find a solution of the governing differential equation for T(x,t) for the component in question. For a solid body of a nonuniform geometry, this is a very difficult task and can only be handled by numerical methods such as finite elements. Exact and approximate solutions for simple geometries (e.g., infinite cylinder, plane wall) are available in most textbooks on heat transfer [4]. Even the approximate solution is tedious and requires tabulated values such as the Bessel functions. Nevertheless, they are relatively easy to implement in an Excel spreadsheet. For example, the solution relevant to the most important practical problem, i.e., steam turbine rotor thermal stress management, is the one for the infinite cylinder. Thus, for Biot numbers > 1, at the surface of a turbine rotor modeled as a cylinder, the temperature difference as a function of time is

$$\frac{T(t) - T_\infty}{T_0 - T_\infty} = C_1 \exp\left(-\zeta_1^2 Fo\right) J_0\left(-\zeta_1\right) \tag{E.16}$$

where the *Fourier* number, Fo, is a dimensionless time:

$$Fo = \frac{\delta t}{\left(D_0/2\right)^2} \tag{E.17}$$

Equation E.16, which is an *approximate* solution for Fo > 0.2, gives the surface temperature of an infinite cylinder with diameter D_0, which is initially (i.e., at time t = 0) at a uniform temperature, T_0, and suddenly subjected to a fluid flow at temperature T_∞.

The *exact* solution of the infinite cylinder problem is similar to Equation E.16 where the right-hand side of the equation is an infinite series of the same form. In dimensionless terms,

$$\theta^* = \sum_{n=1}^{\infty} C_n \exp\left(-\zeta_n^2 Fo\right) J_0\left(-\zeta_n r^*\right) \quad (E.18)$$

$$\theta^* = \frac{T(t) - T_\infty}{T_0 - T_\infty} \quad (E.19)$$

$$r^* = \frac{r}{(D_0 / 2)} \quad (E.20)$$

$$C_n = \frac{2}{\zeta_n} \frac{J_1(\zeta_n)}{J_0^2(\zeta_n) + J_1^2(\zeta_n)} \quad (E.21)$$

where the discrete values of ζ_n are positive roots of the *transcendental* equation

$$Bi = \zeta_n \frac{J_1(\zeta_n)}{J_0(\zeta_n)}. \quad (E.22)$$

The quantities J_1 and J_0 are *Bessel* functions of the first kind and their values can be found using the Microsoft Excel worksheet function BESSELJ. First four roots of the transcendental equation are tabulated in Table E.4.

The thermal stresses in a cylindrical rotor result from the temperature gradient between the surface (in direct contact with steam) and the center (the coldest part). The stress is compressive on the surface and tensile in the interior when heated (i.e., during turbine startup) and vice versa when cooled (i.e., turbine shutdown). Most steam turbine rotors in service have hollow bores since they have been machined at the manufacturing stage to remove material of inferior mechanical and metallurgical properties from the central portion of the original forging.

TABLE E.4
Roots of Transcendental Equation [5]

Bi	ζ_1	ζ_2	ζ_3	ζ_4
0.01	0.1412	3.8343	7.017	10.1745
0.1	0.4417	3.8577	7.0298	10.1833
0.5	0.9408	3.9594	7.0864	10.2225
1	1.2558	4.0795	7.1558	10.271
2	1.5995	4.291	7.2884	10.3658
5	1.9898	4.7131	7.6177	10.6223
10	2.1795	5.0332	7.9569	10.9363
20	2.2881	5.2568	8.2534	11.2677
50	2.3572	5.4112	8.484	11.5621

The thermal stress at the rotor surface can be calculated using Equation E.2, i.e.,

$$\sigma = E\frac{\alpha}{1-\nu}K_T(T_{SURF} - T_{BULK}) \tag{E.23}$$

where the thermal stress concentration factor, K_T. Temperature calculations as a function of time can be done using the "exact" solution for an infinite cylinder as described above, i.e., Equations E.18–E.21, as follows:

- T_{SURF}, at radius $r^* = 1$;
- T_{BORE} at the radius, $r = r^* = 0$.

The average (bulk) rotor temperature is estimated using the following equation based on the assumption of a parabolic temperature profile

$$T_{BULK} = T_{BORE} - 0.359(T_{SURF} - T_{BORE}). \tag{E.24}$$

The thermal stress at the rotor bore is also calculated using Equation E.23 with a K_T of 1. Instead of T_{SURF}, T_{BORE} is used to calculate the temperature difference. Rotor bore stress is a combination of the thermal stress (transient, tensile during startup) and centrifugal stress (steady, tensile). At the rotor bore, which has a very small diameter, the latter is given by the fundamental principles described above which can be combined qualitatively into the steam admission sequence illustrated in Figure E.5.

1. Steam admission starts when the steam–metal temperature difference is at an acceptable value (see Table E.3).
2. High-temperature steam flow heats up the rotor; surface of the rotor heats up faster than the inner regions.
3. As time passes, metal temperature difference between the rotor surface and its bulk average increases. This results in increasing thermal stress in the rotor.
4. When the steam–metal temperature difference reaches an acceptable value, steam temperature and flow are ramped up.
5. This causes an initial increase in thermal stress ramp rate as well.
6. However, due to increased steam flow and HTC, rotor metal temperature also increases at a faster rate and eventually catches up with the steam temperature.

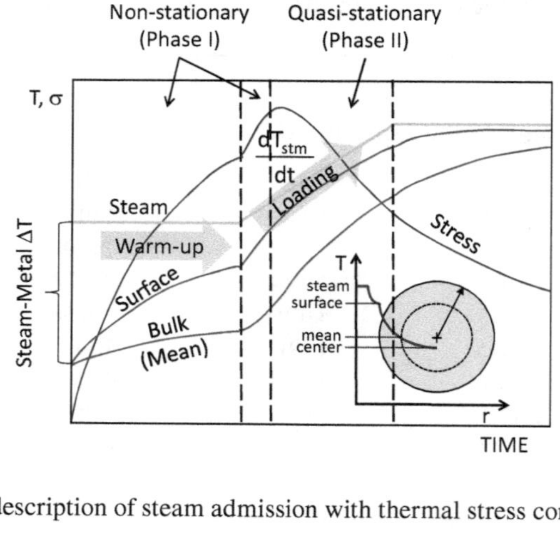

FIGURE E.5 Qualitative description of steam admission with thermal stress control.

7. As a result, thermal stress reaches a peak value and, thereafter, starts decreasing in lock-step with decreasing steam–metal temperature mismatch.
8. Once the steam temperature reaches its rated value, the controller keeps it at that value and rotor warming proceeds in accordance with the lumped capacitance method (LCM) exponential decay formula.

REFERENCES

1. VGB PowerTech Guideline, 1990, *Thermal Behaviour of Steam Turbines*, Revised 2nd Edition, VGB-R105e, VGB PowerTech Service GmbH, Essen, Germany.
2. Viswanathan, R., 1989, *Damage Mechanisms and Life Assessment of High-Temperature Components*, ASM International, Metals Park, OH.
3. Boss, M., 1994, Steam Turbines for STAG™ Combined Cycle Power Systems, GER-3582E, General Electric.
4. Incropera, F.P., Dewitt, D.P., 2002, *Introduction to Heat Transfer*, 4th Edition, John Wiley & Sons, Inc., New York.
5. Özışık, N., 1988, *Conduction Heat Transfer*, John Wiley & Sons, Inc., New York.

Appendix F
Carbon Capture

Min _ Sep _ Work is an Excel VBA function calculating the minimum separation work (MSW) as described in Ref. [2] in Chapter 19. MSW is the equivalent of gas turbine exhaust exergy, which is extensively used in this book.

```
Function Min_Sep_Work(sTemp As Single, sMdot As Single, sCapRate As
Single, sPurity As Single, vComp As Variant) As Single
' 1  'N2
' 2  'O2
' 3  'H2O
' 4  'CO2
' 5  'Ar
' 6  'SO2
' 10  'Air
' 7  'CH4
Dim sGasMolWeight As Single, sMoleFlowRate As Single
' A: Exhaust
' B: Mostly CO2
' C: Rest of Exhaust
Dim n_A_CO2As Single, n_B_CO2 As Single, n_C_CO2 As Single
Dim n_A_Others As Single, n_B_Others As Single, n_C_Others As Single
Dim y_A_CO2 As Single, y_B_CO2 As Single, y_C_CO2 As Single
Dim y_A_Others As Single, y_B_Others As Single, y_C_Others As Single
Dim i As Integer
Dim sTemp_K As Single
Dim G_A As Single, G_B As Single, G_C As Single
Const Rgas As Single = 8.314 'kJ/kmol-K

sTemp_K = (sTemp - 32) / 1.8 + 273.15   'temp. in Kelvins

sGasMolWeight = MWgas(vComp, 10)    'lbm/lbmole or kg/kmole
sMoleFlowRate = (sMdot * 0.4536) / sGasMolWeight 'kmole/s

y_A_CO2 = vComp(4)
y_A_Others = 1 - y_A_CO2
n_A_CO2 = y_A_CO2 * sMoleFlowRate
n_A_Others = sMoleFlowRate - n_A_CO2

n_B_CO2 = sCapRate * n_A_CO2
n_B_Others = n_B_CO2 / sPurity - n_B_CO2
y_B_CO2 = n_B_CO2 / (n_B_CO2 + n_B_Others)
y_B_Others = 1 - y_B_CO2

n_C_CO2 = (1 - sCapRate) * n_A_CO2
n_C_Others = sMoleFlowRate - (n_B_CO2 + n_B_Others) - n_C_CO2
y_C_CO2 = n_C_CO2 / (n_C_CO2 + n_C_Others)
y_C_Others = 1 - y_C_CO2
```

```
G_A = Rgas * sTemp_K * (n_A_CO2 * Log(y_A_CO2) + n_A_Others *
Log(y_A_Others))
G_B = Rgas * sTemp_K * (n_B_CO2 * Log(y_B_CO2) + n_B_Others *
Log(y_B_Others))
G_C = Rgas * sTemp_K * (n_C_CO2 * Log(y_C_CO2) + n_C_Others *
Log(y_C_Others))

Min_Sep_Work = G_B + G_C - G_A  'in kW

End Function
```

Index

A

B

C

D

E

F

I

J

K

L

M

N

O

P

Q

R

S

T

U

V

W

Z

FSC
www.fsc.org
MIX
Papier | Fördert
gute Waldnutzung
FSC® C083411

Kolibri 360 GmbH
Ferdinand-Jühlke-Straße 7
99095 Erfurt, Deutschland
produktsicherheit@kolibri360.de